# DISSENT WIT
# MODIFICATION

## Human Origins, Palaeolithic Archaeology and Evolutionary Anthropology in Britain 1859–1901

John McNabb

Archaeopress

Gordon House
276 Banbury Road
Oxford OX2 7ED

www.archaeopress.com

ISBN 978 1 905739 52 3



Printed in England by Information Press, Oxford

Cover image

*Mesolithic/later Prehistoric tool, with a drawing by W.G. Smith sent to Benjamin Harrison in October 1895. Courtesy of the Maidstone Museum and Bentlif Art Gallery.*

Dedicated to Helena, William, Emily, and Gem
Thanks for everything

# CONTENTS

# PREFACE AND ACKNOWLEDGMENTS

This book has had a long evolution and I describe something of its history in the next chapter. It resulted in eight sets of anonymous comments from peer review. Six were unambiguously supportive, one guardedly so, and one negative. The latter I think from an historian of science who would have preferred me to approach the subject as an historian. However I am an archaeologist, and I have followed the accepted pattern for writing about the history of archaeology adopted by my peers and predecessors. I make no apologies for that.

The first publisher I approached offered a contract on the basis of the comments made by two referees. However, because of their feedback and the advice I had received from colleagues who kindly agreed to read the manuscript, I decided to shift the focus of the book. I would particularly like to thank Professor Mary Orr of the Department of Modern Languages at Southampton University for her extremely helpful advice at this stage of the book's history. So the first half of the book now became the last part of a new book. The next six reviews covered the new sections. I'd like to thank all eight of the anonymous peer reviewers. I thought long and hard about everything you said, and a great many of your comments and critiques are included in the text. I will not refer to these again (as, I believe, their positive influence has made it a better book). I will however briefly touch upon some of the comments I thought were very good, but for various reasons I did not take on board.

Inevitably with so many reviewers there were contradictory views, and negotiating my way around these was sometimes challenging. My writing style was often commented on. Most were more than happy with it, a few less so, and one reviewer hated it. The style of writing I have adopted was specifically tailored to the aims of the work. I wanted to write a book that incorporated original research with synthesis and overview, and at the same time presented original perspectives derived from my overall arrangement of the material. I wanted to target third year undergraduates, Masters students, and PhDs in the earlier phases of their research, while at the same time writing something that would be accessible to the non-academic reader who wanted to know more about the subject. I also wanted to write a book for professional colleagues covering the whole period 1859–1901, which has not been done before. The closest archaeological work to this one is the excellent *Men Among the Mammoths* by van Riper, which stops in the mid-1870s. So I adopted this style deliberately. One of the reviewers disliked the mixing of the first and third person in the text, and another suggested my comments and opinions on particular issues in the first person should be included as textboxes. After much thought I stuck to my guns; in the end one's writing style is a very personal thing. This book has the same approach and target audience as I adopted for my first book *The British Lower Palaeolithic; Stones in Contention*, published by Routledge. I will leave the reader to decide whether THIS ONE has been successful.

One very interesting critique that emerged from the review process focused on my source material. One reviewer, I presume a historian, wanted to know why I wasn't using original letters and manuscripts, as these represented primary sources. I have used, almost exclusively, the original publications from the middle and late Victorian period. For me, as an archaeologist, this is primary data. These published articles are the words and ideas that the people of that time contributed toward the development of debates. While published sources may not reveal someone's innermost thoughts, they nonetheless are the foundation upon which dialogue is based. The written word was what people were willing to be judged by. Although I have used some letters and primary archives, I maintain that published material is a primary resource. Arising from this, two reviewers would have liked to see the book set more within the realm of modern critical scholarship. They asked why I had not included modern explanations for the questions that the Victorians grappled with. Again I thought about this a lot, but in the end I stuck to my original aim. This was to write a history of the early phases of Palaeolithic archaeology and human origins research from the written perspectives of the original protagonists.

Several reviewers felt it was critical to discuss the pre-1859 developments in humanism at the beginning of the book, particularly those in Europe and post-Revolutionary France. After deliberation I decided against this. Not because I disputed the significance of these contributions to what happened after 1859, but rather because I felt the story I wanted to engage with started with the paradigm forming year of 1859 – with the publication of *Origin of Species* and the equally influential but less well known 'Antiquity of Man Debate'. From this point on, time depth, a Prehistoric ancestry and a naturalistic origin for our species, suffused scientific thought to an extent it had not done so before. Excellent introductions to the European pre-1859 period are provided by Rudwick (2007, 2008). I suggest Desmond (1982) for the British material.

Finally on the subject of peer review one referee wanted to see critique developed on the engagement of the Victorian public with the human origins debate, and with origins and ancestry in general. I agree this would be a fascinating topic for further research. Questions about where and in what form the public got its information emerged quite unexpectedly from my research. But that would be a whole new book. Within the aims I set out for this book I decided to identify and describe what those channels of information were. Analysing and critiquing their content will be a research project for the future. Having said this, in the chapters on Victorian science fiction, I have begun to explore this in a little more detail.

A great many people helped in the writing of this book in one way or another. In particular I would like to thank my partner Helena for being supportive throughout the process, and understanding that on a lot of sunny days my mind was elsewhere. Without her forbearance this book would never have been written. I would like to express my appreciation to the staff and the directors of the Maidstone Museum and Bentlif Art Gallery for access to the Benjamin Harrison archives. I would also like to express a huge debt of gratitude to Angela Muthana an assistant curator at the museum whose patient researches into Harrison were a constant source of inspiration. In the British Museum Debbie Buck was, as always, brilliant, and I would like to express my gratitude to all the staff of the Department of Prehistory and Europe at the BM outstation at Franks House for facilitating access to their Harrison archive and for allowing me to photograph artefacts. Thanks to Nick Ashton for photographs and copyright permissions. My thanks go to Rob Kruszynski and the staff of the Palaeontology Department at the Natural History Museum for help in accessing and photographing more artefacts. I am especially grateful to the staff of the Hartley Library at Southampton, in particular Pam Wake, and in inter-library loans, Karin Jazosch whose refusal to give up on finding a resource located a whole journal run on the internet that otherwise I would never have been able to consult. I am especially grateful to Dr S. McLean for reading drafts of the science fiction chapters at short notice.

All of the editors in the various publishing houses I contacted were helpful and very supportive, none more so than the one publisher who, despite five positive reviews, could not accept the book because market conditions had sent his company down a different route in the publishing world. I felt his anguish; his very helpful advice was much appreciated. I would also like to express my deep gratitude to all at Archaeopress for taking the book on, and their patience during its production. You guys are just brilliant. I am especially grateful to Dr John de Vos at Netherlands Centre for Biodiversity Naturalis, Leiden, for a fascinating afternoon looking at the Dubois collection and for allowing me to photograph the reconstruction that Dubois made of *Pithecanthropus*; also for pointing out the significance of the von Max painting. Frances Clarke did a great job in editing and proof reading. I am also very grateful to Penny Copeland and to Dr Susan Hackenbeck for their imaginations and magic pens. Thanks.

This would have been a much better book had my old mentor Roger Jacobi not died prematurely. Somehow I would have persuaded him to read a draft. His copious marginalia, written in his distinctive blocky handwriting, would have improved the text considerably. He would have found all the mistakes I have missed, delivered his trademark admonishing sniff, and then encouraged me to have another go but to pay more attention to the details next time. He had been an enthusiastic eolith hunter when he was yong and we planned to have a long chat in the pub about it all. We put it off because there was plenty of time, and then suddenly there was no time left. You are much missed.

# ILLUSTRATION AND IMAGE CREDITS

The author expresses his grateful thanks to all the following institutions who extended copyright, or in those cases where the copyright had expired, were still happy for the images to be used. In other cases institutions allowed the photographing of their material and its use in this publication.

Maidstone Museum and Bentlif Art Gallery
Figures 9.2, 9.3b, 9.5, 9.12, 9.13, 9.15, 10.1. All images taken by the author.
Imperial College Archives, Imperial College London.
Figure 10.6. The Huxley Papers, 31.158.

Dr Francis Wenban-Smith
Figure 8.1. From Wenban-Smith, F.F. 2009. Henry Stopes (1852–1902): Engineer, Brewer and Anthropologist. In (R.T. Hosfield, F.F. Wenban-Smith & M. Pope, eds). Great Prehistorians: 150 Years of Palaeolithic Research, 1859–2009 (Special Volume 30 of Lithics: Journal of the Lithic Studies Society): 65–85. Lithic Studies Society, London.

Macmillan Publishing
Figures 4.4, 6.7, 7.1,

The Geological Society of London
Figures 4.2, 9.3a, 9.4, 9.6, 9.8, 10.2

The Royal Anthropological Institute of Great Britain and Ireland and John Wiley and Sons Ltd
Figures 4.1, 7.6, 7.8, 9.7, 9.9, 9.10, 9.11

Pearson's Education Limited for Longman's Publishing
Figures 2.7, 4.3

The Society of Antiquaries, London
Figures 2.4, 2.5, 2.6

The Royal Society, London
Figures 2.2 and 2.3

The Director and Trustees of the British Museum
Figure 6.5. The Grays Inn handaxe copyright is held by the Trustees of the British Museum. The Hoxne handaxe copyright is held by the Trustees of the British Museum and the Trustees of the Department of Prehistory and Europe.

The Director and Trustees of the Natural History Museum, London.
Figures 2.1, 9.14. Images taken by the author.

Friedrich-Schiller-Universität Jena, Institut für Geschichte der Medizin, Naturwissenschaft und Technik, 'Ernst-Haeckel-Haus'
Figure 10.5 upper image, Pithecanthropus alalus

Out of Copyright as far as could be ascertained
Figures 5.2, 5.3, 5.4, 5.5, from Lyell 1863 – Hodder & Co. Ltd. for John Murray publishers. Figure 5.6 from Lubbock 1865 – Williams and Norgate Publishers – not traced. Figure 10.3 from Worthington Smith 1894 – Edward Stanford Publishers – no longer in business. Figure 7.2 – privately printed and distributed by Benjamin Harrison. Figure 10.5 lower, image of Pithecanthropus alalus commissioned by Virchow source http://upload.wikimedia.org/wikipedia/commons/1/12/Pithecanthropus_alalus.jpg. Accessed 11.09.2011.

## Copyright for Other Images

- John Lubbock source at http://en.wikipedia.org/wiki/File:John_Lubbock72.jpg. Accessed 11.09.2011.
- Charles Lyell source at http://upload.wikimedia.org/wikipedia/commons/7/7b/ Sir_Charles_Lyell%2C_1st_Bt.jpg. Accessed 11.09.2011.
- Augustus Lane Fox source at http://en.wikipedia.org/wiki/File:A-H-Pitt-Rivers.jpg. Accessed on 11.09.2011.
- Alfred Russel Wallace source at http://en.wikipedia.org/wiki/Alfred_Russel Wallace. Accessed on 13.09.2011
- William Boyd Dawkins source at http://upload.wikimedia.org/wikipedia/en/3/William_Boyd_Dawkins.jpg. Accessed on 11.09.2011.
- E.B. Tylor source at http://upload.wikimedia.org/wikipedia/commons /f/f7/PSM_V41_D080_E_B_Tylor.jpg. Accessed on 11.09.2011.
- T. H. Huxley source at http://upload.wikimedia.org/wikipedia/commons/2/2e /T.H.Huxley%28Woodburytype%29.jpg. Accessed on 11.09.2011.
- J.D. Hooker source at http://upload.wikimedia.org/wikipedia/commons/7/79/Sir_Joseph_Dalton_Hooker.jpg. Accessed 11.09.2011.
- Charles Darwin source at http://en.wikipedia.org/wiki/File:Charles_Darwin_seated_crop.jpg. Accessed on 11.09.2011.
- Édouard Lartet source http://upload.wikimedia.org/wikipedia/commons/2/25/Lartet.jpg. Accessed on 11.09.2011.
- Henry Christy source http://upload.wikimedia.org/wikipedia/commons/2/26/HenryChristyCropped.jpg. Accessed 11.09.2011.
- Boucher de Perthes source http://upload.wikimedia.org/wikipedia/commons/9/90/Boucher_de_Perthes.jpg. Accessed 11.09.2011.

- Sir John Evans source http://upload.wikimedia.org/wikipedia/commons /8/8a/Sir_John_Evans.jpg. Accessed 11.09.2011.
- Hugh Falconer source http://upload.wikimedia.org/wikipedia/commons /9/96/Hugh_Falconer.jpeg. Accessed 11.09.2011.
- Harrison plate image of shop source Maidstone Museum and Bentlif Art Gallery, used with permission. Harrison image copyright to Oxford University Press. Used with permission.
- The Director and Trustees of the Natural History Museum, London for permission to reproduce the Prestwich's 1859 handaxes.
- Images from Creswell Crags are from Boyd Dawkins 1880 'Early Man in Britain'. Out of copyright.
- Rupert Jones from G. Prestwich 1899. 'Sir Joseph Prestwich' Published by William Blackwood and Sons. Not traced.
- Joseph Prestwich from G. Prestwich 1899. 'Sir Joseph Prestwich' Published by William Blackwood and Sons. Not traced.
- Worthington Smith cartoon and Harrison drawing of a body stone are reproduced courtesy of the Trustees of the Department of Prehistory and Europe, British Museum. From Harrison archive, Franks House archive.
- Harrison's sketch cross-section of the Darent Valley is reproduced courtesy of the Trustees of the Department of Prehistory and Europe, British Museum. From Harrison archive, Franks House archive.

# LIST OF FIGURES

Figure 1.1. Modern interpretation of the Pleistocene period. The alternating white and dark blocks represent stages in the Marine Isotope Stage sequence. These are phases of climate (shaded block = cold/glacial; unshaded block = warmer/interglacial) recovered from deep sea cores drilled through Pleistocene sediments. After McNabb 2007.

Figure 1.2. The various sub-divisions of the British Prehistoric period as understood today. After McNabb 2007.

Figure 1.3. Development of ideas and approaches to studying human origins and related disciplines across the Victorian period. Based on Bowler 2003 with additions.

Figure 2.1. Handaxes from Abbeville and St Acheul discovered or bought from quarry men by English archaeologists, notably Prestwich, in the 1860s. Handaxes were more usually called implements by the middle and late Victorian scholars of human antiquity. Sometimes they were also called palaeoliths. A distinguishing feature of their being made by hominins was the evidence of the flake scars. This showed they had been knapped. The scars can be seen clearly on the handaxes depicted. They appear as shallow depressions marked by a clear margin and originating from the implement's edge An example is present on the central pointed implement, in the lower right hand quadrant. Radiating patterns of concentric ripple marks within the bed of the flake scar could be traced to the point of impact where the knapper struck the flake off from the implements edge. When covered by such flake scars it was evident that the handaxes were not the work of nature.

Figure 2.2. Joseph Prestwich's interpretation of the evolution of a river valley. See text for explanation. *Proceedings of the Royal Society* 1862.

Figure 2.3. Joseph Prestwich's illustration of the high and low level valley drifts as clearly demarcated in one of the French valleys. *Proceedings of the Royal Society* 1864.

Figure 2.4. Flint flake illustrated by Evans. *Archaeologia* 1860.

Figure 2.5. Pointed handaxes as illustrated by Evans *Archaeologia* 1860.

Figure 2.6. Ovate/oval handaxes as illustrated by Evans *Archaeologia* 1860. Note the orientation. The ovate is positioned with the wider end uppermost implying Evans thought this to be the tip. Modern convention would suggest the narrow end is the tip and orientate it accordingly.

Figure 2.7. A Neolithic celt or axe as depicted by John Evans in *Ancient Stone Implements* 1872

Figure 3.1. Comparison of proportion of occurrence (as %) of different categories of article and other submission types (reviews, reports etc.) between 1861 and 1871 for the Ethnological (N=381) and Anthropological (N=720) Societies of London. The final category, subject specific papers includes theoretical, methodological, subject overviews and reflective papers.

Figure 4.1. Racial types as suggested by MacKintosh 1866. Faces 1–5 represent the prevailing type in North Wales; includes, long necks, dark brown hair, long narrow faces and sunken eyes. Broad skull, approximately square in shape. Faces 7–9, a second North Wales group. A broad face under the eyes which sinks under the cheek bones Dark complexion and dark brown hair. Skull squarish. Found along the North Wales coast from Mold to Caernarvon. Also present along West Wales coast. Faces 10 and 11, a third type in North Wales. Thick set and large framed. A broad face, often associated with the more prosperous individuals. Possible descendants of the Iron Age Silurian tribes. Types 12–14, allied to the Gaelic peoples. Lower face projects forward. Most extreme type is 15 as seen in law courts in Beaumaris (though whether in the dock or not is not stated). Faces 19–21 Saxon group. A round, short, broad face and very regular features, but with low cheek bones and prominent eyes. Tend to obesity, light brown hair. Anglian type is represented by faces 17, 18, 22, and 24. Like Saxon but with a longer and narrower face. Narrower nose by comparison and more compressed nostrils. Fair complexion and light brown hair. Face 25 is the Jutian group. Narrow head and face, and face very convex in profile. Projecting cheek bones and a long nose. The Danish group were faces 26–28. Long faces with coarse features. High cheek bones with a receding chin. A narrow elongated skull wider at the back. Faces 6, 16, and 23 not mentioned in text. MacKintosh had this to say about Shakespeare. 'I cannot resist the belief that Shakespeare, if not a Welshman, was more allied to the Cymrian type, or one of its lateral variations, than any other type yet classified. In his native district, at least half of the inhabitants differ very little from the Gaelic–British and Cymrian–Welsh. To call Shakespeare a Saxon, would be to show a total ignorance of the science of races; though I should not like to be too confident in asserting that he was not a Dane' MacKintosh 1866, 12.

Figure 4.2. Handaxes from Flower 1872. The two implements on the left hand side of the figure are from Thetford and the gravels of the Little Ouse river. The two implements on the right hand side of the figure are from St Acheul gravels in northern France. The upper two are pointed forms and the lower two are ovates.

Figure 4.3. The Grotto of Aurignac as depicted by Vogt 1864. 1. The inner vault; 2. The rabbit burrow which led to the discovery; 3. Human bones; 4. Rubbish with implements and bones inside the grotto; 5. Rubbish outside the grotto; 6. Deposit of cinders; 7. Rock; 8. Talus of gravel, which concealed the slab of sandstone (10); 9. Slope of the hill covered with gravel. 10. The slab of sandstone erected as a door to seal off the inner grotto. Lartet's excavations actually showed that 4 and 5 were a continuous layer with cave bear mammoth and other Pleistocene fauna. He found fragmentary human remains in 4 and asserted that they were of the same age as the skeletons from the inner grotto directly above that had been removed.

Figure 4.4. The top image shows the cross section of the Cro-Magnon cave as given by Boyd Dawkins in 1874 and 1880. Letters B, D, F, H, and J are accumulations of debris representing occupation horizons. They contain charcoal fragments, flint implements and broken bones. The human bones are represented by lower case letters b and d. The letter a marked the tusk of an elephant, and the bedrock was at

A. The crack in the rock shelter's overhang is clearly visible. The lower image is the rock shelter today with the broken face of the overhang very clear.

Figure 5.1. A schematic illustration of Richard Owen and T.H. Huxley's dispute over the structure of the brain and the hippocampus minor.

Figure 5.2. Redrawn and modified after Lyell 1863, figure 41. His original caption reads as follow. 'Map of part of the north-west of Europe including the British Isles, showing the extent of sea which would become land if there were a general rise of the area to the extent of 600 feet.' The dark line represented the 100 fathom line, the limit of dry land if land levels were to rise.

Figure 5.3. After Lyell 1863 figure 27. His original caption and explanation reads as follows. 'Diagram to illustrate the general succession of the strata in the Norfolk cliffs extending several miles N.W. and S.E. of Cromer. A. Site of Cromer jetty; 1. Upper Chalk with flints in regular stratification; 2. Norwich Crag, rising from low water at Cromer, to the top of the cliffs at Weybourne, seven miles distant; 3. 'Forest Bed' with stumps of trees in situ and remains of Elephas meridionalis, Rhinoceros Etruscus, &c. This bed increases in depth and thickness eastward. No crag (No. 2) known east of Cromer Jetty; 3' Fluvio-marine series. At Cromer and eastward with abundant lignite beds and mammalian remains, and with cones of the Scotch and spruce firs and wood. At Runton, north-west of Cromer, expanding into a thick freshwater deposit, with overlying marine strata, elsewhere consisting of alternating sands and clays, tranquilly deposited, some with marine, others with freshwater shells; 4. Boulder clay of glacial period with far transported erratics, some of them polished and scratched, twenty to eighty feet in thickness; 5. Contorted drift; 6. Superficial gravel and sand with covering of vegetable soil.'

Figure 5.4. Schematic redrawing of Lyell 1863 figure 39. His original caption reads as follows. 'Map of the British Isles and part of the North-West of Europe, showing the great amount of supposed submergence of land beneath the sea during part of the glacial period.' Scotland has been submerged up to 2,000 feet, and other parts of Britain up to 1,300 feet. Isolated islands were too high to be totally submerged. Southern Britain and northern France were interpreted as never having been submerged because of the absence of marine shells and glacial erratics which could only have been emplaced by the melting of icebergs. Whether the whole of the area was submerged at the same time was debateable.

Figure 5.5. Schematic redrawing of Lyell 1863 figure 40. His original caption reads as follows. 'Map showing what parts of the British islands would remain above water after a subsidence of the area to the extent of 600'.'

Figure 5.6. handaxe from Le Moustier as illustrated by Lubbock 1865 figure 131.

Figure 6.1. Proportion of occurrence (as %) of different categories of article and other submission types (reviews, reports etc.) between 1872 and 1880 for the *Journal of the Anthropological Institute of Great Britain and Ireland.*

Figure 6.2. Proportion of occurrence (as %) of different categories of article and other submission types (reviews, reports etc.) between 1863 and 1871 for the

of Holmesdale and the rivers Darent and Shode in the area between the Chalk Escarpment and the Greensand Escarpment. Drawn by Penny Copeland.

Figure 7.5. Map redrawn from Prestwich 1891 and 1892 showing the Chalk Plateau, Ightham, some of the locations of implements found in the fluvial gravels of the area and the patches of Plateau Drift where eoliths were being discovered. Drawn by Susan Hakenbeck.

Figure 7.6. Section through the north London occupation horizon discovered by Worthington Smith, from Smith 1884, plate 11. The upper image shows a south facing section from a pit near Clapton Railway Station. The image below is the section in more detail. It was 11 feet and 6 inches deep, and only reached the top of the basal gravel. Paraphrasing from Smith's text the following is the key to the figure. R is humus; Q is mud belonging to the trail; P is a pocket of London Clay; O is the trail; N is the Palaeolithic sand and loam disturbed by the trail. The Palaeolithic floor was missing from this location. Layers M–B were individual sand and clay units that represent fine grained deposition at the margin of the Thames, today four miles to the south; A was gravel containing in its upper parts 'lustrous and sub-abraded implements of medium age'; at other locations in the area this gravel was exposed to its full depth and at the base of the gravel were '…the oldest class of Palaeolithic implements…they are…greatly abraded, rude in manufacture, and deeply ochreous in colour'

Figure 7.7. Schematic reconstruction of John Allen Brown's interpretation of the Thames river terrace sequence and its relation to older gravel deposits to the north of the Thames and its valley. Drawn by Susan Hakenbeck.

Figure 7.8. A handaxe found during the construction of Chelsea Bridge in 1854. From *Journal of the Anthropological Institute of Great Britain and Ireland,* 1883

Figure 8.1. A shell with an engraved face thought by H. Stopes to be proof of Tertiary man. It is a scallop shell, *Pectunculus glycimeris.* One possible explanation is that it is a Medieval pilgrimage token. Alternatively, it was just a hoax. See Wenban-Smith 2009. Copyright Wenban-Smith.

Figure 9.1. A panoramic (photostich) view of the Chalk Escarpment as it looks today. Taken from a spot to the north of Ightham village.

Figure 9.2. Handaxe found at Rosewood near Ightham Common in 1863 by Benjamin Harrison. This is one of his earliest Palaeolithic finds and long pre-dates the beginning of his interests in the Palaeolithic.

Figure 9.3a. Eolith 464 in Harrison's collection. This was one of the earliest eoliths ever to be discovered, although its significance was not recognised until later. Figure 9.3.b. The corner stone. Another eolith whose significance was not recognised until long after its discovery. Many years later someone drew a figure on the back and varnished it. There is a date that may apply to this of October 1895.

Figure 9.4. Joseph Prestwich's examples of implements and flakes from the hill group as presented in the *Quarterly Journal of the Geological Society of London* for 1889.

Figure 10.3. A section from Cadington as depicted by Worthington G Smith in *Man the Primeval Savage* (1894). A. Surface material; B. Tenacious red-brown clay; C and D and E. Sub angular gravel which has 'ploughed' its way through lower deposits (F, G, H, I, K); C contains brown ochreous handaxes and flakes slightly worn; F is a grey-white clay (possibly boulder clay); G is a gravel with unapraded pale white handaxes and flakes, the flakes refit; H stiff red clay with implements and flakes like in G; I gravel same as G; J–J and O and O, Palaeolithic floor; L and M and N are heaps of flint blocks brought by Palaeolithic humans and stockpiled for knapping into handaxes; these are lying on the floor; K and P are brickearths overlying and underlying the Palaeolithic floor.

Figure 10.4. The contrasting interpretations on the relationship between *Pithecanthropus* and *Homo sapiens* as argued by Eugène Dubois and D J Cunningham and published in *Nature*. Redrawn after *Nature*.

Figure 10.5. Upper image *Pithecanthropus europaeus alalus* painted by Gabriel Cornelius von Max for Ernst Haeckel and presented to him on his 60th birthday in 1894. Lower image a rival interpretation of *Pithecanthropus alalus* commissioned by Rudolph Virchow and also drawn by von Max.

Figure 10.6. Sketch by T.H. Huxley entitled 'Homo Herculei Columnarum' drawn on July 19th 1864 at the Athenaeum Club. The occasion for this sketch may well have been the visit by Hugh Falconer and George Busk to Gibraltar to visit excavations in the Genista Caves. The Gibraltar Neanderthal skull, first found in 1848 was brought to the BAAS in Bath in September of the same year.

Figure 10.7. Sketches of Eugène Dubois' model of *Pithecanthropus erectus* made for the Paris exhibition in 1900. Drawn by Penny Copeland from photographs taken by the author.

# LIST OF TABLES

Table 1.1. Chrono-stratigraphic table of the Earth's geological history. The left hand side shows the modern interpretation. The two sections to the right of this show Victorian interpretations. That on the far right represents a generalised middle and late Victorian outlook.

Table 1.2. Sub-divisions of the Pleistocene, called the drift period by the Victorians. On the left is that supported by William Boyd Dawkins. On the right a generalised middle and late Victorian perspective.

Table 2.1. Discoveries of locations with handaxes as reported by John Evans to the Society of Antiquaries, May 16, 1861, with additions made prior to publication in *Archaeologia* later that year.

Table 3.1. A selection of some of the most important, or best publicised, discoveries of fossil humans found before and after 1859.

# 1

# JOURNEYS IN TIME AND SPACE
## *Introduction and Aims of the Book*

*A man sits alone on a hill side. It is evening and the shadows are lengthening. All around him there are ominous rustlings in the undergrowth, as if creatures are stirring emboldened by the coming night. But he pays them no heed. He stares morosely out over a wide valley. In the distance a great river, burnished like polished metal by the setting sun, flows into an unknown distance. The Time Traveller (or so it will be convenient for us to call him) is consumed by melancholy. He has travelled more than eight hundred thousand years into the future. The scene before him is London, but not the dirty frenetic late-Victorian city of 1895 that he knows so well; nothing could be further from it. In its place is a landscaped garden-metropolis of great palaces, obelisks, and open parks. But this future London is quiet and all but deserted. The great buildings lie in ruins and they have been that way for a long time. What has become of the relentless press of humanity that inhabited the capital of the British Empire, the great super-power of the Victorian age? The London of 1895 was the concrete realization of Victorian optimism; build bigger, aim higher, push the envelope to its very limit – and then push a little more. But this future city is inhabited by a small child-like race. He cannot believe them capable of building anything. They seem to lack even that most basic Victorian quality, curiosity. They idle away their time in frivolous play and do nothing else.*

*The Time Traveller has met the Eloi. He sits on the top of his hill on a bench of strange yellow metal as the twilight gathers. Gloomily he ponders the evolutionary destiny of the human race. Night falls and soon he will meet the Morlocks.*

A scene from H.G. Wells's *The Time Traveller* (2005) first published in book form in 1895, may seem like an odd place to begin a study of Palaeolithic archaeology between the time of Charles Darwin's *Origin of Species*, and the death of Queen Victoria. Actually it is very appropriate. Wells's story is about human evolution, or rather the fear that Victorians had about where evolution was taking our species. By the time that Wells wrote this story most people accepted a 'deep' ancestry for humanity, and one that was rooted in the natural world. Humans could no longer consider themselves as products of special creation. By the middle 1890s embryology and microscopy were beginning to reveal how new life was created, and explore the mysteries of heredity. But with new knowledge came new uncertainties. The Victorian man in the London street intuitively felt he was the pinnacle of the evolutionary process, as was his civilization. But was he? Some scientists claimed otherwise.

Were his comfortable certainties illusory; was his mastery of the Earth not guaranteed after all? Even more worryingly, could the process of evolutionary development actually be reversed? There were also scientists who claimed this could happen. Whether the Morlocks were an atavism, or a completely new evolutionary direction, they were nevertheless a prescient metaphor for Victorian fears about the future. What would have shocked the readers of the *Time Machine* was not the idea of the Morlocks themselves, but that they were an evolutionary possibility.

My original aim in writing this book was to chronicle the story of a very specific debate in human evolutionary studies that took place between the late 1880s and the 1930s. This was the eolith debate (McNabb, 2009; Ellen and Muthana, 2010). Eoliths were small natural stones whose shape and edges suggested to our earliest ancestors their use as tools, either as they were, or with a small amount of chipping to the stone's edge, a process called retouch. These were the most primitive of tools, thought to date to the very beginning of human cultural evolution, and therefore suited to our very earliest ancestors.

The more I researched this topic the more I realised that its explanation was rooted in a number of research questions which today we look at as separate subjects. Yet to the Victorian researcher these were all intimately connected with each other. The links were as follows; a view of time as being inherently linear (at least when viewed from a scale of events that made it seem so), evolutionary change as a process; that process being one of increasing improvement and sophistication over time; and the belief that passing time *should* therefore show progressive development. To the ordinary Victorian these were certainties and they were important; they were at the root of social confidence in the Victorian world view; they had replaced the teleological belief in a special creation. These certainties allowed humans to feel they were still on top of the tree, even if now it was an evolutionary tree and not a divine one. So a book about a forgotten Palaeolithic debate became a book that was just as much about Morlocks, stone tools, racial difference, and the Anthropological Society of London.

Today, the questions surrounding race and the reason why some humans have different skin colours, types of hair or differently shaped eyes, is the preserve of science and DNA. The same goes for the study of heredity, and why we sometimes resemble our parents and grandparents, but are not exact copies of them. Today we draw generalised distinctions between race and ethnicity (always keeping clearly focused the fact that such differences are not statements about the relative worth of one group of humans in comparison with another). The former is often thought of as skin colour and where in the world you were born. The latter is sometimes meant to imply differences between people who share the same skin colour but come from different regions in the same area or country. There is no real connection in modern scholarship between race and Palaeolithic stone tools. Race is not linked to geology, nor is geology often connected with questions of human sterility or inter-racial marriages. For the Victorians this was not the case. All of these were aspects of the same big question. It was not possible to parse one aspect and examine it without it impacting on all the rest.

For the Victorians the question was this. Why was human physical appearance so varied, and why did some groups (races in their terms) appear to be *always* more advanced than others? For many this question highlighted an even more important one. Was the apparent superiority of the European white race a given natural law that would always be guaranteed? Today science has answered that question from a variety of different standpoints, not the least of them genetic studies. All humans are the same, and the physical differences between us are very recent in evolutionary terms. No one group of people can use physical and mental evolution to claim superiority over anyone else. Since DNA and palaeontology show we are all brothers and sisters, and relative newcomers into the world as a species, the physical differences between us should be causes for wonder and delight, something to be celebrated.

But for the Victorians of course this was different. The more I researched eoliths and the background to that debate, the more I felt I needed to understand this huge question of the interconnectedness of the parts in Victorian origins debates. The book I ended up writing was very different from the one I started out to write. I was genuinely surprised at many of the directions that the research took me.

My themes for this book then are as follows:

- apart from interconnectivity itself, I look at the development of Palaeolithic archaeology, its relationship with the study of human physical anthropology in Britain and, to a much lesser extent, on the Continent
- the links between these and the study of race and racial origins
- the question of human origins itself
- the link with geological developments in climate and glacial studies
- the public's perception of the whole origins question
- the public's relationship with race
- how the public got its information on origins related questions and in what form this was served up to them

I end up looking at the opening phase of the eolith debate (1889–1895/1896) as a logical extension of developments in a number of these areas. Victorian science fiction, or at least some of it, discussed at the end of the book, is another aspect of this.

Before going on, it is important to make clear a number of questions concerning Prehistoric chronology, particularly in the relationship between how the Victorians understood Prehistoric time and our appreciation of it today.

## Chronology: Then and Now

It would seem self-evident that the Victorian understanding of the world's geological history was not as detailed or as nuanced as ours is today. Having said that, much of the basic sequence had been worked out by 1859 and still underpins contemporary understanding of the different geological periods in the history of the Earth. The left hand side of Table 1.1 presents the geological periods and their subdivisions as we understand them now. It also shows the dating, in years, for each of these sub-divisions. These dates have been established by a number of different lines of evidence, including radiometric dating techniques applied to the rocks themselves, and the relationship those rocks have to changes in the direction of the Earth's magnetic field (palaeomagnetism). Such precision would have astonished Victorian scientists, as would the duration of Earth's history.

Although there have been name changes, the debt that the modern sequence owes to Victorian (and earlier) science is clear. In their understanding of the post-Cambrian period, three phases of broad geological time could be discerned; these were the Primary, Secondary, and Tertiary periods. Originally these were defined on the stratigraphic super-positioning of certain types of rock (Rudwick, 1992), but by the middle of the nineteenth century they had also become associated with different groups of fossils and so with the

*Table 1.1. Chrono-stratigraphic table of the Earth's geological history. The left hand side shows the modern interpretation. The two sections to the right of this show Victorian interpretations. That on the far right represents a generalised middle and late Victorian outlook.*

| Modern geological timescale | | | | | Lyell 1863 | | Generalised late Victorian stratigraphic column | |
|---|---|---|---|---|---|---|---|---|
| Eon | Era | Period | Epoch | Begins millions of years ago | Roughly equivalent to modern era | Roughly equivalent to modern epoch | Roughly equivalent to modern era | Roughly equivalent to modern epoch |
| Phanerozoic | Cenozoic | Quaternary | Holocene | 0.011 | Post Tertiary | Recent | | Recent |
| | | | Pleistocene | 2.0/1.75 | | Post-Pliocene | | Pleistocene |
| | | Neogene | Pliocene | 5 | Tertiary | Newer and Older Pliocene | Tertiary | Pliocene |
| | | | Miocene | 24 | | | | |
| | | Palaeogene | Oligocene | 34 | | Upper and Lower Miocene | | Miocene |
| | | | Eocene | 55 | | Upper, Middle, Lower Eocene | | Eocene |
| | | | Palaeocene | 65 | | | | |
| | Mesozoic | | Cretaceous | 142 | Undifferentiated Secondary and Primary stratigraphic divisions | Divisions of the Cretaceous, Jurassic and lower strata. | Secondary and Primary stratigraphic divisions | Further divisions of lower strata |
| | | | Jurassic | 206 | | | | |
| | | | Triassic | 248 | | | | |
| | Palaeozoic | | Permian | 290 | | | | |
| | | | Carboniferous | 354 | | | | |
| | | | Devonian | 417 | | | | |
| | | | Silurian | 443 | | | | |
| | | | Ordovician | 495 | | | | |
| | | | Cambrian | 545 | | | | |
| Pre-Cambrian Proterozoic | | | | 2500 | | | | |
| Pre-Cambrian Archaean | | | | 4570 | | | | |

progress of life up the stratigraphic column. There was no overall agreement, but the following characterisation would at least have been recognizable to many students of the time. The Primary (today's Cambrian to Permian, although not all these sub-periods were recognised by the Victorians) was the first phase of geological history in which the Earth was formed, cooled, and a surface crust finally developed; in time this was covered by oceans and by land masses. It was a common belief that the earth had begun as a molten sphere in space and was cooling from the outer surface inwards toward the core.

The various subdivisions of the Primary saw the developments of plants, molluscs, and forests; there were trilobites and fish in the oceans. For some Victorians there were more complicated land animals in the later sub-divisions of the Primary. The Secondary period (very roughly the equivalent of the modern term Mesozoic, although there has been some shifting of the Primary/Secondary boundary in terms of which subdivisions fall within which era in the modern sequence) was the period of the great terrestrial and marine reptiles. This was the age of the dinosaurs. The Tertiary saw the emergence of the mammals, though by the 1830s it was clear that a few smaller mammal species were present in some of the sub-divisions of the Secondary. Later investigators added a Quaternary at the top of the sequence to accommodate the modern fauna and humans. It was Charles Lyell (see Chapter 2) who re-named the subdivisions of the Tertiary as Eocene, Miocene, and Pliocene, based on groupings of snail species (Lyell, 1863a; Gould, 1987). He also coined the term Pleistocene, but applied it to his Newer Pliocene sub-division. Other researchers subsequently applied it to his post-Pliocene period, and the designation eventually stuck. This is also the modern usage of the label Pleistocene. Much of this book will be about the archaeology and geology of this Pleistocene period.

The middle and right hand columns of Table 1.1 show two variations on the Victorian understanding of the sequence. In the middle is Charles Lyell's version from the first edition of the *Antiquity of Man*, the popular name for his 1863 work which is discussed in more detail in later chapters. On the right hand side of the figure is a generic stratigraphic sub-division. Its generalised sub-divisions would have been recognisable to most Victorian scientists.

The modern conception of the Pleistocene as a geological period is depicted in Figure 1.1. It is often labelled as the Ice Age. We now know that the period is characterised by a long succession of glacials (even numbers in shaded blocks) and warmer interglacials (odd numbers in unshaded blocks), and it is against this background of cyclic climatic fluctuation that much of early human history has been played out. These alternating periods of cold and warm climate have been established from analysis of faunal and floral remains in marine deep sea cores, as well as ice cores drilled through the Greenland and Antarctic ice sheets. The information on the right hand side of Figure 1.1 maps the archaeological record of Palaeolithic human occupation of Britain against the geological record of Pleistocene climate

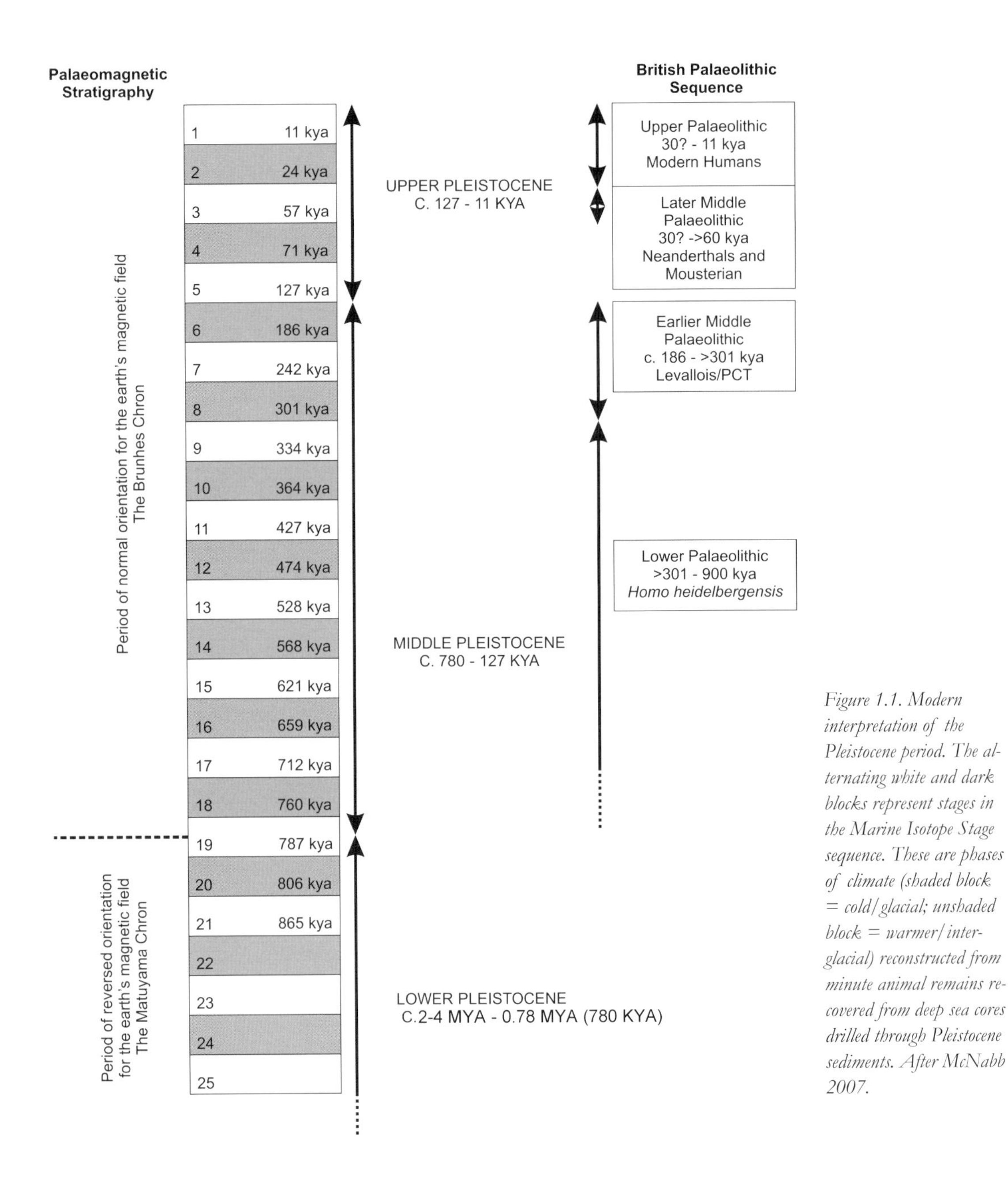


*Figure 1.1. Modern interpretation of the Pleistocene period. The alternating white and dark blocks represent stages in the Marine Isotope Stage sequence. These are phases of climate (shaded block = cold/glacial; unshaded block = warmer/interglacial) reconstructed from minute animal remains recovered from deep sea cores drilled through Pleistocene sediments. After McNabb 2007.*

change. The Palaeolithic (Old Stone Age) is confined to the Pleistocene. It begins, at least in terms of the British sequence, with the earliest hominin occupation of Britain about 900 kya (kya=thousands of years ago). Hominin is a label that includes our ancestors but also incorporates related extinct fossil genera (such as Australopithecines) all of whom evolved after the split with the common hominin–ape ancestor. We do not know which hominin species were the earliest visitors to our shores. By 500 kya Britain was being occupied by a species known as *Homo heidelbergensis*. This earliest phase of the Palaeolithic, the Lower Palaeolithic, ends about 300 kya, and is replaced by the first half of the Middle Palaeolithic. This first part of the Middle Palaeolithic sees the introduction of new stone tools and new ways of making them – the Levallois/prepared core technology. There then follows a long phase in which Britain is abandoned by hominins, possibly lasting more than 125 kya. Sometime before 60 kya the next hominin species to arrive in Britain, the Neanderthals, initiate the second half of the Middle Palaeolithic. At about 30 kya modern humans arrive carrying their Upper Palaeolithic culture. This phase ends about 11 kya. Marine Isotope Stage 1, beginning about this time is the equivalent of our own interglacial, and therefore equates with the Holocene in Table 1.1.

Of course this sequence was not known to the Victorians. As we shall see in the following chapters they (mostly) presumed the handaxes of the drift, what modern geologists would call Pleistocene deposits, were older than the stone tools found in the caves. They were able to further subdivide the different layers within the caves as well, but did not make the association between Middle Palaeolithic/Mousterian tools and the Neanderthals; or the conection between the later Upper Palaeolithic tools and modern humans; to be fair they were very close to it by 1901.

Within our own interglacial, the later Prehistoric period (as I shall define it here) sees the Mesolithic (Middle Stone Age), Neolithic (New Stone Age), and Bronze and Iron Ages, as shown in Figure 1.2. The existence of the Mesolithic only began to gain acceptance toward the end of the Victorian period. The Neolithic with its ground stone axes, domesticated animals, cereals, sedentary life style and ceramics, was therefore the earliest occupation of Britain as far as the early Victorians were concerned. The momentous year of 1859 (see Chapter 2) revealed otherwise. As this book will show, the history of human origins research in Britain was a slowly growing awareness of the Pleistocene as a complex geological and archaeological period. For the Victorians it began with the recognition that the drift, the surface geology of sands, gravels, silts and clays, contained evidence of ancient humans and extinct animals living in the same ice age world.

It is important to understand that geologically, as well as archaeologically, there were very few moments of widespread consensus. This extended to subdividing the Pleistocene as a period. At the beginning of the middle Victorian era (as I have defined it here, 1859/1860–1880), there was a widespread belief that there was only a single glacial phase during the Pleistocene. There was much discussion as to when within the Pleistocene it was, and how long it lasted. Two contrasting views are presented in Table 1.2. On the left hand side are the views of William Boyd Dawkins drawn from his book *Cave Hunting* in 1874. This is an example of one viewpoint on Pleistocene subdivision, in this case based on the presence of suites of mammals preserved in various sedimentary formations. This is discussed in more detail in later chapters. On

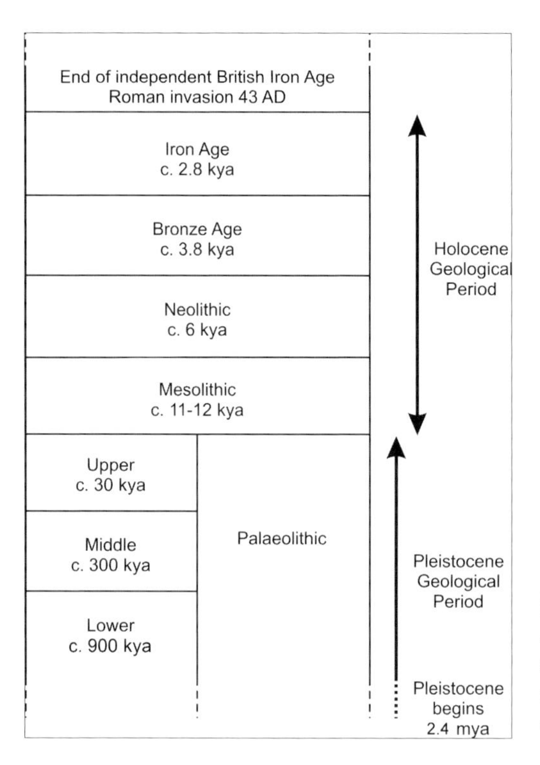


*Figure 1.2. The various sub-divisions of the British Prehistoric period as understood today. After McNabb 2007.*

| Boyd Dawkins 1874 | | A generic view of the Pleistocene in the middle and late Victorian periods |
|---|---|---|
| later Pleistocene | Post-glacial Pleistocene | A post glacial later phase of the Pleistocene |
| | Glacial phase of the Pleistocene | |
| middle Pleistocene | Long pre-glacial phase | A middle glacial phase of the Pleistocene |
| earlier Pleistocene | | An earlier pre-glacial phase of the Pleistocene |

*Table 1.2. Sub-divisions of the Pleistocene, called the drift period by the Victorians. On the left is that supported by William Boyd Dawkins. On the right a generalised middle and late Victorian perspective.*

the right hand side of the table is another suggestion for the way a number of students of this period may have generalised the process of sub-dividing the Pleistocene in the late Victorian period (here taken as 1880–1901). In both of these schemes, Victorian researchers would have placed the Palaeolithic (drift and cave) in the post-glacial phase. Much of the debate examined in this book reflects changing views on these geological and archaeological sub-divisions, in particular whether or not humans were post-glacial in age, or earlier.

*Figure 1.3. Development of ideas and approaches to studying human origins and related disciplines across the Victorian period. Based on Bowler 2003 with additions.*

## The Structure of the Book and the Big Picture

In this section I will present my overall conclusions for the book. These represent original views on my part generated from the order and structure through which I have presented the material, as well as my own interpretation of particular events. This will familiarise the reader with the main themes and how I believe them to be interwoven with each other. However, at the beginning of each chapter, I describe in more detail my conclusions for the topics developed within that chapter. As both an introduction and a description of what I believe are the main lessons to be drawn from the material discussed, I hope this will allow readers to evaluate my conclusions for themselves, and hopefully this will help them to formulate their own opinions on the significance of the data.

The argument I will develop in this book is as follows. It should be read in conjunction with Figure 1.3 which shows the broad development of various research themes over the middle and late Victorian era.

1859 was a momentous year, particularly for Palaeolithic archaeology, as it changed the whole course of debate on the origin of our species. Up until 1859

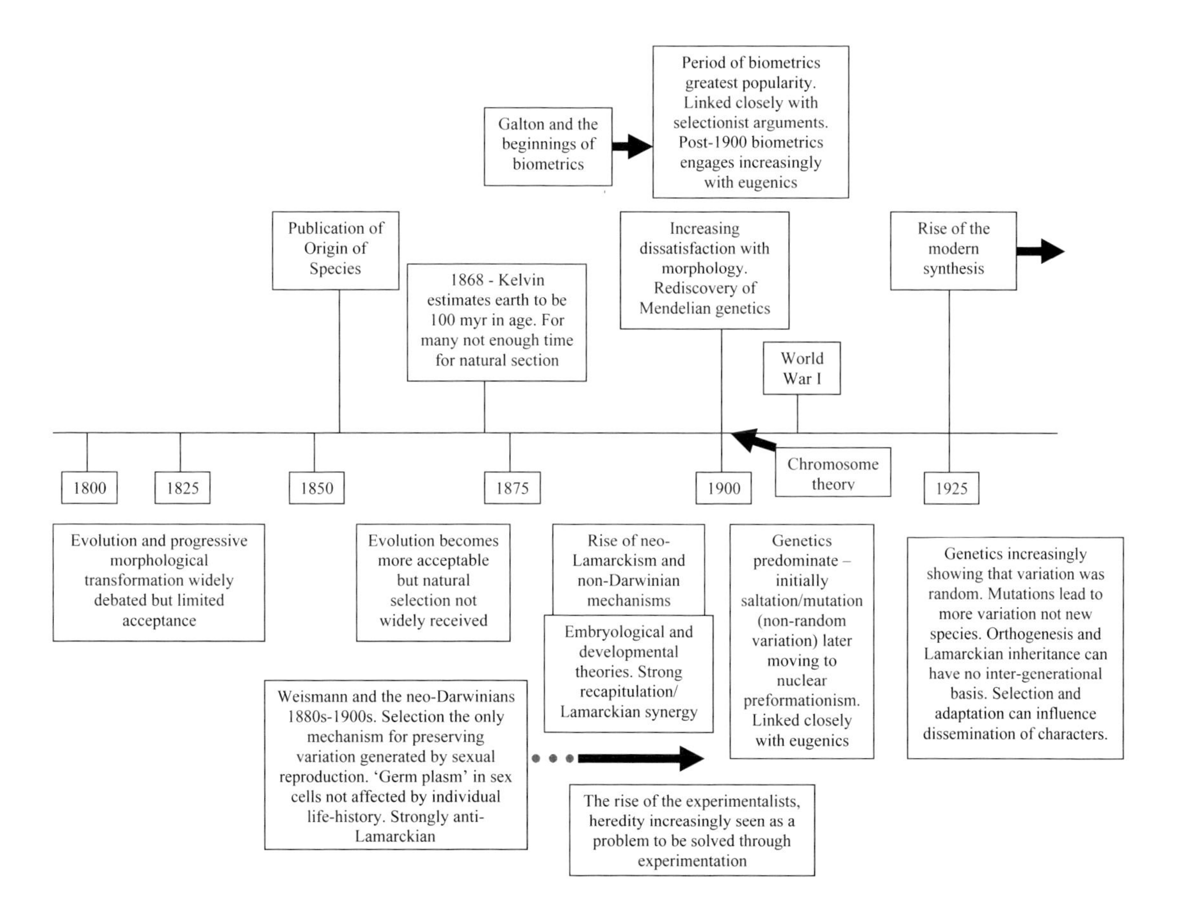

there was no widespread concept of deep time in relation to human antiquity; Prehistory began with the age of the polished stone axes, what today we call the Neolithic. This was the Stone Period for the early Victorians. Estimates on its age varied. The only real discussion on human origins in Britain prior to 1859 was the controversy surrounding the origin of the various human races as the Victorians saw them. Slavery had been the engine that drove this, and debate about the relative 'worth' of the non-white races continued even after 1859. The intellectual scaffolding necessary to conceive of human origins in a different way did not really exist (with the exception of a few far sighted individuals) prior to 1859. But the *annus mirabilis* changed all that. Darwin's work indirectly suggested a natural origin for humanity. If humans were a part of the natural world, like every other animal species, then our present condition was a result of descent with modification over long ages of time. Effectively, Darwin's work predicted a series of human ancestors reflecting differing developmental stages in evolutionary growth. Each stage would have to reflect a world to which that ancestor was successfully adapted and particularly suited.

The Antiquity of Man debate of 1859 revealed the physical reality of a previously unsuspected world. This was the Pleistocene. Although the drift, as it was called, was long known about, the fact that it was inhabited by ancient humans was not really suspected until the middle of the nineteenth century. Geology showed this world to be very ancient, the crucible from which putative human ancestors could have emerged. Others thought this would be earlier still. Palaeontology filled that world with extinct animal species or earlier forms of living ones. Archaeology peopled it by revealing the tools of its human inhabitants, even though they themselves remained elusive. The crudity and simplicity of these Palaeolithic tools suggested the stage of development these humans were at. So, if the pre-1859 debate on racial origins stood for an origins debate because there was no other context within which to set such a dialogue, then Darwin and the Antiquity of Man debate set the question of human origins squarely within the natural world.

Slowly, across the following decades, a true human origins debate began to emerge as racial origins became less and less relevant to physically understanding the human past. This is not to say the question of racial origins simply faded away, it didn't. It was kept alive across the decade and beyond for a number of reasons. These were partly political, and partly the broader Victorian interest in race itself.

After the first flush of excitement, the geological societies who had fronted the Antiquity of Man debate and explored the reality of the newly revealed Pleistocene lost world, returned to debating other matters. Most geologists did not consider the surface deposits and their contents to be real geology.

The two main societies who would have had a key role in exploring human ancestry in the 1860s, the Anthropological Society of London and the Ethnological Society of London, were the focus of much political manoeuvring between different interest groups. In particular the group of rising young scientists who marshalled themselves under the banner of evolutionism, sought to gain greater political control of science through institutions like the Ethnological Society and the British Association for the Advancement of Science.

One unintended consequence of embedding human origins within the older polemics of the Anthropological and Ethnological societies was that although the Antiquity of Man debate began as a geological and archaeological question, in the decade following 1859 it became an anthropological one. This accompanied a shift away from descriptive accounts of stone tools found in ice age river gravels, to a more interpretative approach which tried to anthropomorphise the 'Palaeolithic period' as it was becoming known. The framework for this more theoretical phase was an emerging discipline now labelled 'Evolutionary Anthropology'. 1859 paved the way for this emergence; in a sense 1859 rendered evolutionary anthropology inevitable. The broad evolutionary perspective of progressive time (McNabb, 1996) unintentionally structured ways of thinking that crept into most of the natural sciences. It created synergies between different disciplines, particularly those that incorporated the element of time. Whereas Darwin merely suggested that passing time resulted in more physical variation in populations, evolutionary anthropology explicitly stated that biological evolution was paralleled by a series of incrementally progressive cultural stages, each built on its predecessor, and each characterised by unique signatures in material culture and social institutions. Evolutionary anthropology explicitly linked time with progress, and in so doing allied it to a common misconception about evolution, namely that evolutionary change implied improvement. This was something Darwin had never said. As a discipline, evolutionary anthropology located and contextualised the different periods of Prehistory, placing them in a relative developmental order. It set the Palaeolithic at the very beginning of this sequence.

The anthropologists' use of ethnological parallels to reflect the different archaeological periods automatically lent Prehistory a recognizable and human face. However it was not successful in doing this for the Palaeolithic of the drift. This was too old and too remote. There were no viable analogues here. This partly explains why the Victorians thought of the makers of the handaxes as essentially modern human beings. They didn't really know how else to visualise them. This is a key point. The handaxes and the levels of dexterity required to make them did not suggest some brutish half formed ape-like creature. Quite the opposite, but there was nothing on which to base a visualisation (and hence reconstruction/interpretation) of such a remote ancestor. I use the term human in the text in place of the more technically accurate modern term hominin, in order to reflect this Victorian perception.

But evolutionary anthropology carried its own interpretative difficulties. It couldn't readily explain how and why cultures evolved, merely that they did. This combined with disputes between the learned societies, and a lack of human fossils from the British Palaeolithic, served to hold British Palaeolithic archaeology in something of a stasis in the late 1860s and early 1870s. By comparison the European record was very rich. Belgian and French caves were revealing a whole series of what appeared to be different Palaeolithic cultures, associated with distinct suites of faunal remains that allowed for their placement in a relative chronological sequence of development. A chronology for the Palaeolithic was emerging, and human remains were being increasingly discovered in these deposits. In turn this allowed for the tentative recognition of different Palaeolithic human races. This was the beginning of a racial anthropology for the Palaeolithic that European scholars would attempt to link with the later Prehistoric races thought to be present in the Neolithic burial mounds and Bronze Age tumuli.

Extensive search in British caves and open air sites failed to discover comparable sequences. There was no shortage of British later Prehistoric remains, so British race scientists tended to focus their attentions on this. A gulf between British and Continental Palaeolithic research began to open. This may well have been expressed in a stronger emphasis in Britain on the development of more theoretical perspectives (philosophical as the Victorians would have called it), and evolutionary anthropology may be seen as a reflection of this. Again, it is possible that the lack of British Pleistocene skeletal material led to the emphasis on world ethnology prevalent in the journals of the ethnologists and anthropologists in the 1860s – although this is also accounted for by the interests of administering a global empire.

The 1870s represents a pivotal time in which many things changed. Aspects of human origins research already in flux across the 1860s settled into new directions. With the Darwinians in power in the new Anthropological Institute, human origins finally shed the mantle of racial origins (although individual papers on this topic and its relationship with human origins continued to appear). But the prominence of evolutionary anthropology also began to diminish as the interests of its main advocates were diverted elsewhere. One of them, E.B. Tylor, became more concerned with kinship, mythology, and what would now be considered as straightforward social anthropology. Although he maintained his evolutionary credentials, after the formation of the Folklore Society in 1878 his interests gravitated away from questions immediately relevant to human antiquity. The other main proponent of an evolutionary anthropological approach was John Lubbock. His political career began in earnest during this decade. He was Liberal MP for Maidstone in 1870 and 1874. He maintained a life-long interest in human origins research, but he was increasingly concerned with social reform and natural history. These two men had been at the forefront of the anthropological interpretation of Prehistory. The vacuum that began to open behind Lubbock was filled by John Evans, a hero of the 1859 Antiquity of Man debate.

While Evans engaged with ethnological parallels on occasion, his strongly descriptive and non-interpretative style of archaeology did not lend itself to the broader contextualisation and interpretative stances that evolutionary anthropology offered. With no other theoretical framework available within which to discuss human origins, and Evans unwilling to do so, Palaeolithic archaeology returned to the more geologically orientated descriptive subject it had been between 1859 and 1863. Whether or not this served to encourage the research questions that dominated the 1870s, or *vice versa*, is a moot point. Evans himself became Britain's most senior Prehistoric archaeologist. His prominence afforded his work the status of 'an establishment position' (although to be fair he was unlikely to have been aware of this).

One of the signals of change in the 1870s was the rise of new geological models for the Pleistocene. Alongside Evans, Joseph Prestwich was the other senior player in the Antiquity of Man debate and his interpretation of a post-glacial Pleistocene date for the Palaeolithic (Table 1.2) came as close to being an establishment viewpoint as anything could have done, particularly where chronology was concerned. But it was enmeshed within a particular view of the geological history of the glacial phase of the Pleistocene, that of submergence of large areas of Britain under the glacial ocean. A new glacial/terrestrial interpretation was being increasingly promoted whose developing polemic over the 1870s and 1880s began to offer more credible explanations for the evidence of flint tools in what appeared to be glacial-aged Pleistocene deposits. These were predicated on new views of Pleistocene climate, which offered ages in real years for climatic events. These new ideas frame-worked the potential for discussing inter-glacial and even pre-glacial human occupation in Britain, which the post-glacial hypothesis simply precluded. Inter-glacial and pre-glacial occupation became a key topic whose solution was geological and not anthropological. These debates preface the eolith controversy of the late 1880s and 1890s which were a direct extension of them.

There can be little doubt that the reputation of Joseph Prestwich contributed to the success of the eoliths as an enduring dialogue. In fact it is unlikely that eoliths would have had much of a hearing without his support. The British academic environment was very different to that in Germany, and especially France. Abroad, there was strong institutional and government support for anthropology. The ever growing collection of skeletons from the Reindeer Age caves, and the strong Continental emphasis on probing for long racial lineages lent the subject a contemporary relevance. Continental scholars were not slow in linking race to the political difficulties of European international politics. Race was seen as a motivator in national behaviour. Britain, with its small skeletal database and inability to move back beyond the Neolithic could only look on in parochial envy. There was a real need for an indigenous English origins debate. Most of the work of the British race scientists in the 1870s and 1880s went into analysing current racial health; Francis Galton for example, worried over the dilution of the British as a people.

So part of the success of the eolith controversy as a debate, was that it fulfilled the role of a British origins debate. Prestwich advocated a Palaeolithic date for the eoliths; they were the earliest and first stage in the evolution of stone tools. But this stage was still Pleistocene, probably dating to the early Pleistocene/pre-glacial phase – Table 1.2. A number of his collaborators, like Benjamin Harrison, believed the eoliths were much older, Tertiary in age. The debate raised huge questions. Where had these earliest tool makers come from, and when did they arrive in Britain? Were they evolving in Britain? I characterise this first phase of the eolith debate (the second phase moved to East Anglia in the post-Victorian period) as a 'Second Antiquity of Man Debate' (McNabb, 2009).

But even the British eolith question had a parochial feel about it. The debate was initiated through three papers; two delivered to the Geological Society and one to the Anthropological Institute. It was rooted in Prestwich's geological interpretations of the North Downs and Kentish Chalk Escarpment. But Prestwich's geology was out of date. He was in effect writing up his unpublished notes on sites and sections from previous decades and not taking into account new work. At this time he was suggesting that the Kentish landscape was a result of sculpting by glacial ice. However, no direct evidence for this had ever been discovered in Kent. He was also positing a new glacial episode which post-dated the main glacial Pleistocene (mooted in the second eolith paper). The evidence for this was very controversial. He was even suggesting a major phase of European oceanic submergence in between the end of the Palaeolithic and the beginning of the Neolithic. He would suggest that this was a possible origin for Biblical flood myths. If it hadn't been for Prestwich's prestige, it is to be wondered how far much of this would have been taken. As it was, the eolith debate soon became mired in a lack of new evidence and the entrenched position of the debaters. Old questions could not be answered and new approaches and lines of evidence were not forthcoming. After Prestwich's death in 1896 the debate fossilised.

But eoliths were being discovered everywhere, and interest in them reached an international level. At the heart of this was Benjamin Harrison. He was a central node in a correspondence web which reached out, even to the British dominions overseas. He exemplifies the interconnectedness of Victorian science, in this case spanning the divide between professional and amateur (although the latter was not a label that any Victorian would have recognised). The eoliths even prompted a minor revival in the flagging fortunes of evolutionary anthropology and the use of ethnological parallels. For once, direct connections could be drawn, based on the character of the eoliths. E.B. Tylor revisited earlier work of his which highlighted (as he saw it) the very primitive nature of Tasmanian society. He suggested that the tools these people had made were very similar in concept and limited design to the eoliths of Kent. Conceptually, the eoliths sat at the very root of all evolutionary development in material culture. Archaeologists began to develop long multi-period sequences of typological evolution with eoliths at the base of the tree.

In so doing they highlighted the still poor British Palaeolithic skeletal record. Stone tool development implied that continuity between the Palaeolithic and later periods should be discernable. But where was the skeletal evidence? The discovery of *Pithecanthropus* in Java in the middle 1890s provided a potential maker of the eoliths. The creature was primitive enough for some researchers, but not primitive enough for others. Did such a creature roam the English Downs? Or had the eolith makers evolved from something like Java man on their long journey into the West?

Throughout the book, but particularly after Chapter 5, I have noted the many ways the general public were kept aware of the various arguments concerning human origins. I show that the general public had unprecedented access to detailed knowledge and polemic as scientists fought out their differences of opinion in various journals and periodicals which had wide public readership. The various societies were open to all, and their meetings were another venue for practising scientists and other interested parties to exchange views. Museums ran public lecture series, as did the various education colleges and institutions. The British Association for the Advancement of Science met annually to debate the cutting edge research of the day. All of this meant that there was a widespread availability of information, which I suspect has not been equalled until the advent of the modern internet. Human origins, and the various debates surrounding it, were very much in the public's eye.

I conclude the book with two chapters on science fiction in the late Victorian period. At first this may seem odd, even though I have focused on those works which had a human origins or Palaeolithic theme. Firstly, this allows me to continue developing the concept of public access to information on human evolution. Secondly, it provides me with a vehicle through which to engage in some of the debates of the time in far greater detail than was possible in earlier chapters. Finally it allows me to introduce the question of reconstructing the Palaeolithic, something that has been conspicuous by its absence up to this point.

An objection to all of this this could be that fictional narrative does not reflect scientific interpretation. True, but then I have chosen H.G. Wells as the theme for Chapter 12 as his work did reflect scientifically informed debate. Others who used fiction as a vehicle to engage with science and its presentation to the public, such as Grant Allen, would also have been suitable. But of course not every fiction writer would have been as concerned as these two were with scientific accuracy. These writers are discussed in Chapter 13. Here my point is simple. These story tellers were not really bothered with scientific accuracy, they drew on common understandings (folk psychology) of human antiquity. They gave the public what it expected, and at the same time reinforced that expectation. Effectively, fiction writers were providing the interpretations of what it was like to live in the Palaeolithic, while the scientists who described it, were mostly unwilling or unable to do so. Reconstructions were especially popular in France where a rich tradition of prehistoric fiction grew up around an equally rich Palaeolithic data base. To some extent, the reconstructions of Palaeolithic lifestyles disseminated via folk psychology substituted for evolutionary anthropology after the 1860s, and fictional writing, sparse enough in Britain, fed and reinforced its vision. I suspect this fed back to the scientists and archaeologists, who were in turn influenced by the fiction writers and the public's perception of the Palaeolithic.

Perhaps this is not so surprising at the end of the Victorian period. The evolutionary paradigm continued to exert a powerful influence on how ideas and data were structured. But evolutionary anthropology, once having described something, and allocated its position in the evolutionary sequence, could do little else with it. Anthropology itself had moved to a more social based anthropology in which societies and their institutions became the focus of study. But this could not work for the Palaeolithic, because the data with which to do this was just not there. I get the impression that Palaeolithic archaeologists were stuck with a theoretical framework which seemed relevant, but wasn't particularly helpful – it described but didn't really explain. Not surprising then that it was the fiction writers who seemed to be providing the public with an anthropology of the Palaeolithic.

# 2

## 1859
## *A Brave New World*

In this chapter I will argue that debating human origins in the middle Victorian era (1859/1860–1880) emerged out of much older debates on slavery and on the origins, and relative worth (as then perceived) of the different human races. Race here meant skin colour and where you were born. Prior to 1859 there was no real context in which to embed a genuine human origins debate. Darwin's *Origin of Species* simultaneously provided both a naturalistic background for the emergence of a single human race from successively older and morphologically simpler ancestors, and a sense of the huge timescale necessary for this to happen. The Antiquity of Man debate occurred a few months prior to the publication of the *Origin*. A small group of archaeologists and geologists argued for the existence of a previously unrecognised period in human history revealed through the combination of geology, palaeontology and archaeology. This lost world was immensely old as the crude nature of the humanly made stone tools proved. As such it fitted the requirement of an older and more basic stage of evolutionary development that Darwin's theories required. In the years immediately after 1859, English geologists and archaeologists worked hard to consolidate their position, continuing to rely on the same principle of first hand field work that had proved so effective in opening up the 'vistas of time'.

### 1859

For most people the year 1859 will always belong to the biologists and zoologists.

Charles Darwin's (1809–1882) *On the Origin of Species by Means of Natural Selection, or the Preservation of Favoured Races in the Struggle for Life* was published in November of that year. It was an instant best seller; and the first print run of 1,250 copies sold out almost straight away (Owen, 2008). The core of the book was a theory which explained how new species of plants and animals could develop from pre-existing ones. It was published at a time when most practitioners of science were clergymen or devout Christians who believed the proper way to study the natural sciences was to study nature itself. Within the grandeur and orderliness of nature could be found the proof of God's existence and His presence in the world. This was Natural Theology and for

*Charles Darwin 1809 - 1882*

*Thomas Henry Huxley 1825 - 1895*

*Sir Joseph Dalton Hooker 1817 - 1911*

*Alfred Russel Wallace 1823 - 1913*

many the date of creation was 4004 BC as the Archbishop of Armagh, James Usher (1581–1656) had calculated.

But November 1859 was not the first time Darwin's theory had been unveiled. The idea had received its first public airing in July of the previous year at a meeting of the Linnean Society. This was a joint reading of Darwin's work with that of Alfred Russel Wallace (1823–1913). Wallace was a Welshman born near Usk and a gifted observer of nature. He had hit upon almost the same idea as Darwin whilst island hopping in South East Asia, collecting specimens of wildlife to be purchased by museums and wealthy European collectors (Slotten, 2004). He wrote to Darwin to tell him of his idea and get his opinion. Darwin had always been afraid of the social consequences of publishing his ideas on 'descent with modification'. But now he was afraid of losing that honour. He wrote to friends in a state of panic fearing that Wallace was going to be credited with his discoveries. The celebrated geologist Charles Lyell (1797–1875) and the botanist Joseph Hooker (1817–1911), arranged a joint reading of Darwin and Wallace's papers at the Linnean Society on 1st July 1858. By the middle of the nineteenth century Charles Lyell had become one of the most celebrated geologists of his day with an academic reputation to be reckoned with. Hooker was an old friend and confidant of Darwin's. He would eventually succeed his father as director of Kew Gardens in 1865. In effect these two men bullied the Linnean Society into allowing Darwin and Wallace space on the meeting's agenda by removing several scheduled presenters. Darwin, who thought that his paper would be read as an appendix to Wallace's, intended to write to him renouncing his claim on natural selection (but see Allen 1885 for a different version of events, a hagiographic account aimed at legend building). He abandoned this when he found Lyell and Hooker had arranged a joint and equal reading. Darwin's allies may have deliberately structured the two presentations in such a way as to ensure that Darwin got priority in the honours of who was the first to settle the 'species question' (Desmond, 1998; Desmond and Moore, 1991). They included, in Darwin's material, dated letters proving that he had formulated his views years before. Thomas Henry Huxley (1825–1895) carefully vetted the documents. Darwin need not have worried. By 1857 most of the details of natural and sexual selection, his two great mechanisms explaining change in organisms, had already been worked out. In fact the basis of natural selection had been formulated back in the late 1830s and had been described in two manuscripts in the mid-1840s (Oldroyd, 1983).

Neither scientist was present at the Linnean Society meeting. Wallace was still abroad, and Darwin was mourning the death of his eighteen month old son Charles Waring whose funeral took place on the same day. The papers elicited very little response from the audience at the Linnean. Perhaps Lyell's presence, or the careful stage managing discouraged response (Desmond, 1998). Alternatively, most of the audience may simply not have understood it all. But at least one person did. Ross Slotten, Wallace's biographer, suggests that one speaker who had intended to give a paper on the fixity of species now realised that he would have to abandon his long held views (Slotten, 2004, 156.). This incident also shows that right from the beginning, Darwin was hedged in by a protective band of younger researchers who were willing to fight his battles for him. The clannishness of the Darwinians will appear time and again.

After the joint reading, Darwin went to work to rewrite a huge manuscript he had been working on since the mid-1850s, documenting the proofs of his theory. His original plan had been to publish this as a full exposition of his ideas. Significantly, there had been a whole chapter on humans. But this was now cut out (Desmond and Moore, 2009). He condensed the huge manuscript into a shorter book, what we know today as *The Origin of Species*.

For the Victorians, Darwin's first contribution to the natural sciences was this; he was the man who persuaded the scientific establishment and the general public that evolution was true and that it was a viable explanation for life on earth. The idea of evolution, and even physical modification in animal species, transmutation, was not a new thing (Desmond, 1982; Oldroyd, 1983; Blanckaert, 1988). There had already been a number of attempts by scholars to use evolutionary ideas as explanations for the changing patterns of life seen in the fossil record. Grant Allen (1848–1899), the popular science writer detailed this for the Victorian public in his book describing and contextualising Darwin's life and ideas (Allen, 1885). Allen also described other researchers who had come very close to understanding descent with modification; a Dr Wells in 1813 and the philosopher Herbert Spencer (1820–1903) in the early 1850s among others. But they had lacked the mechanism of natural selection to fully perceive the implications of their ideas. The abolitionist and monogenist James Cowles Prichard (1786–1848; see below) argued in the 1830s and 1840s for a type of transmutation with the effects of climate influencing species. He even believed the earliest human races were black, and that whites were their descendants (Livingstone, 2008). Darwin's own grandfather, Erasmus (1731–1802), had even written poetry and prose with an evolutionary theme (Gillham, 2001) .

Other researchers argued for diachronic change from a teleological perspective. As early as the mid-1830s the Catholic theologian and later cardinal, N.P. Wiseman (1802–1865), argued for divinely inspired monogenism with the different races resulting from natural variation which later became fixed (Livingstone, 2008). At the same time Dean William Buckland (1784–1856) had come to believe that God had initiated a series of distinct and separate creations, each terminated by its own deluge (Sommer, 2007). Some species reappeared in succeeding creations, others were new. This explained the apparent increase in complexity seen in the fossil record which was preserved in the geological strata. Buckland came to distinguish between these successive creations, and that described in *Genesis*. The Biblical creation occurred long after those fossilised in the rocks. The various creations preserved in the geological record had been completed with the advent of the present world. So the Noarchian deluge, in Buckland's view, had been a minor event of more local significance that occurred long after geological history had ended. Richard Owen (1804–1892), the great comparative anatomist, and arguably the most senior palaeontologist of the mid-1850s, was by the end of that decade also promoting a succession of divinely preconceived natural creations. The basic morphology of each living animal was a blueprint (an archetype), and evolutionary change was manifested in progressive modifications to the blueprint. Each blueprint was a complete idea in the mind of God. Desmond notes how similar some of Owen's palaeontological theory building was to the Darwinian notion of the mutability of species (Desmond, 1982).

So the idea of deep time and progressive morphological change in plants and animals was already a part of the Victorian mind set by the time Darwin came to write the *Origin*. Darwin and Owen were both drawing ideas from their times as each sought a mechanism, the one natural, the other teleological, to explain species' mutability.

Darwin's success at convincing Victorian society of the reality of evolution was attributed to his second major contribution to natural science. This was his demonstration of the way in which evolution worked – what he termed descent with modification – again a much debated process in early and middle Victorian times. By showing how effective modern plant and animal breeders were at producing radically new forms of plants and animals from parent stock, he demonstrated that the physical morphology of an organism was not fixed; it could be changed in small ways by careful selective breeding – artificial selection. Different varieties of pigeon, or dog breeds were excellent examples of this artificial selection, carefully manipulated by modern breeders over many generations. In the natural world, living species were also the end result of long trajectories of small incremental changes that had accumulated over time. In effect a modern species was a resting moment in an historical lineage, a family tree of changing descent. Darwin showed how the evidence for these lineages was clearly present in the fossil record of plants and animals. He proved this in the classic Victorian way by meticulously documenting example after example. The process of artificial selection was an analogue for natural selection, and the geological record provided the time depth for the whole concept of descent with modification.

When he died in 1882 these two contributions were enough to earn Charles Darwin a place in Westminster Abbey (with some behind the scenes manoeuvring from T.H. Huxley, John Lubbock (1834–1913), and other friends – see Desmond, 1998). But his third contribution was not so widely accepted. This was the mechanism by which descent with modification, adaptation in modern terms, actually worked – the process of natural selection. If the pigeon or dog breeder was the guiding hand in artificial selection, then this role was taken on by natural selection in the real world. As a theory it was deeply unpopular from the very beginning.

Essentially Darwin argued that all organisms had a tendency to vary in relation to each other. The variation existed between members of the same species, as well as between different species. These variations were not necessarily very great, but they did mean that some individuals had a slight advantage over their neighbours. Some plants may have been slightly taller than others of their kind, or had broader leaves, and so were able to gather more energy than the plant next door; or perhaps they just grew more quickly. So even though two specimens of the same plant species grew next to each other, one might have an advantage over the other. If a third plant grew next to these two, but this time of a different species, its unique set of adaptations may have fitted it to out-compete the first two in making the most of the resources in that small area. Alternatively the first two plants may have been generally more successful at survival than the third, and it suffered as a consequence. The principle applied to all organisms. Whatever the differences, this tendency of organisms to vary gave some individuals a head start in the struggle for existence. In modern terms this would be a positive selective advantage. Darwin argued that variation was a result of environmental influence, and could be passed on to succeeding generations (Oldroyd, 1983).

The political economist and clergyman Thomas Malthus (1766–1834) had argued that there was a universal law at work in all human populations. Population will always grow, and growth will always outstrip resources, in this case the food supply. Population increased in geometric ratio (2, 4, 8, 16 etc.) while the means of subsistence production could only expand in arithmetic ratio (2, 4, 6, 8, etc.; Oldroyd, 1983). The result would be competition for limited resources, and those who were less fitted for success would perish. This was a natural check on human overpopulation. Darwin transferred this to the natural world (Darwin, 1958) and argued that all organisms were engaged in a struggle for life and access to the resources that enabled life. So competitive struggle, whether it was inter-personal, inter-population, or inter-species, was the engine that drove evolution through variation. Those organisms with a slight advantage over their fellows lived long enough to reproduce and to pass on their particular advantages into the next generation more often. Reading Malthus had also been the key for Wallace's independent discovery of natural selection (Darwin, F., 1958, 200). Those less favoured did not reproduce so often or reproduce so successfully, and their adaptations were gradually lost. Over time, relentless competition marshalled advantageous variations together, slowly modifying the bodies and behaviours of a population of organisms to track those conditions that made individuals adaptively successful. With enough time the accumulated changes would modify the members of a population which was separated from its parent stock so much that their physical form would no longer resemble the original population and so a new species would emerge.

*Textbox 2.1. Brief explanations of a number of terms and concepts that appear throughout this book, in particular those relevant to inheritance.*

This was natural selection. It explained in a naturalistic way how new species could develop from pre-existing ones. But many Victorians found this uncomfortable. Many saw a relentless brutality in this vision which was unsettling. For others it was the random character of natural selection that they disliked. Most Victorians believed that change over time had to be progressive. Despite occasional stalls and setbacks (e.g. atavism or stasis) things did evolve and improve, which they saw as the unfolding of a natural law. Moreover, it played to their cultural expectations. But Darwin did not argue that everything had an inherent tendency to progress. He only argued that things had a natural tendency to vary. What is more, he argued that the creation of variability was itself a product of chance. The precise mechanism by which variability occurred, in other words inheritance, was unknown and this rendered his view of the benefits of serendipitous variation unsatisfactory. An animal could not predict what would be useful or when some feature or behaviour would be advantageous, it was just luck. This was too much for many Victorians, it challenged too many deeply ingrained beliefs; Darwin's natural selection was just too messy. Darwin struggled with inheritance and the explanation for variability throughout his career but never satisfactorily explained it (Oldroyd, 1983)

Opposition to the contingent nature of natural selection was fierce and sustained (Bowler, 2003). By the end of the Victorian era very few naturalists accepted this aspect of Darwin's work. Many returned to older forms of selection like Lamarckism to explain heredity, see Textbox 2.1. Even Darwin himself came to doubt the all-powerful efficacy of natural selection. In the first edition of the *Origin* in 1859 he cited natural selection as the major force shaping the emergence of new species with Lamarckism playing a much less significant role. By the sixth and final edition published in 1872, he had down-scaled the importance of natural selection, which was now augmented by other selective mechanisms.

**Theories of Inheritance**

**Lamarckism.** A theory developed by Jean-Baptiste Lamarck variously known as the theory of acquired inherited characteristics, or use inheritance. Constant use of a biological feature or structure in an organism strengthened it and could permanently change it. The external environment therefore had a strong influence on the individual and on inheritance itself, since these changes, so long as they were present in both parents, could be passed on to the next generation. A common explanation of the theory concerned giraffes. They liked to eat the sweeter leaves on the tops of trees. They would stretch their necks to reach up. In time the necks of individuals would become longer and this change would be passed on to their descendants who would repeat the same process. Gradually, as a species, giraffes acquired their long necks. By Darwin's time there was considerable concern with this – whatever process or mechanism allowed phenotypic modification to occur had still not been identified.

**Darwinian Inheritance.** This was the opposite of Lamarck. Although Darwin did admit to a certain degree of use inheritance (more so in later years), and certainly did believe the environment affected the individual, in Darwinian selection the environment had no direct controlling function – it was a filter constantly at work in the background. Populations of organisms were characterised by variability between individuals. This variability arose spontaneously as a result of sexual reproduction. Some of it helped certain organisms to succeed in the competition for resources (1st edition of the *Origin*), or reproduce more effectively and more frequently than others (last few editions of *Origin*). The result was a change in the structure of the individual and the population over long periods of time. Given enough time new species could arise as the aggregation of minor changes had a cumulative effect on the organisms. Commentators were fond of saying that in a Darwinian world giraffes ate the leaves at the tops of the trees just because they could.

**Pangenesis.** A theory first proposed by Charles Darwin in *The Variation of Plants and Animals Under Domestication* (1868) and further refined in *The Descent of Man* (1871). Various structures in the body produced gemmules which were minute particles containing hereditary information about their parent structures. These were released into the body and distributed throughout, eventually aggregating in the germ cells (sex cells) and were then transmitted to offspring. This was a partial explanation for why certain characteristics were persistent across generations. Gemmules or 'undeveloped atoms' as Darwin sometimes called them built copies of the structures that produced them by cell growth and division. Gemmules were transmitted to both sexes but relied on their association with other adjacent cells, as well as factors like timing in the growth cycle of the organism, before they could be switched on. Some gemmules lay dormant during an individual's lifetime, while others could be dormant for a number of generations before becoming active. This explained why some people had features (e.g. nose or jaw line) resembling a grandparent rather than a parent.

**Blending Inheritance.** A widespread belief in the mid-nineteenth century that an offspring will always be an even mix of its parents' characteristics. Folk psychology (untested common sense explanations generated from every day observations) seemed to support it; the children of mixed race parents, black and white for example, appeared to be neither wholly white nor wholly black, rather a mixture of the two.

**August Weismann.** A biologist; his ideas evolved considerably over the late Victorian decades. The following is my interpretation of his developing ideas. Cells could be divided into *soma* or somatic cells which were of the body, and germ plasm which were the reproductive cells. Sexual reproduction involved the passing of germ plasm from both parents to their offspring. The embryo grew from a single cell. Its nucleus contained a combination of germ plasm from both parents. The embryo developed via cell division as the germ plasm controlled different parts of the developing foetus. However, some of the germ plasm was not transferred into the growing embryo in the initial stages of mitosis. It remained in the nucleus of what would eventually become the bodies sex/germ plasm cells. Additionally, a small proportion of the germ plasm from each parent was also retained in the germ plasm cells. It remained unchanged and did not combine. Each new generation's germ plasm therefore contained unchanged portions of its parents, grandparents and great grandparents' germ plasm and so on. Different types of cell division in the egg and embryo prevented a proliferation of ancient germ plasm from dominating the parental material. This was part of the reason for variation in organisms that reproduced sexually and provided the variability for natural selection to work on. Contrary to popular belief Weismann did believe that the external environment could affect germ plasm, but only at a 'molecular level'. These were tiny mutations, which could themselves be passed on to the next generation. This too was a source of phenotypic variation. Weismann was an ardent selectionist. A key element was that the somatic body cells could not contribute any material back to the germ plasm once the initial cell division had begun to form the embryo. It was effectively sealed off from the body and its various components were passed on whole-sale to the next generation. For this reason Weismann denied the existence of use-inheritance. Weismann was a key figure in demonstrating that Lamarckism did not work.

**Herbert Spencer.** A polymath and founding figure in the development of sociology. Spencer was a committed evolutionist who attempted to link physical phenomena of all types under the umbrella of progressive development from the simple to the complex. He was an ardent Lamarckist, firmly convinced that use inheritance governed the development of all organic systems and explained the rise of variation. While not denying natural selection, he focused on Lamarckism as the primary cause of change. An important feature of his later writings and his belief in Lamarckism, rested on the 'correlation of parts'. Organisms are an amalgamation of mutually co-operating structures. One structure cannot be changed without an effect on the others. Darwin had originally argued this. Spencer believed that a Darwinian view would have required each structure to vary independently of the others, and that natural selection would then act on each structure separately, changing them without recourse to the rest. This would preclude the 'harmonious co-operation of the parts' and thus preclude any developmental change in the organism toward increasing complexity. The only solution was Lamarckism where a change in the utility of one structure would force an equally positive change in all the others in order to maintain stability.

**Panmixia.** A theory promoted by August Weismann concerning degeneration of physical structures in organisms. The importance of this was its relevance to Weismann and Spencer's debate in the *Contemporary Review* in the 1890s. Many researchers understood natural selection to mean that every structure in an organism played a vital part in survival and conferred some sort of advantage to the organism. If that was the case, then how did Darwinians explain vestigial structures, or those structures which were clearly retrogressing? How could natural selection support a negative character trait (whether actively declining or simply in stasis)? The enemies of Darwin argued it couldn't, the only thing that could explain the persistence of such a feature was Lamarckian inheritance. Disuse resulted in the feature being ignored while energy was diverted to the replacement character. Weismann thought he could explain this from a selectionist perspective with panmixia. So long as a structure fulfilled a useful role it was maintained by natural selection. Those variants that were not up to the job were weeded out. Once the structure was no longer useful, selection would not bother to weed out the poorer examples. Selection would simply ignore a redundant physical structure. Continuous breeding would do the rest. Individuals in the population with high quality examples of the structure could breed with individuals with poorer quality examples, and slowly the overall quality of that feature in the population would diminish since natural selection was not maintaining its fitness. The result was what would appear to be degeneration in that structure.

**Francis Galton**. Like Huxley, Galton was a saltationist at heart. He formulated a law of ancestral inheritance which asserted that you inherited half your characteristics from your parents, and the remainder was made up of contributions from grandparents and great-grandparents going back in ever decreasing amounts. Consequently he denied any influence from the external environment. He also formulated a law of regression in which he demonstrated, statistically, that individuals born of normal parents, but who were distinctive in some way, would have children and grandchildren who resembled the average of the population. Thus any distinctive or unusual traits (high intelligence for example) would be bred out of the population. He called it his law of regression. For this reason he did not believe natural selection was such a powerful force because after the first generation there would be little for it to work on. Regression did not create variation. Species were stable entities unlike in the Darwinian view where they were temporary pauses in a continuum of developing variability. Real evolutionary change came via big leaps (sports) when individuals were born who were so different from their parents and grandparents that the law of regression simply did not apply.

**Saltation**. A belief that evolutionary change within a species reconstructed from minute animal remains sensitive to ocean temperature, would happen quickly in a single evolutionary jump. T.H. Huxley was inclined to saltation. Lamarckism was a popular mechanism for some saltationists after 1859. Toward the end of the 1880s some geneticists were arguing that each inherited characteristic was controlled by its own genetic particle. A change in these could effect a major change in the organism's offspring. This provided a genetic basis of saltation to work. Hugo de Vries (1848–1935) was arguing that new species arose in a single generation with new forms arising out of pre-existing populations because of changes in the hereditary particles controlling particular physical characteristics. This mutation theory was a brand of saltation, but one that did not allow for considerable environmental input.

In the 1890s the biometricians (generating statistical data from organisms to represent population norms) like W.F.R. Weldon and Karl Pearson, both disciples of Galton, fought an acrimonious debate with saltationists like W. Bateson. The former believed metrical data indicated populations could be described by the normal distribution, i.e. continuous variation. Species change was by a shift in the population mean over time, this could only be done by incremental change – on continuously generated variability in the population – a selectionist approach. Bateson argued for saltation and in the mid-1890s was losing ground to biometrics. But the rediscovery of Mendelian genetics, and experimental work like that of Hugo de Vries, seemed to confirm a saltation/mutation explanation for species formation. After the slow acceptance of Mendel's work , the biometricians found their interests more engaged with eugenics.

## Other Theories Relevant to the Subjects under Discussion

**Uniformitarianism**. A theory proposed by Charles Lyell. All of the process at work in sculpting the surface of the Earth today, are the same as those that operated in the past. There are none at work now that did not exist in the past, and *visa versa.* The physical effects of erosion and deposition can be witnessed today, and these same signatures are also present in rocks and ancient sediments. So the present can be explained by references to forces at work in the past. This will apply to the future also. In addition to this all the physical process which shape the Earth today proceed at a slow and steady (uniform rate) which was also the same in the past.

## Human Origins and the *Origin of Species*

> Inheriting the assumption of the underlying unity of humankind from Greek and Hebrew thought, Western culture has developed in recurrent confrontation with 'others' stereotyped as Asiatic despots, barbarian invaders, Moslem infidels, American savages, African slaves, and unconverted heathen. (Stocking, 1988, 3)

Famously, the *Origin of Species* contained only one line which specifically referred to humans, 'Light will be thrown on the origin of man and his history', although in the large manuscript that Darwin had been preparing in the mid-1850s, he had planned a whole chapter on the 'Races of Man'. Stocking (1988) provides an excellent introduction to the Victorian and post-Victorian study of race and anthropology.

But if Darwin didn't want to discuss human origins, his public did. Desmond and Moore (2009) convincingly show that for many of his readers the *Origin* would have been a book about humans. This worked on two levels. To begin with many of his readers would have automatically assumed that a theory which placed the creation of new species within the natural world would apply to humans as well. The second level concerned those readers engaged in the heated Victorian debate about the origin of the different human 'races'. For them this would have been a book primarily about racial origins.

Not surprisingly there were a number of different definitions and interpretations of what represented a species and what represented a race. Here I will follow the definition adopted by Darwin and a number of his contemporaries, albeit in a simple form. All members of a single species should be able to breed with each other, and importantly, produce viable offspring – in other words fertile offspring whose fertility would remain undiminished over time. Today this is known as the biological species concept. However, not every organism within the same species had to look exactly alike. It was possible to get races, or varieties as some scholars termed them, that looked different from each other. So long as they were able to produce fertile offspring then these different races of plants or animals – or humans – were still part of the same overall species. Conversely, failure to produce persistently fertile offspring meant that individuals actually belonged to distinct species.

For adherents of this perspective inter-racial mixing simply gave rise to new races, not new species. Darwin's belief in ancestral lineages, and their relationship to ever more simple and older common ancestors acted as a metaphor for his belief in the commonality of all humans – that different human races were all part of a single variable species. As with plants and other animals, all living peoples, irrespective of skin colour or racial difference, could trace descent back to a common stock. This belief was underpinned by his passionate hatred of slavery, and his family's long held support for abolitionism.

The racial questions of the Late Victorian era that would become entwined with the questions of human origins, were rooted in the anti-slavery campaigns of the early Victorian period and before (Desmond and Moore, 2009; Stepan, 1982; Stocking, 1987). In 1807 the slave trade was abolished in Britain and throughout her colonies. The right to own slaves, and so the institution of slavery itself, was only made illegal in British dominions in 1834, and full emancipation did not occur until 1837–1838. The abolitionist movement had fought a long and bitter war against slave owners and the propertied classes in Britain. These latter had seen slaves as a justifiable element within the infra-structure of national prosperity and wealth creation, namely their own. The early anti-slavery lobby had centred around a number of key figures, but by the 1830s one individual had come to dominate, James Cowles Prichard. Between 1836 and 1847 his book *Researches into the Physical History of Man* ran through several editions, each larger than the last. (Hodgkin, 1850; Stocking, 1973). Prichard had argued from an ethical and religious belief that all humans had stemmed from one original divine creation. In his view human racial diversification had occurred long after the creation and as a result of migration and linguistic separation.

The Bible only described one creation. Those who were a part of it could be guaranteed to have a soul. For Prichard and the anti-slavery lobby all humans were a part of this Biblical creation. Missionary enterprise was all about saving souls, rooted in the belief that human nature was mutable, it could be changed and improved. In opposition were the pro-slavery campaigners. They argued that humans were the product of a number of separate origin events. For those pro-slavery advocates of a religious outlook, these were separate creations with the white races being the Creator's intended – the descendants of Adam and Eve. As the non-white races were not mentioned in *Genesis* they were not really human, and so did not possess souls. They had been placed on the Earth, like any other of the beasts of the field, for the white man's benefit. Non-whites could be owned and traded as livestock.

But not all who sympathised with slavery accepted a divine creation. For those who believed that humans were a product of nature, the various races of man had arisen separately in different geographical homelands. Progress had favoured the white races because of their inherent superiority. The pre-eminence of European civilization proved this and so afforded whites the licence to enslave others.

For both of these pro-slavery camps there was little point to any missionary work, and no sense in the abolitionist cause, because the non-white races could never be improved. Their nature was fixed and unchangeable. Thus the early debates on race were between the unitarists (anti-slavery, usually religious, single creation, mutable nature) and the pluralists (pro-slavery, sometimes religious, separate creations, immutable nature).

Here I will suggest that the question of racial origins in the early Victorian era stood as a proxy for a human origins debate because the climate of the times was simply not ready to express it in any other way. In the early Victorian period and before, human origins and racial origins were essentially the same thing. They were an ethnological problem (where, who, why and when) more than a biological problem (how).

So from the very start, questions of human origins and racial beginnings were linked, politically charged, and cut to the quick of what new races were, and how they were formed. A parallel origins debate, on the existence and significance of pre-Adamitic races, also reflected this political dimension. This dialogue tracked the shifts in perspective occasioned by the *Origin of Species*. It is described in detail by Livingstone (2008).

For many students of this period, miscegenation, the crossing of the 'blood lines', or what we today we would call racial intermixing, was widely believed to produce either racial hybrids who were sterile, or children who were increasingly so. Many of the pro-slavery lobby were desperate to believe this because hybridism was argued by some of their enemies to be the way new species were formed. A child born of mixed race parents took on many of the physical features and appearances of both parents, but their skin colour would be lighter than one parent and darker than the other. This was visible proof for Victorians that heredity was a blending of the features of both parents – see Textbox 2.1. The desire to believe in the sterility of such children is what Desmond and Moore (2009) call 'planter propaganda' – the untested assertions and folk prejudices of the pro-slavery plantation owners and their allies. The sterility of hybrids would mean that the fixed nature of separate races was left intact. Long term racial intermixing could not endanger the purity of the white race by creating new species.

What made the timing of the *Origin's* publication in 1859 all the more pertinent was its appearance on the eve of the American Civil War (1861–1865). This was a war largely fought around issues of race. With slavery abolished in Britain and its dominions, the Confederate States of America had become the power house for developing intellectual arguments in defence of slavery. These were rooted in anthropological research, designed to show that the non-white races were not actually races, but different and inferior species. Desmond and Moore's (2009) argument makes it clear why Charles Darwin spent so much of the *Origin* on hybrids, and on varieties of pigeons and dogs. He was using artificial selection to address issues of common descent. In the 1840s, pluralists were using dog breeds and pigeons to examine all manner of racial issues including the fertility of hybrids. In concentrating on these in the *Origin*, Darwin was attacking the pluralists on their own ground. As a unitarist he was keen to explore the issue of whether the offspring of mixed racial mating could be fertile. His answer was unambiguously yes. Most of the planter propaganda was unsubstantiated hearsay. Humans of different skin colour, and from different parts of the world, could successfully interbreed and their children and grandchildren could do the same. In the *Origin* races, like different species, were the result of descent from common ancestors through long ages of accumulated incremental changes. What Darwin didn't say, but his work implied, was that all living humans were part of a single species. The different races of humans found around the globe were just varieties of one single human stock. This was the conclusion of his great unfinished work from the middle 1850s. Its substance appeared in the *Origin*, but not overtly framed in terms of human evolution.

But by the late 1850s the unitarist philosophy had a hackneyed feel to it (Stocking, 1971; Stocking, 1982). The pluralists by contrast appeared as the torch bearers of a new and vibrant scientific approach. Some were even linking the different races of humans to (alleged) discoveries of human ancestors – though few of these turned out to be likely candidates. Two American pluralists from the Confederate South, Josiah Nott (1804–1873) and George Gliddon (1809–1857) in their 1857 *Indigenous Races of the Earth* attempted to tie the extant races of humans to different geographical centres of development. Each race had its own long developmental history and ancestral fossil, located in its own unique homeland. Gliddon also defined a new terminology. The unitarists became the monogenists, who were deliberately painted as unscientific, their views tainted with the mark of antiquated religious dogma and unfashionable abolitionism. The pluralists were now re-branded as the polygenists (Desmond and Moore, 2009, 288). They were dynamic and promoted a very modern view of themselves. They used analytical and scientific techniques to back up their claims. They used phrenology, they quantified skull shape (craniology), they objectified skin and eye colour, and hair type, all in an attempt to show that human difference was persistent and deep rooted. These and other approaches supported the image of polygenists as modern researchers in the new science of anthropology.

The timing of the *Origin*'s publication was then doubly important. At a time when the polygenists were gaining ground, Darwin's unambiguously monogenist text struck a powerful note for the commonality of all humans, and in so doing set the agenda for the coming human origins debate.

## 1859: The First 'Antiquity of Man' Debate

But the year 1859 also belongs to Palaeolithic archaeologists.

It was in this momentous year that the first steps toward widespread acceptance of a long ancestry for the human species were made. This was the year that began the 'Antiquity of Man' debate. The story is well known and only a brief outline is necessary here (Spencer, 1990; van Ripper, 1993). Kent's

*J.B.de C. de Perthes 1788 - 1868*

*Hugh Falconer 1808 - 1865*

*Sir Joseph Prestwich 1812 - 1896*

*Sir John Evans 1823 - 1908*

Cavern, in Torquay, Devon, was a famous archaeological cave site (White and Pettitt, 2009) and it had been excavated a number of times during the earlier part of the nineteenth century. There were bones of extinct animals associated with ancient stone implements. However there was always the suspicion that the deposits were disturbed and the bones and stone tools had become accidentally mixed (a common accusation levelled at cave evidence before 1859). Kent's Cavern had been one of the best hopes for those few scholars who, prior to 1859, believed in a long ancestry for the human species.

In May 1858 Hugh Falconer (1808–1865); a palaeontologist with a long standing interest in human antiquity (Anon, 1866–1867); wrote to the Geological Society of London announcing the discovery of a new and undisturbed cave found at Windmill Hill above the small fishing town of Brixham near Torquay, Devon. As a specialist in fossil fauna, and conscious of the problems of association that plagued previously discovered caves, he suggested this was an ideal opportunity to explore the cave systematically. Money was raised and excavations undertaken by the meticulous William Pengelly (1812–1894) who had previously excavated at Kent's Cavern (Anon, 1894a; Dawkins, 1896). A committee of the great and good in Victorian geology oversaw the operations, including Charles Lyell, Richard Owen, and the geologist Joseph Prestwich (1812–1896 – soon to become one of the key figures in the Antiquity of Man debate). Falconer set out the programme of work (Prestwich, 1899). The impact of the excavations is clear in the letters that passed backwards and forwards between Falconer and Prestwich (Prestwich, 1899 *ibid*). For Prestwich it was the systematic nature of observations made within a precise methodological framework that provided growing certainty of the genuine relationship between the bones of extinct animals, and the associated stone tools. The deposits were sealed by an unbroken layer of stalagmite. They had to be very ancient. This was rigorous fieldwork and it was beginning to show the way forward. Even so, Joseph Prestwich still needed more evidence to be fully convinced that humans had been contemporary with the extinct fossil fauna of the Pleistocene period.

Hugh Falconer travelled to Sicily in November of 1858, accompanied by his niece Grace Milne (1832–1899, who was later to marry Prestwich after Falconer died in 1865; Harrison, 1928; Mather and Campbell, 2007). Across the Channel, at Abbeville, the French researcher Boucher de Perthes (1788–1868) had been collecting and publishing his discoveries of stone tools in the gravels of the river Somme for many years. His claims for tools found undisturbed in deep gravels, often with the bones of extinct mammals, were not accepted by most French *savants*. However his fame was enough to induce Falconer and his niece to stop and examine his collection at his Abbeville home. Falconer was convinced that many of the stone tools were genuine. Before continuing to Italy he wrote to Joseph Prestwich (Hicks, 1897; Woodward, 1893; Evans, 1897b) urging him to come and see de Perthes' evidence for himself. Prestwich was at this time a business man involved in the family trade as a wine merchant, but he already had a formidable reputation as a geologist (Prestwich, 1899; Pope and Roberts, 2009).

Prestwich went to Abbeville along with his old friend John Evans (1823–1908), in April of 1859. Evans was also in business. He had joined his uncle's paper making firm at Nash Mills near Hemel Hempstead aged 16 (A.G., 1908; MacGregor, 2008), working his way up to control the business after 1883. By the time of the visit to Abbeville, Evans was an antiquarian, with an interest in ancient British coinage rather than Prehistory. At this time Evans was not even aware that the type of stone tool they would encounter in Abbeville (handaxes, sometimes called implements or palaeoliths by the Victorians) had even been found in Britain (Gamble and Kruszynski, 2009). Figure 2.1 shows a selection of handaxes. Prestwich and Evans saw de Perthes' collection and were indeed convinced that many of the pieces were humanly made tools. More importantly, a handaxe was found *in situ* in gravel deposits at a quarry in near-by St Acheul (Gamble and Kruszynski, 2009; Gamble and Moutsiou, 2011). Prestwich and Evans went to see it dug out of the gravel section. The moment was even photographed. The axe was more than eleven feet below the surface of the ground, and the overlying gravels were clearly undisturbed. This was the same level at which bones of extinct animals had been found by de Perthes. In this case personal observation and field work carried the day. Up to this point both Prestwich and Evans had been sceptical of the evidence for the great antiquity of the human race.

*Figure 2.1. Handaxes from Abbeville and St Acheul discovered or bought from quarry men by English archaeologists, notably Prestwich, in the 1860s. Handaxes were more usually called implements by the middle and late Victorian scholars of human antiquity. Sometimes they were also called palaeoliths. A key distinguishing feature of their being made by hominins was the evidence of the flake scars. This showed they had been knapped. The scars can be seen clearly on the handaxes depicted. They appear as shallow depressions marked by a clear margin and originating from the implement's edge An example is present on the central pointed implement, in the lower right hand quadrant. Radiating patterns of concentric ripple marks within the bed of the flake scar could be traced to the point of impact where the knapper struck the flake off from the implements edge. When covered by such flake scars it was evident that the handaxes were not the work of nature.*

Prestwich and Evans's observations at Abbeville had not been made on the stalagmite-sealed layers of caves. Their conclusions were drawn from the study of open gravel faces in quarries, or in other words at open-air sites. This represented a second and independent line of evidence supporting the cave evidence for the antiquity of humans. The sheer depth of *undisturbed* gravel, with historical remains on top (a Roman cemetery), and the presence of extinct mammals at the same depth as the palaeoliths in the gravel, implied a huge antiquity, the sense of which was often lacking in the cave deposits. On the 26th of May after returning home Prestwich gave a paper on their findings to the Royal Society, and on 2nd June Evans gave a paper to the Society of Antiquaries on the archaeological implications of their discoveries. Prestwich (but not Evans on this occasion) returned to northern France in June with more British geological colleagues for a second look at the Abbeville and St Acheul gravels (Prestwich, 1899, 124). The handaxe found by J.W. Flower (1807–1873) on that occasion has the distinction of being the first *in situ* implement to be discovered in the Somme gravels by an English geologist in person (Flower, 1860). I will return to this below.

Charles Lyell also visited Abbeville. He was a strongly devout man and his antipathy to the concept of the antiquity of the human race was considerable. Yet even he was swayed by the evidence of his own eyes – once more the power of fieldwork and personal observation proved overwhelming. In September 1859 Lyell reported to the British Association for the Advancement of Science (BAAS) meeting at Aberdeen that there had indeed been 'Men Among the Mammoths' to use a phrase coined by T.H. Huxley (van Ripper, 1993). There can be little doubt that the authority of a man like Lyell, a man able to successfully influence the committee of the Linnean Society, contributed much to the acceptance of human antiquity. Many Victorians had a genuinely open mind on the subject as it had been discussed for a number of years prior to 1859 (Desmond, 1982). This was particularly so for those scientifically minded individuals with no strong religious preconceptions. All that had been lacking was demonstrable proof.

There is one thing that is important to keep in mind when thinking about this first Antiquity of Man debate. The main issue was whether or not humanity was old, geologically old, and how to prove it. This question was best solved by careful fieldwork. It could only be proved by finding handaxes, as these were evidence of ancient human craftsmanship. But they had to be found with the bones of extinct mammals as well, because these were evidence of a time long antedating the modern world. The present world was the equivalent of the world described in *Genesis* and only extant animal species were described there. So, extinct fauna clearly belonged to a pre-Biblical earth. But the extinct fauna had to be discovered *in undisturbed contexts,* and that is exactly what Abbeville and Brixham provided; moreover, the English geological archaeologists saw this with their own eyes.

The next obvious question raised by the Abbeville evidence, just *how* old was the human species, was a question for another day. However that day would not be long in coming.

The consequences of 1859 for Joseph Prestwich and John Evans were considerable. Prestwich already had a strong reputation in the geological community, but 1859 made him famous. He retired from the wine trade in 1872, and in 1874 he became Professor of Geology at Oxford University (at the age of 62). John Evans, now committed to the archaeology of early humans, began to research stone tools and their geological context in a more systematic manner, while continuing his other archaeological interests. His position in the Victorian hall of fame was cemented in 1872 with the publication of *The Ancient Stone Implements, Weapons, and Ornaments, of Great Britain.* This was a monumental survey of almost all the known Neolithic and Palaeolithic stone working from Britain that had been discovered up to that time (Evans, 1872). This book came to be viewed by prehistorians as something akin in its authority to the Bible. Like Prestwich, Evans gained the reputation of a meticulous researcher who would only accept clearly proven facts. His obituary in the *Proceedings of the Royal Society* (A.G., 1908), noted

Handaxes acquired by Joseph Prestwich, or given as gifts from the first few visits to Abbeville and St Acheul in 1859 that *Ancient Stone Implements* helped to forge the link between Palaeolithic archaeology and geology by showing how the chrono-stratigraphic record facilitated the division of stone tools into discreet 'varieties' (a term redolent with Darwinian implications) of workmanship. In other words his work led to thinking of different types of stone tool in the same way as biologists thought about discreet animal species.

## 1859 and Onwards: The Geological Implications of Prestwich and Evans's Work

It is difficult today to understand the impact of 1859 on the Victorian mind, even though many of the essential pieces of the puzzle were already in place. Readers will be only too aware that I have not even addressed how the *Origin* and the Antiquity of Man debate affected religious sensibilities. This has been fully discussed by other authors (Desmond and Moore, 1991; Rudwick, 1992; van Ripper, 1993). I will continue to focus only on those aspects I see as relevant to the human origins question.

As a monogenist text, the *Origin* indirectly embedded human evolution within a naturalistic context. Moreover, by association with the rest of the argument, it implied a long history for the human species. This was necessary for descent with modification to work on a series of human ancestors who would have become more primitive the further back one travelled in time. What geology and archaeology provided in the Antiquity of Man debate was the world from which those primitive human ancestors could have emerged. Extinct animals and vanished landscapes clearly showed that this primitive Earth was very ancient, in John Frere's famous expression 'beyond that of the present world' (Frere, 1800).

The Antiquity of Man debate was not only the proof that this ancient world had existed, but that it could be accessed by Victorian empirical science. It was a world they could get to know. In this section I will describe Prestwich and Evan's concept of what this world was like, because their interpretation of the evidence exercised a powerful influence over the succeeding generation's views.

Prestwich gave his lecture on the Abbeville evidence on May 26th 1859, and it was published the following year (Prestwich, 1860; Prestwich, 1859–1860). He followed it up with two more papers in March and June of 1862 (Prestwich, 1864; Prestwich, 1862–1863) and a third with some further observations (Prestwich, 1863–1864). The importance of these papers is that taken together they either influenced understanding of the Pleistocene period for the next quarter of a century, or alternatively, they provided a framework for younger geologists to kick against as the decades passed. In addition they provided the foundation for John Evans's archaeological understanding of the Palaeolithic, one which he maintained until he died.

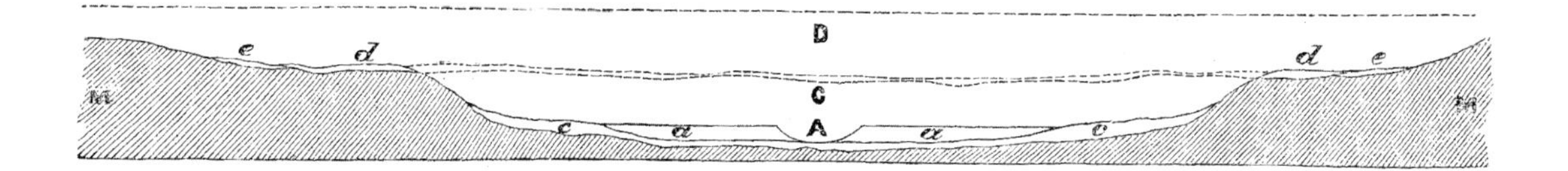


*Figure 2.2. Joseph Prestwich's interpretation of the evolution of a river valley. See text for explanation.* Proceedings of the Royal Society *1862–1863.*

These writings asserted that the gravel deposits of Abbeville and elsewhere were the work of rivers. That the legacy of this originally controversial interpretation was a long one I will show when I look at the earlier phases of the eolith debate in the 1890s (Chapters 9 and 10).

Figure 2.2 gives Prestwich's overall summary of the evolution of major river valleys, drawing on his own and others' research on the rivers of southern England and France. On the base of a broad river valley (D), a river deposited a variety of fluvial residues such as sands and gravels (e and d). Of particular importance was that labelled d. These were Prestwich's high level river gravels or high level valley drifts. This part of the valley represented the higher slopes and often upland areas within the river's broad basin. When the river flowed at this level this had been its flood plain and the river had eroded and deposited gravel at various points along its length. These deposits contained both the bones of extinct mammals as well as flint implements. The pits at Moulin Quignon (see below) and St Acheul were in these high level drift deposits, and there was an emphasis on pointed forms of handaxe (see Figure 2.5). The next phase in the valley's formation was erosive down cutting in one or more stages resulting in the narrower valley C on Figure 2.2. At its base were the low level valley gravels, marked as c on the figure. These were more extensive and more continuous than the high level gravels. The low level gravels were often overlain by the modern alluvium, marked as a on the figure deposited by the modern river, A. The fact that the low level gravels lay beneath alluvium, and the high level gravels could be located on the higher slopes of the valley, or elsewhere within the river's basin, bracketed the time during which the valley could have been formed.

The alluvium was modern, dating to the Recent or Surface period as it was sometimes known. Within this sediment were the celts or polished stone axes of the Neolithic, more often called the Stone Period (see Figure 2.7). In the decade leading up to 1859 the Stone Period was the earliest known period of human existence on Earth. That an entire phase of human history predated this, the Palaeolithic of the drift, was suspected only by a few workers like Hugh Falconer. Clearly then, the formation of the valleys and their gravels had to precede the modern world.

> ...that they...[handaxes]...are found under circumstances which show that, in all probability, the race of men who fashioned them must have

> passed away long before this portion of the earth was occupied by the primitive tribes by whom the more polished forms of stone weapon were fabricated, in what we have hitherto regarded as remote antiquity. (Evans, 1860, 293f, my brackets)

On the other hand the river gravels had to post-date the glacial phase of the Pleistocene, Table 1.2. A sedimentary deposit known as boulder clay or till was characteristic of the Pleistocene ice sheet. Many Victorian geologists of the 1850s believed that glacial boulder clays were left behind when floating glaciers melted. The glacial Pleistocene (see Chapter 1) was accompanied by extensive submergence of the land. Flotillas of sediment laden icebergs choked the glacial sea. Climatic amelioration triggered a re-emergence of the land and the icebergs melted leaving their contained sediment as a mantle of boulder clay over the emerging hills and in the valley bottoms. In East Anglia, the modern topography; the river valleys and the interfluves separating them; was blanketed in boulder clay. In cutting new valleys, the rivers cut through the boulder clay on the higher ground. In reoccupying old valleys, rivers deposited their sediments on top of boulder clay. Rivers and their gravels had to be post-glacial in age.

So the whole period of valley incision and infilling, coincident with the presence of the handaxe makers and extinct mammal faunas as preserved in the drift gravels, was a post-glacial phenomenon, begun and completed by the time the Stone Period and the alluvium with its modern fauna began.

For Victorian geologists the river gravels were termed the drift. While drift encompassed fluvial gravels the term could also be applied to any sediment that sat on top of solid bedrock that had been moved into its current position by some mechanism or other. Boulder clay or any other glacial deposit could have been labelled drift (i.e. various glacial drifts), as could sands, silts, and clays from rivers (fluvial drift). One of Prestwich's major contributions to the 1859 debates was to propose the theory that the high and low valley drifts in the post-glacial valleys were all the product of fluvial deposition; the valleys were cut and infilled by rivers.

Many geologists believed Prestwich's high level gravels were either marine in origin or had been re-sorted by marine submergence during the rise of the glacial ocean, as above. Based on his observations at the gravel pits of Abbeville, St Acheul and Amiens, Prestwich demonstrated the presence of delicate freshwater shells in these upper valley gravels. This meant the gravels could not possibly be marine deposits. Moreover he argued that although the patches of the high level gravels were intermittent, they were always associated with the upper slopes of the valley, or adjacent high ground (the area of d and e on Figure 2.2), and always paralleled the line of the valley along its course. They were never found outside of the river basin. Flow directions from the high level gravels all paralleled the courses of the modern rivers in their valleys below. Additionally the high level drifts in these stretches of

*Figure 2.3. Joseph Prestwich's illustration of the high and low level valley drifts as clearly demarcated in one of the French valleys.* Proceedings of the Royal Society *1864.*

the river, many miles away from the coast, would be too high for post-glacial submergence to reach, and were often located in areas where submergence had never been proved.

The main valley (C on Figure 2.2) was a result of erosive down cutting (erosion was called denudation in Victorian times). The low level gravels were the debris from this. These were more extensive along the valley's sides, again paralleling the line of the river's modern direction of flow. There was often a clear separation between the two sets of gravels, and this was best seen in the French rivers – Figure 2.3. Each gravel train was benched into the valley side at a discreet height. The low level gravels also had delicate freshwater shells within them; Menchecourt, St Roch and other gravel quarries with handaxes in the Amiens and Abbeville area were located in low level gravels. The handaxes in these drifts were different, often more ovate in shape (see Figure 2.6). The French rivers were especially instructive in disproving marine submergence. It could be shown that areas like Amiens and even Paris, both with high and low level drifts, had not been submerged by the sea since the Miocene. Equally important was the proof that these French drifts were not a product of glacial action. They did not contain glacial erratics. The northern French rivers made a signal point in this regard as they contained lithologies *only* found within their own catchment. The French rivers were a very clear example of Prestwich's whole fluvial argument. In Britain the presence of boulder clay with its glacially travelled erratics only confused the issue.

The high and low level drifts contained other information as well. The mammal faunas from the two gravel spreads were indicative of climate. Both showed a colder climate than present; and the upper gravels appeared to be slightly colder by comparison with the lower drift. There were sedimentary

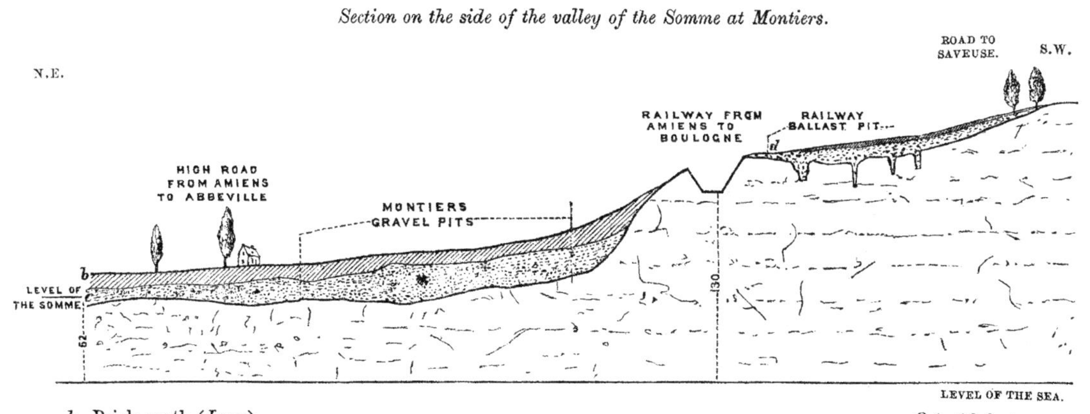


*Section on the side of the valley of the Somme at Montiers.*

*b.* Brick-earth (*Loess*) ........................................................ 3 to 12 feet

*c.* Irregularly stratified sands and gravels (low-level valley-gravel). Remains of Horse, Ox, Elephant, &c. Flint implements of the flake type not rare; one discovered at ✠ ...... 20 to 25 feet.

*d.* Rude mass of coarse gravel (high-level valley-gravel). A few fossil bones; numerous shells 10 to 12 feet.

1. Chalk. (The height at the railway cutting is only approximate.)

structures in these upper gravels, along with large boulders, that implied aggressive river flooding was responsible for depositing the upper drift. The environment of the lower drift was slightly warmer, and the floods somewhat less dynamic. Flooding was a key argument in Prestwich's vision of the Pleistocene river valleys.

By the 1850s Prestwich had rejected Charles Lyell's doctrine of uniformitarianism (see Textbox 2.1), despite it having been a cornerstone of geological thinking for decades. He noted that the modern river channels were tiny by comparison with their Pleistocene predecessors. Modern rivers were too small to carve out the huge valleys they now occupied. This could only mean the erosive power of rivers was different today – less powerful. He noted how modern rivers seemed to erode downwards forming gorges in difficult bedrocks. But the Pleistocene valleys were wide as if erosion had been more effective laterally. In addition gravels (high or low) could be discontinuous along the valley sides, sometimes present on one side of the valley but absent from the other. Only one explanation fitted these observations. He argued that torrential floods were responsible for the emplacement of most of the drift gravels. Although the rivers post-dated the glacial period of the Pleistocene, the de-glaciated final phase of the Pleistocene was still cold as the remains of woolly mammoth and woolly rhinoceros in the drifts proved. The floods were a product of rivers swollen with melt waters and melted permafrost at the beginning of the summer melts. Ice rafting explained the existence of the large boulders within both of the valley gravel deposits. Moreover, ground ice on the river's bed and especially at its banks broke the sediment up when melting occurred, further contributing to the lateral erosion of Pleistocene rivers.

Prestwich was describing what modern geologists would regard as periglacial environments and using the processes associated with them to explain river action and valley incision. He had a nice link between the archaeology and the periglacial landscape too. Why were there more pointed handaxes in the high level gravels? It had been a colder time than that of the low level gravels. There were fewer animals around to hunt. In such circumstances modern ethnologists reported indigenous peoples turned to fishing in rivers to supplement their food supply, so the pointed handaxes were ice chisels used for cutting holes in the ice. No lesser authority than Charles Darwin agreed (Darwin, 1859b).

On one issue Prestwich refused to be drawn, and this was the question of placing dates on his post-glacial/pre-modern period of valley incision. He was happy to place a relative time scale on events but not an absolute one. There simply was not enough information available as yet. Chronologically, there was a tendency for uniformitarianism to stretch different geological events apart because the processes responsible for them were slow and unvarying. In denying uniformitarianism Prestwich opened the door to the opposite view point. Dynamic flood events could erode quickly and at the same time rapidly deposit sands and gravels. In effect a world of violent floods tended to concertina various geological events together. Prestwich was fond of saying his interpretation brought the time of the extinct Pleistocene mammals closer to the present rather than making it seem ever more remote. This was not popular with many of Darwin's supporters who preferred a Lyellian view of time. Darwin once characterised Prestwich as 'too much of ...[a]...catastrophist' for his taste (Darwin, 1859a). Darwin needed time, and bags of it, to make natural selection work. Catastrophist notwithstanding, it didn't stop Darwin from penning an anxious note to Prestwich, concerned about his opinion of the *Origin*.

While Prestwich dealt with the geological interpretations, John Evans concentrated on the archaeological materials, describing the flint palaeoliths to the Society of Antiquaries on June 2nd 1959 (Evans, 1860; Evans, 1861). He provided support for Prestwich's fluvial interpretation of the gravel spreads, emphasising the presence of the delicate freshwater shells in the deposits overlying many of the gravels containing handaxes. He also noted that many of the bones and teeth of the extinct mammals were in good condition, like the handaxes themselves, so they could not have been moved by the river from any great distance. This emphasised the contemporaneity between the makers of the handaxes and the remains of mammoth and woolly rhino. Many of the handaxes were coated with a thin crust of lime-wash and much of the gravel was similarly coated, further emphasising the *in situ* character of the archaeology from the gravels.

Evans also noted that French archaeologists had conducted excavations into the gravels at Amiens (see also Lyell, 1863a) and found palaeoliths *in situ*. This was an important proof in the growing argument for human antiquity, but an element of English national pride crept in here. The French excavations were mentioned in a footnote, but full prominence in the text was given to the first recorded instance of a handaxe found *in* situ. As noted above, this was by J.W. Flower (Flower, 1860), one of the English geologists who accompanied Prestwich back to Amiens in June. He dug a handaxe out of the gravel section at St Acheul. The axe was some eighteen inches into the gravel face, and sixteen feet from the top of the section; Flower dug it out with his own hands and there had been no indication from the section face that it had been there.

Evans briefly addressed the question of human manufacture. He gave good grounds for discounting any arguments which suggested that the palaeoliths were the work of nature. Firstly there was uniformity to their shape – they were made to a design. These shapes could not be repeated by nature so precisely or so regularly. What's more, this repetition of form was found both in England and in France making the argument by design even more likely. Foreshadowing much later research methodologies, Evans had been experimentally replicating the handaxe shapes for himself (Lubbock, 1865b). Using a hard pebble mounted in a handle and used like a modern

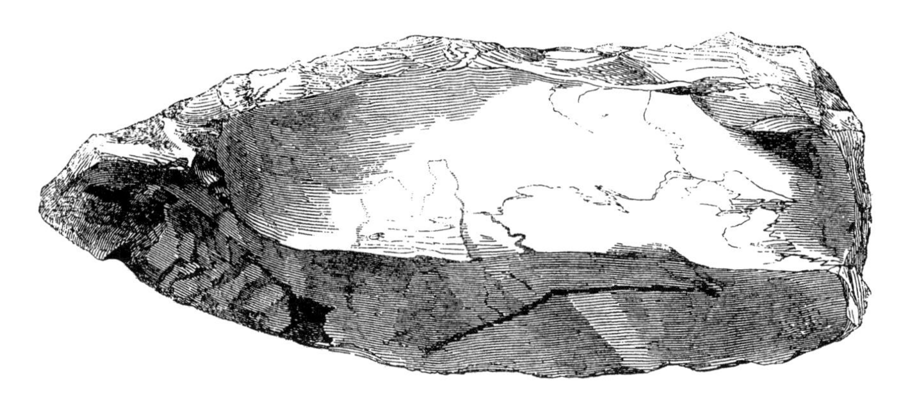

Flint Flake from Menchecourt, Abbeville

*Figure 2.4. Flint flake illustrated by Evans.* Archaeologia *1860.*

hammer he was able to emphasise the importance of the patterns of flake scars (Figure 2.1), detached from the working edges of the tool, and show how the patterns of flaking present on the original tools were best replicated by human action (Lamdin-Whymark, 2009).

He divided the artefacts from the French pits into three broad groups. The flakes and blades, which he interpreted as knives, formed a group of their own. He would later note (Evans, 1861) that these seemed more common amongst the lower gravel spreads in the French valleys. Undoubtedly his own knapping experience informed his ability to identify these artefacts on the basis of their distinctive manufacturing/percussion features, see Figure 2.4. The second group were the pointed handaxes of which there were two sub-divisions; one an elongated handaxe with a very rounded tip, the other more elongated with a sharp pointed tip, often with concave sides, as shown in Figure 2.5. The final group of artefacts were again handaxes; this time ovate with a cutting edge around the circumference, Figure 2.6. Evans noted that individual shape was very variable and examples could be found that connected the three handaxe divisions together.

In 1861 he broadened this first attempt, recognising more types (Evans, 1861) but still felt the original tripartite sub-division was applicable.

There is a real sense of exploring something genuinely new in Prestwich and Evans' early papers, as they sought to convince sceptical geologists and archaeologists (more properly antiquaries as Evans calls them, since there were no real Prehistorians at this time.) of their views. They had opened a window on a whole new world. Public interest was as stimulated about these discoveries as it was about *Origin of Species*. There were articles in the *Times* and in the popular magazines, and Prestwich was defending his fieldwork at Hoxne in the pages of widely circulating periodicals like the *Athenaeum* (Prestwich, 1859a; Prestwich, 1859b).

Prestwich and Evans lost no opportunity to convince their peers of the existence of this new world but there was a lot of disbelief to combat. For

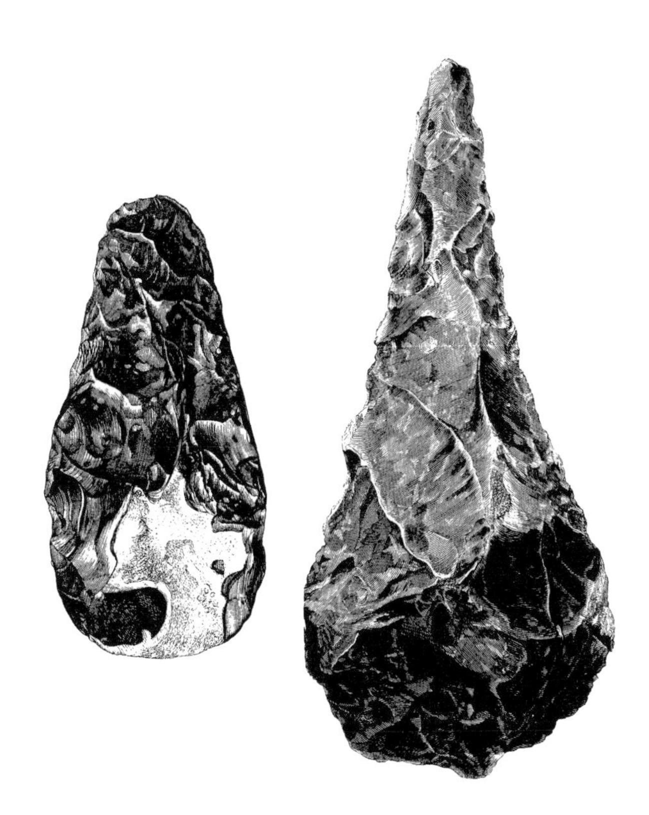

*Figure 2.5. Pointed handaxes as illustrated by Evans.* Archaeologia *1860.*

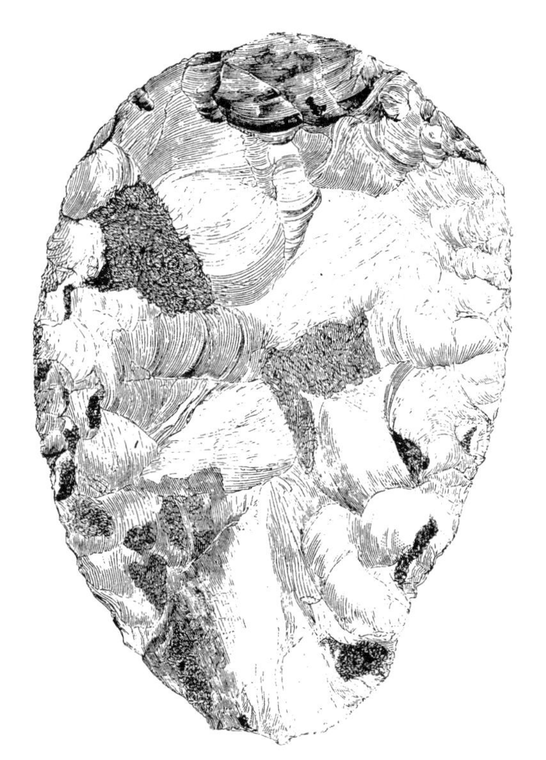

*Figure 2.6. Ovate/ oval handaxes as illustrated by Evans.* Archaeologia *1860. Note the orientation. The ovate is positioned with the wider end uppermost implying Evans thought this to be the tip. Modern convention would suggest the narrow end is the tip and orientate it accordingly.*

| Near | Pit/site/locality |
|---|---|
| France - Abbeville | Porte Marcadé |
| | St Gilles |
| | Moulin Quignon |
| | Menchecourt |
| | Mautort |
| | Drucat |
| France - Amiens | St Acheul |
| | St Roch |
| | Montiers |
| France - Seine gravels near Paris | La MottePiquet (Champ de Mars) x2 pits |
| France - Seine gravels | Clichy (+5 other potential noted localities, in addition to pits near Joinville |
| Other French localities | Pits near Creil |
| | Pits near Clermont |
| | |
| England – East Anglia | Hoxne |
| | Icklingham |
| England – London gravels | Grays inn Lane, gravels of Thames tributaries, also Thames gravels as revealed by drain/sewer extensions and underground railway excavations |
| England – near Bedford | Biddenham |
| | Harrowden |
| | Cardington |
| | Kempston |
| | Gravels of the River Great Ouse –contain fauna |
| England - north Kent coast | Cliff between Herne Bay and Reculvers |
| | Cliff between Swalecliffe and Stud Hill Cliff |
| England – Kent | Near Harten, surface find |
| England – near Guilford | Pease Marsh (an old find) – potential in the gravels of the Wey and Mole rivers |
| England – Hertfordshire | Near Bedmont, surface find |

*Table 2.1. Discoveries of locations with handaxes as reported by John Evans to the Society of Antiquaries, May 16, 1861, with additions made prior to publication in* Archaeologia *later that year.*

example, Darwin found the sheer quantity of the handaxes present in the French gravels difficult to accept (Darwin, 1859c), and many shared this view. It was the growing volume of evidence that was difficult to dispute, and Prestwich and Evans were out field-walking together searching for more palaeoliths, often accompanied by John Lubbock, Charles Lyell, and others. They were also conducting exploratory excavations. Prestwich had sections and test pits dug in the French gravel sites as well as at Hoxne and other British drift localities. They were encouraging colleagues to do the same. Prestwich was lecturing to the Geological Society of London in May 1861 (Prestwich, 1864) detailing new discoveries and deliberately highlighting those areas in Britain where he thought drift gravel could contain palaeoliths. Suitably informed, the audience could now search these gravels on their autumn break. In addition, Evans was exhorting the Society of Antiquaries in May 1861 (Evans, *ibid*) to do the same, as he detailed the latest discoveries. The list of find spots was growing quickly, Table 2.1. Evans was keen to point out in this second paper that the much disputed evidence from Kent's Cavern and from other caves, should be re-examined in the light of the new evidence from the drift. If palaeoliths and the bones of extinct megafauna came from

*Figure 2.7. A Neolithic celt or axe as depicted by John Evans in* Ancient Stone Implements *1872.*

undisturbed open air sites, and were clearly contemporary, then this validated the association of similar artefacts and bones found in the disputed cave deposits; the handaxe makers had lived in the caves too.

Like Prestwich, Evans refused to be drawn on the question of the precise date of this new Palaeolithic world. From a relative dating perspective he could show that the Palaeoliths were different when compared with the Neolithic axes of the Stone Period. The Neolithic axes had a different shape, they were better made, and the wider end formed a cutting edge befitting axes that were mounted in hafts – Figure 2.7. The cruder palaeoliths by comparison were either used in the hand or crudely mounted like the tomahawk of the American Indians.

## The Moulin Quignon Affair and the Triumph of Methodology

The drift gravels of Abbeville had one further surprise for the English geologists. In 1863 a human fossil was found in the drift in the quarry at Moulin Quignon. Prestwich had assigned the gravels in this quarry to his high level drifts. This was the one find that Victorian science had been waiting for (van Ripper, 1993); the maker of the palaeoliths. What made it all the sweeter was its discovery by Boucher de Perthes; a final triumph to crown a glittering career. It became known as *L'affaire Moulin Quignon* (Trinkaus and Shipman, 1993). The general sequence of events is presented in Textbox 2.2 in order to give an overview of what happened. Readers should familiarise themselves with this first. The problem with the jaw was simple, it was fraudulently inserted into the gravels and was not drift-age at all. The jaw and a number of the handaxes were introduced into the section by the workmen in the quarry. One of the more interesting aspects of the whole case was the pivotal role that the handaxes played in uncovering the forgery. I will focus on this aspect.

Boucher de Perthes was informed by a workman of the discovery of a second bone in the section at Moulin Quignon at the end of March 1863. The pit had been known to produce a few palaeoliths of the normal drift varieties found elsewhere. They occurred in the main body of the ochreous stained gravels and their appearance reflected their provenance. However there were not very many of them. The bone turned out to be a partial jaw bone with a few teeth in place. It came from a black manganese stained layer near the base of the section and de Perthes dug it out with his own hands. A friend of his dug out a handaxe from the same level nearby. Others soon followed. Such an important find excited much interest. By the time that John Evans and Joseph Prestwich arrived at Abbeville, others such as the zoologist and anthropologist Armand de Quatrefages (1810–1892) of the Muséum d'Histoire Naturelle in Paris were already there.

The first signs of suspicion emerged on the 13th of April. Evans and Prestwich met with de Perthes who showed them the recently found handaxes from the site, as well as the jaw. Both Englishmen were struck by how different the shape of these handaxes were, when compared with those previously found in the pit. A visit to the pit was unproductive as there had been a section collapse, but while walking to another pit a workman gave them two implements found earlier. These also looked wrong. Prestwich and Evans stopped at the first house they came to and washed the axes. All of the adhering gravel matrix came off, and with it the iron staining that was a tell-tale sign of the artefact having been buried in the drift for untold ages. The flint underneath was black and lustreless, exactly like freshly knapped flint (Prestwich, 1863b). They broached their concerns to de Perthes, but even after he had washed one of his implements to reveal the same black surface he was unconvinced. Neither of the Englishmen were able to stay in Abbeville any longer, and were not present when Falconer examined the jaw

*Textbox 2.2. The chronology of the Moulin Quignon affair.*

**Moulin Quignon**

- 'End of March' 1863. Workman brings Boucher de Perthes a stone tool and a fragment of jaw with a tooth in place from pit at Moulin Quignon.
- 28th March another bone found *in situ*. de Perthes goes to the pit and digs it out himself. It is a jaw bone. A friend digs out stone tools from near by.
- De Perthes announces discovery in local publication on 9th April, but the find is known about before this.
- John Evans, Joseph Prestwich and Hugh Falconer visit Abbeville. They have been on a study tour and were returning home. Evans and Prestwich arrive on 13th April, Falconer day after. Carpenter is also at Abbeville, as is de Quatrefages.
- On 13th Evans and Prestwich are shown the jaw and visit some of the pits in the area as well as Moulin Quignon pit. They are shown handaxes from pit. They become suspicious of the stone tools, believing them to be fakes. No comment is made on the jaw.
- Carpenter is shown jaw and visits pit. His description of section collapse in Athenaeum letter 18th April tallies with Prestwich's letter to Lartet on 5th May, so he may have been in the company with Prestwich and Evans that day.

- Prestwich and Evans return to UK, Falconer stays on for a day or two. He believes the jaw's morphology makes it different from any living race.
- Quatrefages returns to Paris and takes the jaw with him. de Perthes gives Falconer a tooth from the jaw. He returns to London with this, as well as measurements and careful drawings he has made.
- 18th April a letter by W.B. Carpenter is published in Athenaeum. Dated 14th in Abbeville, it announces the find to the general public.
- On 20th April Falconer writes to Charles Darwin, mentions trip to France but no hint of the suspicion of forgery. During this time the tooth was being analysed by George Busk and John Tomes on the request of Falconer.
- 24th April Falconer writes to Darwin asserting the discovery is a forgery perpetrated on de Perthes by the workmen.
- 25th April Falconer writes to the Times. French response is not to accept the accusations. They consider jaw's authenticity to be almost a matter of national pride. de Quatrefages is known to be generally unsympathetic to English scientists with exception of Darwin and Huxley.
- 5th May Falconer and Prestwich respond to French invitation to go to Paris to debate the question of the jaw. Prestwich's tone suggests he unable to attend. Evans also too busy.
- 6th May Falconer tries to recruit people to go with him. Hooker declines, urging him to go alone.
- Falconer, Busk, Carpenter travel to Paris on the 8th. Conference starts on the 9th. Prestwich makes a last minute decision to attend and catches the boat train on the 9th. Arrives next day.
- Conference in Paris 9th to 11th May. French delegates are amongst most senior in French Anthropology including Milne-Edwards (president), de Quatrefages, E. Lartet, Desnoyers, L'Abbé Bourgeois, and Gaudry.
- The jaw is analysed and sectioned and Falconer's tooth discussed. The fresh appearance of the bone and the surface character of staining which washes off, shakes the confidence of some of the French researchers. Prestwich still standing by the fraudulent nature of the implements. Implication of reviewer in Athenaeum 23rd May is that French position has shifted to uncertainty. Conference now moves to Abbeville on 12th on suggestion of Milne-Edwards.
- Lartet at least appears convinced as he spends train journey from Paris to Abbeville worrying about how to word the final report without offending French sensibilities. At Abbeville they are joined by de Perthes.
- At the Moulin Quignon quarry handaxes are recovered *in situ* from the section in the presence of the conference delegates. Workmen repeat their information about the finding of the jaw bone. This all serves to reconfirm the opinions of the French scientists that all is genuine. Prestwich, up to this time is convinced the artefacts are forgeries, now accepts that the implements may be genuine. Busk and Falconer remain committed to the view that the jaw is not old. Both are certain that de Perthes is innocent of any deception and did recover the jaw from the section as he claimed.
- The conference closes on the 13th with Falconer returning to Paris and Prestwich returning home via Boulogne. The majority of the French researchers remain convinced by the discovery. Milne-Edwards and de Quatrefages report extensively in French professional and public circles.
- 21st May Falconer, Prestwich and the others give papers on Moulin Quignon to the Royal Society and speak to its Philosophical Club earlier in the evening. Joseph Hooker reports to Darwin on the evening's events. He thinks the British contingent have made a 'mess' of it, and singles out Prestwich as being a ditherer.
- Evans remains uneasy. That the English delegates have accepted the implements is at odds with his views on them. He travels to Abbeville with Prestwich, Godwin-Austen, J.W. Flower and John

Lubbock on 30th May, reporting on the trip in the Athenaeum on 6th June. Evident forgeries are discovered while they are there and presented to Evans and his group as authentic *in situ* finds from the gravel. Evans has brought with him H. Keeping one of Pengelly's diggers at Brixham. When Evans and party leave he stays on for a few days in early June to observe the continued gravel workings at the Moulin Quignon Pit

- Prestwich presents his findings on June 3rd at the Geological Society. An extensive discussion follows Evans reiterates his opinions as expressed in the Athenaeum. Busk declares that he can experimentally reproduce the implement's staining. He reports on skeletal remains in de Perthes' collection from a site 15 miles south of Abbeville. He and de Quatrefages noted the mineral condition was similar to that in the Moulin Quignon jaw, and a child's jaw from the site showed anatomical features identical to the adult jaw from Moulin Quignon. Both had a limonite coating as well. Moreover, the workman who found these skeletons worked at Moulin Quignon.
- Evan's reports Keeping's evidence (Athenaeum 13th July) on the methodology used by the forgers to insert faked handaxes into the gravel section, and then doctor the section to make it appear that the gravel is undisturbed and the find is *in situ*.
- Athenaeum letter of 13th June now forces Prestwich's to change his mind again. He supports Evan's and Falconer's belief that the handaxes have been faked. He admits this in a short addendum dated October 1863, to Prestwich's June 3rd paper to the Geological Society. The French scientists do not retract their opinion, but over time the Moulin Quignon jaw is referred to less frequently until it disappears from the literature altogether.

with de Quatrefages the following day. Falconer and de Quatrefages visited the pit and dug out some more examples of the 'new type' of handaxes. At this point it is worth detailing what it was that alerted Prestwich and Evans.

- The handaxes were of a different outline form to those from higher up in the section, i.e. the normal forms found in the pit and elsewhere in the drift of the area.
- Their edges were very sharp, they should have been a little more rolled or dulled if they had been in the gravels for any length of time.
- The flake scars were deep and of a rounded concave profile. This was the kind of scar concavity associated with knapping by a metal hammer. Evans as a flint knapper himself, and familiar with the work of the gun flint knappers in Brandon, Suffolk, would have known this. At a later date Evans said he had seen the occasional scratch mark of a metal hammer on the axes themselves.
- There was no sediment adhering to their surfaces, no concretions, or dendritic marks as were often present on genuine implements from the drift gravels.
- Their surfaces were neither patinated white, nor iron stained as was

normal. Rather, they were fresh looking and the same dull black colour of freshly knapped flint.

- Later Evans said that he had been uneasy about the appearance of the surface iron staining (it washed off easily) – it seemed to have been brushed on – there were linear marks such as a brush or a cloth might leave if wet sediment was applied to the surface.

In effect these were a series of criteria against which to measure the authenticity of flint handaxes, a synthesis of archaeological and geological features which, when combined, represented a methodological statement. In one sense what was on trial was the ability to recognise genuine Palaeolithic artefacts. But there was more to it than this. Handaxes were the very cornerstones by which the existence of the newly revealed lost world was actually recognised and validated. If doubt was cast on their authenticity, and on the ability of their supporters to recognise them, it would reflect on the reality of the Palaeolithic period itself. Van Ripper (1993) asserts that this methodological test was a key factor in enhancing the reputations of the English researchers and did much to support the case for human antiquity.

Falconer initially accepted the mandible as genuine, believing it to be unlike the jaw of any extant European race. De Quatrefages thought it resembled Eskimo jaws, but Falconer inclined toward Australian similarities. However, he began to doubt its unique racial characteristics once he was back home (van Ripper, 1993). Colleagues were able to show him modern jaws from London cemeteries that had some of the so-called primitive anatomical features he recognised (Falconer, 1863).

The first public announcement of the discovery was by the zoologist William Carpenter in the *Athenaeum* on the 18th April, accepting the find as genuine. He too had been at Abbeville at the same time as Evans, Prestwich and Falconer (though it is not clear if he was with their party). In the meantime, Falconer showed Prestwich and Evans the handaxes he had found at Moulin Quignon. They were forgeries too. Falconer and two colleagues now cut into the tooth. In section it was white and glistening, and its gelatine content was high. It should have lost most or all of this if it had been a genuine fossil. Falconer wrote to the *Times* on the 25th April outlining his belief that the find was a forgery (Falconer, 1863).

We now skip slightly ahead of the story (see Textbox). The next significant point at which the artefacts played a role was when the conference, called by the French to evaluate the find, moved to Abbeville on the 12th May. Evans was not able to be present at the conference, and Prestwich still believed the stone tools were forgeries. The conference did not announce its arrival at Abbeville ahead of time, and Milne-Edwards, the president, went up early in the morning and set his son on guard to prevent any of the workmen salting the section faces. Sixteen workmen were employed for the day, though whether or not these were part of the normal workforce at the pit is not stated (Anon, 1863b). Five handaxes were recovered that day, one of them a genuine implement of the accepted drift variety, and four of the new shape (Anon, *ibid*). It is not stated whether these came out of the black manganese stained level at the base of the section – the impression given is not.

With little chance of forgeries having been introduced, and new implements being found in the presence of the committee, the whole issue took on a new aspect. Prestwich now came round to the view that the handaxes had to be genuine. He would later argue (Prestwich, 1863a) that one possible explanation for their shape was that they were an earlier type of implement, one not seen up to that point. Moreover, occasional implements of the normal drift shape had been found in a very fresh condition, lacking extensive staining or patination. The metal marks on the implements' surfaces may have been caused by the workmens' picks, or perhaps from objects like keys when the axes were carried around in their pockets. The fact that the apparent artificial staining had striation-like brush marks could be explained by their naturally sliding down the gravel talus. In effect Prestwich backtracked on all the previous arguments he and Evans had put forward. Such was the power of being there in person and seeing these artefacts disinterred. In turn, this of course empowered the authenticity of the jaw bone. Issues like the lack of any deep staining and the fresh un-fossilized character of the bone itself were ignored or brushed aside. It suddenly regained its former status as a genuine drift-aged find.

That evening, the 12th May, the French scientists requested each of the delegates to state their beliefs. The discovery of the handaxes had worked a powerful effect. Prestwich accepted the possible authenticity of the artefacts. Falconer and Busk had to admit that they believed no forgery had been perpetrated, but they stuck to their guns on one aspect. The jaw was not ancient, it was not as old as the gravel in which it had been discovered. It was an unsatisfactory ending to the whole conference. Everyone returned home.

Later, John Evans (Evans, 1863a; 1863b) tactfully tried to suggest that the excitement of the day and the finding of artefacts clouded the judgement of the committee as they stood in front of the section. In fact an anonymous correspondent in *The Reader* gives an entirely different story (Anon, 1863a). The writer claims that all was 'confusion' on the day, and that not one of the five handaxes discovered was actually seen *in situ* by a conference member before it was actually found (implying they were picked up off the ground as if they had just fallen with the gravel that the workmen dislodged). The closest was one seen by Milne Edwards, apparently falling out of the section. This article has such attention to detail that it could have been written by someone who was there. It is interesting to compare this account with that of de Quatrefages concerning his earlier visit (De Quatrefages and Rolph, 1863). The workman undermined a portion of the section, and produced the suspect axe from the talus. De Quatrefages implied that the palaeolith may have already been just under the surface on to which the gravel fell. The ingenuity of the workmen is impressive. With no apparent forewarning of the arrival of the Paris delegates they still managed to introduce fraudulent handaxes under the watchful eye of men who were already suspicious of forgery. A number of contemporary sources comment on the crafty reputation of the Abbeville workmen.

The final act in the drama was initiated by continuing concerns over the implements. For John Evans the new findings of the English delegates were unacceptable. The criteria outlined above clearly established a fraudulent basis for the axes. They had to have been introduced into the section somehow. He travelled to Abbeville with Lubbock and Prestwich, this time taking with them Henry Keeping, one of William Pengelly's diggers at Brixham, and an experienced excavator. He stayed on for a few days after Evans and his party left, working the gravel pit with the workmen. It was his testimony that was the clincher (Evans, 1863a). He revealed the methodology of the forgers, how they could insert the implements into the section and then restore the gravel face to an undisturbed appearance.

Van Ripper asserts that *L'affaire Moulin Quignon* did the English researchers a considerable service. They successfully defended the methodology by which the Antiquity of Man debate had been fought – meticulous personal observation, fieldwork, and a detailed understanding of geological context. Both the English researchers and the subject benefited from this. Furthermore, it was a debate that was conducted right under the public's gaze. Most of the popular magazines and periodicals, as well as the more scientific journals carried articles on it; some, like *The Reader* and the popular *Chambers Journal,* were publishing articles written by individuals whose aim it was to influence public opinion on the authenticity of the jaw. Many of these journals followed the story as it unfolded. *Chambers Journal* for May 30 1863 carried an anonymous article which had this to say.

> Imposture has already been playing its usual tricks in the drifts of Picardy. We mentioned a year ago that a friend of ours, while exploring the gravel beds near Amiens, found a manufactory of ancient flint implements in full activity, preparing relics to sell at a franc each to credulous tourists. (Anon, *Chambers Journal* 491, 351)

The London societies eagerly debated the affair. Not surprisingly the Anthropological Society of London (see Chapter 3) kept up with the debate. It was publishing translations of the French point of view even before the Paris conference (De Quatrefages and Rolph, 1863). The society debated the Moulin Quignon remains on 21st April. Mr Alfred Taylor exhibited one of the fraudulent handaxes supposedly found near the jaw. The meeting agreed it was too sharp to be a genuine palaeolith.

After Evan's paper to the Athenaeum on July 4th, and the meeting at the

Geological Society in June (Prestwich, 1863c), the affair began to lose its impetus. The French never formally accepted the Moulin Quignon jaw as a fake, but gradually they dropped it from their discussions. Perhaps the saddest part of the story was its effect on Boucher de Perthes. He felt betrayed by his English friends and, to the embarrassment of his colleagues, spent the last few years of his life desperately digging at Moulin Quignon to try and find corroborative evidence for the jaw (Anon, 1864b; Trinkaus and Shipman, 1993). He died deeply disillusioned in 1868.

# 3

# THE 1860s
## *Owning, Administering and Populating the Antediluvian World*

> But who shall dwell in these worlds if they be inhabited?...Are we or they Lords of the World?...and how are all things made for man?
>
> (Kepler from *The Anatomy of Melancholy*, quoted by H.G. Wells on front page of *The War of the Worlds*.)

In this chapter it will be suggested that the early phases of the Antiquity of Man debate were characterised by a collection/description mentality as Prestwich, Evans, and their colleagues reinforced the discovery of the Palaeolithic of the drift by more and more handaxe discoveries in undoubted river drift deposits. This was a buttressing phase as they sought to dispel continuing scepticism by the sheer weight of evidence. Informed description was a necessary weapon in this strategy. But as the 1860s progressed the idea of debating human origins began to move away from its purely descriptive beginnings.

The shift toward a more interpretative approach to debating human origins began after 1863 and took two directions. Firstly, Darwin's *Origin of Species* initiated the long process of turning human origins into a question for biologists. Up until this point the anthropologists (*sensu lato*) had debated the origins of the different human races as a philosophical exercise in racial distinction. This move toward a more scientific understanding of the origins of species would take decades, as it was intimately linked with the search for the mechanisms that governed heredity (see Figure 1.3; Bowler, 2003).

The second direction concerned what Gamble and Moutsiou (2011) have termed 'humanizing the past', focusing attention away from the palaeoliths themselves and on to the people who made them. I suggest that the framework for this was the emerging discipline of evolutionary anthropology. The Palaeolithic (drift and cave periods, considered by some to be successive, and by others synchronous) could be slotted comfortably into a relative succession of other Prehistoric periods. These were the kitchen middens of Scandinavia and elsewhere (many of which are now interpreted as Mesolithic); the lake villages of northern Europe (now considered Iron Age); the polished/ground stone axes of the Neolithic and the passage graves; and finally the period of the first metal objects, the Bronze Age, with its round barrows and tumuli. Together they formed a single relative sequence of continuous Prehistoric development. Researchers like E.B. Tylor and John Lubbock began to make the deep past more familiar to their readers by including ethnographic descriptions of indigenous peoples (Lyell, 1863a; Lubbock, 1865b; Gamble and Moutsiou, 2011). There were parallels between the material culture made by contemporary tribal peoples and some of the archaeological artefacts recovered from the sites of these different periods. Evolutionary anthropology strengthened these linkages.

The chapter will also show how the broader evolutionary paradigm opened the way for a number of logical synergies between allied disciplines which further encouraged an anthropological view of Prehistory and the Palaeolithic. Politics, and a pre-disposure to accept monogenist viewpoints, cemented the engagement with anthropology. As part of the process of interpretation, the researchers of the time set out to define and describe the brave new world. In so doing they sought for its limits, and imposed their own upon it. This was akin to an imperial project. After the first flush of discovery came the administrators as the learned societies vied with each other to prove they were the most suitable to debate its history and explore its character. Understanding the inhabitants of a new territory was always the first step towards owning it and controlling them. After 1865 the drift period was increasingly known as the Palaeolithic (Lubbock, 1865b).

## E.B. Tylor and Evolutionary Anthropology

Edward Burnett Tylor (1832–1917) is credited with being one of the founders of the evolutionary school of social anthropology (Stepan, 1982; Stocking, 1982; Stocking, 1987), and one of the founding fathers of anthropology itself (Lowie, 1917). As a younger man in the 1850s he travelled to North America and Mexico. There he met the wealthy banker Henry Christy (1810–1865) with whom he became friendly. They travelled together. Christy would later become more interested in Palaeolithic archaeology and excavate extensively in south-western France. Tylor eventually settled in Oxford to become the archetypal armchair anthropologist. He became Keeper of the University of Oxford Museum in 1883 and was appointed Reader in Anthropology in 1884. In 1896 he became the University's first Professor of Anthropology

Evolutionary anthropology was the study of human society through the framework of evolutionary development. Tylor believed that all humans, irrespective of race or geography shared the same potential for cognitive growth. Whether Australian aboriginal or Zulu, all peoples could in time scale the same social and cultural heights as western Europeans. The human mind was the same everywhere and had the same potential to be improved. However wherever Europeans had gone they had found indigenous peoples in a variety of different developmental stages (as they saw it), some more

*Textbox 3.1. An explanation of the comparative method as used by evolutionary anthropology in the middle and late Victorian period.*

advanced than others, but all very much below those attained by Europeans. This global pattern had to be explained. Through the lens of evolutionary anthropology, and engaging with the broader evolutionary focus on progressive development, these stages reflected some of the phases that all societies had to pass through as they evolved. In effect different cultures were a history of the evolution of the human mind. The universality of the human mind notwithstanding, it was clear that it was nonetheless held in check at different stages of development in different places. The cultures of indigenous people reflected this as well. The differences in detail between geographically distinct social groups sharing the same stage merely reflected local variations on the various developmental phases of cultural evolution. When progress stopped, as it had with most of the people that Europeans encountered (at least in their eyes), it was because some external influence (e.g. presence of a more powerful or more culturally evolved neighbour), or internal force (stranglehold of antiquated social institutions) had held that society back.

The view that a modern hunter-gatherer's life could stand as a proxy for understanding the lives of Prehistoric people was rooted in the broader belief that social institutions and material goods were cross-culturally comparable. The universality of the human mind and the belief in universal stages of cultural evolution empowered what was known as the comparative method – see Textbox 3.1. The label evolutionary anthropology is of course a modern term applied retrospectively to a particular outlook. Few Victorian researchers would have recognised the label. Many would have disagreed with Tylor's particular vision of it (John Lubbock for example emphasised the role of independent invention stimulated by environmental pressure far more than Tylor). Nevertheless, most would have shared its basic precepts.

**John McLennan and The Early History of Man**

McLennan, J. 1869. Article VII. The Early History of Man. *North British Review* 50, 516–549. McLennan's article was a powerful demonstration of the potential of the comparative method to reveal the ordering and structure that underlay earlier societies at the different stages of their development. The piece was predicated on the concept of progressive time, and that development was inherently a part of the human condition.

What he demonstrated was that analogy was able to reveal the historical nature of social systems, its developmental character, and provide insights into the human mind, particularly its earliest times.

Wherever anthropologists looked they always found evidence of development in social institutions. It was human nature for people to organise themselves once they started living in groups. That organisation reflected their developmental status as did the various social structures that contributed to the group's maintenance as a society. Analysis of modern indigenous people could reveal the developmental level they have achieved. But also it revealed the progress they had made. Many societies retained older social practices and customs that were now enshrined in tradition and surrounded by myth. These were windows into their history. Systems of kinship, inheritance, ethics, subsistence, religion, and language, were all a combination of new, old, and borrowed practices. [This was similar to, and draws from E.B. Tylor's work, as McLennan freely acknowledged.] In terms of religions, the most successful were those that proved themselves to be the most adaptable and which appealed to the widest possible spectrum of tastes. McLennan noted the success of religions paralleled the success of businesses in this respect, and likened it to a natural selection on religion. So the comparative method revealed that all societies had evolved from simpler states of social organisation, and were still evolving.

It was also able to reveal other factors in human social evolution. He accepted the Great Man concept whereby individuals of great abilities improved society and induced positive developmental change because of their example and ability. A natural hierarchy existed in all societies. Their leaders, often the Great Men, generated ideas and innovations which filtered down through the various social strata, as the successively lower tiers recognised advantages to themselves. But the further down the social hierarchy you went the greater the numbers of people it was necessary to reach, and the more likely that negative and disruptive forces would derail positive ideas and change. Not all sections of society developed at the same rate. This was a key insight. In quoting John Lubbock, McLennan accepted an age for the Palaeolithic of 20,000 years ago. While the comparative method could not explore the social institutions of such a time, there were some forms of social organisation that it could. Our earliest ancestors were little more than beasts. Many indigenous societies lived at the same level, and others revealed a heritage evolving away from this. But if one looked at the social stratification of cities like London, it was possible to see the survival/re-creation of social organisations and behavioural practices that were little better than that of Palaeolithic beast-men. These were the lowest levels of society in the urban jungles. Innate human nature had not changed that much. Regression, lack of opportunity and social amenities, and natural propensities to differential development, combined to create moral and social systems that reflected Palaeolithic times. Although not clearly stated, McLennan may have tried to imply that under certain environmental conditions humanity would always revert to a Palaeolithic default state, and organise themselves accordingly. Here analogy provided an insight into primitive mental states. So this too was a window on the past. In this sense societies parallel the organisation and histories of races and nations.

**Andrew Lang and The Great Gladstone Myth**

Empowering the comparative method was the underlying assumption of the universality of the human mind. Myths and legends were stories about the world. Although the details were different across time and space many of the myths and legends of indigenous peoples, and those recoded from vanished civilizations, reflected the same stories. This was because the human mind everywhere constructed similar responses to questions and observations about the world. Social institutions and material culture from around the world would also have strong similarities, the same with past societies, because all these represented common solutions to fundamental problems that would face all humans.

This was brilliantly satirised by A. Lang in *The Great Gladstone Myth* (1886. *In the Wrong Paradise*. Kegan Paul, Trench & Co.; first published in Macmillan's Magazine 1886.). Lang sent up the fortuitous similarities and coincidental parallels that he saw underlying many of the comparisons evolutionary anthropologists made using the comparative method.

The story is set in a distant future after some unnamed cataclysm has overtaken England. Society has been re-established, but very little about the earlier history of the UK including the Victorian period is known. It was believed that during this period, known as the post-Christian era, a Teutonic people lived in England. An important figure in the fragmentary archives that remain is known as Gladstone or Mista Gladstone. The future narrator, an academic who has interpreted the Gladstone myth to be a metaphor for the sun and/or the sun god, takes up the story in support of his interpretation.

*Philological evidence.* By the association with elements of Greek, Egyptian, and Indian mythology, it is evident that the name Gladstone is related to a hawk-like deity associated with sun worship. In Victorian England he probably was a name for, or a personification of, the sun or sun god.

*Literary and textual analysis.* Only a few fragments of printed text remain and their content is difficult to properly contextualise. However, and understanding of the link between Gladstone and the sun, clearly empowers the narrators subsequent interpretation of them. Many of the fragments mention his qualities or refer to popular titles associated with his name such as 'popular Bill', and may have been praise songs in greeting of his rising and setting. However rival cults were less respectful. Comments on him in "morning papers" and "evening papers" were evidently curses as they emphasised negative qualities of his nature. Links to a "White Tsar" in the north, when compared with North American mythology, also re-emphasise his association with sun worship. Some fragments which seem to attest to Gladstone's nature being inconsistent and treacherous, are best understood in the context of sun worship in Britain where the weather is never constant.

*Archaeological evidence.* Ninety miles outside of London, on the road to an ancient settlement called Brighton an inscribed stone was discovered. There are a number of interpretations of it. The lettering is difficult to read. Some have reconstructed it as a mile stone on a roadway with information stating the distance to London. This is a ridiculous point of view especially in the face of clear and empowering interpretation that Gladstone was a sun god. Some scholars have reconstructed the stone as a plinth on which a statue of Gladstone once stood. The lettering, they assert, is G.O.M. which stands for Grand Old Man and was an affectionate name for a real historical personality who was a politician. This interpretation is favoured by the Spencerian school and is evidently wrong. The narrator favours the interpretation that the lettering does indicate G.O.M., but that this stands for *Gladstonio Optimo Maximo*, a Latin phrase indicating Gladstone was the chief of the Victorian pantheon of gods.

*Mythological analysis.* These reveal a contradictory picture, difficult to reconcile with a sun god. In a fragment entitled 'Gladstone in Opposition' he apparently made it 'too hot' for his enemies. He also contested with another mythical figure, Huxley, about the nature of creation. Huxley is the Huskley of legend, the brilliant hero of the husk-myth who when released from his husk becomes a luminous and shining being.

Lang filled the text with contemporary references which would have resonated with his readers. He concluded with these words, 'caution, prudence, a tranquil balancing of all available evidence, and an absence of pre-conceived opinions, – these are the guiding stars of comparative mythology.'

Tylor's place amongst the archaeologists was earned through his *Researches into the History of Mankind and the Development of Civilization* (Tylor, 1865). Like Darwin's *Origin of Species* the book was a vast and detailed compendium of observations and facts on almost every aspect of the social life of 'savages'. He tentatively suggested a developmental sequence for stone tool use beginning with naturally sharp stones, possibly grading into fortuitously shaped ones which were unifacially modified by various amounts of retouch as had sometimes been done by Tasmanian knappers. Examples of this kind of simple tool had been recovered from the French drift. This early description is significant in the light of Tylor's invoking of similar tools when commenting on the eoliths described in Chapters 9 and 10.

He suggested there were no significant breaks in Stone Age tool development whether between different modern Stone Age peoples, or between their tools and the tools of the Prehistoric knappers. The binding link was the repetitive nature of the different shapes and forms of tools. There was a persistence of form, despite the clear differences in the skill with which they were made, as befitting the artisans of different phases of Stone Age development. This allowed Tylor to link the drift tools with some of those found in the cave deposits of southern France, and so continue the line of evolutionary development. He made the astute observation that while the makers of the organic tools in the bone caves of the Dordogne were able to work bones or antler to make awls and needles, by grinding and polishing, they were not able to grind their axes like the more advanced Neolithic stone workers. He described some of the tools from Lartet and Christy's discoveries,

> The antiquity of the Drift implements is, as has been said, proved by direct geological evidence [Tylor accepted Prestwich's shorter time scale for the Palaeolithic]. The cave implements, even of the reindeer period, are proved by their fauna to be earlier, as they are seen at a glance to be ruder, than those of the Cromlech period [Neolithic], and of the earliest lake dwellings of Switzerland, both belonging to the Ground Stone Age. To the student who views Human Civilization as in the main an upward development, a more fit starting point could scarcely be offered than this wide and well marked progress from an earlier and lower, to a later and higher, stage of the history of human art. (Tylor, 1865, 197, my brackets)

As already noted, the discoveries of Henry Christy, along with his friend and co-worker Édouard Lartet (1801–1871), will be discussed in later chapters. Here, it is sufficient to note that these two archaeologists, digging in the limestone caves and rock shelters of the Vézère valley, a tributary of the Dordogne in southern France, had revealed another unsuspected prehistoric period. This had become known as the Cave or Reindeer Age.

There is a key point to be made here, and it is this. The comparative method of identifying synergies between the tools of modern tribal peoples, and those from Prehistory, worked well for later Prehistory, and even for the Cave Age remains. However like his friend John Lubbock, Tylor had to admit that when the ethnographic record was pushed back into the drift period the analogies broke down. The handaxes of Britain and northern France had no modern counterpart. Modern native peoples were in the equivalent stage to the Mesolithic, Neolithic or Bronze Age of ancient Europe. In fact no group of stone tool using people had yet been discovered who had not either used metal tools themselves, or who had not been in contact with those who had. Lyell, Lubbock, Tylor and their colleagues humanised these later Prehistoric periods. They were intuitively more comprehensible. But other than trying to link drift and cave tools together in developmental sequences there was little if any attempt to humanise the drift period. As this and the next chapter will show the types of evidence that would make the drift familiar were just not there, it was too old and to remote to humanise.

In fact there was a deeper problem here. Ethnographic parallels, engaged through evolutionary anthropology, actually invited comparison more than they offered direct explanations. Once a society had been slotted into a particular developmental stage, as revealed by its material culture, that was pretty much that. Analogy and the comparative method did not allow more detailed or appropriate contextualised understanding to develop, and, in the absence of another explanatory methodology, it tended to fossilized interpretation. Evolutionary anthropology described and located, and this stood for explanation. This is particularly relevant with the drift period. With no viable analogue to humanize it, it remained unknown territory. Evolutionary anthropology became an interpretative *cul de sac* that the Palaeolithic found it difficult to escape from.

## New Territory: How to Explore the Brave New World

Part of the inability to humanize the drift is simply explained. There was no template for envisioning what a human ancestor would look like; whether more ape than human, more human than ape, or something in between (Moser, 1998). Table 3.1 makes it clear that before 1859 the number of ancient human fossils was desperately small, and there was little support for anyone who made claims for their antiquity. Many scholars would have been comfortable with a progression from some man-like ape creature to something more akin to an ape-like man. But where was the cut-off point? Much would depend on whether you were a monogenist or a polygenist. Reading the literature of the early 1860s there is a real sense that many already thought of the makers of the drift palaeoliths as more recognisably human than anything else, albeit primitive and bestial. Why? Because of the skill and manual dexterity needed to make palaeoliths. No mere beast-man could achieve this. For these scholars the cut-off point would have been much further back in time – into the Pliocene, or even further back into the Tertiary.

All this prompted a further question. Where along the time line were the origins of the different human races to be set? The teasing apart of human origins from racial origins was going to be a long and slow process. Within English scholarship, particularly that centred on the London societies, the historically rooted nature of the racial origins debate ensured that any emergent anthropological or biological perspectives had first to be disentangled from the politics that enmeshed the monogenist versus polygenist debate.

Clear research questions began to emerge after 1859. I have synthesised a number of them in Table 3.2. There can be little doubt that the synergy between the monogenist *Origin of Species* arguing for descent through stages of modification, the acceptance of the Pleistocene world, the Antiquity of Man debate, a natural origin for humans, and the perception of the Palaeolithic as an early stage in a process of natural evolution, encouraged the use of evolutionary anthropology as an explanatory framework through which Prehistory could be made more accessible. Evolutionary anthropology embodied all of these things. I suggest they made the emergence of evolutionary anthropology as a discipline almost inevitable (or at least that particular perspective). The gradual acceptance of a drift period wrapped a diachronic and anthropological framework around these research questions, because a lost world, peopled by ancient stone tool using humans and extinct animals, could be revealed through empirical science.

*Table 3.1. A selection of some of the most important, or best publicised, discoveries of fossil humans found before and after 1859.*

| Site | What found | When and by whom | Comments |
|---|---|---|---|
| Galeinreuth, Bayreuth area of Germany | Mammoth bones, human bones, stone tools | J. F. Esper 1774 | Dismissed by excavators as mixed |
| Natchez in the Mississippi Valley | Fragment of human pelvic bone. Bones of mastodon in close proximity | Dr Dickeson. Post 1812 | Stained black, found in scree. Native American burials similarly stained from top of section. Fauna older and from lower in section. John Lubbock implied it was a mineralised animal bone |
| Maastricht jaw | Jaw bone no other human remains | 1815–1823 elephant remains close by. Found during canal digging. | Not featured in literature until Charles Lyell discussed it in 1863 |
| Bize cavern, Pondres, and other caves of the Languedoc, southern France | Human remains, pottery, extinct animal species, arrow heads | Late 1820s and 1830s by Tournal, de Serres and de Christol | Claims for pre-flood date disputed at time, fauna modern; older fauna probably not contemporary. Neolithic |
| Loess from the valley of the Rhine | Human bones | 1823 Aimé Boué presents bones to Cuvier | Mentioned by M Lartet and L. Figuier |
| Souvignargues, near Nîmes in France | Fossilised human bones | *c.* 1827 | Found with extinct animal bones |
| Engis, Engihoul and other Belgian caves in the valley of the river Meuse | 2 skulls associated with extinct fauna. Some 40 caves examined. | Schmerling 1833 | Adult is now considered modern, child is a Neanderthal |
| Santos, Brazil. | Human bones in solid rock. Published by Charles Lyell in the early 1840s as ancient burials, submerged and indurated by sea water (oyster shells), then subsequently uplifted. | Discovered pre/early 1840s | By 1863 Lyell believed rock was actually tufa and not marine limestone, also oysters were transported up from the coast and not evidence for marine inundation. |

| | | | |
|---|---|---|---|
| Kent's Cavern, Torquay, Devon, UK | Human remains | Early excavations in middle 1820s by MacEnery unpublished. Godwin Austen describes human remains found with stone tools and bones in 1840 | Godwin Austen's excavations are in undisturbed areas of cave and sealed by stalagmite. Bones of extinct animals and stone tools therefore contemporary. Results are dismissed |
| Denise, near Puy-en-Velay, France | Skeleton in volcanic deposits | 1844 by M Aymard | No fauna or stone tools associated. Some believe associated with Pliocene lavas. Suspicion of forgery. Charles Lyell asserts skeleton in Pleistocene lavas. Others claim it is an intrusive burial |
| Gas works excavation near New Orleans | Skeleton of a member of a native American race below 4 separate buried forests | Post 1846 and Charles Lyell's visit to site. Described by B. Dowler | Actually a skull. Dowler suggested an age of 50,000 years. John Lubbock dismissed it on grounds of uncertain association. |
| Gibraltar | Skull | 1847 by Captain Broome | First Neanderthal ever discovered, not recognised as such. Falconer and Busk discuss it at BAAS in 1864 |
| Aurignac, southern France | c. 17 skeletons unearthed in but reburied in an unknown location | Originally discovered in 1852. Site re-excavated by Lartet in 1860. Skeletons never recovered. | Stone tools associated, a sepulchre closed off by a door. Boyd Dawkins asserted burials post-dated Palaeolithic layer. |
| Neander Valley, Germany | Skull cap, ribs, arm and leg bones | Significance recognised by Dr Fuhlrott in 1857. First described by Schaaffhausen | No stone tools. |
| Moulin Quignon | Jaw bone | 1863. Discovered by quarrymen and given to Boucher de Perthes | Handaxes associated. Demonstrated by English researchers to be fraudulent |
| Bruniquel | Skeletons, deliberately buried. | 1863-1864 Vicomte de Lastic, and first described by Richard Owen | Flint and bone implements, art objects, Pleistocene fauna |
| Olmo skull, near Arezzo, Italy | Human skull | I. Cocchi in 1863, 15 metres below surface during railway building | Suggested Pliocene date. Elephant and horse associated and stone tool nearby. Context disturbed. |
| La Solutre | Modern looking human remains | 1866 | Now known to be Upper Palaeolithic modern humans |
| Bone caves in eastern Brazil | Human remains | Lund and Claussen. Reported to London societies in the 1860s. | Examined *c.* 800 caves.many new faunal species.In a cave near Lake Sumidouro found remains of c. 30 individuals. Their condition was identical to bones of extinct animals. |
| Cro-Magnon | 5 or 6 skeletons | 1868. Discovered in Les Eyzies by railway workmen. First described by Broca | Art objects and tools associated. Now known to be Upper Palaeolithic modern humans. De Quatrefages and Hammy use these remains to define their Cro-Magnon race. |
| Calaveras Skull (on a mining claim in Calaveras county), California | Skull at base of deep shaft | 1866 by Mr Matson | Depths vary – from 130 to 153 foot below surface. In gravel, sealed by 5–6 volcanic beds, of 'high antiquity'. Mastodon, pottery, ground stone mortars nearby. Association of skull disputed. |
| Le Trou de la Naulette | Mandible, lower forearms, hand bones | 1867–1868. Dupont | Fauna but no stone tools |
| Mentone, Italy – right on French/Italian border. Riviera coast. | Skeleton lying prone on its side. Found in Caviglione Cave at Balzi Rossi. This is the Man of Mentone. Tall stature. 1873 a triple burial discovered in same cave. All were unusually tall – foundation for a Grimaldi race. Famous double child burial at the Cave des Enfants discovered 1874 | 1872–1873. Dr E. Rivière discovered Man of Mentone skeleton in 1872. Some later authors applied the name Grimaldi to all/some of the Balzi Rossi caves. See Grimaldi below. | Associated with Man of Mentone skeleton were flint implements, shells and pierced deer teeth. Metallic residue near mouth. Date was disputed and thought to be Neolithic; modern fauna. |
| Duruthy | Skull | 1874 L. Lartet and Chaplain-Dupare | Skull badly crushed, associated with a 'necklace' of bear and lions teeth |
| Trinil on Solo river, Java | 2 molars, skull cap, femur, | 1891–1892. Eugene Dubois | *Pithecanthropus erectus* – first *Homo erectus* ever discovered |
| Grimaldi (see Mentone above) | Skeletons in a number of caves at Balzi Rossi from 1872-1895. Post 1895, excavations sponsored by royal house of Monaco – the Grimaldi family.<br>Debate as to whether these later skeletons are of the same race as the earlier discoveries. | 1895 – turn of century | After 1895 the name Grimaldi generally applied to the Balzi Rossi caves, although there had been some application of the name prior to this. |

| | |
|---|---|
| Question 1. | What was the human species' relationship with the animal kingdom? Did a naturalistic mechanism such as progressive transmutation or natural selection explain us (i.e. we were an evolving species), or did progressive evolution have no place (i.e. we were a degraded species)? |
| Question 2. | Was there only one human race, or a number of separate human races?. If the latter, what was their relationship with each other. Were there grounds for considering one was superior to the others?<br><br>If either monogenism or polygenism was correct what was the precise relationship between the races, how were they to be characterised, and where/how did racial separation occur? This was dependent on a resolution to question 3.<br><br>What was the influence of climate and migration on humans? Were such forces more effective in the past?<br><br>Was there a difference in the evolution of the physical body compared with the mental and moral faculties of modern humans? |
| Question 3. | What were the laws of inheritance and how/why/which characters and character traits were transferred from one generation to the next. How did this affect the creation of new species and/or races?<br><br>Could changes in the body be successfully transferred to succeeding generations to become fixed character traits (Lamarckism)? |
| Question 4. | How to link the modern races to the historical/classical races, and then push this back into Prehistory. Was long term continuity really present? Which, if any, of the favourite criteria for exploring racial affinity was the most reliable<br><br>Skeletal anatomy (skull shape, jaw shape, morphology of limbs and other anatomical parts)/material culture /language/ the analysis of myth? |
| Question 5. | Was it possible to build a chrono-stratigraphic sequence of human fossils and material culture that showed one or more Prehistoric races developing over time. Could this be linked to question 4? Was there a link to question 6? |
| Question 6. | What was the implication of the discovery of palaeoliths in other parts of the world, particularly outside of Europe? Had there been a globally significant Palaeolithic period that occurred all over the world at the same time? Or, as evolutionary anthropology suggested, had different peoples gone through a Palaeolithic stage at different times. If the latter, what was the relationship of this to questions 2 and 5? |

The failure of the archaeologists, anthropologists and ethnologists to engage with a biological approach to human origins during the 1860s was the result of a number of things. One major reason was the lack of a viable theory of heredity. Darwin's selectionist theory was not popular, and Lamarckism, in all its variant guises, was unsatisfactory – see Textbox 2.1. Darwin's favoured mechanism of inheritance, pangenesis, was equally unsatisfactory. Without a mechanism to explain inter-generational inheritance, addressing the question of how a new species might arise (and therefore of what a species actually was) remained more of a philosophical discussion than a scientific debate. Freeing the dialogue from the shackles of racial origins was also hampered by the strong historicity of the subject. For many workers (polygenists especially) human origins and racial origins *ought* to have been the same, particularly if they did not accept any real time depth to human antiquity. Such people were particularly resistant to an emergent biological perspective – they would have seen it as a monogenist conspiracy. Time depth was the key.

> In order to abridge the number of centuries which would otherwise be indispensable, a disposition is shown by many to magnify the rate of change in prehistoric times by investing the causes which have modified the animate and inanimate world with extraordinary and excessive energy. (Lyell, 1864, 301)

In this quotation from Charles Lyell's inaugural address to the BAAS at Newcastle in 1864 (Lyell, 1864), the great geologist was alluding to Prestwich's river and flood hypothesis, and his retreat from uniformitarianism. It has already been noted that such an interpretation did away with the need for a considerable age for the Palaeolithic. Since the chronological position and duration of the Pleistocene was relative anyway, it could have been a great deal more recent than people like Lyell were suggesting. In a foreshortened timespan, racial origins and human origins could easily appear to be part of the same question. Shorten the time available and you have to speed up the process of evolution too. If Prestwich and Evans did not wish to discuss the age and duration of the Palaeolithic, others had a vested interest in doing so.

The whole issue was further hampered by a mixture of political intrigue and personality clashes within the two learned societies that would have been most concerned with human and racial origins. These were the Ethnological Society of London (ESL) and its rival, the Anthropological Society of London (ASL). This is often, perhaps unfairly, caricatured as a clash between the monogenists and the polygenists. The majority of the data for this chapter comes from the various periodicals published be these two societies in the 1860s.

*Table 3.2. Some of the major research questions facing Palaeolithic archaeology and human origins research during the 1860s.*

## People and Politics in Human Origins Research in the 1860s

The question of racial origins remained highly politicised in the years just after 1859 as historians of anthropology have noted (Stepan, 1982; Stocking, 1971; Stocking, 1982; Stocking, 1987; Owen, 2008) Much of the intellectual debate on race was developed through the learned London societies. The Aborigines' Protection Society, formed in the late 1830s, split during the 1840s, with the breakaway group forming the ESL in 1843. This was a vehicle for the unitarists, and in the beginning strongly reflected James Cowles Prichard's views (see Chapter 2). In effect it was a polemical and political association, its ethnography dedicated to supporting the missionary and abolitionist cause, and demonstrating that all races were part of the same human family.

By the mid-1850s the ESL was in decline. As its membership began to fall its revenues dropped (Stocking, 1971). It was increasingly driven by more secular outlooks on human origins, while still remaining true to its unitarist/monogenist philosophy. Dissentions between the monogenists and the growing number of polygenists in the ESL hierarchy intensified in the early 1860s, and in 1863 James Hunt (1833–1869) one of its most dynamic secretaries (Beddoe, 1870–1871), broke away to form the ASL, although as Stocking notes (Stocking, 1971) this was as much about personal empire building as it was about academic enquiry. Hunt was a rabid polygenist and in modern terms an out-and-out racist. His new society soon reflected a very active pro-slavery political stance. The Civil War in America was at its height, and issues of race were high octane topics. Confederate sympathisers were working overtime to manufacture the intellectual justification for racial domination. Desmond and Moore (2009) note how the membership of the ASL, and even its governing council, could boast Confederate *agents provocateurs* whose mission was to ensure the polygenist flag continued to fly.

Many students of race in the ASL appropriated evolution to support their polygenist interpretations, using it to show that individual races of humans could emerge naturally and then develop separately while remaining quite distinct. James Hunt advocated such views. Others within the ASL saw a series of separate divine creations but with the white race as God's chosen, and the so-called inferior races destined to serve them. Most of the ASL still remained wedded to the belief that the races could not successfully interbreed for any length of time despite Darwin's counter-proofs in the *Origin*. Hunt himself was fiercely anti-Darwinian, citing natural selection as a new form of the old religious unitarist ideology. But there was a deeper agenda here. Darwinian natural selection focused on the mutability of species arguing that a species was really a pause in a continuous process of change driven by selection. Darwin and other monogenists saw the different human races as inter-fertile varieties within a single species. So the different human races could change, could be improved through the benefits of civilization, and could ultimately give rise to new races under the umbrella of a single human species.

This was a position most polygenists would have found intolerable. They saw the different human races as distinct species. For them, species were insoluble types in the proper typological sense, their central 'essences' unique and not able to be dissolved or mutated by change (Stepan, 1982; Stocking, 1982), even if racial interbreeding (hybridity) could muddy the waters for a short time.

As alluded to already, surrounding Charles Darwin were a number of younger scientists drawn from a variety of physical, biological, and human sciences (Stepan, 1982; Stocking, 1987; Stocking, 1971; van Ripper, 1993; Desmond and Moore, 1991; Desmond, 1998). They have been described as the 'Darwinian aristocracy', a group of like-minded scholars and interested parties, united by a belief in human development and progress. In Chapter 2 we saw how two of them, Charles Lyell (less of a youngster than the others) and Joseph Hooker inveigled the Linnean Society to accept Wallace and Darwin's papers in July 1858. Mostly they were of the younger generation and opposed to the more traditionalist scientific explanations promoted by the older power elites who still held sway in the Royal Society and elsewhere. Most of the old guard were anti-evolution. While not necessarily all Darwinists (i.e. believers in Darwin's non-progressive natural selection mechanism), the members of this Darwinian aristocracy were nonetheless united by a common belief in evolution and the monogenist cause. They promoted ideas of change, and for them change and time were progressive. As a group, and especially motivated by their unelected leader T.H. Huxley, they were committed to the creation of a scientific meritocracy within a professional national scientific framework (Desmond, 1998). Huxley was deliberately forging a weapon to break the stranglehold of the traditionalists in the Royal Society. They looked for vehicles that would allow them to promote their own aims and agendas. The failing ESL, still avowedly monogenist, was ideal.

The disparate nature of the Darwinian group was reflected in its broad church membership, some were part of the core set, the so called X clubbers. Owen (2008) gives their professions as follows; Huxley (biologist/zoologist), Hooker (botanist), Tyndall (1820–1893 philosopher), Hirst (1830–1892 mathematician), Busk (1807–1886 gentleman scientist), Spencer (1820–1903 philosopher), Frankland (1825–1899 chemist), Spottiswoode (1825–1883 mathematician), and John Lubbock (archaeologist and anthropologist). There were others who were not part of this core set but nevertheless would have been counted within the Darwinian circle as they were wholly in sympathy with the group's aims; examples would be Charles Lyell, not admitted to the X Club because of his unitarian religious beliefs, Alfred Russel Wallace co-founder of evolution, and Francis Galton (1822–1911) the sociologist and eugenicist. Right at the centre of the group was the young archaeologist John Lubbock (A.E.S., 1914; Owen, 2008; Duff, 1924), an X clubber and member of the inner core, and of whom we shall hear a great deal more.

Both Stepan (1982) and van Ripper (1993) note how quickly the Darwinians moved in to occupy the ESL. John Lubbock was taking up the presidency as Hunt was defecting to form his new society. Owen (2008) suggests that the ESL became the power base for those in the Darwinian elite whose interests lay in the human sciences and in the origins of the human race. Unfortunately, not all of them could be said to be as committedly monogenist in their racial views as their acknowledged figurehead Charles Darwin. Huxley was only anti-slavery because he felt it demeaned the white slave owners. While an occasional individual of superiority undoubtedly existed (naturally occurring variation), the average black man could never be the equal of the average white. In fact nature (or natural selection) should be allowed to run its course. With no further protection or special favour for the black races they could be left to find their own natural position in nature's hierarchy. He had a similar view about women's emancipation (Huxley, 1902). Effectively natural variation existed between races (within the same species) as it did between genders. This was simply the pattern of nature. Non-interference and an end to special protectionism ensured nature would find its own balance as it was meant to do. A strict selectionist approach to humanity. Equally, Lyell was never convinced of the monogenist belief in racial parity, he toured the American Southern States as the guest of plantation owners (Desmond and Moore, 2009). However, most of the Darwinians were united in their dislike of James Hunt and/or the overt polygenist political agenda of the ASL.

In fact van Ripper (1993, 220f; see also Owen, 2008) asserts that, at least for the anthropologists and archaeologists, a distinct sub-group of this Darwinian aristocracy emerged, which he calls the 'Lubbock circle'. These included archaeologists and anthropologists such as John Evans, E.B. Tylor, Augustus Lane Fox (later General Pitt Rivers 1827–1900), Henry Christy, and A.W. Franks (1826–1897). Prestwich could be included because of his long standing friendship with members of the Lubbock circle as well as the broader Darwinian elite (Prestwich, 1899). However not all of the core Darwinians held Prestwich in as high regard as did his fellow geologists. Joseph Hooker wrote to Darwin about the Moulin Quignon affair, characterising Prestwich as an equivocator, and blaming him and Falconer for the poor impression created by their changes of mind. Sniffily, Hooker conceded Prestwich was a good field-man but did not rate his ability to interpret the bigger geological picture, i.e. as a theoretician (what Hooker called philosophical geology).

John Evans on the other hand seems to come off rather better. Huxley certainly had a high regard as this intriguing quote demonstrates.

> …[Huxley commenting on Evans]…a man who 'knows the wickedness of the world and does not practice it'… (Desmond, 1998, 358, my brackets)

Hooker too had a higher opinion of Evans. He described Evans's July 4th letter on Moulin Quignon to the *Athenaeum* (mentioned in the last chapter) as 'capital'.

*Sir John Lubbock 1834 - 1913*

*Sir Charles Lyell 1797 - 1875*

*Augustus Lane Fox 1827 - 1900*

*Sir Edward Burnett Tylor 1832 - 1917*

John Lubbock, as a member of both groups, provided the immediate connection. The impact of the Darwin/Lubbock Circle's control of the ESL on human origins research, and its troubled relationship with the ASL, will be dealt with in the next section.

Towards the end of the decade there had been a number of attempts at *rapprochement* between the ESL and the ASL, but for various reasons these had not succeeded (see below). It was not until 1871 that the two societies finally united with John Lubbock taking the chair as president of the Anthropological Institute of Great Britain and Ireland (Stocking, 1971), a name suggested by Huxley, as was Lubbock's candidature as its first president. It was not until 1875 that the Lubbock circle finally gained the upper hand, defeating the last attempts by polygenists to seize control of the new society's governing council.

## Human and Racial Origins in the Ethnological and Anthropological Societies

In this section I will give a brief overview of the main trends noted for racial and human origins research within the publications of the ESL and ASL through the 1860s and up to the formation of the Anthropological Institute of Great Britain and Ireland in 1871 (hereinafter AI). A comparison of the proportion of space given to different themes in each of the two societies is presented in Figure 3.1.

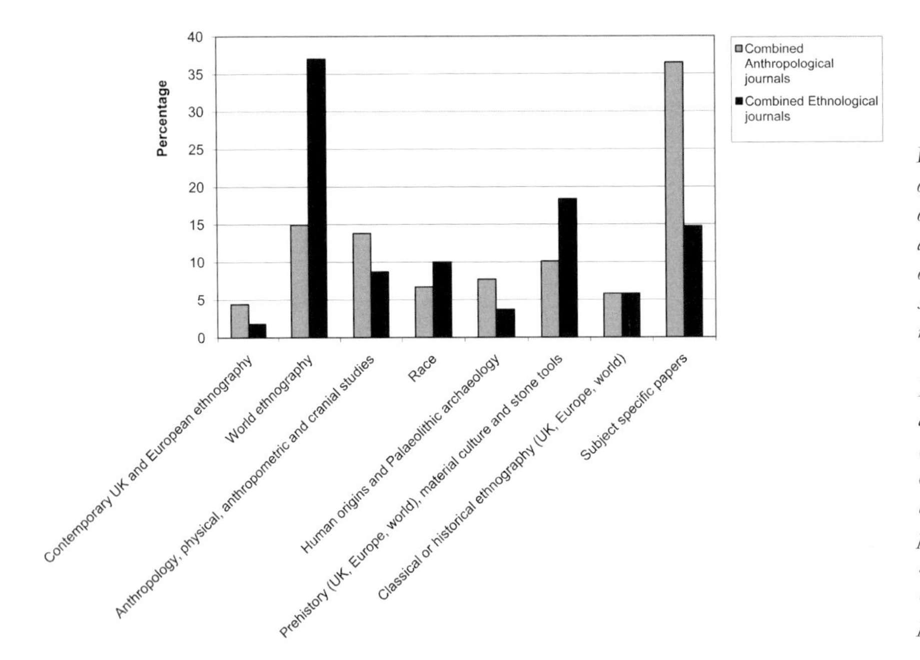


*Figure 3.1. Comparison of proportion of occurrence (as %) of different categories of article and other submission types (reviews, reports etc.) between 1861 and 1871 for the Ethnological (N=381) and Anthropological (N=720) Societies of London. The final category, subject specific papers includes theoretical, methodological, subject overviews and reflective papers.*

Owen (2008) suggests that the Darwinians used the ESL as a power base. If so, then the papers published in its newly revamped journal, the *Transactions of the Ethnological Society of London* suggest they did not make the most of their opportunities. The first two volumes in 1861 and 1863 (see Appendix 1) continued with the same range of material it had delivered in the 1850s. This was despite having John Lubbock as president, and five other Darwinians on ESL's council (three of them were X clubbers). In these first two volumes more space was given over to non-interpretative descriptions of indigenous peoples from around the globe, than any other topic. Effectively these were traveller's accounts of the various tribes that they met. For many people in the ESL (and the ASL) ethnography/ethnology was to be seen as a subject that gathered and compared information about contemporary peoples. The aim was to try and understand what made these people tick. Ethnology was the science of the contemporary races. As such it was an important and useful exercise for a world power whose dominion encompassed many different indigenous peoples. The presence of Lubbock and the other Darwinians in the ESL was not so surprising. The ethnology of the far flung corners of the empire, and that of Europe and even Britain, was intrinsically interesting in its own right. Lubbock and his fellows would have engaged with it as a matter of course. Racial origins at home and abroad were a part of this agenda.

But on the other hand the meetings and publications of the ESL were a vehicle through which human antiquity could have been promoted to the benefit of the Darwinian/evolutionist cause (alongside their short lived publications like the *Natural History Review* of which Lubbock was an editor (Keith, 1924) and part owner). It is curious, however, how few human origins articles there were in the *Transactions*. Human origins, if discussed at all, continued to be embedded within questions of racial origins. Perhaps they were up against a membership who were committed to the ESL's founding principles and were resistant to change (Keith *ibid*).

It is not fair to characterise the ASL as wholly polygenist, or the ESL as a nest of unrepentant monogenists, the situation was more complicated than that. Many members belonged to both societies, and saw no conflict of interest in this. Many lectured on the same topics to both groups, and the same racial and origins questions were of interest to both. The ESL's traditional monogenist and teleological heritage was represented in many contributions published in the *Transactions*, but a variety of polygenist and naturalistic papers were published as well. Representing this diversity in outlook was John Crawfurd (1783–1868), an enthusiastic and very vociferous ethnologist, who'd had a long and successful career as a colonial administrator (Keith, 1917; Livingstone, 2008) in the Far East. He was elected president of the ESL for the second time in 1865 and remained in office until his death in 1868, despite being a committed polygenist. He was publishing voraciously right up to his death. Two of his last papers focused on human antiquity. In 1868, the year of his death, 'On the Antiquity of Man' (Crawfurd, 1868) accepted the extreme antiquity of humans on the basis that many human achievements like metal working, writing, and especially language first appeared in the archaeological record in an already advanced state of development. The human races must have already had a long gestation to achieve that level. In a posthumous paper the following year (Crawfurd, 1869) he confidently predicted that Darwinism would be forgotten within a decade and had little contribution to make to ethnology. James Hunt would have appreciated the sentiments.

Crawfurd's voluminous output touched on common themes explored frequently by the ESL including racial migration (as a favourite monogenist mechanism for dispersing humans around the globe from a central point of origin), and the influence of climate affecting the skin colour and physical features of migrating peoples (as a mechanism to explain how one race could transform into many different races after migration had occurred), and the many ways of objectively quantifying racial distinctiveness.

On the other hand the publications of James Hunt's ASL reflected a very different set of ideals, and were far more overtly polemical. Desmond and Moore (2009) suggest that the ASL was really the only place in London where evolution and Darwinism could be freely discussed. This is curious as most of the ASL's polygenists hated Darwin. At the heart of the society were Hunt's polygenist and anti-Darwinian views. They emerged in his written papers, and in the discussions that occurred after lectures given in the Society's rooms. He was supported by a core group of his own including Charles Carter Blake (*c.* 1840–>1887), a fellow secessionist from the ESL, and the craniologist Joseph Barnard Davis (1801–1881). However Hunt may not have been able, or indeed willing, to totally control his society's agenda. In his writings he certainly presented himself as the unbiased searcher after scientific objectivity and wished others to see the ASL in the same light. The ASL's publications reflected a variety of outlooks, not always in sympathy with its founders' ideas. Hunt and others saw anthropology as being a complete 'science of man', encompassing not only ethnological study but also those researches that looked into the history and origin of humanity, its place in the natural order of life, and the relationship of all ancient and modern humans to each other. In effect anything and everything to do with humans and their place in the world. There was scope within such a widespread brief for any number of contrary points of view.

It is therefore not really surprising that human origins were more visible in the ASL from the very beginning. There were important papers published in each of the *Transactions of the Anthropological Society of London* (1863 volume 1) and *The Anthropological Review* (1863 volume 1, see Appendix 1). Charles Carter Blake (Carter Blake, 1863b) summarised the evidence that had been reported in a number of sources for the existence of humans in the Miocene era. With the concept of a humanised drift period being promoted by Prestwich, Lyell and Evans, this was an appropriate moment to broach such a subject, and it had been debated on the Continent for some time; but it received little favour from the English audience. There were a number of discussions in the *Transactions* on the Moulin Quignon jaw (see Chapter 2).

The *Anthropological Review* was exactly what its title suggested it was – a trawl through a variety of published sources for anything with anthropological relevance. It reported on conferences, papers given at other societies, and it reviewed books. In volume 1 (1863) Lyell's *The Geological Evidences for the Antiquity of Man*, and Huxley's *Evidence as to Man's Place in Nature* were reviewed with disapproval by ASL party loyalists. There was also a very detailed review of the human origins debate up to the early 1860s (Anon, 1863c). The detail in this article made it a very useful overview, providing the society's members with French and German data otherwise difficult to obtain. These papers and others like them in the early editions of the *Review* and the *Transactions* gave the anthropologists good grounds to feel their society (and their science) was a young and progressive one in contrast to the more staid ESL. The *Review* also reported on the ethnographic papers given at the BAAS. There was a further ASL periodical, the *Memoirs of the Anthropological Society of London*, reporting many of its evening lectures in more detail. In presenting large overviews, and the results of other scholarship, the ASL maintained a vivacious public front reflected in its growing membership. In particular the *Review* reported on developments in European scholarship, translated European lectures into English, and especially reported on events in the Anthropological Society of Paris whose leading figure, Paul Broca (1824–1880), was in sympathy with many of Hunt's own views. The ASL took on a programme of translating and publishing in English what it saw as key polemical texts by senior European researchers. In 1864 the first of these was translated by James Hunt, and published under the ASL's auspices. The Swiss anthropologist and naturalist Carl Vogt (1817–1895) was the professor of natural history at Geneva in the 1850s and an outspoken polygenist. His *Lectures on Man* (Vogt, 1864) presented his particular views on man's place in nature, archaeology, anthropology and race science. In effect it was a European polygenist counterpart to John Lubbock's monogenist *Prehistoric Times* (1865b).

The ASL might have temporarily won the public relations war against the ESL, but it was at the BAAS that the ESL's Darwinians fought back. This was where it actually counted.

This annual festival, held in a different city every year, was the showcase for British science, with all the major players attending, and cutting edge work being reported. Hunt once described it as the parliament of British science. Not surprisingly the non-archaeological members of the Darwinian elite were desperate to control the biological sections, and by the early 1860s they more or less did. Ethnology was included as part of Section E, a combined geography and ethnology section (Flower, 1894). Anthropology did not have its own section and sat on the coat tails of the ethnologists. Much of the reporting of the BAAS's discussions in the early years of the *Review* reflected the anthropologists' attempts to get their own section and their belief that the ethnologists were deliberately keeping them out. Hunt's anonymous editorial ranting at the ethnologists in the *Review* is perhaps the best reflection of how much more scientific and modern the anthropologists felt by comparison with the ethnologists of the ESL. One almost apoplectic response to John Lubbock is a case in point (Anon, 1864a). Lubbock believed ethnology was an older and 'prettier' word than anthropology. That the anthropologists should be deliberately denied their own debating section on such grounds was, for Hunt, an affront against science itself. Lubbock was president of the ESL at that time and did not see the need for a separate anthropological section, while at the same time strenuously denying any rivalry between the two societies – it was all in the anthropologists' minds. Or was Lubbock merely being disingenuous? Section D, the biological section, offered to take anthropological papers instead.

The *Anthropological Review* continued as it was, but the second volume of the *Transactions* saw a name change; it now became the *Journal of the Anthropological Society of London*, possibly because the ethnologists were using a similar title for their journal. Appendix 1 charts the shifting patterns in the topics that the two societies debated in the 1860s. There was an increase in the number of ethnological papers in the ASL's publications over the middle years of the decade. Prehistory and allied topics remained popular, but human origins papers were still few and far between; most contributions consisted of papers centred on discussions of the Neanderthal skull that Huxley had described in *Man's Place in Nature* (1863), or more reviews of Lyell and Huxley's books. Casts of the Neanderthal skull, exhibited during meetings on a number of occasions (Carter Blake, 1864b), were variously dismissed as an abnormal but recent Celtic skull (Carter Blake *et al.*, 1864), or a diseased modern skull (Davis, 1865). There is more than a note of satisfaction in the attempts to disprove Huxley's contention that it belonged to a fossil race of humans (*sensu* Darwinian monogenism). Hunt was particularly impressed with Barnard Davis' dismissal. Huxley's views on the Neanderthal skull will be described in more detail in Chapter 5.

The *Anthropological Review* presented a large number of book reviews during 1864–1865, and human origins again featured mostly as part of other origins questions. A distinctive philosophical element began to emerge in these reviews with less of a focus on practical or empirical proofs of humanity's place in the natural world. Perhaps this was a reaction to the materialism implied by Darwin and explicit in Huxley's treatments of mankind.

1866 saw a major triumph for the anthropologists. The BAAS in Nottingham finally recognised the distinction between ethnology and anthropology. While the ethnologists continued in Section E, anthropology became a department in its own right within a revamped biological Section D. However, it would be many years before anthropology would be elevated to the level of a full section – not until 1884. Desmond (1998, 351) suggests that Huxley had pulled strings to achieve this as he and Hunt worked towards a merger between the ESL and ASL. But political events overtook them. A revolt by former slaves in Jamaica was ruthlessly suppressed by the British governor, and the resulting polarisation of opinion at home (Huxley/liberals/monogenist/anti-governor Eyre/ESL vs Hunt/capitol/polygenist/pro-governor Eyre/ASL) derailed negotiations.

Between 1866 and 1868 the *Transactions of the Ethnological Society of London* continued its emphasis on world ethnography. Human origins continued to sporadically appear in papers on racial origins. Another curious lacuna in the ESL's programme was the conspicuous lack of archaeological papers related to the Palaeolithic. This was a productive time for the handaxe hunters who stalked the gravel quarries of southern England, so this is especially surprising for a Darwinian/Lubbock dominated society. The ASL on the other hand developed some persistent research interests around human origins and Palaeolithic archaeology. Impressed by the work of Édouard Dupont (1841–1911) in the Belgian caves (Anon, 1866; Carter Blake, 1867b), the ASL sponsored a visit by Carter Blake and a number of other society members to observe and participate in Dupont's work. This took place in the autumn of 1866 with reports published in all three of the ASL's periodicals, and a long analysis of a jaw bone found in the cave of Trou de Naulette near Dinant, Belgium, published in the *Memoir*, see Table 3.1 (Carter Blake, 1867a). Dupont wrote to Carter Blake announcing further discoveries (Dupont, 1867) including the excavation of the now famous cave of Trou Magrite (Dupont, 1868). What is perhaps most significant about the Belgian evidence was the ready acceptance of the validity of the human fossils and the archaeology. There was none of the doubt that surrounded Kent's Cavern or Brixham. This represented a major shift in attitude in only a few years. Carter Blake, after an exhaustive analysis of the Naulette jaw, and comparison with three thousand other specimens in the company of the Parisian anatomist and polygenist Pruner-Bey (1808–1882; Bey was a title that F.I. Pruner was granted after service to the viceroy of Egypt), concluded that the jaw with its lack of a developed chin, was racially similar to the modern Slavic peoples. Some remains from the Belgian cave of Trou de la Frontal were racially associated with the Calmucks, an indigenous people from Asia (the modern Kalmyk people from the Republic of Kalmykia on the western shore of the Caspian Sea).

These last few years of the 1860s were really the heyday of human origins reporting for the decade. James Hunt died in August 1869 (Beddoe, 1870–1871). He had ceded the presidency of the ASL to John Beddoe (1826–1911), a British racial ethnologist, and a much less controversial figure than Hunt. This may well have been deliberate. There were changes in the character of the papers appearing in the ASL's *Journal* and in the *Review*. Hunt had been the proprietor/editor of the *Anthropological Review*. Although he claimed it technically independent of the ASL, it had been printed, marketed, and distributed through the same network as the ASL's *Journal*. By the end of the decade there were allegations of financial mismanagement on Hunt's part. He was accused of underwriting the *Review* from the profits of the ASL and its official publications, and so running the Society into debt. An attempt by Hunt to oust his chief accuser (Anon, 1868) from the Society failed. Once Hunt died however, things were free to change. The *Review* in its final volume (8, 1870) contained a higher proportion of world ethnology papers than previously, and its metamorphosis into the *Journal of Anthropology* (volume 1, 1870) continued this trend. This was its only edition. The *Journal* did not survive the merger with the ESL.

With the more genial Beddoe as president, the old *Journal of the Anthropological Society of London* continued for two more issues (volume 7, 1869; volume 8, 1870). Volume 7 saw an increase in the number of home grown ethnology papers, possibly reflecting the influence of Beddoe, and the final volume saw a much higher proportion of European and world ethnology, and Prehistoric papers.

In effect the boundaries between the ESL and the ASL were being blurred as each took on more of the mantle of the other prior to the impending merger of the two societies which all members must now have been expecting, if not dreading in some cases. Huxley was president of the ESL by 1869 and actively promoting the merger. The declining fortunes of the ASL were charted by the changes that anthropology as a discipline experienced in the BAAS. Many years later W.H. Flower, in a presidential address to the BAAS (Flower, 1894), implied that after anthropology's status was recognised, it struggled for the next few years to justify its position as an independent department within biology. In 1869 the department was renamed Ethnology when Section E was given over solely to geography. Finally in 1871, not surprisingly the year the two societies merged, it was renamed the Department of Anthropology within Section D.

# 4

# THE 1860s
## *A Growing Sense of Time and Place*

In the previous chapter I concentrated for the most part on the anthropologists and ethnologists, describing the beginning of the process of humanising the past through the theoretical framework of evolutionary anthropology. The Antiquity of Man debate was an object and context orientated polemic, driven by a methodology based on collection and description. It had been a very necessary first step. However, Darwin's *Origin* formalised a synergy between a number of allied disciplines; and from this commonality of understanding evolutionary anthropology emerged to add an interpretative phase to Palaeolithic study. That interpretation primarily involved analogy with modern non-industrial peoples from around the world. Once Prehistoric and modern societies had been located within a unilinear developmental sequence, however, evolutionary anthropology offered little else in the way of explaining those societies. It singularly failed to engage with the Palaeolithic of the drift. It was especially difficult to develop that engagement in Britain, as the drift discoveries did not allow for complex stratigraphic super-positioning of what appeared to be temporally successive cultural remains. This kind of stratigraphy was only really preserved in caves, and the British caves did not preserve the right deposits. However on the continent it was different, particularly in southern France. By the end of the decade a Palaeolithic racial anthropology of the cave age was being discussed. At home, the best that could be achieved was a racial anthropology of later Prehistoric and historic ethnic groups.

But the process of cataloguing and describing the Palaeolithic did not stop, far from it. This project played a key role in expanding the Palaeolithic temporally and spatially, as well as enlarging people's perception of it. In particular, the identification of palaeoliths from abroad began to suggest that the Palaeolithic had been widespread. Theoretically there was an intriguing conundrum here. Did this represent a single contemporary drift/Palaeolithic stage which all humans had passed through at the same time, in the Pleistocene, everywhere, before evolving further? Or, were palaeoliths made by different peoples at different times and in different places as they variously passed through a Palaeolithic stage of evolution?

## Searching for the Evidence: Caves and Skeletons

There can be little doubt that the lack of undisturbed cave sites in Britain affected the ability of the English archaeologists to develop a body of archaeological theory to energise and engage with the emerging archaeological evidence for the Palaeolithic. The evidence was just not there. Caves preserved stratigraphic sequences of different archaeological periods in what E.B. Tylor labelled the 'layer cake' of time. The Belgian caves, and the south-western French caves of the Perigord were increasingly revealing long sequences of stratigraphic complexity which appeared to show a genuine development in material culture over time. European scholars were not slow in trying to create relative sequences of cultural development, and fit the human skeletal remains within them. While the English archaeologists had a strong theoretical platform from which to view evolutionary diachronic change in material culture, they just did not have the evidence to do it with. Their continental colleagues had an abundance of evidence, but were less inclined toward evolutionary theory building (Darwinian ideas were unpopular in France in the middle and late Victorian periods). The English archaeologists, or archaic anthropologists, as the editors of the ASL's journals began to call them, felt the frustration deeply. Table 3.1 listed the principle human skeletal remains discovered up to 1859, and afterwards. Clearly, by the mid-1860s there were still only a handful, none British. The Red Lady of Paviland, discovered by Buckland in 1823, was not formally recognised as a Palaeolithic skeleton until the early 1880s by de Quatrefages and Hamy, although Lartet and Christy suspected as much in the middle 1860s (Sommer, 2007; Lartet and Christy, 1875).

Caves of known later Prehistoric date were excavated to bedrock (cleared out in Victorian terminology) in the hope that lower down drift age deposits might still be preserved. Excavations at the Kirkhead Cave near Ulverstone overlooking Morecambe Bay (Bolton and Roberts, 1864; Morris, 1866) revealed Roman occupation overlying what appeared to be earlier Prehistoric remains, including human bones. However, despite a flint tool, all of the associated animal species were modern. Similarities were suggested between the skull from the site and another recovered from alluvial deposits at Muskham in the Trent Valley (MacKie, 1863). J.S. Mackie, in describing the skull, suggested that certain primitive features implied the skull belonged to a previously unrecognised race inhabiting Britain, whose ability to walk upright had not fully developed. Mackie's audience at the ESL were not particularly impressed.

More promising had been the discoveries from North Wales at the cave sites of Perthi Chwareu and Cefn (Dawkins, 1869–1870) in the then county of Denbighshire. But these too had proved, on excavation, to be more recent in age. Both were Neolithic as thought (Cefn actually did have some Upper Palaeolithic flints but these were not recognised at the time). Interestingly, the human remains from these sites (and other localities in the area) showed a curious feature on the recovered tibias: they were *platycnemic*, which was a flattening of one side of the lower leg bones. This feature was noted by G. Busk (Dawkins *ibid*) to be present on the Cro-Magnon remains from Les Eyzies (see end of this chapter), but not on the remains from the Belgian bone caves, so he was reluctant to assign the condition the status of a Palaeolithic racial characteristic. Elsewhere, the remains from the Heathery Burn cave near Stanhope, Durham, proved on the basis of metallic finds and pottery to be Bronze Age (Mackie, *ibid*). There was, however, sufficient interest in the finds for them to be exhibited to the Society of Antiquaries in June 1862. Near the estuary of the Mersey, a whole skeleton was preserved under a peat bog, sitting on top of glacial boulder clay near Leasowe Castle in Cheshire (Busk, 1866). Its ambiguous context and lack of associated finds made it impossible to date with any more precision.

There were of course known drift age caves, but there were no substantial human remains in them. Apart from Brixham and Kent's Cavern, excavated by the meticulous William Pengelly, most British caves had not been carefully excavated. In general they appeared to lack the stratigraphy of continental caves – there was something frustratingly different about them. A portion of jaw bone and some other human fragments were said to have come from Kent's Cavern (Carter Blake, 1864a), but they were bought from a dealer and so were of no real significance. Carter Blake thought they were probably modern anyway.

Pengelly returned to Kent's in 1865 to resume excavations there (he had previously dug at the site in the 1840s). The project was funded by the BAAS and a steering committee was set up to oversee the project with Pengelly in charge of day to day working. John Evans, John Lubbock and Charles Lyell were also members and so was William Boyd Dawkins (1837–1929), who had published the Neolithic Cefn and Perthi Chwareu cave material. Dawkins was a rising star at this time and he will figure largely in the human origins debates of the 1870s and afterwards. He was a truculent and outspoken Welshman (Woodward, 1931) who began investigating caves in 1859 with the Reverend J Williamson at Wookey Hole, Somerset (see below). He continued his cave researches both while he worked at the Geological Survey and after he left, on the advice of Huxley, in order to become curator of Manchester Museum in 1869 (Woodward, *ibid*). He became professor of geology at Manchester in 1874.

Dawkins' cave experience ensured he was appointed to another committee overseeing a cave excavation, this time that of the Victoria Cave, near Settle in Yorkshire (Anon, 1869–1870; Dawkins, 1874). It was thought that Settle had the potential to be another Kent's Cavern or Brixham Cave, and great things were hoped for it. In the detailed recording required of the excavators by the steering committee, the influence of Pengelly is all too evident. Each artefact recovered was to be individually marked, its layer noted, and its position drawn on a plan of the cave. It will be discussed in more detail in Chapter 6.

## Searching for the Evidence: Links to the Modern Races

It is important to discuss in brief the research conducted on the question of racial origins in later Prehistory and beyond. Human origins was only one aspect of a much broader interest in origins and in racial origins. In many ways, arguments concerning the modern racial history of Britain and its roots in the historical, Roman/classical and Prehistoric periods, were a rerun of the polygenist vs monogenist arguments before and after 1859. Many of the same problems were being debated, just in a different context. Students of historic or Prehistoric race history were equally concerned with hybridism; how, why and especially where new races formed; migration; and the persistence of racial type through physical features. For example the polygenist D. MacKintosh (MacKintosh, 1866) argued that children of mixed racial backgrounds (for example Irish Gaelic and German Teutonic) would resemble one or the other parent, and their children even more so. As far as he was concerned this was a reversion to original stock in successive generations, proving that there was a real fixity of different European racial types (Galton would begin to argue along similar lines in the next few years).

Firstly there were questions concerning the identification and description of modern racial groups within the British population, and the attempt to link them within different regions of the UK. While this could be practised in its own right, it was often associated with a second research agenda: to link the modern British racial groups with those mentioned by the historical and classical authors like Bede, Tacitus or Caesar. This usually involved viewing the various British races as migrants from different parts of Europe and then attempting to link them to their continental homelands. Included in this was how far back in time these original European progenitors could be traced. Yet a third agenda was trying to distinguish racial affiliation for the human remains from the Neolithic and Bronze Age barrows and inhumation burials – the dolichocephalic and brachycephalic skull types. The former referred to skulls that were longer than they were wide, and their owners were characterised as a long headed people. The latter were skulls whose length to breadth ratio was more even. These were characterised as a round headed people. For the braver of the anthropologists there was the possibility that the Prehistoric races could be linked to the classical and historically validated ones through craniology – skull shape.

However on one topic all were agreed. There was little point in attempting to link these back to the British Palaeolithic races. There was too little information, and the actual time depth involved was not known.

For a millennia or more, European peoples had been on the move as they migrated and invaded each other's territories, and new invaders had swept into Europe from the east. So the job of the European ethnologist was to peel off the overlay of racial intermixing and expose the original affiliation underneath. Many attempted to deny that much mixing had occurred

(MacKintosh, 1866), but for other workers it was only too apparent in the people around them. Questions of European ethnology only concerned people with white skins, although for many race scientists, some European people were 'more' white skinned than others. Inter-European racial distinctiveness still applied; for example the Celtic or Gaelic races were distinguished from the Scandinavian and Teutonic races, but the precise biological relationship between these particular groups was not really articulated. This exposes a curious contradiction which was centred on the role of hybridity. In terms of the species concept, races and varieties were the same species if they could interbreed successfully and produce fertile offspring who maintained their fertility. It would have been fairly evident that all these differing white skinned people could successfully inter-marry and raise children and grandchildren. However here many European race scientists were arguing that the essences of the different European peoples should be viewed in the strict typological sense, that of insoluble racial essences. On the other hand it was impossible to deny hybridity as Gaels had bred with Teutons, and Iberians with Britons etc.; there appeared to be a tacit acceptance of hybridity playing a more forceful role in racial variation in these debates. As is so often stated in this book, at the heart of the problem was the lack of a viable theory of heredity that could engage with the origin and definition of race and species.

Given that the ESL was an organization whose mission was to describe the variability present in the peoples of the world, there was very little British ethnology in the *Transactions of the Ethnological Society of London*. There was not that much more in the *Journal of the Anthropological Society of London*, although major pieces on British ethnology were published in the ASL's *Memoir*. But the ASL did appear to maintain a stronger UK ethnology profile than its rival, and this was reflected in Hunt's *Anthropological Review*. There were relevant papers in all but one of its eight year run (appendix 1), and Hunt expressly stated that the *Review* intended to publish individual opinions on British ethnological questions (footnote to MacKintosh, *ibid*).

With the benefit of hindsight it is not really surprising that the anthropologists were more open to racial ethnology. It was they who were reporting on the excavations of later Prehistoric barrows and caves, particularly in the earlier issues of the *Review* and the *Journal*. Between 1855 and 1865 Joseph Barnard Davis (one of Hunt's core set in the ASL) and John Thurnam (1810–1873) published the various sections of the *Crania Britannica* for private subscribers (Davis and Thurnam, 1856–1865), and its results were reviewed and discussed extensively (Anon, 1856; Thurnam, 1864; J.B., 1868). It was a huge compendium of skulls from Prehistoric and historic contexts with attempts to identify racial affiliation and link these to the historical period. Thurnam posited that the long barrows were built by a dolichocephalic people, and that the builders of the later round barrows (now known to be mostly Bronze Age) were brachycephalic. However not all agreed with Thurnam's racial interpretation or his chronology, including apparently his own co-author (J.B., 1868). Thurnam published major articles on ancient British skulls in the ASL's *Memoirs* in 1863–64 (volume 1) and 1867–69 (volume 3). The *Memoirs* contained a number of other articles on British and European craniology.

Publication of raw data like this inspired people like John Lubbock to produce extensive overviews of all this barrow work (Lubbock, 1865a), and any number of related racial overviews were produced (Wilson, 1865; Crawfurd, 1866; MacKintosh, 1866; Anon, 1867; Lewis, 1870–1871), almost all defending contradictory points of view. Some scholars would concentrate on language (Crawfurd, *ibid*), others were adamant language should not be used as a criterion of race (Huxley, 1870). Most agreed that skull shape was a critical variable, but others would fix on physical features such as skin complexion, hair and eye colour (Beddoe, 1866a; Beddoe, 1866b; Beddoe, 1870), or even racial/regional temperament (MacKintosh, 1866). Some of MacKintosh's results are given in Figure 4.1. On one memorable evening at the ASL there was a debate on regional temperament (Pike, 1866a). It led to a spirited discussion with members of the audience commenting in the following vein.

> Dr. Seemann observed that one of the reasons assigned by Mr. Pike why the English were not descended from the Germans was that they had forgotten the use of the fist. But they have not forgotten it, and if Mr. Pike were to express in Germany the opinions he had expressed that evening, he would have practical experience of the fact. (Discussion to Pike, 1866a, cxvi)

In fact L.O. Pike's book *The English and their Origin* (Pike, 1866b) spawned a notorious plagiarism case (Anon, 1869b; Anon, 1870).

The ethnological literature on British racial origins reflected a variety of points of view and approaches. It also encompassed a diverse range of scholarship. For example the empirically based studies of John Beddoe (Beddoe, 1866a; Beddoe, 1870) were grounded in observations of physical appearance from more than 4000 patients in the hospital in Bristol where he worked. Others were more speculative but still data driven (Wilson, 1865; Lewis, 1870–1871), while some were, effectively, unsubstantiated personal opinion (MacKintosh, 1866). There were a number of reasons for this diversity of opinion and apparent lack of rigour. There was no formal professional/amateur divide at this time. There were relatively few people who earned their living from practising science. Men like Prestwich, Evans and Lubbock were gentlemen scientists whose personal wealth allowed them the luxury of following their own interests. Others like Huxley had worked their way up from poverty and needed paid work, so combined many jobs in order to pay the bills. The main arenas for scientific debate and publication were the learned societies whose membership was open to anyone with an interest.

*Figure 4.1. Racial types as suggested by MacKintosh in 1866. Faces 1–5 represent the prevailing type in North Wales; includes, long necks, dark brown hair, long narrow faces and sunken eyes. Broad skull, approximately square in shape. Faces 7–9, a second North Wales group. A broad face under the eyes which sinks under the cheek bones Dark complexion and dark brown hair. Skull squarish. Found along the North Wales coast from Mold to Caernarvon. Also present along West Wales coast. Faces 10 and 11, a third type in North Wales. Thick set and large framed. A broad face, often associated with the more prosperous individuals. Possible descendants of the Iron Age Silurian tribes. Types 12–14, allied to the Gaelic peoples. Lower face projects forward. Most extreme type is 15 as seen in law courts in Beaumaris (though whether in the dock or not is not stated). Faces 19–21 Saxon group. A round, short, broad face and very regular features, but with low cheek bones and prominent eyes. Tend to obesity, light brown hair. Anglian type is represented by faces 17, 18, 22, and 24. Like Saxon but with a longer and narrower face. Narrower nose by comparison and more compressed nostrils. Fair complexion and light brown hair. Face 25 is the Jutian group. Narrow head and face, and face very convex in profile. Projecting cheek bones and a long nose. The Danish group were faces 26–28. Long faces with coarse features. High cheek bones with a receding chin. A narrow elongated skull wider at the back. Faces 6, 16, and 23 not mentioned in text. MacKintosh had this to say about Shakespeare. 'I cannot resist the belief that Shakespeare, if not a Welshman, was more allied to the Cymrian type, or one of its lateral variations, than any other type yet classified. In his native district, at least half of the inhabitants differ very little from the Gaelic-British and Cymrian-Welsh. To call Shakespeare a Saxon, would be to show a total ignorance of the science of races; though I should not like to be too confident in asserting that he was not a Dane' MacKintosh, 1866, 12.*

Another reason was the lack of a commonly accepted methodological practice which would have created an 'industry standard' of quality assurance in what, why, and how things were done. Evolutionary anthropology represented the only unifying theoretical infrastructure there was. Even if many researchers did not buy into it wholesale, or accept Tylor's particular spin on it; they at least were in sympathy with its broader evolutionary ethos.

## The Concept of a Global Palaeolithic Period

At what point the concept of a Palaeolithic period emerged is not too difficult to identify. By 1863 few of the more liberally minded British scholars would have doubted its existence. Van Ripper (1993) asserts that Lyell's 1863 *Antiquity of Man* pretty much ended the debate as to whether or not there had been a Palaeolithic age. The concept of a Palaeolithic period is certainly present in Lubbock's *Pre-Historic Times* and in Tylor's *Researches* both published in 1865. But at what point that idea of a Palaeolithic period translated itself into a clearly conceptualised globally significant stage in human evolution is more difficult to pinpoint. To some extent it was a theoretical necessity required by those who accepted progressive time and evolutionary anthropology.

Immediately after 1859 the geographical extent of the Palaeolithic represented southern and south-eastern England, and the equivalent geological area of northern France. For workers like Flower (Flower, 1866–1867), the similarities in geographical proximity and in the handaxes themselves suggested they were once part of the same region, Figure 4.2 and see below. So it began with chalk and flint. Soon after the limestone country of Britain, Belgium and south-western France was added to the Palaeolithic province. Palaeoliths were recovered in undisturbed fluvial drift sequences, or from the superimposed strata of the caves, but where to look next?

From the earliest reports of mariners returning home from new lands, stone arrow heads and axe-like tools had been part of the European experience of meeting indigenous peoples on distant shores. Imperial conquest usually followed. Europeans were becoming increasingly more engaged with the native peoples they ruled. Relatively few of these were stone tool makers and users. Furthermore, the idea of stone tools from ancient river deposits was still relatively new, even in the years after 1859. In 1865 E.B. Tylor was only able to provide two instances of Palaeolithic/drift implements (as opposed to Prehistoric stone tools) found abroad, one from a cave in Bethlehem, and one from a tell in 'southern Babylonia' (Tylor, 1865). In his seminal *Ancient stone Implements of Great Britain and Ireland* (1872; see Chapter 6), John Evans catalogued the known instances of Palaeolithic tools discovered abroad up to the beginning of that decade. The list was still small. The dates from the original articles Evans cited (1872, footnotes to pages 570–571) show that many of these reports were from later in the 1860s.

*Figure 4.2. Handaxes from Flower 1872. The two implements on the left hand side of the figure are from Thetford and the gravels of the Little Ouse river. The two implements on the right hand side of the figure are from St Acheul gravels in northern France. The upper two are pointed forms and the lower two are ovates.*

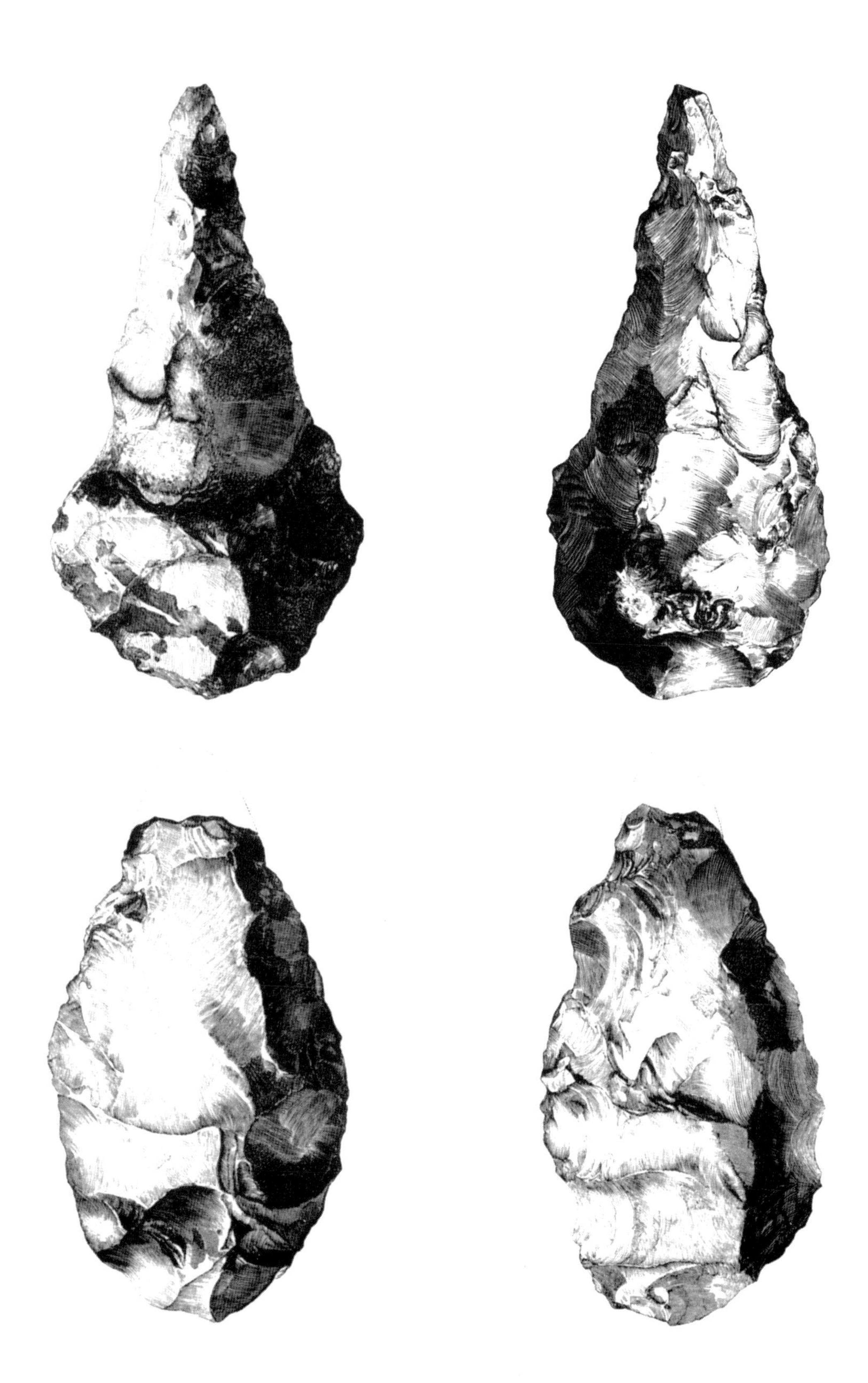

The ESL and ASL had cohorts of corresponding members throughout the world, as well as active members travelling abroad. These frequently sent back examples of material culture to the ASL, and occasionally stone tools. Arrowheads arrived from Canada (Fairbank, 1864), North America (Stirling, 1869) and Portugal, (Evans, 1869) amongst other places, as well as later prehistoric axes from Chile (Carter Blake, 1863a) and other corners of the world. The explorer Richard Burton (a vice president of the ASL in the mid-1860s, and potential president at one point) was a celebrated figure and a good example of the anthropologically informed traveller. Such items had been arriving back in Europe since the beginning of exploration, but now there was an evolutionary and racial spin to their interpretation. The arrowheads described by Fairbank (*ibid*) from plough soil had been made by Native Americans within the last few hundred years, but as they were chipped and not polished, like the Somme Valley finds, Fairbank argued they may have been made by a people living at the same stage of culture as those from Abbeville. Kenneth Mackenzie (Mackenzie, 1867) described stone tools from a location in the Amazon basin, the furthest west they had ever been discovered (i.e. away from the coast). Carter Blake, in commenting on the selection of axes sent from Chile (*ibid*), noted that amongst the more normal polished axes was one which was only partially polished. Boucher de Perthes considered these transitional between the drift palaeoliths and the fully ground/polished Neolithic axes. A commentator on one paper delivered to the ASL (Westropp, 1867) asserted that stone tools had recently been discovered in Burma, while the exhibition of surface collected material from Jubbulpore in central India (probably equivalent of Neolithic material) was rare enough to elicit a comment from Evans who delivered a paper on the artefacts to the Society of Antiquaries in January 1865 (Evans, 1865).

The route by which this material arrived in Britain is a nice illustration not only of the process of material being sent to the learned London societies but also of the relationship between the periphery and the imperial centre. The artefacts had been sent by a lieutenant Swiney to John Lubbock. He in turn passed them on to John Evans. Unfortunately, in the meantime, Swiney had died in India. In this case the fame of Lubbock had acted as a spur for Swiney to send his material on to someone whom he presumably had not met. In other cases many military men and imperial officials were also members of the London societies, further strengthening the direct connections between the public and those who were setting the research agendas of the time. They too would be sending artefacts back to London for identification and discussion. Richard Burton was variously Her Majesty's consul in West Africa, in Santos in Brazil, in Damascus and in Trieste. Like many in his situation he sent objects back to the ASL for discussions and display. Sometimes these included stone tools.

Arguably the most significant finds were those by Bruce Foote (1834–1912) in India (Bruce Foote, 1868, 1869). He presented his researches to the Geological Society of London in June 1868. While surveying stretches of the coastline of eastern peninsular India he noted a series of deposits he termed laterite – a stiff reddish clay with occasional patches of gravel. In some locations the clay matrix had been washed away leaving only the gravel. Bruce Foote demonstrated that the gravels were composed of quartzite clasts eroded from adjacent upland areas. These had once been islands at a time of submergence of the Indian coast when the Bay of Bengal had extended much further inland than today. The handaxe makers had lived on these islands and made their tools out of the quartzite. These had then been washed down into the shallow sea by river action where, with the rest of the quartzite fluvial gravels, they had been incorporated in the developing marine laterite deposits.

These marine laterites had been laid down during a period of submergence, which had drowned the land upwards of 500–600' above current sea level. There followed a period of elevation in which the land was raised up at least 600' and a long period of stasis followed while rivers and erosion began to dissect the laterite formations. Finally there was a second but probably minor phase of marine submergence after which the land re-emerged to its present levels. From his reconstruction of the geological history of the region it was evident that the period of the handaxe makers was very ancient. Further inland, and well above the level the marine submergence, Bruce Foote reported further finds of quartzite implements. Many were surface finds, but their condition was strongly suggestive of their derivation from drift deposits that had only recently been eroded away. Some of his colleagues had actually found implements within drift deposits. He was unable to determine the chronological relationship between the inland artefacts and those from the coastal laterite.

A number of people in the audience who listened to Bruce Foote's paper, including Joseph Prestwich, commented that the implements displayed by Bruce Foote were very similar to the ones from Abbeville 'their fabricators seemed to have been taught in the same school' (Prestwich in Bruce Foote, *ibid*). In a nice link between geology and anthropology, the President of the Geological Society, T.H. Huxley, who was in the chair for that evening, suggested that the makers of the palaeoliths had been the ancestors of the modern hill tribes of southern India, as well as the Australian Aboriginal peoples. The two groups were later separated by 'geological changes'. Both of these groups, he noted, were among the most primitive humans yet discovered (see also Huxley, 1869).

However what is evident in trawling through this literature is how little of it actually referred to Palaeolithic implements. With the exception of the stratified Indian material, by the middle–late 1860s I do not believe the notion of a globally significant Palaeolithic period could have been sustained on the meagre evidence accumulated to that time. The South African material is instructive in this context.

The first mention of a stone tool (at least within the publications of the

ESL and ASL) from South Africa was by a Mr T. Bains, while commenting on a paper given in 1866 at the ASL (Westropp, 1866). He asserted stone tools had been found near the Fish River quite recently and made the point that no known African peoples made stone tools, or had done so since first contact. The indigenous people had always been metal workers, consequently the tools must date back to a remote period and have been made by a different race of people to that inhabiting the district today. In 1869 at a meeting of the ESL John Lubbock (Lubbock, 1869) displayed some surface material from the table lands between False Bay and Table Bay (near Cape Town) which had been sent to him from South Africa. To a modern lithic analyst the collection, as illustrated, appears to contain Levallois points and blades, probably making it a Middle Stone Age assemblage. But the blades were new to Lubbock and there is a slight tone of surprise in his text. He had to ask contemporary flint knappers whether they thought such blades could be made by pressure flaking – they did not. Although Stone Age, none of this material was Palaeolithic as it was understood in the 1860s.

The following year George Grey (Grey, 1870) exhibited more material from South Africa to the ESL and read out letters describing more discoveries (one of his correspondents claimed to have been finding stone tools since 1858). A key point was made by Lane Fox in the subsequent discussion. He noted one quartzite implement was drift-like. He did not say it was a palaeolith, merely that it resembled one, and wondered if drift deposits in South Africa could be of a similar age as those in Europe.

> This type…[the tool resembling a palaeolith]… indeed, is not unknown to us as occurring at the Cape of Good Hope; for one other implement of this form was sent over by Mr. Layard some time ago…As we have now seen two implements of this form amongst the comparatively few specimens that have been sent from the Cape, it is reasonable to assume that it is typical; and being identical with those from the river-drift of this country, it may have been used in the same manner. (Lane Fox in reply to Grey, 1870, 41, my brackets)

But was it the same age? A similarity in tool form was not enough to guarantee a Palaeolithic identification. It required both a similarity of type as well as a secure drift-age context. This is the key to understanding why a global Palaeolithic stage was a difficult theoretical concepts to prove. Under the umbrella of evolutionary anthropology, the concept of discreet stages of cultural development, and the universality of the human mind, a group of people could experience a Palaeolithic stage anywhere in the world and at any time. Some modern native peoples had been at such a developmental stage relatively recently. It was conceivable that artefacts found on the surface were therefore made in recent times; particularly if they were on rock types resistant to weathering and which, unlike flint, did not stain or patinate. However I suspect (but cannot prove) that Evans, and a few other researchers were beginning to suspect that there may have been only one geologically ancient, and globally significant, Palaeolithic/drift period. This would have conflicted with a strict Tylorian view.

So identifying a global Palaeolithic period was not a straightforward matter. Other than Bruce Foote's Indian examples (also quartzite and so identical to the South African implement), Lane Fox did not cite another example of a non-European Palaeolithic site. Boyd Dawkins, also in the audience that night, noted that the shapes of the implements could date to any period; furthermore, since this material had been found on the surface it was probably not drift age.

### Searching for the Evidence: Expanding on the British Sequence, Further Discoveries in the Drift

We have already seen how, upon their return from Abbeville, Joseph Prestwich and John Evans began to search the British drift deposits for palaeoliths, initiating excavations at Hoxne, and encouraging members of the learned societies to get out into the gravel pits and look for themselves. This continued throughout the 1860s with increasing success. The results of these discoveries were not reported to the ASL or ESL to the same extent as they were to the Geological Society of London.

Van Ripper's (1993) data on the contents of the Society of Antiquaries' principle journal *Archaeologia* would suggest that Prehistoric subjects were not of much interest to the country's premier archaeological society anyway. Prehistory represented 7.4% of its published papers between 1849 and 1873 – 2.4% before 1859 and 5% after. Prehistory in this context was taken by van Ripper to be everything up to the Roman invasion of AD55. However, records of its day to day activities, from the *Proceedings of the Society of Antiquaries,* show this was not quite the case. There was a small but significant spread of papers read at meetings throughout the 1860s, they just didn't make it into *Archaeologia.* Perhaps this is not so surprising with John Evans, John Lubbock, A.W. Franks and others of the Lubbock circle on the council of the society. For example in late January 1864 a whole evening's discussion was given over by the Society of Antiquaries to stone tools, with exhibits from Britain and abroad, and even blocks of implementiferous breccia displayed (Haliburton, 1864). In April 1866 a series of skulls from Kew were exhibited at the Antiquaries by Thomas Layton. It resulted in a lively debate between John Crawfurd, arguing that the analysis of skulls provided little data of real worth, and T.H. Huxley who defended their study.

John Evans had predicted that palaeoliths would be found in the drift gravels of the river Ouse (also known as the Great Ouse), which rises in modern Northamptonshire and flows through Bedfordshire into the Wash.

As early as 1861 James Wyatt (1816–1878), a local collector and the founder and editor of the *Bedford Times*, was recovering handaxes from these gravels. One pit at Biddenham was particularly rich, but others along the river's course were also productive. At the base of the gravels were the bones of mammoth, rhino and cave bear (Wyatt, 1862; Wyatt, 1863–1864; Evans, 1872). In Hampshire, at Hillhead and Milford-on-Sea on the South Coast, the gravels of the vanished Pleistocene Solent river were producing palaeoliths, as were drifts near Salisbury which was a fruitful hunting ground for local collectors (Evans, 1863–1864; Blackmore, 1864–1865). In the valley of the Little Ouse in East Anglia, J.W. Flower (Flower, 1866–1867; Flower, 1868–1869) recovered palaeoliths from a series of localities along the river's low lying terrace system. His illustrations of pointed and ovate handaxes from Thetford, next to almost identical examples from St Acheul, see Figure 4.2, were cited by Evans in support of the reality of a widespread Palaeolithic period (Flower, 1866–1867, *ibid*). Remaining within the valley of the Little Ouse, handaxes from Stanton Downham were exhibited to the Society of Antiquaries in April 1866.

As noted above, there was a concerted effort to identify and excavate cave deposits in the hope that Pleistocene sediments would be present in their lower levels. Most were too recent, but at one cave location, Wookey Hole in the Mendip Hills, William Boyd Dawkins hit pay dirt (Dawkins, 1862; Dawkins, 1863). He and his colleagues excavated this former carnivore den between 1859 and 1862. Amongst the fauna from The Hyaena Den at Wookey were characteristic Pleistocene mammals – mammoth, rhino, cave bear, horse, even a cave lion. There were no human remains, but small handaxes, flakes and cores and knife-like flints confirmed they had been there. (Today these are interpreted as a mixture of Neanderthal–Mousterian tools and those of modern humans who arrived a little later.) They were located in the larger open part of the cave. Further in, in the tunnels and passages were bone accumulations from denning hyaenas, their teeth marks abundant on the bones of their prey. On the basis of the unrolled condition of the stone tools and bones, Boyd Dawkins proposed that the cave had not been infilled by a sudden deluge of water. Rather it had been infilled slowly as the cavern had filled with sediment from the natural decomposition of the surrounding limestone. There was some evidence of water action however in the layering of the sediment. This was argued to be a later sorting of the material by gentle water action after the cavern had been abandoned by humans and hyaenas alike and had been infilled.

Boyd Dawkins painted an interesting picture of the broader context of the cave, setting it within one of the earliest examples of Palaeolithic landscape reconstruction. In fact it is one of the earliest examples of Palaeolithic interpretation. He argued that there had been a vast plain in north-western Somerset bordered by hills to the north (the hills of South Wales); the east (Mendips); and the south and south-west (the chalk uplands of the western

Downs and the Quantock Hills). It stretched westwards to the Bristol Channel, and perhaps beyond. It was the home of horse and rhino, while carnivores, in particular bears, hunted in the wooded slopes of the encircling highlands. Humans competed with denning hyaenas for living space in the Hyaena Den cave, and made their tools within its confines. Given that he began excavating in the year of the Antiquity of Man debate it is hardly surprising that he set the discovery in the context of that debate, or that he was keen to show that the cave had not been infilled by flood waters. What is perhaps a little more surprising is the age he assigned to the site. On the basis of the fauna, but especially on the crudity of the implements (by comparison with those from Hoxne and St Acheul which he asserted were more refined), he suggested that it dated to the 'earlier part of the newer Pliocene period', in other words the pre-glacial Pleistocene (see Table 1.2). This was a radical suggestion. Within a few years he would hold very different views (see below). It is however one of the earliest published allusions to pre-glacial humans in Britain.

### Comparisons with the European Evidence: Chronology and Caves on the Continent

While the British Caves were proving to be a disappointment, on the Continent it was the cave sequences of Belgium, Germany and south-western France that were driving the idea of a Palaeolithic as opposed to a drift period, or more specifically a Drift Age, followed by a Cave or Reindeer Age, based on the predominant fauna of the caves. The relationship between the Palaeoliths of the drift and the Reindeer Age remained the subject of much debate.

There is a real sense in the literature of the difficulties of linking the British and northern French drift sequence with the cave evidence. Most researchers (Tylor, Lubbock, Lyell) accepted that the drift was likely to be earlier than the cave evidence, but how much earlier was impossible to assess. Were the two even made by the same race of ancient humans? Faunally, there were similarities between the drift and the caves; cave bear occurred in the river gravels, and mammoth and woolly rhino, common in the drift, were present in some of the caves. A comparison of selected chronological sequences is presented in Table 4.1.

There can be little doubt that the richer cave evidence significantly influenced the attitude of European researchers. Stone tools and human remains were occurring together. There were two ways of frame-working such a sequence. One was using material culture itself – the stone tools. However, the time when sequences of stone tool development could be set against stratigraphic succession was still some years off. The most significant early attempts at this were by de Mortillet in the early years of the next decade (see Table 4.1). Excavation techniques were not sufficiently refined enough to pick out the subtle but important differences between individual cultural layers.

<table>
<tr><th>Period</th><th>J. Lubbock<br>*Prehistoric Times,*<br>1865 1st edit.; + 1869 2nd edit.</th><th>M.E. Dupont 1871<br>*L Figuier 1870*<br>C. Vogt 1867</th><th>E. Lartet 1860s<br>(he became less certain of this chronology later in 1860s)</th><th>de Mortillet for the Universal Exposition in Paris in 1867</th><th>de Mortillet 1872</th><th colspan="2">J. Evans 1872</th></tr>
<tr><td rowspan="4">Palaeolithic</td><td rowspan="2">Reindeer period<br>Cave sites<br>Art works</td><td rowspan="2">Reindeer period<br><br>*Epoch of migrated+existing animals – reindeer*<br><br>Reindeer and auroch periods indistinguishable</td><td>Auroch/bison</td><td>La Madelaine</td><td>Cave deposits of La Madelaine or 'Epoque Magdalenien'. Flint blades, engraved antler and bone, and antler and bone tools.</td><td>Age of La Madelaine</td><td rowspan="4">Drift period in parts contemporary with cave deposits</td></tr>
<tr><td>Reindeer</td><td>Aurignac</td><td>Cave deposits from cave of Solutré or ' Epoque Solutrien'. Leaf shaped lance heads. Upper layers of Creswell.</td><td>Age of Cro-Magnon</td></tr>
<tr><td>Cave period at Le Moustier</td><td rowspan="2">Mammoth period<br><br>*Epoch of extinct animals – cave bear and mammoth*<br><br>Mammoth and cave bear occur together can't be separated</td><td>Mammoth and rhino</td><td>Solutré</td><td>Cave of Le Moustier or 'Epoque Moustérien' Implements worked on one face only. In UK the cave earth of Creswll, Wookey, and Kent's Cavern</td><td>Age of Laugerie Haute</td></tr>
<tr><td>Drift handaxes Archaeolithic?</td><td>Cave bear</td><td>Le Moustier</td><td>River drift age of St Acheul or 'Epoque Acheuléen'</td><td>Age of Le Moustier</td></tr>
</table>

Note – cells that appear next to each other in adjacent columns are not intended to suggest direct equivalence.

*Table 4.1. A selection of different chronological interpretations for the Palaeolithic period including how the drift and cave evidence were related, and the sub divisions of the Cave Age deposits from South West France.*

The other approach was to use fauna. This involved characterising a layer in a cave by the dominant species found. This was a standard methodology, long since familiar to palaeontologists and geologists. Although not everyone thought that even this methodology would apply to British evidence.

> The science of Archaic Anthropology may be divided thus: Subjects which the English understand and the French do not; subjects which are understood by the French and ignored by the English; and subjects which neither of them understand. It might be supposed that the third category would be the largest, but, on careful consideration, most heads are comprised under the second; whilst the first is a mere box for microscopical objects. (Anon, 1869c, 163)

Most caves were thought to comprise broader groupings of layers relating to a single epoch. Indeed the dominance of a group of layers by one species of animal virtually guaranteed the homogenising of any material cultural stratigraphy in a cave. Only when distinct stratigraphic layers showed a difference in fauna would a major break in the chrono-stratigraphic column be recognised. Vogt (1867) could only name four European caves where this clearly occurred. One of these was Aurignac.

The cave of Aurignac had been discovered possibly as early as 1852, but the seventeen skeletons reputedly discovered there had been hastily reburied and were never rediscovered, despite investigations by Édouard Lartet (1801–1871). He excavated at the site in 1860 (Vogt, 1864) and on the basis of this

*Textbox 4.1. Summary of Lartet and Christy's* Reliquiae Aquitanicae *(1875).*

and other work he established a four-fold sequence for the Palaeolithic caves in south western France. A Cave Bear Age was followed by a Mammoth Age, in turn followed by a Reindeer Age. To this period belonged all the artwork that would be discovered by himself and Henry Christy as they pursued their cavern researches throughout the decade. Their results are summarised in Textbox 4.1. They were published in a series of individual fascicules between 1865 and 1874 (Dawkins, 1874), and finally published in book form under the title *Reliquiae Aquitainicae* in 1875 (Lartet and Christy, 1875). Capping the sequence was an Auroch Age. Charles Lyell had suggested (Lubbock, 1865b) that the occurrence of seventeen skeletons buried in what appeared to be a natural tomb deliberately closed off by a stone slab, Figure 4.3, had to mean a more developed race, or at least a more advanced stage of culture, than the makers of the handaxes in the drift.

| Section | Author | Description |
|---|---|---|
| | Lartet, E. and Christy, H. Edited by T.R. Jones | *Reliquiae Aquitanicae; Being Contributions to the Archaeology and Palaeontology of Perigord and Adjoining Provinces of Southern France.* 1865-1875, published as complete volume 1875 by Williams and Norgate, London. |
| | | The plan to publish a single lavish volume of their cave researches was curtailed by the death of Christy in 1875 and Lartet in 1871. Rupert Jones completed the volume as editor. Much of it had been published in French or English journals, or had been delivered as lectures to the French (primarily) and English societies. The final volume took published accounts and lectures and edited them and included letters and comments solicited from various sources. No overall synthesis was attempted. The discovery of the Cro-Magnon remains, and soon after the human remains from La Madelaine and Laugerie Basse, greatly enhanced the volume. Archaeologically, the illustrations/descriptions of plates at the end of the volume were amongst the most useful parts of the volume as they showed in exquisite detail the stone and bone tools and art associated with the Reindeer period. |
| 1 | E. Lartet | Sets the scene describing the Vézère Valley and its limestone rock shelters. Principle sites are Laugerie Basse, Laugerie Haut, La Madelaine, Les Eyzies, and Le Moustier which have all aided in building a relative chronology. Drift race may be same as cave race, and there is a faunal and stone tool break between Palaeolithic and Neolithic |
| 2 | H. Christy | Closely based on a paper in *Transactions of the Ethnological Society of London* for 1865. Strongly monogenist reflecting its date. Embedded in concepts of cultural/evolutionary stages (*sensu* Tylor) which are global. Demonstrated in similarity of tool form in both simple and complex tools when comparisons made between Reindeer period and modern indigenous peoples. Drift and cave peoples may have been separated by long hiatus. Art indicates a lifestyle with free time to explore artistic expression for its own sake. |
| 3 | T.R. Jones | On geology of Vézère valley and region. |
| 4 – 5 | Various | Letters and notes on similarities between Dordogne Reindeer period tools and those used by modern or recent American Indians. |

| | | |
|---|---|---|
| 6 | L. Lartet | Description of Cro-Magnon cave and discovery of skeletons. An old man (CM1 – 'Old Man of Cro-Magnon'), woman with head injury and foetus next to her; 2 other male skeletons. Nearby perforated sea shells (all Atlantic species) and ivory pendant. Worked antler, but no art. There were 4 hearths below sepulchre layer, all sharing identical tools. Dated to Reindeer period (horse predominant species) but before art phase. |
| 7 | Pruner-Bey | Describing Cro-Magnon skeletons. All similar so possibly a family interred together. A tall and muscular people, distinct from remainder of Reindeer Age people who are shorter (e.g. Aurignac). Earlier work by Pruner Bey suggests Reindeer Age race are 'Mongoloid' brachycephalic with 2 groups, Finns (Solutré female and Bruniquel) and Laps. Cro-Magnon definitely Mongoloid but dissimilar to either branch, too dolichocephalic. Crania have abnormalities, and CM1 has rickets which confuses identification. Most similar to modern Esthonian people; palate is low and projecting more suited to producing Finnish languages. |
| 8 | L. Lartet | Fauna from Cro-Magnon |
| 9 | P. Broca | Skeletons from Cro-Magnon. The CM people are earlier than the Reindeer Age people of sites like Les Eyzies who produce art. Not necessarily a different race, possibly ancestral but time depth unknown. CM is Mammoth age. Common belief is only one race from this period, short and stocky, small cranium, prognathous, either Mongoloid or Negroid. But CM are a tall muscular race with a mix of advanced features (skull size, shape, frontal development of brain) and more primitive (broad wide face, big jaw/jaw muscles – evidence of violent lifestyle). Also *contra* Pruner-Bey no evidence of rickets. A dolichocephalic race very different to Belgian cave skeletons. Demonstrates polygenic sympathies by asserting this is a second Palaeolithic race. |
| 10 | A . Quatrefages | Formerly a believer in one Palaeolithic race in Europe to begin with – brachycephalic. Now accepts Broca's evidence for two. Does note so many problems with craniology and the number of exceptions to the rule. Accepts in principle it must be right. Notes all the northern European skulls he has viewed are brachycephalic including the Esthonian ones, as are Belgian Reindeer Age skulls. If Pruner-Bey is right then Estonia must have descendants of both races. |
| 11 | E. Lartet | On needles and sewing in Reindeer period. |
| 12 – 14 | Various | More on fauna |
| 15 | J. Evans | The draft of a lecture to Geological Society in 1864, published now with little updating, based on visit to Vézère in 1863. Does not wish to interpret mode of life of Reindeer Age people, merely establish their antiquity – also refers readers to *Ancient Stone Implements*. Two groups of animals – an older one in very fragmentary condition so probably introduced into cave by men. Later sample is contemporary with Reindeer Age fauna and the cave race. Describes briefly each of major caves. Notes the small handaxes in Le Moustier are more similar to drift handaxes so this is earlier than other reindeer Age sites. Notes the lance heads (in modern terms the Solutrean points) and the arrow heads (antler/bone points and harpoons) and needles indicate a high stage of culture. Possibly these are contemporary with metal users elsewhere. |
| 16 | E. Lartet | Catalogue of fauna by site |
| 17 | T.R. Jones | Lecture to BAAS in 1872 and to AI in same year. Notes pits and linear marks on some bone and antler pieces. Suggests they may indicate gaming pieces, or in other cases tally counts and ownership marks – based on ethnological parallels |
| 18 | T.R. Jones | On flint types and other rocks used in the Reindeer Age caves |

| | | |
|---|---|---|
| 19 - 20 | E. Lartet<br>T.R. Jones | On famous engraving of a mammoth on a tusk from Le Madelaine found in 1864. Jones (section 20) described the engraving of a glutton |
| 21 – 23 | Various | Describing Reindeer, evidence for fishing, and birds |
| 24 | T.R. Jones | Stone objects |
| 25 | E.T. Hamy | CM are the dolichocephalic race of southern France. Because CM1 is an extreme example, secondary characteristics which reflect true racial features can be distinguished from others which just reflect variation between individuals. Quoting from *Crania Ethnica* by Quatrefages and Hamy. 4 individuals have been found at Laugerie Basse, including the 'crushed man' skeleton between 1871–1872, and older finds of a single partial skeleton from Le Madelaine can now be better understood. Stature of LB 4 is similar to CM1 as are peculiarities in the lower leg bones. Skull shape very similar to CM1; also in lower jaw and its robust attachment to skull. Other remains at LB were a male cranium, a female one and a juvenile. La Madelaine and Laugerie Basse are late examples of the Cro Magnon race, while Cro-Magnon itself is an early example. Face and jaw of latter similar to CM1, but not as tall. There is a flattening on one face of the tibia – very characteristic of Cro-Magnon |
| 26-27 | Various | More faunal comparisons |
| | | Plates. Descriptions of the stone tools, antler and bone tools and arte objects from various sites. |

The drift, which most researchers accepted as being characterised by crude handaxes, could have fitted within the Cave Bear or Mammoth Ages of Lartet's faunal succession. Cultural evolution would then have continued to the Reindeer Age with its carved mammoth ivory and antler objects. At the heart of the south western French sequence were the caves and rock shelters of Badegoule, La Madeleine, Le Moustier, Laugerie Haute, Laugerie Basse, Gorge d'Enfer, and the Great Cave at Les Eyzies. These all dated to the Reindeer Age (Evans, 1864). Today these are known to be multi-period sites covering long sequences of separate Neanderthal/Middle Palaeolithic occupation, followed by modern humans with their various Upper Palaeolithic cultures.

Lartet continued to support this four-fold division until late in the decade, certainly as late as 1867 (Vogt, 1867). He died in 1871 and the job of bringing *Reliquiae Aquitainicae* to publication fell to Professor Rupert Jones (1819–1911; Dawkins, 1874). Henry Christy had died in 1865. Christy appears to have differed slightly from his co-worker in that he accepted an early Drift Age dominated by mammoth and rhino, but with a limited presence of horse and reindeer. This was followed at a considerable time later by the cave/reindeer period in which reindeer dominated, with horse and auroch present to a lesser extent. Variations on this simpler scheme were not uncommon (see Table 4.1). Carl Vogt (Vogt, 1864), argued that cave bear and mammoth ought not to be used to distinguish separate periods as they both occurred together in the drift. For Vogt, a mammoth/cave bear/drift epoch was succeeded by a reindeer/cave period. He found little evidence to support a final auroch period.

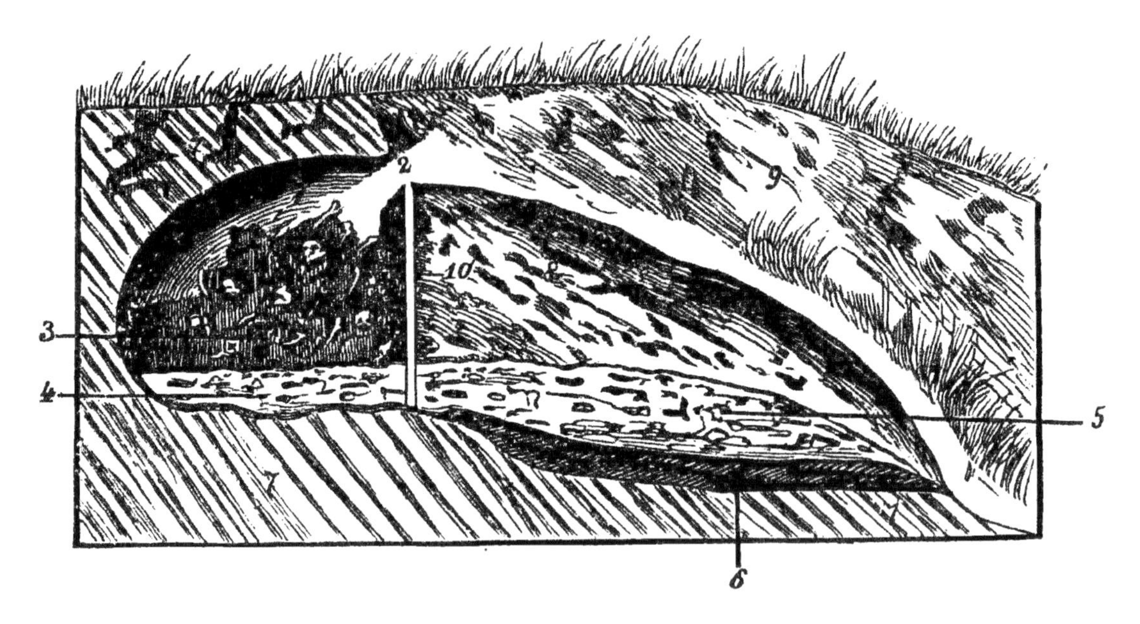


*Figure 4.3. The Grotto of Aurignac as depicted by Vogt 1864. 1. The inner vault; 2. The rabbit burrow which led to the discovery; 3. Human bones; 4. Rubbish with implements and bones inside the grotto; 5. Rubbish outside the grotto; 6. Deposit of cinders; 7. Rock; 8. Talus of gravel, which concealed the slab of sandstone (10); 9. Slope of the hill covered with gravel. 10. The slab of sandstone erected as a door to seal off the inner grotto. Lartet's excavations actually showed that 4 and 5 were a continuous layer with cave bear mammoth and other Pleistocene fauna. He found fragmentary human remains in 4 and asserted that they were of the same age as the skeletons from the inner grotto directly above that had been removed.*

One distinct advantage to chronological subdivision by fauna was that it allowed for the long distance comparison of different sites. By this method, sites from the equivalent of the Dordogne Reindeer Age were identified as far afield as Geneva and Belgium. This was a key point as although by the late 1860s the south-west French caves had not produced any substantial human remains, sites elsewhere had. Here the faunal comparison method came into its own. The interpretation of racial characteristics from one site could be extended to a site of similar faunal age which had not produced any human remains. It allowed various continental researchers to begin to propose what amounted to an anthropology of race for the Palaeolithic.

It will be recalled that Carl Vogt's book *Lectures on Man* (1864) had been translated into English and published by the ASL. His work (Vogt, 1864; Vogt, 1867) epitomises some of these early attempts at reconstructing such a Palaeolithic racial anthropology. He accepted the Moulin Quignon jaw and, because it was associated with mammoth and cave bear, identified it as a specimen of an ancient drift race. Also associated with the cave bear were the skulls from Engis near Liège, and the skull from the Neander Valley. The latter was distinctively dolichocephalic. Vogt (Vogt, 1867) implied that the Neanderthal race, as he defined it, was a later and probably more advanced human race than Moulin Quignon. He drew a vivid portrait of the lifestyle of this Neanderthal cave race. Other Belgian sites such as La Naulette and Trou de Frontal, with their reindeer dominated fauna, contained human fossils which were clearly later in time. The cave race of the Reindeer Age were round headed, a different people altogether, which showed advances on their predecessors. But the sample of skeletal remains was small; two skulls apiece from two caves; Lombrive in southern France, and Furfooz, a cave near

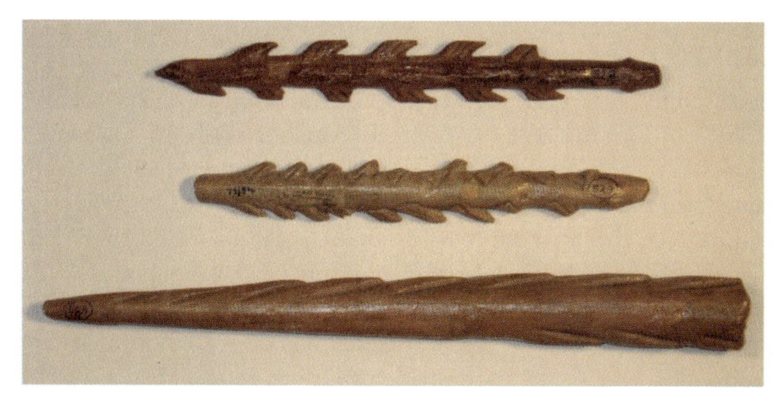

Casts of engravings from Lartet and Christy's excavations in the Dordogne. Top, engraving of mammoth on mammoth bone from La Madelaine. Middle, Ibex like animal from Laugerie Basse. Bottom finished and incomplete harpoons from various sites.

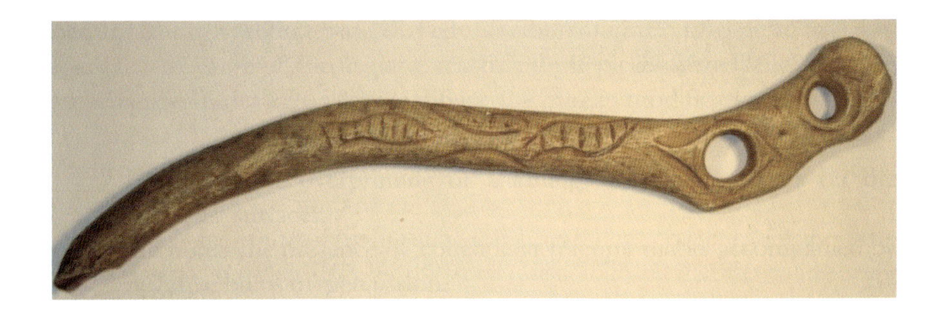

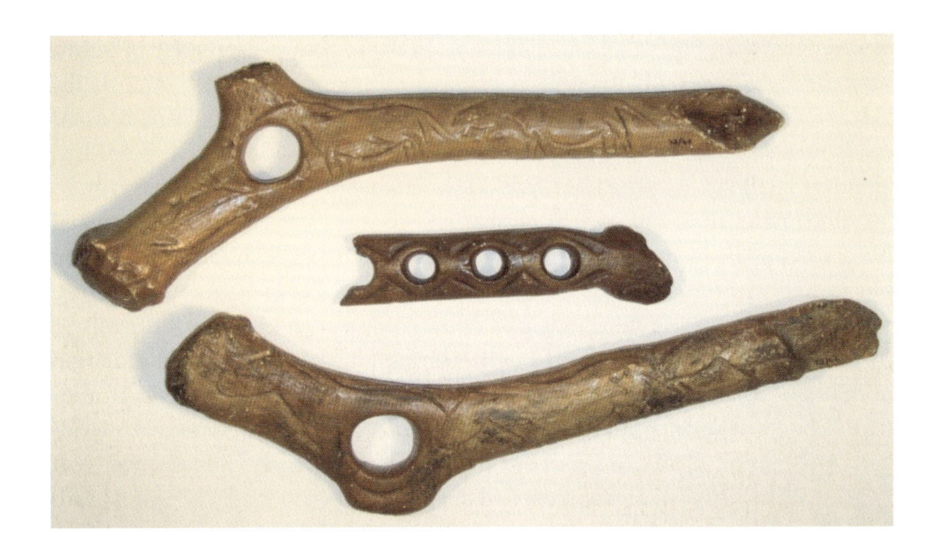

Casts of engraved antler from Lartet and Christy's excavations at La Madelaine

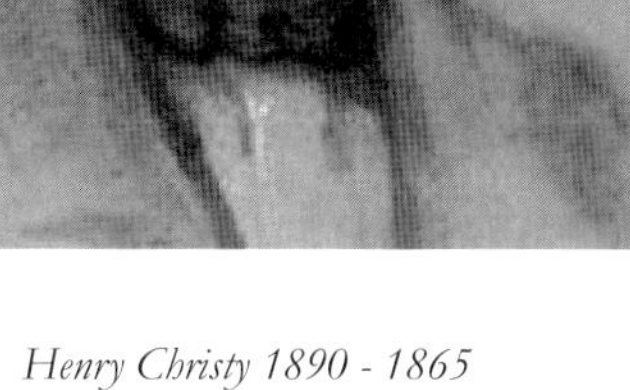

*Henry Christy 1890 - 1865*

*Édouard Lartet 1801 - 1871*

*W. B. Dawkins 1837 - 1929*

*T. Rupert Jones 1819 - 1911*

Dinant in Belgium. In the early 1860s according to Vogt these were the only cranial remains sufficiently well preserved to measure. Other Reindeer Age remains were too fragmentary. (The remains from Furfooz were originally described as occurring with crude pottery and goat, indicating in modern terms a later Prehistoric site. By 1867 Vogt was aware of the problems of dating both this site and Lombrive as well.)

Not all agreed with this dating however. Carter Blake, in reporting to the ASL (Carter Blake, 1867b), asserted that the prognathic (i.e. muzzle-like) La Naulette jaw was associated with elephant and rhino. But the jaw was primitive looking, it more resembled the mandible of an Australian Aboriginal, a people thought at that time to be amongst the most primitive of all the extant human races. In addition to the diagnostic character of skull shape, many Victorian researchers saw prognathism as especially indicative of primitiveness. Herman Schaaffhausen (1816–1893), professor of anatomy at Bonn (Schaaffhausen, 1868) was arguing that the Neanderthal skull and the prognathic La Naulette jaw were more primitive than any living human race. In Schaaffhausen's brand of polygenism, the dolichocephalics originated in Europe and Africa, while the brachycephalics were of Asian derivation. He believed that the brachycephalic roundheads were more amenable to cultural and intellectual development. This was the race producing the exquisite antler and mammoth ivory carvings of the Reindeer Age.

In 1868 railway extensions at Les Eyzies on the Vézère cut through deposits at the base of a rock pinnacle known as Cro-Magnon. The sediments revealed a rock shelter (Dawkins, 1874; Lartet and Christy, 1875) with four layers of hearth debris. Just above the highest hearth level, at the back of the rock shelter were the remains of five individuals, three skulls and the fragmentary remains of another adult and a child, see Figure 4.4, and Textbox 4.1. They were investigated by Louis Lartet (1840–1899), Édouard's son. The detailed anatomical report on the skeletons was by Paul Broca (Broca, 1868; Broca, 1869; Lartet and Christy, 1875) and was later published in *Reliquae Aquitainicae*. Broca, a committed polygenist, was at pains to note how different the Cro-Magnon remains were from any of the Belgian fossils, or from the Neander Valley. Those remains were of a short and stocky people whose build was more like that of modern Laplanders, however the Cro-Magnon were tall and resembled the modern European build. There was little evidence of prognathism. On the other hand there were more primitive features such as the width of the facial area and certain aspects of the jaw that suggested to Broca these people were unlike any he had seen before. Broca was implying these were a new race of Palaeolithic humans. He followed Louis Lartet in assigning the remains to the Mammoth Age (Duncan, 1869), although others asserted the find was Reindeer Age and that the mammoth remains had been brought into the cave by these later hunters.

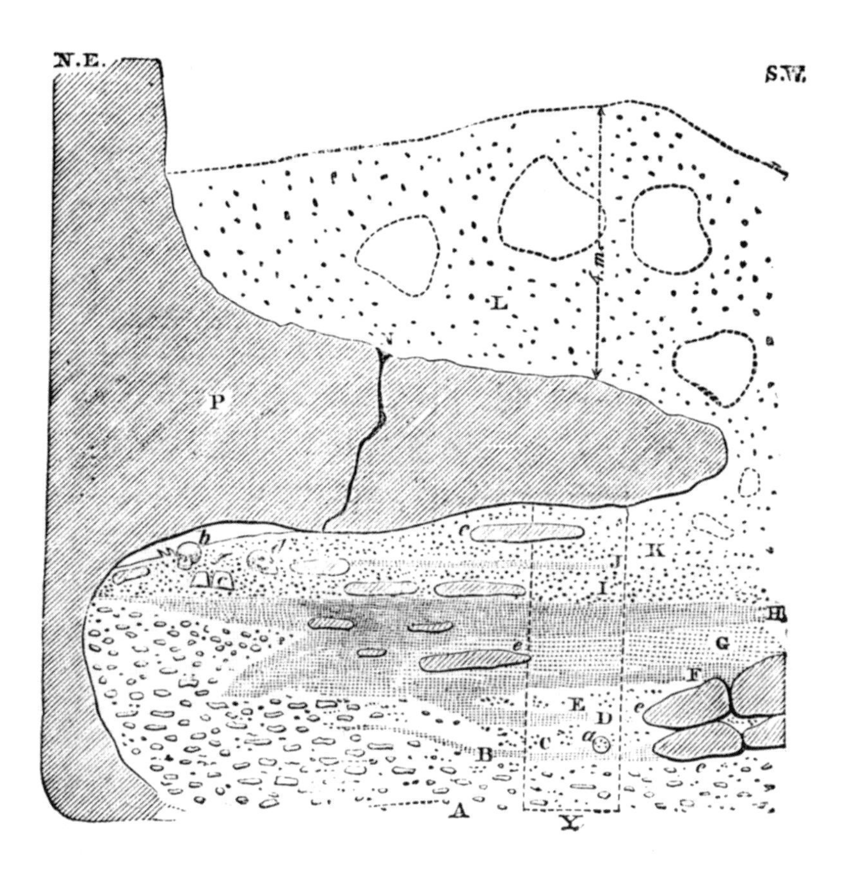

*Figure 4.4. The top image shows the cross section of the Cro-Magnon cave as given by Boyd Dawkins in 1874 and 1880. Letters B, D, F, H, and J are accumulations of debris representing occupation horizons. They contain charcoal fragments, flint implements and broken bones. The human bones are represented by lower case letters b and d. The letter a marked the tusk of an elephant, and the bedrock was at A. The crack in the rock shelter's overhang is clearly visible. The lower image is the rock shelter today with the broken face of the overhang very clear.*

# 5

# PHILOSOPHISING THE PALAEOLITHIC

In a book of this type it would not be normal practice to interrupt the flow of the narrative in order to discuss in detail the contribution of selected works by a few individuals. However, it is the context within which these writings were set that makes it important to review them as individual contributions. The evolutionary perspective gained ground after 1859, as did the belief in human antiquity. While neither concept was necessarily new, the interpretation that developed from their synthesis was. It gave rise to the idea of humanity emerging from the natural world as a result of long ages of progressive transmutation.

This new outlook did not immediately 'level the playing field' for the Darwinians. What it did was establish a new explanatory framework for human origins that sat alongside the more traditional one. The old playing field was the earlier Victorian domination of scientific thinking by the Royal Society, its established social and power elites, and the belief that the natural sciences were the handmaidens of theology. This was the scientific *status quo*. What the Darwinians did was feed a growing view of science as the servant of human betterment and liberalism. Their new playing field existed in setting out their wares in publications, books and scientific papers published in their own journals; their domination of the ESL, the BAAS, and in time the AI, and their presence on the councils of other societies. They created a framework for science that ran independently of the existing power structures (though not exclusively), and whose existence was based on tangible results – the demonstration that their science was providing effective non-teleological explanations for the natural world.

Linking these synergetic institutions were the personal relationships between individuals, their shared goals and aspirations. Here was where the true power of the Darwinian aristocracy lay. They were a new boys club. The links between them ran deeper than mere shared interests. Often it was the independently wealthy middle class background they had in common, or it was the business perspective they shared (Owen, 2008). It especially lay in their commitment to an evolutionary and progressive view of society and human development. In many cases it was a liberal and reforming political stance on social inequalities. For such people academic study provided hard data to support their progressive views. For others of the group what mattered was the establishment of a scientifically informed society in which success was predicated on individual ability. For most of them education was a key factor in building any brave new world. A meritocracy would emerge only from training programmes underpinned by qualified teachers, and especially science teachers (as we shall see in Chapter 12, H.G. Wells was a product of this).

There was no real body of pre-existing evolutionary theory against which to set these new ideas. So although all of the scholar's works discussed in this chapter were influential as individual contributions, we should keep in mind that together they were *creating* the background to a new understanding of humanity's place in the natural world. It is this that makes them important as a group. They drew from each other, and shared with others in their peer group, a set of common values and assumptions that came to underpin the middle and late Victorian view of human evolution. Since there was no theoretical perspective to draw upon or modify, they had to create their own. This was encompassed within the perspective of progressive time (McNabb, 1996) and evolutionary anthropology as discussed in the last few chapters.

Therefore the work of the scholars described in this chapter is important because they literally wrote the subject into being. E.B. Tylor has already been discussed in the last chapter; T.H. Huxley and A.R. Wallace reinforced the natural origins of humans, underlined the monogenist viewpoint, and showed how all humans could be a product of selection; Charles Lyell convinced the public that humans had a long history stretching back into the ice age, and then emphasised that time depth by describing a world very different from our own, and John Lubbock put the people back into Prehistory through his ethnography.

## Theory and Interpretation in the 1860s: Philosophical Evolution

Janet Owen (2008) suggests that while the 1870s were a time of consolidation and the gathering of facts to support the emergent evolutionary consensus, the 1860s were the crucible of dissention from which that consensus emerged (van Ripper, 1993). It was during this decade that the major theoretical positions were laid out. These were underpinned by the common assumption that evolution was almost always progressive and that change usually led to an improvement. The term progressive transmutation began to creep into the ASL's publications during the 1860s. The term may have suited those who were not comfortable with Darwin's non-progressive mechanism of natural selection, but were in sympathy with evolution, descent with modification, progressive time, and a naturalistic origin for humans.

I have grouped the contributions under three headings; biology, geology, and archaeology/anthropology. The divisions are heuristic. It will be fairly evident to most readers that individuals criss-crossed these boundaries with many of their writings, and would not themselves have accepted such simplistic pigeonholing.

It is not my intention to discuss these scientists in detail. Instead, I will take a selection of key early texts; particularly those that dealt with racial origins when viewed through the lens of evolution and evolutionary anthropology; and highlight their relative contributions. It is worth reiterating that these works set the theoretical standard for decades to come.

## The Biologists: Alfred Russel Wallace and T.H. Huxley

As his audiences acknowledged, Alfred Russel Wallace was the first to apply the Darwinian selectionist argument to the origin of race. This he did in 1864 in a lecture to the ASL (Wallace, 1864a). It initiated a dispute with James Hunt that continued over the next few years. Wallace's argument was as follows. Natural selection explained both the emergence of humans from the natural world and the development of racial differences between them. At the same time it explained the development of humanity's unique mental faculties.

At some point in the distant past, possibly the Eocene or Miocene – nine to ten million years ago or more, humans had emerged from a single stock. As with all animals, this stock was infinitely variable in all ways. This included skin colour, facial features, and type of hair. Lacking the basic faculties that rendered us human, especially speech, these early progenitors, more animal than human, spread out over the world from their original tropical home into a variety of different geographic and climatic zones. By chance some combinations of physical variation were better adapted to living in different zones than others. Groups settled where their own range of features best suited them to live. Once settled, the features of the different human groups in their distinct homelands became permanent as these packages of locally advantageous variability were selected for. Isolated from each other the different races began to emerge.

Up to this point natural selection had worked upon early humans as it did on all other animals living in the natural world. Human bodies (and their now racially distinct characteristics) were fixed and would not change anymore. They were by now perfectly adapted to promote survival in each of their regions. Thus humans had a monogenetic origin, but it was very ancient, as were different racial origins in different parts of the world. All of this occurred before the development of human speech and intellect.

Selection now shifted to the brain and natural selection began to work on the mental faculties (which would also be reflected in changes in the skull since mind and head (and face) were all correlated), especially those faculties that allowed an appreciation of the benefits of co-operative behaviour, sociality, forward planning, and mutual concern for others. In effect these were the 'moral' instincts that the Victorians believed had reached their apogee in their own civilization. Geography and climate played a part in determining how variable the development of the moral faculties were in each race – a polygenist slant on the development of individual races. So polygenism was also ancient and could be seen to represent an equally viable explanation for differing human origins simply because racial separation was so old. Wallace suggested that, depending on individual views, you could consider man a biological animal and place our origin in the pre-linguistic migrationist phase (and so be a monogenist), or the term could be restricted to after the time that speech and the mental faculties developed (backing up a polygenist perspective). In emphasising both, Wallace negated the two of them, deliberately suggesting that a monogenist vs polygenist dichotomy was redundant. Natural selection explained everything as part of a single process.

He was adamant that natural selection drove the development of intellect. Those mental abilities that allowed humans to evaluate the benefits as well as the costs of their actions would empower certain individuals and populations who would then gain an advantage in the struggle for existence. Gradually some races would supersede others as natural selection remorselessly drove the intellectual faculties of humans to greater heights. Wallace believed that, in time, human differentiation would diminish as selection perfected the human mind thus reducing the differences between us all, creating a more homogenised human species.

There is some evidence that the Darwinian elite's reaction to Wallace's paper was positive. Wallace had sent Darwin a copy. Hooker wrote to Darwin asking his views, saying how much he had enjoyed it (Hooker, 1864; Darwin, 1864b). Darwin thought it a 'capital' paper. He wrote to congratulate Wallace (Darwin, 1864a). They corresponded further on the matter. Darwin sympathised with Wallace's position though felt his mechanism of sexual selection (a non-adaptive mechanism, see next chapter) was more likely to explain racial development than the adaptive natural selection. This prompted Wallace to provide a strong defence of his ideas in a return letter (Wallace, 1864b).

Not surprisingly the polygenist ASL audience hated Wallace's ideas. Discussions on the paper (Wallace, 1864a pages clxx and following) especially by Carter Blake, and James Hunt, president in the chair for the evening, are instructive. They show the depth of misunderstanding that existed about Darwin's views at that time. Moreover, the notion that all humans would become more alike in the future would have incensed Hunt and the polygenists.

In 1866 James Hunt developed his objections to Wallace's paper in a lecture to the BAAS at Nottingham (Anon, 1866; Hunt, 1866a). The tables were turned now as Wallace was in the chair for that session (the first time anthropology had its own section at the BAAS). Hunt used the opportunity to launch a thinly veiled personal attack on Wallace and T.H. Huxley, arguing that Darwin's disciples were misunderstanding and misrepresenting the views of their master. Desmond and Moore (2009) suggest this may have been a ploy to draw Darwin himself out into open debate; if so it was doomed to fail. Wallace replied to Hunt in the discussions (Anon, 1866 *ibid*), but then wrote to the editor of the *Anthropological Review* (actually Hunt himself) in order to further clarify his views feeling that Hunt had in turn misrepresented him (Wallace, 1866). Hunt, not to be outdone, continued his attack on Wallace and Huxley with another paper (Hunt, 1866b).

Hunt's attacks on two of the most senior Darwinists, combined with the difficulties the polygenists had with the selectionist hypothesis as applied to anthropology, were in fact re-runs of the ASL's difficulties with monogenism in general. On a number of occasions Hunt asserted that Darwinism was a re-branding of the old monogenist view point. What ancient records there were of the different human races, Egyptian tomb paintings, Assyrian bas-reliefs, mention of different races from other Classical sources etc., made it clear that the races were fixed and had been so for a very long time. They were unchanging, and natural selection had not affected them in any way. Hunt pointed out that natural selection should be creating new species of human all the time, but it wasn't. There was no point in going further back in time as there was just no concrete evidence to support one viewpoint or the other (a favourite argument of Hunt's). He also reiterated the polygenist dogma on the creation of new species by racial intermixing. It was a well-known fact (he asserted) that any children born to mixed race parents were either sterile themselves, or had children who were. Hybridism could not threaten the fixed nature of the races. In attacking the old monogenist mechanisms of migration and climate (as explanations for racial differentiation/origin), Hunt asked how was it that climate could change a person's skin colour, or directly affect the type of hair they had? Wallace of course had argued no such thing.

It seems polygenists needed to characterise the opposition as all believers in the doctrine of environmental determinism. Why? Because they already had an excellent response – it couldn't work! Wallace here revealed the real depth of his Darwinian credentials, namely his grasp of the central roles played by variability and chance. Variability was a natural part of a founding population's make up. As people migrated, it just so happened that some of this variability was suitable to life in some of the areas the migrants traversed. So they stayed. Others moved on until they too found areas that suited them; classic Darwinism – natural tendencies to variability conveying fortuitous preferential local advantage. Wallace asserted that climate made further contributions to racial development, but again not in the direct sense the polygenists implied. A tropical climate, lacking harsh winters and providing food all the year round, was not conducive to mental development – there was little need for struggle in an environment where everything was provided for, and in abundance. However, once in the temperate zone, those hominins with lighter skins more suited to this environment would have found it difficult to succeed in an area where seasonality reduced the availability of fruit and plant foods, where the cold could kill you, and where meat was the only stable cross-seasonal food source. Here the struggle for survival would be intense because the challenges to life were intense. Natural selection would begin to work on those individuals with the intellectual abilities to solve the problems of living in a temperate zone. The result would be a selectionist driven development of those faculties described above that guaranteed co-operative survival and success. It's not surprising that the Darwinians liked Wallace's clever synthesis.

T.H. Huxley's *Man's Place in Nature* (Huxley, 1863) drew fire from the anthropologists as much as Wallace's views on natural selection and humans. Huxley was doing what Darwin had not, he was making explicit the naturalistic origin of humans, by revealing from which branch of the animal kingdom we had evolved. It was a monogenist argument, but Huxley was less convinced by natural selection than Wallace. In *Man's Place* he accepted Darwinism, but only because it was a theory that accounted for more facts than any other theory at that time. The book brought together a number of lectures and papers presented since 1860.

What angered Hunt and the polygenists was that Huxley forged the link between humans and the primate world by physically demonstrating their underlying unity. He did it from a zoological and anatomical perspective, but variability was at the heart of the argument. Humans were very variable. So were apes. When put together, there was a great deal of variability between the two groups, but it was all variability in degree, not in kind. For humans the underlying organisational similarities in the arrangement of hard and soft tissues in the body were such that they there could be no question of separate species' of humans. From the primates Huxley chose the gorilla as our closest primate relative, and then examined the hard and soft body parts of these animals. Again not surprisingly there was variation. But when the degree of variability in the range of humans was compared to that in gorillas, there was a strong level of similarity. In fact there was more similarity between the variations in humans and gorillas, than there was when gorillas were compared with the full range of other apes. The conclusion was clear. Humans arose from a similar stock to the higher apes. Our emergence from the natural world also showed in the distinct synergies between the embryonic development of human babies and those of other animals.

Part of Huxley's reasons for publishing *Man's Place* as a book was to make more public a debate he had with Richard Owen, then at the height of his fame. Owen was a powerful advocate for a teleological explanation for the origin of species and the epitome of the establishment that Huxley was at war against. In the book Huxley summarised the position of Owen and his supporters who asserted that humans should be in a separate and unique family in the hierarchy of the animal kingdom. From an anatomical point of view (argued Owen) apes had four hands (hence grouped under the heading of the quadrumana), whereas humans had only two hands. We had two anatomically distinct feet on the lower limbs. However, by following the anatomical construction of both hands and feet, and their differing muscular architecture, Huxley showed quite clearly that the feet of apes were proper feet and not a second pair of hands attached to the hind legs. Once again the similarity in the range of variability between humans and the highest apes was greater than the similarities that existed between the higher and the lower apes.

But the most celebrated part of Huxley's trouncing of Owen was in relation to the brain. Owen argued that the posterior lobe (back portion) of the brain in apes and monkeys extended only as far back as a structure known as the cerebellum, a very ancient part of the brain, Figure 5.1. It did not cover it, so that when the brain was viewed from the top down the cerebellum was visible. But Owen asserted that in humans the posterior lobe extended over the cerebellum and curved downwards covering it. Huxley was able to show that in both humans and all higher primates the posterior lobe extended backwards and did cover the cerebellum. Only in primates whose position in the family tree was below the level of the Lemur was the cerebellum exposed.

Owen continued his argument. Because in humans the posterior lobe did extend backwards and downwards covering the cerebellum, humans possessed a cavity (ventricle) within the posterior lobe which had a particular shape – like an animal's horn, following the downward curve of the posterior lobe. According to Owen this could not be present in chimps or gorillas. Consequently, a structure within this ventricle called a *hippocampus minor;* an upward fold of tissue on the floor of the ventricle was also only present in humans. Huxley showed that both the horn shaped cavity and the *hippocampus minor* were present in humans and all higher apes as well. Effectively, Huxley accused Owen of incompetent dissection and sloppy research (Owen *et al.*, 2009).

In every aspect of the human physiognomy *Man's Place in Nature* showed that humans and the higher apes belonged in the same family, the *Anthropini* or 'Man family' as Huxley characterised it. The gulf between them, and the members of the next primate family, the *Catarhini*, was considerable.

Huxley had to admit that any attempt to forge links between ancient skulls and modern human races was fraught with difficulties. He advocated a standardised methodology for orientating the skull so that its various features could be properly compared. He also made it clear that without the accompanying jaw, any identification of racial characters could only be provisional. He suggested that in terms of the modern world two broad groups of people could be identified on the basis of cranial analysis – those that were prognathous (prominent lower face similar to a muzzle), and those that were orthognathous (more flat faced). A racial 'polar axis' was to be found over the globe. At the southern/western pole, centred on the Gold Coast in western Africa, were prognathous dolichocephalic peoples with curly hair and dark skin. These were the stereotypical black African as far as Europeans were concerned. At the opposite northern/eastern pole, located in the steppes of Tartary (Mongolia), were orthognathous, brachycephalic people with straight hair and lighter yellowish skin colour. These were the

*Figure 5.1. A schematic illustration of Richard Owen and T.H. Huxley's dispute over the structure of the brain and the hippocampus minor.*

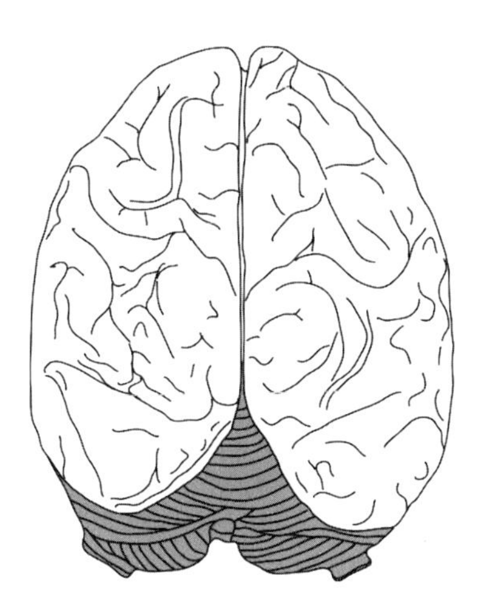

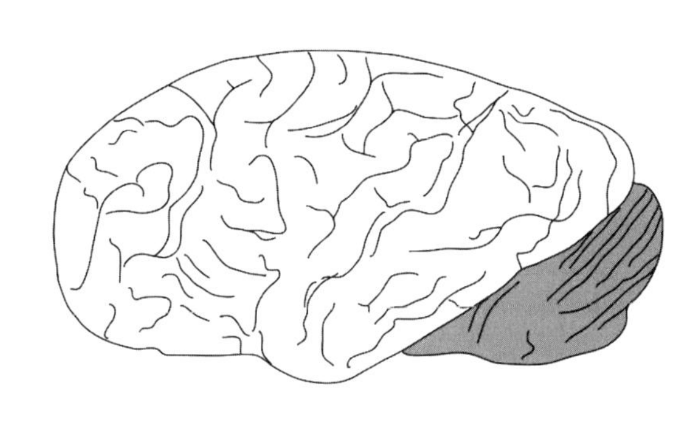

Upper images are schematic illustrations of simian brain from top down (left) and from lateral view (right) as represented by Richard Owen. In this interpretation the cerebellum (dark shading) is clearly visible protruding beyond the occipital lobe (back of the brain). Not to scale.

Lower schematic images are of the human brain. The cerebellum is tucked under the base of the brain. Huxley showed that in both higher apes and humans the posterior lobe did in fact extend backwards to cover the cerebellum. In the right hand image a portion of the lateral ventrical (cavity) of the brain has been cut away to reveal the hippocampus (dark structure) and the hippocampus minor (light structure, now more commonly called *calcar avis*). Huxley demonstrated the hippocampus minor was present in both humans and apes. Not to scale.

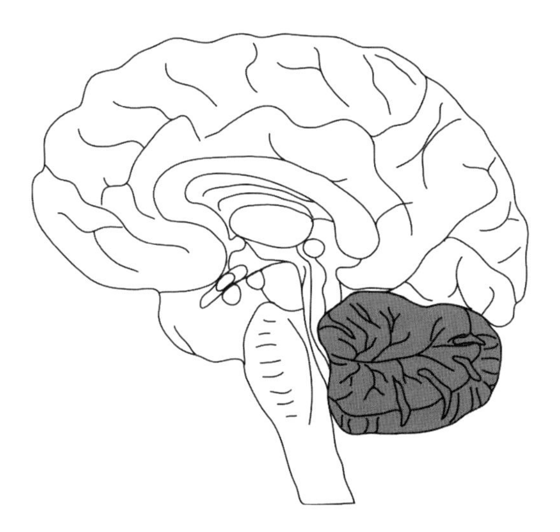

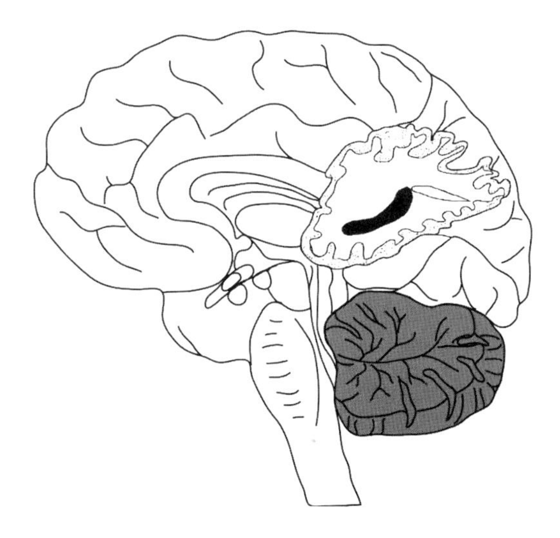

tartars and the Calmucks. Huxley then said if you drew a line at right angles across this polar axis, you would have a racial equator from Europe into India and beyond. Along this equator variations on these two racial extremes would be found, one variation grading into another as you moved from place to place along the equator.

In attempting to link ancient and modern humans on a racial basis there was one particular barrier that was impossible to overcome; there were so few examples of fossilised human remains that were convincingly ancient. Table 3.1 listed some of those found before the publication of *Man's Place*. It's a small list. Huxley was only confident about using two examples, the Neanderthal cranium and the adult skull from the Belgian site of Engis. But linking the Neanderthal and Engis crania to any of the modern racial divisions that Huxley described was difficult, especially without the jaw bones which were missing for each fossil. On the basis of the range of variation seen in humans along his racial poles and equator, Huxley was certain that both skulls belonged within the 'Man family'. Even the primitive characters of the Neanderthal skull fell within observed variation as seen in modern humans. It was nevertheless the most primitive looking he had ever seen. The posterior lobe of the brain must have been flattened out if the shape of the skull was anything to go by. This was similar to some Australian skulls, but again the degree of variability present in indigenous Australian peoples was considerable.

Two years after *Man's Place in Nature* Huxley returned to the question of race in a major article *On the Methods and Results of Ethnology*. Although republished a number of times in later years (Huxley, 1894), it first appeared in 1865 in the popular and widely read *Fortnightly Review*. That a celebrated scientist was publishing on this topic in a magazine with a broad general appeal to the public reflects the currency of these issues at that time. Once again Huxley was unable to link the modern races to the Palaeolithic ones. Although elements of his racial poles and equator remained, this was a much more detailed description of his views on modern racial distribution. Table 5.1 summarises his position. In reviewing the weakness of the monogenist and polygenist arguments, Huxley presented his own particular take on racial origins, albeit in a generalised way. Acknowledging the contribution of Wallace to a Darwinian explanation for race, he too placed variation and chance at the centre of his outlook. Natural selection was the mechanism that would combine these two into a driver for racial development. He approached the issue as a zoological taxonomist. The different races were varieties, they could interbreed, and they were part of the same overall family – a monogenist and naturalistic explanation. But it was clear that the races had been separated for a very long time. The Eskimo, Amphinesians, American Indians, South African Bushmen, Negritos, and Mincopies (in Huxley's terminology) were all relatively recently discovered peoples, but there was no data either on their history or on what time depth these racial groupings actually involved. But

*Table 5.1. T.H. Huxley's views on race and racial distribution across the globe during the middle 1860s (Huxley, 1894).*

the African Negroes, Melanochroi, Xanthochroi and Mongolians were very ancient racial divisions whose history dated back to the earliest records and presumably beyond.

| | Leiotrichi (smooth haired) | | Ulotrichi (woolly or crinkly hair) | |
|---|---|---|---|---|
| | Dolichocephalic (long headed) | Brachycephalic (round headed) | Dolichocephalic (long headed) | Brachycephalic (round headed) |
| Leucous (pale skin colour with yellow or red hair) | Xanthochroi – yellow haired and pale. Overlap in area with Mongolian people but mostly along northern boundary. Westwards into Europe, UK, Iceland, Canary Islands | | | |
| Leucomelanous (pale skin colour with dark hair) | Melanochroi – west of Xanthochroi including north Africa, Canary Islands, western Europe, UK, Syria, Arabia, Persia. Xanthochroi and Melanochroi responsible for European civilization | | | |
| Xanthomelanous (olive, yellowish, or light brown skin colour with dark hair) | Eskimo – group also included Tunguses of eastern Siberia, Japanese, Samoiedes of northern Russia, Tchuktchi of Bering Straits etc. | Mongolians - a huge and diverse peoples occupying land from Manchuria to Lapland. Not continuous across this area. | South African bushmen (San peoples of today). Yellowish brown skins with short stature. Distinctive language | |
| | Amphinessians. The Maori, Tongans, Poynesians, Micronesians, Indonesians | | | |
| | Native Americans from South America, Mexico and up West coast of North America | | | |
| Melanous (black of dark brown skin colour with black or dark hair) | Australian aboriginal peoples | | Negritos Name given to people outside of Africa with dark skin and crinkly hair, including Tasmanians, New Caledonians, New Hebrideans, New Guinea and adjacent islands.<br><br>Negroes/black Africans from sub-Saharan Africa | Mincopies – a people of short stature from the Andaman Islands, considered to be part way between African blacks and Negritos. |

Huxley's explanation was as follows. Geology emphasised the huge stretch of time between our world and its sculpted landmasses, and the Pleistocene world and earlier. Continents, land masses and islands had been submerged below the waves and risen again on a number of occasions, and the earliest 'families' of humanity had to migrate and survive across this dynamic world (see also Huxley's comments on Bruce Foote, 1868; Huxley, 1869). There was huge scope for natural selection to be working on the best adapted. But here chance and natural variation in adaptation came into play. Local circumstances would favour certain groups with particular sets of adaptations to survive in one area. Modern evidence helped refine this, and Huxley gave the example of yellow fever. Certain black African societies were known to have high resistance to this disease. If a group in the past with this trait had happened on an area with yellow fever they would have thrived while other human groups would have perished. Isolation would do the rest.

> Again, how often, by such physical changes, must a stock have been isolated from all others for innumerable generations, and found ample time for the hereditary hardening of its special peculiarities into the enduring characters of a persistent modification. (Huxley, 1894, 166)

In summary, the theory building of two of the most senior of the Darwinians (Huxley the biologist/zoologist; Wallace the biologist/naturalist) sought to emphasise the monogenist origins of all humans, and embed them firmly in the natural world. To varying degrees both advocated a selectionist mechanism to explain human origins, though Huxley was less convinced and he avoided discussing the specifics of racial origins in 1863. Both also shared the conviction that the origin of humans was very ancient, in the Miocene or Eocene (or even earlier for Huxley). For both, this great age explained a key point – why did Pleistocene humans like Neanderthal and Engis (Table 3.1), look so human? Why did they not appear more primitive, more transitional between the apes and man? The answer was they weren't transitional, because the transition had taken place much earlier. Engis and especially the Neanderthal skull were primitive, but undeniably human. It was an intriguing problem. From Huxley and Wallace's monogenist (but racially variable) perspective there had to be more primitive human fossils waiting to be discovered. The Mammoth Age and the Reindeer Age were comparatively recent in the scheme of human evolution and the skeletons from these deposits were recognisably modern.

### The Geologist: Charles Lyell

Lyell's *The Geological Evidences of the Antiquity of Man with Remarks on Theories of the Origin of Species by Variation* (Lyell, 1863a) sought to outline the argument for the antiquity of the human species by showing how the palaeoliths and other stone tools were evidence for 'men among the mammoths'. It was a popular book aimed at a broad public market. The first edition appeared in February to great success. In the previous November at the trade sale, all four thousand printed copies produced by the publishers, John Murray, had been sold. There was a second edition in April, with amendments to the text appearing in appendices, and a third edition in November. Van Ripper (van Ripper, 1993) asserts that it was Lyell's book that pretty much settled the controversy of human antiquity as far as the public was concerned. Most accepted that humans were a genuinely ancient species after 1863.

Lyell's own polemical stance was less clear cut than that of Wallace or Huxley. The book was an attempt to reconcile Lyell's own acceptance of the evidence for human antiquity with his strong religious convictions. He merits his place in a section on the post-1859 theory builders because of the forceful impact his writings had on the public perception of human antiquity. His conversion to this was genuine and he sold that story very effectively. But he evidently had problems accepting a wholly naturalistic origin for humans. In *The Antiquity of Man* as the book was popularly referred to, he equivocated on this key topic, much to the irritation of the Darwinians. Additionally, he was not particularly forthcoming on other issues such as racial origins. Darwin complained to Hooker that he wished Wallace 'had written Lyell's chapter on man' (Darwin, 1864b).

*Figure 5.2. Redrawn and modified after Lyell 1863, figure 41. His original caption reads as follow: "Map of part of the north-west of Europe including the British Isles, showing the extent of sea which would become land if there were a general rise of the area to the extent of 600 feet." The dark line represented the 100 fathom line, the limit of dry land if land levels were to rise.*

The book was divided into three broad sections. The first was on the antiquity of man debate. Lyell described all the major sites in Britain and France, both cave and open air, rehearsing the stratigraphic arguments for artefact location in undisturbed deposits. He accepted both Prestwich's fluvial theory as the best explanation for the drift, and also his interpretations of the cold climate conditions of the Pleistocene gravels. This led him into section two on the geological evidences of the ice age itself. There were long and detailed descriptions of glacial scenery in Europe and numerous examples of the erosive power of land and sea ice. The link between this and the first section of the book was the extent of landscape change that had occurred since the ice age. More so than the extinction of old species and the subsequent rise of new ones, the reshaping of the physical world during the ice age was demonstrable proof of the immensity of time that had passed since the Pleistocene era. The presence of handaxes in the glacial-aged drifts showed that humans lived on the fringes of these events and were witnesses to them. This second section of the book was much admired by the Darwinians. Both Darwin and Hooker commented on the efficacy of Lyell's prose in recreating the lost world of the mammoths.

Graphical depictions of changes to familiar coastlines underscored the time depth involved. Figures 5.2–5.5 show changes in the geography of north-western Europe during the glacial phase of the Pleistocene. Figure 5.2 represents the emergence of the land and the beginning of a long continental period. Throughout the Pliocene climate had been deteriorating. The cliffs of Norfolk and Suffolk were a classic area for showing the stratification of the Pliocene and post-Pliocene/Pleistocene sediments. Based on the mammal and shell remains the glacial period proper began during the series of sediments known as the Norwich Crag, Figure 5.3. The glacial period's first major phase was the equivalent of the Cromer Forest Bed and East Anglia was near the centre of the upraised continental plain depicted in Figure 5.2. Although overall climate was deteriorating, Lyell allowed for the possibility of temperate blips in the gradual cooling. There were elephant, rhinoceros, hippo and horse amongst the faunal remains. Many of the plant species preserved in the Forest Bed were still alive in modern East Anglia. Significantly, in all the years of searching the East Anglian cliffs, not a single stone tool had been found. Humans were not present at this time.

The Forest Bed in East Anglia was replaced by a long series of fluvio-marine sediments, Figure 5.3 marking the beginning of wide spread submergence. Britain would become a series of islands, an archipelago, as shown in Figure 5.4, as the waters of the glacial ocean slowly rose and the land subsided. Just how much of southern England below the Thames/Severn line was actually submerged was open to debate. Certainly north of this line only stretches of high ground would have been above the waves. At its greatest extent Britain would have resembled Figure 5.5. Sea ice was extensive and the boulder clays of the Norfolk coastline (Figure 5.3) would have been laid down by fleets

*Figure 5.3. After Lyell 1863 figure 27. His original caption and explanation reads as follows. 'Diagram to illustrate the general succession of the strata in the Norfolk cliffs extending several miles N.W. and S.E. of Cromer. A. Site of Cromer jetty; 1. Upper Chalk with flints in regular stratification; 2. Norwich Crag, rising from low water at Cromer, to the top of the cliffs at Weybourne, seven miles distant; 3. Forest Bed with stumps of trees in situ and remains of Elephas meridionalis, Rhinoceros Etruscus, &c. This bed increases in depth and thickness eastward. No crag (No. 2) known east of Cromer Jetty; 3' Fluvio-marine series. At Cromer and eastward with abundant lignite beds and mammalian remains, and with cones of the Scotch and spruce firs and wood. At Runton, north-west of Cromer, expanding into a thick freshwater deposit, with overlying marine strata, elsewhere consisting of alternating sands and clays, tranquilly deposited, some with marine, others with freshwater shells; 4. Boulder clay of glacial period with far transported erratics, some of them polished and scratched, twenty to eighty feet in thickness; 5. Contorted drift; 6. Superficial gravel and sand with covering of vegetable soil.'*

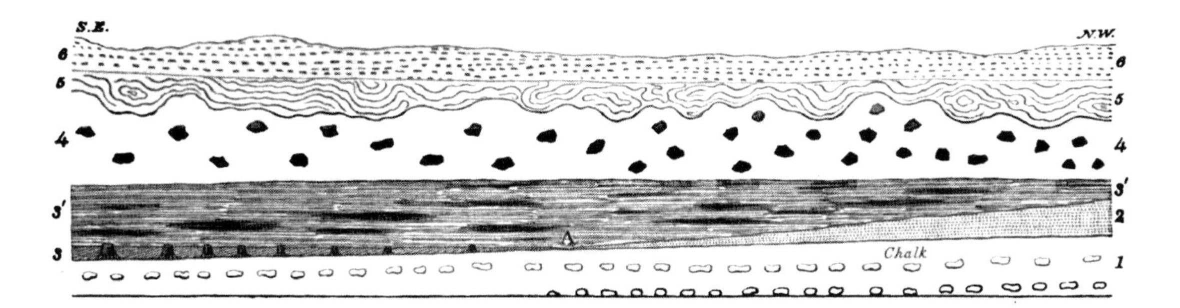


of melting icebergs. The mountains of north Wales showed marine shells at fourteen hundred feet up their flanks, and stratified drifts (believed marine) at over two thousand feet.

A second continental period occurred as the sea level dropped and the land began to re-emerge, reversing the stages shown in the three maps. Since the submergence had been so deep, the re-elevation of the land was considered to be equally as great. It was a time of mountain glaciers when the marine infilling of many valleys was scoured out leaving only isolated remnants of the great submergence. The extent of this continental phase was much greater than that depicted in the land area of Figure 5.2. It was to the end of this second continental phase that Lyell dated the arrival of humans. This was the period of the drift, mammoths, woolly rhinos, hippos, and the makers of the palaeoliths at Hoxne, Bedford, and at Abbeville. The final phase of the geographical history of the Britain was another lowering of the land accompanied by more modest phases of recurrent submergence and uplift until the final configuration of island Britain was achieved.

Lyell shied away from dating any of these events, but in describing continental evidence he quoted the work of Swiss geologists on the deltaic terrace sediments of Lake Geneva studied by Charles Morlot the Swiss archaeologist. The highest terrace at one hundred and fifty feet above the lake contained an extinct fauna similar to that in the Pleistocene gravels of Britain and France. The age of this highest terrace was estimated at one hundred thousand years; by association so were the French and British drift sites. Lyell may have believed the Pleistocene was older, but he was reluctant to push this further.

However, if the second section of the book delivered its message well, the Darwinians were less than impressed with the third section. In this Lyell sought to philosophically explore the place that humans occupied in the natural world. As he later admitted, the religious beliefs of his younger years kept drawing him back from the brink of openly denying any teleological aspect to human evolution (Lyell, 1863b), despite his private conviction that transmutation and natural selection may well have been better explanations. Darwin thought that this was a golden opportunity lost (Darwin, 1863a), especially given Lyell's

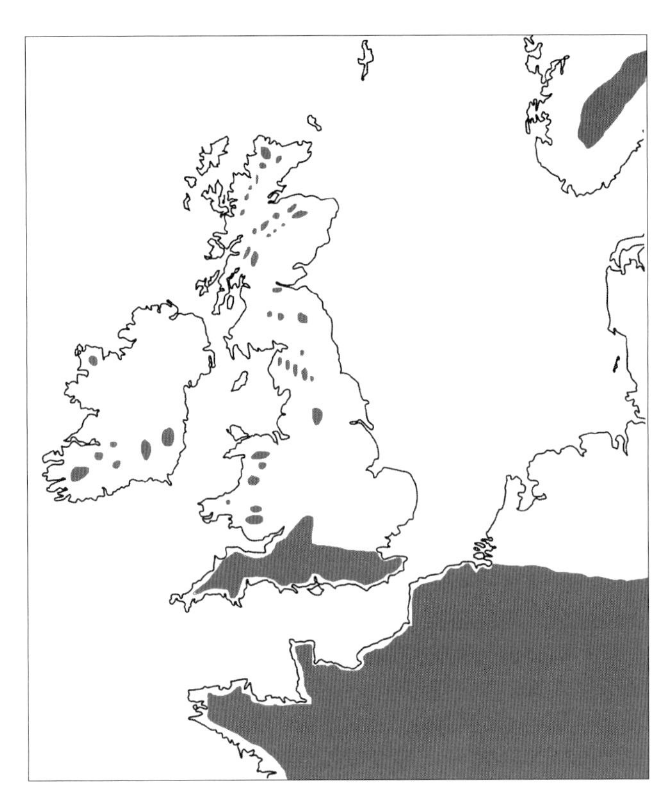

*Figure 5.4. Schematic redrawing of Lyell 1863 figure 39. His original caption reads as follows. 'Map of the British Isles and part of the North-West of Europe, showing the great amount of supposed submergence of land beneath the sea during part of the glacial period.' Scotland has been submerged up to 2,000 feet, and other parts of Britain up to 1,300 feet. Isolated islands were too high to be totally submerged. Southern Britain and northern France were interpreted as never having been submerged because of the absence of marine shells and glacial erratics which could only have been emplaced by the melting of icebergs. Whether the whole of the area was submerged at the same time was debateable.*

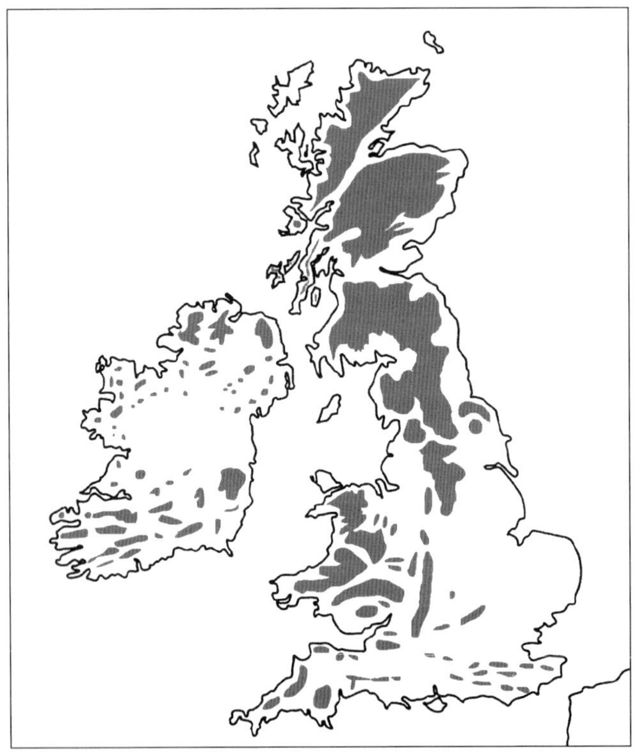

*Figure 5.5. Schematic redrawing of Lyell 1863 figure 40. His original caption reads as follows. 'Map showing what parts of the British islands would remain above water after a subsidence of the area to the extent of 600'.'*

standing in the public eye. Privately both Darwin and Joseph Hooker thought Lyell had stood on the fence too much. The problem was that Lyell did not make his views on human evolution clear. He seemed to accept the principle of progressive transmutation, and could see why Darwinists would argue for a naturalistic origin for humans. He could accept the implication of the ancestor-descendent relationship implied by the physical similarities between humans and higher apes. (At one point he contradicted himself and almost seemed to accept a Richard Owen-like belief in body plans as divinely inspired archetypes.) He noted that humans and animals could share a wide range of emotions and even certain higher faculties such as loyalty and love. But for him the true gulf between the highest form of animal and the 'lowest type of man', was that the mind of the latter was perfectible (an old monogensit perspective), whereas the mind and nature of the brute was fixed for all time. Any progressive transmutation agenda would have had to bridge that gap if it sought to offer a purely naturalistic explanation for humans and their higher faculties. Lyell could not see how transmutation and natural selection could change the one into the other.

Lyell did suggest that the progressive development of human faculties, and thereby human culture as well, could have been through the birth of individuals of great talent and ability whose positive influence served to lift all those around them to higher stages. Since genius was heritable (as Francis Galton was attempting to prove), the children of such gifted individuals would spread positive innovation further. Lyell even intimated that such a progressive saltationist mechanism could explain the jumps that separated but still connected the different stages within evolving species. He asserted it could even explain why some human races seemed to develop while others did not. But when challenged by Darwin on this he claimed it was only a speculation (Lyell, 1863b). He neatly sidestepped the need to commit himself to a non-teleological explanation by offering the reader the commonly accepted choice of three viewpoints to explain God's involvement in the world. On the one hand God was constantly at work tinkering with his creation. Alternatively, He only interfered in the affairs of the world very occasionally, otherwise natural laws were left to determine historical development. In the third option, having set natural laws in motion at the moment of creation, God then withdrew from the world altogether and forces like evolution and progression unfolded naturally. In all three scenarios it was possible to suggest a divine input into natural development, but in the last two it was much easier to reconcile a creator with science's requirement for autochthonous process. This was not what the Darwinists wanted to hear.

But the controversy that enmeshed the book was not a result of outraged public sensibilities, far from it. The outrage came from within scientific circles. In private correspondence Joseph Hooker and Charles Darwin felt the book was too derivative. The first and third sections relied heavily on the work of other people (Darwin, 1863b; Darwin, 1863c; Hooker, 1863). Hugh

Falconer and Joseph Prestwich took Lyell to task on this very subject (van Ripper, 1993; Owen, 2008), claiming that Lyell had not acknowledged their contribution to the human origins debate fully enough in the first section of the book. There were grumblings that the book gave the impression that it was only after a visit from Lyell that the interpretation of a site had the seal of approval. The Darwinian elite, while in some sympathy with this perspective, had little patience with the very public way Falconer in particular pursued Lyell through the press and the pages of the *Athenaeum*. He did not act like a gentleman.

## The Anthropologist: John Lubbock

John Lubbock's *Prehistoric Times as Illustrated by Ancient Remains and the Manners and Customs of Modern Savages* was published in 1865. Mention has been made of it in previous chapters. Inevitably it was to be compared with Lyell's *Antiquity of Man*. Many Darwinians felt this was the book Lyell should have written. *Prehistoric Times* did well overall; a second edition in 1869, eventually running to seven editions (7th edition in 1913), and selling more than twenty thousand copies (O'Connor, 2007; Owen, 2008). As with *Man's Place in Nature* some of the book's chapters had already been published as journal articles in the early 1860s. The ethnographic chapters were all especially written for the volume. Like Lyell, Lubbock situated himself in the mind of his readership by making it clear in the preface that he had personally visited many of the sites he spoke about, often accompanied by the great and good in Victorian science. In so doing he was reinforcing his credentials. He also visited major European museums to view their collections, as well as the private artefact collections of senior archaeologists and geologists on the Continent.

Following Lyell, John Lubbock tracked backwards in time from the Bronze Age and the Neolithic towards the Palaeolithic. Right from the beginning of the book the prose used by Lubbock evoked a strong empathy with Prehistoric humans (Gamble and Moutsiou, 2011), far more so than the writing style of the great geologist – not really surprising for a president of the ESL. Like Falconer and Prestwich, Lubbock too became embroiled in controversy with Charles Lyell over the use of previous publications. Lubbock added a footnote to the preface of the first edition asserting that some of Lyell's information, even wording and sentence structure, had been lifted directly from Lubbock's earlier papers now included as chapters in the book. Lyell had not fairly acknowledged this. Lyell hotly denied it, and the affair ground on for a few months with letters of support and condemnation passing between the Darwinians. Even Darwin himself was drawn into the controversy. The ever gossipy Joseph Hooker informed Darwin that the real reason behind the argument was snobbishness on the part of Sir Charles and his wife toward other members of the Darwinian elite (Hooker, 1865).

One reason *Prehistoric Times* remains highly regarded by archaeologists is that it was here Lubbock first coined the term Palaeolithic, although it only appeared twice in the text of the first edition. The Palaeolithic period was described in three chapters with Lubbock occasionally referring to the drift as the Archaeolithic. He may well have intended that the term apply only to the drift, while Palaeolithic applied to both the drift and caves, but this in not especially clear. Alternatively, it may just have been poor editing.

The river drift represented a period of cold climate with appropriately adapted mammals. He accepted the basics of Prestwich's river theory, and the transporting power of ice. In also accepting Prestwich's post-glacial date for the Palaeolithic, he emphasised that the drift was a time of transition, neither fully glacial nor fully temperate. But some mammals still did not fit comfortably into this scenario. Lions and hippos were, by comparison with their modern counterparts, warm-loving species. Prestwich (Prestwich, 1864) was explaining them by reference to what today we would call non-analogue faunas. Ancient species did not necessarily have the climatic tolerances or behavioural repertoire of their modern equivalents, especially in the case of the hippo remains from the French drift – this was an extinct species. Lubbock now suggested that the few remains of hippo discovered may have been in a different physical condition to other mammal bones and so were derived from earlier deposits (see especially O'Connor, 2007, p47 on this).

By comparison with Lyell's book there was relatively little defence of the antiquity of the drift gravels and especially of the human manufacture of the palaeoliths. These sections in *Prehistoric Times* are brisk and business like. For the most part that battle had already been won. Lubbock identified four types of palaeolith in the drift gravels. An oval shaped handaxe with a rounded point, an oval with a much more pointed tip, the classic heavy butted and long-pointed handaxe, and finally a variant of the last but with a much more rounded tip. Lubbock also gave the astonishing statistic that, up to 1865, some three thousand handaxes had come from the drift deposits of England and France – and not a single polished Neolithic axe. This was a huge number given Palaeolithic archaeology was only six years old.

Lubbock covered much of the ground that Lyell had, especially in attempting to date the Palaeolithic. It should be remembered at this point that there was still some debate as to the precise chronological relationship between the drift and the remains from the Belgian and French caverns. He took advantage of the recent discoveries by Lartet and Christy in the caves of the Dordogne. In particular he described the sequence from the rock shelter of Le Moustier. Handaxes found here looked different and slightly more advanced than those from the drift – see Figure 5.6 (also Chapter 4 and Table 4.1). Lubbock's prose invited the reader to speculate whether these artefacts spanned the gap between the palaeoliths of the drift, and the archaeology of the caves; which appeared in the Dordogne to be mostly reindeer-age (Lartet and Christy considered Le Moustier to be drift-age). This was a developmental sequence for the Palaeolithic rooted in material culture, and critically it was one of the earliest. Placing a more precise age on this sequence was difficult. Lubbock also quoted the evidence of the deltaic fans above Lake Geneva, accepting the oldest as one hundred thousand years old, or more.

However it was with the ethnographic record that Lubbock drew the distinction between his own work and that of Lyell. Whereas the latter had used ethnography in occasional support of some of his arguments, Lubbock devoted three chapters to detailed descriptions of the social habits and material culture of indigenous peoples from all over the world. Many travellers' accounts of a whole variety of social customs were described and compared in detail. It was classic ethnology, worthy of the ESL's president. In the last few chapters he used these data to defend a strongly progressive, positivist, and monogenist interpretation of human evolution.

'Monkeys', he asserted, were known tool users, throwing sticks in self defence and using hammer stones for nut cracking. From here it was a short step to using a naturally sharp edged stone, and then to realising that a hammer stone could also be used to make a sharp-edged stone. The earliest examples of tools would probably be quite similar to those in the drift period. In time improvements would lead to better made axes, pressure flaking replacing percussion, and finally to polishing the surface of the axe to make the Neolithic celt of the Stone Period. Observation and imitation were at the heart of human development. The earliest humans would have observed the many animals that build houses for themselves; they would have recognised that when knapping flint sparks can fly, or in polishing an axe friction develops heat; all would have led to key innovations in human development. The sheer variety of social systems that were present in the contemporary ethnological record deeply impressed Lubbock. It highlighted the ingenuity of the race. Every imaginable adaptation to differing local circumstances had been observed by ethnologists. Similar environments often produced similar societies, but with enough of a local twist to make them distinctive. In this vein, many of the simpler tools for scraping or cutting had been independently re-invented at different times and different places across the globe. Inuit skin scrapers from modern contexts, as well as from graves several centuries old, explained similar artefacts found in the Neolithic of Europe.

Unlike Lyell, Lubbock met the argument about human origins head on. He borrowed a large section of Wallace's 1864 paper to the ASL to explain the origin of the different modern races. Asserting that Wallace was a monogenist at heart, Lubbock adopted Wallace's argument that humans were very ancient and that any diversification of the different human stocks was equally as ancient. Lubbock accepted the possibility of humans in the Pliocene, and even the Miocene, but evolving somewhere away from Europe, probably in the tropics (Lyell had balked at a Miocene progenitor because the fauna was just too different by comparison with that associated with humans in later periods). It was local climate that drove the development of distinctive racial

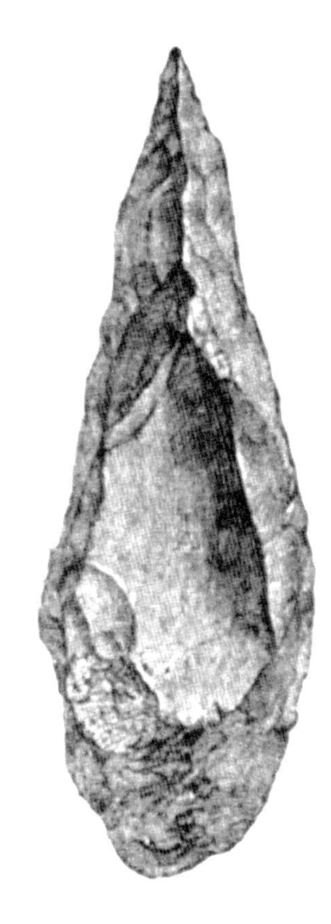

*Figure 5.6. handaxe from Le Moustier as illustrated by Lubbock 1865 figure 131.*

features as well as social customs and technological innovation. Different environments demanded different ways of life and differences in material culture to cope with them: the ethnological record showed that clearly. In their original tropical home humans would have only needed simple tools like sticks and hammer stones to deal with the plentiful resources surrounding them. Once they had slowly moved into new lands the stick and stone-pebble soon became the spear and the club. Culture was therefore as ancient as our original time of racial diversification.

Lubbock developed some intriguing arguments to counter the polygenist staple of a complete lack of change in the races through recorded history. He accepted the polygenist argument that Biblical sources and the tomb paintings of ancient civilizations did demonstrate long term racial stability, but then he asserted that without an external stimulus for change, none should be expected. During the time of recorded history there had been no need for the extant races to have changed. He also conceded the polygenist position that blacks in America had not changed substantially in the several hundred years since they were transported across the Atlantic, or whites grown darker in southern Africa since the Dutch colonists arrived. He argued that such people had only been in new environments for two to three hundred years. This was too short a time for change to begin to manifest itself. In fact some change was apparent. He agreed with those who claimed white Americans were already different in some ways to their European contemporaries, and made the intriguing suggestion that blacks themselves might notice differences between black Americans and black Africans – differences that whites could not yet see.

In 1870 Lubbock published his own take on E.B. Tylor's *Researches.* This was *On the Origin of Civilization and Primitive Condition of Man* (Lubbock, 1870) running to six editions (6th edition 1902). The similarity of the title to Darwin's *Origin of Species* would have been deliberate. He and Darwin were neighbours at Down, and Darwin had been a mentor to the young Lubbock. It was a more ethnological work than either *Researches* or *Prehistoric Times*, lacking Tylor's attempts to use the unity of human mentality to draw links between modern Stone Age peoples and those from Prehistory.

Grant Allen (1885) summarised the importance of both works, but for him their real significance was the use of the ethnology to show that indigenous peoples were not degraded forms of human who had fallen from a higher estate. Civilization, asserted both authors, was not the natural state of our species. It was something we attained through struggle. Our journey up from the animal kingdom, and the many stages of cultural development still seen in the world, was proof of this.

Like Tylor, whom he liberally acknowledged, Lubbock explored the various developmental stages of modern indigenous peoples through their customs and practices. *Researches* and *Origins of Civilization* share strong similarities. Tylor's is, in my opinion, the more innovative because it is an attempt to develop a holistic theoretical stance (although extracting a single integrated theory from the text is difficult). He was more sensitive to theory than Lubbock and understood its limitations better. He realised that similar social institutions in different geographical locations could be a result of independent invention, of contact with other cultures, or of ancestor-descendent cultural relations. Arguing that practices seen in different parts of the world were a product of the universality of the human mind, could only be established when the latter two influences could be discounted. For Tylor similarities in human mentality explained independent invention; for Lubbock environmental pressure led to similar solutions. This probably explains why Tylor was cautious in drawing his linkages between observation and interpretation. Lubbock was less so. Tylor was conscious that in many cases it was just not possible to argue for *in situ* development *because the evidence needed to discount other possibilities simply did not exist.* This highlights another of the drawbacks which haunted evolutionary anthropology. Because it was often impossible to recreate a cultural history for a local group, ascertaining whether their cultural institutions were borrowed or home grown was often impossible. Evolutionary anthropology worked best when its subjects stayed in one place, for a long time, and didn't mix with other peoples; but humans just don't do that.

# 6

# THESIS AND ANTITHESIS, BUT NO SYNTHESIS
## *The 1870s and the Darwinians in Power*

> 'Really, Mr. Holmes, this exceeds anything which I could have imagined', said Dr. Mortimer, gazing at my friend in amazement…'How did you do it?'
>
> 'I presume, Doctor, that you could tell the skull of a negro from that of an Esquimau?'
>
> 'Most certainly.'
>
> 'But how?'
>
> 'Because that is my special hobby. The differences are obvious The supra-orbital crest, the facial angle, the maxillary curve, the '–
>
> (Conan Doyle, August 1901, *The Hound of the Baskervilles,* The Strand Magazine, Chapter 4)

The fallout from the political manoeuvrings of the 1860s affected the study of human origins. The topic was mentioned even less frequently by the end of that decade, but there was also a marked drop in the number of submissions on racial origins as well. With the formation of the AI and its new periodical *The Journal of the Anthropological Institute of Great Britain and Ireland,* human origins reporting took on a somewhat different character. The dynamism and polemical cut and thrust that had characterised the ASL's publications was abandoned, as was its politically intrusive polygenism. The polygenists in control of the ASL had welcomed human origins submissions to their journals because they afforded opportunities to engage with that one question that lay at the root of their viewpoint; racial origins. Gradually their influence was felt less in the hierarchy of the AI. The breadth and variety of reporting within the various publications of the ASL was replaced in the 1870s by a rather staid orthodoxy which emerged as the Darwinian aristocracy and especially the Lubbock circle, settled into uncontested power.

In this chapter I will argue that for British Palaeolithic research in the 1870s there emerged something of a consensus on the Pleistocene occupation of this country by ancient humans. Prestwich's post-glacial time scale (but not his views on the date of the Pleistocene, or its duration) gained a level of acceptance that was facilitated by the emergence of John Evans as the most senior figure in Prehistoric studies. John Lubbock who, arguably, had occupied this position in the public eye up to this time, was beginning to pursue other interests. Between 1859 and 1870 Lubbock published some fourteen major papers and books; in the following decade, thirteen. But between 1880 and 1900 there was only one (data from Keith, 1924). While he maintained a lifelong interest in Prehistory, and was still active in the societies at metropolitan and national level, his interests in other disciplines (particularly natural history), and his political career, overshadowed his archaeological interests. The Lubbock circle mutated into the Lubbock–Evans circle, finally evolving into an Evans circle.

Evans's continued to follow Prestwich's views from the early 1860s, even as Prestwich began to distance himself from his previous opinions. In fact Evans appears not to have shifted from his 1859 position throughout the 1870s and beyond. I suggest that Evans's pre-eminence, and his own descriptive and non-interpretative brand of archaeology, began to adversely affect the development of what the Victorians would have called philosophical archaeology (i.e. interpretation and reconstruction of the past). The Tylor–Lubbock approach, where material culture and archaeology were synthesised with ethnology in an interpretative humanisation, stalled somewhat, and the opportunity for philosophical archaeology stalled with it. Under the influence of Evans, the 1870s saw a return to the more descriptive archaeology of 1859.

Yet almost as soon as the fledgling post-glacial consensus appeared it was challenged. This was the beginning of a more concerted investigation into the allied questions of intra-glacial, pre-glacial, and even pre-Pleistocene man. Moreover, a younger generation of geologists was questioning the very character of the ice age as the middle Victorian geologists had conceived it. This geological challenge to consensus highlighted the shift away from the evolutionary anthropology of the previous decade.

*Figure 6.1. Proportion of occurrence (as %) of different categories of article and other submission types (reviews, reports etc.) between 1872 and 1880 for the* Journal of the Anthropological Institute of Great Britain and Ireland.

## Under New Management

John Lubbock assumed the presidency of the new Anthropological Institute for Great Britain and Ireland in mid-1871 and held office for two years. He was replaced by the anatomist G. Busk (1807–1886) in 1873 who also served for two years. This period of tenure was standard for the AI in the 1870s. A glance at any list of vice presidents for this decade reveals a domination by Darwinians and the Lubbock–Evans circle, and it was from these that future presidents were chosen. Their control over anthropology at the BAAS was equally evident. By the 1870s the anthropological department of the BAAS was under the aegis of Section D, Biology, itself dominated by Darwinians and manipulated by Huxley, who, for example, attended the 1871 BAAS at Edinburgh in his capacity as a vice president of the AI, standing in for Lubbock who was absent. The following year Lubbock was the president of Section D at Brighton, and Augustus Lane Fox was his vice president. Lane Fox was also chairman of the Anthropology Section for that year.

*Figure 6.2. Proportion of occurrence (as %) of different categories of article and other submission types (reviews, reports etc.) between 1863 and 1871 for the Anthropological Society of London as plotted by year. Data from* Anthropological Review *and the* Transactions of the Anthropological Society of London.

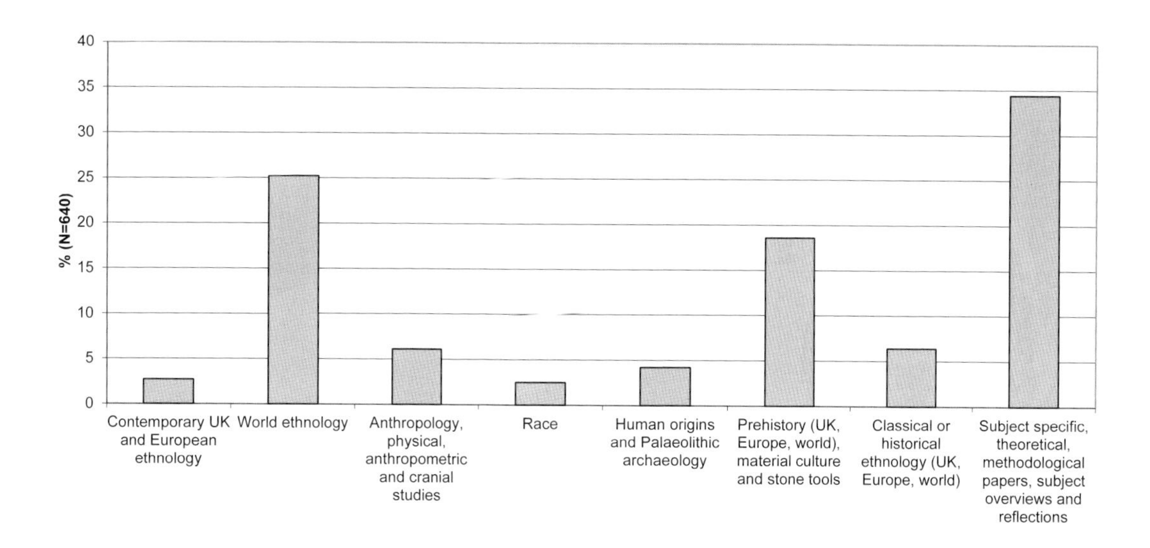


Figure 3.1 compared the pattern of published articles between the ASL and ESL for the 1860s, and Figure 6.1 does the same for the 1870s. There are evident differences even between the ASL and the AI across the twenty year period, with a gradual drop in physical anthropology, as well as those articles discussing racial origins and racial differentiation. Proportionally, there was a decrease in contributions on Palaeolithic archaeology and human origins, but an overall increase in Prehistory and Prehistoric material culture studies once the new society had been formed.

Equally instructive are the data in Figures 6.2 and 6.3 which show patterns of article publication by year, and compare the ASL, as represented by the *Anthropological Review* and the *Transactions of the Anthropological Society of London*,

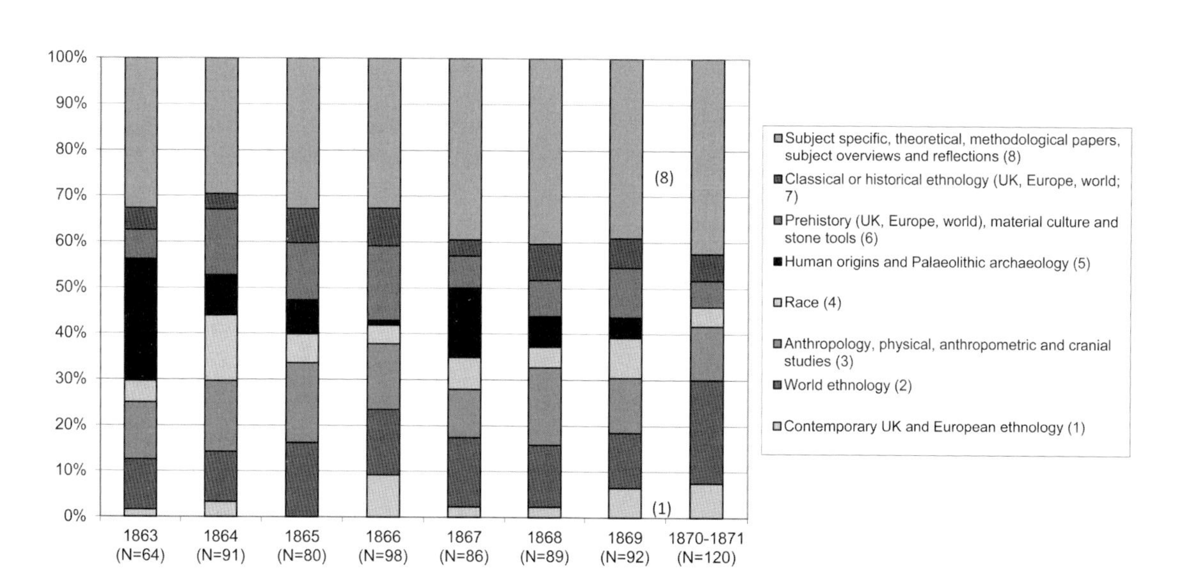

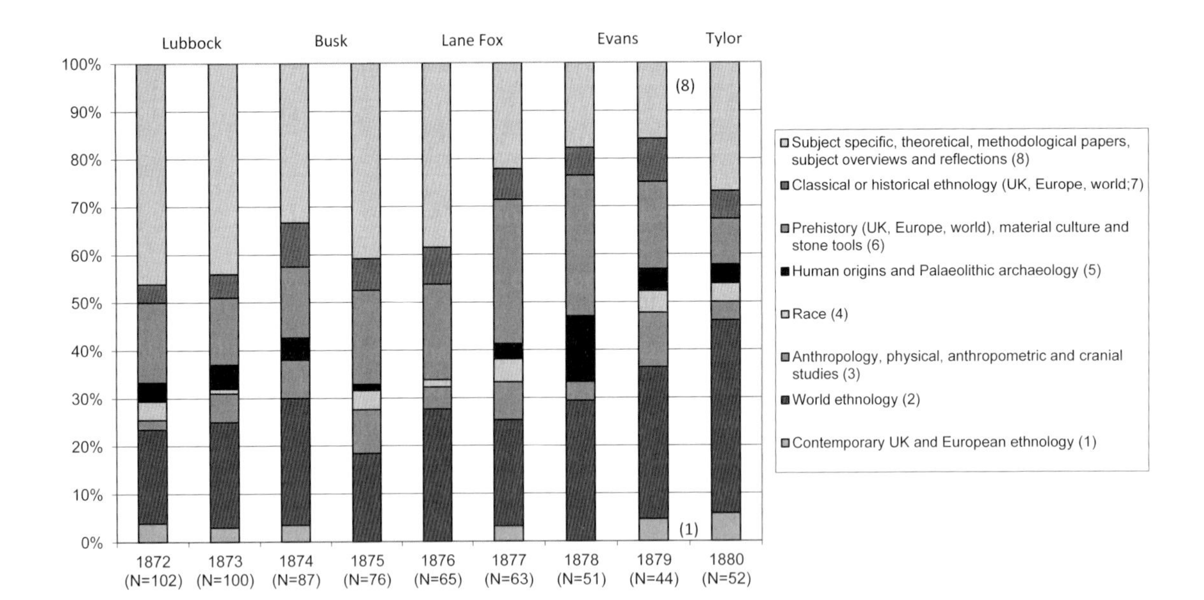


*Figure 6.3. Proportion of occurrence (as %) of different categories of article and other submission types (reviews, reports etc.) between 1872 and 1880 for the Anthropological Institute of Great Britain and Ireland as plotted by year. Approximate term of office of presidents of the AI indicated.*

with the AI as represented by the *Journal of the Anthropological Institute of Great Britain and Ireland.* Across the 1860s articles on world ethnology (descriptive pieces on aspects of the lifestyles of indigenous peoples etc.) were allotted a broadly similar amount of space to those on race, racial origins and racial parity, and various aspects of physical anthropology. In fact the scaling down of these themes after 1871, and the increase in world ethnology, is rather surprising given the increase in interest in anthropometrics during this period. The eugenicist Francis Galton (Gillham, 2001) makes his first appearance in the publications of the anthropologists after the merger. He already moved within the right social circles – he was a cousin to Charles Darwin – and his scientific interests in heredity aligned him squarely with the Darwinians. He became a council member of the AI, a vice president, and eventually president in the mid-1880s. Galton will reappear in later chapters. In fact it was his brand of anthropometrics, allied to his interests in heredity that replaced the previous focus on physical anthropology as the proving ground for racial distinctiveness.

If anything, human origins and Palaeolithic archaeology fared better under the ASL than it did in the AI, a reflection of the amount of shelf space afforded it in the *Anthropological Review.* Human origins and Palaeolithic archaeology were a persistent if small contribution to the combined societies' journal after 1872, and reach their prominence in the *Journal* for 1878, when John Evans was president between mid-1877 and mid-1879. The increase in general and later prehistoric articles after the merger is evident from Figures 6.2 and 6.3. Within this category are included material culture articles incorporating aspects of prehistoric weaponry amongst other things. Material culture achieved a greater degree of prominence under the AI, and this reflects the increasing visibility of Augustus Lane Fox within the newly formed society. Interestingly one subject that retains its position across both decades is classical and historical ethnography, which would include the racial origins of the modern European peoples. More than likely, this was too contemporary to fall out of fashion.

If Owen and colleagues (Owen, 2008) characterise the 1870s as a decade of consolidation, it was not one without significant discoveries. New prehistoric sites were constantly reported to the BAAS and to the AI, including a slow but steady trickle of Palaeolithic ones. On-going cave research was paying dividends. Boyd-Dawkins read a paper at the Brighton BAAS in 1872 describing the latest discoveries at the Victoria Cave, Settle. Beneath later Prehistoric levels with burials a Pleistocene horizon had been discovered at the base of the excavation (Lane Fox, 1873). This is what everyone had been hoping for. The signs were good – hyaena, cave bear, mammoth and reindeer bones were present. He led an excursion to the site in 1873 from the BAAS meeting at Bradford (Rudler, 1874) . The site will be discussed in more detail below. A new Palaeolithic cave was reported from the Wye Valley (Wake, 1872) on the Welsh borderlands, King Arthur's Cave (now known to contain Late Upper Palaeolithic), but little was made of it at the time. William Pengelly (Rudler, 1874) was reporting to the BAAS on the continuing excavations at Kent's Cavern. He was now convinced there were two periods of occupation in the cave. Below a stalagmite level was a breccia which contained cruder looking palaeoliths (though he did not commit himself to associating these with the drift). Above the stalagmite was what we would now interpret as a mixture of middle Palaeolithic flake tools and Upper Palaeolithic blade tools. Pengelly recognised them as distinctive and argued that their stratigraphic position implied they were later than the handaxes. Additionally, the long awaited final report on the Brixham cave was published, edited by Joseph Prestwich (Prestwich, 1874).

Professor Thomas McKenny Hughes (1832–1917), a Welshman from Aberystwyth, who had risen to become Woodwardian Professor of Geology at Cambridge University, was excavating in a cave in the valley of the river Elwy near St Asaph in North Wales (Hughes and Thomas, 1874). Pontnewydd cave was close to the cave of Cefn mentioned in Chapter 4. The stratigraphic profile of Pontnewydd was revealing. A gravel unit dated to the period of time when the cave was sealed and it was the route of an underground river. The river originated on the surface where it was draining across glacial boulder clay. This was the only possible source of the cobbles in the cave's gravel because, lithologically, they were not native to the Elwy valley. In the breccia that overlay the gravel were implements similar to those found by Prestwich and Evans in the Somme valley, and scrapers similar to those found by Lartet and Christy in the French cave of Le Moustier in the Dordogne.

They had been made on a rock type known as felstone, another lithology not natural to the Elwy's catchment. It too was found in the local drift deposits which in this part of North Wales were considered glacial. Since the artefacts were fresh and sharp, but the original nodules they were made from were worn as if by transport, McKenny Hughes argued the tools had been made on erratics brought into the valley by glacial action. Combining the geological evidence, the stone tools had to be post-glacial in age. Pontnewydd provided a strong argument in support of Prestwich's post-glacial date for the human occupation of Britain.

If anything after 1859 approached an orthodox point of view – this was it. The Palaeolithic dated to the post-glacial phase of the Pleistocene. This had been Prestwich's conclusion in 1859 and in the following years. Evans threw his weight behind this position and as will be seen below his monumental *Ancient Stone Implements* (1872) was a supporting testament to this very orthodoxy.

Open air drift sites were not neglected, but were few and far between. Lane Fox had been studying the gravel terraces of west London near Acton and Ealing since 1869, and reported his findings to the Geological Society in 1872. As an example of just how quickly things were changing, he based his geological appreciation of the Thames terraces on the work of William Whitaker (1836–1925), an officer of the Geological Survey, who had been mapping the area. Prestwich had identified two terraces, his high and low drifts. He often subsumed other high level gravel trains beyond the confines of the valley itself (see Chapter 2) within his high level drift, because he believed that these were the remains of the former high level flood plain. However Whitaker mapped three terraces (low, middle high) within the valley itself. Any higher drifts beyond the *immediate* confines of the valley were considered high-level drifts. These were not the same as his third high terrace, or the Plateau gravels on the interfluves between river basins. A much more complex history of the river was being suggested. Lane Fox (Lane Fox, 1872) described sections within the high and middle terraces. Handaxes and flakes were discovered on what in some circumstances today would be described as an ancient land surface. Implements were not as common as in the valley of the Somme and their spatial distribution within the valley appeared uneven. His researches were crowned by the first discovery of a mammoth in the high terrace (*sensu* Whitaker). He also suggested that the shape of the handaxes was more a reflection of the individual flint nodule they were made on.

John Brent, exhibiting Palaeoliths to the AI, believed some from near Canterbury could be drift in age (Brent, 1875).

Worthington George Smith (1835–1916) is a luminary of Palaeolithic archaeology (Dyer, 1959; Dyer, 1978; Roe, 2009). His later work at Caddington in Bedfordshire, published in the volume *Man the Primeval Savage* (1894), is a classic contribution, rarely bettered even by modern standards. This work will be discussed in more detail in a later chapter. Between 1878 and 1884

Worthington Smith, then a resident of Highbury in north London, was observing sections in the drift capping the terraces of the central and eastern areas of north London. He was visiting sand and gravel quarries in both the metropolis and in the more rural areas of Hertfordshire to the north of the city. This was a busy time of urban expansion for London (Conway, 1996; Harrison, 1963) with the building of residential housing whereever possible. Roads were being metalled by gravel extracted from quarries cut into the Thames' ancient terraces. In the middle and late 1870s Smith pioneered a unique method of investigation. He would keep a sharp eye out for roads with a newly spread gravel surface, and then examine them carefully for handaxes. If discovered, he would enquire from the carters where the gravel originated and then visit the pit for further investigation. In time he taught himself to be able to recognise from which pit new road-metal originated. During the late 1870s, and following this strategy, he researched the valleys of two tributaries of the Thames, the Lea in north-east London, and the Brent in the north-west (Smith, 1879; Smith, 1880). He suggested implement collectors keep even the roughest of handaxes. He argued that these were the everyday field tools of the Palaeolithic, made in a few moments for one job and then discarded. They had been neglected by collectors. Researching them and not just the fine specimens would pay dividends in terms of understanding tool behaviour.

John Evans (Evans, 1878a) also reported the discovery of a large number of handaxes from the vicinity of Broom and Chard Junction in the valley of the Axe river, Dorset. These were the furthest west implements had yet been found in Britain, and they were away from the chalk country. These were made on chert and Evans noted that while they seemed to be dominated by flattish ovoid forms, these were in no way different from the classic St Acheul implements, proving that the Palaeolithic knappers could make anything they wanted to.

Abroad, there was a stream of new discoveries throughout the 1870s with Prehistoric stone tools being found pretty much everywhere. There were further discoveries from southern and western Africa (Chapter 4), Egypt, the Sinai Peninsula, Syria, Patagonia, Honduras, New Zealand, and even Newfoundland. Lubbock thought some of those from Egypt closely resembled Palaeolithic forms, with a St Acheul type from Deir-el-Bahari (Lubbock, 1875). The amount of Palaeolithic/drift material discovered was still negligible by comparison with what today would be interpreted as Mesolithic, Neolithic and other later Prehistoric stone tools. The publication of Lartet and Christy's *Reliquiae Aquitanicae* was eagerly anticipated, with Rupert Jones completing the task of bringing the volume to publication (Lartet and Christy, 1875), see Textbox 4.1. He also described incisions on bone and antler finds from the Dordogne caves (Rupert Jones, 1873) thinking they were tally marks, signs of ownership, or magical symbols. Much of this was rooted in ethnological comparison.

I noted earlier that the discovery of Cro-Magnon presaged the beginning of a Palaeolithic racial anthropology of the kind British race scientists were desperate to develop. A good example is the translated work of Jean-Louis Armand de Quatrefages, Professor of Anthropology at the Museum of Natural History, Paris, and a colleague of Paul Broca and Pruner Bey (Chapter 4). Two of his works (De Quatrefages, 1879; De Quatrefages, 1875) are summarised in Textbox 6.1. The wealth of continental specimens that were being discovered allowed French *savants* to forge any number of individual takes on race and racial origins. De Quatrefages was only one such scholar, but his work exemplifies the richness and detail with which links were being made between the Palaeolithic races of continental Europe and the Prehistoric interments of the Neolithic and Bronze Age. They were also describing racial continuities and admixtures between later Prehistoric races and modern ethnic populations. Whereas British scholars could, and did, do this for later Prehistory, they could not venture further back in time. Undoubtedly the lack of institutional and government support in Britain, by comparison say with France, and a weaker anthropological tradition in general, contributed to this. It is perhaps little wonder that human origins research returned to its roots in the 1870s.

de Quatrefages, A. 1875. *The Natural History of Man: A Course of Elementary Lectures.* D Appleton and Company. New York

de Quatrefages, A. 1879. *The Human Species.* D. Appleton and Company. New York

Both these texts were published in America. The former was based on a course of public lectures to French working men given by de Quatrefages. Although these publications reflect one individual's views, they present a very complete exposition of the potential inherent in the science of Anthropology as practiced in France. The detail and breadth of the subject matter contrasted sharply with the impoverished character of British anthropology and race science in terms of Quaternary studies. This was even reflected in the degree of government financial support given to public funded anthropological foundations. Germany was similar; there were no comparable institutions in the UK.

Humanity originates from the central Asian plateau to the north of the Himalayas. Here a pale-yellow skinned race, possibly with red hair had its homeland. This was a human race and recognisably so. Humans did not descend from the same lineages as higher primates. Natural selection was not able to create new species. Adaptation via competition for resources creates individuals and groups suited to living in the conditions of one location. Natural selection could adapt groups to new conditions either in the same place or elsewhere. It did so by minute changes over time. In so doing new races (all of which are inter-fertile with each other and the parent stock) are formed. But these are new races, not new species. Gradual change by incremental iterations cannot separate parent and daughter/conspecific races sufficiently for new species to form. de Quatrefages places racial origin and racial intermixing right at the heart of his theories. His is a monogenist position, although he does not support the monogenist vs polygenist polarization.

From this homeland men moved out and adapted to new conditions. From this migration three major human races emerged. The pale-yellow skinned race remained in the core area for a long time. However they also moved eastwards where their descendants, intermixed with other pale races, are still found. A black and dark skinned group emerged to the south of the old homeland, and a pale-white skinned group are now found to the west. These three groups represent the racial groupings from which all other modern 'pure' and mixed race groups have developed. There was continuous migration from this core area. Tertiary occupation in western Europe represents an early migration. As the Pleistocene ice age grew in severity the heart land became depopulated because of further migrations. The Quaternary/Prehistoric populations are larger than Tertiary ones and there is considerable racial intermixing.

Europe has 6 Palaeolithic races that can be related to those of modern Europe. There are two dolichocephalic races. The first is the oldest recognised racial group. These were the Neanderthals, known as the Canstdat (Canstatt) race. Their on-going contribution to the European racial heritage was demonstrated in occasional examples of atavism. These occur everywhere throughout Europe, and in all races, consequently the Neanderthals were widely distributed. The second race were the Cro-Magnons. From their heartland in the Dordogne they sent out racial colonists across western Europe. These intermixed with other races later on, but their physical features are often seen as part of contemporary European people. A developmental sequence could be traced for this race from Le Moustier through a number of stages to La Madelaine. The race of La Truchère was represented by only one quaternary specimen. It approached the brachycephalic condition, but was very different from Cro-Magnon. The sub-brachycephalic Furfooz race was split into two groups distinguished by the roundness of the skull in profile. Finally the 6th race was that of Grenelle, a site near Paris. This race was contemporary with Cro-Magnon as examples of both races were found at Solutré. These too have descendants in the modern European population.

*Textbox 6.1. Two selected texts by A. de Quatrefages from the 1870s showing the type of detailed Palaeolithic racial anthropology that continental scholars were able to develop given the richness of the Pleistocene skeletal sample available to them.*

## *Ancient Stone Implements* and *Flint Chips*

Arguably, the most significant contribution to the process of consolidating the gains of the preceding decade was John Evans's *Ancient Stone Implements*. It forcefully promoted a consensus view, the authority of which was enhanced by the reputation of its author. In this section I will discuss this book and compare it to another book that has been unfairly overshadowed by the Evans volume.

*The Ancient Stone Implements, Weapons, and Ornaments, of Great Britain* (Evans, 1872) is still regarded as a classic volume. It represented the first systematic national survey of Palaeolithic artefacts in Britain. Its impact bridged the divide between the middle and late Victorian periods, and its influence was wide ranging. As one of the most senior archaeologists of his time, the word of John Evans carried an appropriate weight of authority (MacGregor, 2008; Owen, 2008). He was a careful and conscientious worker with a reputation for scepticism (i.e. proper scientific prudence). He was nick-named the 'Doubting Thomas' of the archaeological world, accepting nothing unless it was supported by a mass of factual and observational data (White, 2001). Throughout his life he continued to follow the methodology so successfully emplaced by himself and Prestwich at Abbeville – fieldwork, personal observation, and undisturbed deposits with *in situ* finds. His visits were often made to verify the context of a particular find (Roberts and Barton, 2008). In this Evans's work was in keeping with that of his peers, and in amassing a huge compendium of discoveries, he did for Prehistory and the Palaeolithic what E.B. Tylor had done for evolutionary anthropology.

By the 1870s Evans's Palaeolithic reputation (A.G., 1908), and middle class, entrepreneurial liberal values, had positioned him with the social set that included the Darwinian elite (Stocking, 1987; Stocking, 1971; Owen, 2008). If not a member of Huxley's inner circle of Darwinians he was nevertheless respected by them, and had a close association with many of them through his friendship with John Lubbock, of whose circle of intimate associates and allies he was definitely a core member. Owen (*ibid*) asserts that by the mid-1870s Lubbock and Evans were already household names to the middle classes.

Palaeolithic archaeology was too young a subject in the two decades following 1859 for a 'Palaeolithic official viewpoint' to emerge. There was no institutionalised focus that could reflect or promote such an outlook anyway. Overall too little was known and the basic information came from too many other disciplines, each of which had its own established reporting structures. The Palaeolithic, and even Prehistory, never did figure very highly in the more established archaeological journals such as *Archaeologia* (published by the Society of Antiquaries), or *The Archaeological Journal* (published by the Archaeological Institute), even after 1859 – see Figure 6.4. These periodicals concentrated

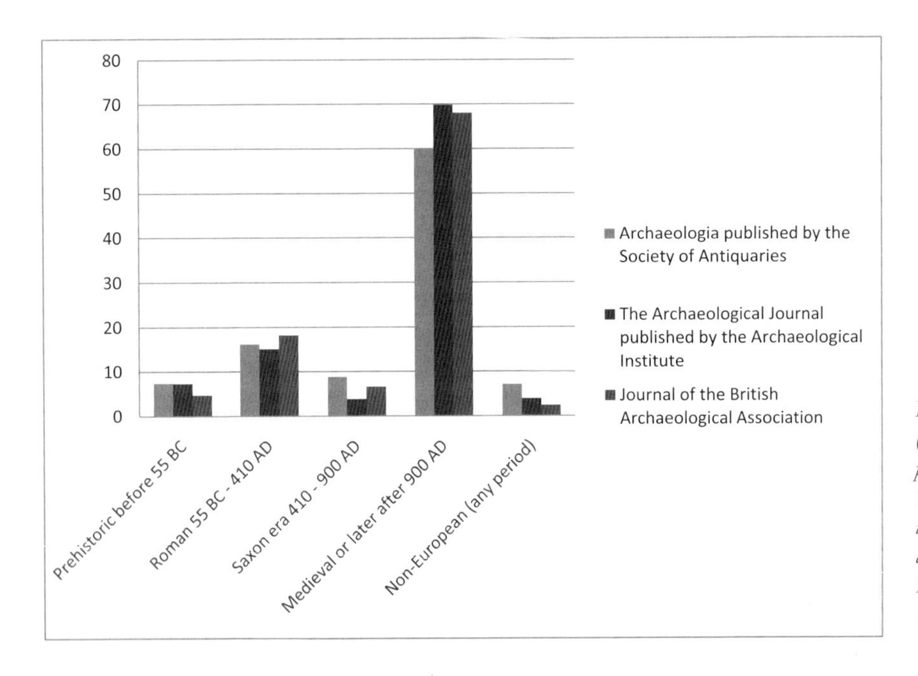


*Figure 6.4. Frequency (%) of archaeological papers published between 1849/1850 and 1875 in Britain's leading archaeological journals. Data from van Ripper 1993.*

mostly on traditional Roman and Medieval archaeology. Members of these societies would have considered the Palaeolithic archaeologists as geologists more than anything else. As we have seen, human origins was relegated to the anthropological and more geologically orientated periodicals.

Evans was already an established figure in a number of different historical and archaeological fields, especially numismatics, and capitalising on the fame of 1859, this allowed him to move between different disciplines and assume the role of Palaeolithic spokesman. In effect his word became the Palaeolithic establishment view, and he came to personify that role.

Only four of the twenty-five chapters in *Ancient Stone Implements* were dedicated to the Palaeolithic, the remainder focusing on later Prehistoric artefacts. This fairly reflected the state of knowledge at the time. The Palaeolithic section came after the Neolithic, actively mirroring the declining level of knowledge the further one travelled back in time. This was accepted practice, and Evans was only following the format of Lubbock and Lyell in this. The structure of the Palaeolithic section was simple. It began with a chapter on caves and their contents, and then moved on to a chapter on palaeoliths and the drift. There was a third chapter on other artefacts from the drift such as cores, flakes, and retouched tools (i.e. small flake-tools such as scrapers). In this chapter the similarity of the drift and cave implements to each other, and to those found abroad was also briefly described. The alleged Pliocene artefacts from St Prest, and the Miocene evidence from Thenay, were not accepted as genuine (this is discussed in more detail below), although the potential for pre-glacial/pre-Pleistocene artefacts was readily acknowledged. Evans, like most of his contemporaries, accepted the necessity of a long ancestry for human kind, but believed that the evidence for it would be found outside of Europe.

The final chapter in the Palaeolithic section was a long defence of the drift gravels as being of fluvial origin. Even in the 1870s there was still opposition to this, and to Prestwich's model of dynamic end-of-glaciation flood events (see Lubbock, 1873; Flower in reply to Bruce Foote, 1869; Flower, 1872). For many geologists drift gravels were still interpreted as beach shingle laid down in one of the numerous periods of submersion beneath the glacial ocean. For others the valleys had been excavated almost to their present depths by floods, often up to eighty feet or more in height. These had filled up the valleys spreading a thick mantle of drift over the valley sides. Today's river terraces were simply eroded through this thick infilling of sands and gravels. Even for those who accepted the fluvial hypothesis there was still the difficulty of whether the rivers that cut their ancient valleys, like the Thames and the Solent (Lane Fox, 1878), were similar in erosive powers to modern rivers (Lyell, uniformitarianism), or had been more powerful in the past (Prestwich, dynamic Pleistocene floods).

Evans drew on the authority of Prestwich in contextualising his chrono-stratigraphic and archaeological interpretations. For Prestwich a key geological insight from 1859 had been the recognition that implements in the valley drifts were in the same stratigraphic position in both France and Britain. For many geologists of the 1850s and 1860s the evidence supported only one glacial period within the Pleistocene (though this was the subject of increasing controversy – see below and Table 1.2), and the Palaeolithic post-dated it; human occupation of Britain was a post-glacial phenomenon.

It is worth reiterating that the book was published only thirteen years after the *Origin of Species* and the first Antiquity of Man debate. By the 1870s the idea of ancient humans was no longer new or desperately controversial, neither was the notion of a distinctive Palaeolithic period; Lyell and Lubbock had merely formalised what most people already accepted by the middle 1860s. However as I noted in Chapter 4 the idea of a *global Palaeolithic period* (as opposed to a Palaeolithic stage that could appear anywhere in the world at different times) was still something new, and difficult to reconcile with the prevailing theories of the time. I believe some of the book's polemical stance reflected an exploration of this possibility, although Evans never actually articulated this openly as he was far too cautious. Certainly within a decade or so other researchers were comfortable with the notion (Allen, 1885).

In recalling that it was less than a decade and a half since the visit to Abbeville and Amiens, the first edition of *Ancient Stone Implements* was a testament to just how much had been discovered in a relatively short time. When discussing 1859 Evans would often remind his audience that he had not known at the time that implements like those in the Somme valley had been found in England. He had been profoundly surprised to see examples from Hoxne on display in the Museum of the Society of Antiquaries on his return from France. The first edition of *Ancient Stone Implements* details only two instances of pre-1859 discoveries of drift handaxes, other than the celebrated Grays Inn Lane and Hoxne examples, see Figure 6.5. (These were Peasemarsh on the river Wey found in the early 1840s (Evans, p. 528; see Table 2.1; Lubbock had also detailed this find in Prehistoric Times), and a discovery near Salisbury in 1846 (Evans, p. 548; also Stevens, 1870)). Evans also stated that the bulk (if not all, p. 605) of the handaxes found in the southern English drift deposits were found after 1862. Owen (2008) notes that Evans began work on the book in 1863, and we have already noted that Lubbock asserted that by 1865 some three thousand handaxes (Lubbock, 1865, 280) had already been found in the British and French drifts. So, in less than a decade, the ubiquity of Palaeolithic implements within the river drifts of England and northern France had been amply demonstrated.

If Evans' opinion stood as proxy for an official view, then the ideas he expressed in *Ancient Stone Implements* were the party line.

To begin with, and unlike many scholars, Evans did not draw a clear line between the makers of the implements from the caves and those from the drift. In many cases he believed they were contemporary but it was also clear that some cave sequences post-dated the drift. For Evans they were all

*Figure 6.5. Handaxes from Hoxne on the right, and Grays Inn Lane, London, on the left. The Grays Inn handaxe was found in 1690, and those from Hoxne reported by Frere in 1800.*

made by the same race of ancient humans and there was no apparent break between drift and cave. Furthermore, there was no convincing evidence for any cultural development within the drift, but there was some support for development within the cave sequences of southern France. He disputed the successive faunal sub-divisions of Lartet, making the salient point that the faunal remains in caves did not necessarily reflect natural changes in animal communities. Rather, they reflected the preferences of the humans who lived in them. Palaeolithic hunters were bringing home prey species. But he did have some sympathy with an earlier cave period dominated by mammoth faunas, and a later cave phase dominated by reindeer. This was also being proposed by Dupont for the Belgian caves – see Table 4.1.

Like Lubbock, Evans chose to accept a material culture based sub-division for the Cave Age, adapting that proposed by Gabriel de Mortillet – Table 4.1, (Roberts and Barton, 2008; Monnier, 2006). Evans' succession for the caves was as follows: first came the age of Le Moustier in which mammoth and hyaena dominated the fauna, and the handaxes bore similarities to those in the drift. The handaxes were broad as a rule, and scrapers frequent. Today much of this archaeology would be characterised as Neanderthal/Mousterian. This was followed by the age of Laugerie Haut. Mammoth and some of the cave species were still present, but faunally horse and reindeer predominated.

Handaxes were slender and elongated, and there were leaf-shaped points and arrow-like points, all delicately made. There was a small amount of bone working. Today at least some of this would be classified as Solutrean – an Upper Palaeolithic culture made by modern humans. Then came the age of Cro-Magnon people. There were few, if any, drift-like implements. End scrapers predominated (scraping edges made on the ends of long blades of flint), and the horse was the key element in the fauna. There were more bone and ivory tools. Today we would label this as being mostly Aurignacian in age – an Upper Palaeolithic/modern human culture that we now know pre-dated the Solutrean. Finally there was the age of La Madelaine. These people made long blades and blade-cores, and often crafted their tools in bone or ivory – needles were common. Art objects were also common. Reindeer were the predominant animal. Today these would be termed the Magdalenian people, the last major European culture of the Upper Palaeolithic.

Despite the currency of discussion there were two topics on which Evans refused to be drawn, and would not speculate. These were the age of the glacial period, and the amount of time necessary to excavate the river valleys (thus providing a date for their occupation by ancient humans). There were not enough facts available to make a clear interpretation, and the various geological methods he described for elucidating a date for the ice age were too contradictory and too insubstantial. Evans did however believe a considerable hiatus existed between the Palaeolithic and the Neolithic. They were two different races of people. Part of his reasoning was that no intermediary set of implements existed which graded from Palaeolithic axes into Neolithic ones. This served to highlight the distinctiveness and separateness (and antiquity) of the river drift era.

In reading *Ancient Stone Implements* there is a strong sense of Evan's monolithic view of the Palaeolithic. His writings convey a largely unchanging and undifferentiated block of time in which ancient humans showed very little digression from the normal routine of hunting, eating, making handaxes, and then moving on. No biological or cultural development was proposed although, as stated above, I do suspect he was beginning to contemplate the idea of a globally synchronous Palaeolithic/drift period. E.B. Tylor (1865) had only been able to quote two Palaeolithic sites outside Europe but, as seen in the previous chapter, toward the end of that decade reports of drift like implements were becoming increasingly more common.

Two years earlier, E.T. Stevens the honorary curator of the Blackmore Museum in Salisbury had written and published *Flint Chips. A Guide to Prehistoric Archaeology as Illustrated by the Collection in the Blackmore Museum, Salisbury* (Stevens, 1870). The Blackmore was a private museum based on an enormous collection made by Mr William Blackmore 'of Liverpool and London' (W.B.K., 1870), which opened in 1867. Today *Flint Chips* is all but forgotten overshadowed by Evans's more famous work yet it was an important book in its time as it presented the public with a highly readable yet still detailed opportunity to directly engage with the material culture of Prehistory, and it appeared first. Evans's fame clearly ensured that *Ancient Stone Implements* would have seemed more authoritative – more academic. After all *Flint Chips* was only a museum guide. But this was its very strength. Evans's work described a lot of material held in private collections, often his own. Much of it would never have been seen by the public. Stevens's book described a large and comprehensive collection of European and British material culture that was readily accessible – the museum was open five days a week and entry was free. One contemporary reviewer of the book was in little doubt as to its true significance.

> We may, however, in conclusion, state our opinion, that the author has done himself scant justice in the choice of his title, since it conveys no adequate notion of either the intention or the execution of the work. From both points of view, it must be admitted that Mr. Stevens has presented us with a veritable Guide to Prehistoric Archaeology, and as such we commend it to the study of all who are interested in the topics of which it treats... (W.B.K., 1870, 169f)

The book was indeed a comprehensive guide to Prehistoric archaeology since it covered everything from the drift to the end of the Bronze Age, with a small amount of Iron Age thrown in for good measure. It also incorporated a substantial collection of modern ethnological objects. New World ethnology and archaeology was particularly well represented. Whereas Evans' remit was largely British, what Stevens wrote was, effectively, an introduction to world Prehistory. In many respects *Flint Chips* scored over *Ancient Stone Implements*. At its heart was the clearly stated view that human history was a progression, and the museum's cases were laid out to reflect that. From the entrance the museum – which was a single room – opened up before the visitor. Turning to the right and walking around the outside of the room there were cases of mammal bones and drift implements until half way around then the visitor progressed on to the Neolithic. Continuing to follow the wall cases around, until back at the entrance, the Neolithic gave onto the Bronze Age. Later European Prehistoric artefacts, and others from abroad, occupied strategically placed cases toward the centre of the room.

Interestingly, Stevens embedded his clear progressive agenda within a teleological explanation. The divinely inspired natural law of survival of the fittest and strongest, ensured that the balances between the species seen in nature today had operated in the Pleistocene (a term Stevens was one of the first to use) as well. His reviewer (W.B.K., 1870) cocked a literary eyebrow at that. Unlike Evans, Stevens saw no great break between the Palaeolithic and the Neolithic, yet believed that there was something different about the very thought processes of the Palaeolithic knappers. In most periods of Prehistory the dominant artefact forms of a later phase were occasionally present in

| Evans *Ancient Stone Implements,* 1872 | Stevens *Flint Chips,* 1870 |
|---|---|
| Flakes | Flakes |
| Trimmed Flakes | Scrapers |
| Handaxes – pointed<br>tongue shaped | Handaxe – pointed/pear shaped<br>shoe shaped |
| Handaxes – sharp rimmed including<br>discoidal<br>ovates/ovals<br>heart shaped<br>sub-triangular<br>almond shaped<br>lozenge shaped<br>lunate shaped<br>perch backed; | Handaxes –<br>discoidal (crude)<br>ovates/oval<br>heart shaped |

*Table 6.1. A comparison of two early stone tool typologies published in Britain in the early 1870s.*

the preceding phase, even if only in an embryonic form. This was seen throughout the world, and explained the similarities between many ancient and modern Stone Age societies scattered across the globe. Not so with the people of the drift. None of their tools survived into the Neolithic proving that there '…must have been also something wholly different in the drift people themselves…' (Stevens, *ibid*). Table 6.1 compares the typology that Stevens used with that of Evans. Stevens asserted that since Evans published his earliest descriptions of the drift implements in the 1860s, new forms had been identified (e.g. the heart and shoe-shaped implements). Stevens awaited the master's pronouncement on how these would be incorporated into the original scheme. The table shows that in fact Evans changed very little, although he did accept the newly identified implement shapes.

While Evans was willing to explain a Prehistoric artefact's use by reference to its similarity to a modern ethnographic type, he was unwilling to make the broader social comparisons that Lubbock and Tylor had done. Stevens sat well within the latter camp, using not only ethnological artefacts from the Blackmore collection to underscore his points, but also drawing on the writings of travellers and explorers – particularly when discussing the Perigord caves. As an educational work aimed at the public (which perhaps in fairness Evans' wasn't to the same extent) *Flint Chips* is somehow a much more rounded book than *Ancient Stone Implements*; it reads better, yet remains more archaeological than either of Lubbock's two books discussed earlier (Lubbock, 1865b; Lubbock, 1870). *Flint Chips* should be rescued from under the shadow of *Ancient Stone Implements*. It deserves to be better known.

*Table 6.2. A selection of some of the sites mentioned in the Tertiary man debate. The evidence was a cross-section of supposed stone tools, bones said to have cut-marks resulting from stone tool use, and human skeletal remains in supposed ancient contexts.*

## Challenges to Consensus and Consolidation

One topic that remained contentious was the existence of Tertiary man in Europe, cutting to the heart of the human origins question in Britain. Brief mention of this subject has been made in previous chapters. In Chapter 1 the popular belief in a pre-glacial Pleistocene, followed by the glacial phase itself, and then a post-glacial Pleistocene was briefly introduced. As we have seen Prestwich's view after 1859 favoured a post-glacial age for the Palaeolithic, but there were those like Boyd Dawkins (Chapter 4) who had suggested an intra- or even pre-glacial presence of humans in Britain. This debate began to open up in Britain in the 1870s. But part of this question revolved around where and when humanity had first developed. Most people believed this had occurred somewhere 'out east', and humans had migrated into Europe during the Palaeolithic – but beyond this the question could not be resolved. A few researchers, notably Huxley, believed that humans were very ancient, originating in one of the three periods in the Tertiary (see Chapter 1, Table 1.1). They were open to the possibility of Tertiary man in Europe and Britain.

On the Continent this subject had been debated more seriously for some time, Table 6.2. Cut-marked bones from the Pliocene deposits of St Prest, France, were discovered by Desnoyers, and claimed to be evidence of the earliest occupation of Europe (Carter Blake, 1863b). Later the Abbé Bourgeois asserted that stone tools had been discovered as well. Bourgeois also claimed to have discovered Miocene implements at the French site of Thenay at a congress in Paris in 1867 (Stirrup, 1885).

| Site | Discovered or promoted by | What was found | Suggested age | Difficulties and associated material |
|---|---|---|---|---|
| Thenay near Pontlevoy, in France | Abbé Bourgeois | Flint implements (described as splinters) | Miocene | Flints either out of context or natural |
| Pouancé in France | Delaunay | Cut marked rib of extinct species on manatee (*Halitherium*) | Miocene | Cut marks probably natural – bite marks from carnivorous fish |
| St. Prest near Chatres in France | Desnoyers<br><br>Abbé Bourgeois and others | Cut marked bone<br><br>Flint Implements | Pliocene | Cut marks probably natural<br>Flints either out of context or natural |
| Unnamed site near Sienna, Tuscany, in Italy | Discovered by a Mr Lawley, figured and discussed by Prof. Capellini | Whale bone with cut marks (Evans believed they were bite marks from a shark) | Pliocene | Layer may be more recent. Cut-marks natural. Flint flakes and pottery |
| Crano dell' Olmo near Arezzo in Italy | I. Cocchi | Human skull | Pliocene | Context disturbed Neolithic axe |
| Wetzikon near Zurich in Switzerland | Rütimeyer and Schwendener | Shaped wooden staves some with shavings wrapped around the base | Interglacial | Natural staves showing differential preservation, with knots at base |
| Castelnedolo | Described by Sergi | Skeletons | Pliocene, family of shipwrecked mariners | |
| Otta in Portugal | | Implements | Unknown but early | Surface finds which could be natural percussion flakes |
| Unnamed site in Auverne, France | Laussedat | Cut marked rhino jaw | Late Eocene-early Miocene | No details |

Both Lubbock and Busk referred to the question in their Presidential addresses to the AI (Lubbock, 1873; Busk, 1875), the former in reference to Evans' rejection of the evidence for Miocene man in *Ancient Stone Implements*, and the latter on the basis of discussions at the Anthropological Society of Paris. The Abbé Bourgeois had made further discoveries at Thenay which Gabriel de Mortillet accepted as being genuine artefacts, but now explained as the product of an extinct genus of hominin, and not a 'man'. Since all other species from the deposits belonged to extinct genera, why should the maker of these tools be different? The AI spent the best part of a whole evening (Charlesworth, 1873) discussing whether some shark's teeth from the Red Crag (Tertiary deposit exposed in the cliffs of East Anglia) were perforated by Tertiary humans for personal adornment. Most felt they were natural. Abroad, Frank Calvert, the co-discoverer of Troy with Heinrich Schliemann, wrote to the AI to announce the discovery of stone tools and bones in possible Miocene deposits in the Dardanelles (Calvert, 1874). At one location at a 'geological depth' of eight hundred feet below the surface, he had removed, from the base of a cliff, a bone belonging to an elephant-like animal. It had incised artwork on one face including the image of a 'horned quadruped'. Modestly, Calvert assumed the question of Tertiary man was now settled!

Toward the end of the decade Evans served on a committee whose purpose was to oversee investigations which explored the possibility that the cradle of the human race was in the Far East, in particular the caves of Sarawak, Borneo (Everett, *et al.* 1879–1880). An eastern origin for humanity was a widespread belief at this time. Evans had often made it clear that he was not against ancient humans, merely the quality of the evidence for their presence in Europe. This was a chance to test that theory. Private subscription, and grants from the Royal Society and the BAAS, allowed A.H. Everett to investigate 32 caves, including the famous Niah Cave, and excavate in 12 of them. Unfortunately he found no evidence for the origins of humans in this part of Asia. All the human remains and material culture found were recent. This accorded with the idea that marine submergence in the area was recent and the infill of the caves all post-dated this. The oldest thing discovered was an axe identified by Charles Lyell as Neolithic.

The evidence for Tertiary man was one form of challenge to Prestwich and Evans' post-glacial date for the Palaeolithic. Another was the evidence for humans in the pre-glacial Pleistocene (the period of time after the first appearance of Pleistocene mammals but before the glaciers arrived, Table 1.2), or, equally as controversial, in an interglacial.

An interglacial was a warm period, possibly many thousands of years in duration, which occurred *within* a glacial period. Richard Hill Tiddeman, an officer in the Geological Survey, was one of the excavators of the Victoria Cave near Settle, in Yorkshire for the BAAS. In the journal *Nature* (Tiddeman, 1873) he announced the discovery of a human fibula in a drift deposit containing

| | |
|---|---|
| 1a Cromer till and contorted drift<br>1b Gravel, sand, clay – **interglacial**<br>1c Great chalky boulder clay of eastern counties etc. Northern Drift of Midlands and southern counties; unfossiliferous bottom boulder clay of Cheshire and Somerset etc;<br>1d Gravel, sand &c; Bridlington &c; shell bed – **interglacial**<br>1e Purple boulder clay of Yorkshire; lower shelly boulder clay of Lancashire and Cheshire &c. | Disturbed climate, a series of oscillating glacial and interglacial conditions. Earliest cold period (Cromer beds) not as intense as chalky boulder clay period which is as cold as it gets; later boulder clays result of glaciations that are less severe. Northern Drift + chalky boulder clay are only evidence of the southern most extent of this 1c phase, all else destroyed by later glaciers. Hoxne boulder clay = 1c |
| 2 Hessle sands and gravels with mild fauna and marine littoral shells. **Last interglacial**. Middle sands of Cheshire and Lancashire, these are the only convincing remains of this phase even though it was found everywhere | Temperate open fauna (mammoth and horse) in Yorkshire. An extensive landscape develops with rivers and soil formation. Valley drifts of Southern England, with their implements, date to the earlier and middle part of this period. At end of interglacial the land gradually becomes submerged, in Wales this reaches 1300 ft. Climate deteriorating at end of interglacial. Southern England not flooded. |
| 3 Hessle boulder clay; younger boulder clay of Durham and Northumberland; upper shelly boulder clay of Lancashire and Cheshire &c | Re-elevation of land and return to intense glacial conditions, but not so severe as chalky boulder clay times. South of England not glaciated |
| 4 Evidence for glacial action high up on mountain slopes in Wales and Lake District; flood gravels &c | Glaciers on retreat, 'great floods scouring the drifts of the low grounds' glaciers still large, but local and not confluent with each other |
| 5 Nar Valley beds | Cooler climate than present and partial submergence |
| 6 Valley moraines | Glaciers vanish |

*Table 6.3. British Pleistocene geological sequence according to James Geikie in his book The Great Ice Age (1877), second edition.*

Pleistocene mammal bones beneath a deposit of glacial boulder clay. Its human status was suggested by Professor Busk (Busk, 1874). The fibula was unusually thick walled, and if genuinely human, was from the extreme end of the human range. W.H. Flower of the Hunterian Museum had already suggested it was an elephant bone, and Boyd Dawkins thought it a bear. Both the bone-bed which the fibula came from, and the overlying glacial boulder clay, dipped into the cave at an angle as if they had been emplaced there by external forces. In the case of the boulder clay this would have been either an ice sheet, or the melt from icebergs. These two contradictory explanations introduce a long running dispute within the geological fraternity.

Tiddeman (Tiddeman, 1876) was using the chrono-stratigraphic framework of James Geikie (1839–1915). Geikie was at this point working for the Scottish Geological Survey, mapping drift deposits in Scotland. His chrono-stratigraphic sequence is presented in Table 6.3, from the second edition of his book *The Great Ice Age* (1877b). The English river drifts, and their Palaeolithic implements dated to an extensive interglacial phase, the third and last such period within the ice age. The valleys were all cut and infilled during the earlier part of the interglacial when the climate was still cold and summer melt would release huge quantities of water into the nascent valleys. There was a cold 'arctic-like' fauna present in the high drifts/terraces. In time the cold fauna was replaced by a temperate one, and finally as the interglacial became fully warm there was a southern, warmth-loving fauna in the lower valley drifts. One point to note in Table 6.3 is that Geikie promoted a major phase of marine submergence for much of Britain at the end of the last interglacial. Later he would abandon this (Tiddeman had already done so), and argue that all marine features were actually glacial.

In effect the British geological community was split into two. In the north and Scotland, the terrestrial glaciologists were arguing that boulder clays and other glacial deposits were emplaced by the movement of extensive ice sheets. In hindsight their position was unsurprising considering the glacially moulded mountainous terrain that was their 'patch'. But it did set them against the southern English submergence school, and the likes of Joseph Prestwich and the majority of the Geological Survey.

The problem at the Victoria Cave was that there was only a small remnant of this boulder clay, perched on one side of the cave mouth. McKenny Hughes (Anon, 1876) suggested that the deposits had fallen in from above, an interpretation Tiddeman hotly disputed. Additionally there were no Palaeolithic implements in the faunal bed, indeed there were none in Scotland or the north of England as Evans had pointed out in *Ancient Stone Implements*. This made the human fibula from Settle appear anomalous, despite the presence of a rich interglacial-like fauna including elephant and rhinoceros.

The Settle Cave contributed to the glacial/interglacial debate. The fibula was stratified below the boulder clay. Tiddeman suggested that the great ice sheet that had emplaced the boulder clay had wiped out all evidence of humans, leaving caves like that in Settle as isolated time capsules. Consequently, humans were of interglacial age in the north of England. How did this affect Prestwich and Evan's post-glacial age for the river drift implements in the south? According to Tiddeman, and the interglacial supporters, there wasn't one glacial boulder clay, but many, deposited at different times within the ice age. This complicated the more straightforward framework of the southern submergence school. For example an ice advance into northern England could overlie river drifts with handaxes. So the humans there would be pre-glacial in relation to the local glacial sediments. Elsewhere, to the south, human occupation could have continued uninterrupted until the advent of a later and more extensive ice sheet. In terms of a relative stratigraphic relationship, both deposits would appear pre-glacial and contemporary, whereas in fact they were of different dates – the southern occupation was post-glacial in terms of the northern deposits but pre-glacial in terms of its own local stratigraphic succession. What the northern school were claiming was that a pre- or post-glacial age for any deposit stratified in relation to boulder clay was only of local significance.

Equally as controversial, and potentially damaging to the post-glacial age bracket for the Palaeolithic, was work being done toward the end of the 1870s in East Anglia. Sydney B.J. Skertchly (1850–1926) was an officer of the Geological Survey. He had spent eight years mapping the East Anglian chalky boulder clay, a very distinctive till/glacial deposit. He estimated he had mapped some two thousand square miles of the formation. This experience gave him the confidence to overcome what some termed his 'temerity' in suggesting an interglacial age for humans in East Anglia, and then to face his critics (Miller and Skertchly, 1878; Skertchly, 1876a; Skertchly, 1876b). In 1876 he wrote to the journal *Nature* announcing that he had clear proof of the presence of interglacial humans.

In deposits of brickearth, stratified beneath the East Anglian chalky boulder clay he had found two sites with a number of ovate shaped handaxes. There were many more such deposits without palaeoliths elsewhere in the region. The implements were rather crude looking, and very similar to other crude ovates found in drift deposits that overlay the chalky boulder clay. This glacial unit, unfortunately called the 'upper boulder clay' by some geologists, was stratigraphically lower than other boulder clays/glacial deposits in the Midlands, Lincolnshire, Yorkshire, and in Scotland. Skertchly supported an analogous argument to that just described above. Inter-glacial (i.e. period with warm fauna) in one area, could be post-glacial elsewhere. This argument is summarised in Figure 6.6.

Skertchly's first letter led to a run of correspondence on the subject in the pages of *Nature* and elsewhere. In the second edition of *The Great Ice Age* (Geikie, 1877b), Geikie included a chapter written from notes that Skertchly provided and he elaborated further on the cruder ovate handaxes that were found in the drifts above the chalky boulder clay. There were two groups here, but one group was more refined looking and less damaged than the other.

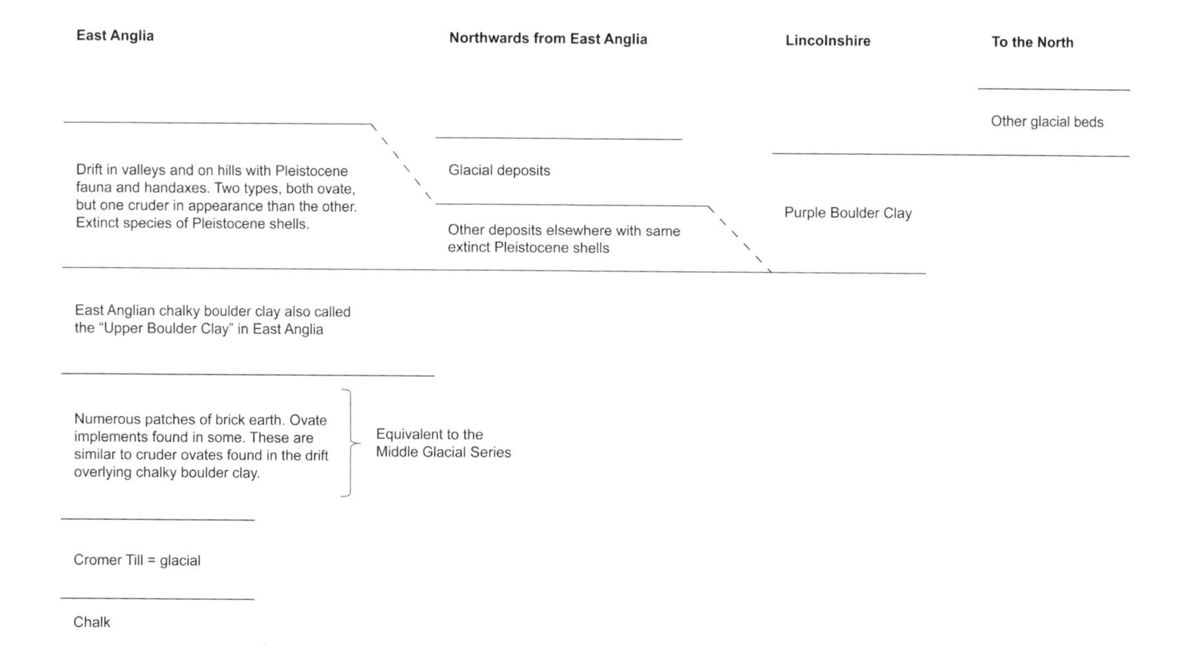


*Figure 6.6. Schematic reconstruction of J.B. Skertchly's stratigraphic arguments for East Anglia.*

He implied the cruder and more worn palaeoliths had been derived from the destruction of implementiferous deposits below the boulder clay, possibly from floods as the ice sheet melted. The two variants (crude handaxes in lower and upper deposits, and the more advanced looking handaxes in the upper deposits) represented an evolution in tool manufacture over an unknown time period. This flew directly in the face of Evans's views as he saw no evidence for an internal evolution of handaxes within the Palaeolithic period at all.

Other challenges to a post-glacial consensus focused more on theoretical concerns. An excellent example of this is a critique from J.W. Flower (Flower, 1872; and discussion to Bruce Foote, 1869). It was made all the more important by the fact that it came from someone within the Lubbock–Evans circle. Flower had accompanied Prestwich on his return trip to Abbeville in 1859 (Chapter 2) and gained the honour of being the first of the English academics to find a handaxe *in situ*. Flower argued that there was a long hiatus between the Palaeolithic of the drift and the Palaeolithic of the bone caves. There was another equally long one between the end of the Palaeolithic and the Neolithic. He began from the same position as Stevens in *Flint Chips*. Whereas most cultural phases of the Neolithic and later Prehistoric periods could be seen to have been repeated in historical times or by modern indigenous peoples, nothing like the Palaeolithic of the drift had ever been recorded other than in the Pleistocene. It was a unique period in human history.

Much of what Flower said (the paper was originally delivered at the BAAS in Edinburgh in 1871) he had already noted in earlier papers. He believed southern England and northern France formed a single geological (chalk) province connected by dry land across the English Channel at the time of the drift. The drift deposits themselves, as well as overlying peat and loess deposits had also been continuous across the land bridge – thereby dating the formation of the Channel to quite a late period in Prehistory. This was why the mammals of the drift, and the implements in the two countries were so similar but the rivers that drained this landscape were not the same as those today. Prestwich and Evans's belief that the high level drifts were the ancestral flood plains of the modern rivers was wrong. The drift deposits were too high up and often too far back from the valley's sides. There were drift deposits on interfluves. All of this implied that the higher drifts were laid down by different rivers, flowing in different courses, and draining highlands now long vanished. Furthermore, Flower disputed Prestwich's interpretation of the gravels as the product of normal Pleistocene river action. Explaining away the presence of shells in the gravels, he asserted that the nature of the drift was more in keeping with short periods of violent flooding – as the French geologists believed (Flower ignored Prestwich's argument about dynamic flooding, presumably he felt this was flooding at a different scale). He then argued that the juxtaposition of mammal fossils with flint implements did not prove their contemporaneity, in fact the Pleistocene mammals were older and dated to a time before humans arrived. The implements themselves were too fresh to have been transported far by the rivers, so they were found pretty much where they were used and dropped.

Flower then went on to argue that the more recent age of the bone caves was demonstrable on a number of grounds, fauna being the most important. In the drift there were about seven species of mammal continuously recovered compared to twenty seven in the caves. Since this must reflect the controlling influence of climate, and climate change, it was evident that much time must have elapsed between the two periods. Any similarity in implement forms between the drift and the caves (he implied such similarities had been exaggerated) was accidental, a function of the limited number of shapes that could be made out of flint. Defending himself against criticism from Lane Fox, he argued that he believed there were more handaxe types in the drift gravels than accepted by Evans (or by implication Stevens), and that this distinctive suite went to prove just how different the Palaeolithic was. The discussion between Lane Fox and Flower after the reading of Flower's paper must represent one of the earliest debates on handaxe typology – whether the shape was specifically predetermined (Flower), or whether the original flint nodule shape could influence the axe's final form (Lane Fox); the same debate continues today (Lane Fox, 1872).

## Debating the Challenges to Consensus

John Evans, by this time the head of the committee overseeing the Victoria Cave excavations, had his own way of rising to these challenges: he met them head on.

In 1876–1877 John Evans was President of the AI. He used the opportunity to initiate a mini conference on the 'Antiquity of Man', which took place on the evening of 22nd May 1877. The programme is interesting, and the line-up of speakers deliberately chosen. They were supporters of the orthodox view point as represented in *Ancient Stone Implements* – Professor William Boyd Dawkins, and Professor T. McKenny Hughes. Both of these men, while disagreeing with Evans on some points, supported a post-glacial age for humans and the drift terraces. Neither were supporters of the Victoria Cave fibula or its interglacial age, and both had been members of the steering committee appointed to oversee excavations at the cave. Boyd Dawkins had directed the excavations for a while. Although Skertchly was not on the programme, Tiddeman was present to defend the case for inter-glacial humans.

Evans's opening address (Evans, 1878b) focused on the key elements of context and identification. It was a reiteration of the Abbeville methodology. In the first instance, the interpretation of the nature and stratigraphic position of a layer was the province of the geologist. Evans may have been making an oblique reference to problems with Tiddeman's stratigraphic interpretations by referring him to the distinguished geologists in the audience – Prestwich and McKenny Hughes. The job of correctly identifying a stone tool belonged to an archaeologist, and a human bone to a physical anthropologist. Evans made explicit reference to the discoveries from Thenay, and other European pre-glacial/pre-Pleistocene sites (Table 6.2) where evidence of human action had been inferred from supposed cut-marked bone. The incised whale bones from near Sienna, he said, were probably the result of shark bites. In this he reiterated his position from 1872, but again made it clear that he did not dispute the possibility of pre-Pleistocene humans, merely the quality of the current evidence for them being in Europe (Evans, 1878b; Evans, 1878c). In referring to cut-marked bone in his opening address, it is tempting to suggest that Evans was setting up Tiddeman as this was a topic that would be central to his presentation. Evans finished his introduction with his usual rallying cry of 'caution, caution, caution' (Evans, 1878b).

Boyd Dawkins stood up next (Dawkins, 1878). He delivered a no-holds barred attack on James Geikie and the interglacial supporters. Their proofs for the existence of a temperate phase in the glacial were simply untenable in Dawkins's opinion. He also took Tiddeman to task for his poor judgement in bone identification and stratigraphic interpretation. O'Connor (2007, 62) suggests that Boyd Dawkins was actually accusing Tiddeman of poor excavating. At the very least he accused him of sloppy interpretation. Tiddeman had suggested there was a reindeer horizon (i.e. cold fauna), overlying a warmer fauna with hippopotamus. Boyd Dawkins said that when he had been in charge of excavations in 1872 they had found both together, and reported it at the BAAS at Brighton in 1872, but Tiddeman had overlooked this. Promoting a wholly zoological approach, Dawkins demonstrated that humans had to be post-glacial, their association with a reindeer-dominated fauna, of northern origin, placed them in the last of three consecutive phases of mammal development in Britain. He also alluded to McKenny Hughes' ssite of Pontnewydd. Here was the proof of post-glacial humans. Dawkins reiterated his belief, already expressed to the Geological Society, that the Victoria Cave fibula was from a bear.

He also alluded to his latest excavations at the site of Creswell Crags. This was a deeply significant locality. He had noted a stratigraphic succession of human occupation (Dawkins, 1878; Dawkins, 1880a). At the base were cruder stone tools (later identified as flakes). In the middle layers a second industry was present (later equated with handaxes). At the top of the sequence were the most refined stone tools, accompanied by bone or antler tools, and even an example of art to match the evidence of Lartet and Christy in the Dordogne. This was a horse's head engraved on a fragment of rib.

> A sequence of Palaeolithic remains of this sort has not, as far as I know, been obtained from any other bone cave either in this country or on the continent. To my mind at least it implies a distinct progress in the arts among the cave-dwellers, while the fauna…remained on the whole

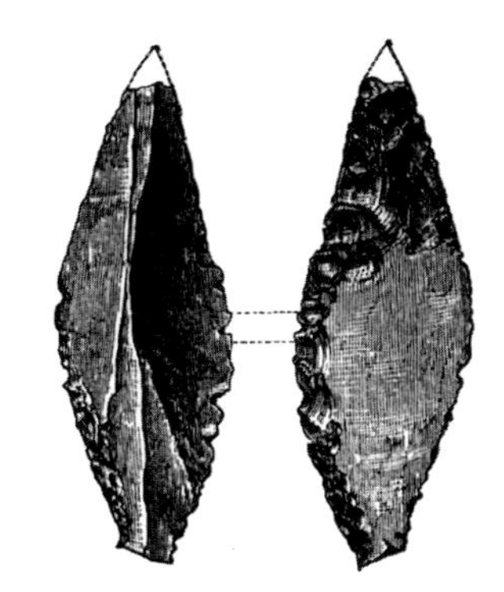
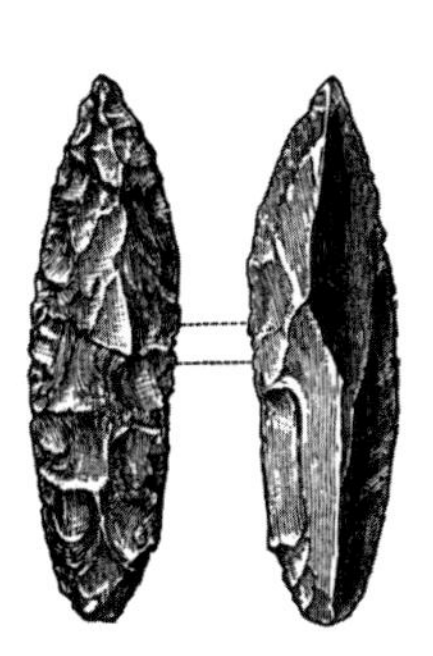

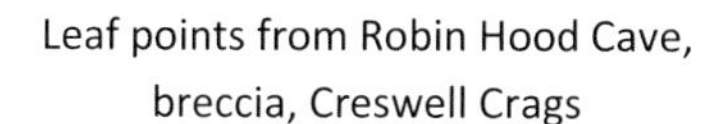

Leaf points from Robin Hood Cave, breccia, Creswell Crags

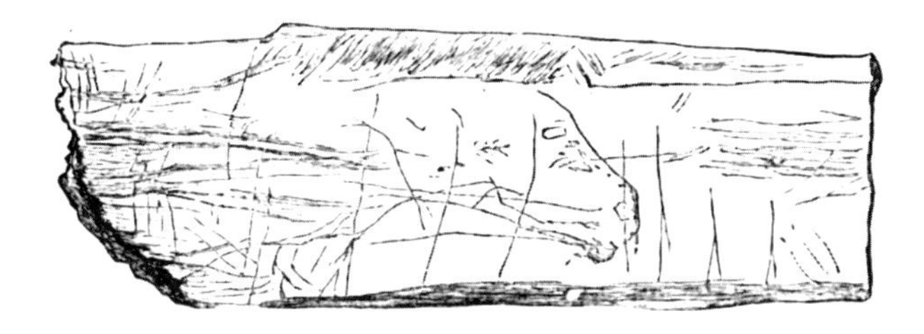

Engraving of a horse on fragment of rib, upper cave earth, Robin Hood Cave, Creswell Crags

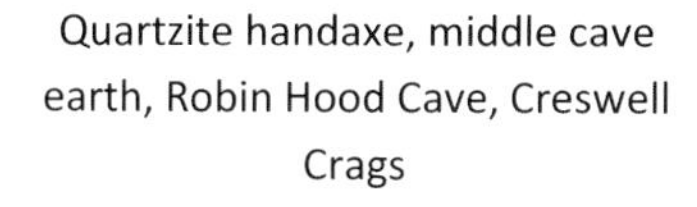

Quartzite handaxe, middle cave earth, Robin Hood Cave, Creswell Crags

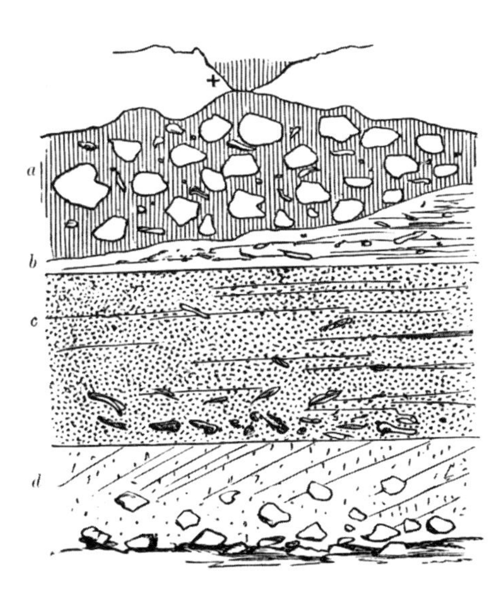


Stratigraphy of Robin Hood Cave, Creswell Crags.

+ represents stalagmite connecting roof to breccia

a) Stalagmite breccia with bones and implements
b) Cave earth with bones and implements
c) Red clayey sand with a few hammer stones and primitive quartzite flakes
d) Light coloured sand with limestone blocks. No archaeology

> without change. In Kent's Hole also, the implements found in the breccia at the bottom are of a ruder form than those which have been met with in the cave-earth above. (Dawkins, 1878, 154).

Switching to stratigraphic geology, McKenny Hughes (Hughes, 1878) took on the interglacial arguments of Skertchly. The focus was now the open air sites. He attempted to show (although his evidence was not convincing and I find his logic difficult to follow) that the East Anglian fenland sites with implements were not overlain by real glacial boulder clays, and that they belonged, stratigraphically, to a group of sediments which actually post-dated the glaciation of the area.

Tiddeman, in his paper (Tiddeman, 1878), described new bones from the Settle cave which showed anthropomorphic modification. He argued (backtracking somewhat) that if bone identifications were ambiguous (i.e. the fibula), then the presence of anthropogenic modifications in the form of incisions deliberately cut into bone could not be. They were cut and dried proof of a human presence. His paper is a fine example of the tail wagging the dog. He actually noted that the cut marks on the bones were sharp and modern-looking, and that the bone itself was in a fresh condition and looked recent. Moreover, the bones were from a goat. What could this mean? Clearly, he asserted, physical condition as a criteria for age should now be thrown out because the Victoria Cave bone showed it to be unreliable. He also tried to suggest that goats were a natural component of the Pleistocene fauna. With hindsight, it is perhaps clear why Evans enjoined his audience to scepticism. Boyd Dawkins in the discussion left the audience in little doubt as to his opinion on Tiddeman's interpretations. The one good point that Tiddeman did raise, which concerned the difficulty in determining the contemporaneity of glacial tills between the north and south of England, was rather lost. In addition to the three papers mentioned, a letter accompanied by photographs of the recently published Brixham cave artefacts was read out, as was another letter from an Italian anthropologist describing later prehistoric remains from the caves of Italy.

Tiddeman put up a spirited defence of his interpretations in the ensuing discussion. Unfortunately no mention was made of the East Anglian evidence and Skertchly's ideas were not discussed further. Whether the bulk of the audience left the AI's meeting room that night convinced of the continuing post-glacial age of the river drift is not possible to ascertain. James Geikie, not present at the meeting, read the transcripts later. He at least was unimpressed (Geikie, 1877a). In defending the orthodox interpretation Evans noted that some implements from the river drifts were made out of gravel clasts derived from glacial drifts. Arguing by association (the basis of much of Evans's polemic), similar implements should also be post-glacial. However, in one respect the evening was successful. Evans appeared to accept Boyd Dawkins's evidence of progressive development within the stone tool assemblages of

Creswell Crags. This was a key moment. It presaged official sanction for an evolutionary Palaeolithic sequence in Britain based on material culture. Frustratingly, Prestwich suggested the simplistic tools in the lowest layer were not really more primitive, they just reflected their manufacture on quartzite.

The most telling comments from the evening were by Prestwich. He accepted the possibility of the Victoria Cave being inter/pre-glacial, though wanted more conclusive evidence. He noted the mammals in the northern pre-glacial deposits were the same as those in both the southern pre- and post-glacial deposits, so there was no specific reason why humans should not be present before the boulder clays. Prestwich then went on to provide much more conclusive evidence. He noted that in the Thames Valley palaeoliths were only present in the high terrace at one location, at Reculver in the Lower Thames estuary. Everywhere else in the valley the implements were confined to the lower terraces. His explanation was climatic. At the time of the deposition of the upper Thames terrace it was too cold for humans in the upper and middle parts of the valley (he described glacial features underlying the terrace gravels that he had seen himself). Conditions were just warm enough for humans to survive in the lower part of the valley near Reculver. The point here was that the high terrace at Reculver was a downstream continuation of the archaeologically sterile high terrace in the upper and middle sections of the valley and contemporary with them. This meant there were humans in the south of England in the glacial phase of the Pleistocene. This was a variant on the argument of the ages of the different boulder clays in north and southern England.

> Thus the data for carrying man back to the boulder-clay period, may be considered as an account audited and passed. (Prestwich in discussion to Tiddeman, 1878, 177)

Although not strictly contemporary with the glaciers, the humans in the higher terrace at Reculver existed in a cold phase that was definitely older than the conventional understanding of the term post-glacial. By definition then, in southern England, humans were both 'glacial' and post-glacial. He added, offhandedly, that he was even prepared to accept the French evidence for pre-glacial humans at Thenay and elsewhere.

## Theory Building in the 1870s

Unlike the 1860s this decade did not see an outpouring of articles and books on human origins, but then there was little to compare with that first flush of excitement immediately after 1859, or the energising effect of the Moulin Quignon affair. With the rise of the new anthropological society and its domination by the Lubbock–Evans circle, the older debates on racial origins declined markedly in frequency.

Charles Darwin's own position on race and human origins had been eagerly awaited since 1859. It saw publication in 1871 – *The Descent of Man and Selection in Relation to Sex*. Much of the core theory had been worked out in the 1850s – part of the big book that he abandoned in favour of writing the *Origin of Species*. He maintained his belief that hybrids of truly different species could not successfully breed viable (i.e. fertile) offspring. If individuals from different races of humans could conceive children and grandchildren, then by definition all humans were part of one single species. His ideas on sexual selection had also been largely fleshed out in the 1850s. He was certainly writing to Wallace in the early 1860s suggesting that sexual selection was a better explanation for racial origin than any other mechanism (Darwin, 1864a). Natural selection adapted organisms to succeed in the battle for life, so it was utilitarian (Stepan, 1982).

However sexual selection was non-utilitarian, it did not adapt organisms for success in the struggle of 'all against all'. Sexual selection worked on a sense of aesthetic value which varied from place to place with different peoples. It was selection for characteristics and features that had no specific adaptive value in survival. As humans spread out into the world spontaneous variation would give rise to a variety of different physical traits (including skin colour of every shade, hair and eye colour, hair type, etc.) in differing local populations. In humans, women would chose those traits of physical appearance that more closely approached the ideal of male attractiveness for that group, and later on, those features that were advantageous in a male (hunting prowess, strength, wisdom etc.). In this way discreet racial features, as well as particular traits of behaviour, specific to distinct geographical locations, would slowly become fixed in a population. Men, in particular chiefs and other powerful tribal males, would be selecting for themselves those women who were the most beautiful according to the ideals of the group. Over time, sexual selection would continue to emphasise the most desirable traits in men and women, eventually fixing them into the population. They would become hereditary. Darwin believed that sexual selection had been a powerful force during the evolution of early humans, but that by modern times it had ceased to play a distinctive role as racial differentiation was now fixed. Primitive humans had made their choices on the basis of physical desirability alone. By modern times people were choosing their mates for reasons other than just attractiveness, so sexual selection was no longer as effective.

By any standard of comparison the *Descent of Man* is a remarkable book. It provided in a single volume a complete theory of the biological evolution of humans explained through two types of complementary selection, natural and sexual. It provided an explanation for the development of the modern races, and at the same time explained the evolution of culture and material culture. The thread that seamlessly joined all these disparate elements was selection. Like Wallace, Darwin emphasised the evolution of human mental faculties, in particular dwelling on altruism and sociality. He posited that their benefit to the survival of a group, and the benefits that would accrue to altruistic individuals within a group, would not have been lost on early man. Like many of his generation Darwin believed that habitual behaviour, especially if enacted to evident benefit, would eventually become instinctual and therefore hereditary (i.e. a Lamarckian component). Through mechanisms like praise and blame the power of positive sociality would be reinforced. The superior mental faculties of some individuals, as well as superior social morals (the traits that made for effective social cooperation), would slowly accrue within a society as more individuals with these characteristics were born to each generation. Consequently, natural selection would create some societies that were inherently superior to others. Just as there was competition between individuals, so there was competition between all social groups and societies, irrespective of size and scale. This was because population at any level would always come to outstrip resources. So Darwin used selectionism to link competition between individuals and groups together to explain how certain societies could succeed whereas others would fail.

He went still further. Competition now existed between the races. Variation resulting from different individual, physical, moral and social characteristics collectively ensured that some races were superior to others. The Victorian imperial world-view seemed proof enough of that. Darwin suggested cultural institutions were not shared by all humans (*contra* Tylor). Different geographical centres/races would develop their own unique cultures. By definition, the superior races would have had superior cultural institutions, and these contributed directly to the dominant position enjoyed by those races. Race, origins, and culture were therefore indivisible.

The book did well (Moore and Desmond, 2004). The two volume first edition of 1871, priced at twenty four shillings, was reprinted twice that year. A second edition, with rewrites now condensed into one volume, was published in 1874 at nine shillings to help boost sales. It was reprinted the following year and again in 1876 with a new section. This was reprinted again in 1879. So the book spanned the whole of the decade and, as ever with Darwin, rewrites and new editions were used to correct mistakes and answer critics. In effect this holistic interpretation of human origins and cultural evolution was before the public eye across all of the 1870s. Why *Descent of Man* did not become popular with the anthropologists and archaeologists therefore invites speculation. Perhaps its focus on racial origins was seen as slightly out of date. This was becoming more of a question for biologists and those concentrating on the development of a viable theory of heredity. Theories on heredity such as Darwin's pangenesis lacked empirical support at a time when experimental methodologies were beginning to become more popular. For the anthropologists research questions had moved on and Prehistory was more the vogue; it had data to work with. Additionally, selectionist interpretations remained unpopular.

## Augustus Lane Fox and Material Culture

As a distinct strand within the area of theory building, I will now look at studies on material culture, ancient and modern. Like Darwin, Augustus Lane Fox believed that habitual action could become inherent and so hereditary. An innovation in the use or design of a tool or weapon represented the insight of an individual and would, over time, become a routine part of that object's manufacture. Lane Fox argued that each innovation in material culture that was made was actually an *idea* preserved in the materiality of the object. A progressive succession of such objects and ideas was a demonstration of the evolution of thought. The value of a collection of material culture from differing modern indigenous peoples was that they therefore preserved different stages in cognitive evolution. This drew significantly on the currency of evolutionary anthropology. Theoretically it was possible to show the links between one stage of cultural development and the next, because a people's level of development was reflected in the things they made. However, these links would not often be preserved in Prehistoric examples; and collections of Prehistoric artefacts were not complete enough to show the sequence of progression. On the other hand with the material culture of modern peoples, it was (Lane Fox, 1875).

Perhaps more than anyone in the Lubbock–Evans circle, it was Lane Fox who explored the link between evolutionary theory and material culture in the modern world and in the past. One of his biographers (Thompson, 1977) suggests that he was heavily influenced by a crude form of Darwinism during the 1860s (probably best read as a belief in progressive transmutation), and by a saltationist approach which he took from T.H. Huxley. In the first half of the following decade he was more influenced by the philosopher and sociologist Herbert Spencer (1820–1903). It led to a more gradualist approach (Bowden, 1991). While this may be true, his evolutionary credentials remained intact in the 1870s, and Darwinism certainly crept into his writings. In explaining his belief that the forms of Palaeolithic handaxes had arisen by chance, and in many cases simply reflected the natural variation of the nodules of flint they were flaked from (Lane Fox, 1875), he demonstrated a solid grasp of the role of random variation.

Another key point made by ethnographic collections was in highlighting the danger of judging innovations in sequences of tools from a modern perspective – automorphism (a term Lane Fox borrowed from Spencer). To European eyes many of the innovations appeared insignificant and rather trivial. Yet to the peoples who had made them they were significant advances and the product of many generations of thought and application. Like Lubbock and his contemporaries, Lane Fox began his developmental sequences with comparisons to the natural world. The highest examples of tool use in nature were chimpanzees cracking nuts. This was the starting analogy for the earliest human progenitors. Initially tools would have been of perishable materials (Lane Fox, 1875; Lane Fox, 1977). His ethnographic specimens showed an evolution from mere sticks used for offence and defence, to clubs, shields, and eventually even boomerangs in Australia. Many Australian aboriginal peoples reflected these stages. Borrowing from Lubbock he suggested that nut cracking led to the use of stones as tools when accidental breakage revealed the utility of a sharp edge. Utility was an important concept in Lane Fox's theorizing. It was recognition of the utility of function that was the driver for change in tool use and morphology over time. He postulated an early stage of stone working, not dissimilar to that of Stevens, in suggesting that prior to the bifacial and multidirectional flaking patterns seen on Palaeolithic implements, there would have been a unifacial phase, where single flakes were knocked off flints and just used as they were. Tylor (1865) had argued the same in *Researches*.

Unlike most ethnographic museums at the time, which were arranged by geographical areas, such as Henry Christy's collection or the Blackmore in Salisbury, Lane Fox arranged his more by classes of artefacts, for example, grouping disparate collections of spears or clubs together, and in this way made another important point about the value of ethnography. The sequences of development showed that indigenous peoples were not degenerate cultures, their material culture illustrated clear progress in social evolution. His collection was on public display between 1874 and 1878 at the Bethnal Green Museum in north London, a satellite of the South Kensington Museum. The museum was deliberately located in a working class area of east London, and this would have suited Lane Fox's passion for public education and in particular museums as instructional institutions (Bowden, 1991). The displays were also laid out to encourage visitors to believe that gradualism and slow incremental change was the natural order of things (Bowden, 1991). This was the venue for an important paper in the development of his theoretical outlook on material culture which was delivered to the AI at the museum's opening (*ibid,* 1875). The collection was moved to South Kensington in 1878, and remained there on public display until 1884 when it was acquired by the Oxford University Museum. A few years later, it was displayed in a special annex of the museum.

In 1880 Lane Fox inherited the estate of Cranborne Chase from a distant relative, and took up the name of Pitt-Rivers. But by this time he was no longer active in evolutionary studies. From the mid-1870s onwards he had already been at work at the later Prehistoric excavations for which he would become famous. After Lane Fox, the theoretical link between evolution and material culture was never pursued to the same intensity; a second interpretative strand to fall by the wayside in the 1870s.

I will end this section by looking at a book by William Boyd Dawkins, his first. *Cave Hunting, Researches on the Evidence of Caves Respecting The Early Inhabitants of Europe* was published in 1874. As the title suggested it was a comprehensive review of the British cave evidence compared to selected caves from the Continent. Dawkins merits a place in a section on theory building because his views challenged many of the orthodox assumptions that underpinned the geological interpretations of the Pleistocene, and which were supported by the type of material culture studies practised by Lane Fox and others. For Boyd Dawkins, it was fauna that was the most important type of evidence, particularly for climate change, as animals responded directly to the climates they lived in, and reflected those where they had evolved.

Refuting the evidence of European scholarship, Dawkins asserted that all the major European finds of human remains associated with Palaeolithic artefacts and/or Pleistocene mammals were later intrusive burials. A dolichocephalic Neolithic race used caves for their interments. This was the case at Aurignac, Cro Magnon, Paviland, and all of the celebrated Palaeolithic caves. He went to great lengths to highlight inconsistencies in their stratigraphic profiles, and was often able to imply that modern mammals were amongst the fauna from the Pleistocene layers, confirming these were modern intrusions into much older deposits. In fact there were only two caves containing Palaeolithic human remains that passed muster, one was the Belgian site of Trou de Naulette, and the other was the Victoria Cave at Settle, Yorkshire, the excavations for which he had directed for a time. The Victoria Cave remains represented, as Boyd Dawkins believed at this stage in his career, evidence for a pre-glacial human. Later, as seen, he came to believe the fibula was from a bear. Other British caves clearly contained Palaeolithic artefacts in association with Pleistocene mammals, these were King Arthur's Cave in the Wye valley, Wales; Pontnewydd (Chapter 4) and Hoyle's Mouth Cave, Dyfed, both in Wales; Brixham and Kent's Cavern, and Wookey Hole.

These long headed Neolithic burials from the caves, and the tumuli were part of an extensive Neolithic race which Boyd Dawkins allied to a non-Aryan people who inhabited Europe before the round headed brachycephalic people of the Bronze Age arrived, and alongside whom they lived for a while. This Neolithic race represented the forefathers of the modern Basque and Berber peoples. But of the Palaeolithic race/races there were no survivors in Europe. However, the cave race of Palaeolithic Europe did have descendants, part of an unbroken bloodline; these were the Eskimos of the Arctic, today called the Inuit people. Their life style and behavioural traits, but especially their material culture allied them closely to the race who had made the artwork in the south-western French caves. Boyd Dawkins had been arguing this since the middle 1860s, as had others. The similarity was not so much in individual items of Inuit culture which could be matched to Palaeolithic artefacts (although he described and illustrated many), it was in the *set* as he described it. The match between whole groups of related tools and weapons meant that the Inuit were continuing to live a Palaeolithic Ice Age lifestyle.

The relationship between the Palaeolithic cave race and the race whose handaxes were preserved in the drift was more complicated. Boyd Dawkins accepted the stratigraphic implications of Kent's Cavern. The drift-like

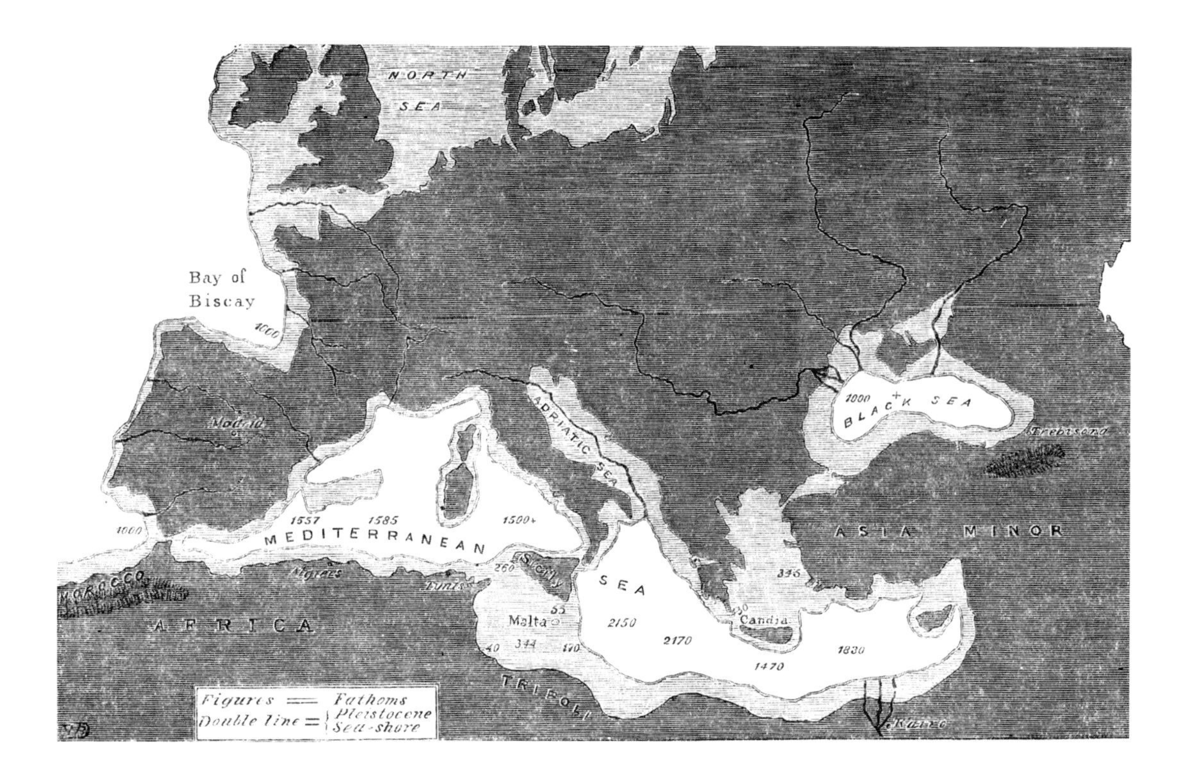


Figure 6.7. The connections between Britain and the Continent as revealed by the lowering of sea level during the Pleistocene. After Boyd Dawkins 1874

implements in the breccia which were sealed in by stalagmite were earlier in date than the handaxes, flake tools, blades, and objects of carved antler found in the overlying cave earth. This in turn supported the view that the drift was probably earlier than the cave period. He also accepted the idea that Kent's and Brixham had very similar sequences. The breccia with its drift-like implements had originally filled both caves, but its upper portions had been eroded away. The succeeding stalagmite and cave earth implied a long hiatus between the breccia and the later cave implements. Later in the book he suggested there may have been distinct provinces with Palaeolithic peoples respecting their own territorial boundaries. The occurrence of cave implements in the south and west of Britain, implied a connection between the south and west of France where similar discoveries had been found.

On the other hand the handaxes from the gravels of Hampshire, Sussex and Kent were similar to those directly across the channel in the drift of the Somme and the Seine. They were similar to others in the Thames basin and to the north-east in East Anglia. This was a distinct Palaeolithic province or culture area. The implication is they were broadly contemporary. It may be that Boyd Dawkins was referring to the handaxe-like component found in the cave deposits and not necessarily to the drift of Prestwich and Evans, but this is not clear from the text.

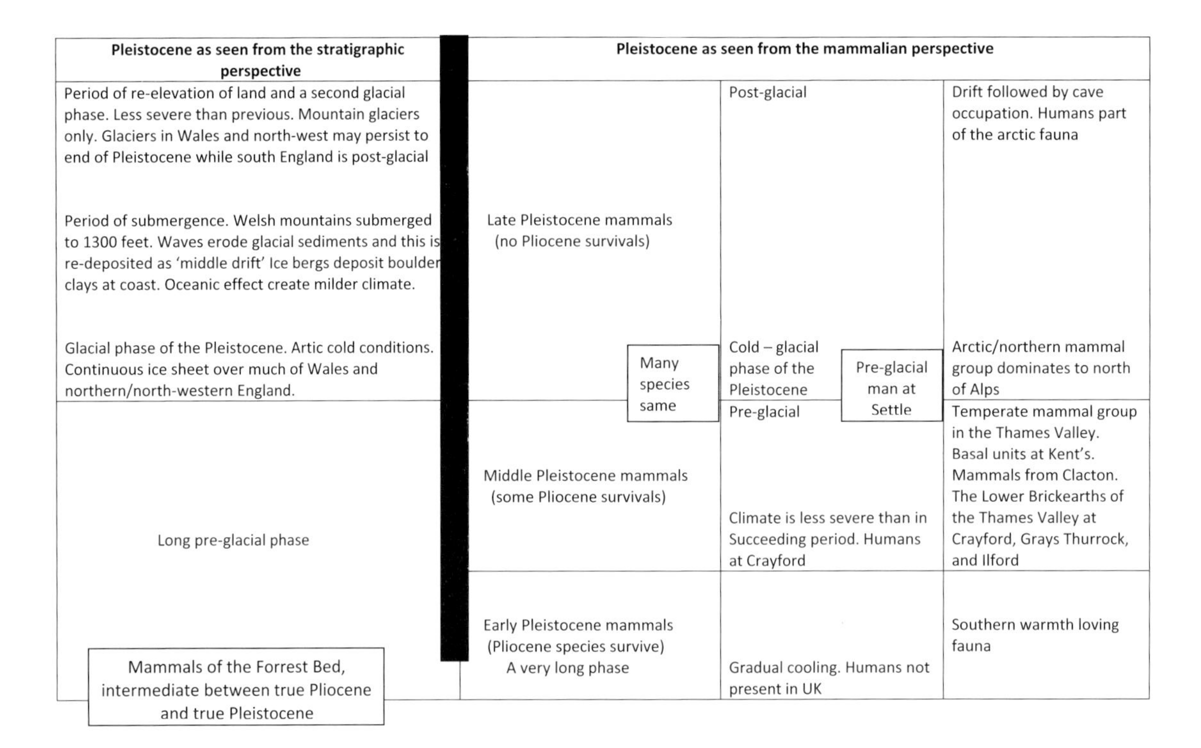

| Pleistocene as seen from the stratigraphic perspective | Pleistocene as seen from the mammalian perspective | | |
|---|---|---|---|
| Period of re-elevation of land and a second glacial phase. Less severe than previous. Mountain glaciers only. Glaciers in Wales and north-west may persist to end of Pleistocene while south England is post-glacial | | Post-glacial | Drift followed by cave occupation. Humans part of the arctic fauna |
| Period of submergence. Welsh mountains submerged to 1300 feet. Waves erode glacial sediments and this is re-deposited as 'middle drift' Ice bergs deposit boulder clays at coast. Oceanic effect create milder climate. | Late Pleistocene mammals (no Pliocene survivals) | | |
| Glacial phase of the Pleistocene. Artic cold conditions. Continuous ice sheet over much of Wales and northern/north-western England. | Many species same | Cold – glacial phase of the Pleistocene<br>Pre-glacial man at Settle | Arctic/northern mammal group dominates to north of Alps |
| Long pre-glacial phase | Middle Pleistocene mammals (some Pliocene survivals) | Pre-glacial<br>Climate is less severe than in Succeeding period. Humans at Crayford | Temperate mammal group in the Thames Valley. Basal units at Kent's. Mammals from Clacton. The Lower Brickearths of the Thames Valley at Crayford, Grays Thurrock, and Ilford |
| Mammals of the Forrest Bed, intermediate between true Pliocene and true Pleistocene | Early Pleistocene mammals (Pliocene species survive) A very long phase | Gradual cooling. Humans not present in UK | Southern warmth loving fauna |

*Table 6.4. William Boyd Dawkins's interpretation of the chrono-stratigraphic sequence for the Pleistocene in the British Isles as he viewed it in the early and middle 1870s.*

How were such connections possible? This was down to Pleistocene geography, and Boyd Dawkins's view of Ice Age Europe is shown in Figure 6.7. The lowering of the sea levels during the Pleistocene created land bridges that connected up the eastern English Channel with the southern North Sea and the East Anglian coast; and between south-western Britain and south-western France. This was not that different to the view held by Lubbock or Lyell a decade earlier. My interpretation of Boyd Dawkins's – views on Pleistocene chronology is given in Table 6.4. Two phases of cold were separated by a more temperate period, but not one that could be called an interglacial.

He was keen to emphasise that the left hand side of the table, the geological succession, did not map directly onto the faunal succession shown on the right hand side. There was overlap in the species and genera present, and this highlighted a clear problem with his belief that fauna was the only true barometer of climate change. On the face of it, his reasons for disputing other lines of evidence appeared sound. He disagreed with researchers like Dupont, and especially Lartet, who used the dominant mammal at a site to date it. To the north of the Alps, animals like the mammoth and reindeer were part of the same ecology, and their presence, absence and frequency in one locality was dependent on many factors. But he also took people like Lyell and de Mortillet to task for their dating of sites based on sequences of material culture. It was a similar argument to the last. Differences between tool types at discrete sites, or their presence and absence locally, could reflect different contemporary groups expressing their own distinctive tribal character. It did not necessarily reflect diachronic change. This was seen in the modern ethnographic record. Also crudeness was no test of age, it just reflect the variable skill base present in any group of humans.

Yet his preferred line of evidence, mammal communities, allied to extinction and speciation events over time, was equally problematic. He identified three groups of mammals that reflected changes in Pleistocene climate. The problem was that the groups shared common species between them, and they were found in the same sites over wide areas. In general, a southern group of warmth loving fauna occupied Britain earlier in the Pleistocene, some species being survivals from the warmer Pliocene climates. The home of most of the non-Pliocene species was Africa and the Mediterranean area. They migrated northwards during warm summers and returned south in colder winters. This migratory pattern was facilitated by the land bridges in Figure 6.7. In time a northern faunal group gradually appeared. These were adapted to colder climates, and as a faunal community, at its most developed, it was arctic in character. This was the mammal fauna of the glacial period and it dominated the later Pleistocene. Humans were a part of this fauna in post-glacial times.

The only way Boyd Dawkins was able to reconcile the appearance of species from these two groups in the same cave deposits was to argue for annual migrations in response to seasonal changes. With the sea-level lower and more land exposed, continentality in terms of climate was greater and the difference between summer and winter more extreme. His reason for not wanting to have the mammal groups mapped on to the geo-chronological sequence was because of this permeable boundary. The third mammal group, the temperate mammals, were present in the Thames Valley and elsewhere during the period characterised by the middle Pleistocene mammal phase, Table 6.4. As individual species, many of the temperate mammal group were not particularly useful as diagnostic indicators.

In one instance association with mammals of the temperate group was very revealing. In the brickearths of Crayford the Reverend Osmond Fisher, whom we will meet later, had found a flint flake embedded in the undisturbed deposits. Boyd Dawkins was present when it was found and accepted the find and its context as genuine. This was the only evidence for humans in the middle Pleistocene mammal period. Like the Victoria Cave fibula, this was evidence of a human presence prior to the glacial, or at least the coldest glacial phase of the Pleistocene.

Boyd Dawkins' views were highly individual and contradicted many of the tenets of accepted wisdom He disagreed with Evans on a number of points, but overall his opinions meshed in with the post-glacial consensus view. Like Evans, he was not one for changing his mind much. He stuck to the views outlined in *Cave Hunting* for many years, and as his influence grew, so did his effect on the progress of the human origins debate, particularly in the later eolith phase.

# 7

# THE EVE OF THE WAR

## *The 1880s: Questioning Post-Glacial Man at the Start of the Late Victorian Period*

In 1879 E.B. Tylor, president of the AI and of the Department of Anthropology at the BAAS in Sheffield, delivered his president's address (Tylor, 1880). It was a succinct summary of the state of anthropology twenty years after *Origin of Species* and the Antiquity of Man Debate. He reflected on how far the dialogues surrounding human origins had resolved themselves. His commitment to evolutionary anthropology, as a viable alternative to geology in revealing a deep human past, was clear in the emphasis he placed on philology, myth, and material culture studies. In other words, subjects which utilised a sense of progressive time. These were an independent check on archaeology. Accepting language and race were not the same thing, Tylor asserted that linguistic connections did provide evidence of cultural connections despite contrary opinions. But in listing the broad ability of the 'science of man' to answer many questions, it was clear that the Palaeolithic, and especially the drift, remained too remote. It continued to defy connection with later Prehistory.

The evolution of culture and the early history of civilization served to reinforce the outlook that was by now being presented by linguistic research; that human differentiation lay in the black box between the Neolithic and the Palaeolithic. Trawling back into the Miocene for evidence on origins was pointless in the light of the contemporary understandings of the time. The geological jury was still out on Skertchly's glacial handaxes. Not surprisingly then, chronology was an increasingly important subject; but whether the Pleistocene should be made more recent (an allusion to Prestwich) or older remained unclear. Tylor expressed a common opinion; that the implements of the drift were not the most primitive imaginable. Humans must have had a long period of evolution outside of Europe. Locating where and when was a major issue.

An important focus of this chapter is the continuation of debate concerning the post-glacial date for the occupation of Britain by humans, or in other words, the date of the Palaeolithic itself. The discussions of the 1880s were to some extent the preamble to the eolith controversy of the succeeding decade. The question concerned the presence of humans within the glacial phase of the Pleistocene, and, equally as important, the precise character of the glacial Pleistocene itself. Was it a single monolithic phase or a period of multiple ice advances separated by one or more warm interglacials? The 1870s had produced a number of challenges to post-glacial man, and the 1880s would continue the trend; but they were still delivered piecemeal, on a site by site basis. They appeared as cracks in the post-glacial orthodoxy, but the cracks were not joined up. As such they could be dealt with on an individual basis.

In part this reflected geological uncertainty, but in part it also reflected the lack of a consistent origins debate in Britain which could have engaged with intra- or pre-glacial humans. Ideas about a single monolithic glacial period with human occupation confined to a post-glacial Pleistocene did not provide much scope for expanding a debate. John Evans's monolithic view of an unchanging Palaeolithic was equally inhibiting. The researches of men like Spurrell, Worthington Smith and John Allen Brown (see below) could have been engaged within debates on the timing of human occupation in Britain, but they weren't. To be fair, they had different agendas, but my point here is that others did not pick up on these aspects of their work either.

The pre- or intra-glacial evidence was both geological as well as archaeological. Questions concerning the meaning of flint artefacts in drift deposits had become more subtle. This applied to fauna as well. Ancient and extinct species no longer just indicated extreme antiquity. More sophisticated understanding led to more sophisticated questions asked of the data, which in turn exposed contradictions and discrepancies between the main data sets of differing disciplines. But the debates were focused only on Pleistocene occupation. The continental evidence for Tertiary Man still did not impress most English archaeologists, and the equivalent evidence in this country was decidedly lacking. This will be discussed further in Chapter 8.

## The Geological Background

The 1880s opened with a major work, William Boyd Dawkins's *Early Man in Britain and His Place in the Tertiary Period* (Dawkins, 1880a). The book was really a sequel to and an update of *Cave Hunting*: the basic geological and palaeontological sequence was a continuation of that laid there and fauna remained the primary driver for his geological interpretations. A summary of his views of the Pleistocene are given in Table 7.1. In the years after *Ancient Stone Implements,* with John Evans still unwilling to broaden his descriptive archaeology to encompass explanation, it was Boyd Dawkins who took up the challenge.

> The geologist…tells us of continents submerged, and of ocean bottoms lifted up to become mountains; and he points out to us that side by side with the ever changing conditions of life there were corresponding changes in the living forms. Group after group of animals and plants pass

over the field of vision, each connected with that which preceded it, and each becoming more and more highly organised, until man appears the last born as well as the highest and the noblest creature in the realm of geology. (Dawkins, 1880a, 1)

| **Temperature** | **Relationship of land to sea** | **Geography and geology** | **Human presence** | **Fauna**<br>Early/Middle/Late Pleistocene identified on faunal groupings. Relationship of fauna to specific phases difficult to isolate. Suggested interpretation of B.D.'s meaning here. |
|---|---|---|---|---|
| Temperate but becoming cooler<br><br>Cold. Glaciers form | Uplifted<br><br>Still high. | Forest covers much of Britain which is connected to continent as part of a single late Pliocene land area<br><br>North of Britain and Midlands covered by ice. Ice margin along Midlands/southern England border | No evidence to date | Latest Pliocene and Early Pleistocene. A temperate fauna migrates from Asia in response to gradually declining conditions in higher latitudes. Intermingles in Britain with small remnant Pliocene fauna. Forest bed and overlying fluvio-marine/lignite beds (cooling) in East Anglia<br>Low lying forests of Norfolk and Suffolk drowned as next climatic phase begins |
| Cold | High in North and Midlands. Southern England depressed. | Southern England flooded. Icebergs from ice margin melt and form the lower boulder clay in flooded East Anglia. Britain as a whole is an archipelago. | ---------- ? ----------<br>Yes associated with one of these phases, but not able to be more precise. | ---------- ? ---------- |
| More temperate (but not interglacial) | Depression of land continues. | Wide spread inundation. Britain still an archipelago. East Anglian 'middle drift sands and gravels' form on top of boulder clay. | Would represent 'pre-glacial' or 'inter-glacial' presence. These | Middle Pleistocene. Migration from Asia continues, and new species arrive. Pliocene relict species die out. The 'river-drift men' are associated with the Middle Pleistocene fauna. |
| Cold | Still archipelago | Glaciers on higher ground and icebergs float over drowned lower lying areas. When melt form the East Anglian upper boulder clay. | are the 'river drift men'. Many beds with temperate | Earliest evidence of occupation, if genuine, is flakes from Erith and Crayford. Described as simplest and earliest technology. |
| Becoming more temperate<br><br>(Passing into modern equivalent of post-glacial and Mesolithic) | Uplift. | Initially northern Britain, Wales and Ireland inaccessible due to remnant glaciers and high sea barriers<br><br>Eventually all of Britain will emerge from sea. Britain becomes part of extensive late Pleistocene North Sea dry-land area continuous with continent. | fauna overlain by 'trail' – glacial disturbance<br>---------- ? ----------<br>Yes. Post glacial. River drift men arrive, unrelated to any living races. Asian in origin but migrate from south. Relationship to Mid-Pleistocene hominins uncertain. Later and distinct 'cave race' also Asian but via northern route. Equivalent of modern Eskimos. | Late in period a few northern/arctic forms appear<br>---------- ? ----------<br>Late Pleistocene. Cold Artic-like fauna predominates. Final phase of Asiatic migration. Present in caves and river drift. Initially fauna contemporary with ice and sea barriers in north and west. So this fauna is both glacial and post-glacial.<br>Hominins in Europe in glacial, contemporary with fauna; move north and south as ice barrier waxes and wanes (often seasonally). Hominins present north of Thames only in post-glacial. In post-glacial the river drift men precede the cave men |

The table makes a number of significant points. It was the same basic pattern as had been suggested six years earlier (Table 6.4), but now there was more detail. The early Pleistocene period began with an influx into Britain of fauna from the central Asian plateaux, driven south and westwards as climate began to deteriorate there. So the break with the preceding Pliocene period was geologically recognizable through a change in mammal, communities occurring in the layers within river and cave sediments. It was not a complete break because a few Pliocene survivals lingered alongside the new early Pleistocene temperate arrivals. The middle Pleistocene was distinguished from the early phase by the continuation of the temperate fauna, as well as the arrival of some new temperate species, and the final extinction of all of the surviving Pliocene species. At the end of the middle Pleistocene a new faunal group would begin to appear, this was the northern or arctic fauna, reflecting the cold of the end of the middle Pleistocene and the early part of the late Pleistocene. These northern mammals would dominate the late Pleistocene: 28 early and middle Pleistocene species survived into the late Pleistocene, with 17 new arrivals in the artic group including Arctic and Norwegian lemmings, Arctic foxes and reindeer. Humans were still primarily a post-glacial species.

The southern mammals were found alongside animals of the arctic late Pleistocene and temperate early–middle Pleistocene groups in the cave and river deposits of the late Pleistocene. The overlap and geographical range of these three faunal groups is shown in Figure 7.1. Boyd Dawkins was at pains to point out that these alternating cold and warm phases were not proper interglacials as argued by Geikie and Tiddeman because they were not warm enough. Summer and winter migrations, as well as those occasioned by more widespread climatic change, remained the only explanation for the occurrence of species from these different groups side by side in the same cave layers.

The Scottish geologists, whose data was terrestrial and driven by a focus on glacial features in the mountainous heartland of Scotland, remained unimpressed by the extent of marine inundation proposed by their southern colleagues. Unofficially, the Scots were led by James Geikie whom we met briefly in the last chapter. The second edition of his *The Great Ice Age* (Geikie, 1877b) was published three years before *Early Man in Britain*, and the two books could not have been more different. Compare Table 7.1 with Table 6.3. Geikie's work is closer to modern geology in its reconstruction of Pleistocene climates. Significantly, it supported a cyclical pattern of climate change with full interglacials succeeding intensely cold glacial phases. The major agent of landscape change was land ice, although as noted previously, at this stage Geikie still accepted some role for marine inundation and icebergs.

*Table 7.1. William Boyd Dawkins's interpretation of the Pleistocene in Britain from his 1880 book* Early Man in Britain.

This was the latest phase in a longstanding antagonism between the two geological communities that had traversed a number of debates on different topics. An echo of this Scottish–English national division may appear in H.G. Wells's short story *The Moth* originally published in the *Pall Mall Gazette*

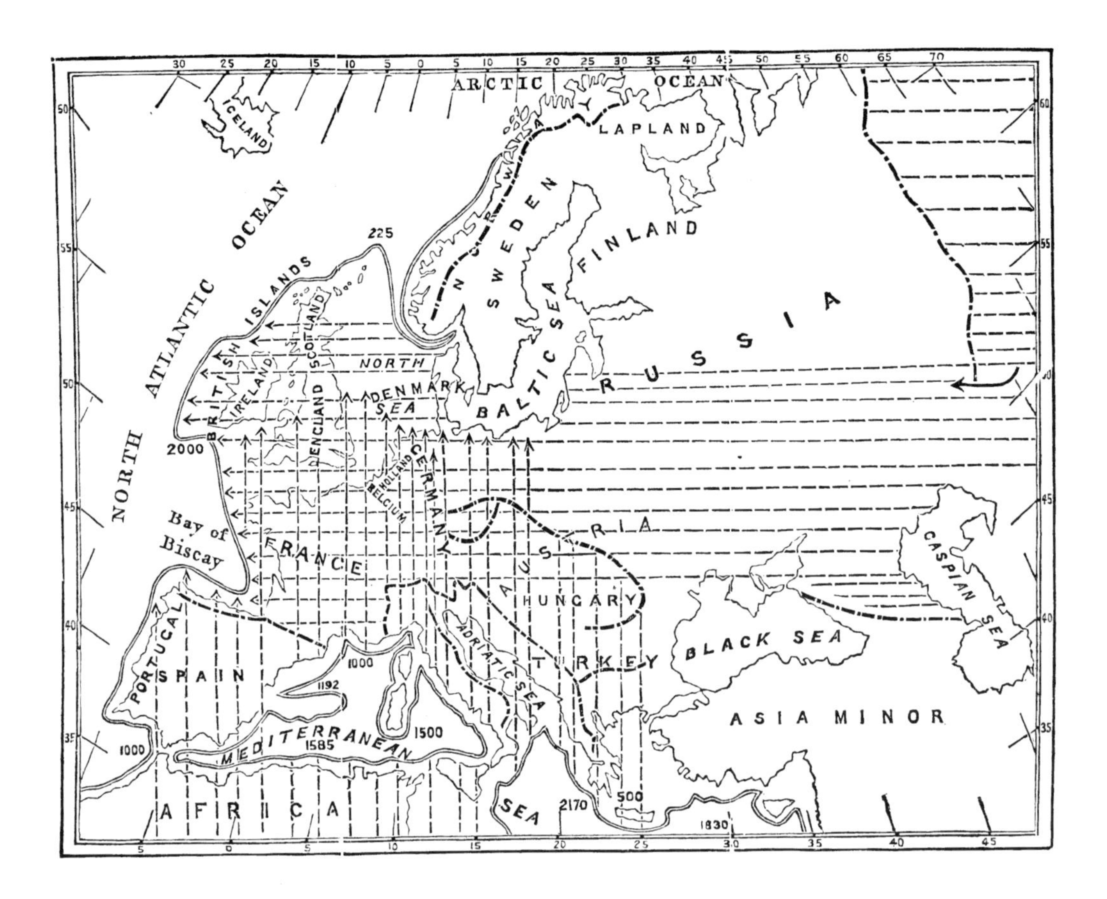


for 1895. The story refers to the 'convulsions' in the Geological Society and asserts that the 'great hate of the English and Scotch geologists has lasted now half a century' (Wells, 2000; Rudler, 1888; Geikie, 1889). Anne O'Connor (2007) details the disintegrating personal relationship between Geikie and Boyd Dawkins during the 1870s and into the early 1880s.

The work of Geikie and the glaciologists was embedded in a fundamentally different concept to that held by the English submergence school. It concerned ideas on how and why glacial periods occurred, and engaged with the theories of James Croll (1821–1890). Croll was a self-educated Scotsman (Fleming, 2006) who, after turning down a position as a geologist for the Geological Survey of Scotland, worked for the Scottish Survey as a secretary and an accountant. Croll was a brilliant mathematician and from the mid-1860s onwards he developed the Astronomical Theory of Ice Ages, drawing on the work of earlier researchers. Calculating the effects of the motions of the Sun and planets on the Earth's orbit, and on the precession of the equinoxes, he predicted that there had been a number of ice ages in the past, and that these had been interspersed with periods of more genial climate. The ice ages had coincided with periods when the orbital eccentricity of the

*Figure 7.1. William Boyd Dawkins's interpretation of the overlap and origin points of the three major Pleistocene faunal groups in Europe as depicted in* Early Man in Britain *in 1880. The southern fauna are represented by the vertical arrows. The temperate group are represented by the horizontal arrows which originate from eastern Europe and the Russian steppes, and the northern/arctic group originate from further north and their entry into Europe is suggested by the black arrow on the right of the figure.*

Earth had been at its greatest (i.e. the orbit was an ellipse and not a circle), and the winter solstice in the northern hemisphere occurred when the angle of the planet's axis was furthest away from the Sun. At such times a tipping point occurred and the build-up of polar ice and mountain glaciers could not be stopped. Much of Croll's work is embedded in today's explanations of glacial/interglacial climate change. He further argued that the astronomical effects merely triggered a series of terrestrial phenomena such as the diversion of warm ocean currents, and the build-up of ice and snow reflecting the Sun's heat back into space rather than absorbing it at ground level. He argued that it was the cumulative effect of these feedback loops, initiated by orbital and precessional events, that finally tipped the balance and initiated glacial advance.

The 'ice ages' were themselves groupings of glacial and interglacial events (Ball, 1892), although those terms were not used at this time. Croll demonstrated that when one hemisphere was in the grip of a glaciation, the other would be ice free. He was able to map the orbital eccentricity of the Earth three million years into the past, and a million years into the future (Fisher, 1892). He calculated that the last period of maximum elliptical orbit had begun two hundred and forty thousand years ago, and ended eighty thousand years ago. This he equated with the last glacial period of the geologists. In effect he was placing a real year timeframe on the ice age and its human inhabitants.

His work was widely cited, and he further developed it until his death in 1890. Its reception was mixed however. Astronomers and mathematicians with a strong geological interest like Sir Robert Ball (1840–1913) and the Reverend Osmond Fisher (1817–1914) were supportive, as were geologists like Geikie and Lyell. Even biologists and naturalists like Charles Darwin (Fleming, 2006) were sympathetic to the Astronomical Theory. In Darwin's case it gave him the time depth necessary for incremental change in species to occur. Moreover, Croll had provided a wholly independent proof of the repetitive patterns of glacial and interglacial climate that the record of the rocks was beginning to reveal. Other geologists such as Prestwich and members of the English submergence school were dismissive (Prestwich, 1887; Prestwich, 1888). Prestwich was reducing the length of time that he believed humans had spent living in the Pleistocene, only twenty thousand years. So his opposition to Croll's views were no surprise. Boyd Dawkins, as an ally of Evans was not a fan either. He and other opponents of Croll's work focused on what they saw as key problems, for example the lack of independent dating to demonstrate the alternating of climate between the northern and southern hemispheres, and the physical evidence for previous and more ancient ice ages.

As noted above Boyd Dawkins identified the beginning of the Pleistocene with the appearance of a specific set of mammals, and it was on this basis that he was convinced that humans could not have had a deep evolutionary history in Western Europe. Our ancestors, he argued, were one element within a 'mammalian package'. For Boyd Dawkins the different geological epochs (Tertiary, Pleistocene, etc.) were each characterised by their own unique animal community. As these broad geological periods succeeded each other, moving slowly toward the present, so the various faunal communities in each successive period came to resemble more closely the animals of the present day. New groups of animals, genera, and ultimately new species would appear and would then continue into modern times. The mammalian community was gradually developing into the one we know today with each successive geological period.

He made the key point that humans are a highly specialised type of animal. However, there were no traces of any ancestors for them in the Eocene or the Miocene, nor were there any other animals that showed extreme specialisation indicating that such adaptations were a normal part of mammalian adaptive strategies at this time. Humans would have stood out as anomalous. He also speculated that given the immense amount of time that must have passed, had humans or their ancestors been present, it was astonishing that they had not become extinct – all other Eocene and Miocene species had. A number of modern species had evolved by the Pliocene, some of them higher placental mammals, but Dawkins argued that there were too few of them to expect humans to be a part of the Pliocene mammal community. Evolution would surely have changed humans more than it had if they had been present since earlier times (Dawkins, 1880b; Dawkins, 1882). Other animals had evolved and changed, would humans be any different?

In short, he found no evidence for humans in the Tertiary in Britain.

Only in the late Pleistocene was there evidence for a distinct human presence in Britain (palaeoliths), and this was associated with the Arctic faunal group, essentially modern in character, but lacking the domesticated species. Pre-glacial occupation by humans remained problematic. By 1880 Boyd Dawkins had long since abandoned the Victoria cave fibula, it was from a bear but he continued to support a possibly limited phase of occupation as suggested by the flake from Crayford (see last chapter and Tables 6.4 and 7.1) and a few further finds in a similar situation from nearby Erith (Dawkins, 1880a; Scott and Shaw, 2009).

Reaction to *Early Man in Britain* was guardedly positive. While some took the author to task on points of detail (Woodward, 1880), others cautiously praised him for producing a work which bridged the gap between the ever increasingly specialist literature inaccessible to most readers, and the need to inform the broader reading public (Anon, 1881). Boyd Dawkins's views changed very little throughout the 1870s and 1880s. In his presidential address to the geological section of the BAAS (Dawkins, 1888) he maintained his opinion that fauna provided the most appropriate framework on which to hang Pleistocene chrono-stratigraphy. From the opposite point of view, and in the following year, Geikie defended the terrestrial ice position in his address as president of the section in September 1889 (Geikie, 1889). He confidently asserted that the Pleistocene map of Europe could be explained far more effectively by glacial land ice than by submergence and deposition by icebergs. He embedded his arguments in a detailed review of the European evidence thus attempting to convince his audience that independent continental research was amply confirming not only the land-ice hypothesis, but the cyclic glacial-interglacial nature of Pleistocene climate. He implied that continental geologists were not concerned with whether interglacials existed, they argued about how many there were. By appealing to the evidence of the European glaciologists he cast the English submergence school in rather a parochial light.

> I have often thought that whilst politically we are happy in having the sea all around us, geologically we should have gained perhaps by its greater distance. (Geikie, 1889, 466)

In his presidential address Geikie also made a clear appeal for stratigraphy to come before the palaeontological content of individual geological layers. In this he cut straight to the heart of Boyd Dawkins's methodology, and its major flaw. If Pleistocene faunas lived in successive phases of the period, as the cyclical glacial–interglacial hypothesis required, then a deposit could only be safely placed in its proper chronological position once its true stratigraphic place had been determined. If a deposit was found in isolation, and contained species found in a number of successive periods, how could its real position be assessed? It was a good point.

But despite Geikie's ebullience, the English inundation and iceberg school remained a force to be reckoned with, and during the 1880s a number of archaeological sites were interpreted with reference to theories of marine inundation. While Boyd Dawkins and his collaborator the Reverend J. Magens Mello (1836–1903) continued to excavate at Creswell Crags in the late 1870s, and publish their findings in the *Quarterly Journal of the Geological Society of London*, a new phase of cave hunting opened up in Wales. Henry Hicks (1837–1899), a medical doctor from Pembrokeshire, and a well-respected amateur geologist (T.G.B., 1904–1905), was investigating the limestone caves of south and north Wales (Anon, 1882; Jones, 1882). But it was the Vale of Clwyd, in the north, that received most of the attention. The caves of Cefn and Pontnewydd in the same area had already proved their worth. Hicks and colleagues were excavating new caves that seemed to offer evidence of an interglacial, and just possibly a pre-glacial, age for humans. At Cefn and Pontnewydd, McKenny Hughes (Hughes and Thomas, 1874; Hughes and Wynn, 1881) had issued a stern geological warning. The stone tools found were on lithologies that originated beyond the Vale of Clwyd. They could only have been brought there as erratics carried by ice. This meant that human use of them as raw material sources had to post-date the glacial period.

At two new cave sites near St Asaph, Cae Gwyn and Fynnon Beuno, Henry Hicks believed the stratigraphic record told a different story. He was at pains to establish the undisturbed nature of the stratigraphy. At the base of Cae Gwyn there was a gravel of local origin. It filled the cave when the river in the valley flowed at nearly the same height as the cave's mouth. Later, the valley was cut down and the cave became isolated. Hyena and humans competed for living space leaving bones and stone tools. Then a stalagmite layer formed over these deposits. It was smashed by violent marine action when the cave and its surrounding countryside were submerged by rising sea levels and a fall in the level of the land. This was the intra-glacial submergence of Boyd Dawkins and others. Later, as the land was re-emerging, icebergs deposited erratics from the north. Finally, as the land continued to uplift, the local glacial boulder clay was deposited in the cavern and around the local area (Hicks, 1885; Anon, 1886b). The glacial deposits were therefore *in situ* and sealed the cave.

For Hicks the case was simple; humans occupied the cave in the mid-glacial period. A committee was formed to proceed with excavations at Cae Gwyn and Fynnon Beuno, and it reported back to the BAAS in 1886 (Hicks, 1886). Both Hicks (as secretary) and McKenny Hughes were members. Hicks opened up a new area of the cave at Cae Gwyn for investigation. This allowed him to confirm the basic sequence he originally proposed, stressing that the sequence showed undisturbed layering within the cave. The 'bone earth', containing the remains of Pleistocene mammals and the broken stalagmite floor, was overlain by what was interpreted as a thin layer of undisturbed marine sand, which implied that the cave sediments had been disturbed by wave action before the marine sediment was laid over them. The succeeding glacial deposits, sands and gravels and boulder clay, must have been laid down by sea-ice as the land was emerging from the ocean. The report then broached the tantalising possibility that hominin occupation was early Pleistocene, because the marine sands overlying the bone earth were laid down by the middle Pleistocene submergence. Even more tantalising was the discovery of a flint flake *in situ* in the bone earth while Hicks and others visited the excavation.

Not surprisingly McKenny Hughes did not accept Hicks's findings, but the stratigraphic argument was a good one. McKenny Hughes had to resort to an unusual line to counter it. In 1880 Joseph Prestwich reported evidence to the BAAS for what he believed was a period of marine submergence that post-dated the Palaeolithic (see Chapter 1), and even post-dated a set of Pleistocene deposits variously labelled as 'warp' and/or 'trail'. These were geological layers found at various places in Britain. They overlay the *latest* of the valley drift gravels and the river terraces with Palaeolithic tools (Prestwich, 1880). They showed every evidence of having undergone extensive disturbance (Jones, 1884). In effect, the warp and trail were the last evidence for glacial action affecting the landscape. Prestwich's hypothesised final marine submergence was stratigraphically later even than the warp and trail, so chronologically it was final or even post-Pleistocene. Many geologists and archaeologists, including John Evans, believed a hiatus existed between the disappearance of Palaeolithic humans and the arrival of Neolithic people. Prestwich was fitting in a last marine submergence into this period, whose duration was open to speculation.

It was to the idea of this final post-Pleistocene inundation that McKenny Hughes seemed to have appealed (Hughes and Wynn, 1881; Hughes, 1886) in support of his own post-glacial chronology for the northern Welsh caves. This was a convoluted attempt to refute Hicks's pre-glacial/early Pleistocene chronology. (I have to admit I am not certain of this as I find his argument difficult to follow.) If Hicks's marine inundation was not middle Pleistocene but post-Palaeolithic (*sensu* Prestwich), then all of the archaeology contained within the cave could be comfortably inserted within the late Pleistocene period when, according to researchers like Evans and Boyd Dawkins, Palaeolithic handaxe making humans had been present in Britain. Hughes was also arguing that the glacial boulder clays and tills were not *in situ*, they were re-deposited by solifluction and therefore could not be used as a chronological datum.

How many geologists accepted Prestwich's evidence for a final marine inundation is a moot point. A.J. Jukes-Brown took Hughes to task for confusing his glacial and post-glacial terminologies, and asserting that his version of the submergence was not a view 'usually held' (Jukes-Brown, 1887).

Hicks was only too aware of the importance of the Cae Gwyn site for the date of the earliest occupation of Britain. He noted that there was no robust evidence anywhere else in Britain for the presence of humans before the glacial period. Humans were associated with the northern or Arctic fauna (*sensu* Boyd Dawkins), and co-occurred in caves and terrace deposits with mammoths, woolly rhinos, and especially reindeer. There were no examples of tools occurring with any pre-northern fauna, with the exception of the few flakes from Crayford and Erith (Anon, 1888a). On the evidence of Cae Gwyn, Hicks was now suggesting that the northern fauna must have appeared earlier in the Pleistocene than Boyd Dawkins had suggested. In defending his interpretation, Hicks alluded to many of the problems with Boyd Dawkins's palaeontological approach that Geikie had highlighted (Hicks, 1887). Only stratigraphy was a reliable guide to chrono-stratigraphic position. Even more worrying for Boyd Dawkins's position, animals migrated and humans migrated with them. A pre-glacial interpretation for a site in southern Britain would be predicated on the lack of northern species. This was a faunal variant of the argument about the contemporaneity of glacial deposits noted earlier. Northern arctic fauna could be present in one part of the country, possibly already starting to migrate southwards, while a cave deposit in the south was filling up with temperate mammal remains.

Cae Gwyn presented a seemingly substantial case for humans in the middle glacial or even pre-glacial. So much so, that Joseph Prestwich now withdrew from his post-glacial position on the age of the drift river terraces. As we saw in Chapter 6, he had expressed his openness to such an interpretation in the 1877 meeting at the AI. It is not difficult to see why; it was pure stratigraphic observation generated by fieldwork; it was the same kind of argument that had prevailed in 1859 (Prestwich, 1887; Anon, 1888a).

In addition to his paper on the late Pleistocene/post-glacial submergence; in the early 1880s Prestwich had begun a remarkable series of investigations which led to a re-interpretation of much of the late Pliocene and glacial-Pleistocene chrono-stratigraphic framework of southern England (Prestwich, 1881). He embedded much of this thinking in a major geological text book (Prestwich, 1886; Prestwich, 1888; Prestwich, 1899), and then developed further aspects in a series of papers delivered to the scientific societies, principally The Geological Society of London. It is important to look at his ideas in some detail because of their relevance to the succeeding eolith debate. These interpretations were the background against which the geological context of the eolith controversy would be set.

Following many geologists of his era, Prestwich believed in the existence of a vanished range of hills called the Wealden range (technically the Wealden anticline). This low mountain range was located in the area of the modern Weald of Sussex and Kent, see Figure 7.2. As conceived by Victorian and Edwardian geologists, it stretched across the English Channel as far as the Belgian Ardennes. Today the English Weald is a low lying stretch of land between the chalk dowlands of the North and South Downs – Figure 7.2 (Prestwich, 1890a; Prestwich, 1890b; Prestwich, 1890c). These Wealden hills had been folded upwards by earth movements in the Pliocene to form a low mountain range 2000–3000 feet in height. Composed mostly of marine sands and other soft lithologies at their core, the heights had been subject to erosion. Characteristic rocks had therefore been carried down by rivers from the higher slopes as gravel clasts. These were cherts, and pebbles of ragstone that originated in the Lower Greensand part of the core of the Wealden hills. The rivers flowing off the northern slopes of the hills, in the direction of what would later be the Thames Valley, carried clasts of these indicator lithologies onto the flat-lands to the north of the Wealden range. These drifts with their characteristic Wealden rock types were termed by Prestwich, the Southern Drift (Prestwich, 1890c). This was one of a number of key drift deposits widespread over the south of England that contributed to the overall interpretation of the geological history of the region in the Pliocene and Pleistocene.

Another set of drifts were termed the Westleton and Mundesley Beds (Anon, 1889; Anon, 1890b; Anon, 1890c), and in particular a band of shingle that was part of these beds. They were exposed in the sea cliffs of Suffolk, and lateral extensions were found in Norfolk (Prestwich, 1890a). They lay below the glacial beds in the cliffs, and were therefore pre-glacial. The

*Figure 7.2. Benjamin Harrison's visual depiction of the history of the Weald from his privately printed and circulated booklet of 1904. His original caption reads as follows. 'In [the figure] is shown a geological section across the Weald of Kent and Sussex, from the river Thames to the English Channel, a distance of about fifty miles from north to south. Section A represents the country as it now exists; Section C is the same section as A, with the ancient Wealden dome reconstructed over the present landscape of Kent and Sussex; and section B represents a conjectural intermediate stage between A and C, showing the Wealden hills in a partly denuded condition as they may possibly have existed when the eolithic implements were made. The actual amount of denudation that has taken place when Eolithic Man dwelt in the land is unknown, and [B] must not be taken as equivalent to a pronouncement on the point. In the sections: 1 represents Chalk; 2. represents Upper Greensand and Gault; 3. represents Lower Greensand; 4. represents Wealden Beds; T. represents Tertiary deposits; O. represents Plateau Drift containing Eolithic Implements; X. represents the ancient land surface whereon the implements are supposed to have been made. The vertical scale of the sections is exaggerated.'*

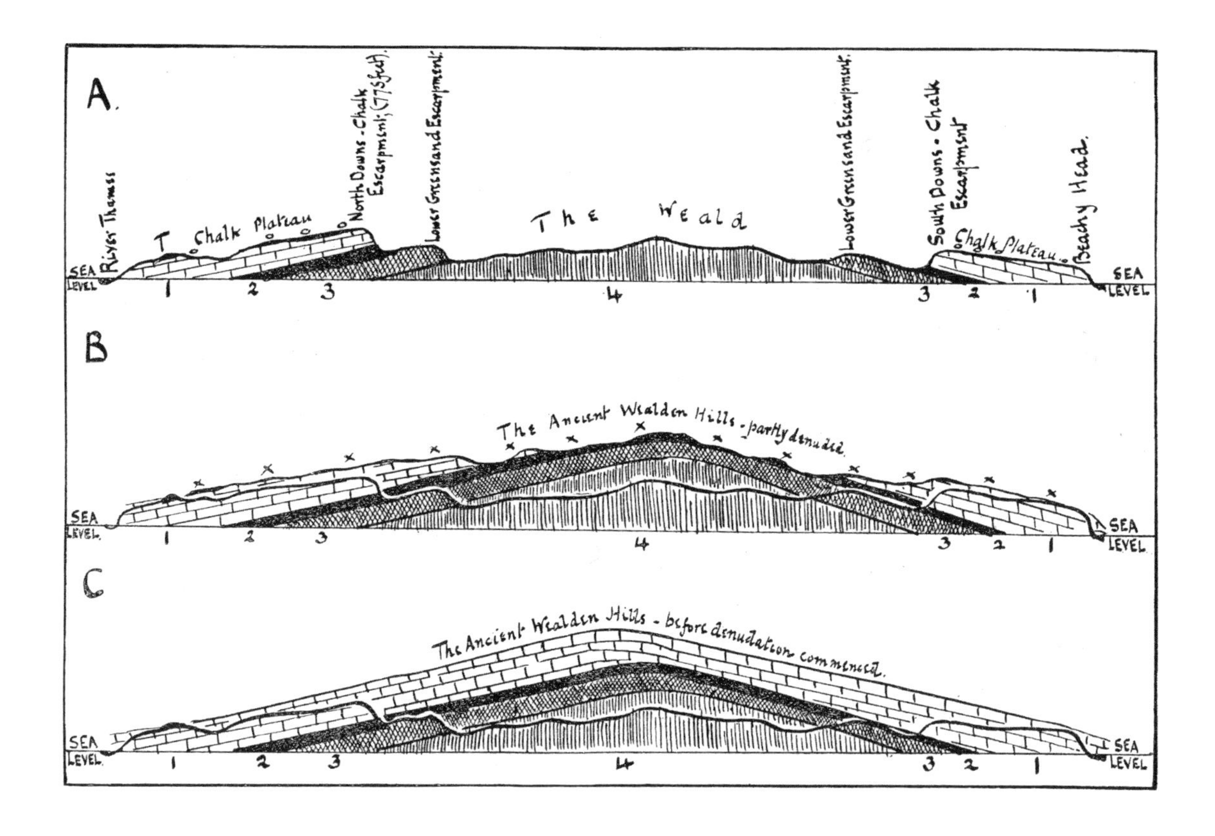


Westleton shingle could be traced inland from the East Anglian coastline into Oxfordshire, Hertfordshire, and south Buckinghamshire; south into the Thames Valley, and then westwards almost to Somerset. The implication was that this vast spread of drift pre-dated the cutting of the Thames Valley itself. For Prestwich these beds were marine, and they marked out the extent of the Pliocene marine submergence that occurred before the Pleistocene see Figure 7.3. The Westleton Sea could not penetrate further south because of the Wealden–Ardennes mountains. These had begun to uplift in the early Pliocene, constraining the sea to the north of them. The uplift continued throughout the (unknown) duration of the Pliocene. The Westleton beds were thus a component of the Pliocene sea floor, the waves of which lapped the edge of the plain which fronted the Wealden heights and onto which the Southern Drift was deposited. This meant that the Wealden heights were being eroded and the Southern Drift formed from at least the middle of the Pliocene onwards. This plain would one day become the Chalk Plateau of North Kent, upon which the eoliths would be discovered (Figure 7.2).

During the later Pliocene/early Pleistocene, before the glaciers arrived, the area of the Westleton Sea was uplifted and remained dry land long enough for river valleys to be begin to developing. On the plain (future Chalk Plateau) in front of the Wealden heights, north-south valleys were cut. Northward

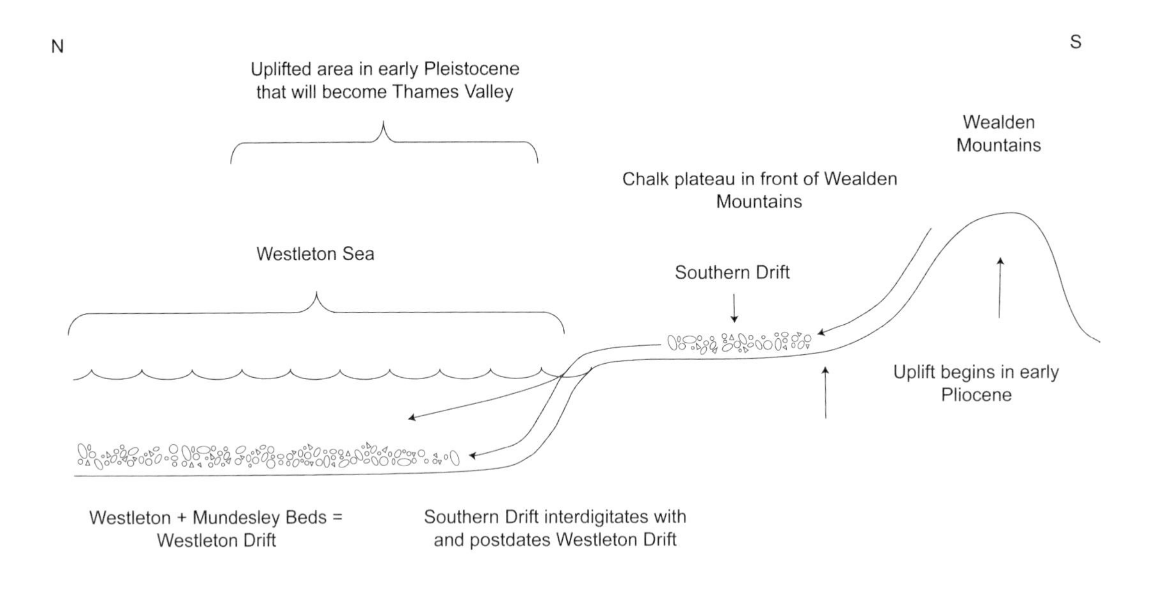


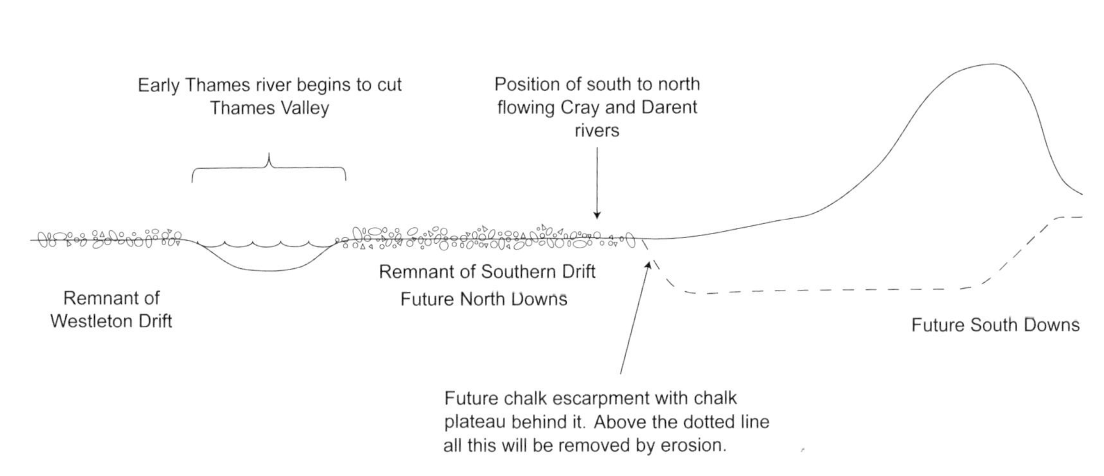


*Figure 7.3a and b. Schematic reconstruction of Prestwich's interpretation of the relationship between the Westleton Sea, the Westleton beds, and the Southern Drift of the Chalk Plateau. 7.3a is prior to uplift, and 7.3b is later in time, after the uplift has occurred. Drawn by Susan Hakenbeck.*

flowing rivers like the Cray and the Darent began to be cut through the Southern Drift; their earliest phases dating to some time in the Pliocene. Later post-glacial erosion greatly deepened and extended these valleys. The Thames valley, once the bed of the Westleton Sea (and having a small component of Southern Drift along its southern margin where Wealden rivers had drained into it), was raised as part of the early Pleistocene phase of uplift. Part of the river's later course was determined by an early west-east flowing Thames, running across the freshly uplifted land surface. However the modern valley, in its extent from upper to lower reaches, was cut and connected by glacial action after the arrival of the mid-Pleistocene glaciers.

The Vale of Holmesdale in front of the Chalk Plateau, Figures 7.3, 7.4 and 7.5, was also glacially eroded. So, at the same time as mid-Pleistocene glaciers were carving the Thames, glaciers flowed out from the Wealden mountains and moved downwards into the Holmesdale valley and began to erode what is now the Chalk escarpment, thus creating the Chalk Plateau behind it. This is a key point for the eolith debate which was to begin at the end of the decade (Chapter 9). The Chalk Escarpment of the North Downs, and the Chalk Plateau behind it, as depicted on Figures 7.2–7.4, was the area where Benjamin Harrison and his fellow eolith hunters would make their discoveries. Prestwich's geological interpretation would provide both an explanation and a date for the presence of the eoliths on the chalk landscape.

Prestwich's 'big picture' interpretations were dramatic. But not all geologists accepted them. The problem was that Prestwich was now in his 80s, and had constructed this framework from notes he made on field trips many years earlier. Much work on the stratigraphic relationship between different geological beds had been done since then by other, younger workers, but Prestwich did not include their work (Anon, 1888c; Geikie, 1899). The result was an interpretation that was considered by many rather 'old fashioned'. I have no doubt that it was the prestige of this hero of 1859 that lent these increasingly anachronistic views authority. Nevertheless, these were the interpretations that would inform thinking on the geological age of the eoliths.

In the early 1880s there was still very little physical evidence (Anon, 1883b) for the existence of interglacials in southern Britain. But as the 1880s progressed this would change. There was a growing belief that at least two boulder clays could be identified in the south of England (reminiscent of Skertchly at Brandon). In South Nottinghamshire and Derbyshire stratigraphic evidence appeared particularly convincing. Here there were actually three deposits of boulder clay. Valleys cut through the middle unit

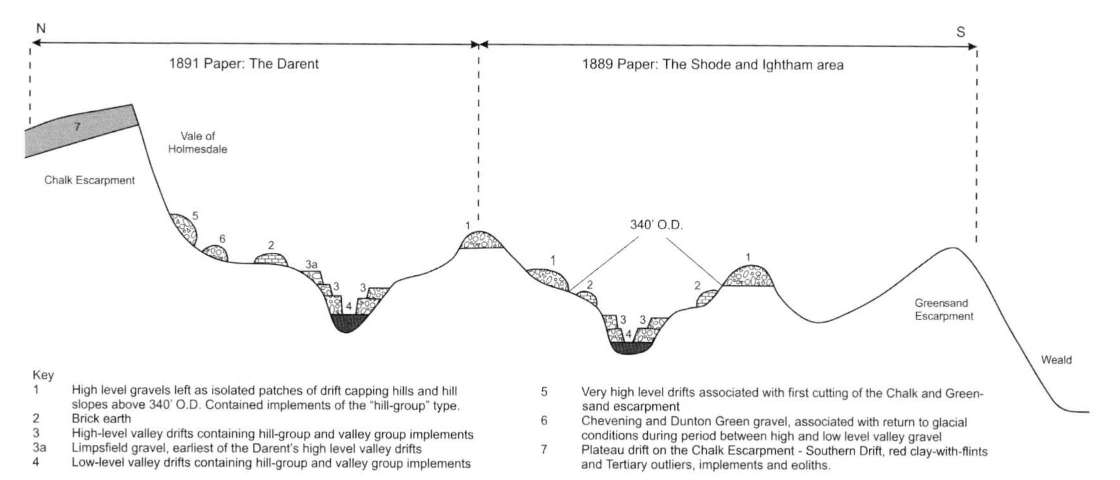
N
S
1891 Paper: The Darent
1889 Paper: The Shode and Ightham area
Vale of Holmesdale
Chalk Escarpment
340' O.D.
Greensand Escarpment
Weald
Key
1 High level gravels left as isolated patches of drift capping hills and hill slopes above 340' O.D. Contained implements of the "hill-group" type.
2 Brick earth
3 High-level valley drifts containing hill-group and valley group implements
3a Limpsfield gravel, earliest of the Darent's high level valley drifts
4 Low-level valley drifts containing hill-group and valley group implements
5 Very high level drifts associated with first cutting of the Chalk and Greensand escarpment
6 Chevening and Dunton Green gravel, associated with return to glacial conditions during period between high and low level valley gravel
7 Plateau drift on the Chalk Escarpment - Southern Drift, red clay-with-flints and Tertiary outliers, implements and eoliths.

*Figure 7.4. Schematic cross-section reconstruction of the topography of the Vale of Holmesdale and the rivers Darent and Shode in the area between the Chalk Escarpment and the Greensand Escarpment. Drawn by Penny Copeland.*

had river terraces developed on them, and these in turn had boulder clay covering them. The terraces were described as inter-glacial. Evidence from the Midlands, North Wales and Lincolnshire led A.J. Jukes-Brown to argue for an extensive interglacial land surface in Britain, based on purely geological grounds (Fisher, 1887; Jukes-Brown, 1887)

## Palaeolithic Archaeology in the Late Victorian Period

As we saw in Chapter 2, the first 'Antiquity of Man' debate had focused on one question: was humanity an ancient species or not? The answer had been yes. In 1859 it had not been possible to answer the follow up question of just how old was the human species? At this time it had been enough to win the first battle and then consolidate the new perspective. But it had not been long before questions on the origin and age of the human species had emerged.

By the late Victorian period some archaeologists thought these questions could now be answered. Although no agreement was in sight, the very fact that there was debate concerning the nature and timing of the glacial Pleistocene, opened the subject up for discussion about the first arrival of humans within this period. The concept of a global Palaeolithic period with humans covering much of the Old World was something John Evans and others could now begin to more readily appreciate.

Some of the more pressing questions being asked were as follows:

1. Were humans present in any numbers during the glacial period itself?
2. Were there genuine interglacials within the overarching glacial period?
3. Linking to question 2 were land surfaces that humans and the great mammals could have roamed across present in the temperate phases?
4. Could a time frame in real years be placed on this occupation?
5. How good was the evidence for humans in the pre-glacial/early Pleistocene?
6. Could the geological evidences for submergence and uplift, or successive glacials and inter-glacials, be used to provide an intra-glacial chronology for hominin occupation?

Palaeolithic archaeology had made slow but steady progress during the middle Victorian period, and this continued after 1880 into late Victorian times. The local societies reported the occasional discoveries of Palaeolithic sites or isolated finds of palaeoliths, and their news filtered up to the bigger national societies centred in London. As noted elsewhere, the link was the societies' membership base. Many of the members of the national societies in London were also members of the smaller regional scientific societies. Members would have reported back to their colleagues in the country what was being talked about in Town. For those from the Home Counties, efficient rail transport made attendance at the weekly meetings in London relatively straight forward. Many of the regional societies reported on the papers presented at The Geological Society, the Anthropological Institute and others. Subscribers to *Nature* were kept regularly informed of scientific meetings at local and national level. Table 7.2 gives a breakdown of the frequency with which articles specifically on human origins, and after 1889, eoliths, appeared in the pages of the national societies' journals. It is clear that these topics

| **Journal Title** | **Years in which volumes appear** | | | |
|---|---|---|---|---|
| *Quarterly Journal of the Geological Society of London* | 1880 - 1884 | 1885 – 1889 | 1890 - 1894 | 1895 - 1901 |
| | 0<br>(260) | 3.0%<br>(269) | 4.6%<br>(219) | 1.5%<br>(324) |
| *Geological Magazine* | 1880 – 1884 | 1885 – 1889 | 1890 - 1894 | 1895 - 1901 |
| | 2.2%<br>(1100) | 2.6%<br>(1097) | 1.6%<br>(1238) | 2.3%<br>(1556) |
| Reports of the British Association for the Advancement of Science | 1880 – 1884 (Sections C+D+H) | 1885 – 1889 (Sections C+D+H) | 1890 – 1894 (Sections C+D+H) | 1895 – 1901 (Sections C+D+H) |
| | 5.8%<br>(549) | 3.0%<br>(830) | 2.8%<br>(713) | 3.6%<br>(923) |
| *Proceedings of the Geologists' Association* | 1879-1884 | 1885-1890 | 1891-1894 | 1895-1902 |
| | 2.8%<br>(211) | 6.9%<br>(246) | 5.2%<br>(153) | 4.6%<br>(323) |
| *The Nineteenth Century* (including '*The Nineteenth Century and After*') | 1880 – 1884 | 1885 – 1889 | 1890 - 1894 | 1895 - 1901 |
| | 0.3%<br>(649) | 0.6%<br>(678) | 1.0%<br>(789) | 0.7%<br>(1229) |
| *Journal of the Anthropological Institute of Great Britain and Ireland* | 1880 – 1884 | 1885 – 1889 | 1890 - 1894 | 1895 - 1901 |
| | 6.8%<br>(206) | 4.6%<br>(218) | 5.4%<br>(204) | 2.7%<br>(451) |
| *Archaeologia* | 1880 – mid 1885 | mid 1885 - 1890 | mid 1892 - 1897 (no 1891 or earlier 1892) | 1898 - 1901 |
| | 0<br>(72) | 1.1%<br>(94) | 1.5%<br>(68) | 0<br>(40) |
| *Man* | No volumes prior to 1901 | | | 1901 |
| | | | | 0.7%<br>(154) |

*Table 7.2. The frequency of articles on human origins, Palaeolithic archaeology, its relationship to Pleistocene stratigraphy, or any topic with a relevance to human origins expressed as a percentage of the total number of articles for those years covered. Major national journals only.*

did not achieve any degree of prominence. I have included in the table data for the AI, the society that might reasonably be expected to have been more interested in human origins and Palaeolithic archaeology. The percentages show a limited coverage for the twenty one years analysed. The total even remains low for 1890–1901, the decade that saw the most discussion on the eoliths of Benjamin Harrison and Joseph Prestwich.

The pattern is difficult to explain. Certainly in the case of the number of articles directly addressing human origins, part of the answer must be the lack of primary data. A selection of the best known Pleistocene (and reputedly earlier) skeletal remains known by the early 1880s was presented in Tables 3.1 and 6.2. These were, effectively a few Neanderthal specimens, and examples of the earliest modern humans. While both of these groups were contested in various ways, with the exception of the Neanderthal specimens the remainder were modern human, or at least what could pass muster as such. There were still no examples of any creature that could be argued to be a truly early or transitional form of human.

During the 1860s the framework of evolutionary anthropology invited ethnological comparison between modern indigenous peoples and Prehistoric ones. But during the 1870s Palaeolithic archaeology had returned to something of a cataloguing exercise. Admittedly most of the archaeological discoveries were isolated finds (Evans, 1872). At this point in the history of Palaeolithic archaeology it was not common for individual artefacts, or even assemblages, to be used to reconstruct aspects of ancient behaviour. Ethnological parallels had been a proxy for that. Even using different artefact types to interpret the chrono-stratigraphic position of geological deposits was in its infancy. The observation–description (cataloguing?) outlook continued into the 1880s. The on-going lack of any suitable ethnological parallels with which to invite comparison with Palaeolithic humans contributed to this. It will be recalled that Lubbock and Tylor's ethnological comparisons were applicable only to later Prehistory; they could not be pushed back in time to the Palaeolithic.

In March 1880 Flaxman C.J. Spurrell (Scott and Shaw, 2009) made a spectacular discovery in the brickearths of Crayford, Kent, reporting his finds to The Geological Society of London (Anon, 1880), and to the BAAS meeting in Swansea in August and September 1880 (Spurrell, 1880). These were the same brickearths that Boyd Dawkins had accepted as containing evidence of mid-glacial humans in *Early Man in Britain*. What was novel about Crayford was Spurrell's realisation that here was a factory site that lay totally undisturbed after being abandoned by its makers. He refitted many of the flakes back on to each other and on to their parent cores, anticipating an important analytical technique used by modern Palaeolithic archaeologists by almost a century. Spurrell believed it was a handaxe making factory, and indeed he found two such implements at the site to support his claim. Discovered too late to be included in the book, Boyd Dawkins acknowledged the new Crayford finds as evidence of mid-glacial occupation of Britain in later publications (Dawkins,

1882). In 1880 he had assigned the brickearths at Crayford and Erith to the mid-Pleistocene period, on faunal grounds. Spurrell's site was now grouped in with the other Crayford and Erith finds, but Dawkins was less clear about how the site fitted into the glacial chronology, shown in Table 7.1. Crayford had no glacial deposits in the vicinity, and the mammals were a combination of warm and temperate species that were common throughout the earlier and middle Pleistocene. If Crayford was faunally middle Pleistocene, it could equally be pre- or intra-glacial, depending on when the glaciers arrived. When Boyd Dawkins gave his presidential address to the anthropology section of the BAAS at Southampton in 1882 (Dawkins, *ibid*) he asserted:

> 'Nor am I able to form an opinion about their relation...[humans]...to the submergence of Middle or Northern Britain under the waves of the glacial sea. They are quite likely to be pre- as post-glacial.' (Dawkins, 1882, 601)

This is a good example of the limitations of Boyd Dawkins's method, and it is not difficult to understand why James Geikie and the terrestrial glaciologists felt this approach to be so flawed. The early 1880s were the height of the Dawkins–Geikie feud (O'Connor, 2007).

Spurrell's site attained a well-deserved level of prominence because it was carefully positioned in high-status journals such as *The Archaeological Journal*, (published by the Archaeological Institute), *The Quarterly Journal of the Geological Society of London,* and *The Journal of the Anthropological Institute of Great Britain and Ireland*, and discussed at the meetings of these and other London Societies. A good example of the type of general Palaeolithic article discussed at the less prestigious county level is again by Spurrell (Spurrell, 1883). Spurrell's paper, given to the Kent Archaeological Society, focused on handaxes from the collections of Benjamin Harrison as well as others. The paper's significance is that many of Harrison's artefacts were from the valley of the river Shode in Kent and Harrison's home at Ightham, Figure 7.5. This is the same valley that, in the late 1880s, Prestwich and Harrison would focus on for the first of their eolith papers (although Spurrell gave the valley a number of alternate local names). Spurrell had been exhibiting handaxes found by Harrison since the early 1880s (Harrison, 1928, 89), using them to illustrate talks to local societies (for example at Blackheath), as well as the more prestigious London societies (such as The Archaeological Institute).

The valley of the Shode, according to Spurrell, once contained an ancient course of the Darent, Figure 7.5, its higher drift terraces preserving evidence of a northward flowing river that curved westwards near Ightham to join the modern Darent valley (*contra* Prestwich, 1889 who interpreted it as a tributary of the Medway which flowed to the south). Spurrell attempted to give relative ages to the implements based on how worn they were; grouping them into five divisions, the first three of which were the oldest and most worn, and so

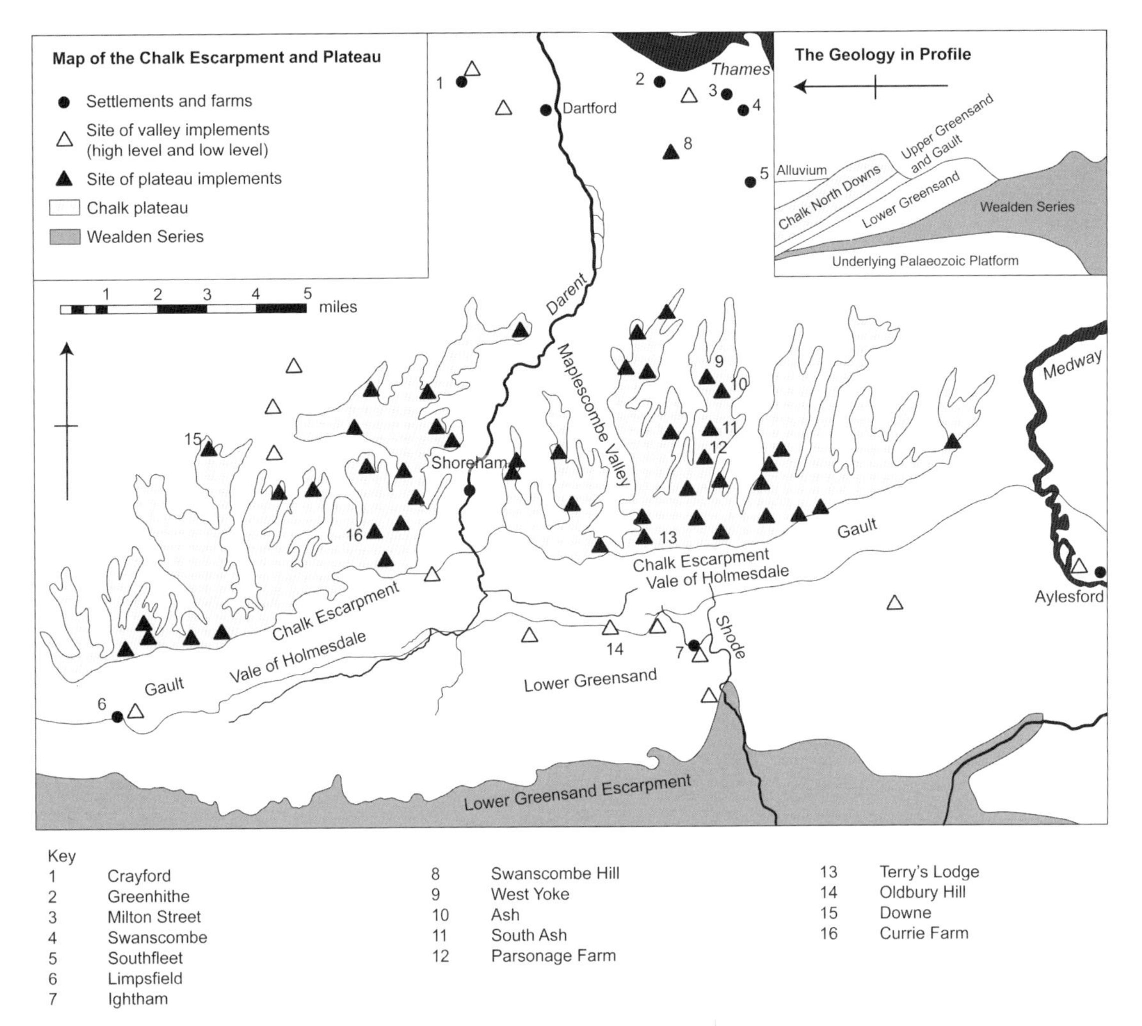


must have originated from the highest terraces in the valley. He was making the point that the age of an implement, as indicated by the wear, supported a derivation from older and higher deposits than those they were found in. He further noted that Harrison's collection contained handaxes which had been recovered from deposits described as warp and trail (see above). Another group of Harrison's palaeoliths showed the kind of wear that suggested they should derive from the warp and trail layer, but were found in low lying post-glacial drift terraces. The implication (though not explicitly stated) was that these latter palaeoliths must be older than the glacial warp and trail.

Here is the key point; on both a national (Cae Gwyn and the Anthropological Institute, British Association, Geological Society etc.), and a regional level (e.g. Spurrell at the Kent Archaeological Society), the post-glacial occupation of Britain by Palaeolithic humans was being questioned.

Similar inferences can be drawn from Worthington Smith's work in north London on the river Lea and other tributaries of the Thames (Smith,

*Figure 7.5. Map redrawn from Prestwich 1891 and 1892 showing the Chalk Plateau, Ightham, some of the locations of implements found in the fluvial gravels of the area and the patches of Plateau Drift where eoliths were being discovered. Drawn by Susan Hakenbeck.*

1880; Smith, 1884a; Smith, 1884b). Worthington Smith (see Chapter 6) was observing the foundation trenches for houses and even newly-dug graves in the London cemeteries near his home in Highbury. After his work on the Lea and the Brent, he turned his attention to part of the area in between the two rivers, in particular the area around Stoke Newington. He produced his definitive report in 1884 (Smith, 1884b). Like Spurrell, he discovered a number of localities where *in situ* Palaeolithic archaeology was revealed. He described an extensive 'Palaeolithic floor', in effect a land surface, perfectly preserved, with the evidence of human occupation undisturbed across this extensive feature. Taking the lead from his old friend Spurrell, he too was able to conjoin flakes and refit them onto their parent cores. The condition of material on the floor was pristine.

But at Stoke Newington he was able to reconstruct a geological sequence that contextualised the floor and its palaeoliths – see Figure 7.6. The lowermost unit was gravel, varying in depth from 8 to 18 feet. At its base were crude and worn palaeoliths which Smith described as the 'oldest class'. They were crudely made, worn, and stained a deep ochereous colour. At this basal level of the gravel there were enormous blocks of sandstone, too big to have been emplaced by river action; they had been dropped from floating icebergs (though this was only implied for this level). At the top of this gravel was another level of handaxes. These were of 'medium age'. They were damaged by transport in the gravels from their place of manufacture, but not nearly as much as the oldest class. They had a lustrous sheen to their surface appearance. They too were accompanied by gigantic blocks of sandstone, some up to 5cwt. These erratics in the upper part of the gravel sometimes showed glacial striations, and Smith was clear that these had been dropped from passing icebergs in a later and separate ice advance.

In Figure 7.6 there are a series of bands of clay and sands running up to the warp and trail. The Palaeolithic floor, with its conjoining artefacts, would have been more or less at the height of the warp and trail in this section. However the passage of the ice, or the solifluction resulting from torrential snow melt (which had created the warp and trail here) had swept away the Palaeolithic floor in this vicinity. Smith assured his readers it was present a few yards further on. So the *in situ* Palaeolithic floor that Smith recognised in north London was a third and later phase of Palaeolithic occupation. Just like Spurrell's data, here was a relative chronology established on the basis of the stratigraphic position of artefacts, and their condition. Smith noted that the knappers of the Palaeolithic floor occasionally found palaeoliths of the oldest class and re-flaked them for their own purposes. The warp and trail itself, evidently glacial, also contained implements and palaeoliths scoured up from deposits further up slope and carried along by the 'muds' of the warp.

This was powerful evidence, and the sequence was robust – it was extensive over a wide area of north London and beyond. It was capped by the warp and trail so it all pre-dated the end of the Pleistocene. It was based

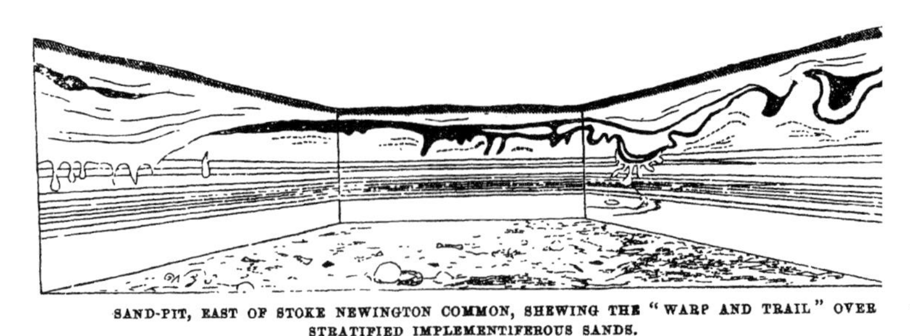

SAND-PIT, EAST OF STOKE NEWINGTON COMMON, SHEWING THE "WARP AND TRAIL" OVER STRATIFIED IMPLEMENTIFEROUS SANDS.


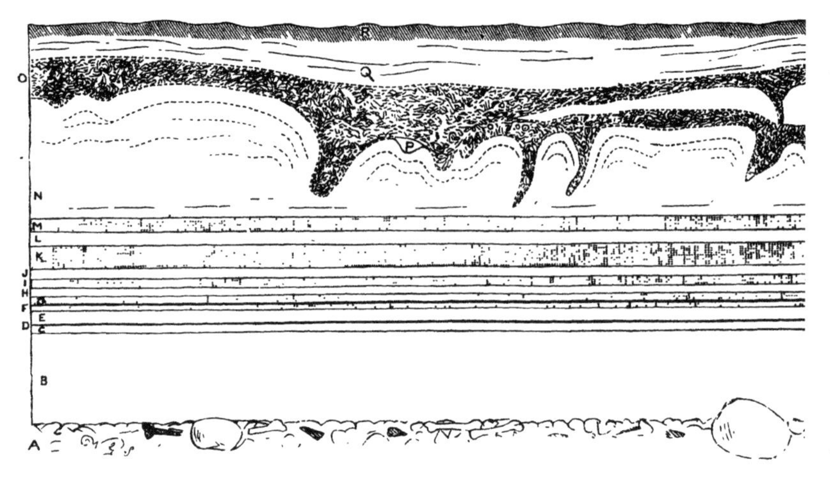


*Figure 7.6. Section through the north London occupation horizon discovered by Worthington Smith, from Smith 1884b, plate 11. The upper image shows a south facing section from a pit near Clapton Railway Station. The image below is the section in more detail. It was 11 feet and 6 inches deep, and only reached the top of the basal gravel. Paraphrasing from Smith's text the following is the key to the figure. R is humus; Q is mud belonging to the trail; P is a pocket of London Clay; O is the trail; N is the Palaeolithic sand and loam disturbed by the trail. The Palaeolithic floor was missing from this location. Layers M–B were individual sand and clay units that represent fine grained deposition at the margin of the Thames, today four miles to the south; A was gravel containing in its upper parts 'lustrous and sub-abraded implements of medium age'; at other locations in the area this gravel was exposed to its full depth and at the base of the gravel were '…the oldest class of Palaeolithic implements…they are…greatly abraded, rude in manufacture, and deeply ochreous in colour'.*

on direct observations made during fieldwork, so it was difficult evidence to dispute. It also set up an interesting conundrum. Smith's Palaeolithic floor was well within the Thames terrace sequence as were the middle and older classes of palaeolith from his basal drift. With evidence of icebergs, and of the final glacial event in the form of the warp and trail sealing the deposits in, it would be difficult to argue that such terraces were post-glacial in age. This was something that was not picked up at the time.

Worthington Smith was not alone in exploring the terraces of London. John Allen Brown (1831–1903) was placing the archaeology of west London into a chrono-stratigraphic sequence. Brown was a Victorian business man who retired early and devoted himself to geological work – a 'useful local man' (Rudler, 1903). His work differed from Smith's. Whilst Smith focused on his Palaeolithic floor and its terrace context; Brown moved beyond the terraces of the Thames Valley and attempted to adduce the relationship between them and the drifts on the hills to the north of the valley. His book *Early Man in N.W. Middlesex* (Brown, 1887), while not as well known as Smith's, is equally a classic. The cross section in Figure 7.7 shows Brown's attempt to fit his findings into the bigger geological picture which Smith (1884a; 1884b) did not do. The oldest drift in the area was marine, located on

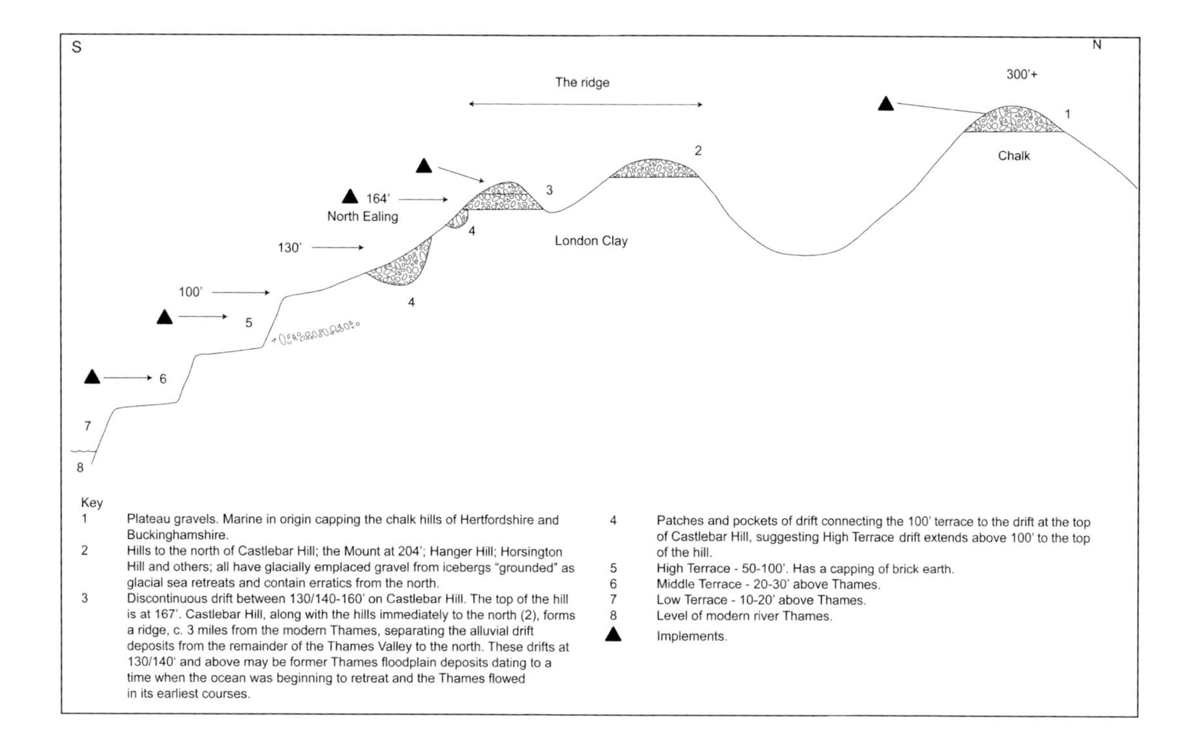


*Figure 7.7. Schematic reconstruction of John Allen Brown's interpretation of the Thames river terrace sequence and its relation to older gravel deposits to the north of the Thames and its valley. Drawn by Susan Hakenbeck.*

the chalk hills of Hertfordshire at heights of up to 300 feet above sea level. These dated to a time of widespread submergence. In time, falling sea levels resulted in the grounding of icebergs on what later became high ground, explaining the presence of glacial gravels at The Mount (204 feet above sea level) and elsewhere. At Castlebar Hill in Ealing, north London, there was drift present at the very top of the hill, 167 feet above modern sea level. Some 37 foot lower on the slopes of Castlebar Hill the drift was fluvial and Brown suggested that it was fluvial at the top of the hill too. What he was implying was a gradual lowering of the Thames flood plain from the earliest levels at which it flowed (167 feet) as the land slowly re-emerged from the sea, down the sides of the valley to the prominent High Terrace with its drift at 100 feet, and the succeeding Middle and Low Terraces.

Brown accepted a pre-glacial date for humans, suggesting that they had lived in the north of Britain before the glaciers arrived (Cae Gwyn and Cefn), but migrated south as the glaciers advanced. When the land began to uplift again humans migrated across a dry English Channel and reoccupied the emerging land surface. At best, Brown's Thames evidence could have been considered 'late glacial', contemporary with the melting of the ice, the falling of sea levels, and the slow uplift of the land. Brown did not accept differences in colour staining (i.e. differing degrees of ochreous staining) as indicators of relative age differences on implements, but he did accept physical condition

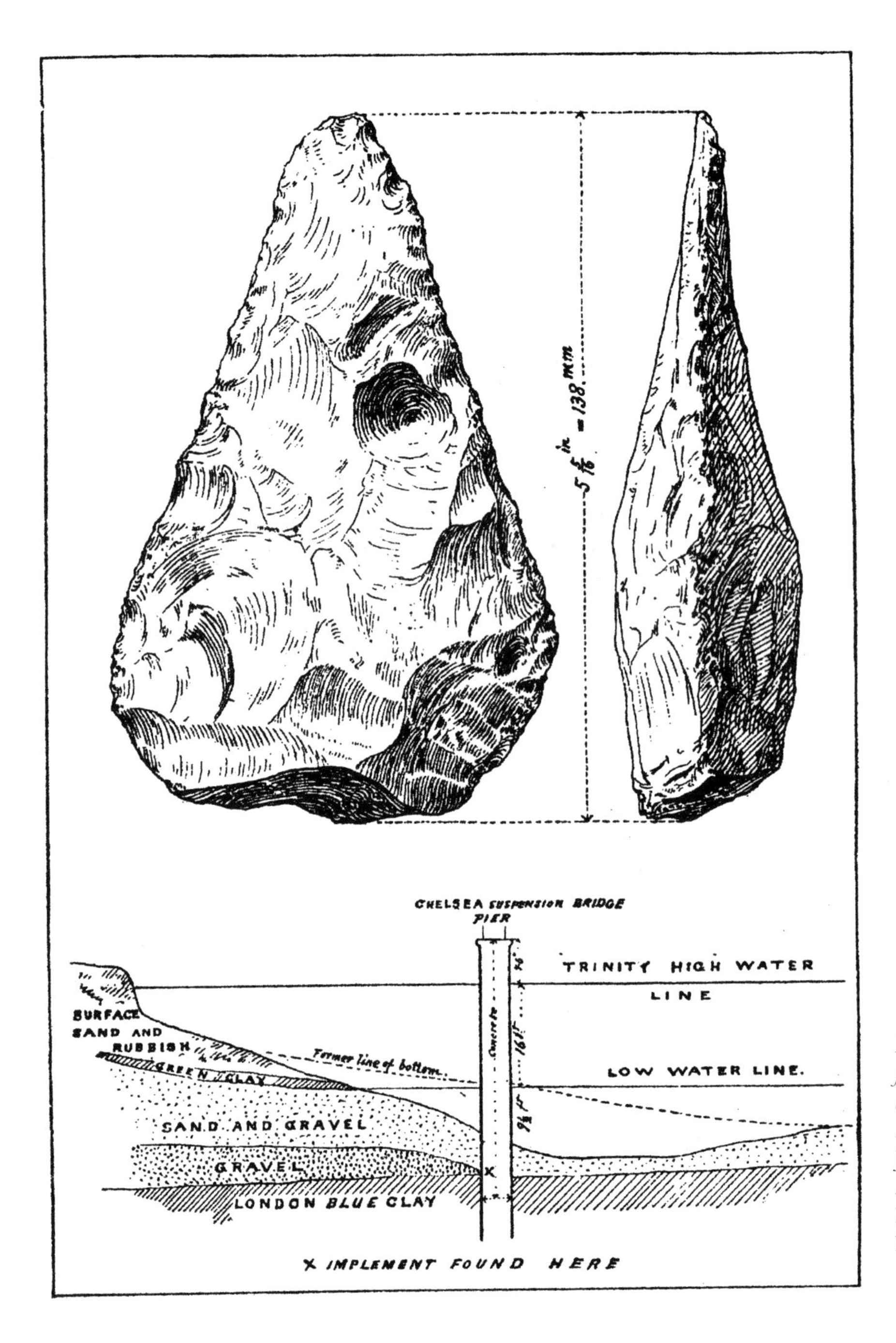


*Figure 7.8. A handaxe found during the construction of Chelsea Bridge in 1854. From* Journal of the Anthropological Institute of Great Britain and Ireland, *1883*

and wear. Some of his implements he noted to be similar to Worthington Smith's oldest class, and a few were in so damaged a condition that it suggested they predated the submergence; he was however reticent on this point. Much of the archaeology chapters from *Palaeolithic Man in N.W. Middlesex* were dedicated to the *in situ* floor that Brown discovered at Creffield Road, Acton, in west London, in the High Terrace deposits.

There was a steady trickle of Palaeolithic research published in the journals of the local and national societies. Much to his irritation Worthington Smith's success in the valley of the Lea and the Brent, and at Stoke Newington, brought others who invaded his territory (Greenhill, 1884). The results were disastrous with forgers, and canny workmen selling implements imported from elsewhere (Smith, 1884b).

Elsewhere, J.A. Brown noted the discovery of a mammoth skeleton with a stone tool in direct association (in this case a Levallois point very similar to those on the Palaeolithic floor at Creffield Road). As in 1859, the interest in new finds energised a review of older discoveries. In 1854 a palaeolith was discovered while dredging the Thames to build the Victoria Bridge, Chelsea (Anon, 1883a). It was exhibited at the AI in 1882, and its context discussed the following year. Like the Hoxne and Grays Inn handaxes, and those from Salisbury and Peasemarsh, this pre-1859 artefact was found too early in the discipline's history to have been recognised for its true significance when first discovered, see Figure 7.8.

The Thames Valley was a fertile area for Palaeolithic exploration (Ashton and Lewis, 2002) as Victorian urban and industrial expansion swallowed up more and more available countryside (Conway, 1996; McNabb, 1996; Harrison, 1963). Professor Rupert Jones reviewed the situation for the Middle Thames (as Spurrell's 1883 paper could be said to be a partial review for the Lower Thames), on the occasion of Worthington Smith's exhibition of finds to the Geologists' Association (Jones, 1884). Jones discussed the plateau drift or plateau gravels, a name first coined by Prestwich, and which would have equated with the high hills of Hertfordshire (300 feet) that John Allen Brown described in his book. Rupert Jones suggested that the plateau drift was either deposited on the base of the sea bed when Britain was an island chain (thus making it similar in context to the Westleton shingles of Prestwich), or the gravel was deposited by rivers at sea level, and then distributed across the shallow ocean floor by currents, after submersion and further deposition from ice flows. At the end of the glacial period the Thames Valley itself was subsequently incised by retreating sea water and river action cutting through the plateau drifts, as the land re-emerged from the sea. For Jones, this explained the relationship between the river terraces of the Thames Valley and the plateau gravels capping the hills beyond the valley at Acton (see also J.A. Brown above).

Rather than identify a high and low level river terrace within the valley at Acton, following Prestwich's basic river terrace model, Rupert Jones followed Whitaker's remapping of the Thames terraces in London (see Chapter 6) into High, Middle and Low. J.A. Brown had also followed this scheme in 1887. In the vicinity of Reading, further up the valley, three such terraces were also visible, picked out by drift gravels at specific heights. In the highest terrace, the quarry of Grovelands had a possible land surface near the base of the deposit where the majority of the artefacts were found. Jones noted that there were artefacts in all three terraces at Reading and inferred a presence for humans at each successive period. In describing the localities of Worthington Smith in the Stoke Newington neighbourhood, Rupert Jones alluded to earlier work which raised the possibility of linking the brickearth containing Smith's floor, with that in which Spurrell's workshop at Crayford had been found.

O.A. Shrubsole (Shrubsole, 1885) also explored the Reading gravels, and he too noted that at the Grovelands pit the artefacts occurred near the bottom of the deposit. But he argued for a glacial or pre-glacial date for the gravels. This was on the basis that the implements at Grovelands were covered by a thick body of archaeologically sterile gravel. Such a depth of gravel could only be generated by the erosive forces of a glaciation (although many geologists would have taken exception to this). But Shrubsole went further. He described a number of objects which were very unlike the palaeoliths from the site (which he noted were ovate and thus different from the more common pointed forms elsewhere). These artefacts had pronounced concavities, some picked out by chipping, which he claimed would have been useful for making spear-shafts and dealing with organic materials. While these were not eoliths in the 'Harrisonian' sense, they were conceptually very similar. Given his geological views and his eye for unusual artefacts, it is not surprising that Shrubsole became an early supporter of eoliths after 1889, and one of those researchers who sought them outside their original Kentish homeland.

Elsewhere, other reports of glacial or pre-glacial archaeology met with a mixed reception. William Pengelly was taken to task by the editor of the *Geological Magazine* (Woodward, 1884). A canoe, discovered above the modern river valley of the Bovey in Devon, was entirely covered by 'head' deposits. This was a term used by some members of the Geological Survey to indicate glacial deposits such as boulder clay. In fact it refers to solifluction deposits – water saturated sediments (often under cold conditions but not necessarily glacial) which have slipped downhill under the pull of gravity. Pengelly and his colleagues were claiming the canoe was at least of glacial age. Woodward reminded them that the term 'head' was open to different interpretations by different researchers. This type of geological process was not necessarily glacial in age. This was the same argument put forward by McKenny Hughes at Cae Gwyn, where a head deposit was, he asserted, being interpreted by Hicks as the equivalent of the local upper boulder clay.

W.J. Knowles (Knowles, 1883) read a paper to the BAAS meeting in Southport in 1883. This is an interesting paper and unpublished elsewhere, to the best of my knowledge. It attempted to establish the existence of a pre-Palaeolithic industry on the basis of artefact typology. At Larne, in north-eastern Ireland, Knowles discovered a series of implements whose shape were cylindrical. They had points, and he claimed the butts of the tools were shaped to the hand. Knowles asserted that the cylindrical tool-shape would precede the various Palaeolithic forms with their focus on flat axes. He had also recovered pear shaped natural stones that, to his mind, must have suggested to these earliest knappers the transition from cylindrical to flat shaped tools. There was already a history of research on the large Neolithic axe-making centres in this part of Ireland. He contrasted the flat palaeoliths with the Neolithic axes (see Figure 2.7). On the Neolithic celt the wide end was the business end. On Palaeolithic axes it was the point. So for Knowles, each prehistoric period had its own fixed idea of what an axe should be, and they were evolving over time. This is an intriguing article as it argues for what is, effectively, an eolithic culture and cultural stage, characterised by its own signature tool type – in this case natural stones with a minimum of shaping, pre-dating the Palaeolithic.

More plausible suggestions were made by Worthington Smith, who had moved to Dunstable, where he was resuming his collecting activities. He was finding handaxes on the Dunstable Downs at heights considerably above the river valleys there, and in brickearth deposits unrelated to the modern river valleys (Smith, 1889). These were the interfluves between the rivers, the equivalent of what Rupert Jones had described as plateau drifts in the Thames.

# 8

# THE SHAPE OF THINGS TO COME
## *Anthropology, Heredity, and Race in the 1880s and early 1890s*

> 'You interest me very much, Mr. Holmes. I had hardly expected so dolichocephalic skull or such well-marked supra-orbital development. Would you have any objection to my running my finger along your parietal fissure? A cast of your skull, sir, until the original is available, would be an ornament to any anthropological museum. It is not my intention to be fulsome, but I confess that I covet your skull.'
>
> (Conan Doyle, August 1901, *The Hound of the Baskervilles,* The Strand Magazine, Chapter 1)

In this chapter I will primarily focus on the anthropological background to the eolith question as it stood in the beginning of the late Victorian period. The biologists were slowly transforming a once anthropological debate into a proper scientific study, and in so doing they increasingly excluded the anthropologists. They could only look on and concentrate on the results of inheritance; so they continued to measure and quantify race in all its outward manifestations. From this quantified perspective, they could address its qualitative consequences. Anthropologists like Francis Galton (see below) could take the same kind of statistical data as race scientists were using and apply it to questions of racial health and the continuity of positive racial development (or more worryingly for the Victorians, the opposite).

In this sense human origins remained connected to race, but without the emphasis on racial origins. In earlier chapters I suggested that evolutionary anthropology had managed to humanise later Prehistory (Gamble and Moutsiou, 2011), but not the Palaeolithic. There were no viable analogues for the drift. This situation persisted into the late Victorian period as the inadequate British skeletal database continued to prevent either an effective humanizing of the Palaeolithic, or the linking of the later Prehistoric races to the Palaeolithic ones. British Palaeolithic researchers had to engage with the racial lineages suggested by their continental colleagues, using them as proxies to explain British evidence. While continental, and particularly French, anthropologists were forging direct links between the Palaeolithic and Neolithic races, English scholars had to rely on the archaeological evidence to approach this question, but the conflicting data surrounding questions of archaeological continuity or replacement generated a diversity of opinions.

Once again, the peculiar combination of circumstances that had precluded the development of a lively and indigenous origins debate in Britain continued to influence what was debated. The belief that humans in the European Palaeolithic were essentially modern contributed to this. The expanding evidence for a global Palaeolithic seemed to imply that, by analogy with the western European drift, Palaeolithic humans beyond Europe would be human-like as well. But what would an earlier progenitor or transitional creature look like? Where and when were the man-like apes turning into the ape-like men, and, if human evolution was more rapid (*sensu* Prestwich) than Darwin had envisaged, how long had the process actually taken? During the 1880s and early 1890s it was difficult to see where an answer could even come from.

The eolith question, discussed in the next two chapters, was an early salvo in the process of formulating those questions. It was a direct outgrowth from the unanswered questions outlined in this and the last chapter and turned the attention of British archaeologists and anthropologists away from the end of the Palaeolithic toward its beginning, thus providing a sustained and British origins debate for the British academic establishment.

## Tertiary Man

By the 1880s the former Darwinian elite were no longer young rebels banging on the door of the establishment. They were the establishment. Stocking (1987) notes that by the middle (and later) Victorian period the prosperous middle class were distancing themselves from the lower. They now had a vested interest in maintaining the *status quo*. Evolutionary gradualism as opposed to turbulent revolution had a stronger appeal. Rapid change through upheaval would not be good for business or science. Both Bowden (Bowden, 1991) and Stocking (*ibid*) suggest that Pitt Rivers even laid out the artefacts in his museum deliberately to convince the working classes of the propriety of gradual evolutionary change.

While giving his presidential address to the Geologists' Association in 1888, its president, F.W. Rudler (Rudler, 1888), used the occasion to look back over 50 years of the Association, to the year Victoria came to the throne in 1837. In referring back to 1859 and what he called the 'anthropological aspect of geology' he noted,

> '...from these modern researches a new department of science has sprung into being – the science of prehistoric archaeology...'

and in describing archaeology he thought it

> '...Janus headed, with one face turned towards geology, the other towards history.' (Rudler, 1888, 250).

He was alluding to the difficulties of distinguishing between the disciplines of Palaeolithic archaeology, anthropology, and the geological study of the Ice Age. As a geologist Rudler was sceptical about any evidence for Tertiary man in Britain, and he was not convinced by the inter-glacial interpretations of Skertchly, or Hicks in the Welsh caves either. This was serious opposition, as Rudler was president of a national scientific society. He claimed the right of proper scientific scepticism. John Evans, serving as vice-president of the Anthropological Institute in the early 1880s (Evans, 1883), used the occasion of the president's unavoidable absence one evening to launch an impromptu attack on the inter-glacial and pre-glacial evidence for human occupation in Britain. He continued to dispute Skertchly's evidence and neatly sidestepped Boyd Dawkins's middle Pleistocene age for the finds from Erith and Crayford. Evans drew a very clear distinction between the Tertiary evidence for human antiquity, which was all continental, and the pre-glacial/inter-glacial evidence in Britain.

Evans was not alone in disputing the European Tertiary evidence for humans on the continent (Stirrup, 1885), but then Tertiary man had his supporters as well. It is not known what the immediate reaction was to a paper by Henry Stopes (1850–1902) to the BAAS at York in 1881, announcing the discovery of a shell with a human face on it from the marine Pliocene Walton Crag (Stopes, 1881), see Figure 8.1. Stopes thought the crudity of the face emphasised its extreme age. Its discoverer had passed it on to Stopes, uncomfortable with the implications of its potential age. The geological and archaeological community apparently received the news with derision (Wenban-Smith, 2009). Wenban-Smith provides an intriguing answer to its discovery; it was a Medieval pilgrimage token. This type of

*Figure 8.1. A shell with an engraved face thought by H. Stopes to be proof of Tertiary man. It is a scallop shell,* Pectunculus glycimeris. *One possible explanation is that it is a Medieval pilgrimage token. Alternatively, it was just a hoax. See Wenban-Smith, 2009. Copyright Wenban-Smith.*

'evidence' would only have served to strengthen Evan's opinion against Tertiary Man.

On theoretical grounds, the concept of Tertiary humans was being given a new airing during the 1880s. W.S. Duncan (Duncan, 1882; Duncan, 1883) argued for the origin of humanity along an arc from southern Europe to the Himalayas, following the line of the great mountain chains. His premise was that some higher apes displaced southwards by climate change would come under extreme selective pressure when they hit barriers such as mountain ranges or seas, both of which would impede their progress further south. Those species which were already man-like apes gradually transformed into ape-like men as climate altered their diet, and forced new behavioural patterns under direct and intense competition. Here Duncan was using a specific methodological approach, that of taking the modern distributions of apes, and so-called primitive peoples, to map out the likely areas of evolution for the human species. The chairman of the session at the AI where Duncan gave his paper dryly noted that few people believed that men were descended from monkeys. However, in passing, he did allude to the polygenist connotations in the paper showing that there was still life in the old debate yet.

A similar approach was adopted by E.S. Morse (Morse, 1884) in his vice-presidential address to the American Association for the Advancement of Science in 1884. This was a more focused piece than Duncan's, and confronted some of the major theoretical arguments offered whenever human origins was discussed. He noted the global distribution of humans compared to the more limited spread of the apes, and further asserted there were no very early human remains because they were unlikely to survive from such a remote period. Furthermore, if humans were ancient, why was there no evidence of their undergoing further evolutionary change (Morse, 1884)? In effect he was taking a very similar line to that of Wallace and Darwin years before. Like them, he argued that the apparent stasis in human physical development was a result of co-operative social living and culture. Although this had derailed natural selection on the human body, the human brain had developed out of all proportion, and had done so because intelligence was a more effective weapon in survival than physical adaptation.

This idea of the lack of further evolutionary development in humans was a key concept, one which cropped up time and again in both geological and anthropological discussions on the Tertiary age of humanity. All other mammals with which we were associated showed evidence of adaptation and evolutionary change; but humans appeared largely unchanged since the Palaeolithic. As already noted, Victorian anthropologists conceived of Palaeolithic humans as being essentially modern in physical appearance. There had been no discoveries of men-like apes or ape-like men. Whether this belief was a survival from the pre-1859 era, when humanity was almost always conceived of and depicted as human (Rudwick, 1992), or whether it was because there was no template to conceptualise what a transitional ancestral form might actually look like, is difficult to say; possibly a bit of both. Certainly for many archaeologists the best of the palaeoliths showed considerable dexterity in their making, something that only a human might be capable of.

In an article in the popular *Contemporary Review* for 1882, Grant Allen (Textbox 8.1) addressed the issue of just how physically modern-looking Palaeolithic man ought to have been. The article is an important one as it broached a number of key issues, including that of the existence of Tertiary man, on a very public stage (Allen, 1882). Allen, a committed evolutionist, effectively argued for a continuum of evolving forms. It was not possible to identify one ancestor and say that after this point all descendants would be human. He charted the potential trajectory that an evolving man-like creature could take from its earliest form in the Eocene through the various radiations and branching events that would lead to a separate line of evolving early human. The modern appearance of Palaeolithic man meant that a long history *was* required. Evolutionary theory predicted earlier forms, and the tertiary history of the mammals and anthropoid apes provided the evolutionary context for them. The evidence of Thenay and St Prest was the proof. The logic of Darwinian prediction was presented as evidence in itself. This countered Boyd Dawkins' arguments that human ancestors were not present in the Miocene and Pliocene because the other mammals they were associated with were not there either. Allen also provided one of the few published physical descriptions of how the man-like Tertiary ancestor would look (Textbox 8.1), controversially warning the public that they might be descended from a black skinned ancestor. He suggested that even Palaeolithic man, while being physically similar to modern humans, would have retained certain primitive traits not present in modern so-called primitive people. On the other hand the cave races certainly showed an advance in culture compared to many indigenous tribes.

*Textbox 81. Summary of an article by Grant Allen from the popular Fortnightly Review magazine summarising human evolution and presenting a case for Tertiary Man to the general public.*

**Allen, G. 1882. Who Was Primitive Man? *Fortnightly Review* 32, 308-322**

- Human evolution was a slow and gradual process with no sudden leaps. Identifying one point at which it was possible to say that before this humans were not really human, but afterwards they were is impossible. But it is important to stop thinking of humanity as a single species, and think of it as a Tertiary genus with only one surviving species remaining.
- During the Eocene our ancestors were a small and generalised mammal, a lemur-like creature, not that different from the ancestors of many of the major animal groups of today. They could use all four feet as hands (quadrumana), were small brained and possibly nocturnal fruit eaters. It was in this period that the main branches of the modern mammals began to develop.
- Lower Miocene the lemuroid split into two groups, one branch will ultimately lead to the lemurs themselves, the others to the modern monkeys. Monkeys are common and specialised in the mid-Miocene, so must have been present in earlier Miocene; none yet found however.

- Middle Miocene sees the monkey branch undergoes further diversification with the splitting off of the anthropoid branch and a separate man-like branch. Allen paints a vivid word picture of the creatures appearance; tall, hairy, 'black faced', a slouching gait, beetle-browed and fierce looking like a gorilla. This was a fire user. This is also the period of the Thenay flints and Allen attributes them to this creature.
- Pliocene evidence is largely missing, but this is an artifice of the geological record and a common one, so the absence of evidence is not a problem. There are the worked flints from St Prest and various examples of cut marked bone from a number of European sites. It is possible that the Miocene and Pliocene races of man-like animals could be classed as different species within genus *Homo.*
- Little evidence in the earlier Pleistocene, but again this reflects the inconsistencies of the geological record. First clear flint flakes are found in the middle Pleistocene period. The later Pleistocene is the period of the drift and its handaxes. Clear evidence of evolutionary development from the preceding middle Pleistocene or the various Tertiary tools. Allen agrees with Boyd Dawkins that this race is recognisable as men, but more primitive, probably retaining a few atavistic features to emphasise their being closer to earlier progenitors than they are to ourselves. The river drift men have no parallels amongst modern peoples, though some of the cave races may be represented in indigenous tribes today.
- The cave race was later in time than the drift race. We are descended, in all likelihood, from a cave race of this period. They may have not been white skinned or Aryan-looking. They may have been black skinned with skulls like Neanderthals.In many cases they may have been more intelligent than many modern savages, or even people in isolated rural communities of the West (though Allen was careful to distinguish between what he called backwardness, and a generic low level of intelligence common to a race as a whole).
- Dating this period was difficult. The Palaeolithic may have been from 200kya to 80kya, with the cave races being forced southwards by the last advance of the glaciers at this later date. Allen very cleverly evoked a sense of the length of time involved by alluding to de Mortillet's division of the Palaeolithic into three periods; Le Moustier, Solutré and La Madelaine. Based on the heights of these rock shelters above the Vézère river, the river had eroded down 27 metres between the occupation of Le Moustier and that of La Madelaine. However, between the period of La Madelaine and the present, it had only eroded 5 metres.

## The Biological Study of Heredity in the Late Victorian Period

The Victorians believed that there was strong evidence for the survival of ancient racial features into modern populations, and that in studying modern groups of people their real racial origins could be identified. In other words they were hereditary. This line of thinking would have applied to Palaeolithic evidence, had there been any, as much as to prehistoric and historic data. It must have been a matter of deep frustration that the French anthropologists were successfully demonstrating the persistence of physical traits from the Palaeolithic skeletons they had recovered.

But over the millennia racial intermixing (miscegenation) had occurred and the physical features originally unique to each race had been watered down. However, the essence of each race, those features that originally defined them, could not be completely bred out. These essences were immortal. Even within mixed blood individuals, at some generations removed from parent stock, aspects of this essence would shine through. The most common signposts to true racial affiliation were usually facial. The recognition of racial identity by observation was one of a number of methodologies available to the Victorian race scientist. What was actually being discussed was heredity, or rather the outward physical manifestation of a process that remained unexplained.

In part, it was to explore racial origins that Darwin had written the *Origin of Species* in 1859. He had written *The Descent of Man* 1871 to try and understand the persistence of different races (Desmond and Moore, 2009), but since the early 1870s there had been advances in biological science, and these questions could now be approached from a more scientific basis. The biological study of heredity and inheritance was beginning to address the question of race formation, the persistence of racial types, and the true role of hybrids. I will return to the anthropological view of the persistence of physical features in the next section.

The competing scientific disciplines which were engaged in debating heredity during this period are characterised in Figure 1.3 which gives a brief overview of their shifting fortunes during the late Victorian and Edwardian eras. The study of the biological basis of heredity had made considerable progress. In his presidential address to the anthropology section of the BAAS at Newcastle in 1889, Professor Sir William Turner gave a short and business-like summation of the state of the subject (Turner, 1889). Despite some serious theoretical lacunae, the piece had a thoroughly modern feel to it. Ever more powerful microscopes were allowing individual structures, such as the chromatin filaments (chromosomes) to be identified within sex cells (then called germ plasm). In turn, these allowed greater insights into the process of genetic inheritance. With advances in microscopy came advances in the important field of embryology (Gould, 1977; Bowler, 2003). The egg and the embryo were increasingly seen as providing the potential to solve questions of inheritance. Clearly, practical science had a better chance of solving the riddle of inheritance than armchair anthropology, or the theoretical biology of Darwin.

Anthropology's questions were the same as those of biology. How did change occur in a species? How were modifications transferred to the next generation? How much of an influence was the external environment? How did a physical or structural modification become fixed in a population? The answers would apply equally to the formation of new animal species as they would to understanding how different races of humans had originated and evolved. An excellent review of the state of research into heredity in the late Victorian period was published by McKendrick (1887–1888). From an origins perspective, if the how and why of the transmission of individual characters down the generations could be understood, then the persistence of racial features could be explained. At the same time the true nature of racial intermixing could be ascertained.

In addition to biological science, and anthropological observation, there was a third aspect to studying heredity. If culture and human intelligence had derailed natural selection, as many researchers like E. Morse and Alfred Russel Wallace believed, then what was happening to the human race now? In particular what was happening to white western Europeans, removed from nature, and living in huge urban conurbations cushioned from the influence of natural selection? Was the white race showing signs of degeneration? The best way to study this was via the collation of metrical data on modern people. This was achieved through anthropometrics. I will return to this below.

A fundamental question was how much the external environment affected an individual organism; how changes made within the lifetime of an organism were passed on to the germ plasm. This was Lamarckism, and its adherents believed that new variants (i.e. individual organisms with new physical features that distinguished them from the remainder of their group) could potentially arise in populations fairly quickly. New races could develop in a relatively short space of time – a few generations. For the creation of new species, or human races, a population would not need a large quota of naturally occurring variation as was necessary in strict Darwinian terms. The raw material for change was built into the malleable nature of the individual and his or her germ plasm. If, on the other hand, the germ plasm was not influenced by the body, and could not be easily changed, then any alterations that did occur (however this was affected) would be occasional, infrequent, and the formation of new species/races would proceed by infinitesimally slow incremental change on a geological time scale. This perception was more suited to Darwin's original belief in the role of random variation and natural selection. It required large quantities of heritable variation to be present in a population, and new variants being spontaneously created all the time. New races/species could arise only from those variants already present within that population. Useful variants would persist and the less useful ones die out.

These two opposing ideas were encapsulated in two very specific schools of thought. The inability of the body to influence the germ plasm was an idea promoted by August Weismann (1834–1914) and his students at Freiburg-im-Breisgau, Germany. (Bowler, 2003; Poulton, 1917). His works were translated into English during the 1880s, and were debated in popular magazines. More detailed discussion of his work will be presented in later chapters (see also Textbox 2.1). For Weismann the germ plasm carried the unchanging code for each generation, which was faithfully copied from parent to offspring. It could not be affected by anything happening in the body. Variation within the reproductive germ plasm occurred very occasionally by what we today would call random copying errors.

But the germ plasm also fed information into the embryo (a one way transfer only). This information was a fusion of a portion of the germ plasm inherited from both parents, and contributed to the physical development of the embryo. This fusion was another source of individual variation. This was why children had some of the characteristics of their parents, but at the same time were not identical copies. Sexual reproduction and parental input was the key to variation in the next generation. Weismann's hypothesis was Darwinian because those offspring who had inherited useful variations gained a preferential advantage, and their variations persisted. Selection was essential to Weisman's vision of an immortal germ plasm. Useful lineages of germ plasm survived, and hence began new races.

The opposing school sometimes went under the name of the neo-Lamarckians (Bowler, 2003, 236f). Lamarckism was the theory of acquired inherited characteristics (Textbox 2.1). A major change in an organism's behaviour or morphology could be the root of a new species. The neo-Lamarckists were a disparate group. There was no single Lamarckian methodology that applied to all practitioners. What they did share was a common understanding that somehow the external environment influenced body and behaviour, and that this influence was heritable.

Lamarckism was popular in late Victorian biology because it could be adapted to other hypotheses of species/lineage generation. For example, saltation (favoured by Huxley) was the belief that changes in species could occur quickly and new species arise very rapidly requiring only a few major structural changes within an organism and its immediate descendants. Many of the early geneticists were saltationists. As the understanding of genetics grew, many saltationists turned to genetic mutations (where a mutation occurred in a single gene) to explain morphological change and the rise of new species. A significant change in behaviour leading to a change in the organism itself was one way of explaining a variety of saltationist perspectives.

On the other hand the orthogenesists believed that developmental constraints built into an organism channelled its life history, and consequently its species' history, down a restricted number of developmental pathways. An organism started with a pre-determined set of developmental options and, depending on which options were taken, these further channelled the direction taken by subsequent morphological restructuring. Again, Lamarckian changes could be invoked here to interpret which pathways an organism took and why.

Recapitulation suggested that the evolutionary history of a species was revealed in the stages of development an embryo went through prior to birth. This was a widespread and popular belief in Victorian biology. The influential German evolutionist Ernst Haeckel (1834–1919) popularised his own brand of recapitulation and coined the phrase 'ontogeny recapitulates phylogeny' (Gould, 1977).

What linked all these different theories was a common belief that internal somatic change was a result of external pressures from the environment. This is why they were all amenable, to differing degrees, to Lamarckian interpretation. They were not necessarily mutually incompatible viewpoints either. A saltationist could also accept recapitulation. This mutuality strengthened the appeal of the shared outlook, and lent it the feel of being 'right'.

Recapitulation was particularly important and its appeal to the Lamarckists was clear (Bowler, 2003, 191f). As a theory of acquired inherited characteristics, each major species-forming adaptation was a stage added on to the life history of the lineage, almost like a bolt-on. Change was linear and therefore directional. These stages were reflected in the embryonic development of an organism. The embryo's growth (ontogeny) passed through (recapitulation) the adult stage of earlier phases of the species' evolutionary history (phylogeny). The notion of successive stages of development meant there was a natural synergy between Lamarckism and recapitulation theory. Again, this would have appealed to many in the Darwinian elite (although not Darwin himself); a directional evolution with developmental stages already mapped out – the inevitability of progress – with a Lamarckian potential for the individual to preferentially alter their own circumstances for the better.

Conversely it is not difficult to see why a selectionist agenda contained problems for anthropologists. Geologists like Prestwich (Prestwich, 1887), were arguing for a shortened time scale to human occupation in Europe – less than twenty thousand years. For others, such as Lyell, even a longer chronology in the order of one hundred thousand years still served to concertina all of human evolutionary history into a relatively short space of time. These spans fell far short of what Darwin had originally conceived of as necessary for evolution to take place. In the *Origin* he had calculated the Earth was nearly three hundred million years in age, but by the late 1860s Lord Kelvin (1824–1907) had calculated the earth could be no greater than one hundred million years old based upon the laws of thermodynamics (Bowler, 2003). Such timescales did not provide enough time for Darwinian natural selection to work. Kelvin refined his dating over the years and eventually reduced this to sixty million years. The scientific nature of Kelvin's work, and his personal reputation, leant such pronouncements a powerful authority.

As we have seen, most archaeologists and anthropologists believed that the earliest phases of human evolution had occurred somewhere other than in Europe, although down-dating the age of the Earth, and thus shortening the time span of geological periods, seemed to preclude any long phase of evolutionary development. By the late Victorian era there was a feeling that evolutionary change had to be a lot more rapid than Darwin had predicted and Lamarckian explanations seemed to facilitate such a view.

I will conclude this section by returning to William Turner's presidential address to the BAAS in 1889. What I find significant about his introductory talk is that it was an early attempt to position research on biological inheritance within the broader anthropological debate as it affected human evolution and specifically race. His audience would have been those anthropologists whose work will be described in the next section – the more traditional, observation-driven practitioners. Turner and colleagues sought fundamental laws governing change and believed that one observation with a microscope could reveal such a law. Turner expressed his neo-Lamarckian credentials:

> 'By accepting the theory that somatogenic characters …are transmitted we obtain a more ready explanation, how men belonging to a race living in one climate or part of the globe can adapt themselves to a climate of a different kind.' (Turner, 1889, 770)

And slightly later:

> 'We know, however, that this process of the dying out of the weakest and the selection of the strongest…[Darwinism]…is not necessary to produce a race which possesses well recognisable physical characters. For most of us can, I think, distinguish the nationality of a citizen of the United States by his personal appearance, without being under the necessity of waiting to hear his speech and intonation.' (Turner, 1889, 770–771)

A lighter moment at the conclusion of his address, but implicit in his speech was the belief that the biologists would soon provide anthropologists with an answer to how new races would form, and at the same time explain the individual variability which lay at the heart of racial anthropology.

## Heredity and Race, the Anthropology of Observation

As the foregoing suggests, the job of determining the laws of inheritance and how they operated was slowly being ceded to the biologists. However, anthropologists retained a strong interest in the results of inheritance and in heredity itself. The primary methodology for this was observation of physical characteristics, particularly those of the head and face. Equally as important to the anthropologists was the question of exactly how such observations should be made.

Paul Topinard (1830–1911), Professor of the School of Anthropology in Paris, presented a paper to the AI in 1881 (Topinard, 1881) on the science of anthropometrics. This was a Victorian obsession, and it is easy enough to understand why.

> Anthropometry…means the measurement of the entire human body (living or upon the dissection table) with the view to determine the respective proportions of its parts: 1st, at different ages, in order to learn

> the law of relative growth of the parts; 2nd, in the races, so as to distinguish them and establish their relations to each other; 3rd, in all the conditions of surrounding circumstances, in order to find out their influence upon the ascertained variations. (Topinard, 1881, 212).

The data thus generated were the means by which humanity's position in the 'scale of living beings' could be fixed, as well as the position of the different races relative to each other.

Topinard's paper was an appeal for uniformity of practice in measuring the human body. He was attempting to influence the Anthropometric Committee of the BAAS (see below), trying to make them accept the measurement system he and the Paris school favoured. With the benefit of hindsight the paper epitomised the shortcomings of the whole anthropometric project. There was an obsession with process and the people being measured were soon lost in the systematizing of practice. But no one could agree on what measurements to take or on how to take them. Consequently different regional and national schools of anthropometrics were producing reams of incompatible data. Data in the field was often collected in a slipshod manner so that much of what was generated outside the laboratory was useless. It was not only physical measurements of body parts that were of interest. Colour of skin, hair, and the eyes were equally important, as was hair type and the shape of certain parts of the body such as the nose, ears and lips. All of these were variously taken as key indicators of racial heritage.

What Topinard and others sought were the insoluble essences of race that underlay the melange created by miscegenation. It was the collection of objective observations and measurements, designed to reveal those true essences, that lent the whole enterprise a scientific air. This reinforced the link between the different branches of anthropology, and the study of heredity.

A classic example of this was John Beddoe's *The Races of Britain* (Beddoe, 1885). He devised his own methodology for distinguishing differences in the colour of hair and eyes which were, for him, the only true persistent racial indicators. Unlike many anthropometric analysts he did not view cranial measurements as being definitive, noting that they failed to distinguish between the Flemish (German racial root) and Walloon (Belgic racial root) populations of Switzerland, Belgium and Germany. However, hair colour did. He claimed hair colour formed a genuine racial frontier in northern Europe (Beddoe, 1880). He mapped out the various racial groupings within Britain (*ibid,* 1885), and related them to the known historic population movements. There were possible Palaeolithic survivals in certain parts of the country; he noted the occurrence of 'Mongoloid' racial features such as oblique 'Chinese' eyes amongst particular populations. In hindsight what Beddoe couldn't explain was why it was hair colour which was a racial indicator; what made it persistent, and impervious to change? For that, a viable theory of heredity was required.

Beddoe did not entirely neglect skull shape. He coined the term 'coffin-shaped' for the majority of the skulls of the population of Britain in the Bronze Age and later, suggesting that one explanation for this skull shape might be a fusion of the round headed and long headed peoples. *The Races of Britain* was well received (H.W.H., 1886), and continued to be a standard work. Beddoes' obituary in 1912 noted it remained the only monograph on the subject, possibly implying it was the only extensive work on British racial groupings (A.C.H., 1912).

Naturally there was no consensus. If Beddoe believed hair or eye colour was more significant, there were many others who did not. Paul Broca (Dallas, 1886), Topinard's mentor in Paris, believed the nasal index (breadth/length nasal cavity) was a good racial indicator, while W.H. Flower believed tooth size was better (Flower, 1885). For linguists like A.H. Sayce, (1846–1933) a presidential address to the anthropological section of the BAAS in Manchester was an opportunity to expound on the importance of language in determining racial origin (Sayce, 1888).

J.P. Harrison gave a paper to the BAAS meeting at York in 1881. For him, the classic methodology of skull shape was the only persistent feature of racial origins. He quoted European researchers who asserted that hair and eye colour, and other soft tissue features, actually disappeared early in racial intermixing (Harrison, 1883a; Harrison, 1883b). Their data, said Harrison, suggested that new races arose from the intermixing of old ones. He concentrated on the facial profile of the skull, believing that this was a clear mirror of racial history. He focused on the Anglo-Saxon/Teutonic groups in British burials, contrasting them with the older indigenous racial group that preceded the Saxons, namely the descendants of the brachycephalic Bronze Age round-barrow people. He produced average profiles from the skulls of known prehistoric and historic cemeteries, and then compared the two 'smoothed-out' out profiles, finding the Teutonic (specifically more Saxon than Angle or Jute) face to have smooth brows and moderately projecting nasal bones, while the older British population had more prominent brows and a higher bridged nose. He associated this with specific facial features on modern living peoples; for example the round-barrow descendants, when they had more rounded skulls, were noted for thin lips, average stature, and pear shaped ears as well as fair hair.

Significantly, Harrison noted that at the York meeting the Anthropometric Committee of the BAAS had exhibited photographs of people from different parts of the country. The regional differences between the displayed faces was clear proof that persistence of racial features was a fact. The research of Harrison, Beddoe, the Anthropometric Committee, and many others, was embedded within the belief that there *had* to be a link between Prehistory and modern European ethnography, one that would be revealed by following the methodological route of anthropological observation. Harrison even cited portraits of 'ethnic Germans' from Trajan's Column and the human studies of Renaissance portrait painters to support the identification of persistent racial features. From a modern perspective his approach now seems ludicrous. He generated averages on what were tiny sample numbers (4–6 skulls in some cases), and drew his primary data from examples of known 'ethnic' status, thereby revealing the differences he expected to find. But his work fitted in well with the ethos of the time. John Beddoe was in the audience when Harrison presented his data to the AI in 1883, and complemented him on his approach.

My favourite example of the belief in the persistence of racial characters is again by Harrison (Harrison, 1880). The continuing tradition of flint knapping at Brandon, a small village in Norfolk, suggested to a number of researchers that a continuous population had been present in this isolated rural village from the Neolithic onwards. Since it was likely such traditions were kept alive within families, Neolithic racial features could well be preserved amongst a group of people for whom isolation meant 'the march of civilization was less rapid'. The BAAS visited the village and confirmed,

> At Brandon, there can scarcely be a doubt that we still possess examples of an early British race. (Harrison, 1880, 627).

They possessed dark hair and eyes, and a 'depressed' nasal bone. When compared to army recruits from across Norfolk they appeared very different.

Discussions of this kind were not just confined to the learned societies in London. Books like those of Beddoe (1885) or Boyd Dawkins (1880) were intended for consumption by the wider reading public, who also made up a portion of the membership of the various London societies (Beddoe himself was a practising GP and only held honorary academic positions). There was a wide public interest and access to these debates. T.H. Huxley is a good case in point. His involvement with racial anthropology was a fairly brief one, mostly confined to the late 1860s and early 1870s as the Darwinians sought control over the emergent AI. As already noted, Huxley's writings on race and racial origins were published in widely read magazines such as the *Nineteenth Century* and the *Fortnightly Review*. A celebrated scientist of his standing would ensure a wide readership for any scientific article. In later life, H.G. Wells recalled the excitement amongst his fellow students at the Normal School of Science when one of Huxley's articles appeared in the immensely influential *Nineteenth Century* (Desmond, 1998). I will return to this aspect in a later chapter. Here it is sufficient to note that the general public were only too aware of the racial debates of the day.

There were political dimensions as well, notably when racial anthropology was drawn in to contemporary discourses. John Lubbock's views on the racial make-up of Britain were published in the AI's journal (Lubbock, 1887), but were originally published in the *Times* newspaper. The journal had offered him the opportunity to reprint the articles and answer subsequent critics.

Using the data generated by Beddoe amongst others, he rejected the common assumption that the United Kingdom contained three or four basic racial groups (roughly corresponding to Welsh, Irish, Scots, and English) each with their own historic national boundaries. He marshalled his evidence, and showed that the Victorian population of Britain was a mish-mash of any number of ancestral and historical racial groups and their subsequent intermixing. No one area of Britain was a pure racial homeland. He was responding to Irish nationalists who were claiming a historical legitimacy for their 'Ireland for the Irish' campaign. Lubbock was no disinterested bystander. His position reflected his own membership of the Liberal Unionists, a breakaway group from the Liberal Party who split with Gladstone in 1886 over his insistence on Irish Home Rule. Lubbock was keen to show there were no real racial homelands in Britain. A mongrel nation would be more likely to unite behind the particular vision of a United Kingdom that the Liberal Unionists and Conservatives supported.

An earlier example of the political currency of racial anthropology is provided by de Quatrefages (1872). France was decisively defeated in the Franco-Prussian War of 1870–1871. Paris had been besieged, and the Museum of Natural History where de Quatrefages worked had been shelled. He wrote a damning indictment of the Prussians arguing that their unique racial heritage led to an aggressive militarist racial attitude. He contrasted that with the French and the French elements in their mongrel-like national character. These had been the cultured and artistic traits that had been swamped by other racial characteristics. This piece was a classic example of the influence of miscegenation tempered with racial persistence, while trying to directly influence the reader. One of its overt political dimensions was to try and drive a wedge between the Prussians and the ethnic Germans who formed a confederation of independent German states supporting the Prussians. They were racially different and did not share the nationalist and racial characters of the Prussians. He was giving ordinary Germans an historical excuse not support Prussian imperialism. (Another victim of the Franco-Prussian War was Édouard Lartet. He fled Paris to his country residence but depressed by events he succumbed to apoplexy (Lartet and Christy, 1875)).

### Francis Galton and the Anthropometric Laboratory

If anthropometric data could reveal a person's true racial origins, it could equally be applied to determining a race's future. This section continues the thread of Victorian racial research that entwined itself around both the study of human origins and ethnic affiliation, but its relevance to the former is more indirect. My aim is to highlight aspects of observational anthropology in its broader context. Prehistoric and Palaeolithic racial studies were a part of this bigger picture. Through the person of Francis Galton (Gillham, 2001), whom we met in earlier chapters, I will briefly outline the project of racial health as revealed by anthropometrics. There is a second reason for highlighting Galton. I briefly alluded to the popular appeal of Huxley's writings on race in the preceding section. Galton is an excellent vehicle for showing that the awareness of anthropometric research, and therefore of the implications of racial issues in general, was widespread among the scientific community and the general public. Galton's researches reflected a much wider public concern over race, racial health, and whether or not the British were beginning to degenerate physically and mentally.

Stocking (1987, 233) suggests that Galton's primary concern lay with the relationship between biological and social evolution. It was widely held that the two ought to march hand in hand, yet somehow, as far as Galton was concerned, they had become uncoupled. The leading classes of society restrained their natural urges to procreate, while the working classes in the industrial slums seemed to do little else. The result was what Stocking labelled an unnatural selection, a fear that everything that was positive in society would be swamped by the mediocre and the inferior.

In earlier life Galton had been an explorer, sporting gentleman, and a meteorologist (he was the first to recognize the existence of the anticyclone, see Gillham, 2001). Amongst his many achievements was the development in Britain of the use of fingerprints in criminal detection (G.H.D., 1912). He also founded psychometrics, the measurement of intelligence, another anthropometric-style endeavour. He came from an illustrious family and was independently wealthy. His mother was a daughter of Erasmus Darwin, Charles's grandfather. But he is best known for his studies on heredity, and later on eugenics, a discipline for which he is credited with being one of the founders, as well as coining the name. For Galton anthropometric data on the contemporary British population was the key to understanding what would be the likely fate of the British as a people. He was president of the AI in 1885–1886, and of the anthropological section of the BAAS in 1877 and again in 1885. Galton's presidency of the AI came at a propitious moment. Stocking (1987) asserts that the AI went into a phase of decline, after a series of archaeological presidents and archaeological domination of the council. W.H. Flower was president of the anthropological section of the BAAS at York in 1881 (Flower, 1881). His address certainly painted a bleak picture. The study of anthropology was sadly underfunded, ignored by government (unlike the competition in France and Germany) and public alike, and the membership was falling.

> This certainly does not from any want of good management in the Society itself. Its affairs have been presided over and administered by some of the most eminent and able men the country has produced. Huxley, Lubbock, Busk, Evans, Tylor, and Pitt-Rivers have in succession given their energies to its service, and yet the number of its members is falling away, its usefulness is crippled, and its very existence seems precarious. (Flower, 1881, 688)

Stocking (1987) suggests the AI had lost its way, and needed to return to its roots, to the study of race. Bowden (1991) implies it was the data driven archaeologists who, in alienating the ethnologists, had precipitated the decline. The implication was there had been too much Prehistory and too many prehistorians. During the 1880s it got back to basics and its fortunes began to revive. Data was still on the agenda but now it was linked to race again. Galton was one of the more anthropologically orientated presidents who spearheaded the return. In 1894, when Flower was again president of the anthropological section of the BAAS (Flower, 1894), he happily reported a clean bill of health for the Institute. Perhaps part of the success of Galton's presidency was that he was addressing topics of *immediate* concern to the middle and upper classes.

Like many of his contemporaries Galton wrote books which had a wide circulation amongst the specialist and non-specialist reading public. He published frequently in both the scientific journals and in the popular magazines and reviews (eg. Galton, 1882 in the *Fortnightly Review*; or his 1873 article in *Fraser's Magazine*, in which he laid out in detail his beliefs on racial health and improvement, for the appreciation of the reading public). All of these had a broad public appeal. He opened an anthropometric laboratory in 1888 at the Imperial Institute (later Imperial College), South Kensington, for which he paid out of his own pocket. He had been advocating the importance of such facilities for many years prior to this (Galton, 1882). It was an early version of a 'drop-in centre' where members of the public were encouraged to volunteer for anthropometric measurement. When building plans forced the closure of the laboratory at Imperial, the South Kensington Museum found the laboratory a new home. Galton's very brief description of the laboratory's successes (Galton, 1892) clearly shows its relevance to the first and third of Topinard's criteria for anthropometric study quoted above. Over three and a half thousand people were measured in the first three years of its existence, and it had generated more than seven thousand measurements by 1894 (Galton, 1892; Flower, 1894). Temporary anthropometric laboratories were a useful way of gathering data. He set one up at the 1884 International Health Exhibition and collected data from the visiting public on a variety of physical and mental traits. An annual laboratory was opened every year at the BAAS meetings from the early 1890s onwards. It was organised on Galton's lines and its data published in the annual reports of the Association's meetings with the deliberate aim of increasing awareness of the importance of anthropometric study.

Galton's work was cited in the *Times* newspaper and mention made of the success of his presidencies of the anthropological section of the BAAS, and the AI (Anon, 1886a). He gave public lectures on heredity at the South Kensington Museum (Anon, 1888b). These three Saturday afternoon lectures offered in November 1887 were on diversity in physical and mental characteristics, recapitulation, the inheritance of specific traits (including peculiarities), and the influence of environment through lifestyle and nurture. It was Galton who initiated the 'nature versus nurture' debate.

Although his primary interest in anthropometrics was inheritance and the transmission of individual characters from one generation to the next, his anthropometric data could be used to determine the long term effects on heredity of both life styles and environments. While this was not specifically focused on the creation of new species/races, it was complementary to this project. It was about how the external environment could influence heritable characteristics and so would have had a more immediate relevance to ethnological studies comparing different European races and miscegenation. Galton promoted a non-Darwinian brand of selectionism which was anti-Lamarckian, preferring to see inheritance as a product of deep-seated laws. There is more than a hint of orthogenesis here. For example, in one of his famous studies he deduced an anthropometric cross-generational pattern that he formulated as a law of regression. Individuals in a population who departed from the average (say were more intelligent, or taller) and thus formed a sub-set of that population, would witness their children and grandchildren return toward the more common condition (Galton, 1886). A selectionist type of approach was the only way in which the departure from the norm could be fixed into a population. The problem was that Galton's own law seemed to show that such departures from an average tended not to persist in populations. It was the bio-statistician Karl Pearson (1857–1936), Galton's avid disciple, who began to translate his master's ideas into statistically grounded hypotheses on inheritance and evolution in the 1890s. It was he who demonstrated that departures from an average could become fixed in the population, thus creating the basis for new species (Bowler, 2003).

Galton's influence is very clear in the subject matter presented to the AI during his presidency (1885–1886). The full range of anthropometrics was on display. The mental development of school children was tested by their teachers on his request (Bryant, 1886). Not surprisingly race was widely debated. Papers from a conference on the native races of the British Empire were presented in volume 16 (1887) of the AI's journal. The proceedings from another conference, this time on the native races of America, also appeared in this same volume. As president for the year, Galton introduced both.

Nor were European racial differences ignored. During his presidency a short series of papers on the Jewish peoples of Europe was presented, using historical and anthropometric data, to define and discuss 'Jewishness', a subject with a long history of debate in European racial research (Jacobs, 1886a; Jacobs, 1886b; Neubauer, 1886). The anthropometric data was presented by Joseph Jacobs, and the debt to Galton and his methods was made explicit. At Jacob's request Galton applied a photographic technique, which he had previously pioneered, to Jewish school children. By taking photographs of individual faces, and then producing composites of them, an approximation of the 'average' face could be produced. Individual features tended to blur and cancel each other out, but strongly developed 'persistent' racial features (the undying typological essences) present in each subject would be clearly emphasised. In effect, the photographs depicted a racial standard. Jacobs was seeking the average Jewish expression. To modern tastes this is a ludicrous (and offensive) procedure, but the audience (which included a number of senior figures in the Jewish community) were impressed by the scientific data and accepted much of what they heard. Galton seems to have had little sympathy with his subjects. He described some of them as sharing a 'cold scanning gaze'. Jacobs, to his credit, disputed the master's word. In discussing the face on one of the photographs he asserted, 'There is something more like the dreamer and thinker than the merchant in ...[composite photograph number]...A.' (Jacobs, 1886b, 55). Jacobs reminded his audience in the AI that if one or two of the older lads did have a cold stare, it was to be recalled their young lives in the East End of London were not easy.

I will return to the public aspect of these discussions in Chapter 11. My aim here has been to show that even though these anthropological concerns were not directly associated with human origins, they had a clear link to them in that they were about heredity and race. They were also very much on the public radar.

## Evolutionary Anthropology, Race, Material Culture and the Universality of the Human Mind, 1880s to Early 1890s

In an essay in the *Fortnightly Review*, originally published in 1865, T.H. Huxley had taken his readers on a whistle stop tour of the 'primitive' peoples of the world (Huxley, 1894). It was a continuation of the views he had presented in *Man's Place in Nature* (1863, see Chapter 5).

As the late Victorian period progressed and more evidence accumulated, Huxley was able to do in 1890, what he had been unable to in the early 1860s, which was to link a Palaeolithic skull to a modern racial type (Huxley, 1890). The Friesians, of the north German coast, were an isolated long-headed xanthochroid group. They were also rather isolated from their long-headed neighbours. Drawing on the researches of the influential German anatomist Rudolf Virchow (1821–1902), Huxley suggested certain cranial features on Frisian skulls were reproduced on those of the Neanderthals. Either they were direct descendants of the 'men' of Spy, Canstatt, and the Neander valley, or their ancestors had mixed with them. Either way there was a direct connection here. Huxley noted that Neanderthal-like features also cropped up in other races, but how to explain them?

> If these characters belong to a stage in the development of the human species, antecedent to the differentiation of any of the existing races, we may expect to find them in the lowest of these races, all over the world, and in the early stages of all races. I have already referred to the

> remarkable similarity of the skulls of certain tribes of native Australians to the Neanderthal skull;...Neanderthaloid features are to be met with, not only in ancient long skulls; those of the ancient broad headed people entombed at Borreby in Denmark have often been noted. (Huxley, 1890, 328).

Here Huxley was following the route laid by the continental scholars, forging a link between a Palaeolithic race and a later Prehistoric one. The link was based on structural similarities. But a British flavour was introduced by the inclusion of ethnological references. Australian aboriginal people were often singled out as exemplars of physical primitiveness. They had protruding jaws and heavy overhanging brow ridges (in Victorian terminology beetle-browed). Huxley only noted this feature in the native Australians. But here he forged a direct link. If the Neanderthals were primitive humans who antedated the division of the races, then some of their primitive features would persist in some later races once the division had occurred, and possibly be included within the most primitive human races still living. With the Neanderthal remains it was possible to do this because these demonstrated substantial differences when compared to modern humans (although not enough in Huxley's opinion to preclude them from being human). With the Cro-Magnon and other remains from the bone caves, this was not possible. The overall similarities were too striking. But the Australians were also thought to be culturally backward because their society had no knowledge of pottery, metals, weaving or agriculture. For John Allen Brown (Brown, 1887) one of the criteria of their primitiveness was a matrilineal society. Other indigenous groups had some or all of these features.

In earlier chapters I noted that while E.B. Tylor's interests had moved away from those aspects of anthropology that could be directly linked with human origins, he had nevertheless maintained his interest in the topic. During the 1890s he returned to it and made an important contribution to debates on early Palaeolithic culture by amplifying an idea presented in *Researches* in 1865. He asserted that the aboriginal groups of Tasmania, by that time hunted to extinction by whites, represented an equivalent of the earliest Palaeolithic societies that could be imagined, almost pre-Palaeolithic; as their material culture was more primitive even than that of Australian Aboriginals, and certainly less developed than that of the handaxe makers of the drift. He described the use of simple flakes, occasionally with a minimum of unifacial retouch. Boyd Dawkins in 1880 (Dawkins, 1880a) had noted that such flakes appeared to be the earliest form of stone tool used by humans. Tylor went to great pains (Tylor, 1894a; Tylor, 1894b; Tylor, 1895) to demonstrate that Tasmanian aboriginal people had only possessed flakes as tools, they had never developed the bifacial working applied to the Pleistocene handaxe, had never polished or ground their axes and had never developed the concept of hafting stone tools. When such tools were found, they were clearly a result of cultural contact with the native groups of the Australian mainland or elsewhere. There was a specimen of such a simple Tasmanian tool in the Taunton museum and a comparable artefact had been found in the French drift. Tylor's work in the 1890s went on to postulate that Australia was once peopled by Tasmanians, and that this culture had been everywhere throughout the continent. Only later did a people from the Torres Straits, carrying a Neolithic culture with ground stone axes, displace the older population, marginalising them in isolated areas (Tylor, 1898; Tylor, 1900). In Tasmania they were left behind by the march of progress. Tylor's white correspondents in Australia assured him that in the remote and isolated western part of the continent Aboriginal groups had been discovered whose material culture was just as simple.

Tylor's polemic on the Tasmanians may be seen as part of an attempt to shore up the 'Tylorian edifice' as new ideas and younger workers keen on anthropological fieldwork were beginning an increasingly hostile critique of armchair evolutionary anthropology (Stocking, 1995; Stocking, 1987). Even old allies like Andrew Lang (1844–1912), the mythologist, were turning on him (Stocking, 1995). The impetus for promoting the Tasmanian evidence is likely to have been the eolith debate which began at the end of the 1880s. There was a strong synergy between these simple, sometimes unifacially retouched, natural stones and the Tasmanian flake tools; the supposed stage of culture involved also invited comparison. Theory and interpretation here validated each other. A contemporary people stuck in a pre-Palaeolithic cultural state, making eolith-like tools validated the eoliths themselves. However, as far as Tylor was concerned, they were also a living proof of the reality of his universal stages as well – much need support in times of difficulty.

Physically all of the human races discovered by the Victorians were anatomically modern, as were the Palaeolithic cave races. Moreover, there was no evidence for subsequent evolutionary change in humans since that time. Tylor, like many of his colleagues, explained the Tasmanians and other so-called primitive people as cultures in stasis. They occupied lands where there was no impetus to change. There was no competition for resources, and so no need for innovation and evolution. Biologically then, it was difficult to envisage indigenous peoples as reversions or atavisms, as some people claimed. The evidence did not fit. John Lubbock entertained a long public debate with the Bishop of Dublin and then the Duke of Argyll on this topic (Lubbock, 1868; Anon, 1869a). Their Graces took a religious stance; arguing that archaeology and anthropology showed man to be a fallen species. Lubbock would have none of it.

## The Limited Palaeolithic Skeletal Data Set

As Table 7.2 shows. The AI remained the mainstay of journal publications for human origins-related subjects throughout the 1880s and 1890s, although the Geological Society of London continued to maintain an interest in

Palaeolithic topics where Pleistocene stratigraphy was involved. Even so, the frequency of Palaeolithic and human origins-related articles did not change much from the levels of previous decades. Tables 3.1 and 6.2 show that the Palaeolithic human skeletal database remained very small well into the late Victorian period. The major British find for this period was the Galley Hill skeleton; internationally it was du Bois' *Pithecanthropus* from Java. These will be discussed later.

Support for the status of the Neanderthal skull being Palaeolithic in age was growing, fuelled by subsequent discoveries such as at Grimaldi in Italy and at Spy in Belgium, Table 3.1. Hermann Schaaffhausen, who had collaborated in identifying the Neanderthal skull, exhibited the original at the BAAS in Swansea in 1880, and supported Huxley's diagnosis that this was a primitive human form (Schaaffhausen, 1880). Huxley (1890) noted that the discovery of the Neanderthal skeletons from Spy had fully confirmed his interpretation of the original remains, and on the basis of associated fauna and stone tools firmly established their Pleistocene status. But not everyone was convinced. William Boyd Dawkins continued to assert that all evidence of Palaeolithic interments were actually the result of a Neolithic long headed people with a propensity for burying their dead deep within the strata of caves. Since he had argued that the Cro Magnon remains were also Neolithic (1874, 1880), then by definition any skeletons that were similar would be later Prehistoric as well. Removing a large sample of comparative material from the debate made it difficult to establish a baseline for comparison. By the publication *of Early Man in Britain* in 1880 Dawkins was willing to accept a few discoveries as being possibly the equivalent of Palaeolithic cave men. He conceded that the jaw bones at Naulette in Belgium, and those from Arcy-sur-Cure in southern France, were similar, and probably Palaeolithic. The latter had been found with stone tools and a classic Pleistocene fauna. He was ambivalent about fragmentary remains from the Dordogne caves of Laugerie Basse and La Madelaine. He was more positive about the British discovery of a tooth from Pontnewydd.

His antipathy to interment led him to reject one of the best of the south west French examples, the Cro-Magnon-like remains from the cave site at Duruthy in the French Western Pyrenees. Here a series of Neolithic interments occurred in a layer above a Palaeolithic skull. The skull was sealed by a thick undisturbed layer, clearly separating it from the later Prehistoric remains. The French excavators claimed there was a strong link between the long headed Palaeolithic and Neolithic skull shapes. They were consequently of the same race and so showed a racial continuity between the two archaeological periods. Unconvincingly, Boyd Dawkins denied the implied racial continuity, and adopted his default position; everything from the cave was Neolithic.

The British record of Pleistocene skeletal material remained as poor as ever. Apart from the tooth from Pontnewydd there were few enough remains let alone new discoveries. A skull was reported to the BAAS from Southport when the Association held its annual conference in that city in 1883 (Barron, 1883). G.B. Barron asserted the skull had been found in association with the skull and antlers of an Irish Elk. It lay in a peaty deposit, a landward extension of the drowned forest. The prognathic, dolichocephalic skull was reminiscent of the skulls of modern Inuit people and dated to the 'Cave Man' era. However T.M. Reade, an engineer present when the drainage ditch in which it was discovered was being dug out (Reade, 1883), emphatically denied it was associated with anything but fragmentary remains. At the very most it was a few thousand years old as its stratigraphic context proved. It did not date to the 'Cave Man' period.

Only one other find of a skull was reported in the scientific literature of this period and it is a sad tale (Prigg, 1885; Smith, 1894). A portion of a skull cap was found in 1882 near Bury St Edmunds in pockets of loam cut naturally into chalk. In an adjacent pocket was a handaxe, and in another, what appears to be the description of a Levallois core. Tradition had it that a complete skeleton had been discovered some years earlier in another pocket along with a tusk. It had disintegrated almost immediately. The skull cap, considered Palaeolithic in age, was duly dispatched to Worthington Smith for illustration, and then returned by railway parcel post. However at this time railway stations were targets for terrorist bombs planted by supporters of Irish Home Rule, and the authorities opened the parcel, breaking one of the stone tools and the skull, and then repackaging the whole lot in a careless manner. When the package was finally returned, the skull was smashed beyond repair. Smith ruefully remarked that bad fortune appeared to attend the discovery of human remains.

# 9

# THE BRITISH EOLITH CONTROVERSY
## *A Home-Grown Human Origins Debate*

> One hot afternoon, after a long walk over the hills together, they [Benjamin Harrison and Colonel W.C, Underwood] arrived at Shoreham village, weary and thirsty. Making their way to an inn they asked for tea. But it was the day of the local flower show, and the innkeeper's wife professed her inability to spare time to get ready the meal.
>
> 'Damn!' ejaculated Colonel Underwood with vigour, as the disappointed travellers turned to leave the inn. They had proceeded only a few yards when a little girl overtook them.
>
> 'Please sir,' she asked, 'are you Colonel Underwood?'
>
> 'Yes I am', was the answer, 'how did you know me?'
>
> 'Oh, please sir, my father was in the bar when you said "damn", and he was once in your regiment, and he said, "That's Colonel Underwood's damn", and so mother said if you will come back she will be very pleased to get you some tea'.
>
> (Harrison, 1928, 214f, my brackets)

I have already alluded to my opinion that the British eolith question represented a 'Second Antiquity of Man Debate' (McNabb, 2009). The reason for this is that it focused on a series of primitive tools whose very nature and provenance set them at the beginning of any evolutionary sequence of material culture development. Other than organic tools, it was impossible to imagine a more primordial stage. What is more, they occurred on the chalk down land of north Kent. This was as close to an indigenous origins debate as British scholars were likely to get, short of finding the remains of a transitional creature in an as yet undiscovered cave. Eoliths prompted questions like who made them and where did the makers come from? If they came from somewhere other than Britain or Europe, then when did they arrive? What was their relationship to the handaxe makers of the drift and the cave races?

Curiously these questions were not aired as much as they could have been by the British archaeological and anthropological community. Rather they focused on the physicality of the tools themselves. This fitted with the largely non-interpretative framework within which human origins and Palaeolithic archaeology operated at the end of the Victorian period. This Second Antiquity of Man debate was fought very much on the same grounds as the first – geological context and the reality of the tools themselves. This time however two of the main players, Joseph Prestwich and John Evans, were on opposite sides of the barricades.

## *Dramatis Personae*

Joseph Prestwich's contribution to late Victorian geology will be apparent from the foregoing pages. So far I have only focused on his interests in drift geology. He also made important contributions to the geology of the British coalfields, the water bearing strata beneath London, the Tertiary stratigraphy of Britain and northern France (he identified the Thanet Sands, and the proper chrono-stratigraphic position of the London Clay), and even plotted out the safest strata through which to dig a channel tunnel (Hicks, 1897; Woodward, 1893; Evans, 1897b). H. Woodward, in describing Prestwich in 1893, suggested he belonged to the second generation of great British geologists; men like Darwin, Lyell, Owen, and Godwin-Austen (1808–1884); men who had known members of the first generation such as Dean Buckland, Henry De la Beche (1896–1855), and Roderick Impey Murchison (1792–1871). He also said Prestwich was the last of his generation.

Until 1872 Prestwich was only a part-time geologist, practising in the evenings, weekends and on holidays. Otherwise he was a full-time businessman, following the family business as a wine importer. His daily commute from Shoreham in Kent to London took four hours a day, leaving him tired and often with little energy for geology (Mather and Campbell, 2007). In 1872 he sold the business and from that point on devoted himself wholly to geology. In 1874 he took up the position of Professor of Geology at Oxford University (at the age of 62) which he held until 1888. His retirement was split between his home at Darent Hulme in Shoreham, and, during the 'scientific season' when the various metropolitan societies met, a house belonging to relatives in London. The Prestwichs also spent a lot of time travelling. Contemporary summaries of his life and work were presented in his wife Grace's, *Life and Letters* (Prestwich, 1899) and by Archibald Geikie (brother of James and Director General of the Geological Survey; (Geikie, 1899)). A modern assessment is presented by Pope and Roberts (2009). His wife Grace was herself a geologist of note, as well as a successful novelist and writer (Mather and Campbell, 2007). She was the niece of Hugh Falconer, and travelled extensively with her uncle throughout Britain and Europe. She accompanied Falconer on his geological excursions acting as assistant, often sketching and describing fossils and sections. She worked on the cave of Cefn with him producing, among other things, sketch sections of the stratigraphy. Her role (and Falconer's) have been somewhat neglected in the 1859 Antiquity of Man debate. She was both eye witness and full participant in the opening stages of the events of 1859. She left a record of that crucial visit to Abbeville which is rarely acknowledged nowadays (G.A. Prestwich, 1895). This was recognised

*Figure 9.1. A panoramic (photostich) view of the Chalk Escarpment as it looks today. Taken from a spot to the north of Ightham village.*

by Benjamin Harrison commenting on an occasion in 1894 when she had mentioned those far-off days. 'Here was a supreme moment, to be in contact with the first persons who had examined and recognised the artificiality of Boucher de Perthes' finds.' (Harrison, 1928, 192).

We also know a great deal about Benjamin Harrison. If anything his life can be reconstructed in greater detail than Prestwich's. His son, Edward Harrison, produced a 'Life and Letters' volume (Harrison, 1928) focusing on his father's archaeological interests and the campaign to establish the authenticity of the eoliths. The two books, although very similar in style and presentation, could not be more different. Grace Prestwich wrote a eulogy to her husband's memory, a catalogue of honours won and famous friends. Towards the end (as Lady Prestwich seemed to become more obsessed with death) Prestwich the man disappears beneath Prestwich the legend. Not so with Harrison. His son draws out a warm and very human character from the voluminous letters, diaries, and notebooks that Benjamin Harrison kept throughout his life. *Harrison of Ightham* is an insight into the life of a remarkable person – it is also a fascinating glimpse into the practice of late Victorian science from an outsider's point of view.

By trade Harrison was a shopkeeper, running the village shop at Ightham. In 1852 he left school at age 14 to assist his father in the shop, taking over the business in 1867 and running it until 1905 when it folded. Harrison was not a business man, he spent too much time field-walking. He also appears to have been a serial insomniac and a workaholic (at least on those subjects that engaged him). He would rise early, and think nothing of a walk of several miles up the scarp slope of the North Downs, see Figure 9.1, and then across the plateau 'geologizing' and hunting for artefacts, before returning for breakfast to open the shop. He was a prodigious walker and his knowledge of the physical character, flora and fauna of the local countryside was profound. His education had been basic but he read voraciously to feed an insatiable curiosity about the natural world around him. He was a shy and diffident man, although I sense a bullish streak of stubbornness in him. He became increasingly deaf in both ears, particularly after the middle 1860s. His son blamed Harrison's refusal to join scientific societies, attend their meetings, and give public lectures on his deafness. He was a man who was only really comfortable when on home territory (A. Muthana, pers. comm.). But since he was happy to give informal lectures either in his museum (a room above the shop where he kept his artefact collections and his notebooks), or in his garden to visiting field trips, perhaps his shyness is more to blame. Deafness did not prevent him from leading field trips to visit archaeological sites, or conducting parties of visiting scientists around the area.

One aspect that is worth examining from the early years of Harrison's life is just how much access he would have had to knowledge regarding his various interests, including archaeology. Ightham was a small rural village off the beaten track. In his early days as a tradesman (and not a particularly successful one) he may not have been able to afford regular journeys to London to attend the meetings of the scientific societies, or to take their journals. Regular subscriptions to book catalogues would also have been difficult to afford. Certainly Harrison himself felt the intellectual isolation. But I suspect that this became less of a difficulty over time. In the late 1870s he took *Chamber's Journal,* which regularly reported the proceedings and debates of the BAAS. Edward Harrison notes that it was from 1878 that his father began to take a deeper interest in Palaeolithic archaeology, so he would have been familiar with the big questions of the time, if not their detail. Harrison also subscribed to the *Cornhill Magazine* from its first issue in 1860. This too had occasional articles on archaeology. It was through its pages that he first made the acquaintance of Grant Allen (writing as GA) the popular science writer, evolutionist and science fiction author.

If Harrison was reluctant to appear in person at scientific meetings, he certainly had no fear of entering into correspondence with people, no matter who they were. Harrison is the one who began the correspondence with Allen when he wrote to congratulate him on an article in the *Cornhill.* Similarly, in 1864, he initiated a long correspondence with Professor Rupert Jones, the geologist, which developed into a lifelong friendship. Jones was at that time co-editor of the influential *Geological Magazine*, a sister publication to the prestigious *Quarterly Journal of the Geological Society of London.*

Perhaps more than anything else it is as a correspondent that Harrison would have received most of his up to date information, particularly as his reputation and fame grew. Even from the 1870s onwards this was the case. In typically Victorian fashion he represented a node within a network of similar individuals. Informative letters, news, gossip and opinions on other researchers' ideas would be regularly exchanged. Attendances and performances at public lectures would be reported on. The reactions of the audiences and the pronouncements of the great and good would be picked

ABOVE IS THE CORNER OF BENJAMIN HARRISON'S SHOP (LEFT HAND SIDE), AND PART OF THE MAIN STREET AT IGHTHAM. DATE UNKNOWN. BELOW IS BENJAMIN HARRISON AGED 74

HARRISON UNDER THE BEECH TREE
1912

apart by letter or post card. Published off-prints would be often exchanged – Harrison's notebooks have many pasted-in articles received from friends and co-workers. Books would be loaned, sometimes from distant and distinguished correspondents such as Rupert Jones or even Gabriel de Mortillet, and relevant passages copied out by hand into the notebooks. There was a lively trade in artefact exchange. A parcel of specimens from Worthington Smith or Flaxman J. Spurrell would be accompanied by opinions and explanations, requiring comment and interpretation from Harrison by return of post. His long field walks accompanying visiting scientists of all calibres were opportunities for debate and gathering more up to date information.

For those readers interested in knowing more about Harrison I heartily recommend reading *Harrison of Ightham*, as it is as informative as it is charming. It is a unique glimpse into the connections between two parallel worlds, both now vanished. One is the world of late Victorian/Edwardian scientific anthropology and archaeology. The other is the life of a scientifically-curious, lower middle-class village shopkeeper in rural southern England. Benjamin Harrison had appalling handwriting which I confess I am mostly unable to read. The remainder of this chapter draws heavily from his notebooks and diaries, curated in the Maidstone Museum and Bentlif Art Gallery, and I would like to acknowledge the enormous debt I owe to Angela Muthana, a curator/researcher at the museum, who has patiently transcribed his writings.

*Table 9.1. Brief sketch of a selection of the major stages by which Benjamin Harrison arrived at his beliefs in eoliths.*

## Prestwich's Geological Views Just Prior to the Beginning of the Eolith Debate

Table 9.1 summarises some of the main events in the journey that Prestwich and Harrison took in the lead-up to the reading of their first paper on eoliths and the Chalk Plateau at The Geological Society of London on 6th February 1889 (Prestwich, 1889). As we have seen, by the late 1880s Prestwich had accepted that humans pre-dated the post-glacial period. In the discussions that followed John Evans' mini-conference on human antiquity in 1877 (Chapter 7), Prestwich had argued that the evidence from Reculver demonstrated that human occupation of Britain dated to the glacial, not the post-glacial phase of the Pleistocene. For Prestwich this was an 'account audited and passed'. He backed this up a decade later with a paper directly addressing the relationship of human antiquity and the glacial period (Prestwich, 1887). This paper, read to The Geological Society of London in May 1887, is worth focusing on as it is the theoretical background onto which Prestwich would graft his acceptance of the eoliths.

To begin with he disputed the evidence for multiple glaciations as predicted in Croll's astronomical theory, which underpinned Geikie's terrestrial glacial theories. He preferred to believe there had only been one great glacial epoch. True there was evidence for minor warm and cold fluctuations within it, but the warm phases were not warm enough to be full interglacials. Returning to

| Date | Basic methodology | Collecting activity | Influences on Harrison's opinion of the date of artefacts and drifts |
|---|---|---|---|
| pre-1880 | Based on context and topographic location. Handaxes and implements had to be located in drift deposits. Harrison is aware of post-glacial date of river valley drifts. Also aware of greater antiquity of drifts located higher up or outside of the valleys.<br><br>Later Harrison begins to concentrate on the drift deposits found capping interfluves between the modern river valleys. These preserve older drift not related to modern valley formation | Mostly concentrating on Neolithic material. Occasional Palaeolithic finds such as Rosewood handaxe, figure 9.2.<br>August 1879 Harrison meets Prestwich. Pointing to Darent Valley Prestwich explains handaxes in Somme gravels half way up valley side. Harrison realises he has palaeoliths from higher up – i.e. older | Presumably influence of Prestwich and Evans at this time was strong. Evans committed to post-glacial date. Prestwich less so, but not published on the subject for some time |
| 1880-1881 | | Discovery of a handaxe (September 1880) at Highfield, to the east of Ightham village, initiates a phase of collection focused on Palaeolithic. Attention is on the valley gravels of the river Shode. Prestwich is kept up to date with discoveries and visits some of the sites, sometimes with Harrison. Evans is also informed of discoveries, and also visits. | |
| 1882 | | Extends area of investigation to the Upper Darent, but continues with higher level gravels associated with Shode. | Corresponds with James Geikie who opens his eyes to possibilities of interglacial and pre-glacial humans. Visits British Museum to look at Skertchly's inter-glacial material |
| 1883 | | Expands his search area to the gravels of Medway valley and to Aylesford. Continues with higher level gravels of the Shode | Visits British Museum and A.W. Franks shows him large numbers of Neolithic axes. Considers his implements from 'base of chalk hills' are likely to be much older. Opinion based on typology and assessment of crudity/workmanship |
| 1884 | | Particularly interested in high-level gravels on hills above 400' O.D. Drifts on watersheds are searched. All of these are below the chalk escarpment.<br><br>Locates a drift patch on the top of the Chalk Escarpment at Parsonage Farm at 520' (or 530') O.D.. The gravel is worn, stained, and has quartzite in it. Eoliths are first noted at this patch of gravel but not recognised as humanly worked at this point. | Begins to take notice of presence of quartzite in high level drifts, particularly in the drifts on the interfluves and watersheds. Quartzite pebbles not native to chalk. Implication is they are brought south in glacial drifts.<br><br>Harrison believes a spread of gravel on Oldbury Hill is of glacial age. It contains a few Palaeoliths. |
| 1885-1888 | From 1885 onwards focus shifts away from the Shode and its higher level deposits to the Chalk Plateau itself.<br><br>Key elements in methodology for recognizing Plateau palaeoliths were colour (ocherous or deep brown/red staining); wear (more worn are older); appearance (assessments of crudity – less refined looking means older). | The first handaxe/implement is found on the Chalk Plateau 19/11/1885. In a deposit of gravel – a Tertiary outlier at Ash, to the north of the church. Between 1885 and 1888 a further 20 will be found. Prestwich accepts the context of this palaeolith as evidence of interglacial or pre-glacial humans (probably late 1885). Harrison begins to believe that the eoliths from Parsonage Farm found previous year, and kept, are also genuine artefacts. But he shares this with no one. Included in sample was piece 464. April 1885 Henry Walker, geologist, looks at 464 and declares edge chipping human but astonished at topographical position. From this point on Harrison begins to explore Chalk Plateau drifts with clear idea of the eoliths and palaeoliths being evidence of interglacial or pre-glacial humans.<br><br>But Harrison still harboured some doubts. His own final convincer (i.e. eolith) was found on 15/2/1886.<br><br>Prestwich accepts that artificial retouch is present on some natural stones in July 1888 and 464 was included in the material that persuaded him. His final conversion to eoliths as a class of tool was in September 1890 | The Chalk Plateau/North Downs are a remnant of a land surface to the north of the Wealden anticline. Tertiary deposits at Friston, East Dean in Sussex, represented the southern fore-plain of Wealden anticline (i.e. the South Downs). A palaeolith found in these outliers served to reinforce Harrison's belief in ancient date of Tertiary outliers on the North Downs.<br><br>Prestwich is convinced of glacial deposits in area (for example Park Farm Brick pit, and Highlands, near Wrotham). Striations on the palaeoliths and other natural stones also suggest glacial activity |

a favourite theme of his, he then argued that any attempt to estimate how long ago the glacial had been was impossible if it followed uniformitarian arguments. Such interpretations would imply that the process of erosion in the Pleistocene occurred at the same rate as today – by analogy with modern rivers. In other words, it was infinitesimally slow. Many geologists worked on a formula of one foot of soil removed from the surface of the earth every six thousand years. Measuring the depth of river valleys thus gave a rough estimate of age. But Prestwich (followed by Evans) had long believed that the river valleys were cut by torrential Pleistocene floods unlike anything in the modern world. These meant that the river valleys could have been formed relatively quickly (Prestwich, 1886; Prestwich, 1888).

Prestwich reminded his readers that in 1859, as the first Antiquity of Man debate was being fought, the pendulum of human antiquity had swung from very recent, to very old. However, as the Noachian chronology had collapsed, the pendulum had continued to swing making humans ever older. Estimates widely believed at the time suggested that the glacial began a million years ago, and lasted for eight hundred thousand years, giving a post-glacial age for humans of two hundred thousand years (closer to the view held by Lyell). For Prestwich this was far too long as humans had shown no evolutionary change in that time.

Prestwich (1887), abandoning his former stance, now calculated that the length of time it would have taken glaciers to grow, reach their maximum distribution, and then retreat, was between fifteen to twenty five thousand years. This therefore represented the duration of the glacial epoch. The post-glacial period he estimated at between eight and ten thousand years in duration. This gave a span to the length of time humans had been in Britain of twenty to thirty thousand years at most if you believed they were pre-glacial (see below) or intra-glacial in age. If post-glacial, then they had been here for less than ten thousand years.

Then Prestwich dropped a mini bombshell.

> My first impressions with regard to the Valley of the Somme were: that the high-level gravels originated in early Glacial times; that the intermediate stages and terraces were formed during the excavation of the valley as a consequence of the great glacial and post-glacial floods; and that the low level gravels formed the concluding stage of those conditions. But in the absence of data, since acquired, the strong prepossessions then existing, and the novelty of the subject, I was then led to conclude that the whole might be Postglacial. So much evidence has, however, since been brought forward with respect to the so-called Preglacial Man, that I feel I am now justified in reverting in great part to my original position. (Prestwich, 1887, 406)

He had always believed that humans were present in the glacial even in 1859. (The expression on John Evans's face can be imagined.) Prestwich then drew the threads together. Humans were pre-glacial in the sense of being in southern Britain before the ice sheets arrived, although not of being here before the glacial epoch, i.e. before the Pleistocene. The evidence of Hicks in the North Welsh bone caves, of Tiddeman at Settle, and Skertchly in East Anglia demonstrated this. Furthermore, there were palaeoliths in what he interpreted as glacial gravels in Mildenhall and Lakenheath, in Norfolk. These palaeoliths had to predate the glacial too. When the glaciers arrived, humans fled south to parts of the Lower Thames Valley (Reculver) and to the Somme and Seine river valleys which were never glaciated. Some valley incision began at this time, or just before it, and high level drifts were formed. Then, when the glaciers began to melt, humans moved back northwards which was why the lower drifts in all the major valleys, now deepened by floods from melt water and torrential rain storms, contained implements. Prestwich would now use the second Antiquity of Man Debate to reposition himself and his views away from the post-glacial interpretations still adhered to by John Evans, Boyd Dawkins and others.

## The Eolith Debate Begins

It was during the 1880s that Harrison was actively collecting Palaeoliths from the Ightham area and further afield. It is legitimate to ask how much Harrison's discoveries contributed to Prestwich's 1887 perspective, even though they were not mentioned. The answer requires that we understand something of Harrison's own journey toward his belief in eoliths.

Harrison turned his attention to the Palaeolithic in the late 1870s. His first meeting with Prestwich (21st August 1879) was a significant one. He asked the Professor about the position of the high level Somme gravels.

> Using the Valley of the Darent as an illustration, Prestwich pointed from his window to the little river, saying, 'If we take the Darent to be the Somme, the gravels would lie at about the level of the railway station.' (Harrison, 1928, 84).

Harrison realised that a number of the palaeoliths he had found already were from situations that were higher. Edward Harrison suggests this triggered a realization in his father's mind that these handaxes, such as the Rosewood specimen found in 1863 at 500' OD, Figure 9.2, were older than those from the Somme valley. This was reinforced by the discovery in September 1880 of a handaxe from Highfield, less than half a mile to the east of Ightham village. The palaeolith came from drift gravel on the field. But its position was significantly above the river Shode which at this point ran through a deep gorge. The implication was that this must be an implement from a drift unconnected with the modern river's course, and located well above it. This find stimulated a phase of collecting between 1880 and 1885 focused on

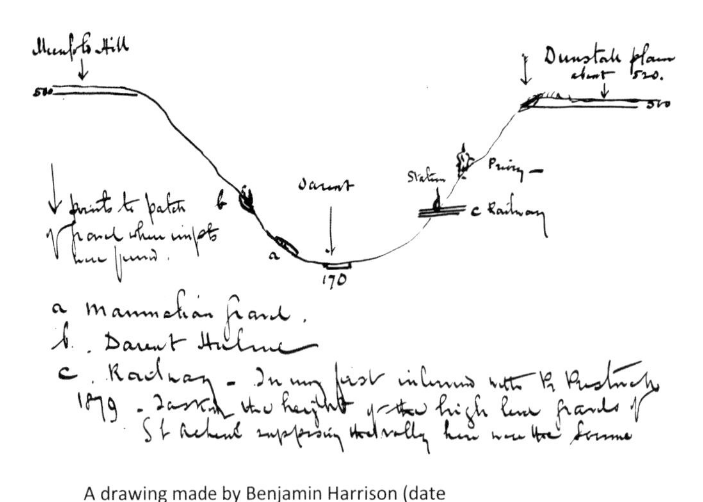


A drawing made by Benjamin Harrison (date unknown) many years after his first meeting with Joseph Prestwich and describing the event

Harrison's Annotations.

Dunstall Plain about 520 (i.e. feet ASL)

Meenfold Hill 500 (i.e. feet ASL. On modern maps a wood at this location is marked Meenfield)

a – mammalian gravel

b – Darent Hulme (Prestwich's house)

c – Railway . In my first interview with Professor Prestwich 1879 I asked the height of the high level gravels of St Acheul supposing the valley here were the Somme

d- Station

Priory (above station)

Darent 170 (i.e. river at 170 ASL)

↓ points to patch of gravel where implements were found

implements from various drifts/terraces associated with the Shode valley (see Figure 7.5). Towards the end of this period Harrison began to concentrate on deposits higher than the Shode's terraces. Although he made occasional forays up the Chalk Plateau (after 1881) he was mainly concentrating on the area below and in front of the North Downs escarpment, and at heights above 400'OD, (Figures 7.2 and 7.4 and 9.1).

Throughout this time Harrison was in communication with Prestwich, and wrote occasionally to John Evans apprising him of his latest finds. Prestwich would sometimes visit the localities with Harrison, and on rare occasions bring Evans along with him. Both visited Harrison's museum during this period.

By late 1884 Harrison was beginning to visit the Chalk Plateau on a more regular basis, and he stimulated Prestwich's interest too. There were quartzite pebbles in these high drifts. These were not native to the chalk landscape and one of the ways they may have been introduced was through glacial activity. The district around Ash, South Ash, and Parsonage Farm (Figure 7.5) began to attract Harrison's attention. The latter location was at 520/530' OD and the stained and ochreous gravel there appeared to be a remnant of drift; a Tertiary outlier (i.e. an isolated remnant deposit of Tertiary drift from which any contemporary or later overlying deposits and any other later deposits

*Figure 9.2. Handaxe found at Rosewood near Ightham Common in 1863 by Benjamin Harrison. This is one of his earliest Palaeolithic finds and long pre-dates the beginning of his interests in the Palaeolithic.*

around it had been eroded away). Harrison collected a number of pieces from Parsonage Farm. They appeared retouched, but he was uncertain and kept them for later examination. However in November of the following year he found a handaxe in a similar gravel-spread to the north of the church at Ash. This was the first implement from the plateau and Prestwich, who saw it soon after its discovery, accepted its provenance. As Tertiary gravel, this spread was pre-Pleistocene. Prestwich must have immediately conceded its pre-glacial date in the sense of it antedating the arrival of the glaciers.

Harrison now switched his attention to the plateau and began collecting from there in earnest. More palaeoliths from Tertiary outliers followed.

Harrison was also intrigued by the retouch on the stones recovered from the Tertiary outlier at Parsonage Farm. One of these, piece 464, Figure 9.3a, was shown to a visiting geologist, Henry Walker, in April 1885. Walker pronounced the retouch human but expressed surprise at its location on the Plateau. Possibly it was this which had induced Harrison to collect in the area and which led to the November palaeolith. In fact 464 was an eolith, amongst the earliest to be discovered, but as yet its real significance was not recognised. An eolith called the Corner Stone, Figure 9.3b, was shown to Prestwich and Evans in 1881, again long before Harrison accepted the eoliths as genuine. At that point he believed it was a handaxe, but Evans, not surprisingly, rejected it and Harrison threw it away. He rescued it from his waste heap some years later when he began to recognize eoliths in greater numbers. What I find interesting about the Corner Stone is its provenance. It was from drift deposits on the watershed between the Darent and the Shode, in high-level gravels above the valley drifts of these two rivers (see Figures 7.4, 7.5), so clearly not on the plateau, but not a river drift deposit either. Such a key specimen should have been showcased more, yet apart from the anecdote of its rejection, it is not mentioned further in the published literature. Quite possibly this is because its location was embarrassing. As the eolith controversy warmed up, Prestwich and Harrison argued that eoliths in a fresh state were not found below the Plateau – only derived ones were, and they were keen to keep the numbers of these low.

The eolith that finally swept Harrison's doubts away was discovered in February 1886, but we know little else about it. From this point on he accepted their existence as a class of tool; he had long suspected their existence but he had never been sure. He had also not mentioned these tools to Prestwich or Evans. He continued to remain silent.

To return to the question posed above: 'how much did Harrison's discoveries influence Prestwich's views in 1887?'. At the time when Prestwich gave his lecture in 1887 he would have been aware of palaeoliths from high level gravels below the chalk escarpment that could not be tied to the post-glacial drainage, and of palaeoliths from drifts on top of the chalk plateau that were, in all probability, glacial or pre-glacial given their location. Why he did not mention them is curious, but they must have helped to reinforce the opinions he ventured in his 1887 lecture.

By the 1887 lecture, Harrison was a believer in eoliths, but Prestwich was unaware of them. The first stage of his conversion came in 1888. Harrison included piece 464 in a batch of flints from the Plateau which he sent to Prestwich for consideration and illustration as Prestwich was preparing the first lecture on the high level palaeoliths for the Geological Society. Prestwich accepted the retouch on this and a number of other flints as being human, but at this stage did not accept that they represented a particular class of artefacts

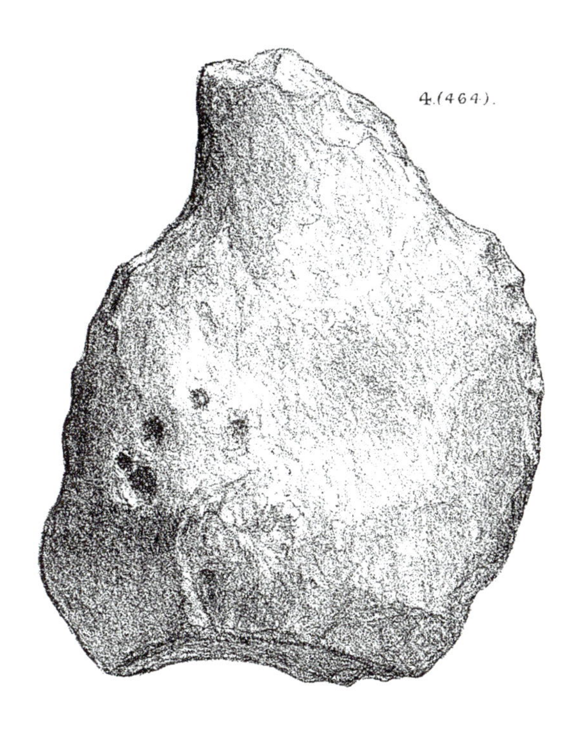

*Figure 9.3a. Eolith 464 in Harrison's collection. This was one of the earliest eoliths ever to be discovered, although its significance was not recognised until later. Figure 9.3.b. The corner stone. Another eolith whose significance was not recognised until long after its discovery. Many years later someone drew a figure on the back and varnished it. There is a date that may apply to this of October 1895.*

in their own right. This came later in September 1890 when Prestwich was researching the second of the trio of papers, this one on the Darent valley. He asked Harrison to keep a number of specimens he had found from an ochreous drift at West Yoke.

> It was the dawn of the era of the eoliths [Harrison claimed this for Prestwich's acceptance of 464 also], for on this day he pressed me to take home specimens that only a few months earlier he would have regarded as too doubtful to be preserved. (Harrison, 1928, 135–136, my brackets)

*Figure 9.4. Joseph Prestwich's examples of implements and flakes from the hill group as presented in the* Quarterly Journal of the Geological Society of London *for 1889.*

## 1889–1892: The Fighting Begins! The Geological Society and The Anthropological Institute

The first of Prestwich and Harrison's eolith papers was read before the Geological Society on February 6th 1889, and published the same year (Prestwich, 1889). The paper was actually read by the secretary of the society following normal practice, but Prestwich sat at the front with a pointer and indicated various items on diagrams set up for the audience to see. Harrison sat next to him, but his deafness prevented him from hearing very much. Edward Harrison suggests it was the greatest day of his father's life. Prestwich impressed upon the audience that up until now there were relatively few handaxes found at elevations outside the confines of the river valleys. This lent extra significance to the sheer number that Harrison was finding. These two points alone would have alerted the audience to the fact that something uncommon was being revealed.

*Figure 9.5. The artefacts from Figure 9.4 that can today be relocated in the Maidstone Museum and Bintlif Art Gallery from the Harrison collection.*

The geological argument is shown in the schematic diagram in Figure 7.4. A high and low terrace of the Shode (3 and 4) represented the post-glacial drifts. Implements and other artefacts were present in these, some worn and ochreous in appearance suggesting they had been derived. Others were sharp, contemporary with the deposits.

But above the 340' O.D. level there were drifts that Prestwich could not relate to the modern drainage patterns. These were either older courses of the Shode and other extant rivers in the area, or they belonged to an older and very different drainage network. These were the hill drifts/gravels or hill group. They were similar to the implements of the succeeding river drift only a little 'ruder', and long pointed handaxes were less frequent. Prestwich's illustration of them is reproduced in Figure 9.4 and the artefacts themselves shown in Figure 9.5.

To modern eyes this is a very heterogeneous group. Most of the artefacts are not Lower Palaeolithic. Numbers 4, 5, and 6 are probably Mousterian, the stone tool culture of the Neanderthals (4 and 5 are small handaxes; 4 has a twisted profile; 6 is a small Mousterian disc core – A. Muthana, pers. comm.) and number 3 is probably a Levallois flake, another Mousterian-like artefact. Number 8 is a classic large Mousterian handaxe, in this case found by a fellow

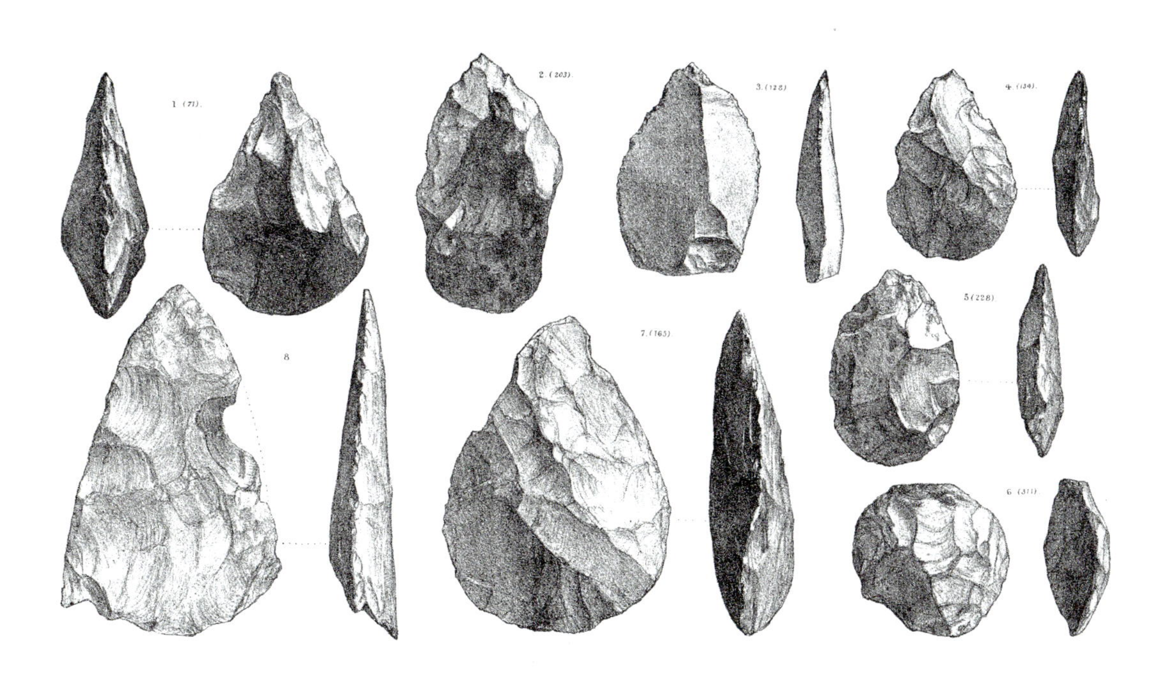
1. (71).
2. (203).
3. (128)
4. (134).
5 (228)
6. (311).
7. (165)
8

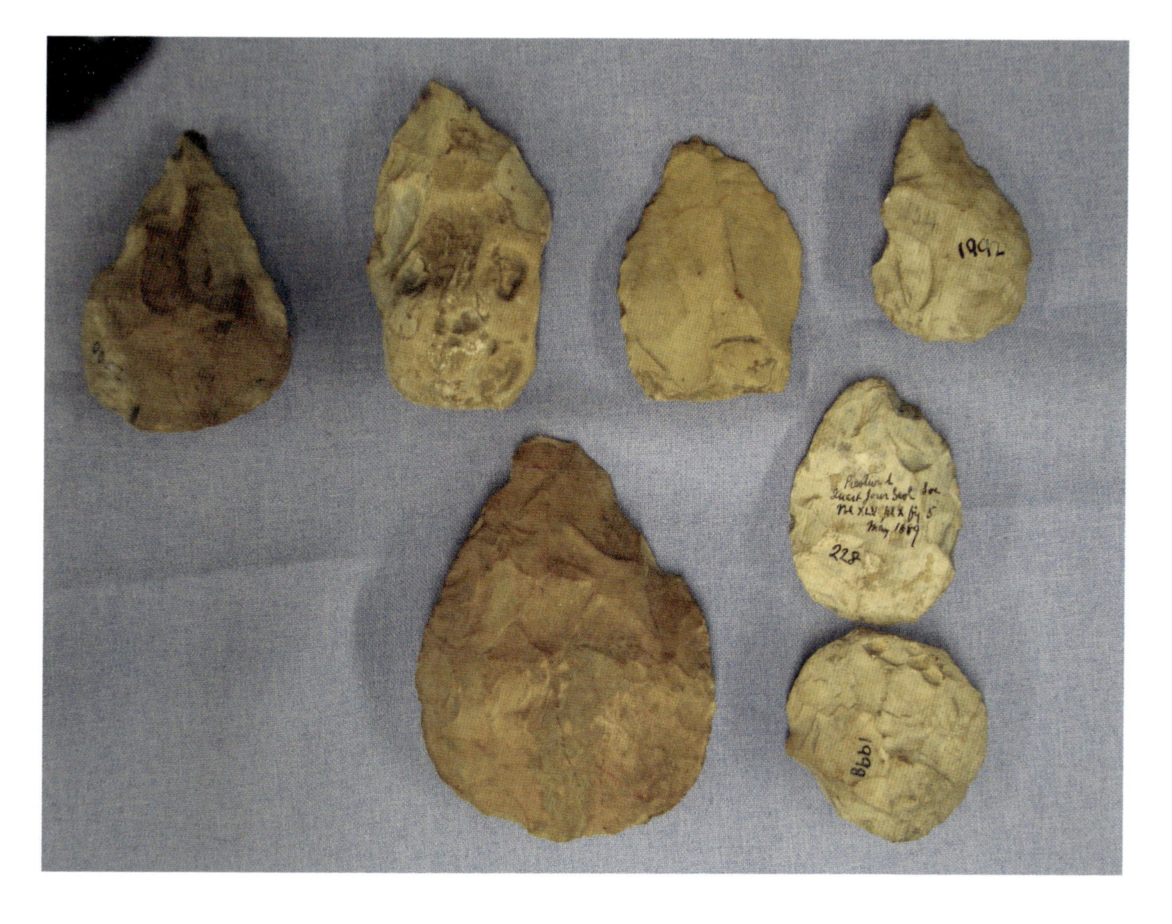
228

flint collector of Harrison's, De Barri Crawshay. Numbers 1 and 2 may or may not be Lower Palaeolithic; number 2 is either a small retouched point, or a convergent scraper on a genuine flake. Number 7 is Lower Palaeolithic. It is in sharp condition and shows a marked twist to the profile. This is the only one of the extant set with a deep ochreous stain, the others are either patinated white or bluish-white.

In terms of what these tools were supposed to be (i.e. evidence of the earliest occupants of Britain), the group should be unambiguously Lower Palaeolithic, yet, to modern eyes it appears dominated by Neanderthal/ Mousterian tools of the later Middle Palaeolithic. It is ironic that one of the later arguments levelled against the eoliths, that they were surface finds dropped by accident by later Palaeolithic humans, was in all probability the real explanation as to why so many of Harrison's hill group implements were Mousterian. These were Neanderthal discards and losses. Although not relevant to this book, and therefore not detailed here, Benjamin Harrison discovered and excavated a primary context Mousterian encampment at Mount Pleasant on Oldbury Hill. This links to much of his collection of surface artefacts from the area around Ightham, and from the Plateau, which include previously unrecognised Mousterian artefacts. Together with the Oldbury site this suggests a hitherto unsuspected wider Mousterian landscape.

There was a second body of gravel identified in this hill-drift group. Archaeologically sterile, a drift of white unstratified flints at Park Farm brick pit to the east of Ightham, suggested the passage of ice in the area. Both the implementiferous hill-group and the glacially modified white gravels were difficult to date, but Prestwich suggested they might both be glacial, by which he meant equivalent with the time of the glacier's greatest extent and so broadly contemporary with the Reculvers in the Thames, and the high terrace of the Somme at Abbeville and Amiens.

Finally there was the Plateau group or Ash group as they were called at this stage. Prestwich's figure is reproduced here as Figure 9.6. Only number 5 has been relocated to date, and this is a rolled natural cortical flake. These artefacts appear to be a mix of genuine artefacts such as number 1; what appear to be natural pieces, suggestive of an implement's shape, such as 2 and 3; and other pieces that may or may not be proper artefacts. The archetypal eolith, piece 464 (see above) is present – number 4 on the figure. Number 5 is a classic eolith with natural edge chipping resembling retouch. The mix is interesting and a good illustration of Prestwich and Harrisons' opinions on the more crude character of the Plateau group. In the descriptions of the plates he labelled these 'presumed pre-glacial' in date.

Prestwich skirted around the issue of explaining the drifts, this was to be left to the second paper. He did however note that on the Chalk Plateau were lithologies such as ragstone from the Lower Greensand, Oldbury stone (a waxy chert from Lower Greensand beds found on Oldbury Hill and adjacent hills), and other rocks that indicated the presence of ancient river drifts.

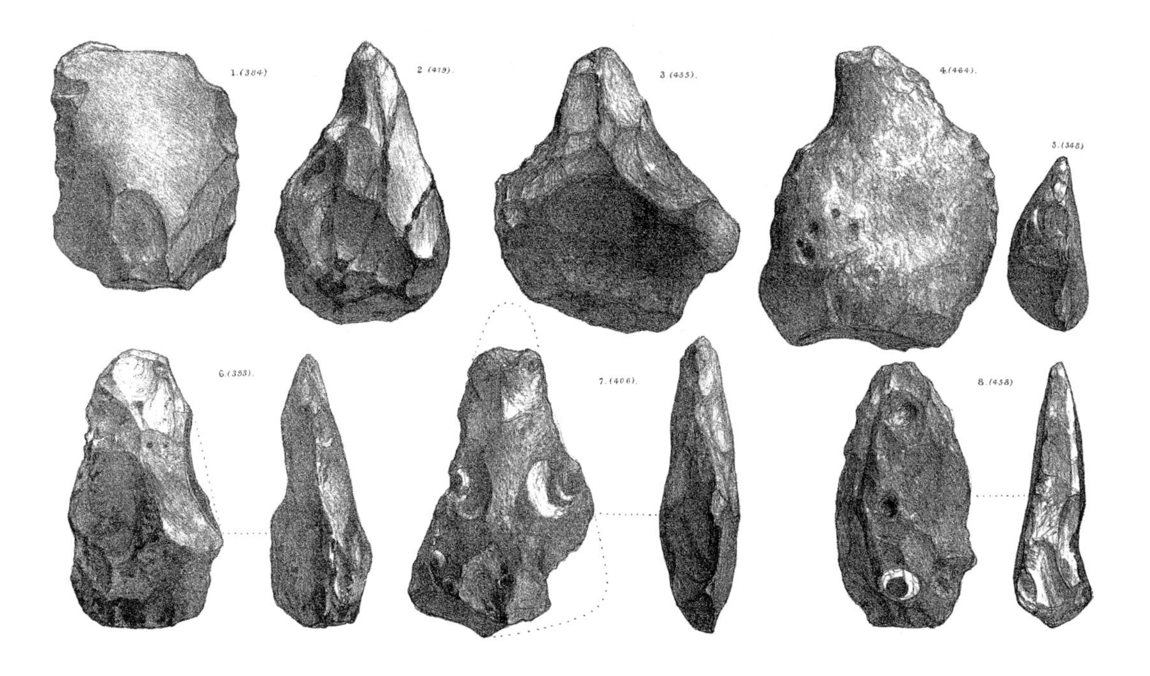


*Figure 9.6. Prestwich's Plateau or Ash group of artefacts provenanced to the Chalk Plateau. From the* Quarterly Journal of the Geological Society of London *for 1889.*

Prestwich and Harrison were at pains to point out (though this point was missed by many in the audience) that the ancient drifts on the Plateau were denuded/eroded. They were the remnants of once thicker and more spatially-extensive deposits that had covered the Chalk Plateau; but erosion had removed these, and only a few isolated patches (the outliers) now remained. The artefacts from the Plateau were always associated with these patches and were not found anywhere else. The artefacts from them showed every sign of originating from within drift deposits, but they were on the surface now because of the erosion. These were Prestwich's Southern Drift as described in Chapter 7.

I have tabulated the most significant objections raised by John Evans and William Whitaker of the Geological Survey in the discussions that followed the lecture, Table 9.2, and Prestwich and Harrison's responses to them. It is worth reiterating here that although Prestwich had at this point accepted eolithic-like retouch, he had not as yet accepted eoliths as a separate group of tools. Everything figured and discussed in the paper was Palaeolithic as far as Prestwich was concerned, he was simply stretching the length of time that the Palaeolithic could be equated with (Plateau-group+hill-group+normal river drift-group), while not extending the actual age of humans in Britain (i.e. pushing them into the early Pleistocene). His 1887 paper allowed for a long Palaeolithic, but he still set this in a short Pleistocene timescale of 20–30 thousand years.

| Objections to the paper | Response by eophiles |
|---|---|
| **The Geological Society read February 6 1889, published May 1889.** | |
| Evans. Given huge length of time, and huge landscape changes, we can't know whether drifts are really pre or post-glacial. The lithologies of high level hill drifts, and high and low level valley drifts can be explained by other drainage patterns now destroyed – the escarpment was further to the south in the Pleistocene and rivers ran from the Greensand Escarpment, which was higher, northwards. | The escarpment was cut by glacial action *isolating* the Plateau drifts. Nothing else really explains the escarpment. There are Lower Greensand lithologies on the Plateau because rivers carried eroded debris south when the Greensand and the Chalk formed a continuous plain. There are Plateau lithologies in the drifts below the Chalk Escarpment because glaciers cut the escarpment and then deposited their debris at the base of the escarpment. Rivers have then carried these south as they flowed toward the Weald. |
| Evans. Presence of three condition groups only reflects sediment burial (i.e. not original provenance). Implication is one group of rather disparate handaxes | Discreet location is important. Crude implements in a very worn state are restricted to plateau. If in Hill or Valley drift – clearly derived (this was also used to support the erosion argument above). Explanation below is also directly relevant here. |
| Whitaker. They reflect a general problem of surface finds (there are so many of them) not related to their parent gravel deposits | The implements are always found in gravely patches. This problem was addressed in the paper. Concretions on artefact's surfaces', and on natural flints, imply that both originate from same ferruginous drift. The artefacts and natural flints are deeply stained, brown and ochreous, and heavily rolled/worn. So flints and implements must come from the same drift. They can't come from hill or valley drifts as contemporary implements in these are less worn and less ochreous. |
| Whitaker . There is no evidence of glacial deposits south of the Thames | Not really addressed, by implication Prestwich's glacial erosion of the escarpment as most parsimonious explanation serves as an answer. |
| **The Geological Society read January 21 1891, published 1891** | |
| Topley and Le Neve Foster. Not satisfied with the arguments for provenance of the artefacts on the Plateau. Want to know, specifically, if any eoliths or implements have been found *in-situ* within Plateau drift deposits.<br>He also disputed the glacial formation of the escarpment, and thought that the Darent at that time flowed further south so that the Limpsfield Common gravel probably belongs to a different/earlier river. Foster added that had there been evidence of extensive glacial activity, there should be glacially striated stones and Prestwich had not mentioned these. | Lack of sections and excavations on Plateau does hinder identifying them *in-situ*, but Harrison has found a small number of implements/handaxes *in-situ*. The *in-situ* find from South Ash, discovered in spoil from digging a fence post was a handaxe 'St Acheul'-like and unworn! Prestwich restates his point about surface modification and encrustation that could only be a result of burial within a drift. They are on the surface because of ploughing. Crawshay intended excavating on the Plateau in order to demonstrate this very point. Ducked the question of the attribution of the Limpsfield gravel to a different river. Didn't rule out a glacial component to this gravel as well. Reiterated the evidence for glacial derivation of Chevening gravel and Limpsfield brickearths.<br>There were a few striated stones, but could easily be a result of fluvial transport |
| Archibald Geikie (president of the Geological Society) also dubious about the glacial cutting of the escarpment. Made the comment that no evidence for a glaciation in the south of England (no boulder clay or ice scratched stones) | In a general reply on glacial origin of the Escarpment, Prestwich agreed that there was no extension of northern ice sheets southwards. But a southern 'ice-area' could have existed in the Wealden heights and glaciers have proceeded in all directions from here. Ice and snow in the valleys therefore of possible local origin. |
| **The Anthropological Institute read 23 June 1891, published February 1892** | |
| John Evans. The retouch on eoliths was natural. Moreover it blunted the edges not sharpened them, so producing tools that were of no practical value.<br>No need to argue palaeoliths (or eoliths) were made by a different race on the basis of primitive appearance, since *all* drift deposits contained a *range* of appearances.<br>The argument that eolithic retouch was always marginal because at this early evolutionary stage invasive flaking had not been invented, did not work. Marginal and invasive working were *both* present in many river valley drifts. This continued the previous point about range. | Prestwich accepted that a few of the displayed eoliths might be ambiguous, but a few days examining these tools would be enough for Evans to develop a practiced eye for recognizing the right features.<br><br>Prestwich had always admitted there was a range of finishes in all drifts. The point was a) the *proportion* of ruder forms was so much greater on the Plateau, b) the *geological circumstances* of their discovery, restricted to ancient drifts, on an ancient plateau, c) the *constant repetition* of particular forms/types. |
| Pitt Rivers acknowledged genuineness of eoliths as tools but raised concerns over their surface derivation. If they were that common on the surface, proving their provenance by excavation would be simple. | No direct response to this |
| Boyd Dawkins. The presence of finer and cruder palaeoliths on the Plateau just reflects river drift hunters moving across high ground. The absence of the cruder palaeoliths in the valley drifts was because they were never looked for.<br>Also made the point about a range of appearances occurring in an assemblage of tools; this is seen in river drift assemblages, in his own cave assemblages, and in those of modern Native American people. Surface provenance not prove age, or help to clarify it. | Prestwich dismissed Boyd Dawkins' objections in a very off-hand manner. He said he had missed the point and not understood the geological arguments properly. Possibly making an intentional pun, Prestwich asserted he failed to see the drift of their argument! |

| 'Nature and Art', letter from Prestwich to the editor of the Geological Magazine, August 1895 | |
|---|---|
| *Assuming* that some of the 'more highly finished' implements (implication here seems to be eoliths, but it may well refer to crude/proto-handaxes as well) are genuine artefacts, then if they are found on the same surface as unambiguous palaeoliths they must date to the Palaeolithic period, not earlier. | No, this is a spurious argument. Eoliths, Palaeolithic implements and Neolithic axes are all found on the same surface. Their association is accidental. This proves nothing. In the Parsonage Farm excavation there were only eoliths found in the deeper layers. |
| Wave action (transport by waves, whether on the sea shore, or in a river, is identical) can simulate the type of edge damage seen on particular forms of eolith. They are therefore natural. John Evans' take on this was more subtle; since humans and waves can potentially make the same retouch we can never tell the difference, so we should err on the side of caution and assume they are natural till proven otherwise. | Rolling in a stream, or on a shore, will round off edges and tend to transform flints into pebbles. It will not produce retouched edges and retouched points. You do not see these in natural beach and river accumulations – only rounded pebbles and shingle. |
| Arising from the previous point, the Aldborough (spelling differs from Evans) flint from the Suffolk beach proves that the eoliths are natural | Admittedly this is a good example of a genuine similarity between a natural flint and an eolith. However it is the *only one* found to date that does look same. There are also important differences. The edges of the natural piece are not suitable for scraping, whereas those of eoliths are. The natural pebble's edges are also blunted and rounded because of the wave action, this is not present in eoliths. |

*Table 9.2. A catalogue of some of the more important objections raised against the eoliths as tools, as well as Prestwich's geological interpretation of their context.*

The remainder of 1889 and 1890 was a period of intense field walking for Harrison, as Prestwich began to prepare for the next paper on the River Darent. Harrison was now focused on the Plateau and on eoliths, searching both the downland block to the west of Shoreham for Prestwich, as well as that to the east which was his own stamping ground. Eoliths, which were now being called the 'old olds', were everywhere. Harrison was even beginning to make comparisons between the different patches of drift and notice differences and similarities in artefact appearance and types, as well as the character of the gravel patches. For example artefacts from Snag Hill, to the west of the Shoreham Gap, were similar to those of the hill-group around Ightham. Harrison, conscious of Whittaker's point about the surface finds, (Table 9.2), began to look for sites and sections where implements and eoliths could be found *in situ* within the Tertiary drifts.

One example, drawn from the Harrison archive at Maidstone, will suffice to show the energy and enthusiasm of the chase. This period is covered in Harrison's notebooks numbers 15 and 8. In January 1890 Prestwich wrote to Harrison to request he retain some eoliths from Ash that Harrison had sent him. He wished to show them to John Evans. In February two palaeoliths were discovered at West Yoke within a short space of time. One was number 534, the other 537 is illustrated in Figure 9.7. Both were deeply stained (Harrison, 1928, 153). Harrison sent 534 to Prestwich who, in turn, showed it and the eoliths to Evans at a dinner party. Evans wrote to Harrison on March 4$^{th}$ accepting 534 as an implement (Harrison, n.d.-g) but rejecting the Ash eoliths. Harrison informed his friend A.J. Montgomerie Bell (1845–1920) of the finds, and possibly sent one or both of the palaeoliths to him. Montgomerie Bell was a classical scholar who taught at Limpsfield, although his home was Oxford where he was a very active member of the Ashmolean Natural History Society of Oxfordshire (Nicholas, n.d.). He was also a graduate of

Oxford University. Today he is mostly known for his later archaeological work at the Lower Palaeolithic site of Wolvercote, Oxfordshire. He began teaching in Limpsfield in 1877 (Nicholas, *ibid*). Presumably his collecting in the Limpsfield area, and on the slopes above the town, dates from around this time. How he met Harrison and why he was so receptive to the concept of eolithic humans is not known. He had been a keen field walker for many years, and Harrison was visiting him and his artefact collection in the early 1880s (Harrison, 1928, 107). By 1891 Bell was convinced that the eoliths were genuine, and older than the plateau palaeoliths (Harrison, 1928).

Bell now took it upon himself to inform various members of the Oxford elite about the discovery, including E.B. Tylor (who was by now Reader in Anthropology at Oxford University), Arthur Evans of the Ashmolean Museum (John Evans's son), and H. Balfour who was an assistant at the museum (Harrison, n.d.-g). The importance of 534, and its relationship with the deposit in which it was found was impressed upon Tylor. Again in early March John Evans had another look at the flints, confirming his earlier opinion that 534 was an implement but the eoliths were not.

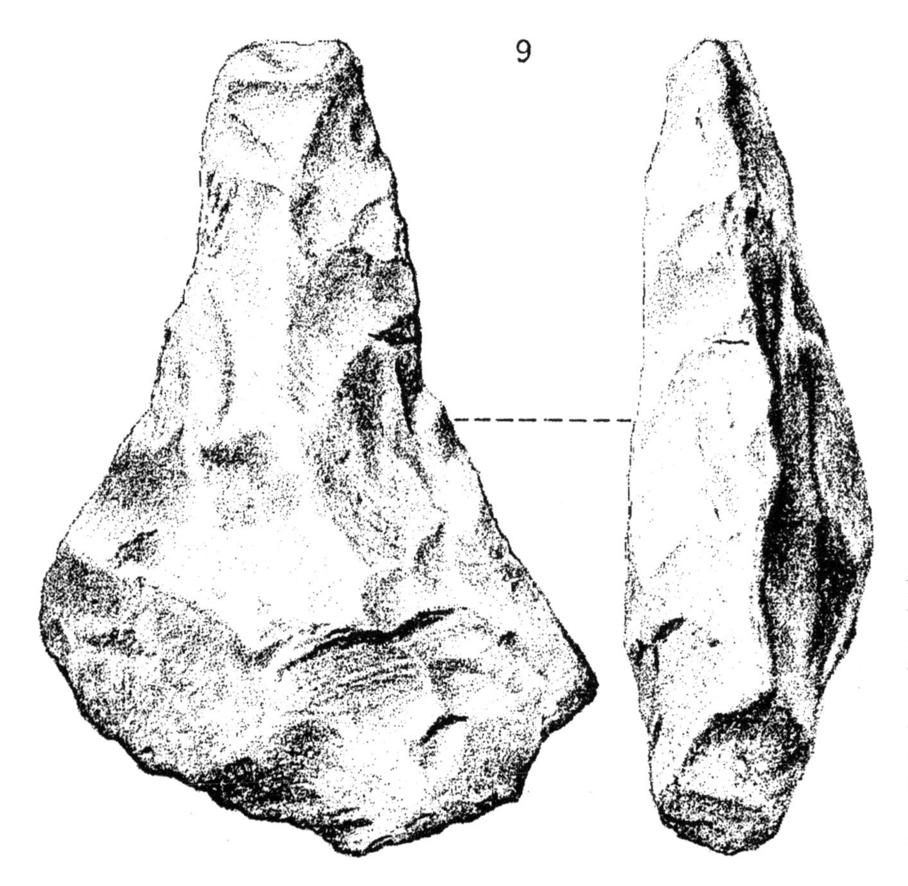


*Figure 9.7. Handaxe 537 found in February 1890 at West Yoke. This was one of two handaxes (the other being number 534) that were found in the vicinity and shown to Evans. Crucially he accepted their authenticity and their provenance.*

On the 12th of March, Montgomerie Bell wrote to Harrison having heard that Evans had accepted the authenticity of 534. It seems he had not been as successful with Tylor as he thought and wanted to hang on to 534 and 537 (not clearly stated but assumed by me to be so) in order to have a second go at persuading the great anthropologist as well as trying to persuade the geologist Professor Green, who was Prestwich's replacement at Oxford (Harrison, n.d.-g). Interestingly he refers to one of the artefacts as a 'regular St Acheulian' and the other as an 'Upper Chellean' – probably a reference to one of de Mortillet's typologies.

Bell admitted these handaxes were a difficulty to their eolith theory, presumably implying that accepted later forms of Lower Palaeolithic handaxe should not really be found on the Plateau. It seems that Tylor remained sceptical, though Balfour and Green were won over (Harrison, n.d.-c). Tylor did write to Harrison to request a selection of eoliths for the Oxford Museum and Bell urged Harrison to accept – but make him pay (Harrison, n.d.-g)! The peripatetic Bell was on the case again in Oxford in November, still trying to win over the museum and its anthropologists. He also button-holed Edwin Ray Lankester (1847–1929; Harrison, n.d.-c).

This rather long example is instructive (although reconstructing the exact series of events from Harrison's notebook is difficult) because it shows that genuine palaeoliths/handaxes were being recovered in a slow but steady trickle from the Plateau. What is more they were, as Evans had implied in the discussion to the first paper, ordinary palaeoliths, a fact the eophiles seemed to accept amongst themselves. Secondly, this is an excellent example of the web of interconnections mentioned earlier. Harrison was writing to a variety of people, in turn receiving letters and forwarding others, accompanied by specimens, designed to convince the recipients of the veracity of the eoliths. His allies, like Montgomerie Bell, were taking the fight to the front line and confronting sceptics with specimens they hoped would win them over. Then they reported back to base. There was a constant stream of information exchange comparable to the modern internet and e-mail. This served to heighten awareness of various issues central to the eolith debate as letters and specimens were posted in order to address specific problems, as well as reinforce accepted points. So, away from the society meeting rooms, there was a concerted if uncoordinated effort to keep the whole issue firmly on the radar.

The second paper was read before the Geological Society on 21st January 1891 and published in the *Quarterly Journal of the Geological Society of London* for the same year (Prestwich, 1891). Whereas the first paper concentrated on a broad strip of land from the Plateau down into the Vale of Holmesdale and southwards into the Shode basin (a north-south orientation), this paper moved the focus westwards (an east-west orientation). Prestwich now focused on the formation and drift geology of the Darent Valley and the Plateau above and to the west of the gap at Shoreham through which the

Darent flowed, see Figure 7.5. The geological story is summarised in figures 7.2–7.4, and described in Chapter 7. He noted three prominent sedimentary deposits on the Plateau. The first was the reddish clay-with-flints which was of local origin, a result of sediments and chalk decomposing on the surface *in situ*. It was laterally extensive and it surrounded patches of Tertiary drift (the second drift identified on the Plateau) which were older in date (often as sandy patches). Third was the Southern Drift. This was associated with the clay-with-flints but younger, presumably lying in patches on the surface of it. It is implied, though never clearly stated as such, that the implements and eoliths were associated with the ochreous and worn flints of the Southern Drift. As ever Prestwich was cautious, leaving a door open for a potentially earlier date for the implements by associating them with pre-Southern Drift deposits.

> In several cases it is noticeable that the implements occur on or near small Tertiary outliers, as though they might have preceded the Red Clay with flints. (Prestwich, 1891, 133)

The cutting of the valleys, including the nascent Darent Gap, through the Escarpment (Figures 7.4 and 7.5) and the Vale of Holmesdale and thus the Chalk Escarpment itself, were either late pre-glacial or very early glacial (*sensu* Prestwich 1887). The Southern Drift pre-dated the arrival of the glaciers as the drift on the Plateau contained elements from the Lower Greensand and Oldbury cherts which could only have got to their present positions through fluvial transport as rivers flowed across a *continuous* gently sloping plain from the Wealden range southwards, across the as yet uneroded Vale of Holmesdale and on to the Thames Valley to the north. The Escarpment was cut, and thus the Plateau formed, by glaciers moving across the area. This was the interpretative framework Prestwich had presented to the Geological Society the year before (Prestwich, 1890b; Prestwich, 1890a; Prestwich, 1890c) which is described in Chapter 7. In terms of the chronology of the artefacts on the Plateau they had to be either broadly contemporary with, or pre-date a late pre-glacial or very early glacial age bracket.

Importantly, Prestwich was still not passing outside the frame of chronological reference set up by the 1887 and 1889 papers, as late pre-glacial and early glacial still did not fall outside of the Pleistocene period. The erosion of the Wealden heights and the initial incision of the valleys, including the Darent, were all broadly contemporary. However, the Wealden erosion had to begin somewhat earlier for the rivers to carry the eroded debris southwards and deposit it on the plain that would one day become the Chalk Plateau.

The Darent was clearly a post-glacial river in that it flowed at the base of the Chalk Escarpment. The topography in this western area was rather different from the Shode basin and adjacent land to the north. The hilly landscape with its drift-capped heights was only present in the south-east corner of this western area as covered by Prestwich in 1891, although possible contemporary deposits in the Darent Valley may have been present (Figure 7.4). The low and high level valley drifts (3 and 4 on Figure 7.4) were present in isolated patches along the Darent's course and these contained the usual mixture of rolled and unrolled palaeoliths. A few eoliths were present, but they were clearly derived.

Prestwich described a new and higher terrace. It was present on the watershed between the Darent and the eastward flowing Oxted Stream at Limpsfield Common. This Limpsfield gravel (3a on Figure 7.4) could be traced as isolated drift patches along the course of the Darent, through the Shoreham Gap and into the Thames Valley where its equivalent downstream deposits were present at Dartford Heath and at Milton Street, Swanscombe. In 1877 Prestwich had asserted there were no implements at Swanscombe but by 1891, this was not the case. This Limpsfield terrace was clearly included within the high-level valley drifts but it was the oldest of them. In this gravel at Limpsfield Common, Montgomerie Bell had found a number of palaeoliths similar to those of the hill-group from nearer to Ightham (and another eolith hunter, E. Lewis, had found a 'derived' eolith as well). Elsewhere implements were scarcer, though Harrison found a broken implement and an eolith in an equivalent gravel deposit in the eastern tributary of the Darent at West Yaldham/St Clere. Prestwich suggested the eolith was probably derived from the Plateau.

Outside the Darent Valley, a number of drift units existed that were not associated with the river. There were some high level gravels (5 on Figure 7.4) that may have been associated with the cutting of the Escarpment itself and may possibly have been contemporary with the hill group around Ightham. Prestwich believed that there was a temporary return to glacial conditions after the glaciers of the main glacial phase had retreated. He believed he found evidence of this in the Darent (2 and 6 on Figure 7.4). There were brickearth deposits in the region of Limpsfield which showed deformation structures as if a body of ice and frozen snow had passed over them. The same was true of an extensive spread of isolated gravel patches he termed the Chevening and Dunton gravels. They occurred along the northern bank of the Darent, positioned between the Escarpment and the river, and they looked like a series of eroded terrace fragments. A number of these gravel units seemed to have been pushed into the underlying sediments, with the soft bedrock deformed by the passage of ice over the top. Positioning them in the relative chrono-stratigraphic sequence of the valley's development was difficult; Prestwich was only confident that they post-dated the Limpsfield gravels.

This second paper was primarily on the geology of the Darent Valley but Prestwich went into detail on the artefacts to answer critics of the first paper. He noted once more that the staining and appearance of the artefacts made it clear that although they were found on the surface of the drift, they had in fact originated from within it. Benjamin Harrison was beginning to find artefacts *in situ* within the drift outliers. Tackling Evans' arguments, Prestwich said he did not think it appropriate to judge a whole body of artefacts by reference to the very few that appeared to be better made than the overwhelming majority which were more poorly made. That just implied there were a few gifted artisans. The problem addressed here was the finding of unambiguous handaxes in the drifts on the Plateau. In the 1889 paper he had illustrated a number of 'proto-handaxe' like implements (see Figure 9.4, and Prestwich, 1899 Plate 11). Since then, however, much better examples (e.g. the handaxes noted above – Harrison find numbers 534 and 537, Figure 9.7) had been found. Evans was aware of these. They clearly confused the issue (see A.M. Bell's 10/8/1894 letter to Harrison quoted in Chapter 10). He reiterated that no eoliths had been found in the valley gravels that were contemporary with those gravels and that the few that were found were evidently derived. Prestwich had instructed Harrison to diligently search a number of gravel beds in river drifts, including Aylesford and Milton Street, Swanscombe, to address this very issue.

It will be recalled that by 1891 Prestwich was a convert to the eoliths, believing not only in the artificial character of the retouch but also in their existence as a tool-group. At this time he produced the first attempt at a typology of eoliths, with clear groupings of specific forms (types) within the group as a whole, see Figure 9.8. The repetition of different forms strengthened the interpretation that the retouch could not be natural. Nature could not reproduce a variety of consistent shapes made by retouch (a repetition of the 1859 methodology). Emphasising the importance of type recognition, Prestwich asserted that some of the eolith forms were at the root of much later tool types indicating a clear evolutionary trajectory from these early tools. For example, number 4 in Figure 9.8 developed into the pointed handaxes of the valley drifts. This was exactly the kind of thing you would expect from an industry at this very early stage in human development. It is perhaps no accident that Prestwich chose not to illustrate any of the genuine handaxes from the Plateau, nor any of the ruder implement forms that were prominent in the 1899 paper. Only eoliths were presented.

The second of the two meetings at the Geological Society appears to have been a rather muted one. Harrison does not make much of it in his notes despite being present and displaying a large number of specimens. John Evans was away, holidaying in Sicily (Prestwich, 1899). The objections raised in discussion were primarily geological, as was to be expected. The real battle was to come when Prestwich and Harrison were to lay their evidence before the AI.

The question of the age of the 'old olds' was of paramount importance. However, the eophile camp was divided. A.M. Bell wrote to Prestwich in May 1891 (Harrison, 1928), and to Harrison not long afterwards on this subject. Bell believed, thinking Harrison was of a similar mind, that the eoliths were

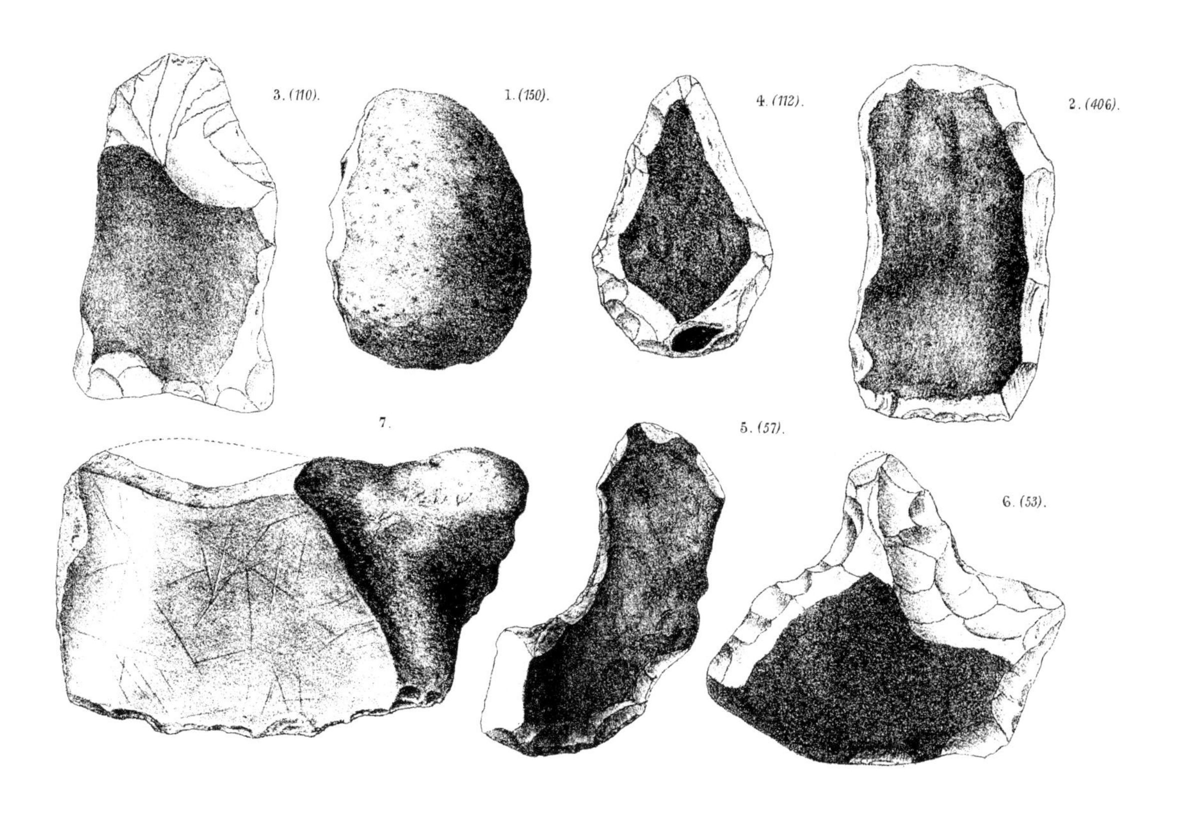


*Figure 9.8. Eoliths from the Chalk Plateau as illustrated by Prestwich in his second paper to the Geological Society delivered in 1891.*

much older than the palaeoliths (of any type) from the Plateau. He thought there might have been geological evidence to show that the drift spreads at South Ash and West Yoke (major eolith localities) were older than the Ash spread which contained palaeoliths. He noted that Prestwich had asserted that all the drifts were pre-glacial. In my opinion Bell was assuming Prestwich meant pre-Pleistocene; which is unlikely. Prestwich's reply on the 27th May 1891 was ambiguous. There were no geological grounds for distinguishing between different spreads of drift on the Plateau, and geology had to take the primary position if such an interpretation were to be suggested; artefact 'form and shape alone are not sufficient'. Nevertheless, he admitted he found it difficult to believe that the eoliths and palaeoliths from the Plateau were made by the 'same race of men'. The two types of artefact were so different.

Harrison's reply to Bell explained that he too had suggested to Prestwich that the eoliths were older and that Prestwich had agreed that they were while continuing to believe they were still Palaeolithic, but they were just the first stage in the evolutionary history of Palaeolithic tools. Edward Harrison implies that his father was already convinced of the greater antiquity of the eoliths in 1889 (Harrison, 1928, 145). But just how old were they? The notebooks and letters are silent on this point. In notebook 15, there are a number of entries discussing the Thenay flints (see Table 6.2) discovered by the Abbé Bourgeois. These were considered of Miocene age by some researchers. Harrison copied either an article or letter from Samuel Laing (E. Harrison, n.d.) in which the human characteristics of the Thenay artefacts were confirmed by ethnographic comparison with the tools of the Andaman Islanders. Another undated extract copied by Harrison into notebook 15 on the subject of the Thenay flints, this time by J.A. Gaudry (1827–1908) a French palaeontologist and geologist at Paris, suggests the flints were made by the ape-like creature *Dryopithecus*. Whether Harrison believed the 'old olds' were really this age is not known. A.M. Bell certainly did.

The reading of the paper at the Anthropological Institute of Great Britain and Ireland was on the 23rd June 1891, although it was not published until February the following year (Prestwich, 1892). The meeting was a full one. Prestwich was second on the agenda after a talk from H. Balfour of Oxford on an exhibition of Tasmanian implements (see last chapter). E.B. Tylor (Balfour's boss), president of the AI for that year, was in the chair. Harrison and De Barri Crawshay exhibited some three to four hundred specimens to support their case. Both Harrison and Crawshay stood to speak after the main paper was delivered (Crawshay, 1892; Harrison, 1892). Flaxman Spurrell wrote to Harrison a few days later to compliment him on the exhibited pieces which were much more convincing than those he had seen on previous occasions (Harrison, n.d.-h).

Prestwich went over the old ground once again to ensure his audience understood the significance of the artefacts and their geological context. In particular he reiterated an argument developed in the Darent paper, namely that eoliths and cruder implements did not occur in the valley drifts unless derived. Recent investigations of well-known, valley drift, implement-rich sites like Aylesford and Milton Street at Swanscombe (by now a well-known collecting locality), showed that their absence was not a case of eoliths having been missed by collectors in the days before they were recognised as tools. They were genuinely not there. Similar searches in the gravels of the Shode and elsewhere by Harrison and the other eolith hunters had confirmed this. Picking up on the theme of his correspondence with Bell earlier in May, Prestwich had this to say:

> But the special question I wish to bring before you this evening, is whether, taken as a whole, the plateau implements exhibit distinct characters and types, such as would denote them to be the work of a more primitive and ruder race than those fabricated by palaeolithic man of the valley-drift times. (Prestwich, 1892, 254)

The seeds were beginning to take root. In a footnote to the Darent paper (p134) Prestwich had speculated that the 'old olds' (and by implication the cruder worn palaeoliths as well) might have been washed down from the old Wealden heights by the rivers draining the slopes before the glaciation set

*Table 9.3. Joseph Prestwich's (left hand side of the table) and Benjamin Harrison's (right hand side of the table) contrasting typologies of eoliths. The column marked '1892 plate number' represents Prestwich's illustrations. The numbers refer to individual specimens from Figures 9.9, 9.10, and 9.11. The numbering on the figures is the same as Prestwich's originals – see text. Entries on either side of the figure number therefore show the two contrasting interpretations of an eolith's type for which that illustrated example is representative.*

in. He reasserted this to the AI, implying that if this was the case then the tools would have to be 'relegated to a still earlier period than I have named'. I suggest this was a hint at a Pliocene date. Not surprisingly he took this no further.

Prestwich flagged up the continuing difficulties with interpreting the eoliths; 1) some of the eoliths had retouch lighter in colour than the rest of the tool, 2) there were more and more discoveries of unworn and lighter coloured handaxes from the Plateau which were consistently found with the eoliths on the surface, 3) there were still relatively few eoliths found *in situ* within the gravels, 4) the retouch could sometimes be difficult to distinguish from natural chipping. He admitted all of these points were difficult to explain. Cleverly, he played to his anthropological audience on the last point. Many tools made by Australian Aboriginal people showed the same limited degree of retouch. These were also difficult to distinguish as human. If such obviously humanly made tools were not easy to identify, it was not fair to dismiss eoliths on similar criteria.

Prestwich's paper was shorter than usual, giving Harrison and Crawshay a chance to present their material. In the Darent paper Prestwich had presented

| Joseph Prestwich's categorisation of the forms of implement found on the Chalk Plateau as presented to the Anthropological Society in June 1891 (Prestwich 1892.) | | | | Ben Harrison's categorisation of eoliths from the Chalk Plateau given at same meeting |
|---|---|---|---|---|
| Group | Description | % | 1892 Plate No. | Description (B.H's categories illustrated by same figures as Prestwich) |
| 1. Natural flint blanks. Shape of the stone determines form. Little purposeful shaping by retouch. These are described as 'implements' on page 260 | a) Thin flat pieces or natural flakes, no shape, sides retouched, sometimes into concavities or points | 40% probably an under-estimate | 19.1<br>19.2<br>19.3<br>19.4 | Single curve scraper<br>Combination tool (a point + concavity + an end scraper)<br>Semi-circular tool<br>Tool with retouch all around |
| | b) Split half of tertiary pebble, marginal retouch | | 19.5<br>19.6 | Split pebble group with retouch on one face, found everywhere on plateau (19.5 and 19.6) |
| | c) Larger stones, some shaping by retouch, used mainly in percussion activities like hammering or trimming bone or wood; also could be used to shape other stone tools | | 19.7<br>19.8<br>19.9 | |
| 2. Natural stone still used as blank. However retouch shows more design, intent to shape tools for particular tasks. More evidence of attempts to reproduce specific forms or types | a) Scraper - ordinary on flattish blank or natural flake | 54% | 20.1 | |
| | b) Scraper – two types, knob headed and shoe shaped. Plano-convex split pebble, cortical, working mostly at one end | | 20.2<br>20.3 | |
| | c) Scraper – massive, thick plano-convex, retouched on one or two edges | | 20.3 | |
| | d) Scraper – square or chisel shaped ends | | | |
| | e) Scraper – crescent shaped (drawshave), concavity formed by retouch, very common eolithic form | | 20.4<br>20.5 | Drawshave or hollow scraper – most common form of all (20.4 and 20.5), found everywhere on plateau |
| | f) Scraper – double, like e) but two adjacent concavities formed by retouch, point in middle where concavities meet. Two forms – ordinary has elongated point (20.7, 20.8), 'depressed' has short point (20.6, 20.9).<br>Very common eolithic form | | 20.6<br>20.7<br>20.8<br>20.9 | Double curve scraper, common, many much worn down |
| | g) Scraper – double in form of hour glass, infrequent | | | |
| | h) Scraper – beak shaped '*implements*', projecting pointed corner (beak) formed by meeting of natural/retouched concavity with a retouched edge. Common eolithic form | | 20.10<br>20.11 | Crook point tool, found everywhere on plateau |
| | i) Crook shaped '*implements*', elongated concavity | | 20.12 | |
| 3. Flaking of surfaces more extensive, making definite types. Often both surfaces wholly flaked. Common in valley drifts, but rare on Plateau | a) Percussion flake, unretouched, used for cutting | 6% | 21.1 | |
| | b) Percussion flake, wide and marginally retouched | | 21.2 | |
| | c) Ovate handaxes, common in valley group, but twisted forms frequent in valley drifts (and Shode Valley) are absent from plateau group | | 21.4<br>21.5<br>21.6 | |
| | d) Pointed handaxes, smaller by comparison with those in valley group | | 21.8<br>21.9 | |
| | e) Pointed handaxes with point slightly curved | | 21.7 | |
| 4. Infrequent types or single forms | Circular trimmed 'fling-stones'<br>Rod shaped and short<br>Crook shaped<br>Drill shaped<br>Triangular point | | <br><br>20.12<br>21.12<br>21.3 | (Harrison also described the presence of 'rude choppers'<br>found in abundance but defying classification because of variety of form. Not illustrated) |

(Figures 21.10 and 21.11 were illustrated, but were not listed in Prestwich or Harrison's text)

a very simple typology of the eoliths (only six types were described) in order to demonstrate to his geological audience that there were regularly reoccurring shapes. This was now expanded greatly for his archaeological and anthropological audience. His types are given in Table 9.3, and illustrated in Figures 9.9 to 9.11. The figure numbers in the illustrations refer to the original numbering for ease of recognition and comparison with Table 9.3. They were divided into three broad groups on the basis of how invasive the retouching and shaping was and how much it imposed a clear form on the tool. Whereas Prestwich's groups were based on form and shape, Harrison's typology was, Prestwich claimed, more influenced by assumptions of use. My reading of the two typologies does not support such a distinction. Harrison's descriptions are no more assumptive than Prestwich's were – Table 9.3.

The artefacts were, as previously, a mixture of natural and genuine ones. It has not been possible to relocate all of them, but for those which have been identified we may conclude the following. Prestwich's specimens 19.2, 19.3, 19.4, 19.6, 19.8, 19.9, 20.1, 20.3, 20.5, 20.11 and 20.12, are all natural. In some cases (such as 20.1) they would appear from the illustrations to be genuine tools but the illustrator has exaggerated apparent flake scars. What unites them is a wholly fortuitous shape that appears to have been deliberately accentuated by purposeful retouch – sometimes steep.

Figure 9.11 contains a mixture of real artefacts and natural pieces, but again the mix is not always clear cut. Specimens 21.6 and 21.9 were not rediscovered

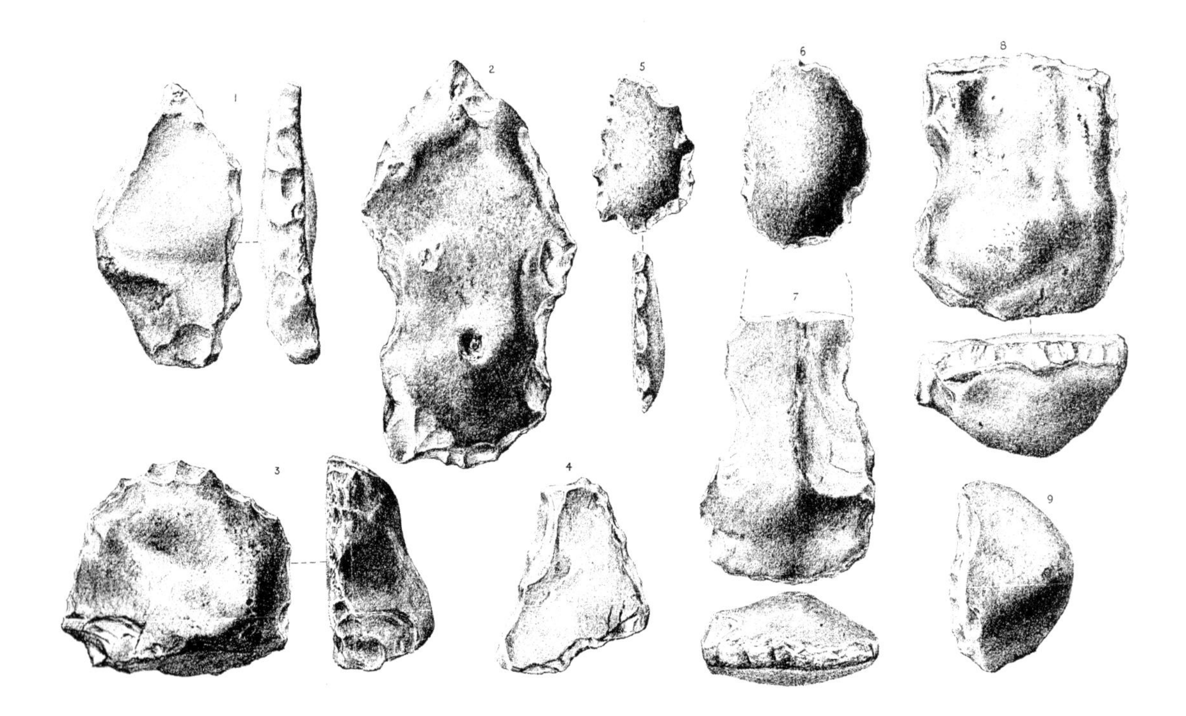


*Figure 9.9. Artefacts from the Chalk Plateau as illustrated by Joseph Prestwich from the* Journal of the Anthropological Society of Great Britain and Ireland *for 1892. This was Prestwich's plate 19, so the artefact numbers here refer to pieces 19.1–19.9.*

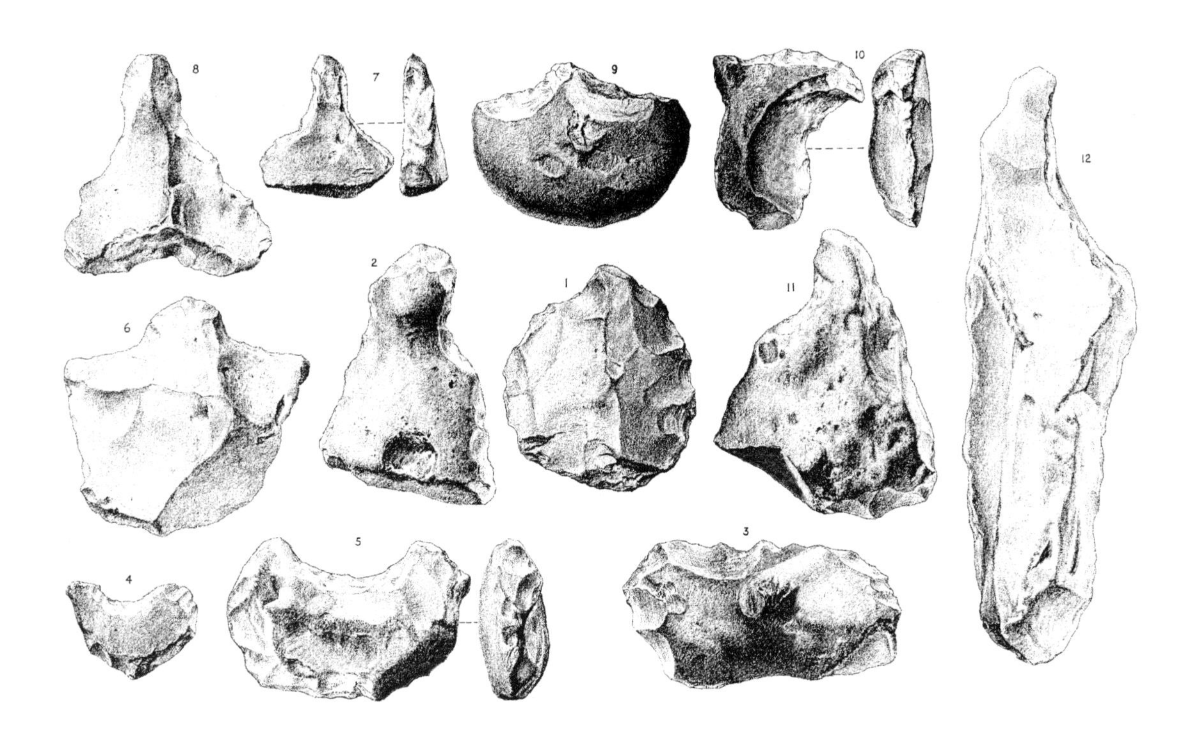


*Figure 9.10. Artefacts from the Chalk Plateau as illustrated by Joseph Prestwich from the* Journal of the Anthropological Society of Great Britain and Ireland *for 1892. This was Prestwich's Plate 20, so the artefact numbers here refer to pieces 20.1–20.12.*

but their illustrations leave little doubt they are genuine palaeoliths. Specimen 21.9 is number 537, the palaeolith from West Yoke on the Plateau that Evans accepted as genuine (see above). Additionally, 21.7 appears to be a small iron stained hand axe and 21.5 may well be a small discoidal core, also iron stained. Whereas the former two handaxes are liable to be Lower Palaeolithic, the small handaxe and core are more difficult to interpret; it is possible they are Middle Palaeolithic/Mousterian. 21.12 is a struck but damaged flake. Specimens 21.4 and 21.8 are natural. The former, again, has had its surface features exaggerated by the illustrator until it appears to resemble a flake's dorsal surface. The illustrator, W.S. Tomkin was a nephew of Harrison. To anyone unfamiliar with the actual pieces, these illustrations would have lent the tools real credibility.

Prestwich concluded his talk to the AI as follows:

> '...[the eoliths and plateau implements]...constitute characters so essentially different from those which typify the latter implements [i.e. valley drift] that by those characters alone they might be attributed to a more primitive race of men; and as this view accords with the geological evidence which shows that the drift beds on the chalk plateau, with which the implements are associated, are older than the valley drifts, I do not see how we are to avoid the conclusion that not only was the plateau race not

> contemporary with the valley men, but also that the former belonged to a period considerably anterior to the latter – either an early glacial or a pre-glacial period.' (Prestwich, 1892, 261, my brackets)

Finally, the old campaigner had nailed his colours to the mast. The Plateau material was fashioned by an older race than the post-glacial valley drift, though he did not (in my opinion) alter his belief that humanity was confined to the Pleistocene. Neither Harrison nor Crawshay ventured this far at the meeting. Their contributions were straight forward artefact presentations. Harrison gave some background on his discoveries and then presented his own typological divisions, which are in the right hand column of Table 9.3. Prestwich was evidently a splitter and Harrison a lumper, and the most obvious difference between the two is the third group of Prestwich's which Harrison does not engage with at all; a clear reflection of his belief in the distinct age difference between eoliths and plateau palaeoliths.

The discussion following the paper was a lively one and Prestwich's biographer notes that he was somewhat taken aback by the strength of the negative reaction to the paper (G. Prestwich, 1899). John Evans and Boyd Dawkins were sceptical, the latter particularly so. Given the tone of comment and response between Prestwich and Boyd Dawkins there may well have been some personal friction here. Angela Muthana's transcription of volume 8 of Harrison's notebooks includes this comment on the evening's proceedings.

*Figure 9.11. Artefacts from the Chalk Plateau as illustrated by Joseph Prestwich from the* Journal of the Anthropological Society of Great Britain and Ireland *for 1892. This was Prestwich's Plate 21, so the artefact numbers here refer to pieces 21.1–21.12.*

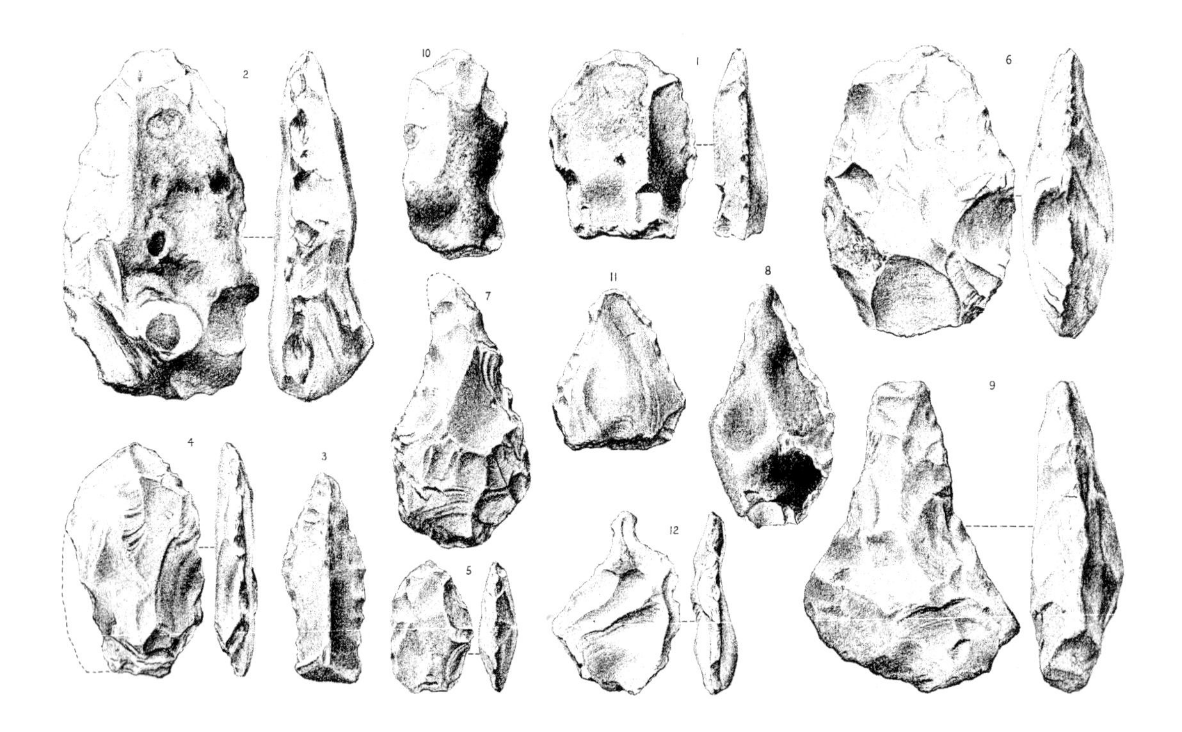

> In red ink … Sep 25 1891 from Mr Crawshay. The remark that Pr Prestwich whispered to me when Sir John Evans and Pr Boyd Dawkins spoke and immediately left the room at the Anthropological meeting of June 1891 'Their objections are puerile and they don't even wait to have them replied to'. (Harrison, n.d c., Notebook 8, page 18 and following, A. Muthana transcription)

Harrison was rather put out by Evan's behaviour at the meeting (Harrison, n.d.-a), and the manner of his rather abrupt departure. Evans had been very dismissive of Harrison's exhibited specimens before the meeting began.

> 'His behaviour and his Esquire… [Professor Boyd Dawkins]…was certainly not in good form as on my getting up to read my paper out had only read two paragraphs my coat tail was pulled and I was informed Dr Evans and Pr Boyd Dawkins wish to speak as they go by train. An eminent geologist waved to me to say he did no approve of this behaviour.' (Harrison, n.d. a, Autobiography volume 2. Transcribed by A. Muthana, transcription direct from notebook, my brackets)

The responses of Evans and Boyd Dawkins are presented in Table 9.2. Evans' comments were significant in that he admitted in public the genuineness of the palaeoliths from the plateau. The background work of Prestwich and Harrison in trying to convince Evans was beginning to pay off, although Evans saw no reason to argue they were the product of an older race than the valley drift knappers. John Allen Brown made a significant point. Supportive of Prestwich and Harrison's position, he stated that he too was beginning to find plateau type implements (handaxes at least) on the high hills to the north of the Thames (Figure 7.7). He had been unable to explain them until now, and postulated the existence of a long eroded higher land surface, comparable with the Kentish Plateau, only to the north of the river in the softer and easily eroded Tertiary deposits found there.

The search for eoliths outside their core area had begun. This will be explored further in the next chapter.

## 1892–1896: In the Storm – the Battle for the High Ground

In summarising the issues raised by the three papers (Table 9.2), the key archaeological objections to the eoliths were as follows:

- The lack of evidence for eoliths being *in situ* within the gravels
- The presence of well-made valley drift-like palaeoliths on the Plateau
- The character of the retouch on the eoliths
- The reality of specific types
- Their supposed presence in valley drifts and cave deposits as unrecognised background noise

- The eoliths represented one end of a natural continuum of variability found everywhere and at all times.

To which list, the following geological concerns could also be added:

- The ambiguous evidence for the glacial formation of the Chalk Escarpment
- The total lack of any other evidence for the presence of glaciers south of the Thames Valley

The years between 1893 and 1896 are largely covered by the 13$^{th}$ volume of Harrison's notebooks (Harrison, n.d.-e). It is a clear illustration of the level of interest that Harrison and Prestwich's discoveries generated. Harrison received a constant stream of visitors, both amateur and professional; he was guiding tours, as well as lecturing to visiting groups in his garden, contrary to his supposed dislike of such engagements. They ranged from the Geologists' Association and the London Geological Field Class, to the National Society of Cyclists. But as much as anything, and like all the other notebooks, volume 13 shows the extent of the correspondence web within which Harrison was enmeshed, and how powerful a tool it was for maintaining heightened awareness of the eoliths in both scientific and public circles.

It is evident from the interchange of letters between the eophiles from 1889 to 1892, that John Evans was an important target in their campaign. As Britain's most senior Palaeolithic archaeologist, his opinion carried weight and had the power to influence people and institutions. As suggested in Chapter 6, the Lubbock circle had gradually evolved into an Evans circle, and he epitomised the establishment viewpoint. Volume 12 (Harrison, n.d.-d) of Harrison's notebooks details an important incident in the campaign to convert Sir John (he was knighted that year; as usual it is difficult to reconstruct the exact sequence of events from Harrison's notes). On the 13$^{th}$ October Harrison went to Darent Hulme to see Evans, who was visiting Prestwich. An opportunity was taken to lay out eoliths for Evan's examination. Typically, Evans was unconvinced. Sometime later, Prestwich sent Evans some eoliths (possibly the same ones) and informed Harrison that he had done so, chiding Harrison that he had not brought over the most convincing examples when he came over. In a panic Harrison wrote a long and rambling letter to Evans at 4.30 in the morning of the 26th. He wished to let the evidence (eoliths) speak for itself, and not present any particular theory which would establish a case for accepting the eoliths as genuine. The draft of the letter seems to detail a number of instances when Harrison had met with Evans and had wished to broach his feelings on the eoliths and describe them in detail to Evans, but his diffidence, and the presence of others, had prevented him from speaking up. The wording clearly reflects a man all too conscious of his social status and a sense of inferiority in the presence of Evans. The following day, the 27$^{th}$ of October, Harrison dispatched a large sample of eoliths to

Evans (Harrison, 1928, p183f; Harrison, n.d.-d, p42f). Evans replied on the 29th (Harrison, 1928) reiterating many of the points that he had raised at the AI. He specifically addressed Prestwich's theory of a different race of early humans making the Plateau artefacts, asserting he saw no need for this. In effect, he stated that he could not conceive of north-western Europe as being the home of a race of humans at such an early date. He did however reassert his belief that there were genuine early implements (i.e. handaxes) on the North Downs. He even implied that some of the cruder implements (i.e. the proto-handaxe like examples from the 1899 paper) might have been occasional tools or quickly made field-tools, but these were exceptions and the majority he maintained were natural. He insisted the eoliths were all natural.

Edward Harrison (Harrison, 1928, 185) suggests that there was correspondence between Prestwich and Evans concerning Harrison. Possibly Evans was complaining about the deluge of letters and specimens he was subjected to. Harrison had replied to Evans' letter of the 29th October in a rather tongue in cheek manner; and perhaps Evans felt it was a little disrespectful. Harrison opined; 'Perhaps some day you may not be jealous of the elevation of distinguished commoners to sit in the upper house.' (Harrison, n.d.-d, 49). He was referring to Sir John including eoliths from Harrisons' museum (which he likened to the House of Commons), in Evans' own museum at Nash Mills (the House of Lords). Others too felt the pressure of the eophiles enthusiasm. Montgomerie Bell's attempts to convert the Oxford anthropologists has already been noted. Henry Lewis, an eophile from London and friend of Harrison's, took specimens to show Augustus W. Franks of the British Museum at Bloomsbury who was a senior member of the Evans circle in February 1890 (Harrison, n.d.-g, 63). He received short shrift. Lewis was interested in scratch marks on flints, taking them to be glacial striations, and so evidence of glacial age. Franks looked at the eolith in question and immediately pointed out that the striations did not continue onto the edge chipping. Undeterred, Lewis showed him another eolith pointing out the retouch on the side.

> 'I can see no chippings at all'...[said Franks]... 'All along the side there Sir and slightly around the point.' [replied Lewis]. He...[Franks]...carefully laid down the specimen and said 'Dear me is there now, well I am a bad judge of chipping, thank you very much for showing them to me, good morning'. (Harrison, n.d. g, Notebook 15, p63, brackets and quotation marks are mine, A. Muthana transcription).

Presciently, Lewis finished his explanatory letter to Harrison with the observation that it was Evans who had to be convinced. Benjamin Harrison himself was cold shouldered by Franks when he visited him some years later (E. Harrison, n.d.). Harrison was also busy trying to recruit allies abroad. He wrote to the Gabriel de Mortillet who had first coined the term 'eolith'.

On receipt of an encouraging letter from the great man, Harrison wrote to Worthington Smith, who replied on 7th December 1892 in his usual caustic but nonetheless friendly manner:

> Am truly sorry that your ideas agree with that of Mortillet, am also truly sorry that you fancy you can see anything in the Thenay things either from the illus[trations] or the descriptions. You see, I do not believe in Tertiary Man but I expect you will be going in for Cretaceous man soon; with a race of lithocephalic or some other hideous nightmare. (Harrison, n.d. d., Notebook 12, p64–66, A. Muthana transcription)

By this time Professor Gaudry, who had once suggested that the Miocene man-like ape *Dryopithecus* might be the maker of the Thenay flints, had gone back on his original suggestion (R.L., 1890).

While Worthington Smith was another target the eophiles had in their sights, so was Flaxman Spurrell. Their refitting work at their respective sites had earned them solid reputations in the anthropological and archaeological communities. Spurrell was undecided, while Smith remained a fervent eophobe. He made his point clear (Harrison, 1928, 175f) to Harrison in letters in March and April 1892. If eoliths were indeed tools, and they occurred with obvious implements, then it was more likely they dated to the time of the implements. Only if Harrison found the eoliths separately would the case for their greater antiquity be recognised.

This was a significant point and both Harrison and Prestwich were only too aware of this. There had been a handful of instances were eoliths or implements had been found on the plateau *in situ* but none under controlled conditions. In the spring of 1894 the farmer at Parsonage Farm near Ash, a Mr Pink, dug a soak-away pit. One of Harrison's scouts (a group of labourers and lads trained by him to recognise eoliths and implements) brought some eoliths down from the spoil. On visiting the site, by that time filled in, Harrison saw more evidence that intrigued him. The opportunity to excavate came later in the year. At the BAAS meeting in Oxford there had been a special session on the eoliths (discussed in the next chapter). Given the currency of the debate, the committee who had supervised Harrison's excavations of the Oldbury Rock Shelters and Mount Pleasant, awarded the remaining funds, £10, to the investigation of the Plateau drifts. Harrison's notes imply it was Evans who actually proposed this. Work began in mid-October and appears to have continued for a few months.

Between the rather uninformative report (Harrison, 1895), and Harrison's own observation at the time (Harrison, n.d.-e), the following can be reconstructed. Close to farmer Pink's original pit, a twelve foot by six foot pit was opened. Further digging was prevented by torrential rain, and a second similar pit opened nearby. The story was the same in both. The top two and a half feet was a flinty drift with white angular and rounded ochreous flints in a loamy matrix. There were broken palaeoliths, Neolithic implements (possibly many flakes of Neolithic origin) and a few eoliths. The deposit's character was similar to the Southern Drift (Oldbury chert, chert from the Greensand, and Ragstone). Below this was three and a half feet of loam with occasional eoliths, and then a foot or so of gravel. Within the gravel there were many eoliths but no sign of palaeoliths or of Neolithic material. Below that, twelve to thirteen feet of fine grained Tertiary deposits with no artefacts at all.

On face value Harrison should have been delighted. The eoliths were coming out at depth, and there were no palaeoliths at the same level. This seemed to vindicate his ideas that the eoliths were older than the cruder palaeoliths, as well as the finer palaeoliths found on the surface. The written sources, however, appear quite muted. Evans, characteristically was unimpressed by the results (Harrison, 1928, 198), suggesting the retouch on the specimens was more impressive in Harrison's report-sketches than in reality (something already noted for the illustrations of the published papers above). In fact at the BAAS meeting in Ipswich in 1895 the Parsonage Farm report was very nearly not read out at all. John Allen Brown lobbied for its inclusion on the programme after Flinders Petrie, the section president, suggested it should not be read out because Harrison was not present. (This may have been prompted more by considerations of time in a busy schedule than anything more sinister; that and the fact that most of the delegates at the BAAS meetings no longer found human origins and the Palaeolithic that exciting).

In the summer of 1895 the excavated eoliths were displayed at the Royal Society at a conversazione. Prestwich coached Harrison on which pieces to send, emphasising the importance of groups as well as both excavated and surface specimens. Sadly Harrison did not take his leader's advice and sent large examples which did not sell the message very effectively. Prestwich wrote afterwards to say that it was an opportunity missed. The programme was pasted into Harrison's notebooks, see Figure 9.12, and in pencil along the side was written 'smaller specimens would have looked less clumsy and would have been more impressive' in Harrison's hand, echoing Prestwich's letter (Harrison, 1928, 197f). He did however sketch an eolith over the programme.

In April 1893 Prestwich grumbled to Harrison that 'In the meantime neither the geological nor the anthropological questions seem to advance beyond where I left them 2 years ago' (Harrison, n.d.-e, 6). In fact the debate had reached a deadlock. Neither side gave ground. Prestwich and Harrison were unable to produce new evidence to further their own position. Since the conservative scientific establishment believed the eoliths were natural, no amount of carefully controlled provenance information from excavations made the slightest difference. On the 15th April 1894 Evans wrote to Harrison to comment on another parcel of flints Harrison had sent him. Some of the pieces reminded him of a natural flint he had previously sent to Harrison.

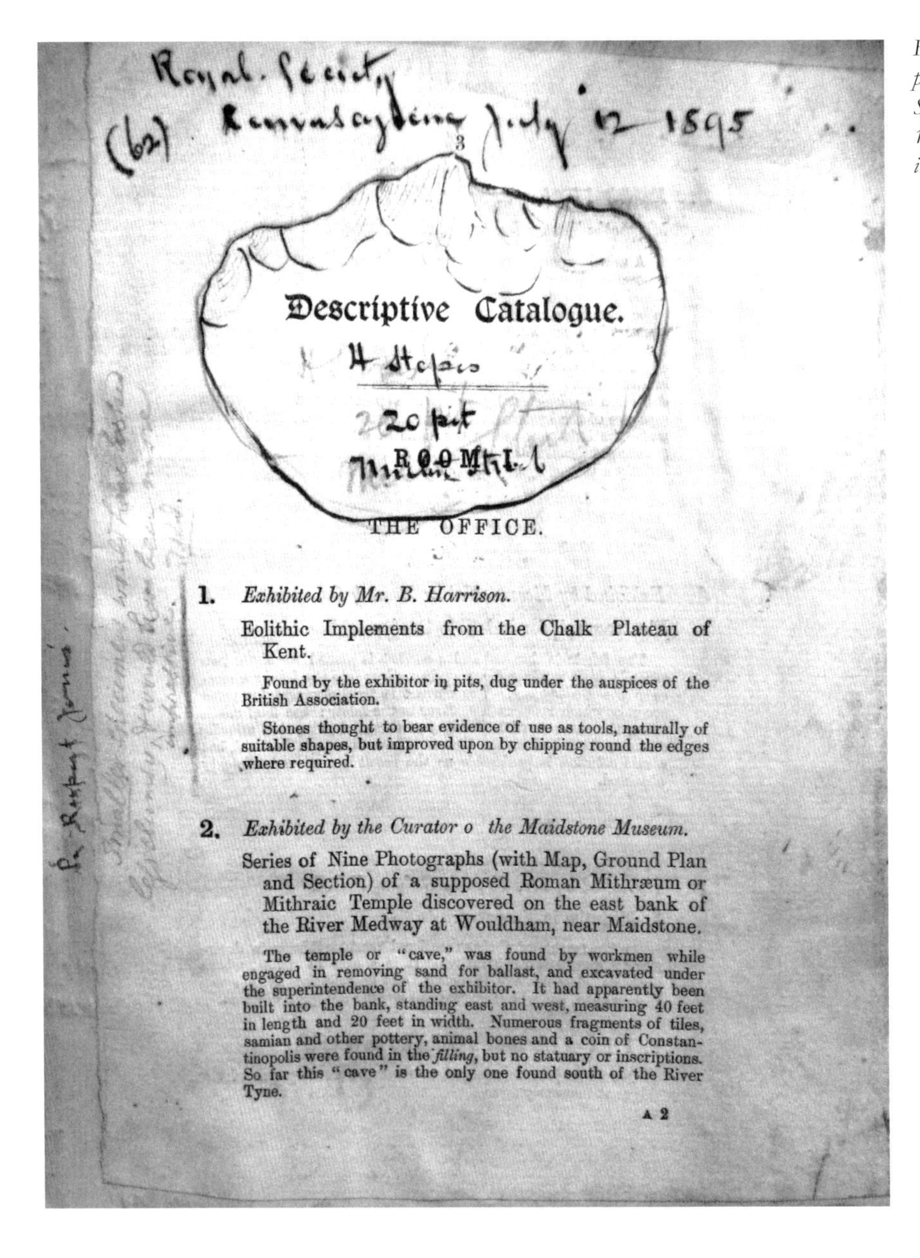
Royal. Society
(62) Conversazione July 12 1895
3
Descriptive Catalogue.
ROOM No. 1
THE OFFICE.

1. *Exhibited by Mr. B. Harrison.*

Eolithic Implements from the Chalk Plateau of Kent.

Found by the exhibitor in pits, dug under the auspices of the British Association.

Stones thought to bear evidence of use as tools, naturally of suitable shapes, but improved upon by chipping round the edges where required.

2. *Exhibited by the Curator o the Maidstone Museum.*

Series of Nine Photographs (with Map, Ground Plan and Section) of a supposed Roman Mithræum or Mithraic Temple discovered on the east bank of the River Medway at Wouldham, near Maidstone.

The temple or "cave," was found by workmen while engaged in removing sand for ballast, and excavated under the superintendence of the exhibitor. It had apparently been built into the bank, standing east and west, measuring 40 feet in length and 20 feet in width. Numerous fragments of tiles, samian and other pottery, animal bones and a coin of Constantinopolis were found in the *filling*, but no statuary or inscriptions. So far this "cave" is the only one found south of the River Tyne.

A 2

*Figure 9.12. The programme from the Royal Society conversazione in 1895. Harrison pasted it into his notebook.*

> I found it and several others of the same kind in the shingle on the coast at Aldeburgh in Suffolk by the sea, and I regard it as a typical example of the manner in which the sea or other water in violent action rolls a split pebble among the stones around it. It is therefore very instructive! (Harrison, n.d. e, Notebook 13, p60, A. Muthana transcription)

A year later the implications of these sea-shore stones was still troubling Harrison. Prestwich wrote in March 1895 (Prestwich, 1899) to reassure him.

The occasional naturally retouched piece was bound to resemble some eoliths. The key point was that eoliths could be grouped into clearly repetitive types. Nature might be able to simulate one type (in this case the naturally split Tertiary pebble), but not all of the types.

In fact, these sea-shore specimens represented the only real methodological development in the eolith argument in the years after the paper to the AI. Prestwich offered a free copy of his geology text book to anyone who could bring him a naturally fabricated eolith from a verifiable context, although he was concerned enough to write a short piece to the *Geological Magazine* (Prestwich, 1895a) in order to stave off the seaside challenge. The arguments are presented in Table 9.2, and the type of eolith that the sea-shore specimens resembled is shown in Figure 9.13. Curiously, this is not covered in *Harrison of Ightham,* but the subject clearly rattled the eophiles. However, a letter from Prestwich to Evans is reproduced in *Harrison of Ightham* (p227, actually reproduced from Grace Prestwich's *Life and Letters*) in which Prestwich gloats over the *Geological Magazine* letter, claiming it will force Evans to recant his eophobe ways. Unsurprisingly it didn't. The seriousness of the sea-shore threat is only too evident in other letters from Prestwich to Evans (G. Prestwich, 1899, 388 and 394). Evans later claimed that he never intended to imply marine action, merely the effect of turbulent water flow. This is likely given his reliance on Prestwich's 1859 explanations of the drift.

Prestwich wrote two more pieces on eoliths and human antiquity (Prestwich, 1895b; Prestwich, 1895c). Additionally in his *Collected Papers*

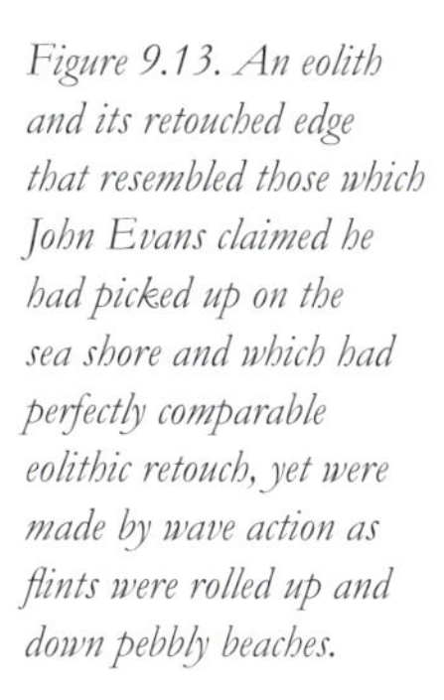

*Figure 9.13. An eolith and its retouched edge that resembled those which John Evans claimed he had picked up on the sea shore and which had perfectly comparable eolithic retouch, yet were made by wave action as flints were rolled up and down pebbly beaches.*

*Figure 9.14. Handaxes (Figure 9.14a top left and 9.14b top right) and a flake (Figure 9.14c bottom) from the Plateau as identified in Prstwich's 1895 volume* Collected Papers on Some Controverted Questions of Geology.

volume (Prestwich, 1895d), the AI lecture was updated and reprinted with a number of amendments and some new observations. Of more significance was the inclusion of new plates illustrating more examples of the eoliths and the crude palaeoliths that comprised the Plateau group. The handaxes in Figure 9.14 are important as they represent the last clear statement that Joseph Prestwich made on what Plateau implements/handaxes were. These are now curated by the Natural History Museum in London. Again they appear to be a mixture of worn and stained Lower Palaeolithic-like pieces such as Figure 9.14a, and small, potentially Mousterian handaxes in a similar state, 9.14b (the true nature of this example is uncertain). Interestingly Figure 9.14c is a genuine Palaeolithic flake, which was clear from its original illustration. Evidently Prestwich maintained his belief that the Plateau artefacts, while dating from an earlier phase of the Palaeolithic, remained unambiguously a part of the Palaeolithic stage.

Arguably the more important of the two new papers was the one for the influential popular magazine *Nineteenth Century* in April 1895. In effect it was a review of work done and arguments made, and as such did not advance the debate. However it succinctly summarised and then refuted each of the archaeological and geological criticisms levelled against the eoliths and their chrono-stratigraphic context. (A Pliocene date for the eoliths was ruled out because the Plateau drifts sat on top of Pliocene deposits which was also noted in the *Collected Papers* reprint – a reference to the results from the Parsonage Farm excavations). In choosing the *Nineteenth Century* the piece was clearly an exercise in heightening public awareness. Prestwich's 1859 credentials were liberally cited. So as far as the public were concerned, here was a hero of Victorian science doing what he did best – thinking outside the box and confounding the establishment; and he was doing it all over again. The simple format and clarity of style sold the message very effectively. Importantly, it brought into the wider public domain the concept of an earlier race of humans than those who had made the palaeoliths of the valley drifts – only now in a purely British context. In this it was quite emphatic. Prestwich was embedding the academic debate about eoliths within the broader context of human origins in the minds of the general public. He reminded his readers that what little evidence there was for the Palaeolithic race showed them to be similar to modern people. He speculated that the eolithic race might have been more arboreal because their eolithic tools were smaller and designed to be used in the hand, while the pointed and ovate palaeoliths seemed to be designed for mounting as spears possibly for use against the large Pleistocene mammals like the mammoth, bear, or rhino. The implication was of a simpler way of life for a less advanced human type.

## 1896–1901: Trying to Hold the High Ground

Sir Joseph Prestwich died on 23 June 1896, only a few months after being named in the New Year's Honours List. Edward Harrison notes that his father had lost his leader, and the years that followed were a time of consolidation rather than fresh discoveries. Benjamin Harrison's relationship with Prestwich had been a warm one. Edward suggests that Prestwich's character could be imperious at times, and he certainly could be impatient with the 'staff', but I sense a genuine affection on the part of Prestwich. When Edward Harrison achieved honours, Prestwich wrote to congratulate. On occasions when Harrison failed to take his leader's advice, to the detriment of the eolith cause, Prestwich would express his disappointment but finish with supportive reassurances that all would be well in the end. He was unfailingly supportive of Harrison, who was a frequent visitor to Darent Hulme, often staying over for lunch or spending the afternoon there to cheer up Prestwich when the older man was convalescing. My favourite image of the two of them is having tea in the snug of a public house on the afternoon in 1890 that Prestwich accepted the eoliths from West Yoke (Harrison, 1928, 156). Here was one of the greatest geologists of his age, together with a partially deaf obsessive-compulsive local shopkeeper, reminiscing about how biscuits used to be much nicer when they were younger.

Harrison continued to collect from the Ightham area and especially across the Chalk Plateau. He continued to send specimens to his friends and supporters. Evans was in receipt of constant parcels of eoliths and implements. Harrison excavated and watched diggings at a number of localities on the Plateau confirming the findings of the Parsonage Farm excavations. His notebooks and his son's biography of him amply demonstrate the continuation, if not expansion, of his role as a focal node in the correspondence web. Parties and field trips continued to seek his advice, request a viewing of his museum, or have him guide them across the Plateau. He achieved a measure of public fame and interested parties from other walks of life would visit; for example Alfred Russel Wallace in 1891 (Ellen, 2011), and Kier Hardy in 1901. Senior figures in the archaeological, anthropological and geological world wrote and visited, and he continued to provide specimens to illustrate public talks and museum displays. All of this kept eoliths very much on the radar, but although there was no end of supporters ready to champion the cause at the BAAS or at the London societies' meetings, there were none who could command the same respect and attention that Prestwich had done. There was no one who, after penning a hasty note, could confidently expect to be included on the speaker's platform at the next BAAS meeting, or at the Geological Society. There is a telling comment by F.W. Knowles writing to Harrison from the BAAS meeting in Liverpool in September 1896:

> I do not doubt but your view of Sir John is correct. He is too old now to have an open mind and he will I have no doubt hold firmly to the doctrines he has found regarding flints during his life. The study of the Egyptian Mycenean and other early art culture has taken such a hold on the mind of the younger archaeologists that the subject of the early stone culture sticks in their noses and they are impatient if anything of that kind is brought up… (Harrison, n.d. f, Notebook 14, p29, A. Muthana transcription).

The president of the section was Arthur Evans, Sir John's son.

Public visibility was maintained at a number of prestigious events at which the eoliths figured prominently. Harrison was requested by the Archaeological Association to exhibit at the Guildhall London in September 1896 (see next chapter for a fuller discussion). As with Prestwich's *Nineteenth Century* article, this was an opportunity to bring the debate directly to the public and allow the eoliths to speak for themselves. That the association made this request was a mark of the impact the eolith debate had made. Some years later, in the summer of 1899, and rather out of character, Harrison gave a public lecture at Rochester (Harrison, 1899) to the South Eastern Union of Scientific Societies (a regular joint conference of all the various scientific societies and clubs in the south-east of England). The lecture was rather hurried as Harrison had to present it almost as soon as he arrived. Although he concentrated on only one type of eolith, the double scraper (see below), 'the general opinion was favourable to Mr Harrison's views' (Harrison, *ibid*). Rupert Jones requested a set of eoliths from Harrison for an exhibition at the BAAS meeting at Dover in 1899. Jones later wrote to Harrison (Harrison, 1928, 231) to report that it was the smaller eoliths that had attracted the most attention, because these were of such a size that wave action could not be responsible for making them.

Although the years after 1896 were not characterised by any major theoretical or methodological advances in the Kentish eolith debate, Harrison did make two further important contributions, but they were not widely discussed. On a number of occasions; John Evans had asked Harrison whether or not the 'utter uselessness' of the eoliths had ever struck him. He was referring to the blunt, near right angle retouch or chipping on their edges. John Lubbock had also once asked him if he cared to speculate on the function of the eoliths. Harrison had declined. But clearly the idea intrigued him. A solution for at least one type of eolith was suggested by ethnographic parallels. From the middle 1890s onwards, and before Prestwich died, Harrison promoted the idea that the double scrapers (Harrison figure numbers 20.6–20.9 on Figure 9.10) were better described as body stones or body scrapers, for removing patches of rough skin. The idea appears to have come from a Dr Thompson, an international traveller (Harrison, n.d.-b), who had seen similar stones used by indigenous peoples. Blunt chipping would be much better for such a job as it would not cut into the skin. In particular Harrison's sources emphasised that local people would use them to rub the rough skin on the soles of their feet which puzzled Harrison until later informants (Harrison, 1899) suggested an answer. Constant barefoot walking produced a tough layer of outer skin which was prone to corns and deep cracks in hot dry environments. The result was lameness and much pain. Keeping the layer of toughened outer skin down to manageable proportions with an abrading 'body stone' was the solution. What Prestwich thought of this is unrecorded, if Harrison ever told him.

From the middle of the 1890s Harrison's notebooks and sketch books began to mention occasional finds of artefacts that he described as transitionals, pieces that bridged the gap between the eoliths and the palaeoliths. Descriptions and illustrations suggest a number of different varieties. In some cases roughly worked pieces that resembled, through

Top. A drawing by Benjamin Harrison of a body stone

Bottom. A humorous cartoon drawn by G. Worthington Smith expressing his opinion of Harrison's ideas

the edge working, the overall shape of a later palaeolith-type. Alternatively transitionals could be an actual shape resembling a later palaeolith but which had characteristically blunt eolithic retouch. Harrison was finding them at a number of localities on the Plateau, but one of the most prolific areas was what he called the Maplescombe Valley (see Figure 7.5). Here he believed transitionals were coming out of drifts at heights within the valley that were consistently below the eoliths which were restricted to patches of higher ground. These transitional tools would have provided another important link in the chain of plausibility which linked eoliths to the palaeoliths of the valley drifts. Harrison was well aware of the importance of transitional specimens. If a complete evolutionary sequence could be established, with a seamless transition from one major type to the next, then eoliths would be strengthened in their position at the beginning of the sequence. Their distinctiveness would appear less incongruent by comparison with other later tool types.

Academically, Harrison's visibility in the eolith debate diminished after the turn of the century. The Kentish eoliths, with no new champion and no fresh impetus to reanimate the debate, fell into an interpretative limbo where they remained. They were further overshadowed by a new debate. This was the question of an East-Anglian pre-Palaeolithic phase, chronologically situated somewhere between the eoliths and the Palaeolithic proper. Harrison played little part in these developments. He continued to collect enthusiastically until his death in 1921.

> Harrison was one of those humble workers for science who, in the face of great difficulties, rise superior to their surroundings by strength of character and industry, and leave an imperishable name behind. (Anon, 1921, 251)

His obituary noted dryly that the eolithic controversy had not yet 'received its final solution'. One is left wondering whether the eolithic debate would ever have risen to prominence if Joseph Prestwich had not retired to Shoreham, or been a less sympathetic character.

# 10

## MANY MEETINGS
### *Eoliths and Palaeoliths in the 1890s*

Against the unfolding eolith debate, Palaeolithic and anthropological research continued, and important contributions were made throughout the decade. Inevitably the eoliths dominated many of the debates. Despite being an origins debate for British scholars, on their own territory, it was one that did not transcend its very narrow lithic focus. John Evans remained sceptical throughout the decade. He was once again applying the 1859 methodology; rigorous adherence to a set of archaeological criteria (especially clear percussion features on the eoliths – which were missing), and a sound geological context (found *in situ* – which the eoliths were not). With hindsight Evans was proved right. The much awaited second edition of *Ancient Stone Implements* was published toward the end of the decade. It was a continued testament to the object-centred way of thinking, and remained as resolutely non-theoretical as the first edition. Evans also ignored the more complicated picture of Pleistocene climate and ice age geology that was emerging.

In fairness to Evans, no single body of theory had emerged during the late Victorian period which could have contextualised the Palaeolithic. The idea of evolution and progressive change still held sway, but the limitations of the evolutionary paradigm (and evolutionary anthropology) were becoming more evident. How to explain change within cultures, and how to show that cultures evolved from one stage to the next, was difficult within the very static evolutionary framework. The functionalist explanations of the post-Edwardian years were still more than a decade away. Most British anthropologists had moved on to studying individual elements of social systems such as kinship, myth and other social institutions, laying the foundation for the later British school of social anthropology (of which functionalism was a part). As a major figure in anthropology, Tylor began to decline from the mid-1890s, as it seems did his mental health (Stocking, 1995). Stocking refers to this period as an 'extended moment of slowly waning paradigmatic potency' (*ibid*, 125). Part of the vacuum left by Tylor's slow demise was filled by James Frazer (1854–1951) and the *Golden Bough* (1st edition 1890; 2nd 1900; 3rd edition of 13 volumes published between 1906 and 1915), but it had little direct relevance for the Palaeolithic.

The lack of skeletons from the British cave and drift sequences continued to have an adverse influence on British scholarship. A number of archaeologists were arguing for unbroken sequences of stone tool development from the eoliths to the Neolithic, and in so doing implying that there ought to be a comparable racial/physical continuity; others, like Evans, disputed this. The supposed hiatus between the Palaeolithic/Pleistocene, and the Neolithic/Recent became increasingly more important in the 1890s. It is perhaps not surprising that scholarship focused on the beginning and ending of stages when concern was raised about how cultures could effect change.

Eoliths fitted any developmental sequence perfectly. They anchored the earliest ends of progressive sequences by playing to expectations of what the earliest tools should look like. They remained objects of British origins debate because they continued to prompt the question of why such early tools were present in Britain. Perhaps the most important discovery of the decade was *Pithecanthropus erectus,* Eugene Dubois' upright man of Java. Here was a fossilised creature which, if genuine, might have made eoliths.

## Eoliths on Display: Academia, the Public, and the Press

In September 1890 the BAAS was in Leeds, and the president of the anthropology section (H) was John Evans (Evans, 1890). His presidential address dwelt on the origins of the Aryan language family (he scathingly dismissed comparative philology as not very scientific) and the state of ethnological research in Britain. Not surprisingly his first topic was human origins, and he reiterated the position he had held since the 1870s. No new data had come to light to change his mind on the big issues, and he made the usual appeal to 1859 to legitimise his remarks. The palaeoliths were still dated to the post-glacial river valleys, and the European evidence for pre-glacial Pleistocene or Pliocene humans was still unconvincing. He did not dispute the possibility of an early phase of human evolution, indeed it was likely, but the evidence for it had not yet been discovered, and it was not to be found in western Europe. Significantly, he did not mention Harrison and Prestwich's Plateau implements or the eoliths. In 1897 the BAAS met in Toronto, and Evans was a speaker there too. This time he was the president of the whole Association for that year (Evans, 1897d). With a few important exceptions, noted below, it was a rerun of 1890 and earlier. Also in 1897, the second edition of *Ancient Stone Implements* was published – he was still touting his own party line.

Elsewhere there was enthusiastic support from a small but dedicated group of eophiles. In August 1892 the BAAS was in Edinburgh, and despite a non-committal mention of eoliths by the section H president for that year, A. J. Montgomerie Bell delivered a spirited defence of the eoliths accompanied by an exhibition of specimens provided by Harrison (Bell, 1892). He wrote to Harrison with news of the meeting. He concluded with an anecdote that pleased Harrison very much, in which he recalled an occasion when Barry De Crawshay showed one of Charles Darwin's sons a collection of eoliths. The son had remarked. 'Oh that my father had been alive: how he would have entered into this' (Harrison, 1928). Bell lost few opportunities to promote the eolithic cause. At some point during the Edinburgh meeting he had dinner with the Geikies, using the opportunity to spread the word (Harrison, n.d.-d). Geikie was open to the arguments and specimens presented by Bell (E. Harrison, n.d.).

In 1894 Bell was back in action at the BAAS in Oxford. This time eoliths were formally on the agenda. A joint discussion was arranged between the geological section (C) and the anthropological section (H), whose president that year was W. H. Flower (Flower, 1894), by that time director of the Natural History Museum (which had replaced the South Kensington Museum). Flower gave a brilliant speech on the past history and current wellbeing of British anthropology in the late Victorian era. In fact the 1894 meeting was a return to form for Palaeolithic archaeology at the Association, with a number of Palaeolithic and human origins' related papers other than those featured in the joint discussion. The topic had been poorly represented in the meetings of previous years, quite possibly reflecting the resurgence of race and world ethnology as an issue in the AI (see last chapter). The Oxford debate of 1894 is significant. To have a major debate between two important sections from the social sciences highlights how seriously eoliths were being taken. Moreover, its results were reported in detail in the national press, which meant that eoliths were receiving the widest possible coverage. A number of the eoliths sent by Harrison for exhibition at the meeting are shown in Figure 10.1. The auditorium for Bell's address was full, with up to three hundred people attending. The joint meeting opened with a defence of the eoliths by Professor T. Rupert Jones. His paper (Jones, 1894a; Jones, 1894b) was a basic rehearsal of Prestwich and Harrison's views as given to the Geological Society and the AI.

Jones was followed on the platform by William Whitaker of the Geological Survey (Whitaker, 1894) who injected what he called 'wholesome doubt' into the proceedings. He had remained a wholesome doubter since Prestwich's first paper to the Geological Society in 1889 (Table 9.2). He made the point that the Plateau Drifts were not actually drifts *sensu stricto*. They were the result of *in situ* erosion of the surface of the chalk, and whatever deposits lay on top of it. So any palaeoliths (he freely admitted the reality of the handaxes from these deposits) found in those deposits had to be at least as old as the deposits were. However, the surface deposits had been forming for a very long time, and they were continuing to form. There was no way of knowing just when, in their history, the palaeoliths had been introduced. He then sidestepped the eoliths, but noted the retouch on them looked like 'very ordinary chippings'. The plateau deposits and their artefacts were clearly earlier than the true drifts and their implements in the river valleys. Other than that there was little to say. A pre-glacial or glacial date for them could not be established because of the utter lack of any glacial deposits in the vicinity, and there were

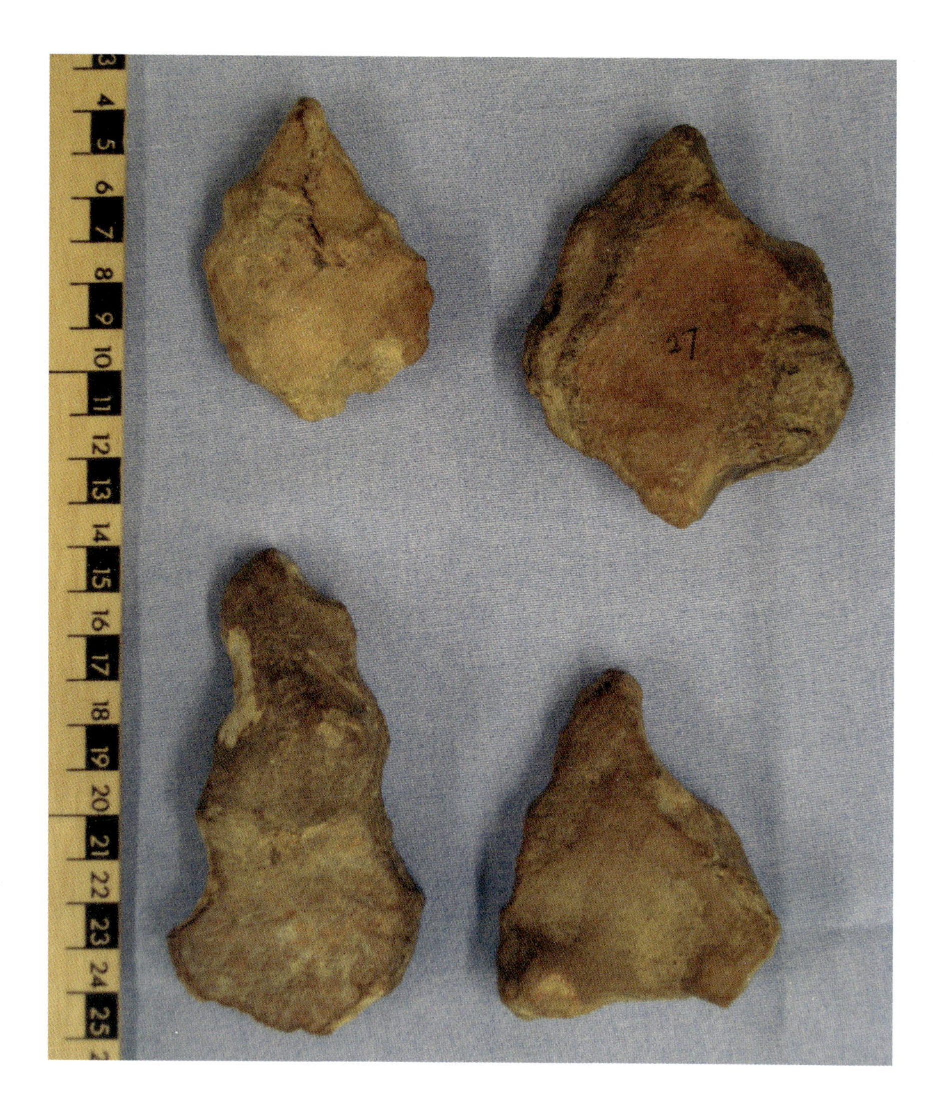

*Figure 10.1. A small selection of the large collection of eoliths displayed at the* British Association for the Advancement of Science *meeting in Oxford in 1894. The eoliths were supplied by Benjamin Harrison.*

no stratigraphic relationships to overlying deposits that could help narrow down their chrono-stratigraphic position. On the face of it Whitaker's purely geological objections were simple, but difficult to refute.

Bell, in a letter to Harrison (10/8/1894) continued the story of the Oxford meeting. This is an interesting document as it represents the opinions of a participant at the meeting so it is worth quoting in full:

> The debate…is over…It is not a triumph, it is not a defeat, but leaves things much as they were. Rupert Jones opened, and has been firm as a rock throughout. Whitaker followed, and I could not understand his drift. I came on, and spoke on the anthropological side only…Evans followed, admitting the great age of the drifts, and accepting them as belonging to the *palaeolithic* stage. Would not admit one of the primitive type; saw no

> reason to seek to find some new kind of implement when the Palaeolithic hâche was there; could not persuade himself that man was born in Kent... Hicks followed, to prove man of glacial age. Boyd Dawkins next, chiefly geological; *thought he could match all our tools*...Pitt-Rivers next, who said he agreed with me there was a long advance before palaeolithic man came on...Then Clement Reid spoke: saw no sign of palaeolithic man.

Rupert Jones replied very firmly, said that everyone had been on his side.

> Last, Sir William Flower said that the discussion had been useful, but he did not think any opinions had been changed by it. There are, as you know, some obscurities both geological and anthropological. It was always a difficulty to me when the worn palaeolith appeared with the other types. *It did complicate the question.* (Harrison, 1928, 193; original emphasis)

The report of the Oxford meeting, and the original of Bell's letter are preserved in the Harrison archive at Maidstone (Harrison, n.d.-e; E. Harrison, n.d.). In fact the *Times* newspaper reported fully on the meeting, and the individual lectures were detailed in the 11th August edition. Harrison cut them out and pasted them into a scrap book. Edward Harrison did some editing of Bell's letter for *Harrison of Ightham*. Evans had been gently heckled as he asserted that Kent could not be the home of the human race; cries of 'no, no'. The original letter also included the astonishing statement that Boyd Dawkins admitted to Bell, in private, that he accepted the authenticity of all the specimens that Bell displayed, but could not admit it in public (Harrison, n.d.-e, 87). Bell finished by reassuring Harrison. 'I am sorry for it, old chap but believe me with Prestwich that all will come right in the end.'

The Oxford debate, as depicted in Bell's letter perfectly sums up the status of the eolith question in the middle 1890s. The usual suspects were on parade, and none of them were inclined to alter their long-held positions, although Evans was by now admitting the authenticity of the worn palaeoliths from the Plateau but insisting they were Palaeolithic nonetheless, a position not too dissimilar to that of Prestwich. It was an entrenched stalemate. In response to a paper by Boyd Dawkins, given to the AI early in 1894, Montgomerie Bell was once again defending the eoliths (Bell, 1894). In fact the paper had been written some years earlier as a response to Boyd Dawkins' denials of Prestwich's original papers at the Geological Society and the AI. Bell added to and updated his Ms and presented another vigorous defence of Harrison and his views on the tools. A summary of Bell's arguments is given in Table 10.1. In essence these points encompass the range and scope of almost the whole of the eolith debate up to the end of the Victorian period.

For the most part A.M. Bell confined his defence of the eoliths to the material from the North Downs, focusing on Harrison and Prestwich's arguments and the geology of the Chalk Plateau. This is not surprising as

| Objections of William Boyd Dawkins | A.M. Bells Points in Defence of Eoliths from Bell 1894. |
|---|---|
| You find palaeoliths and other tools all over the downland and in the valleys below because Palaeolithic humans wandered over the whole landscape | True. But there is a clear distinction. The Palaeoliths from the Plateau are old and worn. Any more recent ones (contemporary with the river drifts below) are in a very different condition – much fresher and easier to distinguish. Anyway, the frequency of palaeoliths on the Plateau, of any sort, is very limited.<br>Eoliths are in patches or worn ochreous gravel. The worn Palaeoliths are from other surface Plateau deposits. |
| Plateau eoliths are the waste from ordinary knapping scatters. Any axes made during these knapping events have been removed. | This applies to Neolithic workshops and there is nothing like this on the Chalk Plateau. Moreover there is a real difference in the character of the flints from these kind of deposits. Would workshop debris be rolled as are the eoliths. Palaeolithic workshops such as Caddington and Crayford were not. |
| Neolithic axe makers made implements of the same shape as those made by Palaeolithic knappers, often by accident. All would be found on the surface. This means that anything found on the surface resembling a palaeolith can only be dated with confidence to the Neolithic. | Agrees with the logic. But the condition and physical context (in ochreous gravel) does allow surface material to be distinguished into different groups. Groupings of Neolithic material are visually distinctive.<br>As a case study he has a group of material from the surface of the Limpsfield gravel that Boyd Dawkins would identify as Neolithic. But he has contextual evidence that proves it older. |
| Patches of ochreous abraded gravel are of local significance, they have no wider importance. | When different types occur in frequency together, and are all stained in a similar way this cannot be accidental. When you get a number of similar patches over a wider area then it is not locally insignificant. |
| Objections raised by John Evans on the eoliths in general | Handaxes cannot be thought to be the earliest tools, too advanced in their manufacture. Recently acquired material from Tasmania by E.B. Tylor shows what the earliest tools may have been like, and the link between these Tasmanian tools and the eoliths is clear |
| | Types of artefact found below Plateau very different from those on it. But if you put a true flake next to an eolith, and the retouch on both is identical, surely this validates the retouch on the eolith, even though it lacks Evan's criteria of acceptability – a bulb of percussion. Also now are beginning to find transitional tools between eolithic forms and palaeoliths. |
| | Palaeolithic people chose surface weathered flint to work. Neolithic people mined their flint from chalk. The patina and cortex is a clear distinguisher. |
| | Valley drift palaeoliths have long scars that cover the face of the artefact, but don't bite deeply into it. This is a learned technique as long scars are naturally deep. Similar thin, but not quite so long scars are present on Plateau palaeoliths. Some eoliths approximate this. But nature cannot replicate it. |
| | While the shape of a pebble can influence the shape of the final tool, the presence of clear types in the eoliths, and the frequency with which these types occur cannot be precludes nature just enhancing the shape of pebbles with natural retouch |
| Evans' arguments about the utility of the blunt edges. | He has no good answer to this. Reiterates the importance of the arguments about different types of eoliths and that nature cant reproduce these clear cut shapes in exact detail and in any frequency. He also adds that the eoliths are not found in every gravel deposit everywhere. |
| | While originally he didn't agree that worn palaeoliths did occur on the Plateau, he now accepts this. He does however believe they occur in different drifts to the worn Palaeoliths – something that Prestwich doesn't agree with. |

*Table 10.1. A selection of William Boyd Dawkins's objections to the eoliths and the response to those objections by A.M. Bell (from Bell 1894).*

this was home territory for him. He was a friend of Benjamin Harrison (they collected and field walked together) and this was the landscape he knew best. When he did stray out of his patch he continued to apply the North Downs logic to other sites. As late as the 1900 BAAS meeting at Bradford, he described a river gravel in a channel at Wolvercote, near Oxford (Bell, 1900), and he had spoken of the site at the 1894 meeting as well. Normal palaeoliths were contained within the channel gravel. However these channel deposits cut through an older land surface; this one containing palaeoliths, some flakes, but also a few eoliths. All these older artefacts were deeply stained and markedly different in appearance when compared with the (presumably) post-glacial Wolvercote channel implements. Bell asserted these were the earliest human tools from the Thames Valley, the logic being pure Harrison.

Other workers were also taking the debate away from its core area. O. A. Shrubsole discussed gravels in and around Reading (Shrubsole, 1890; Shrubsole, 1893). The highest gravels in the region did not contain evidence of humans, but lower down, in drifts at *c.* 235 feet above sea level, implements were discovered. Moreover, implements from the lower drift terraces were identical to those described by Prestwich from the higher terraces of the Shode valley. In his 1893 paper, Shrubsole described simple 'hollowed out scrapers' (i.e. with a distinct concavity created by retouch) and other very primitive tools which nevertheless conformed to distinct types. Although not mentioning eoliths by name, he was clearly implying the discovery of similar tools from above the river valleys. Shrubsole suggested that these gravels capping the interfluves between the local river valleys may have been laid down by rivers draining the Wealden mountains, and later reworked and redistributed by a phase of marine subsidence.

In a later paper (Shrubsole, 1895) he argued that no handaxes had ever been found in these plateau gravels, nor flakes, implying that only these very ancient tools were present. The Berkshire plateau implements were of three types; hollow scrapers were made on natural fragments of flint, and linked directly to the hollow scrapers made on flakes found in the valley drifts; large sized implements with rounded butts (only one found); and pointed tools which were elongated nodules with 'chipping' at one end to make a point.

Shrubsole supported his interpretations by recourse to the standard arguments made by the eophiles:

- Yes, they looked primitive and natural, but that's what you would expect of such a primitive cultural stage.
- The Abbeville and Amiens palaeoliths had not been believed at first either.
- No true contemporary palaeoliths had been found on the plateaux above river valleys.
- The tools could be placed in clear groups (and in the case of the hollow scrapers linked as the beginning of a series that would lead to those in the lower drifts).
- The gravels were associated by Prestwich with his Southern Drift and therefore of pre-glacial age.

While Shrubsole was rather reticent about the identification and naming of his primitive tools at the Geological Society meeting (Shrubsole, 1893), one member of his audience was in little doubt. W.J. Lewis Abbott (1863–1933) pronounced them human. Abbott was a tireless but uncritical champion of Tertiary man, with a tendency to write himself into the story of other peoples' discoveries (compare Abbott's 1894 version of the discovery of the Basted Fissure (see below) with Benjamin Harrison's (Harrison 1928); or Spencer (Spencer, 1990) on Abbott's appreciation of his own role in the Piltdown discoveries). At the same time he could be unstinting in his praise for those whom he admired. He wrote supportive articles in the 1890s praising Prestwich and especially Harrison in the popular scientific press (Abbott,

1894a; Abbott, 1894c). In one article at least Abbott seemed to be indulging in some deliberate myth making. In a piece aimed at the general public he cast Harrison into the mould of a lone hero engaged on a noble quest. In this case the holy grail was the proof of Tertiary man's existence. In generalising the events, Abbott gave the impression that Harrison's discoveries were the inevitable result of a dedicated search to reveal Tertiary man to a disbelieving world; Harrison, battling against the odds, finally won through to see truth conquer. Harrison's own record of events (in Harrison, 1928) reveals a more haphazard and organic journey toward eoliths and enlightenment. Abbott, conscious of the power of 1859, may well have been trying to craft a similar legend for the 'Second Antiquity of Man' debate, at the same time as position himself at the very beginning of the story.

Lewis Abbott first accepted Harrison's eoliths as evidence of Tertiary man in spring or early summer 1892 (Harrison, 1928). Not long after, he began to work on the Basted Fissure. This was an ancient vertical crack in the local rock which was Kentish Rag (ragstone is a name for a type of limestone) into which Pleistocene animals had either fallen or their bones had been washed in by surface water, before the fissure was sealed. The fissure, an important time capsule for Pleistocene fauna (Abbott, 1894a; Abbott, 1894b; Anon, 1894c; Anon, 1894b), was a Harrison discovery and was close to his home. Sadly no implements were ever recovered from it, only a few rather unconvincing flakes. But for Abbott, the fauna provided a key insight into early human occupation in Pleistocene Britain. The faunal record from the site was prodigious and included many new species. It bore a distinctly cold aspect to it, reminiscent of Boyd Dawkins' arctic fauna (1874; 1880). For Boyd Dawkins these were post-glacial animals. Abbott reversed the logic of the standard Evans/Boyd Dawkins argument. If the fauna was associated with glacial conditions as Abbott now asserted, then so must the humans be who were associated with these animals. The post-glacial age of the valley drifts was wrong.

> Day by day the discoveries of relics of our pre-glacial and eolithic progenitors are thundering from various parts of the Kingdom, which must, ere long, wake the last slumberer from his post-glacial-man nightmare. (Abbott, 1894a, 172)

The sleeper to be awoken here was likely to be Evans. Abbott was another researcher who was looking for eoliths outside of north Kent. Although much of his later work would be focused on Sussex, he was among the first to seriously claim to have discovered eolith-like tools in the Forest Bed of the East Anglian coast line, in this case to the west of East Runton (Abbott, 1897). This locality had a thick deposit of glacial beds, marked by twisted and contorted strata, overlying the Forest Bed itself. Within the latter was an iron pan layer noted for its rich faunal remains, so much so it was known as the elephant bed. The artefacts came from here, as well as being found on the beach, although Abbott was satisfied these had originated from the iron pan too. He was aware of the negative reaction these pieces were likely to evoke, so he applied strict criteria in identifying the human modification.

- He noted the overprinting of thinning and shaping scars on the artefacts' surfaces.
- He noted the presence of conchoidal fracture marks accompanied by negative bulbs of percussion.
- He identified a small scar type on one face of the tools known as an éraillure scar. He taught himself to flint knap and his experimental work suggested to him that this was a more reliable hallmark of human knapping than even a cone and bulb of percussion.

In this he was reproducing a methodological approach that many eophiles would employ, that is amassing an impressive catalogue of *individual* features which, when listed, seemed to confirm deliberate knapping by humans. Unfortunately it was not a list that would sway the likes of John Evans. He and the other eophobes were only too aware that it was the *combination of sets* of these features found together, which was indicative of artificial retouch. Isolated scars, even with conchoidal ripple marks and a negative bulb of percussion, did not prove anything. Even a few scars close together on otherwise unshaped nodules (usually on prominent extremities which would be the most prone to natural percussion damage) were not sufficient to prove human workmanship. Abbott showed the artefacts from the Forest Bed to John Evans. He was not convinced. Nothing more was ever made of them in the professional literature. Suggestions of cut-marked wood from the Forest Bed fared equally poorly (Longe, 1901); the face cut into a Forest Bed shell that Henry Stopes thought was human has already been discussed.

In retrospect, the significance of these and similar finds was they were part of an uncoordinated programme aimed at recognizing a widespread pre-Palaeolithic stage in Britain. It would be another decade before there was a sustained effort to identify pre-Palaeolithic tools in this key geological location.

H.P. Blackmore of the Salisbury Museum wrote to Harrison in July 1894 to inform him that he had discovered eolith-like tools in drift deposits at Alderbury, near Salisbury, in Wiltshire (Harrison, n.d.-e). These were outliers of Prestwich's Southern Drift. The eoliths from here were rather fresh in appearance. Significantly, there also seemed to be evidence of tools which were much more worn. These, Blackmore implied, had been derived from even older beds, now long since destroyed, and for which no evidence remained. A Mr Lasham of Guildford was also finding eoliths and Palaeoliths in the gravels of Farnham (E. Harrison, n.d.). To keep matters well within the family, Edward Harrison informed his father that he had discovered eoliths in Yorkshire (Harrison, 1928).

The eoliths themselves were also on the move during the 1890s. They were being regularly exhibited at the meetings of the major scientific societies. When a paper on artefacts was read out at the Geologists' Association, the Geological Society or the AI, there would be a display of implements and eoliths to accompany the talk. Benjamin Harrison was regularly sending eoliths to many of his correspondents and providing small collections for people to take to the various meetings. Eoliths were also attracting the attention of museum curators and they were now on display at the Blackmore Museum at Salisbury. The British Museum was negotiating to buy a representative sample as early as 1893, but Prestwich was unhappy with the specimens that Harrison sent them (Harrison, 1928). The professor thought there were too many plateau palaeoliths and not enough convincing looking eoliths. Interestingly, Worthington Smith (1894) implies a small collection of eoliths was on display at the British Museum before his book *Man the Primeval Savage* was published. He noted there were few of them and they were not a very representative sample, suggesting these were the very ones Prestwich was complaining about. There had been some haggling about the price too (Harrison, n.d.-e). The British Museum also bought the specimens from the Guildhall exhibition (see below) for £100, and later bought six of those that Harrison excavated from Parsonage Farm for the BAAS in 1895. At the Natural History Museum, South Kensington, they were on display in 1896, and Harrison arranged for more specimens to be sent to augment the display of eoliths at the Royal College of Science (formerly the Normal School of Science which changed its name in 1890) which already had them on show (Harrison, n.d.-e). There were displays at the Museum of the Geological Survey and elsewhere (Jones, 1901).

The provinces were not neglected either. Many individuals gave papers and exhibited collections of eoliths (March, 1898; Anon, 1901). John Ward of the Cardiff Museum and Fine Art Gallery wrote in April 1894 enquiring about purchasing palaeoliths and eoliths from the Plateau (Harrison, n.d.-e). Thomas Wilson of the United States National Museum wrote as early as 1891 hoping that implements could be bought or exchanged (Harrison, n.d.-c). In May 1899 a Captain Hutton, the curator of Canterbury Museum in Christchurch, New Zealand, wrote to thank Harrison for his donation of eoliths. He noted they showed more evidence of workmanship than did the tools which had been made by the aboriginal people of Tasmania (E. Harrison, n.d.).

In fact delegates to the BAAS meetings saw a number of exhibits of eoliths during the 1890s. A regular attendee who often displayed substantial collections of stone tools and eoliths at these meetings was Henry Stopes (Wenban-Smith, 2009). Stopes was keen to display his material in chronological order to reveal the evolutionary progression in tool form. He was an early convert to Harrison's eoliths (c. 1893; Harrison, 1928), perhaps because he believed the evolution of human mental capacities could be tracked through ever more sophisticated manufacturing skills. For Henry Stopes, like Lane Fox, stone tools were fossilized thoughts, and the eoliths were the perfect starting point for a sequence of development. A well ordered chronological series of artefacts which showed the progression from early/crude (and therefore primitive), to late/refined (and therefore sophisticated) was a particularly powerful way of visually displaying the link between mental evolution and the development of practical skills (Stopes, 1893; Stopes, 1894; Stopes, 1895; Stopes, 1900).

The idea of displaying a single ordered series of artefacts was a powerful weapon in the arsenal of any archaeologist who wished to show evolutionary development. Lane Fox had employed it to sell the very same message in his museum. Eophiles were quick to appreciate the efficacy of this display method. Their insistence on grouping tools in series and highlighting their transformation into well-known later Palaeolithic and Neolithic forms through transitional specimens, was particularly empowered by a linear visual approach. It would have immediately appealed to the Victorian public's understanding of progressive evolution. At the same time it would have reinforced their existing prejudices – the simplicity of the visual display would merely have served to emphasise how right it felt to them.

This is highlighted in articles sent to the *Times* newspaper in September 1896. Harrison achieved a small degree of notoriety at this time as his eoliths were prominently displayed in the London Guildhall in a large public exhibition accompanying a meeting of the British Archaeological Association (Harrison, n.d.–f; this notebook contains the text of a speech that Harrison may well have given on the specimens at the Guildhall). Correspondents singled out Harrison and Ightham for their combined contribution to prehistoric studies. One, a Mr Henry Walker of Kensington Park Road (Harrison, n.d.-f), wrote to urge the public to visit the displays. On the day of the British Archaeological Association's meeting the displays were unordered,

> ...and therefore unintelligible to most of the visitors,... [they are now]... arranged in...historical sequence. Instead of a medley of all ages, the implements are now arranged, as far as may be, chronologically, so that the sequence and evolution of most of the forms is visible at a glance. (Walker, 1896. Letter to the *Times*, September 29, my brackets)

Although the 1894 BAAS meeting at Oxford reflected the deadlock in the eolith debate, general interest in the eoliths continued throughout the 1890s. On 26th March 1896 (Harrison, 1928) Harrison received a distinguished visitor to his shop, a man with an unimpeachable archaeological pedigree; William Cunnington (1813–1906) the distinguished Wiltshire palaeontologist and antiquarian. He had an equally famous grandfather of the same name (William Cunnington 1754–1810) who had excavated hundreds of Wiltshire barrows with Sir Richard Colt-Hoare (1758–1838), as well as conducting the first recorded excavation in Stonehenge. Their work had done much to establish the character of later prehistoric studies in southern England. Cunnington was initially persuaded by the eoliths Harrison showed him but quite quickly changed his mind. He focused on their retouch. In a paper in the journal *Natural Science* (Cunnington, 1897) he took a small number of eoliths that Harrison had presented to him as being representative examples. He showed (very convincingly) how the natural damage to the stones was effectively a biography of their post-depositional history. There were six phases of damage or alteration to the surfaces of these flints including staining, striations as a result of glacial activity, edge chipping and infilling of the various striations with (as he thought) white silica. These episodes of damage had occurred at very different periods. The so-called retouch was not all contemporary, but dated to different periods of surface modification. Eophiles required the flints be retouched once and then used and abandoned. He presented other arguments as well, questioning why the tools seemed to appear in such profusion, even from excavated contexts (i.e. Harrison's excavations in 1894 and 1896 at Parsonage Farm).

In my opinion Cunnington's paper in *Natural Science*, and the subsequent one to the Geological Society (Cunnington, 1898), represent some of the most devastatingly cogent critiques of the eoliths ever made. The eophiles were not slow in responding (Kennard, 1897; Abbott, 1898; Bullen, 1898). It was the outspoken Alfred Santer Kennard (1870–1948; *ibid*) the Natural History Museum's malacologist who recognised the true danger in Cunnington's views. Re-use of the tools on a number of occasions explained why the retouch appeared to date to different periods. It was common enough in the Palaeolithic and Neolithic. He gently criticised collectors (possibly a sideswipe at John Evans) and museum experts for holding negative opinions based on the few unrepresentative pieces they held in their collections. Only field men could perceive the truth as they understood the whole range of eoliths types from the best to the worst. Curiously he asserted that Clement Reid (1853–1916) of the Geological Survey had decided that many of the gravel outliers on the North Downs had been adversly affected by glacial climate. Therefore, this extreme weathering had rendered the surface flints highly susceptible to edge damage. Reid was able to explain away the eoliths and their edge chipping as natural. Kennard noted with satisfaction that Benjamin Harrison had flint knapped eoliths into replica palaeoliths to prove the flint was not so affected!

Cunnington's Geological Society paper was particularly strong. He was not able to be at the meeting and it was read for him by a Dr Gregory. Cunnington accepted as genuine the worn and stained handaxes from the chalk plateau, and their contemporaneity with the drifts they were found in, and took *these* as proof of 'plateau-man'. In the paper he illustrated a broken palaeolith, Figure 10.2, from the plateau, probably broken by frost fracture. On the break surface was retouch, and it was *identical* to that seen on eoliths. From this he then made two inferences:

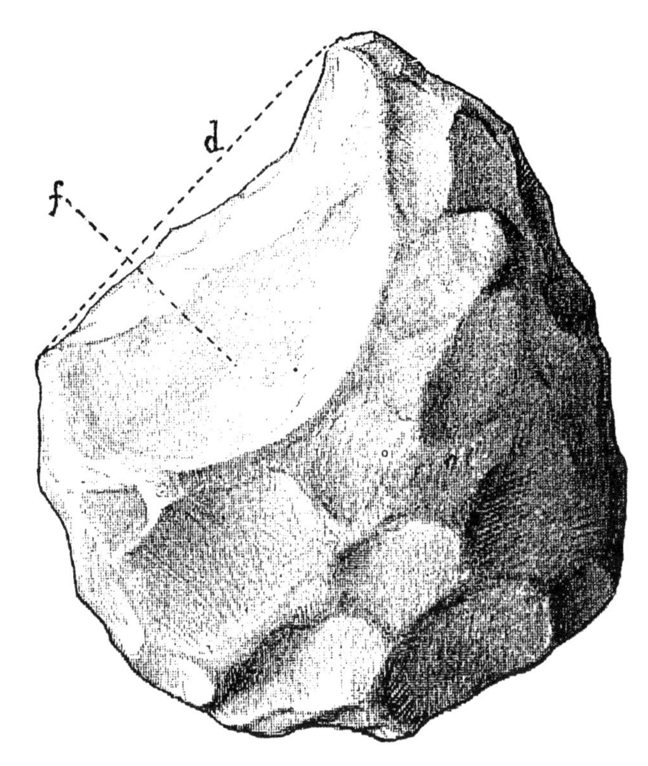


*Figure 10.2. Broken handaxe illustrated by William Cunnington in the* Quarterly Journal of the Geological Society of London *for 1898. Paraphrasing Cunnington's explanation for the artefact, he identified 7 stages in its life history. 1. Handaxe knapped by Palaeolithic humans; 2. worn down and abraded; 3) frost fracture breaks the axe in its upper left hand quadrant and the remains of that fracture surface is f on the Figure; 4) the outer edge of this natural fracture surface, d on the figure, is very fragile and it has become steeply retouched; this retouch is identical to that on Harrison's eoliths; 5) the whole axe becomes stained dark brown, characteristic of Plateau flints; 6) next the implement is marked by a series of 'glacial' striations on its surface; 7) thin layer of silica is deposited over the surface which when infilling the striae make them appear white.*

a) since the retouch post-dated the manufacture and breakage of a true palaeolith (albeit a very old one), the retouch could not be eolithic *sensu stricto* – most likely it was natural.
b) alternatively, if the retouch was humanly made, it occurred on a true palaeolith and so must post-date the palaeolith itself and consequently could not be eolithic. By implication, the eoliths themselves had to date to the Palaeolithic or later.

Simple but insightful. In both papers Cunnington noted that the striations on the eoliths' surfaces need not imply the presence of glaciers as was normally supposed. Slippage in frozen ground, with stones sliding across each other, could easily explain the surface scratches (the process today is termed solifluction), so this need not be taken as evidence of a glaciation or the passage of glacier ice. This point too was ignored. Gregory's presentation of Cunnington's paper that evening was accompanied by an exhibit of eoliths and implements provided by Harrison. Gregory, in fielding questions after the paper, wryly noted:

> Those who believed in these questions [i.e. eoliths] still could not agree as to which are genuine and which not. Prof. Rupert Jones…[eophile]…had

> that evening doubted one which Mr Harrison selected as a fine example.
> (Gregory in discussion on Cunnington's paper, my brackets)

Combine William Whittaker's geological arguments with William Cunnington's observations on artefact condition, and the eolith question was settled. But of course the eophiles weren't listening.

On occasion the eolith debate seemed to descend into parody. Sir Henry Howorth (1842–1923) was a dyed-in-the-wool Victorian polymath and controversialist. He wrote voluminously, on anthropology, zoology, archaeology, church history, Asia, gypsies, slavery, the Germans, mammoths and many other subjects. His works were characterised by a maniacal attention to detail which, if his eolith writings were anything to go by (Howorth, 1901), were usually wrong. Boyd Dawkins, writing Howorth's obituary (Dawkins, 1923), noted that his efforts to reinstate the concept of the Biblical flood in north-western Europe convinced few.

Howorth's eolith paper in the *Geological Magazine* attempted to show that in Europe, the gap between the end of the Palaeolithic, and the beginning of the Neolithic corresponded with just such a deluge (Prestwich had argued something similar). The gravels of the plateau drift as well as the valley drifts were 'distributed' by the flood waters. Howorth's work always seemed to excite controversy and the eolith paper was no exception. F.W. Bennett (Bennett, 1901) noted that Howarth's assertion that there were palaeoliths in the Forest Bed was wrong, as was his belief that palaeoliths were never recovered with eoliths. Both Rupert Jones and A.R. Hunt took Howorth to task for asserting that T.H. Huxley had deliberately suppressed McEnery's manuscript on Kent's Cavern at the Royal Society (Hunt, 1901; Jones, 1901). They noted that Huxley was one year old when the Ms was completed, and only 16 when McEnery died; remarking with admirable restraint that Huxley's influence with the society was probably limited at that time. Anyway, the Royal Society had never owned the manuscript. R.A. Bullen (Bullen, 1901) noted that Howorth had mixed up Lewis Abbott with another amateur collector, Harry Lewis (I notice that Howorth also includes arrow heads amongst a short catalogue of Palaeolithic tools).

### Progress of Palaeolithic Research during the 1890s

Palaeolithic sites and artefacts continued to be discovered, hand in hand with the discovery of eoliths, as researchers continued to scour the river valleys and gravel pits of southern England. The Shrubsoles, Bells, Abbotts and Harrisons of the world were equally as pleased with the discovery of a fine post-glacial handaxe as they were with an eolith. Palaeoliths were claimed to be present in Scotland (Smith, 1892) which Evans doubted, while Irish Sea beach evidence (Coffey, 1901) was used to show how eolithic forms could be easily reproduced by waves. An early paper by Samuel Hazzledine Warren

(1872–1958), an ardent eophobe, reported the discovery of eolith-like flints on the Isle of Wight, along with palaeoliths (Warren, 1900). Like William Cunnington, Warren was sceptical about the edge chipping on eoliths, finding it on later implements, including Neolithic axes. Warren would become one of the key players in the later eolith debates (Oakley, 1959).

Abroad, H.W. Seton-Karr was reporting the discovery of palaeoliths in Somaliland (Seton-Karr, 1895), discovered while lion hunting. Rupert Jones (Jones, 1899) described artefacts from Swaziland. He could not resist a brief plug for Harrison and the eoliths, hinting broadly that eoliths had been discovered near Pretoria. (The note following Jones' paper is by W.H. Penning (Jones, *ibid*), and actually suggests that handaxes from localities near modern diamond mines, including the famous open air site of Canteen Koppie near Kimberley, were used by primitive people in search of diamonds!). Rupert Jones' eophile credentials were never more apparent than in a review he wrote of a publication by Franz Noetling of the Geological Survey of India in 1894 on the discovery of chipped flints in Miocene strata in Burma (Jones, 1894c). He used the discovery to attempt to validate the Thenay flints. He expressed surprise that the discoverer had added a question mark to his title as it implied that the author was not sure if the flints were really chipped by ancient humans.

John Evans (Evans, 1897a) described the many new instances of Palaeolithic discoveries in Africa, the Near East and the Indian sub-continent, all discovered since the first edition of *Ancient Stone Implements* in 1872. As in the early edition, he noted the marked similarity in the character of the palaeoliths from all these areas. But the remarkable similarities between the Indian handaxes, those of Seton Karr in Somaliland and the western European examples now prompted him to suggest they were either all made by the same race, or by separate peoples who were in close contact with each other. In the second edition I get the strong impression that Evans accepted a globally significant Palaeolithic period. Indeed, the African discoveries stimulated him to suggest that Africa could be a key to determining the route by which the Palaeolithic race travelled from their eastern homeland into the west. Quite possibly they were clues which would help in identifying the exact location of the original home of our species (Evans, 1897c; Evans, 1900). Common consensus still held the human cradle to be in the 'east', but occasional dissenting voices located it elsewhere. S. Waddington, for example, suggested in the *Nineteenth Century* that the northern Palaearctic was a more likely area as it explained Palaeolithic humans' association with the arctic fauna and the appearance of early humans in America (Waddington, 1900).

The second edition of *Ancient Stone Implements* (1897a) was eagerly awaited. Harrison was chomping at the bit to find out what Evans had written about the eoliths. His son notes he was disappointed but he could hardly have been surprised. In the first edition the only mention of artefacts in the area had been of a handaxe found at Currie Wood, in drifts on top of the North Downs to the west of Shoreham. This palaeolith's anomalous position on the interfluve between the Darent and the Cray had been noted. In the second edition there had been some changes. Evans had followed Prestwich's basic sequence of events (as always). He accepted the worn and stained Plateau palaeoliths as dating to the period before the Vale of Holmesdale was cut (Figures 7.2 and 7.4), but exactly how old, and just when, the deep Holmesdale valley was excavated was, he maintained, anybody's guess – 'beyond all ordinary means of calculation'.

Benjamin Harrison, A. Montgomerie Bell, and De Barri Crawshay were singled out for praise for their collecting activities, but the eoliths were not accepted as proper human artefacts. Although Evans had rewritten much of the catalogue sections for the second edition, not a great deal else had changed. The river drift was still post-glacial, and he effectively side stepped the implications of sites like Hitchin and Hoxne, Caddington and Stoke Newington, and the Palaeoliths from the top of the North Downs. The final chapter still focused on proving the drifts were fluvial rather than marine (at this point, a fact surely accepted by most workers, apart from the occasional 'left-field' controversialist like Howorth). Evans used the Plateau deposits to assert that they supported the concept of fluvial drifts because they could be explained as being part of the former drainage from the Wealden heights before the scarp slope was cut. He still argued that river valleys were cut by periodic floods from higher elevation, and he made no attempt to engage with the newer glacial theories. Although today it is rather heretical to criticise *Ancient Stone Implements*, I must confess I think the second edition was a disappointment. Very little of the interpretative sections had been rewritten, and Evans' ideas were similar to what they had been a quarter of a century earlier.

The eolith debate progressed against a background of on-going Palaeolithic research, and represented a dialogue about the earliest occupants of Britain. I will only concentrate on a few examples here in order to focus on some of the key points in the debates, and to illustrate the character of mainstream Palaeolithic research against which the eolith debate was to be increasingly compared.

Hoxne had been a key site in the original Antiquity of Man Debate. Evans and Prestwich had cut and recorded sections there in 1859 and 1860. It retained its importance in the second edition of *Ancient Stone Implements* (Evans, 1897a). New work supervised by Clement Reid, under the auspices of the BAAS, had provided a much more detailed stratigraphic picture of the site. A stream, flowing over a mantle of boulder clay (glacial), gradually became a lake, which filled up with sediment, in time becoming an alder carr. Renewed lacustrine conditions followed, but now under arctic or near arctic conditions. This was demonstrated by the discovery of leaves of the arctic willow. This deposit was covered in flood loams (brickearth), containing the palaeoliths, also reflecting cold conditions, which in turn were covered by a sandy deposit. The implication was for two distinct periods of cold climate, the second one post-dating the boulder clay which was the traditional deposit associated with the glacial period. The timing of the human occupation did not present difficulties for Evans's post-glacial Palaeolithic, but he did not expand on the implications of two discreet cold climate events.

Clement Reid then looked at Hitchin (Anon, 1897), a site which produced an almost identical sedimentary sequence with boulder clay overlain by alluvial deposits and then a Palaeolithic brickearth with implements confined to its stony base. At the close of the cold, represented by the boulder clay, the land had been at a greater elevation and this had resulted in the cutting of deep valleys. Then the level of the land had fallen and the alluvial/lacustrine silting up of the valleys had begun, in a temperate environment, and continued across the transition to colder conditions and the final deposition of the implementiferous brickearths. As at Hoxne, the Hitchin brickearths were deposited in a cold steppe-like environment. Once again the post-glacial date of the Palaeolithic implements was clear, because the glacial itself was associated with the boulder clay. The later cold could have been a minor post-glacial reversion to colder conditions. Nevertheless, there were two distinct cold episodes, and humans were associated with one of them. Evans bypassed this as well.

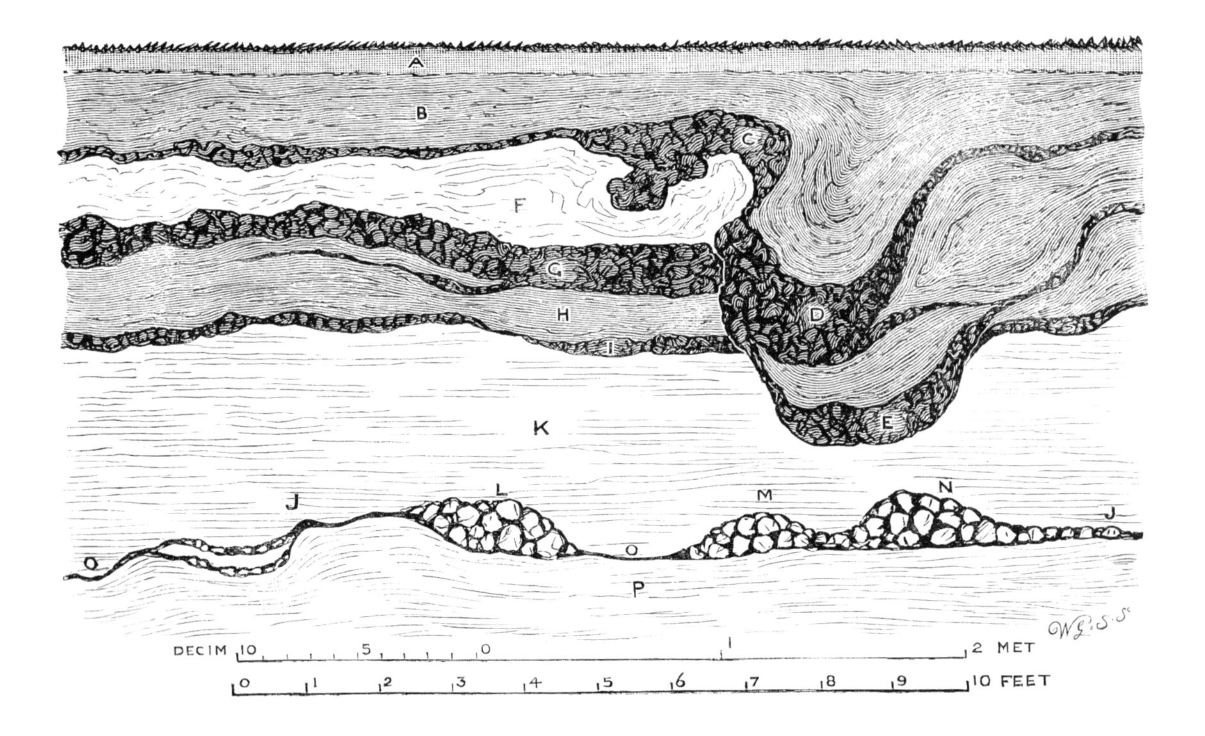

Worthington Smith moved from London to Dunstable in 1885. As noted in a previous chapter, he soon recognised the presence of palaeoliths on interfluves in positions well above the current valley systems, undoubtedly in response to Harrison and Prestwich's work on the Chalk Plateau (Smith, 1889). Smith had been friendly with Harrison since 1878 (Harrison, 1928), but was no eophile. The work for which he is best known, *Man the Primeval Savage*, was published in 1894, and dealt with his discovery of *in situ* Palaeolithic knapping floors at Caddington and elsewhere on the Dunstable Downs. Worthington Smith discerned a sequence in the archaeology of Caddington. Figure 10.3 reproduces one of his sections and gives a good impression of what the locally variable stratigraphy was like. The archaeological–geological sequence was as follows. Originally there was boulder clay capping the hill tops of the Dunstable downs and adjacent areas of the Chiltern Hills. This was glacial. Sometime later humans arrived. They made rather crude handaxes which could now be identified by their distinctive ochreous staining. Later, at Caddington, at a somewhat lower elevation in the landscape, there was an extensive occupation surface and knapping floor – the Palaeolithic floor. The Palaeoliths from here were similar to the others except that these were unstained, and a little more advanced in technique. These people had scrapers and other flake tools not present in the higher and older deposits. The handaxes from the *in situ* floor were identical to the normal river drift handaxes found in the valley gravels, so the Caddington floor evidently dated to the main drift period. The whole sequence was technically post-glacial, and there was not that much difference between the stained and ochreous handaxes and those from the floor. The contorted drift-like deposit that overlay the capping brickearth (Figure 10.3) was a return to cold conditions as gravels and clays sludged down from higher ground. These deposits contained the older ochreous palaeoliths, solifluction explaining how slightly cruder and older handaxes ended up as stratigraphically higher than later better made ones.

*Figure 10.3. A section from Cadington as depicted by Worthington G Smith in* Man the Primeval Savage *(1894). A. Surface material; B. Tenacious red-brown clay; C and D and E. Sub angular gravel which has 'ploughed' its way through lower deposits (F, G, H, I, K); C contains brown ochreous handaxes and flakes slightly worn; F is a grey-white clay (possibly boulder clay); G is a gravel with unapraded pale white handaxes and flakes, the flakes refit; H stiff red clay with implements and flakes like in G; I gravel same as G; J–J and O and O, Palaeolithic floor; L and M and N are heaps of flint blocks brought by Palaeolithic humans and stock-piled for knapping into handaxes; these are lying on the floor; K and P are brickearths overlying and underlying the Palaeolithic floor.*

Exactly when the more ancient land surface was destroyed is not stated, but Worthington Smith's work (as well as Hitchin and Hoxne) highlights a developing contradiction. The contorted drift-like sediments (C, D, E, on Figure 10.3) at Caddington (these had been present in Smith's London sites, see Figure 7.6) were also a cold climate deposit, and since the worn palaeoliths in them pre-dated this deposit (because they were eroded and emplaced by whatever mechanism laid it down), yet post-dated the boulder clay/glacial, they must technically have been *intra-glacial* in age. (This same argument would have applied to Hitchin and Hoxne). Did intra-glacial mean interglacial? For those who accepted Geikie's multiple glacial/interglacial theories there was no real contradiction here. But for the southern English submergence school, particularly those who believed in a single glacial event, how did you explain this end of Pleistocene cold event? Was it glacial? Prestwich, as we have seen, was arguing for a post-Pleistocene submergence as well as a return to glacial conditions at the end of the Pleistocene, although I suspect not many agreed with him.

Smith did not clearly resolve this in *Man the Primeval Savage*, but he did offer the following explanation:

Professor G.H. Darwin (Charles's son) had suggested that the last glacial period occurred 100 thousand years ago. However, the Reverend Osmond Fisher (Davison, 1900), whom we met earlier, a noted geologist interested in the age of the earth and mathematical calculations on the date of the ice age, suggested 110 thousand years ago for the glacial layer labelled the 'trail' (implying that the boulder clay was therefore older again). But there was a similar layer called the 'warp'. These labels, warp and trail, were introduced in earlier chapters, and were used to describe thin contorted horizons usually overlying sediments with handaxes in them. At some sites one or the other was present, at a few, both. They were evidently episodes of intense disturbance, often thought to date to the end of the Pleistocene.

Fisher considered the warp to be have formed 30 thousand years ago as a result of late Pleistocene glacial-like conditions. Smith thought the contorted drift at Caddington, as well as that in Stoke Newington, the equivalent of the warp. The warp was present at both sites because the Palaeolithic floors at each location were part of the same extensive Palaeolithic land surface, continuous between the London and the Hertfordshire downs. If the warp was a glacial deposit, a result of glacial conditions predating the final melting of the ice sheets, then the human occupation had by definition to be at least intra-glacial.

Evans, in the second edition of *Ancient Stone Implements*, continued to promote a monolithic view of the river drift in the post-glacial Palaeolithic. There was little room for sequences of progressive development in his outlook. The Palaeolithic of the river drift showed no evolution. He glossed over the Caddington data implying that the ochreous implements, with their lack of associated scrapers were merely the sweepings of contemporary living floors located away from the implements' manufacturing floors. The older palaeoliths in the lower gravel at Stoke Newington, Skertchley's evidence in Norfolk, and John Allen Brown's carefully mapped out sequence (Figure 7.7) of palaeoliths in gravel units above the high terrace of the Thames valley (Brown, 1887) were not engaged with.

But it was becoming increasingly difficult to ignore such evidence. Although John Allen Brown's linking of the river drifts to various stages of marine submergence and re-emergence was becoming dated by the late 1890s, it was nevertheless embedded within the concept of a climatically complicated ice age. The cave evidence from Creswell Crags and Kent's Cavern was the stratigraphic proof of that complexity. Hoxne, Hitchin, Stoke Newington and Caddington were equally persuasive. Times were changing. The Scottish terrestrial glaciologists were gaining ground (Deeley, 1893), and the English submergence school were appearing ever more parochial. The third edition of James Geikie's '*The Great Ice Age*' (1894) was now promoting six glacial periods and five interglacials. The second edition in 1877 (G.J.H., 1895) had argued for only two or three interglacials.

Workers like Henry Stopes (see above) were taking a somewhat different approach. He largely ignored the geological evidence, preferring to turn to the developmental stages present in the evolution of material culture (Stopes, 1893; Stopes, 1894). Human technological skills, as well as tool uses and applications, marched in lockstep with cultural and biological progress. His outlook was more in keeping with Lane Fox's from the 1860s and 1870s. Comparing Stopes' work with that of Evans, the two were methodologically worlds apart. Evans worked by observation and analogy; he argued that similarities in tool form and manufacture implied parity. He largely ignored interpretations of function. But Stopes used a more inductive approach (Stopes, 1900). He assumed outline form masked a variety of very different uses and consequently classified stone tools on edge damage, as well as on the character of their edges' potential for a variety of uses. In effect he produced a functionalist classification based upon the uses he *thought* a tool would be suitable for (an approach liable to have driven the cautious Evans insane with frustration). This allowed him to develop long sequences tracking the evolution of particular tool types, often into the historical period and later.

Stopes spoke frequently at the BAAS meetings and at the AI, exhibiting collections to support his case. His utilitarian outlook predisposed him to accept eoliths. For Stopes, before humans had shaped tools they had picked up natural stones and used them. Then, they picked natural stones they judged suitably shaped for particular jobs, (Harrison's eoliths were fitted in to Stopes sequence here), later graduating to the limited modification of an edge to enhance suitability (actually, it is here that Stopes should have placed Harrison's eoliths). Stopes characterised this stage as transitional between 'selected and used', and 'worked and used.' These transitionals, reminiscent of Shrubsole's examples from the Berkshire plateaux, were often large and roughly worked pieces, and for some reason commonly left-handed. The palaeoliths succeeded these with their variety of different forms related to different tasks.

In the judgement of most of his peers, Stopes's work was uncritical and based on unwarranted assumptions about function. Yet I believe there was a sound contemporary logic to it. He grounded it in a shared evolutionary belief in progress where physical skills, expressed through material culture, would be evolving as humans did. Lane Fox had argued much the same decades before. In this sense Stopes anticipated the culture historical approach that was still years away, when developmental sequences of material culture could be discussed independently of stratigraphy, and even used to date particular geological layers (McNabb, 2007; McNabb, 1996).

Stopes is also significant in that he was the first researcher to promote the significance of the drift deposits at Swanscombe in a stratigraphic context (Wenban-Smith, 2009). The great chalk pit at Milton Street, Swanscombe, was known as a prolific source of handaxes (McNabb, 1996), but Evans in 1872 and 1897 largely ignored it. In 1899 Stopes discovered the Greenhithe shell bed (Stopes, 1899), a layer within the higher terrace of the Thames' valley drift with a very rich molluscan fauna and associated with palaeoliths. The precise location is now unknown, but it was close to the site of Dierden's Yard on the west side of the narrow valley known locally as the Ingress Vale. The Galley Hill skeleton described below was discovered only 500 yards to the north-east in the same gravel, and the Milton Street pits were a stone's throw away. Significantly there were eoliths within the deposit, but no mention was made of their being in a different condition to the palaeoliths (Newton, 1901).

I will close this section by reference to a debate that we have already come across very briefly, and which continued throughout the late Victorian period, that of the relationship between the Palaeolithic and the Neolithic (Shone, 1894; Smith, 1894; Howorth, 1901; Watson, 1901). In 1893 John Allan Brown gave a very detailed paper to the AI on the transition between the two periods (Brown, 1893). Evans, in both editions of *Ancient Stone Implements,* had denied that a bridging series of artefacts existed to fill the gap between the two quite distinct series of tools, and Boyd Dawkins had consistently maintained that the faunal record of the two periods was quite separate. In the former phase the great Pleistocene mammals had thrived and then become extinct. In the latter period, although a number of lesser species continued into the Neolithic, the key elements in the fauna were the domesticated species which never occurred in the drift. However, Dawkins had been unsure as to how long the hiatus between the two periods was (a limitation of the faunal method).

Brown took on both the faunal and the archaeological record and, in reference to the latter, provided a series of distinctive implement types that did indeed appear to show a development from palaeoliths of the drift through to the Neolithic axes. He defined these as Mesolithic after a term first coined by John Lubbock in the 1860s. In fact he argued for an unbroken sequence of artefact development from Harrison's eoliths all the way through to the Neolithic. He also provided a geological context for the Mesolithic. This was in the head and solifluction deposits, and brickearths, in the dry valleys on the chalk downs. They were exactly where, topographically, such transitional assemblages should be located in the landscape, mid-way between the Palaeolithic/drift and the Neolithic/surface sites. He amassed an array of evidence using British and European cave data to show that a Mesolithic phase characterised by its own unique tool forms was widespread both in Britain and abroad (Brown, 1889).

The following year Boyd Dawkins responded (1894a; 1894b), restating his usual case on the fauna. The two periods were characterised by suites of different animals, with domesticates present only in the Neolithic. However, his denunciation of the archaeology (Dawkins, 1894a) took an interesting line. At the AI he exhibited a large set of implements in three groups. The first were from a flint mining site on the chalk downs, Cissbury, near Worthing, which Brown had asserted contained clear evidence of Mesolithic axes. Dawkins showed that a number of Palaeolithic-like axes were present, as well as a number of Mesolithic-like examples, all in the flint heaps from the site. They were in fact all failed or rejected attempts to make Neolithic axes. The site was a Neolithic flint mine and axe factory. Boyd Dawkins made the point (Dawkins, 1894b) that the Mesolithic axes from the dry valleys of the South Downs were unstratified, the implication being they too were probably Neolithic discards.

His second group of artefacts came from the Creswell Crags cave sites. He stated that in the *in situ* cave earths associated with handaxes, were artefacts which resembled eoliths. He had made similar remarks before to the puzzlement of his audiences. He was arguing that these were the debris from the manufacture of handaxes. The two types of flints, eoliths and manufacturing waste, were visually very similar. This, he said, proved the eoliths were merely the 'wasters' of knapping episodes which dated to the river drift period. Their celebrated edge damage and ochreous colouring was a result of local conditions and quite incidental. This opinion explains comments such as that quoted above by A.M. Bell in his 1894 letter to Benjamin Harrison, that Dawkins could 'match' the eoliths with examples of his own. Finally, he exhibited a group of implements from America which he had acquired on a visit in 1880. He noted that Palaeolithic drift handaxes had never been found in any cave deposits, and that when similar forms were discovered it was on the surface and usually in association with Native American material culture. The identification of the implements as true handaxes was therefore dubious, and there was no clear example of an uncontaminated association between an undoubted handaxe and a North American Pleistocene river drift deposit. His argument was prescient. Within three years a sea change would occur on the whole question of the Palaeolithic occupation of the eastern United States, prompting D.G. Brinton to launch a devastating critique of the evidence supporting it (Brinton, 1897).

The whole of Boyd Dawkins's argument really centres on the use of analogy and association. He was asking that only artefacts from *within* secure geological contexts be used in establishing cultural comparisons, and that the safest type of data for determining developmental sequences was that which came from stratified localities. This cut to the very heart of the eolith debate as the majority of them were surface finds and therefore from insecure geological contexts (their status as eroded *in situ* remnants was not accepted by the eophobes). Their stratigraphic context depended on validating a geological hypothesis – Prestwich's interpretation of the erosion of the Vale of Holmesdale. However, Boyd Dawkins's argument did rather fall down on the cave evidence as he had clearly misunderstood the natural character of the eoliths – they were not waste flakes from knapping, as Montgomerie Bell had pointed out in no uncertain terms (Bell, 1894). The force of Boyd Dawkins's observations were lost on his audience as John Allan Brown (Boyd Dawkins, 1894a) showed when he argued that the presence of Native American artefacts alongside supposed Palaeolithic implements in the caves just proved how long-lived many of the simple tool forms really were. Henry Stopes would have appreciated such an opinion.

## Eoliths and Anthropology

In terms of physical anthropology there were significant developments during the 1890s. In the late 1880s at Galley Hill, Northfleet, northern Kent; a partial skeleton was discovered. It was found in gravels of the 100 foot or higher valley terrace, at a locality which was known for producing palaeoliths (Swanscombe and the rich Milton Street pits were close by – see Figure 7.5). Stopes (Stopes, 1899) stated that it came from near the mouth of the Ingress Vale. The report on the skeleton (Newton, 1895) mentions eoliths being found in the same gravels, but as they were deeply stained, the implication is they were derived (Stopes's paper on these (1900) made no such claim).

It will be recalled that by general expectation Palaeolithic humans would be similar to extant modern peoples, a bit more robust or hirsute perhaps, but recognizably human. They would not be man-apes, or ape-men, this was a stage in human development that had occurred outside of Europe, long before Palaeolithic man had arrived. The Galley Hill skeleton conformed to type, it was clearly human. In comparing it to living races of modern humans, and to fossil skeletons, E.T. Newton (1895) had difficulties in assigning it comfortably to a particular group. The skull was dolichocephalic, but more so than the majority of the Neolithic long barrow people. The jaw was not heavily built as in the Neanderthals from Spy, and it had a projecting chin which Neanderthal specimens lacked. However the face had been quite prognathic and the body size, as reconstructed from the femur, short and squat. Galley Hill was unlike the tall Cro-Magnon cave race. The skeleton did not really fit comfortably anywhere, either with living peoples, or extinct ones. Newton's preferred interpretation was that it was a Palaeolithic ancestor to the Neolithic peoples of Britain, but an ancestral race which had developed differently in the north-western corner of Europe, and for which no other examples had yet been discovered. Wisely, Newton left the door open on this one.

Right from the very beginning there was doubt about the skeleton. It had only been possible to make very preliminary observations on the exact site of the discovery before it was quarried away. The remains were described in detail at a meeting of The Geological Society of London on May 22nd 1895, seven years after they had been found. John Evans and Boyd Dawkins, in the audience that night, believed the skeleton to be a later intrusive burial into the Pleistocene gravels (which in fact it was). The remaining four discussants in the audience all accepted its authenticity, including John Allen Brown and Lewis Abbott. The discussion also engaged with the relationship between the Palaeolithic and Neolithic races, the archaeology of which has already been mentioned.

While some archaeologists were happy to create long typological sequences of artefact progression from the Pliocene to the Neolithic, they were entirely divorced from any corroborative data from physical anthropology, and so not very helpful to questions of racial origin. The small size of the British skeletal data set continued to hamper research. On the other hand such long archaeological sequences did serve to remind researchers that connections between cultural stages could, and perhaps more importantly *should* exist. The argument about the Mesolithic exemplifies this. I think that by the 1890s the question of a continuity between the periods was increasingly being debated: quite simply, its time had come. Bridging the gap would open the door to the possibility of demonstrating much deeper racial roots. Here, archaeology was a proxy for physical anthropology.

Controversy also surrounded another partial skeleton, that of the Javanese remains of *Pithecanthropus erectus* (now *Homo erectus*) from the banks of the Solo river, near Trinil in Java. It was discovered by the Dutch medical doctor Eugène Dubois (Dubois, 1896; Shipman, 2001). The remains were two teeth of large size, a skull cap, and a thick femur. They were mineralised and in the same condition as extinct Pliocene faunal remains found close by. During the middle 1890s Dubois took the remains on a European tour visiting the learned societies in the major European capitals in order to persuade people of its transitional character, part way between apes and humans (Leakey and Slikkerveer, 1993). Not surprisingly opinions were divided. For some authorities the remains were too human-like (especially the femur) to be transitional, others argued that the remains were mixed, the skull cap was simian and the thigh bone human (C.W.A., 1895). In Britain, the general public were treated to a discussion on *Pithecanthropus* in the pages of the *Nineteenth Century* by no less a person than the émigré Russian prince and exiled revolutionary Peter Kropotkin (standing in for T.H. Huxley in the magazine's recent advances in science section; see Desmond 1998 for further details on Huxley and Kropotkin). He provided a thumbnail sketch of the shifting fortunes of *Pithecanthropus*' journey across Europe (Kropotkin, 1896). In Berlin the reception had been hostile, in Dublin supportive.

A paper on the remains by Professor D.J. Cunningham of Trinity College Dublin, to the Royal Dublin Society in January 1895, had been sympathetically received but with the usual concerns about the integrity of the remains (Anon, 1895a; Anon, 1895b). When Dubois himself spoke in the Irish capital in November 1895 he seemed to convince his audience that the remains all belonged to the same creature. Much of the discussion was on the relationship of *Pithecanthropus* to modern humans and Neanderthals. One of the principal discussants was Professor Cunningham himself who neatly summarised the issues in a diagram, see Figure 10.4, which contrasted his position with that of Dubois. Cunningham's interpretation was reported thus:

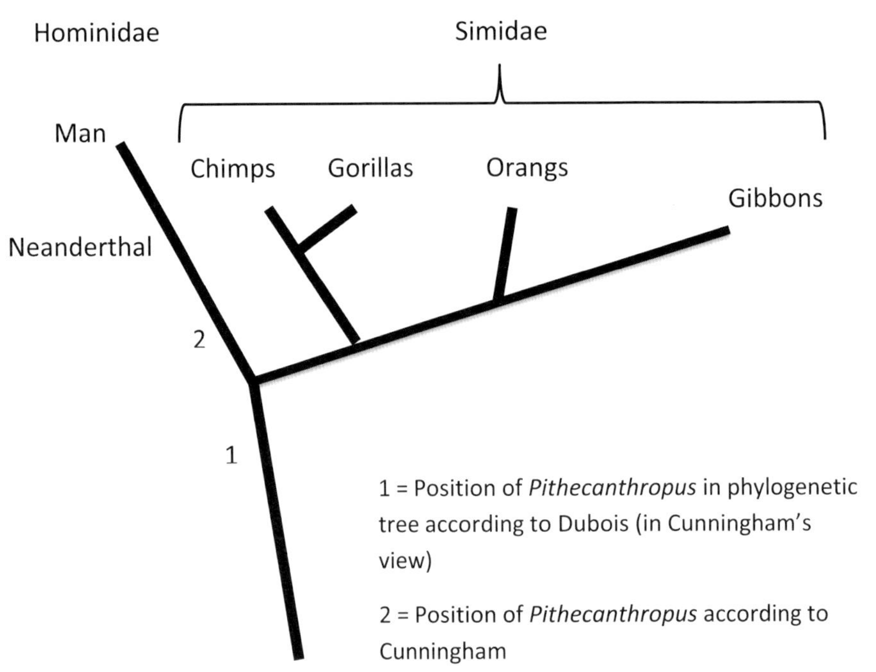


*Figure 10.4. The contrasting interpretations on the relationship between* Pithecanthropus *and* Homo sapiens *as argued by Eugène Dubois and D J Cunningham and published in* Nature. *Redrawn after* Nature.

> …he could not believe an ape-form with a cranial capacity of 1000 [i.e. internal skull volume measured in cubic centimetres as a proxy measure of brain size] could be the progenitor of the man-like apes, the largest of which had a capacity of only 500. Such a supposition would necessarily involve the assumption that the anthropoid apes were a degenerated branch from the common stem. (Anon, 1895b, 115f, my brackets)

For Cunningham the question revolved around *Pithecanthropus*'s relation to the modern apes, while Dubois was focused more on its relationship to humans. Professor William Sollas (1849–1936), who chaired the meeting, suggested that the deposits in which *Pithecanthropus* was found might equate with the Forest Bed in Norfolk. He considered the physical differences between the Javanese finds and Neanderthals, and those between Neanderthals and humans, as 'proportional'. A young Arthur Keith (1866–1955), speaking in London when Dubois exhibited the remains before the AI, supported a similar age for the remains. Keith was destined to become one of the most senior anthropological figures of the early twentieth century, and was intimately associated with Piltdown. For Dubois, *Pithecanthropus* was a representative of the human race in the Tertiary, albeit a transitional one. But for Keith *Pithecanthropus* was, quite simply, 'Pliocene Man'.

Keith's position would have raised intriguing questions for the audience. If *Pithecanthropus* was a progenitor of humanity, and if it was of a similar age to the Pliocene Forest Bed, how long would its evolutionary journey

*Figure 10.5. Upper image* Pithecanthropus europaeus alalus *painted by Gabriel Cornelius von Max for Ernst Haeckel and presented to him on his 60th birthday in 1894. Lower image a rival interpretation of* Pithecanthropus alalus *commissioned by Rudolph Virchow and also drawn by Max.*

have been between Asia and Western Europe? Where and when was the transition from *Pithecanthropus* to *Homo*? Were Harrison's eoliths made by a *Pithecanthropus* like creature? Did Pithecanthropus sit at the root of a single evolving human branch, or was it related to other branches (other races), and if so which ones?

A significant part of the problem was that the late Victorians did not really have any concept of what a transitional creature of the Tertiary age might really look like. In 1894, to celebrate Ernst Haeckel's sixtieth birthday, the artist Gabriel Cornelius von Max (1840–1915) had produced a painting of Haeckel's *Pithecanthropus,* which was entitled *Pithecanthropus europaeus alalus.* Haeckel had named this putative ancestor in the 1870s, long before any evidence of it existed, on the basis of it being a necessary stage in an evolutionary sequence. In so doing he popularised the concept of the missing link. This had inspired Dubois to go out and search for it.

Von Max's rendition of speechless man showed a creature more ape-like than human, apart from the nascent humanity implied in the direct gaze of the female as it stares out at the observer

– Figure 10.5. Professor E.P. Evans, commenting on the painting, believed that humanity was best demonstrated in the single tear running down the mother's cheek (Evans, 1894). However, this wasn't a creature in the process of evolving into something else.

A section of Nicholas Ruddick's *The Fire in the Stone* (Ruddick, 2009) is a critical analysis of Prehistoric fiction in the late Victorian era. He reproduces a number of illustrations of so-called cave men from a variety of fictional stories, both British and European (see Chapter 13). While evidently not for academic consumption, these images reflect broader public perceptions of Palaeolithic humans or their Tertiary ancestors. These are what the public thought (or were being told to think) human ancestors would have looked like.

*Figure 10.6. Sketch by T.H. Huxley entitled 'Homo Herculei Columnarum' drawn on July 19th 1864 at the Athenaeum Club. The occasion for this sketch may well have been the visit by Hugh Falconer and George Busk to Gibraltar to visit excavations in the Genista Caves. The Gibraltar Neanderthal skull, first found in 1848 was brought to the BAAS in Bath in September of the same year.*

I suspect many academics would have been sympathetic to these perceptions. They all reflect one of two themes – either a modern human is depicted wielding an axe or club and wearing skins or, alternatively, an ape-like creature is shown. These latter illustrations are, effectively, renditions of apes, but they have been drawn upright and bipedal, or almost so. They carry axes and tools to further emphasise their nascent human status. A light-hearted attempt from 1864 by T.H. Huxley, Figure 10.6, has a muzzle, bent knees and splayed divergent big toes; it seems to have the vestige of a tail. All very ape-like, yet the sloping forehead is Neanderthal and the bipedal stance human, but the hunched shoulders and stoop suggest the bestial and primitive. On the other hand it carries a handaxe hafted in the way many Victorians believed they were used – like a tomahawk.

The difficulty of how to realise a transitional creature is superbly demonstrated in a reconstruction that Dubois had made of *Pithecanthropus* for the Paris Exhibition in 1900, shown in Figure 10.7. It is neither ape nor human. It provides a useful insight into the difficulties that Victorians had in visualising transitional forms. It is depicted as a mixture of elements from both (like Huxley's image). This is more like Baron Frankenstein's monster; an accommodation of body parts; rather than an animal shaped by evolution. Dubois explained the reasoning behind his image of *Pithecanthropus* in a pamphlet written to accompany the reconstruction (Dubois, 1900). The Trinil femur indicated an upright bipedal animal, but certain muscle attachments suggested a more ape like condition which led Dubois to suggest that *Pithecanthropus* was still a tree climber. This lay behind his reconstruction of the big toe as divergent, and the shorter lower leg by comparison with the longer forearm, all necessary for even a partially arboreal life style. The hand, clutching the deer antler, is also proportioned mid-way between humans and apes, the thumb shorter and the fingers longer. Originally the left hand held the other part of the antler rack (ready to throw), and at its feet was a flint flake with which it had chopped off the antler tine. The use of an antler tine rather than a handaxe, or even the stone flake, served to further emphasise its primitive ancestral status. The face was a careful reconstruction based on features drawn from both apes and humans. Surprisingly, Dubois also used a fragment of a fossilised chin found in the same layer as the Trinil finds, but from a locality some kilometres away. Facial features of modern indigenous peoples were used to model an appearance part way between those of the 'lowest' extant human races and the anthropoids. As with Gabriel Max's painting, however, it is the contemplative gaze that reveals its evolutionary potential.

The only other remains discovered during this period were in 1900 (Layard, 1901). A skull was dredged up from a peat layer at the base of the river Orwell in Ipswich, Suffolk. As such it was likely to be Neolithic, though identification was not helped by it being stuck on the end of a pole adorning the dredger for nine months before it was rescued by Nina Layard (1853–1935) whose excavations at Foxhall Road, Ipswich, after the turn of the century would reveal what was in all probability an *in situ* Palaeolithic occupation site (White and Plunkett, 2005).

As Worthington Smith had so presciently noted, bad fortune seemed to attend the discovery of Palaeolithic human remains in Britain.

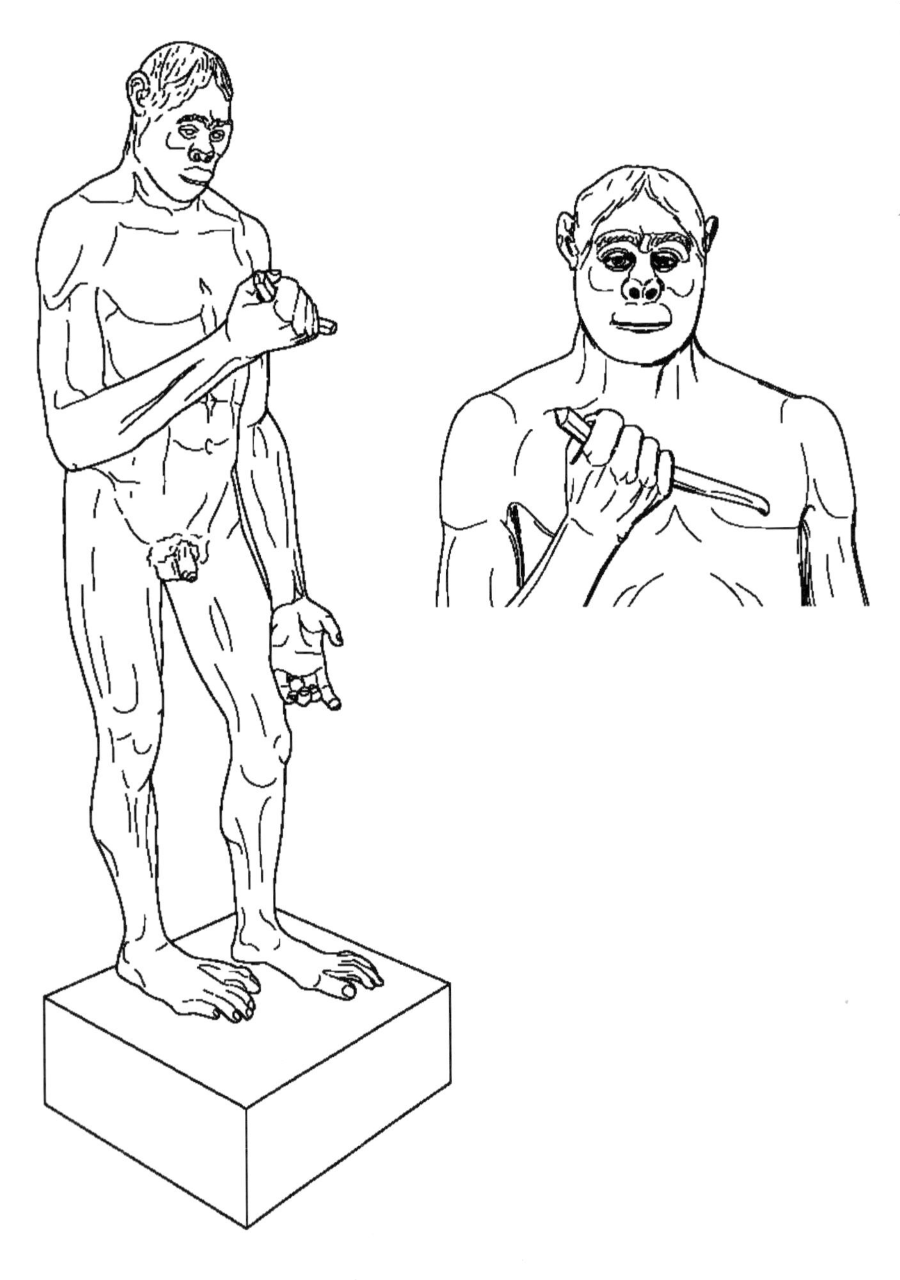

*Figure 10.7. Sketches of Eugène Dubois' model of* Pithecanthropus erectus *made for the Paris exhibition in 1900. Drawn by Penny Copeland from photographs taken by the author.*

# 11

# THE VIEW FROM THE HILL
## *Human Evolution in the 1890s*

In Chapter 1 we left the Time Traveller sitting on a hillside gloomily contemplating the ruins of a future London. As the preceding chapters have shown, there was enough doubt and contradiction in evolutionary science in the late Victorian period to prepare a man of the mid-1890s for the twin eventualities of the Eloi and the Morlocks. I will begin by reviewing the key points from the preceding chapters. This will represent the Time Traveller's view of human evolution (or at least my interpretation of it). The Time Traveller was a late Victorian gentleman scientist, but he can also represent the well-educated and scientifically informed Victorian man in the street. This short chapter is about what he knew and where he got that information from.

### The story so far

The earliest phases of human evolution had occurred 'elsewhere' and, despite claims on the continent for an early human presence, there was no substantial British evidence to fuel debate. The earliest origin and home of humanity was somewhere 'in the east'.

- Consequently, British archaeological disputes concerning human origins were confined to questions of the timing of the first occupation of Britain. The eoliths of the Chalk Plateau were very old and very primitive. a local phenomenon that could be engaged with. The evidence was on the doorstep so the eolith question became a British archaeology of origins – autochthonous or allochthonous.
- It was commonly held that the palaeoliths of the river drift had arrived with a people who reoccupied Britain after the glacial period (or as some believed during, or possibly even before). They too came from elsewhere and they brought the knowledge of palaeoliths and their manufacture with them. By the late Victorian period the archaeology of the river drift had long since ceased to be an origins debate.
- The makers of the palaeoliths were considered to be essentially modern human in appearance. When the Galley Hill skull was found, despite it actually being of Neolithic age, no one expressed much surprise. However the eoliths were older by several orders of magnitude. What

would their maker have looked like? The Neanderthals were essentially human looking as well, and had large brains similar to modern humans. *Pithecanthropus* was a puzzle. Its brain size was larger than an ape's, but smaller than a modern human's. Dubois claimed it was a genuine transitional animal, but its relevance to human origins depended upon which evolving lineage it was a part of and where along that line it should be located. With nothing else like it and not enough information to situate it, *Pithecanthropus* was an intriguing anomaly.

- The prevailing theoretical paradigm of the time was evolution. Evolution was largely seen as progressive and linear. Darwin's own version of selection was unpopular, but the more generally conceived notions of progressive transmutation and progressive time were widely accepted. They translated comfortably into social and cultural anthropology. While not every anthropologist or archaeologist would have accepted all of the theories of E.B. Tylor and his particular brand of evolutionary anthropology, the basic approach would have been familiar to most. Palaeolithic archaeology had no particular theoretical position of its own at this time. It drew on anthropology for its theories of origins and development; particularly in the early years.
- No one had as yet come up with a rival interpretation to Prestwich's explanation of the formation of the Chalk Plateau and the North Downs. Although the mechanisms he proposed may have been disputed, the relative sequence of events was accepted by many (e.g. Evans, 1897a). This was therefore the only chrono-stratigraphic framework within which to explain the eoliths. This constrained the debate as well as directed it.
- The late Victorian archaeologists had no real conception of the spans of time they were dealing with or of the temporal distances between different events. There was widespread engagement with the dates offered by people like Croll, G.H. Darwin or Fisher, and disputes over whether parts or all of the Pleistocene should be contracted or expanded by so many thousands of years, but the *time separating different events* was anybody's guess. The occasional fixed date for an event did not really help; there were too few to make them meaningful. However, what the relative chronology did convey was the immensity of time involved. Paul Broca had once noted that the Vézère river had eroded its valley by 27 metres between the time that the Le Moustier rock shelter was occupied and people inhabited La Madelaine (Allen, 1882). But it had only eroded down another 5 meters between the time of La Madelaine and the present day.
- There was still no viable theory of inheritance that could be comfortably linked with a mechanism that explained how inherited traits were moulded into new species and races. A fixed chronology in real years would have helped as it would have set limits within which

particular biological events in the broader stratigraphic record could have occurred. It would have also helped in refining the geographical and temporal relationships of the historic and Prehistoric races. The chronological mind set of the Victorian scientist was a relative one.

- This had a knock on effect that brings us back to the first three bullet points. Because time was still conceived relatively, the amount of time by which eoliths predated the palaeoliths from the Chalk Plateau could not be gauged; the same with the amount of time separating the eoliths from the palaeoliths in the valley drifts. I believe this served to conceptually separate the eoliths, to cut them adrift from the main course of interpretable events. *Pithecanthropus* couldn't bridge the gap between eoliths and palaeoliths because nobody understood the fossil's relationship with the European evidence. For some researchers the archaeology itself showed that a link was present between the eoliths and all later tool forms, so physical anthropology needed to catch up. At least the plateau palaeoliths, and those from the river drift, were still recognizably palaeoliths. John Evans had admitted them all into a Palaeolithic stage, but not the eoliths. The eoliths were too remote to be accommodated in any meaningful sequence despite the claims of a few like Harrison and Stopes.
- But prevailing theory came to the rescue. Evolutionary anthropology required a Tasmanian-like stage of development, and all evolving lineages need a beginning. So the eoliths perplexing dissimilarity to everything, and their temporal remoteness were used to situate them at the beginning of the evolutionary sequence in Britain. For British scientists the eolith debate became an archaeology of origins because of their anomalous position. Moreover, because the links to later archaeology were still tenuous it could be debated in isolation.
- A final point on evolutionary perspectives in late Victorian science: there was a strong synergy between different branches of science imparted by the broader paradigm of evolution itself. This was a function of the perception of linearity inherent in both geological time and evolutionary progress. We have seen a number of examples of it in preceding chapters. It was present in the recapitulation theories in biology, in the orthogenesis theory, and in the more popular Lamarckian outlook. It is little wonder that when a museum display, or exhibition at the BAAS, showed a sequence of stone tools with eoliths at one end, the overall impression would have 'seemed right' to the audience. Elsewhere (McNabb, 1996) I have applied the label 'progressive time' to this whole concept.

## The Time Traveller's View of Evolution and Human Origins: the Informed Opinion of a Gentleman of Science

As a man of science, the Time Traveller would be aware of much of this, and receptive to the broader debates of his time, even if he would not necessarily have constructed it in quite the same way. I characterise him as a good representative of the late Victorian scientific gentleman. Another good example, this time more anthropological, is Dr Mortimer from Conan-Doyle's *Hound of the Baskervilles*. The Traveller was most likely a reader of *The Nineteenth Century* and a number of other popular magazines and journals which carried scientific articles. I have no doubt that he was a regular attendee of the BAAS meetings and although he would have spent most of his time in the science sections, he would probably have attended some of the anthropology debates in section H and possibly some of the geological debates in section C. I also have no doubt that the Time Traveller was an evolutionist and probably familiar with the works of Tylor and Lubbock and others. I would not be surprised to learn that he was a member of the AI as well, possibly acquainted with John Evans, Joseph Prestwich and others. Very few Victorian scientists were specialists in only one field of study. This luxury belongs to a later period when the sciences had been professionalised and individuals lacking independent financial means had to follow paid careers in only one branch of research.

The Time Traveller experienced no difficulty in rationalising the world of the future. As he sat on his seat of strange yellow metal surveying the Thames Valley he came to the conclusion that he was witnessing a moment at the end of human history.

> It seemed to me that I had happened upon humanity on the wane. The ruddy sunset set me thinking of the sunset of mankind. (Wells, 2005, 31).

And as he thought more about the garden city before him with its palaces and domes, he realised it was set in a perfectly controlled environment. Everywhere beautiful flowers bloomed, but the pollinating insects had been eradicated. There came another realisation; the world of the Eloi lacked any form of competition. Struggle, for whatever reason, had been eliminated. There were resources for all, adequate to the needs of all. The perfect balance of humanity in nature had been achieved. The Time Traveller then reasoned:

> But with this change in condition comes inevitably adaptations to the change. What, unless biological change is a mass of errors, is the cause of human intelligence and vigour? Hardship and freedom: conditions under which the active, strong, and subtle survive and the weaker go to the wall; (Wells, 2005, 32)

In contemplating the child-like Eloi, the Time Traveller's thoughts were beginning to turn to ideas of degeneration and reversion, concepts that he would have been familiar with through anthropology. They were also concepts that fascinated late Victorian society in general. For example, readers of the *Nineteenth Century* in February 1895 would have encountered an article by B. Kidd entitled 'Social Evolution' (Kidd, 1895), which was a response to critics of his celebrated book of the same name. In the article he asserted:

> Except on one condition, the inherent tendency of all the higher forms of life appears to be toward retrogression and degeneration. There is only one way, to all appearance, in which this tendency has ever been held in check; namely, by prevalence of conditions in which selection can prevail, and each form be enabled to carry on its kind to a greater degree from individuals rising above the average to the exclusion of others falling below it. (Kidd, 1895, 228)

The implication of the quote was that degeneration was a default condition but that natural selection acting through progressive time could hold it at bay. H.G. Wells was at odds with the vision of progress that Kidd, and other sociologists were promoting, and debated the point with them in magazines and various periodicals (Pearson, 2007). On the other hand, in August of the previous year, H.P. Dunn had confidently asserted that there was no tendency toward degeneration (Dunn, 1894). Also writing for the general public in *Nineteenth Century*, he offered as proof of this the continuing attractiveness of blonde Anglo-Saxon women (brunets and redheads it appears did not get a look in). More seriously, E.B. Tylor had also asserted that there was no natural inclination for races to decay (Tylor, 1878). He was adamant that the Tasmanian culture was not a result of degeneracy (Tylor, 1894b), it was an Eloi-like stasis. When degeneration was observed, the explanation was almost always a result of a detrimental external influence. In cases where the causes appeared to come from within the society, then the retarding influences of outmoded social traditions were at fault. There were even statistics to back this up. The Anthropometrics Committee of the BAAS had sat between 1875 and 1883, reporting annually in the *British Association Reports*. In collecting and collating information on the physical characteristics of the British population, such as height, weight, breathing capacity, shape of head etc., they addressed the popular conception that degeneration was evident within the population, particularly in the 'manufacturing towns of northern England'. The stultifying effects of urban life were a long term concern for Victorian liberal thinkers. Over the years the committee sat, they found no evidence for any such degeneration – perhaps to their surprise. The data showed that the English 'professional classes' came out near the top of the biological league table when compared to other races – although the Scandinavians were dangerously close to the English in their national averages.

Regression and degeneration were aspects of a broader angst within society that expressed itself in art and in literature, as the Time Traveller would have been aware. It is sometimes called *fin de siècle*:

> In its widest sense *fin de siècle* was simply the expression of a prevalent mood: the feeling that the nineteenth century – which had contained more events, more history than any other – had gone on too long, and that sensitive souls were growing weary of it. (Bergonzi, 1961, 3)

Bergonzi has argued that H.G. Wells's early writings, his romances, were steeped in this late Victorian phenomenon. (The Time Traveller may well have been familiar with some of the more scientific of Mr Wells's stories) and it pervaded archaeology and anthropology as well as the arts (Müller, 1892). Had the Traveller been at the BAAS in Leeds in 1890, he would have heard papers on degeneration by Henry Stopes (1890), and Nina Layard (1890). She argued that true reversion to a more primitive state was unlikely because the proportion of the 'ancestral blood' retained after many generations would be minimal. Her paper underlined the continuing need for a workable theory of inheritance. Layard articulated a key notion; that the 'why' and the 'how' of regression were intimately related. Similar regressive trends were observed in Victorian biology too (Bowler, 2003, p244, 248, 307). Wells's friend Edwin Ray Lankester, a tireless populariser of science, wrote extensively on the topic for professional scientists and the public alike. Wells himself wrote a piece on 'Zoological Retrogression' for the *Gentleman's Magazine* in September 1891 (Wells, 1891). Unfurling his recapitulationist colours he asserted:

> There is, therefore, no guarantee in scientific knowledge of man's permanence or permanent ascendancy...The presumption is that before him lies a long future of profound modification, but whether that will be, according to present ideals, upward or downward, no one can forecast. (Wells, 1891, 253)

Implicit in the notion of 'ancestral blood' (Layard *ibid*) was the notion of an ancestral condition (*viz*. recapitulation theory), so the potential for reversion was something we may well have all carried within us. Not surprisingly atavism was a popular theme in literature and scientific study. Microcephalics were commonly thought to have brains that were half way between apes and humens in size and structurally midway between the two. They were often seen as a throwback to an ancestoral condition as well as proof of its existence. It all reflected the lack of a meaningful theory of heredity.

> The subject of Heredity, if I may say so, is in the air at the present time. The journals and magazines, both scientific and literary, are continually

> discussing it, and valuable treatises on the subject are appearing at frequent intervals. (Turner, 1889, 757)

Sir William Turner, president of the BAAS section H in Newcastle, was probably referring to the public and scientific reaction to August Weismann's germ plasm theory, aspects of which were discussed, or translated into English (see following chapters), during the late 1880s and early 1890s. This would be another debate the Time Traveller as a gentleman of science would have been aware of.

The Time Traveller may have been bitterly disappointed by the direction human evolution had taken, but he cannot have been completely unprepared for some of its outcomes. Francis Galton and the race scientists had been predicting doom for some time. The Morlocks could have been a product of degeneration, a return to an ancestral condition, or equally, a new evolutionary direction dictated by changing circumstances. Ethnology might have suggested parallels with some of the most 'primitive' indigenous peoples encountered by Victorian imperial expansion. For the Victorians, cannibalism exerted a horrific fascination. Time and again in literature of all sorts this was the barrier that separated the beast from the civilized man. Although revolted by the Morlocks' eating habits, the Time Traveller could call on anthropology and evolution to account for them in the far future.

Ethnology and evolutionary anthropology could find parallels for the Eloi as well. Their childlike simplicity and lack of curiosity was reported in some of the indigenous peoples encountered by Victorian explorers. E.B. Tylor and other anthropologists drew on analogies between some tribal societies and stages in the development of European children. Alternatively, isolated on their island home the modern day Tasmanians had undergone a cultural atrophy in the face of adequate resources. The model could be applied to other groups as well. Wallace, Darwin and others had repeatedly asserted that struggle and competition were the incentive for progress, without them stagnation resulted. The future races of the Earth resembled actual and potential stages in human development – possibility recapitulating human ontogeny.

## The Public's Engagement with Evolutionary Issues: Getting the Story Out. Or, how the Time Traveller Would Have Acquired his Understanding of Human Evolution

The extent of public interest in issues of heredity, indeed on anything that reflected on origins questions, is shown in the amount of space it was given in print. As the quote from William Turner (see above) implied, articles were frequently appearing in magazines aimed at the general public as well as at the practising scientist (Allen, 1882 and Textbox 7.1 for example; or Büchner, 1894 and Textbox 12.1).

A further insight into this can be seen in the data presented in Table 11.1. The table lists articles that appeared between July 1893 and October 1896 in a variety of popular magazines, and which were reviewed in the journal *Nature*. George Romanes (1848–1894), Weismann, Pearson, Galton, Spencer, Wallace, Huxley and many other leading thinkers of the time are represented. It is worth reiterating that these were periodicals aimed at the non-specialist, yet written by practicing scientists who were engaging in detailed debate with their detractors. In terms of public understanding of evolutionary and origins debates, these publications provided detailed critiques as well as contextualised

| Year/volume of *Nature* | Month | Page in *Nature* | Magazine name | Contributor | Subject |
|---|---|---|---|---|---|
| 1893/48 | July | 249 | *New Review* | Spearman | Criminal detection/anthropometrics |
| | | | *Humanitarian* | Bertillon | Anthropology in France |
| | | | *Contemporary Review* | Romanes | Supporting August Weismann in his argument with Herbert Spencer |
| | September | 443 | *Contemporary Review* | Weismann | On Natural Selection |
| | | | *Contemporary Review* | Weismann | Replying to Herbert Spenser |
| | | | *Contemporary Review* | Weismann | Replying to Herbert Spenser |
| 1893/49 | November | 31 | *Fortnightly Review* | Wallace | On the ice age |
| | December | 155 | *Fortnightly Review* | Wallace | Continuation of the Ice Age article from previous |
| | | | *Contemporary Review* | Spenser | Reply to Weismann |
| | | | *Longman's Magazine* | Evans | Forging antiquities |
| | | | *Blackwood's Magazine* | Seth | Critique of Huxley's Romanes lecture (evolution and ethics) |
| 1894/49 | January | 235 | *Fortnightly Review* | Buechner | Origins of humankind |
| | | | *Quarterly Review* | Anon | Review of Howorth's *Glacial Nightmare* |
| | | | | | |
| 1894/50 | May | 66 | *Fortnightly Review* | Grant Allen | Origins of cultivation |
| | | | *Longmans Magazine* | Grant Allen | Origins of language |
| | July | 234 | *Fortnightly Review* | Pearson | Socialism and natural selection |
| | | | *Contemporary Review* | Bonney | On glacial erosion a reply to Alfred Russel Wallace |
| | August | 419 | *National Review* | Galton | Response to Kidd's *Social Evolution* |
| | | | *Fortnightly Review* | Pearson | Responding to Lord Salisbury |
| | | | *Fortnightly Review* | Linton | Review (devastating!) of Drummond's *Ascent of Man* |
| | October | 585 | *Contemporary Review* | Spenser | Reply to Weismann |
| 1894/51 | November | 44 | *Humanitarian* | Mivart | Heredity |
| | December | 162 | *Contemporary Review* | Caillard | Critique of Huxley |
| | | | *Century* | du Bois | Critique of Huxley |
| 1895/51 | January | 259 | *Century* | Maxim | Discussing his own flying machine |
| | | | *New Review* | Anon | Comment on first instalment of Wells' '*Time Machine*' |
| | | | *Humanitarian* | Mivart | Heredity |
| | | | *Fortnightly Review* | Wallace | Evolution and variation, attack on Bateson |
| | | | *Monist* | Lloyd Morgan | Natural selection |
| | March | 450 | *Fortnightly Review* | Wallace | Evolution and variation attacking Galton |
| 1895/52 | May | 45 | *Contemporary Review* | Fogazzuro | History of the evolutionary debate |
| | | | *Humanitarian* | Thompson | Heredity |
| | | | *Contemporary Review* | Spenser | One of a series of articles ultimately combined in *Principles of Sociology* |
| | June | 159 | *Contemporary Review* | Spenser | Another in series of articles ultimately combined in *Principles of Sociology* |
| | July | 257 | *Contemporary Review* | Spenser | One of a series of articles ultimately combined in *Principles of Sociology* |
| | | | *Fortnightly Review* | Grant Allen | On August Weismann |
| | | | *Science Progress* | Keith | *Pithecanthropus* and human evolution |
| | August | 355 | *Fortnightly Review* | Brodrick *et al.* | Reminiscences on T.H. Huxley |
| | | | *Contemporary Review* | Spenser | One of a series of articles ultimately combined in *Principles of Sociology* |
| | September | 451 | *Fortnightly Review* | Mivart | Attack on Weismann, Haeckel and Pearson |
| | | | *Contemporary Review* | Spenser | One of a series of articles ultimately combined in *Principles of Sociology* |

| | | | | | |
|---|---|---|---|---|---|
| | | | *Contemporary Review* | Spenser | One of a series of articles ultimately combined in *Principles of Sociology* |
| | | | *National* | Hill | on Huxley, Virchow and agnosticism |
| | October | 586 | *Contemporary Review* | Spenser | One of a series of articles ultimately combined in *Principles of Sociology* |
| | | | *Fortnightly Review* | Wallace | On speech |
| 1895/53 | November | 18 | *Fortnightly Review* | Pearson | Reply to Mivart |
| | December | 116 | *Contemporary Review* | Spenser | One of a series of articles ultimately combined in *Principles of Sociology* |
| | | | *Popular Science Monthly* | Marriott | Fossil skeletons from Mentone |
| | | | *Popular Science Monthly* | Wright | Glacial man in Ohio |
| 1896/53 | February | 331 | *Contemporary Review* | Spenser | One of a series of articles ultimately combined in *Principles of Sociology* |
| | | | *Fortnightly Review* | Reid | Chapter from forthcoming *The Present Evolution of Man* |
| | March | 428 | *Contemporary Review* | Spenser | One of a series of articles ultimately combined in *Principles of Sociology* |
| | April | 548 | *Contemporary Review* | Spenser | Final in the series of articles ultimately combined in *Principles of Sociology* |
| 1896/54 | July | 260 | *Science Progress* | Beddoe | Complexion and disease |

*Table 11.1. A selective list of magazines and periodicals appearing between 1893 and 1896 in which subjects of a Prehistoric, evolutionary, or human evolutionary theme were published, as they were reviewed by the journal* Nature *across that period.*

polemic. The public had access to the 'facts', in detail, and could make their own minds up as to which side of the debate they favoured. (I would venture to suggest that there are few publishers of popular science magazines today who would dare to print prolonged debate between scientists and engage with such detail).

As both a practicing scientist and a member of the reading public, the Time Traveller would have had access to a number of sources through which anthropological information was disseminated to the general public. There have already been a number of instances in this book where I have noted that texts on specialist topics received a surprisingly wide circulation. The following list shows some of the more important routes by which science was made accessible:

- Encyclopaedias and science based periodicals, often built up through weekly instalments. This is a much underrated source of polemical argument. Scientists frequently contributed entries on their various specialist subjects but made no attempt to remain neutral. They promoted their own viewpoints at the expense of their critics. For example, a sharp critique was aimed at Edwin Ray Lankester because of the republication in a single volume of a series of outdated biological articles by himself and others from earlier editions of the *Encyclopaedia Britannica* (Anon, 1891). Lankester was a frequent contributor to encyclopaedias and popular science journals Others, such as Worthington Smith, wrote the Prehistoric archaeology sections in the Victoria County History (Bedfordshire in Smith's case); John Evans wrote the section for Hertfordshire.
- Popular magazines and journals, especially those focusing on topical issues. Table 11.1 is a good yardstick of the frequency with which anthropological information was presented and of the variety of magazine outlets that carried such material. A quick trawl through

the titles suggests there was a stronger presence of 'human' related pieces than geological. Grant Allen in the *Contemporary Review* (1882) and Ludwig Büchner in the *Fortnightly Review* (1894) have already been mentioned, and any number or articles by H.G. Wells could serve as representative examples – see next chapter.

- Newspapers. Reporting issues of the day, they also reported events in the BAAS meetings, in scientific societies, and were a useful advertisement for upcoming lectures, exhibitions and the like. Benjamin Harrison's archive in Maidstone Museum is an excellent record of the frequency with which Palaeolithic and human origins material was reported in the national press and in a variety of local and regional newspapers as well.
- Books
  - Introductory texts and simplified primers aimed at the general public. I will mention two here as examples. Edward Clodd's *The Story of Primitive Man* (Clodd, 1895; see also Pearson, 2007) was a small book intended as an introduction to prehistory in Britain, setting the Palaeolithic in its broader evolutionary and prehistoric context. It was published by George Newnes (1851–1910; publisher of the *Strand Magazine*) in his *Library of Useful Stories* series. Thus Clodd's book was embedded within a wide distribution network managed by a publishing house specialising in reaching the general public. It is also dedicated, with permission, to John Evans. This is a situating device that positions the author and his writings. The reader knows that what they will read is 'authorised and approved'. Evans was a well-known authority by this period. The book promotes what is, I suggest, the standard viewpoint for the public on evolution; red in tooth and claw and driven by relentless competition. Eoliths are not mentioned but are noted in the bibliography, which lists books by the leaders of the day (Darwin, Huxley, Evans, Boyd Dawkins etc., a number of whom in turn referenced Clodd in their papers and spoken presentations). Interestingly, Pearson (*ibid*) notes that H.G. Wells and Edward Clodd were friends and correspondents, and Clodd's work, with which Wells must have been familiar, may have directly influenced Wells's views on human evolution. The book was popular. It sold 2000 copies in the first few weeks and 5000 copies in the first few months (Pearson, 2007). Sir Robert Ball's *The Cause of an Ice Age* (Ball, 1892) was a slightly more upmarket text, probably aimed at a more informed general reader. It too was part of a science series, in this case *Modern Science*, and it too had its validating and positioning device. In this case the *Modern Science* series was edited by Sir John Lubbock, who wrote a general introduction to the series which was included in the book. Lubbock's name, with its archaeological, liberal and political connections, would have immediately reassured the readers of the

quality of the work. Many readers may have read the book on their bank holiday, the introduction of these holidays being Lubbock's doing (Harrison, 1963; Benjamin Harrison called them Lubbock days). The book was cautiously but favourably reviewed by other professionals (Fisher, 1892).

- ◦ Polemical books and subject specific texts. Scientists presented their researchers and polemical writings in books written for a market that included both specialist and non-specialist readers. In other cases books were written to span the divide. Texts written with the specialist in mind would have included Weismann's *Essays Upon Heredity and Kindred Biological Problems* 1891–1892, or his *The Germ-Plasm. A Theory of Heredity* 1893. In other instances subject specific reviews (with personal interpretation) such as A.R. Wallace's *Darwinism* (1889, and subsequent editions) spanned the professional/non-professional line, presenting detailed argument which answered critics on the broadest scale. Books like these have been mentioned in detail in preceding chapters. There was a lively market for such volumes, and a great many went into numerous editions which were then frequently revised to a greater or lesser extent. Darwin used the five subsequent editions of *Origin of Species* as opportunities to answer specific objections raised by his critics; the same for the *Descent of Man*. They were a useful vehicle for updating theoretical and factual positions. Subsequent editions of popular or topical volumes, could be published within months in some cases, and frequently within a year or two of their first editions.
- ◦ Volumes of collected essays and lectures. These could refocus public interest on a series of topics through the re-issue of older papers and lectures on topics still in current debate. Relevant rewriting would add to the interest. Excellent examples already noted are Huxley's republication in the early 1890s of many of his earlier writings in the *Collected Essays* series and Prestwich's *Controverted Questions* volume, with its update on the eolith paper to the AI. (Prestwich, 1895d). The Victorian practice of publishing posthumous volumes of 'Life and Letters' are appropriate here too.

- Scientific societies and their meetings, and the society journals. These were not only the national societies in the capital, but those in the major provincial cities and towns as well. The public were free to join such societies and attend their meetings. A variety of topical issues of local and national significance were presented. The London societies tended to reflect national and international interests.
- Public lectures. Galton's free lectures on heredity and anthropometrics at the South Kensington Museum have been noted. Many associations and institutions ran such lecture series for the benefit of public

education. Huxley's famous working men's lectures are another example. In 1868 he was the co-founder of The Working Men's College where public lectures from a variety of visiting speakers were available to all for a few pence. On the other hand, for fifteen guineas the AI would arrange for a series of six popular lectures distilling the most up to date scholarship in the various branches of anthropology. They could be delivered at local institutions, or even in a drawing room for groups of private students (Anon, 1890a).

- British Association for the Advancement of Science meetings, annual, and open to all.
- Public lending libraries.
- Museums. Eolith displays were featured in a number of museums, local and national, along-side collections of Palaeolithic and Prehistoric material culture. Specialist anthropological museums such as the Pitt Rivers Museum in Oxford were also open to the public.
- Specialist attractions, such as the various anthropometric laboratories set up at the BAAS meetings, at various other scientific events, and by Galton in South Kensington; or the eoliths exhibitions at the Guildhall noted earlier.

# 12

# SCIENCE FICTION AND H.G. WELLS
## *Resolving the Conflict Between Past and Future*

> Culturally, in the mid-1890s, prehistoric man and concepts of primitivism became bound up with notions of the place of science in society, the development of man, human intellect, and modern consciousness and identity. (Pearson, 2007, 60–61)

Victorian evolutionary research and race science had predicted a set of circumstances that could explain the Eloi. Remove competition and the struggle for survival, and stagnation follows. The Morlocks had been anticipated also, whether interpreted as atavistic metaphors for the Palaeolithic beast within, or as a warning that evolutionary change is dictated by circumstances and could be reversed. As McLean (2009) observes, in the world of the far-off future, ironically, the man from the past was the pinnacle of evolution.

But why break the narrative of human evolutionary studies to look at science fiction at all; how can it be relevant? I have no intention of reviewing science fiction's history, or of exploring its various definitions. That is a highly specialist project with a vibrant critique of its own (Alkon, 1994; Roberts, 2005a; Roberts, 2005b; Ashley, 2011). I will deliberately avoid anything other than common understandings (folk psychology) of what this subject is.

Fiction is an exploration of the human condition. It achieves its aim by placing imaginary characters in a variety of testing situations, and in so doing achieves a better understanding of human nature through the reactions of those characters. Thus authors invite readers to examine what their own reactions might have been. Science fiction simply pushes the envelope further. These writers define what it is to be human by testing their characters' humanity against ever more exotic challenges and extra-ordinary backgrounds. It is in such a field of literature that human origins are most likely to be discussed. Science fiction often demands that its readers journey beyond the limits of their own humanity and in taking a backward look, engage with other possibilities. I noted in the last chapter that the 1890s, the Victorian *fin de siècle* decade, was a time of great concern with what it meant to be human, in both literature and science (Pearson, 2007). Much of the work described in this chapter and the next reflect this.

Just like fiction, science fiction mirrors what the public wants and believes, as well as telling it what it should or could expect. As a part of the broader project of literature, science fiction can therefore reach, influence and in turn be shaped by a wide public audience. It is an appropriate place to look for what the Victorian public understood about human evolution.

Just as the evolutionary anthropologists in the 1860s humanised Prehistory by inserting images of indigenous peoples as proxies for Prehistoric hunters, so the Victorian story writers humanised Prehistory by inviting readers to share experiences of the lost world by identifying with the characters in the stories. The deep past became a place the reader could actually visit. This was a powerful device for shaping public understanding and opinion. I made the point earlier that the Victorians had difficulty in conceiving of a transitional ancestor, citing as an example, Dubois' impossible *Pithecanthropus* in Figure 10.7. But this was confined to the 'professional' view. There were very few scientific reconstructions of Palaeolithic humans made by scientists. Most thought the handaxe makers would be recognisably human. Actually it was the fiction writers and their illustrators who shaped the public's impression of what Palaeolithic man might look like. An example of this is given in Textbox 12.1.

*Textbox 12.1. A summary of an article published in a popular magazine which reflects a popular view of human evolution and represents one form of behavioural reconstruction.*

Büchner, L. 1894. The Origin of Mankind. *Fortnightly Review* 55, 74-82.

Chosen as a relatively representative example of the general public's view of indigenous people, the lives of Prehistoric people, and the relationship between ethnology and Prehistory. The implications drawn about the state of Prehistoric people, on the basis of ethnological parallels are an example of the comparative method. Büchner's article is a review of a French pamphlet written by Abel Hovelacque and published in 1882. The overwhelming similarity between Prehistoric lifestyles and those of modern primitive peoples empowers the use of ethnological parallels. The need is the more pressing as many of the tribes discussed by Hovelacque are becoming extinct. Some like the Andaman islanders appear to be a transitional race between Indians and Australians.

- There is a closer relation between modern civilized man and modern primitive man, than there is between extant primitive peoples and the earliest ape-like ancestor.
- Physically many ape 'races' (various species of chimp, gorilla etc.) resemble primitive people in terms of stature and physical features. Beetling brows, prognathism, flattened noses, divergent big toe etc., all mark the modern primitive and the Prehistoric ancestor as being physically undeveloped. Even the lack of a chin in indigenous peoples is matched in the La Naulette jaw.
- Intellectually, non-Europeans are deficient as their brain sizes are smaller. A gorilla has a brain size of 530 cubic centimetres, a modern primitive has between 1100 and 1400, while the European has 1400–1500cc.
- In behavioural terms the modern primitive is more animal-like, and this must have be the state of the Prehistoric peoples as well.
- Knowledge of Prehistoric man makes it clear that the ethnological parallels described are valid in terms of reconstructing ancient human life-ways, but why some of these ancient peoples subsequently developed and others did not, as is seen in the stagnation of modern primitive

peoples, is a question only future anthropology can solve. This shows there is no inherent drive to progress in our species. There have to be external and/or internal stimuli before progress can be initiated. This applies to psychological and cognitive evolution as well.

- In the lack of moral and social/cultural sensibilities indigenous people are actually below the level of animals as Büchner's two books *Intellectual Life in the Animal World* and *Love Life in the Animal World* show.
- There is a long catalogue of specific behaviours, customs and institutions from different tribal groups. All are designed to highlight a bestial or primitive state of existence.
    - Nudity allied to indifference to personal morality and concern for public decency and the rights of others. Treatment of women and girls.
    - Living in trees and holes in the ground. If they build huts they are the most basic.
    - They will eat anything and show no discernment in their diet. Cannibalism is rife and both the young and the old will be eaten in times of hardship. Cooking is a relatively recent invention. Fire has only been discovered by some tribes, though it seems most have learnt its use from contact with others.
    - Hunting and fishing are practiced with only the most rudimentary tools which primitive groups have invented for themselves. Any advanced tools or practices were learnt by contact with more advanced peoples.
    - Pottery is another late invention, again not developed by a number of tribal peoples.
    - Indigenous people are capable of artistic practices but they are well below the standard of Palaeolithic cave art. Music is simplistic and monotonous. For example, Australian aboriginal music and dancing is an imitation of the behaviour of kangaroos.
    - The mobile lifestyle results in a society with no sense of property or personal ownership. Possessions are few and transitory. There are few social traditions and laws and moral propriety sanctified by tradition is limited. This suits the 'fickle' nature of the primitive man's character. No injunctions against theft, murder or deceit, and pity and gratitude are unknown. Büchner asserts that this proves that morality and a social conscience can only be achieved through education.
    - No sense of God or of religion, and where present belief in a supernatural is very basic and restricted to a belief in spirits.
    - Language is basic, often with no proper names for things, and augmented by gesture.
    - Modern primitive peoples cannot be changed or improved. When they come into contact with higher civilizations they only acquire their vices and not their virtues.

## Science Fiction in Victorian England

The term science fiction was first coined in 1851 and then promptly forgotten until 1929 when the legendary American publisher Hugo Gernsback reinvented it (Alkon, 1994) for his landmark magazine *Amazing Stories*. This was the first science fiction magazine – a pulp. Prior to this, these kinds of stories went under a number of labels in Britain and America. H.G. Wells, beginning with the *Time Machine* in 1895, called his own stories, scientific romances. They celebrated the romance of scientific discovery, or warned against its excesses. They could be tales of the present or the future. Jules Verne's stories had been labelled the, 'Fantastic Voyages' beginning with *Five Weeks in a Balloon* in 1863. Nor was there such a thing as a science fiction writer. Men like Wells, George Griffith, whose work will be described in the next chapter and their contemporaries were jobbing writers and they wrote all kinds of stories for all kinds of magazines. The most profitable were then turned into books. Some magazines paid by the word, others by the type of story. Some storylines could command higher prices than others, but fame could cross-cut this. Well known writers could charge much higher prices for their work and place it in the most prestigious magazines, or target those that paid the most.

Story writers of the late Victorian era were helped by two very different phenomena. The first was the increase in public education. Not only could more people read, but reading was becoming more widespread in the lower–middle and lower classes, so reading tastes were expanding too. Once demand was satisfied, the next problem was supply. The first magazine was the *Gentleman's Magazine* published in 1731. It was followed by a host of imitators. Most of these provided contemporary comment and criticism (Ashley, 1977). It was not until 1817 that *Blackwood's Magazine* began introducing fiction, two years after the first infants' schools were founded (Ashley, *ibid*). By the mid-century there was an ever increasing demand for magazines. By January 1865 there were 544 magazines being published in Britain (*ibid*). The railway boom also helped, as reading was considered an excellent way to pass the time on a train journey. William H. Smith secured the first contract for selling magazines on railway stations in 1849 and W.H. Smith is still there. Some authors founded their own magazines. The celebrated novelist William Thackeray founded the *Cornhill Magazine* which was the first to reach a circulation of one hundred thousand copies. But it was priced at a shilling which was not cheap (Moskowitz, 1974a; Ashley, 1977). This was a significant outlay for the likes of Benjamin Harrison on the lower/middle class border who took the *Cornhill* from its first issue.

So there was a gap in the market. Weekly and monthly periodicals were largely aimed at either the low budget and low quality children's market, or the much higher priced better educated readership who bought *Cassell's Magazine* or the *Cornhill*. Neither of these were printed on good quality paper. There was a need for a reasonably priced quality general magazine with a mixture of fiction and factual pieces, well illustrated, and aimed at the better educated of the working and lower middle classes. In fact the target market was the likes of Harrison. It was the innovative publisher George Newnes who saw the niche and filled it, kicking off the late Victorian publishing boom in the process (Moskowitz, 1974a; Moskowitz, 1974b). At the same time he created a legend. The first issue of the *Strand Magazine* hit the news-stands in January 1891 and sold for sixpence. Within months there was a flurry of imitators and competitors, but none ever bettered it. *Pearson's Magazine* launched in 1896 was the only one that came close.

There is no doubt that authors like Wells and Griffith benefited from the new markets. So did the scientific romance story line. By modern acclaim (Moskowitz, 1974a; Moskowitz, 1974b; Ashley, 1977), it was *Blackwood's Edinburgh Magazine* that published the first science fiction story in the magazines. This was G.T. Chesney's *The Battle of Dorking: Reminiscences of a Volunteer* (May, 1871). It was actually the beginning of the immensely popular 'Future War' storyline. The story served as a warning to an unprepared British military establishment that the Prussian war machine, fresh from its success in the Franco–Prussian war might turn its eyes on Britain next. As we have seen in earlier chapters Paul Broca, A. de Quatrefages and Édouard Lartet had first-hand experience of the effectiveness of Prussian militarism.

## H.G. Wells's Science Fiction between 1888 and 1901

One morning in September 1884 a young man walked down Exhibition Road in London's borough of South Kensington (Mackenzie and Mackenzie, 1987). He paused to look up at an imposing red brick building. All his youthful hopes and dreams were focused on this one institution. Herbert George Wells (1866–1946) stood before the Normal School of Science. Perhaps he could become a science teacher, or even a practicing scientist. One thing was certain, he did not want to be a draper's assistant. There had already been two failed attempts at this encouraged by his mother. For the young eighteen year old H.G. Wells, with a prestigious scholarship in his pocket, education was the way up and his ticket out.

Wells did well in his first year. He studied biology under T.H. Huxley, whose charismatic brilliance ensnared the young aspirant. Despite their closest contact being a terse good morning as Wells once held a door open for him (Mackenzie and Mackenzie, *ibid*), he later stated that Huxley was probably the greatest man he would ever know. But the old bulldog was tired and ill, exhausted after a lifetime of battle and he left at the end of Wells's first year (Desmond, 1998). The two years that followed were not as inspiring for the young scholar, and lacking motivation his work gradually declined. He left in 1887 with no qualifications (Mackenzie and Mackenzie, 1987; Foot, 1995), worn out and in poor health. A series of unfulfilling and badly paid teaching/tutoring posts followed. Gradually things improved as did the teaching jobs. He passed his Intermediary Examinations in 1889 with a second class in zoology, and graduated in 1890 with a first class BSc in the same subject. The jobs continued to get better and pay more.

While still a student at South Kensington Wells had begun writing. The first version of what would one day become *The Time Machine* was published in the *Science Schools Journal* in 1888. The story was called the *Chronic Argonauts*. But he was not just writing fiction. He had begun to write and sell scientific pieces to the magazines. He co-authored *Honours Physiography* with R.A. Gregory, and was sole author of *Textbook of Biology*, both published in 1893. Right from the very beginning he was both an educator and a polemicist (McLean, 2009). His works reflected his own developing views as well as his concerns; his writing could enlighten as well as warn. With his science training, and an appetite for sweeping up the science articles in the journals and magazines, he was well placed to provide his readers with contemporary and informed scientific commentary (Wagar, 1964; Partington, 2008). Most of his science writing followed this pattern. He commented on the scientific topics of the day that interested him, subjects which were being actively debated.

*Table 12.1. A selection of H.G. Wells's scientific writings up to 1901 which reflect evolutionary themes, some of which include Palaeolithic archaeology and human origins.*

In dealing with the public his method was to instil doubt by offering scientifically plausible alternatives to comfortable or cherished beliefs. Huxley's influence was manifest from early on (Mackenzie and Mackenzie, 1987; Alder, 2008; McLean, 2009), something that Wells acknowledged all through his life. Wellsian analysts often cite the similarity between Huxley's dark and occasionally brutal view of evolution (for example his views on blacks and women, Huxley, 1902) and Wells's interpretations of it as expressed in the *Time Machine* and the *Island of Doctor Moreau* (Philmus and Hughes, 1975). But Wells didn't simply reflect the master's views. As McLean notes (2009) Huxley had never emphasised the competitive aspect of Darwinian theory, but Wells placed this at the centre of his thinking in some of his early work (also Pearson, 2007). One thing Wells did learn from Huxley was the power of scientific research (Draper, 1987). It could change the world, and the failing old man had proved it.

Table 12.1 represents a personal choice of scientific writings and stories which seem to me to reflect Wells's views on evolutionary theory and human evolution. I have grouped them into four themes, some of which have already been identified by Wells scholars. It will be apparent that a number of stories and articles cross-cut these themes. Between 1888 and 1901 he wrote well over a hundred books, articles and stories; 1894 to 1896 were particularly productive years. The table therefore reflects only a small proportion of this body of work. In his earlier writings, at least up until the publication of *The Time Machine* in 1895, Wells frequently focused on the public's view of evolution as a process. His biographers all agree that within this, anthropocentrism and the general complacency that accompanied it were important concerns.

In *Zoological Retrogression* (*Gentleman's Magazine* September 1891; see also last chapter) Wells confronted the public's perception of evolution head on. Anthropocentrism was fuelled by a belief that evolution was a linear and progressive process. This was a commonly held view at this time. Herbert Spencer was a strong advocate of the view that evolution (in everything) progressed from simple and homogenous to complex and heterogeneous (McLean pers. comm.). Many of Spencer's earlier papers were republished in a collected volume *Essays: Scientific, Political and Speculative* in 1891 which McLean (*ibid*) suggests may have underlain the writing of *Zoological Retrogression*. The linear and progressive outlook led to the view that humanity was either already perfect, or if perfection was impossible, then physical evolution could

| Year of Publication | H.G. Wells (F = fictional story; S = scientific article) | | | | Other Relevant Publications |
|---|---|---|---|---|---|
| | Anthropocentrism | Progressive nature of evolution | Man's place in nature | Evolutionary implications | |
| 1887 – 1893 | *A Talk With Gryllotalpa.* F. 1887;<br>*A Vision of the Past.* F. 1887;<br>*Ancient Experiments in Co-operation.* S. 1892;<br>*On Extinction.* S. 1893; | *Zoological Retrogression.* F. 1891; | *Zoological Retrogression.* F. 1891;<br>*On Extinction.* S 1893; | *The Rediscovery of the Unique.* S. 1891;<br>*Man of the Year Million.* S. 1893;<br>*Ancient Experiments in Co-operation.* S. 1892; | Weismann 1889 *Essays on Heredity.* 1st English edition; 2nd edition in 1892<br>Weismann 1893 *Germ Plasm. 1st English edition.*<br>Huxley – 1893 Romanes Lecture<br>Spencer-Weismann-Romanes debate 1893 |
| 1894 | *Time Machine* in National Observer. F. March-June;<br>*Lord of the Dynamos.* F September; | *Time Machine* in National Observer. F. March-June; | *Life in the Abyss.* S. February;<br>*The Stolen Bacillus.* F. June;<br>*In the Avu Observatory.* F. August;<br>*The Strange Orchid.* F. August;<br>*Aepyornis Island.* F. December;<br>*The Rate of Change in Species.* S. December; | *Time Machine* in National Observer. F. March-June;<br>*Province of Pain.* S.;<br>*The Biological Problem of Today.* S. December;<br>*The Rate of Change in Species.* S. December; | Weismann – 1894 Romanes Lecture |
| 1895 | *Time Machine* in New Review. F. January -June;<br>*Time Machine.* Novelization. F. May 1895; | *Time Machine* in New Review. F. January - June;<br>*Time Machine.* Novelization. F. May 1895 | *Island of Doctor Moreau.* F. Draft finished in February; | *Time Machine* in New Review. F. January -June;<br>*Limits of Individual Plasticity.* S. January;<br>*Bye-products of Evolution.* S. February;<br>*Island of Doctor Moreau.* F. Draft finished in February;<br>*The Duration of Life.* S. February;<br>*Death.* S. March;<br>*Our Little Neighbour.* F. April;<br>*Time Machine.* Novelization. F. May 1895 | |
| 1896 | *Intelligence on Mars.* S. April ; | *Bio-optimism.* S. August; | *Under the Knife.* F. January;<br>*Island of Doctor Moreau.* Novelization. F. April;<br>*Intelligence on Mars.* S. April;<br>*Bio-optimism.* S. August;<br>*In the Abyss.* F. August; | *Under the Knife.* F. January;<br>*Island of Doctor Moreau.* Novelization. F. April;<br>*Intelligence on Mars.* S. April;<br>*Bio-optimism.* S. August;<br>*Human Evolution an Artificial Process.* S. October; | |
| 1897-1901 | *War of the Worlds* for Pearson's Magazine. F. 1897;<br>*War of the Worlds.* Novelization. F. 1898; | *When the Sleeper Wakes.* F. Novelization. 1899;<br>*First Men in the Moon.* F. Novelization 1901; | *War of the Worlds* for Pearson's Magazine. F. 1897;<br>*A Story of the Stone Age.* F. 1897;<br>*War of the Worlds.* Novelization. F. 1898; | *War of the Worlds* for Pearson's Magazine. F. 1897;<br>*Morals and Civilization.* S. 1897;<br>*A Story of the Stone Age.* F. 1897;<br>*War of the Worlds.* Novelization. F. 1898;<br>*When the Sleeper Wakes.* F. Novelization. 1899;<br>*A Story of the Days to Come.* F. 1899;<br>*First Men in the Moon.* F. Novelization 1901; | |

only continue to endlessly improve us. Technological and social development marched hand in hand with the inevitability of continued human achievement.

In *Zoological Retrogression*, Wells gave examples to show that nature could sometimes take a backward turn. Retrogression, or the degeneration of species, was an important concept in biology at this period and subject to intense debate as to its true meaning in a progressive evolutionary framework – see last chapter. For many it provided proof of evolutions' power to effect change. Wells showed that retrogression could be a strategic move on the part of the organism (Morlocks?), at other times an inevitable response to environment (Eloi?). Natural history, he argued, taught that retrogression was more prevalent than many of the public assumed. Wells drove his point home, there was no guarantee that natural selection would continue to act upon us in the future as it had in the past. The 'Coming Man', might be a far cry from the public's expectations. His reasoning was that population increase would ensure that intense selective pressure continued to affect our species, so circumstances could arise that would force human evolution to take a backwards step. This was straightforward Malthusian Darwinism.

The theme of the public's expectations vs evolutionary reality continued in *Man of the Year Million* (*Pall Mall Budget* November 1893), originally a student talk at South Kensington. Wells drew a picture of what human evolution might ultimately look like. We become unemotional creatures; nothing but enormous brains, our bodies having long ago atrophied through disuse as machines and technological advancement replaced the need for organic bodies. The only other aspect of our physical bodies that would continue to evolve was one of our hands, now greatly enlarged, on which we would hop around, and which sufficed for all manual tasks. Our lives would be lived out in deep caverns, the surface of the Earth having become too cold for life under a dying sun. A number of Wells scholars have noted the parallel with the Martians from *The War of the Worlds*.

If our evolutionary progression was not to be taken for granted, then our very dominance of the Earth was equally questionable. Nature may already have had a 'Coming Beast' waiting to replace us (Bergonzi, 1961; Philmus and Hughes, 1975). Later stories like *In the Abyss* (*Pearson's Magazine* August 1896) and the *Sea Raiders* (*Weekly Sun Literary Supplement* December 1896) both dealt explicitly with these themes. In the former an unsuspected civilization is discovered on the sea bed worshipping the 'gifts' that the gods from up above occasionally allow to descend to their realm. In the latter tale a sea creature begins to terrorize the shores of the English coast line (Hammond, 1998). The fragility both of human life in the face of unknown nature, and, by implication, human dominance of the world is being examined in these stories. This was also explored in stories published a few years earlier, see Table 12.1. *In the Avu Observatory* (*Pall Mall Budget* August 1894) and *The Flowering of the Strange Orchid* (*Pall Mall Budget* August 1894) both have people attacked from unsuspected quarters when seemingly at their ease. In the former story a legendary creature in Borneo, while in the latter the attack is by a new species of orchid.

In *Aepyornis Island* (*Pall Mall Budget* December 1894) a castaway hatches the egg of an extinct flightless bird commonly known as the Madagascar Elephant Bird. He rears it as a pet but on reaching adulthood it turns on him and he has to kill it before it kills him. The story nicely reflects the anthropocentrism of the times. Here an extinct creature, possibly hunted to extinction by humans in the seventeenth century, comes back to life and threatens a human being – the 'heir of the ages and all that'. The man is physically unequipped to deal with the giant bird. He manages to capture it with a bolas and then kill it. Paradoxically only with ancient hunting technology is the human able to triumph over larger nature. But only just – it could have gone either way. *The Rate of Change in Species* (*Saturday Review* December 1894) also addressed this same issue. Humans might succumb to disease or a bacillus; they may encounter rapid temperature increase or even a new Ice Age, but because we are slow to breed as a species our ability to resist and survive change is not as great as those species which breed more quickly – another biological warning against complacency. *On Extinction* (*Chamber's Journal* September 1893) tried to warn the public that the extinction of species was a natural part of life.

In these stories and scientific articles Wells was speaking directly to the general public on issues that had currency in contemporary scientific debate. Using both fiction and scientific speculation in a variety of magazines and journals, he reached a very wide audience. In some cases he used satirical humour to make his point. In *Vision of the Past* (*Science Schools Journal* June 1887) Wells played on the famous cartoon 'Awful Changes' from the 1830s by Henry de la Beche (McCartney, 1977; Rudwick, 1992) in which a school of young ichthyosaurs are being given a lesson in history by Professor Ichthyosaurus (de la Beche was actually satirising Charles Lyell with this character). An allusion to Byron's poem *The Dream* strengthens the association. In Wells's story a traveller falls asleep on a sunny day. He dreams he sees extinct dinosaur-like creatures. One of them is congratulating the others on being the very pinnacle of creation. They are the Něm of Dnalgne (Men of England) and the world was made for them. The sleeper attempts to correct the lecturer's mistake. He points out that these arguments actually apply to humans as it is they who are the pinnacle of evolution not an extinct species of reptile. The Něm become angry and the sleeper realises he has to wake up quickly in order to escape. Whether it was a religious or a secular view that was being satirised is irrelevant, the story works on either level by exposing hubris.

But Wells's fiction before 1895 did not just dwell on the fragility of man's place in nature and anthropocentric arrogance. Dangers arising from cultural and technological evolution paralleled the threats that nature posed. *The Lord of the Dynamos* (*Pall Mall Budget* September 1894) remains today an unsettling tale of racial and personal intolerance. A brutal and uneducated white overseer is murdered by an equally uneducated native of the Straights

Settlements, part of modern day Malaysia. Both had been employed in an electricity generating station providing power for the London underground railways. The Malaysian had gradually come to see the dynamo as a god, while the bullying racist overseer treated the dynamo as if it actually was one, certainly believing it to be more important than other humans. After the murder, a white official from the power company arrives to investigate but all he really cares about is ensuring there is no interruption to service. The Malaysian is killed by the dynamo while trying to murder the official.

Three different types of person are presented in this story. Each in his way is subservient to the machine, an obvious metaphor for the modern world, and this is juxtaposed with the inability of the three to rise above their own primitive inner nature. The uneducated (in Western terms) Malaysian has lived in the West for some time, but cannot shed his primitive beliefs. The white overseer cannot overcome his brutish character. The replacement manager, who ought to know better, can not rise above his selfish indifference to his fellow humanity. On the one hand he has a Palaeolithic indifference to other people (see *Story of the Stone Age* and other Prehistoric fiction stories below). Wells's treatment of the principle characters, especially of the Malaysian, is unsympathetic and today the story has a disturbing after taste.

Our technological culture lifts us away from the primitive, but doesn't replace it as it still hides in all of us. We rely on our culture to maintain our control of our environment, but at what price? In *The Stolen Bacillus* (*Pall Mall Budget* June 1894) an anarchist steals a vial of cholera intending to poison London's water supply. Instead he swallows it hoping to spread the infection by human contact, but it turns out he has been tricked and it is merely a harmless dye. Here again the illusion of our mastery over the world is highlighted. Wells tapped a rich vein of Victorian anxiety here. Terrorists appeared in a number of his stories and were part of the literary fabric of thrillers of the time, as will be seen in the next chapter. In Chapter 8, I described how the Bury-St-Edmonds skull was destroyed by police searching for hidden explosives. A literary device like this, utilising familiarity but with the element of danger, was one more way of bringing home to the reading public the fallacy inherent in anthropocentric complacency. By appeal to the familiar set in exotic circumstances, or alternatively the unusual set in more mundane surroundings, Wells invited his readers to examine their own position and ask questions of themselves.

During the early 1890s Wells was thinking about a problem common to all the themes suggested in table 12.1, and which was interwoven into much of his fiction and science writing of this time – plasticity (Philmus and Hughes, 1975). Or put another way, where does human nature come from, and can it be changed. This concerned to what extent human physical morphology and human behaviour were fixed and immutable. It was a theme being hotly debated by scientists in the professional journals as well as in the more widely read popular magazines. Pearson (2007) notes that the 1890s were a time when human nature, primitiveness and atavism, were all under particular scrutiny. Popular literature was tracking scientific debate. Francis Galton had set the ball rolling decades earlier when he had coined the phrase 'nature versus nurture'. As we have seen, Palaeolithic archaeology had already made a contribution to the discussion. Humans had not changed physically since the Neolithic, and even since the end of the Pleistocene. Every indication suggested Ice Age humans were similar to modern humans; Edward Clodd informed the public of such in his *The Story of Primitive Man* (1895; see last chapter). Wells too came to believe this (see below). Only *Pithecanthropus* was different, and how old that was, was anybody's guess. At the core of the debate was the mechanism by which inheritance worked. Darwin's pangenesis remained unpopular. Lamarck's concept of the inheritance of acquired characteristics, or use inheritance, experienced a resurgence in the late Victorian period, despite there being no substantive evidence that it actually worked.

Lamarckism persisted because no one had come up with anything better. August Weismann's work has been mentioned in earlier chapters (see Textbox 2.1, and Chapters 8 and 11). Throughout the 1880s and 1890s he developed and promoted a view of heredity that challenged use-inheritance. Weismann's work was widely discussed in the journals and magazines especially in the part-professional and part-popular journal *Natural Science*. He was an embryologist and a gifted microscopist – both were important scientific credentials in the late Victorian world. His works were critiqued and reviewed in English publications throughout the 1880s. However a milestone occurred in 1888 when some of his most influential essays were translated into English. There was a second edition a few years later (Weismann, 1891–1892). His seminal work *Germ-Plasm* was published in English the following year (Weismann, 1893). He was presenting a new theory of heredity and it was strongly Darwinian. Variation in populations was crucial and it arose through what we would characterise today as random mutations in the germ-plasm. These created the necessary variation for sexual reproduction to maintain variability in populations. Natural selection did the rest.

During the time Weismann's works were becoming widely available, Wells was still a Lamarckian (Philmus and Hughes, 1975). The appeal was clear, and it was one that many others like Wells (and Spenser?) wanted to believe in. Plasticity was inherent in the human condition. Humans could improve themselves, behaviourally and physically. Individuals could make a positive contribution to the race because their own improvements could be passed on to subsequent generations and then built on. It had a powerful appeal for anyone with a positivist social outlook. In an unpublished essay *The Universe Rigid* (Philmus and Hughes, 1975) and in *The Rediscovery of the Unique* (*Fortnightly Review* 1891) Wells argued that despite what appeared to be immutable laws governing the progress of life and which gave the impression that nothing could be left to chance, when one took a bottom up approach variability was present in everything. No two things could be absolutely identical; organisms were infinitely varied, as for example were colours (no two shades of red were ever absolutely the same). Darwinian variation was proof of this for organic life. For Wells the Lamarckian, this lack of universal laws working on changeable characters meant that neither nature or nurture were fixed from birth, both could be adapted by selection working through use-inheritance. These themes were combined in his article the *Province of Pain* (*Science and Art February* 1894). Plasticity would underpin future changes in the human body as cultural and technological advances affected our lives. The article defended an evolutionary explanation for the existence of pain, usually debated as an ethical and philosophical issue. It had begun as a necessary warning from the body to take note of injury and sickness. But Wells posited that as humans evolved further there would be little need for pain. Our cultural and technological sophistication would slowly remove its causes over time (as in the world of the Eloi). Use inheritance facilitated the decrease in the body's sensitivity as technological aids took over more of the human body's functions.

*Table 12.2. The main points of the debate between George Romanes, Herbert Spencer, and August Weismann from the magazine* Contemporary Review *1893-1894. The themes they debated were inheritance and heredity.*

Philmus and Hughes (1975) suggest that Wells first encountered Weismann's theories late in 1894, and converted in March 1895. This seems unlikely given the increasing exposure that Weismann was achieving in the late 1880s and early 1890s, not to mention the many articles published by others debating his theories. Wells, in keeping up with the popular journals and magazines, may even have started to come across some of the German geneticist's work second hand, while still a student. In an early version of the *Time Machine* published in the *National Observer* between March and June 1894, Weismann's theory of panmixia is mentioned (see Textbox 2.1). It is fair to assume that Wells may have been working on the draft late in 1893, or in early 1894. So his acquaintance with Weismann must at least date from then. I am certain that he would have been aware of a celebrated debate between the biologist George Romanes, Herbert Spencer and August Weismann. The main issues are summarised in table 12.2. The debate was pursued through the pages of the *Contemporary Review* for almost the whole of 1893. There was no convincing victory for one side or the other. Weismann gave the prestigious Romanes lecture in 1894, on his view of heredity and Spencer concluded the debate with a somewhat peevish finale in October of 1894. I suspect that the debate is actually alluded to in the novelization of the *Time Machine*. On first encountering the Morlocks, the Time Traveller is reminded of the white fish of the Kentucky caves. This is an allusion to the bleached fish and amphibians that live in the permanent darkness of deep caves. These fish formed a key element in the Romanes-Spencer section of the *Contemporary Review* debate.

In this debate Spencer robustly defended Lamarckian inheritance across a range of topics, while Romanes, also a Lamarckian, criticised Spencer for what he believed was a poor interpretation of the theory. Spencer's broader agenda concerned intervention by the state in social matters. He was vehemently opposed to any help for what he saw as the evolutionary less-fit sections of

| Name | Article Title and Source | |
|---|---|---|
| Spencer, H. February 1893. | The Inadequacy of Natural Selection. Part 1. *Contemporary Review* 63, 153-166. | Natural selection can only be invoked to explain those characteristics which make a significant contribution to survival and reproduction. But many characters cannot be explained by either. Only the inheritance of acquired characters explains them. Degeneration of formerly useful structures in the body is a good example of the difficulties inherent in the theory of natural selection – how and why are they maintained at all. Weismann's explanation, *panmixia,*is unacceptable. Selection cannot take a trait and completely modify it to advantage an organism. It can only work on pre-existing traits that already show potential for materially advantaging survival and reproduction. Natural selection cannot be creative. |
| Spenser, H. March 1893. | The Inadequacy of Natural Selection. Part 2. *Contemporary Review* 63, 439-456. | Darwinian law of correlation of parts illogical. A structure cannot develop/modify by itself – it will require supporting structures to develop as well. But they can't develop as quickly as the primary one. Consequently, selection on individual traits/structures cannot occur. Only Lamarckism can explain a large structural change, followed by supporting changes in succeeding generations. He provides proofs of acquired changes being passed on to succeeding generations thus disproving Weismann. |
| Romanes, G.J. April 1893. | Mr Herbert Spenser on "Natural Selection". *Contemporary Review* 63, 499-517. | Darwin always accepted that acquired character inheritance occurred alongside natural selection. So does present author. However he disagrees with Spencer's interpretation of Lamarckian inheritance. Argued that Spenser had misunderstood how selection worked on degenerating structures and the inter-relatedness of structures, so his use of them as a proof against natural selection was equally invalid. |
| Spenser, H. May 1893. | Professor Weismann's Theories. *Contemporary Review* 63, 743-760. | The first half of the paper deals with Weismann's theories. He focuses on the difference between reproductive germ cells and body cells. Firstly he denies this is a valid distinction. Different parts of the body are mutually co-operative, and so are their cells. He then denies the idea of somatic cells 'dying' and the germ plasm being immortal. The second half of the paper answers specific criticisms of his earlier work by Romanes, Wallace and Lankester. An interesting set of second hand anecdotal data is presented as evidence that white women with white husbands can produce children who show the influence of previous black husbands. This is support for the influence of somatic cells on the germ-plasm. He finishes by asserting that a proper understanding of the mechanism of heredity is the basis for understanding education, ethics, politics, biology and psychology – in other words everything about the individual and their society. |
| Romanes, G.J. July 1893. | The Spenser-Weismann Controversy. *Contemporary Review* 64, 50-54. | Replying to Spencer's claim that he couldn't understand Romane's drift in his previous article. Romanes gives a lovely example to illustrate panmixia/cessation of selection, and elaborates further on whether the offspring of parents will display characters inherited from their mother's previous husband. He still disputes this, but makes it clearer he believes in aspects of Lamarckism (use inheritance in his terms), but not in Spencer's version of it. |
| Spencer, H. July 1893 | Note. Controversy. *Contemporary Review* 64, 54. | A Reply to Romanes last. |
| Hartog, M. July 1893. | *II. Contemporary Review* 64, 54-59. | A none too simple over view of Weismann's theories. Hartog is not a supporter. Does note the significance of Weismann's latest pronouncements that climate and environment can influence structures within the germ plasm. |
| Stephen, L. August 1893. | Ethics and the Struggle for Existence. *Contemporary Review* 64, 157-170. | Inspired by Huxley's Romanes lecture. The reality of struggle for existence is accepted and the question is how do we map morality and ethics onto such a concept. What are responsibilities of superior societies to those perceived as inferior. |
| Weismann, A. September 1893. | The All-Sufficiency of Natural Selection. *Contemporary Review* 64, 309-338. | A detailed response to Spencer and Romanes. Weismann concentrates on ants and shows how new behaviours and physical structures emerge in the worker and soldier castes. Since they are persistent traits in neuters, then their continuation is not a result of use-inheritance. They have to develop by natural selection. Robustly defends Darwinian explanations and panmixia. |
| Weismann, A. October 1893. | The All-Sufficiency of Natural Selection II. *Contemporary Review* 64, 596-610. | Weismann continues to refute Spencer's objections to his work. This is a more technical paper, focussing for the most part on cell division and Spencer's criticism of the inability of the somatic cells to influence the germ-plasm. |
| Spencer, H. December 1893. | A Rejoinder to Professor Weismann. *Contemporary Review* 64, 893-912 | Responding to Weismann, Spencer tries to show that the patterns in the insect data are the result of either anciently evolved behaviours, or patterns of ontogeny and life history, particularly affected by nutrition. Weismann's germ plasm explanation for the inter-connectedness of parts responding to modification in a major trait is criticised. |
| Spencer, H. October 1894. | Weismann Once More. *Contemporary Review* 66, 592-608 | A final shot by Spencer. He reviews his main conclusions and asserts that Weismann has failed to answer his critique or just failed to answer. Includes some new arguments concerning the persistence of bone modifications which arose as a result of non-European peoples' habit of squatting. European pre-historic ancestors had the same modifications, but they have lost them. This is clear evidence for use-inheritance. He reasserts that a proper appreciation of the truth of Lamarckism underpins a proper appreciation of society and the wisdom of social reform. |

society. If allowed to breed they would weaken the race irreparably. Wells on the other hand believed in positive intervention The opportunities a scholarship had given him while still a poor working class boy, had deeply affected his world view (Draper, 1987).

Spencer made it clear at the end of several of his articles in the *Contemporary Review* that a proper understanding of inheritance (i.e. Lamarck) not only underpinned the correct attitude to social reform, but also politics, religion and ethics; in fact everything. For Spencer the appeal of Lamarckism was its ability to manipulate the future through direct modification of the present. Another aspect of the appeal of use-inheritance for Spencer was that it was quick. Changes affected in the current generation could begin to have immediate effect in the next. He believed that only 'superior' elements in society should be encouraged to produce more children because their superiority could quickly shift the national average upwards. Societies evolved and they followed the same rules that human biological evolution followed. Societies, like humans grew because they could build on their own achievements. The negative side of Lamarckism was that poor traits could be dispersed within the population as quickly as beneficial ones.

Wells was certainly still supporting a Lamarckian interpretation of heredity at the end of 1894. An article in December's *Saturday Review* appropriately entitled *The Biological Problem of Today*, was a strong critique of Weismann's views, it also showed Wells keeping up with the latest literature. He asserted that research into mitosis showed that the daughter cells resulting from cell division each shared half of the original cell's contents. But Weismann was arguing that germ plasm divided unequally with some cells containing more heritable particles than others. So Weismann's theories were untenable, according to Wells, because they did not map onto the latest research on cells division.

*The Time Machine* was a great success when it was published as a novel (in May of 1895) by William Heinemann. Up to this time Wells had been a relatively unknown short story writer and scientific commentator. With some critics now referring to him as a genius, H.G. Wells had become a name to be reckoned with; his opinions carried weight. The storyline of *The Time Machine* will be familiar to many readers and was briefly introduced in the last chapter. A brilliant and unorthodox scientist invents a machine that can travel in time, but not through space. The Time Traveller moves more than eight hundred thousand years into the future of London where he encounters humanity's descendants. The Eloi are a child-like race who live above the ground. The subterranean Morlocks prey upon them for their food. Human evolution had taken two different pathways as two groups of humans became separated and ever more distinct from each other; the result was physical, mental and behavioural adaptation that created two new species. Humanity had adapted itself out of existence – another warning against believing that evolution would continue to perfect the human species.

In the Eloi and the Morlocks, McLean (2009; pers. comm.) sees a clear critique of Spencerian-like progressionism. Time Traveller muses on what circumstances could have led to this future world. As human society conquered both the human condition and the physical world, population was brought into a controlled balance. Institutions such as the family became unnecessary. In later tales *A Story of the Days to Come* (1899 *Pall Mall Magazine* June–October), and *When the Sleeper Wakes* (1899 Harper), the beginnings of such a society were sketched out. The Time Traveller asserts that the social differentiation that would one day lead to the underworld of toilers and the pleasure seekers of the over-world, began in Victorian times.

The influence of Darwinian thinking is also clear. It is the removal of struggle and competition that is the root of the problem. Create a paradise, and then remove the need for struggle within it and the result is stagnation. This was a common belief in the late Victorian period. It is a frequent point in the Romanes–Weismann–Spencer debate. Domestic breeds only showed variation when human breeders interfered. Otherwise a constant supply of food and protection of the young resulted in stable blood lines. The Eloi are the result of this process. In fact, Wells asserted, through the character of his scientist, that intelligence itself was a product of competition which was also necessary for its maintenance. This is very Darwin and echoes *The Descent of Man* (Darwin, 1871).

Both the Eloi and the Morlocks are described in the novel as degenerate, but their degeneration had evidently taken different pathways. The Eloi were descendants of those who mastered the physical and biological life of the over world (see Chapter 11), and it is easy to imagine them as the consequences of a controlled environment without struggle. Their garden-like world may have been suggested to Wells by the Botanical Gardens at Kew (Draper, 1987). Pamboukian debates whether the Time Traveller was romantically attached to the Eloi girl Weena, but concludes not, as he is more emotional about his time machine. She suggests Wells was asking his audience to debate their reaction to future evolution through the Time Traveller's very ambiguous relationship with Weena (Pamboukian, 2008).

It is less clear how the lack of competition led to the Morlocks. They continue to maintain their ancient machinery as well as provide for the Eloi (the Morlocks make the Eloi's clothes), partly out of habit and an instinctual memory of their former servitude. It is even implied that the Morlocks control the Eloi's breeding. But they still possess the cunning of the hunter. They still retain enough intelligence, curiosity and mechanical skill to take apart and then reassemble the time machine itself. Their care for the effete Eloi is also a practical one; the Eloi are their food supply. This adaptation was clearly learned, and had to have developed after the human race had split. A simpler explanation of the Morlocks is that they are a result of increasing competition as the dwindling resources of the underworld forced this branch of humanity down a darker pathway. They either reverted to an inherent primordial state, or developed a new adaptation that suited their changed circumstances. The Eloi and the Morlocks are two sides of the same problem. If this is correct, then once more Wells was showing how adaptation need not necessarily imply progressive perfection.

It is also interesting that the physical description of the Morlocks is couched in terms of an ape-like appearance. There is a strong hint of atavism here. Despite their pale skin, scanty body hair and their large luminous eyes (so linking them to the cave dwelling fish noted above), there is also a suggestion that their limb proportion and their gait is more ape-like than human. Their bestiality invites us to think of a degenerative path that includes elements of a dark human past; Wells sometimes calls them 'white-apes' – adding the suspicion of racial degeneration as well (McLean pers. comm.).

Philmus and Hughes (1975) suggest that the degeneration of the Eloi and the Morlocks implies a Lamarckian explanation. I disagree. It is strongly implied in the *National Observer* serialisation of the story in March–June 1894 (*ibid*, 79), but not in the later novelisation. In the novel Wells does not ask the reader to engage with heredity and inheritance, rather he deals with the consequences of the evolutionary process itself. Nevertheless, the laws of inheritance were the key to plasticity and adaptability, two central themes in the story. The Eloi and the Morlocks would have forced readers to consider exactly how the two species had formed, while taking on one of Wells's many warnings – this could happen to us one day. For Philmus and Hughes, Wells was highlighting the ability of human intervention to make a difference. Pamboukian (*ibid*) suggests the opposite. When the Time Traveller witnesses the end of the world thirty million years hence, the narrative implies that the power of human technology to influence evolution is shown to be false by its ultimate futility. In fact this is very Huxley. We can't derail the cosmic process of evolution in the long run, but we can and should make a difference now. Why? Because we can.

The first half of 1895 saw the publication of a series of scientific articles by Wells that possibly represented his coming to terms with the implications of Weismann. It will be recalled that the appeal of Lamarckian inheritance would have been the ability of individuals, as a collective, to pass on mental and physical improvements to succeeding generations. These improvements would then become instinctive as time passed. Human plasticity vs. the fixed character of nature and/or nurture lies at the heart of *The Limits of Individual Plasticity* (*The Saturday Review* January 1895) and *Bye Products in Evolution* (*The Saturday Review* February 1895). In the latter Wells put forward the still controversial view that not every structure in an organism's body necessarily had a selective purpose. Darwin had posited the interconnectedness of parts – that individual structures in an organism were inextricably linked to each other through function. Change one by adaptation, and there would be changes in others. The new adaptations may then themselves confer adaptive advantages. Wells noted that pleasure and the artistic sensibilities were the result of chemical activities in the brain. They were also features of human cultural interaction. They must have arisen fortuitously because of other chemical reactions that generated pleasure from the sense of security that group living brought to our Palaeolithic forefathers. This correlation of parts was another strand that threaded through the Weismann–Spencer–Romanes debate.

It was *The Limits of Individual Plasticity* that nailed Wells's colours to Weismann's mast. He stated clearly that it is now possible for science to alter the physical character of a body, alter its chemical balance, and, through hypnotism (the main form of education in *When the Sleeper Wakes*), also affect behaviour. Almost every aspect of what makes up a human being could be manipulated and changed; pugnacity could be transformed into self-sacrifice, or sexuality into religious emotion. Here moral education was a key player – it was the moulding of primal instinct. But crucially Wells also asserted that these characteristics would not be passed on to any offspring.

Many Wellsian commentators point out that before his acceptance of Weismann, Wells had believed in the plasticity of humans both mentally and physically (as in *The Time Machine* and his critique in *The Biological Problem of Today*). Afterwards, the implications of the inability of the germ-plasm to be influenced by the somatic cells, meant that only extra-somatic culture could make a difference to humanity's future. Society and education would now have to be the shapers of our future. It would be fascinating to know what Wells's views on inheritance really were when he finalised the text for the publication of *The Time Machine* as a novel.

The emphasis on culture comes out in the next group of scientific articles he wrote. In *The Duration of Life* (*The Saturday Review* February 1895) and its sequel *Death* (*The Saturday Review* March 1895), Wells took themes and titles directly from Weismann's works (Weismann, 1891–1892). In the views of Darwin and Wallace reproduction was a key factor in how long an animal lived. For many species death would occur naturally after successful mating or egg-laying had occurred. Life was longer for species such as our own with a parental investment in feeding and rearing. Fortuitously, this created enough time for the cultural element to be emplaced and develop in individuals. In this sense human culture was an unintended consequence of the adaptation of human life to its own reproductive cycle (e.g. *Bye products of Evolution*). But adaptations like this could take many forms, and there were some which conferred clear benefits over others. Monogamy had given the most scope for cultural knowledge to grow and be transmitted because it conferred long term social stability. The alternative would have led to chaos. The implication was that this explained the success of western European society over others.

The nature vs nurture debate and its relationship with inheritance sat at the heart of the *Island of Doctor Moreau* published as a novel in April 1896. A draft was ready more than a year earlier, so the story's gestation will have been across the time that Wells was thinking about the implications of inheritance. As one of Wells's most celebrated works it will be familiar to many readers. Prendick, a shipwreck victim, winds up on a Pacific island where Moreau, a notorious vivisectionist, has continued his experiments on animals, attempting to turn them into humans. His life's interest has been the plasticity inherent in our species, but all his experiments are failures and his Beast-Men gradually revert back to animals. Moreau and his assistant, the only other European on the island, are killed by the creatures and Prendick escapes but only after becoming partly beast-like himself in order to survive. Many Victorians would have seen a parallel between the Moreau's vivisection process and the transforming nature of evolution itself. Philmus and Hughes (1975) note how Moreau's callous disregard for the pain and suffering he causes his victims, could stand for pitiless nature and the seeming brutality of natural selection (Draper, 1987). I have noted elsewhere that many people had rejected Darwin's vision of natural selection for this very reason. Draper also notes that Moreau's first real success was turning a gorilla into a black man. It took him a week, and then he rested!

In *The Island of Doctor Moreau* I suspect that the failure to create a true human, and the reversion of the Beast-Men back to animals, is a reflection of Wells's conversion to Weismann, or at least his rejection of Lamarckism. The animal nature always shines through and eventually reclaims the creature (as in the *Lord of the Dynamos* above). Moreau's assistant openly asserts that when the Beast-Men reproduce, their human characters do not appear in the offspring, he says there is no 'inheritance of their acquired human characteristics'. It is only nurture that can now make a difference.

Critiques of the novel often cite the conflict between the Beast-Men's nature, and the 'Law' – a series of social rules given to them by Moreau designed to emphasise their newly acquired humanity (McLean, 2008). The Law emphasises they are separate from nature. Increasingly the Beast-Men find it difficult to keep these laws. The novel invites the reader to ask what it is about humans that allows us to keep our laws which are the corner stones of our ethically informed culture, while the Beast-Men fail. Simple recitation of the laws is not enough. It has to be reinforced by comprehension. Although Wells does not provide the answer in the novel, the key to this comprehension is obviously education.

After the conversion Wells quickly came to believe that human bodies were no longer affected by natural selection, and had not been since the Pleistocene. Lacking the need to change, natural selection simply maintained the physical form of the race keeping it healthy through variation. Instead selection now focused on the brain and on a consequence of its development, human culture. The latter, in turn, frame-worked and contributed to enhanced intellectual growth. Wells seems to have had no coherent personal outlook on human evolution prior to 1895 other than a vaguely Darwinian belief that humans had evolved from the primitive to the more complex. But after rejecting Lamarckism a clearer vision emerged. In *Human Evolution an Artificial*

*Process* (*Fortnightly Review* October 1896) and its sequel *Morals and Civilization* (*Fortnightly Review* February 1897) this vision was explored. Just how much else of Weismann's bigger vision of heredity he accepted is less clear from his writings. He certainly did not accept that the somatic cells died as Weismann was claiming – see *Death* above. *Human Evolution* and *Morals and Civilization* were exceptional articles. They drew in many of the motifs of Wells's earlier scientific and fictional writings, and engaged them in a new perspective that was broader and more coherent. Plasticity, slow breeding rates, monogamy, the manipulation of behaviour and a number of other themes were intertwined.

Wells had come to believe that humans were composed of two parts. First there is the Palaeolithic animalistic creature of instinct. In modern humans this is the Culminating Ape fashioned by natural selection. This was the 'inherent' nature that Moreau could not change. Atavism was a powerful proof of the continued existence of this darker Palaeolithic nature. *Our Little Neighbour* (*New Budget* April 1895) is an unsettling tale about a young couple who move into a new house only to find that the brother of their neighbour demonstrates an extreme degree of atavism. The implication is that his condition predates the Palaeolithic and is more like a missing link part way between human and ape. Chapter 10 showed that *Pithecanthropus* was a hotly debated topic at this time. The creature's bestial nature comes to the surface with tragic results. The second part of a human is that influenced by the Acquired Factor; culture; 'a fabric of ideas and habits' which overlies, softens and redirects the primal instincts. It is this which is now seen as plastic and is subject to change in Wells's post-Lamarckian outlook. In such a perception, the role of education became clearer. It alone was the mechanism which intercalated the Acquired Factor into the Culminating Ape to produce the moral and ethical man of the modern era. Societies establish their own ethics and moralities as relevant to their needs and circumstances.

There is much in these two papers that modern evolutionary psychologists would find familiar. There is also a nice paradox. In relinquishing Lamarckian explanations, but accepting the empowering role of a malleable culture, Wells did not notice that culture is actually a Lamarckian phenomenon. Social advances made in one generation can be passed on to the next. In *Morals and Civilization*, successive cultural phases were built on each other, improving on E.B. Tylor's vision of stages of development because of the inclusion of the transformative power of cultural enhancement to shape each succeeding stage.

For many readers The *War of the Worlds* is Wells's most famous book. It was first serialised in *Pearson's Magazine* between April and December 1897 and then published as a novel in 1898 by William Heinemann. It is a grand tale of human vulnerability and is another of those stories that ticks most of the thematic boxes in table 12.1. Not surprisingly it has engendered a large body of scholarship (Philmus and Hughes, 1975; Draper, 1987; Williams, 2008; McLean, 2009) on the various layers of meaning in the text. I will not discuss this book in detail. The only two things I wish to point out here are the unusual role played by chance and luck and the evolutionary reasons underlying the Martian invasion.

One thing about the book that shocked Victorian England was how quickly the 'empire of man' was swept away. A technologically more sophisticated species, millions of years more advanced than ourselves, defeated us in a matter of weeks. The Martians were driven to it by a lack of resources and fierce competition for survival on their own world. It is a classic tale of survival of the fittest but with humanity as the loser. The Martians had already triumphed over other species on their own world. Humanoids had evolved on Mars too, but they were now food for the superior species and humanity was destined for the same fate. But because evolution had taken a radically different direction on Mars they had no knowledge of bacteria and so were unaware of the threat they posed. Pure chance saves humanity from the fate of so many other species. This type of large scale derailment of process by contingent events is more in keeping with much later views of evolution (Gould, 1989), but does demonstrate a continuing Darwinian substrate to Wells's views on evolution. *Bio-Optimism* (*Nature* August 1895), and *Intelligence on Mars* (*Saturday Review* April 1896) both emphasise Wells's commitment to competition and selection. In the latter the Martian context of the story suggests that selection is a universal law.

For students of human origins the most important story by H.G. Wells from this period is *A Story of the Stone Age* (*Idler* May–September 1897). The tale is set in southern England in what will one day be the county of Surrey (Hammond, 1998, Pearson, 2007). Wells's marvellously evocative prose conjures up a rich natural world of forests and rivers, as the reader is transported back into the world of fifty thousand years ago. The Wealden range (Chapter 7 and following) provides the backdrop for the unfolding of the story of Ugh-lomi and his mate Eudena. She is coveted by the leader of the tribe Uya the Cunning. The two flee. While in exile Ugh-lomi observes a stick accidentally forced into a sharp flint with a hole in the middle. The tribe already have carefully shaped handaxes. With some flint knapping he now invents the first hafted axe. He returns to the tribe and kills Uya. More innovations follow; horse riding and a club with lions' claws as teeth – the first composite weapon of different materials. The society of the tribe is brutal and merciless. Here is Huxley's 'nature red in tooth and claw' as applied to human evolution (though the phrase was first used by Tennyson). Ugh-lomi and Eudena are the only humans who show affection and concern for each other. The remainder of the tribe are motivated by an almost Spencerian self-sufficiency and ruthless individualism. There is a callous disregard for the suffering of others. In fact they are more like the Martians in their unemotional indifference to other's misfortune, and like the Martians, evolution has made them that way. Wells concluded *A Story of the Stone Age* with the following lines, and their introduction into the text in an almost offhand manner was clearly designed to underscore the brutality of the time:

> Thereafter for many moons Ugh-lomi was master and had his will in peace. And on the fullness of time he was killed and eaten even as Uya had been slain. (Hammond, 1998, 332)

In the physical appearance of the Stone Age people there is plenty of variability for natural selection to be working on. But they are nevertheless recognisably human. These are not ape-like Morlocks, nor are they *Pithecanthropus*. Our heroes appear to be from the better end of the physical range, while at the opposite end, Uya's brutality is in part explained by his atavism (beetling brows and prognathism).

In my opinion, this one story epitomises the general public's view of what life was like in the Palaeolithic. It has all the right elements; Huxley's brutal visions of nature, as well as progress, innovation, the Victorian view of the male/female hierarchy, domestication, and the dominant position of humans in nature. Darwinian competition and chance are present, but culturally embedded innovation drives progress forward rather than the indifferent self sufficiency of Ugh-lomi's defeated enemies. It provides an interesting alternative to Worthington Smith's (1894) view of Palaeolithic life which contained many of the same elements, yet lacked the callous brutality of Wells. Pearson (2007) speculates that Wells might have drawn some of his views on Palaeolithic society from Edward Clodd's *The Story of Primitive Man* (1895). If so that would be a nice link between the archaeological and anthropological community, and the general public.

Wells's post-Lamarckian views were not without their critics. He defended his position in the journal *Natural Science* (Perry Coste, 1897a; Perry Coste, 1897b; Wells, 1897). Amongst a number of issues, F.H. Perry Coste was concerned by one of the ideas from *Human Evolution* and *Morals and Civilization*. This was the 'Great-Man' theory'. Wells had argued that culture might advance by individuals learning from their own experiences, or through education instilling an appropriate sense of what was good and right. Equally important, education could be emplaced by emulating the actions of superior individuals who advanced various aspects of cultural life. This process of reflection and emulation would carry innovation into the next generation (Lamarckian cultural evolution). The idea was not new. Wallace and Lyell had mooted as much in the 1860s and it had been implicit in Lane Fox's views on innovation in material culture. Grant Allen (1885) suggested Darwin was such a man in his life of the great scientist (Allen's text is a classic example of Victorian hagiography and 'Great-Man' myth building.) Wells needed a very specific mechanism that would ensure the cultural transmission of positive characters now that biological use-inheritance was no longer tenable. Emulation fitted the bill. Perry Coste conceded that there may have been gifted individuals who made practical contributions, but he denied that moral or ethical genius could have arisen in the Palaeolithic. The comments are particularly relevant to *A Story of the Stone Age* because Ugh-lomi is evidently one of these practical innovators. Wells's reply suggested that he sympathised with this view. Most innovations and ideas developed by individuals would only really take off once settled community life, speech and the family unit were part of the common human condition. In effect this would be the Neolithic. *A Story of the Stone Age* predates that time, but it would nevertheless have served to illustrate to the public the concept of how individuals could influence cultural evolution; how education through innovation and dissemination by Great Men could lift the species from its brutish nature. Furthermore, in Wells's reply to Perry Coste, there are direct links with the emerging cultural philosophy that he was developing in *The Duration of Life*, *Death*, *Bye-Products of Evolution*, *Human Evolution* and *Morals and Civilization.*

Unsurprisingly, speech was a key element in Wells's emerging vision of human evolution. McLean (2008; 2009) placed it at the centre of his critique of *The Island of Doctor Moreau.* In *A Story of the Stone Age* it is in a very rudimentary state of development. Thoughts and concepts cannot be articulated as vocabulary is not sufficiently developed. The animals appear to have more developed communicative powers than the humans do, they can communicate on an intra- and inter-species level, reminiscent of Edgar Rice Burroughs's Tarzan. In *Rediscovery of the Unique* and *The Island of Doctor Moreau* speech had been highlighted as a bench mark in cognitive evolution. In the former our loss of the awareness of the uniqueness of all things dated to the origins of speech, when a limited vocabulary enforced descriptive and conceptual pigeonholing. The idea of the limiting power of nascent language resurfaced when Moreau explained to Prendick what the real difference was between ourselves and our nearest relatives:

> And the great difference between man and monkey is in the larynx, he continued, – in the incapacity to frame delicately different sound symbols by which thought could be sustained. (Wells, 1896, quoted in Wells, 2007, 229)

The chapter in *The Island of Doctor Moreau* that this quote comes from is called 'Doctor Moreau Explains', and was a rewrite of *The Limits of Human Plasticity* (Philmus and Hughes, 1975). The implication is that the capacity for versatile speech must precede complex thought. I will return to this in the next chapter

There is a huge body of scholarship surrounding Wells, and here I have only dipped into and borrowed from some of it. As the Victorian *fin de siècle* decade came to an end and the Edwardian period began, Wells too was undergoing his own evolution as a polemicist. The period of the scientific romances was over. He was becoming more of a prophet as he looked for human salvation in social reform. He had pursued his fiction and his scientific journalism in the public domain, making the most of the late Victorian publishing boom and the demand for 'light journalism and fiction' (Draper, 1987). In this sense his late Victorian writings were very much a part of the contemporary debates He provided informed opinion and scientific accuracy on the biological and evolutionary issues of the day and his background gave him a privileged insight into the workings of evolution and its possibilities. His reputation ensured his work would be noticed. Through his work he forced his readers to confront the reality of evolution, and the physical truth of human origins. Wells emphasized ambiguity and uncertainty and in so doing reflected the understandings of many practising scientists. Because he was a widely-read writer he was able to engage the public with the actuality of origins debates and not just what they wanted to believe.

# 13

# AMAZING STORIES

In this chapter I wish to look at the opposite end of the scientific fiction spectrum. With the exception of Grant Allen and Andrew Lang, the authors discussed here were not scientifically informed in their writing. They wrote for money as this was their livelihood. They gave the public what it wanted and what it expected. In doing so they reinforced stereotypical views of evolution and human Prehistory. I will divide the material into three groups; those stories that could loosely be grouped under the broad umbrella of scientific romance (which we would today call science fiction), prehistoric fiction, and lost worlds/lost races fiction. All, to varying degrees, reflected the biological and social evolutionary themes that have been discussed for H.G. Wells and human evolution. But I will begin by discussing in detail the work of one writer who perhaps more than any other epitomised the Victorian *fin de siècle* decade.

## Scientific Romances: George Griffith

George Chetwynd Griffith-Jones (1857–1906) wrote under the name of George Griffith, George Chetwynd, Chetwynd Griffith, G.G. or Levin Carnac. In the early stages of his career he had occasionally used the name Lara, and once toward the end of his life he called himself Mrs Stanton Morich to appeal to a religious audience. Wells's fame has ensured that his own life has been well documented. We know little about Griffith by comparison (Moskowitz, 1974a; Moskowitz, 1974b). Before he was eighteen he had shipped before the mast to Australia, jumped ship and then spent several years in the bush. By twenty he was back in England. Like Wells he spent some time as a teacher/ tutor although he had no formal teaching qualifications. He then tried his hand at journalism and publishing. Facing starvation he took a job with the London publishing group headed by C.A. Pearson (1866–1921) who later founded the *Daily Express*. From that point on until the late 1890s he was pretty much a company man for Pearson's diverse publishing interests.

Griffith wrote action-adventure stories, poetry, travel yarns and factual pieces on transport, foreign capitals, or on prisons – a favourite theme of his. His contribution to science fiction, if it is remembered at all today, was to the 'Future War' story begun by Chesney's *The Battle of Dorking*. Griffith was conscious of the possibilities of science and his technological descriptions were informed by contemporary debate (McLean, pers. comm.), but ultimately he wrote for money; he was a storyteller pure and simple – and a popular one at that. During the middle 1890s some of his fiction outsold that of Wells, much to the latter's frustration. During the early period of Wells's writing the two were often compared, which also annoyed Wells. He was aware of Griffith and admired some of his work (Moskowitz, 1974a; Moskowitz, 1974b). Griffith never really achieved the literary recognition he felt he deserved. Wells gradually overtook him in sales and public recognition. In truth Wells was the better writer, but what Griffith lacked in literary style, he made up for in imaginative and exuberant story telling. He was the British Edgar Rice Burroughs. He died in 1906 on the Isle of Man. His biographer summarised him as follows:

> George Griffith was obviously a popular, admired and influential writer of his time. He has not survived because his literary output was for the most part a reflection, not a shaper, of the feelings of the period. He danced to the beat of the nearest drummer. (Moskowitz, 1974b, 47)

Today it is difficult to locate Griffith's work except in anthologies (Moskowitz, 1974a; Moskowitz, 1974b; Evans and Evans, 1976) and there are few enough of those. However, the internet and out-of-copyright e-books have made access easier. Of the work that is available, three of his better-known pieces contain a sustained viewpoint on human and social evolution. *The Angel of the Revolution* (*Pearson's Weekly* January–October 1893) was published in book form under the same title by Tower Publishing in 1893. This was his first book and launched his career as a writer. Equally successful was its sequel *The Syren of the Skies* (*Pearson's Weekly* December 1893 to August 1894). This was published as *Olga Romanoff; Or The Syren of the Skies* by Tower Publishing in 1894. The third is *Stories of Other Worlds* (*Pearson's Magazine* January–June 1900) which appeared as *A Honeymoon in Space*, published in 1901 by C. Arthur Pearson.

The first two books played on the threat of terrorism that has already been alluded to, and secured for Griffith his justly deserved accolade as one of the greats of the Future War story.

In *Angel of the Revolution,* Europe teeters on the brink of war. The Franco–Slavic Dual Alliance (France and Russia) is ranged against the Triple Alliance (Great Britain, Germany–Austria, and Italy). But another power is at work undermining the political map of Europe, a secret society known as The Terror. With enormous wealth and secret agents placed at every level of European society and government, the Terror is poised to overthrow European civilization. Its aim is to sweep away an unjust and despotic capitalist system supported by corrupt financiers and hereditary monarchies. But they lack a weapon which will give them superiority over the military forces of Europe. A disillusioned and destitute young inventor called Richard Arnold gives it to them. He has conquered the problem of 'aerial navigation'

– powered controlled flight. He builds a fleet of airships for the Terror, and as the equivalent of the First World War unfolds, the Terror gradually gains the upper hand until by the end of the novel, it rules the world from its newly-founded country in Africa. (The irony of wanting to banish war and end tyranny by precipitating a war that costs tens of millions of lives and devastates Europe appears lost on the Terrorists.) The Terror calls its new land Aeria. Warfare is now banished, and property and ownership are restructured along egalitarian lines. (Pleasingly, the legal profession is scrapped and much social progress inevitably flows from this one act!) The Terrorists allow the nation states a certain freedom to rule themselves, but under their watchful eye. They enforce their will by a total control over all forms of air power which is the exclusive inheritance of the Terrorists and their descendants.

There is little biological evolution in this first novel. In their new state of Aeria, a geographically- and geologically-isolated part of Africa, the Terrorists find primates that had been evolving and now were 'closer to man'. The classic Darwinian mechanism of isolation drives this accelerated evolution. Apart from this, there is no other connection to physical evolution.

Nevertheless the novel is relevant to this study. It becomes apparent very quickly that it is a novel about race war and social Darwinism. While Griffith's political map of European alliances was fictitious, his Franco–Slavonic Dual Alliance versus the Anglo-Saxon League explored the racial fault lines that Beddoe, Galton, Ripley, and Huxley had been debating in the AI and elsewhere. (Huxley, 1870; Beddoe, 1885). In this sense Griffith's novel played to the fears of race and race war that were endemic in Victorian society. This is not surprising as race scientists/anthropologists in every European country sought the empirical evidence for the racial divisions they already knew to be present amongst the peoples of Europe.

Throughout the story, Britain, as the mother country of the Anglo-Saxon race has an almost mystical importance. It has been the bulwark of freedom and justice throughout history and as such its people have earned their destiny as the pre-eminent race in Europe. Underlying this is the creed that, inevitably, certain individuals, groups and societies or nations, manifest an inherent superiority. Such individuals or collectives will always win through. This is a version of the 'Great-Man' theory described in the last chapter. The members of the Terror are such individuals, and they form the nucleus of the new Aerian state.

Other racial concerns are drawn into the novel. Quite early on the Terrorists learn of a new Buddha having arisen in the East. He has formed his followers into an army which may soon threaten the West. The Terrorists fear that as a race, whites will be fatally weakened by the coming European struggle leaving the west open to successful attack. The Terror cannot allow superior white races to fall to 'yellow barbarism'. However the Buddhists are defeated by a Moslem army, which is in turn defeated by the Terrorist aerial navy. In contrast to this there is none of the casual (or overt) anti-Semitism that often accompanied this kind of story. In fact Jews are dealt with sympathetically. Natas, the master of the Terror, is in fact Natas the Jew. His Nihilist plot against Europe was inspired by the cruelty and crippling torture he received at the hands of the Tsars' agents who also raped and murdered his young Gentile wife.

Racial anthropology therefore sits at the heart of *Angel of the Revolution.* It is developed further in *Syren of the Skies*, but now the social Darwinist subtext really surfaces.

It is one hundred and twenty five years since the events of the first novel. The world is at peace and the Aerians have policed it for some four generations. Under their over-lordship the nations of the world have removed all social inequalities. Competition is recognised as a key factor in life because it is a natural part of our human heritage, but as in the over-world of the Eloi, competition in nature has been harnessed. In this brave new world individual ability and innate worth will allow anyone to succeed who has the talent to do so, and from a fair and equal footing with all others. This is equitable competitiveness.

Following the last wishes of Natas, the Aerians return self-determination to the nations of Europe. They retain air power solely for themselves however. But there is a shadow on this bright future in the shape of Olga Romanoff. She is a descendent of the Russian Tsars. She passionately hates the Aerians and all they stand for, and is consumed by one thought; revenge for her murdered father and her fallen dynasty. She too believes in innate superiority, but for her it is the bloodlines of the aristocracy who have the ancient right to rule. Democracy in her eyes (and Griffith's too possibly) is government by 'counting heads' – mob rule. Drugging a young Aerian who is one of the heroes of the novel she transforms him into her slave. He reveals all the secrets of the Aerians and Olga Romanoff builds a vast aerial battle fleet. Along the way she ensnares the Moslem ruler of the East and together they set about the conquest of the world. But at the moment when the final battle is to be joined, the Aerian fleet turns tail and runs away. They have received alarming information from Martian astronomers with whom the Aerians have been in contact for many years. The world will pass through a toxic cloud and all life will be destroyed. After many adventures, and the destruction of the Earth, the Aerian chosen elite emerge from their underground shelters to begin rebuilding a devastated world.

Central to both novels is the premise that innate superiority is a finite quality. In the second novel the Aerians, by carefully controlled marriages, have almost become a superior species of human. This is practical eugenics at its most successful. Much of this would have appealed to Francis Galton and Herbert Spencer. Griffith presents imperialism as the perfect vehicle for the deployment of innate superiority, allied with a martial outlook that maintains the health of the bloodlines. There are good reasons (according to Griffith) for this. Human nature is seen as naturally brutal and our default condition is violence and anarchy. Democracy is dangerous as it gives free rein to these primitive instincts. There is clear proof of this. With some help from Olga Romanoff, the nations of Europe and elsewhere quickly revert to barbarism and the worship of Moloch. The speed with which social change occurs, either upwards as with the Aerians, or downwards as Europe descends into anarchy, begs the question of the inheritance of characteristics, although heredity is not directly addressed in either novel.

Griffith contrasts the Aerian enlightened despotic socialism with the society that it replaced (i.e. Victorian society). It is implied that the old social pattern was based on Darwinian principles of unbridled competition and this led to many inequalities. The strong and financially secure preyed on the poor and the weak. While competition between individuals and groups is still present in Aerian society it is rooted in a strong moral philosophy which is accepted and self-policed by all citizens – pre-empting Wells's insistence on education after his rejection of use-inheritance. This negates the downside of competition but accentuates the positive side with wealth being placed at the service of both the individual and the state. The Aerian society is a foil for the social pattern of the Martians who are the classic Wellsian super intelligences, socially-engineered, and devoid of emotion. It will be recalled that *Man of the Year Million* was published in 1893 also. The contrast is therefore with Aerian society where a perfection has been achieved which did not involve engineering the humanity out of being human; a perfection through mutual appreciation of self evident superiority.

Recalling that Griffith had no scientific training, and was not trying to educate or warn his readers, we may ask what would the Victorian public have got out of these two novels concerning evolution?

1. Both texts reflect a strong social Darwinist stance developed through a utopian perspective on social evolution. Both reflect the belief that life is a struggle and the weak will go to the wall (usually civilians and the innocent) unless protected by morally superior rulers.
2. Certain individuals and their bloodlines, as well as races, show a hereditary superiority.
3. Careful social control, especially over procreation (through social peer pressure) can protect and enhance bloodlines. But I don't get the impression this is Galton-like eugenics, i.e. direct, state-approved, social control. This seems to lead to the type of dehumanized society found on Mars. As practiced by the Aerians it appears to be the recognition and acceptance by a society of intelligent individuals, that certain of them are especially gifted (Darwinian variation?). By common consent these individuals become social leaders. They act for the benefit of all, a fact that will be recognised and accepted by all. The naturally superior will automatically recognise the logic of their own superiors.
4. Human nature is instinctive and brutal, and unless controlled (by

society) reverts to savagery, barbarism and idolatry, through incessant warfare.

5. Technological superiority (i.e. technical evolution) confers a clear advantage to superior societies, especially in weapons of warfare.
6. Time is progressive. The society of the Aerians is an evolutionary progressive next step.

While Griffith's thinly-veiled political views would not have been acceptable to all, many of his core beliefs on social development would have found widespread acceptance. For many Victorians it would not have been difficult to blur the boundaries between their own society and that of the superior Aerians, even if Griffith intended the opposite. The superiority of English Victorian society was self-evident to the Victorians: the very complacency and arrogance that H.G. Wells was criticising. It is worth recalling that both of the novels were best sellers – they gave the public what it already knew to be true.

What of biological evolution? Griffith tackled this in *Stories of Other Worlds*. The plot is simple enough. An English aristocrat, the Earl of Redgrave, has invented an interplanetary ship, the *Astronef*, powered by the mysterious R-force. Accompanied by his new bride Zaidie, daughter of the discoverer of R-force, and the taciturn engineer Murgatroyd, he takes the *Astronef* on a voyage of discovery across the solar system. They have many adventures and only return home safely after nearly being dragged into the sun.

There is a strong Darwinian subtext explaining the nature of the civilizations they encounter. The Moon and Mars both have similar evolutionary histories – they are both old worlds now in decline. Environmental decay has led to intensified competition for resources and habitable living space. On the moon this process is far advanced. Great cities lie in ruins on the beds of the lunar seas, tracking the shores of the ever diminishing oceans. In the centre of the dried up sea beds the last habitations are shanty towns. For a time, life had survived in the craters where oxygen had still been plentiful. But gradually these settlements had succumbed. In the crater, Tycho, they find one city with the skeletons of its last inhabitants littering the streets. They were humanoid but large and barrel chested. Redgrave explains that natural selection would have favoured those in the population with bigger lung capacity – natural selection working on variability. This last city was inhabited by the descendants of those who, by chance, had been able to breath better in the dwindling Lunar atmosphere. Later, at the very base of the crater, Newton, they find water and just enough of an atmosphere to sustain life. The last descendants of the moon now live here in cold empty blackness. They have degenerated to become blind, pale, ape-like creatures, not dissimilar to the Morlocks.

On Mars, decay is not as far advanced as the Moon but it has led to the most technologically-advanced race defeating all others and dominating the planet. These are (probably) the Martians of *Syren of the Skies*, themselves on the way to becoming the Martians of H.G. Wells. The Martians appear to have bred out of themselves any trace of their animal ancestry. However, at the sight of Zadie they become consumed with lust as their primitive passions resurface and the *Astronef* has to make a quick getaway. Griffith's Martians are similar to his humans – the beast lies close to the surface.

The story is the same for both these worlds. The physical character of natural selection is driven by competition, fuelled by diminishing resources. Only the strongest (where strength=best adapted) survive. Selection works on the presence of variation in the population. Retrogression and decay appear to be the inevitable results of evolutionary change under these circumstances. 'Human' nature cannot be wholly subsumed by nurture. Aggression is a default state for any form of humanity under difficult or competitive circumstances. Moreover there are dangers in social engineering since the baser passions cannot be wholly bred out – even if they should be. Inadvertently Griffith is re-running the lessons of the Eloi and the Morlocks (and prefiguring Edgar Rice Burroughs's Mars by nearly twenty years).

Various evolutionary explanations are used to framework the peoples of Venus and Ganymede (one of Jupiter's moons). Everywhere, survival of the fittest is the universal law. Often technological superiority confers, or reinforces, that status. Finally on Saturn they witness the whole spectrum of evolutionary development paralleling the history of life on Earth as they travel from the Saturnian equator to the south pole. At the equator they discover the reptiles of the oceans (Earth's Secondary epoch). As they travel further south they find the great mammals of the plains (Earth's Tertiary epoch). Finally on mountain slopes near the southern pole they find anthropoids living in caves (Tertiary–Quaternary). The journey is reminiscent of Borough's *The Land that Time Forgot* trilogy, and like that work there is a strong hint of recapitulation in the idea of stages of progression.

A comment by Zaidie will have struck a chord with a number of Victorian readers. While happy to accept that Darwinian evolution was a fact, she objected to Darwin's title *Descent of Man* (implied in the story but not stated). She preferred to think of the ascent of man, especially when applied to women.

As a mirror of the times Griffith's view of biological evolution was, I suspect, the one that most people held. Nature 'red in tooth and claw' set the agenda, and relentless competition was the engine that drove progressive evolution. It will be recalled that Wells's *Experiments in Co-operation* was written as a direct counter to this very same outlook.

## Prehistoric Fiction

> Prehistoric fiction, hereafter abbreviated 'pf' is a speculative literary genre dependent on extrapolations from scientific or quasi-scientific discourse. In the densely branching genealogical tree of popular literary genres, pf is very closely allied to science fiction… (Ruddick, 2009, 2)

In his informative and original book *The Fire in the Stone*, Nicholas Ruddick sees pf and science fiction as compatible but not identical projects. The former asks us to consider what it means to be human by plausibly speculating on our origins. The latter uses imagination to question what it may be like to be human in the future – a very Wellsian concern.

In tracing the history of pf, Ruddick notes the disparity between English and continental pf written before 1901 (Ruddick, 2009). That the French invented pf will come as no surprise given the wealth of archaeological evidence that was discovered there after 1859. Writing fiction was a logical next step. The first such story was *Paris avant les hommes* published in 1861 by Pierre Boitard, and many more such tales soon followed. But it was not just fiction writers who were visualising France's Stone Age past. One of the discoverers and excavators of the French Upper Palaeolithic site of Solutré, Adrien Arcelin, wrote a novel in 1872 about its original inhabitants. Under the pseudonym Adrien Cranile, he wrote a tragic tale of race war at this celebrated horse-kill site. Reading Ruddick's descriptions of the various plot lines in this and other stories it is clear that race was a central concern of this early French literature. Ruddick notes that in Arcelin's 1872 novel, the incoming blond-haired Aryans employ their technological superiority to besiege and slaughter the indigenous Solutreans. He posits the likely impact of the Franco–Prussian war and the siege of Paris on this storyline.

Such was not the case in Britain. Reflecting a much poorer data set, British pf did not begin until 1880, with Andrew Lang's *The Romance of the First Radical; A Prehistoric Apologue* published in *Fraser's Magazine* in September of that year. There were few successors. Wells's *Story of the Stone Age* was one, and Stanley Waterloo's *The Story of Ab: A Tale of the Time of the Cave Man* published in America in 1897 was another. Waterloo was an American, so strictly speaking his book should not be included in a list of British pf publications. However he set *The Story of Ab* on the banks of the Pleistocene Thames, and the ideas underpinning his story are of relevance to the themes already highlighted by Wells's and Griffith's work. The same applies to his short story *Christmas 200,000 B.C.* (from *The Wolf's Long Howl* published in America in 1899). *Zit and Xoe: Their Early Experiences* by Henry Curwen is one of the few other notable examples of British Late Victorian pf (*Blackwoods Edinburgh Magazine* for April and May 1886). I will begin with this.

Ruddick considers *Zit and Xoe* to be satire, but admits it is difficult to pinpoint just what is being satirised. Zit is born into a family of apes but he is different. He is hairless, lacks a tail, and is much more intelligent than his siblings or parents. He is thrown out of the family and journeys down river to the sea (a classic metaphor for personal development). On the way he comes to understand how different he is from the other animals around him, which leads him to mistreat them. As he travels he invents a series of 'manly' things such as weapons. At the sea shore he meets a girl, Xoe, whose story is similar to his. She too has made a number of 'womanly' discoveries on her journey down another river. As the only two humans on earth, they join forces and slowly come to love each other. In learning how to live with one another they invent the idea of Victorian family values. They also learn about sexual politics and the innate differences between male and female psychologies (at least from a Victorian perspective). It seems that human nature and gender stereotypes stretch back to the dawn of time. Zit also becomes a father. He leaves the house one morning and returns to find he has a son. Eventually the animals of the forest attack their home so they flee to a nearby island, where they learn animal domestication and husbandry. Finally they leave in search of a new home. The story ends with Zit as an old man writing his memoirs. He has taken on the mantle of Biblical patriarch. Their village is full of their great-great-grandchildren.

The story bears no relation to any Prehistoric scholarship or serious interpretation of the past. Evolution is not mentioned, and discussions of origins are assiduously avoided, although the concept of progressive time is present. I suspect the real message of the story, if it has one, is along these lines: Humans are clearly distinct from the natural world, special, even if they evolved out of it. Apart from our intelligence, the positive influence of human love distinguishes us from brute nature. Consequently, our distinctiveness should be celebrated for its own sake. Xoe is unhappy that Zit is writing their story down and declares:

> 'Why should we degrade them (their descendants) so terribly with the tale of our mean origin?' (Curwen, 1886, 634, May issue, my brackets)

Many Victorian readers would have sympathised with such sentiments, they echo Zaidie's discomfort with a brutish and naturalistic origin. I suspect a key reaction to human evolutionary heritage was being expressed here. By the late 1890s it was difficult for most rationally-minded individuals to deny humanity a deep Prehistoric past, but this conflicted with Victorian hubris. It was this very sentiment that H.G. Wells was so keen to expose through his fiction and his science writing. Many traditionally-minded Victorians would have preferred the following storyline.'Despite a wholly natural ancestry, humans are clearly different and special now. We have transcended our own origins.' Consequently, there is no point in re-exposing human ancestry – to do so was in bad taste. The writing of the story is tongue-in-cheek with a certain arrogance emplaced through patronising humour. This attitude is not uncommon in the stories described here.

Stanley Waterloo's *Christmas 200,000 B.C.* describes a day in the life of a Palaeolithic girl called Red Lips. Her bullying father, Fang, gives her to a hunter called Wolf whose atavism ran to a tail and a propensity for howling. She escapes and encounters her preferred suitor, a young man called Yellow Hair. They kill Wolf, and just for good measure they kill her father too. The day ends with them occupying her parents' cave, accepting her mother as a slave, and having a celebratory feast. This very short story lacks any subtlety but would have fed straight into the public's expectation of what a day in the Palaeolithic would have been like. In particular Wolf's courtship appears to have been an assumed standard of behaviour; he caught Red Lips by the hair (with her father's permission) and dragged her off to a cave. Seemingly ludicrous today, there was a serious body of anthropological scholarship from the 1860s that saw this as the root of all later forms of marriage relations (McLennan, 1865) as well as exogamy, the widespread human practice of seeking wives from outside one's own social group. This Palaeolithic courtship, today a stereotypic image for cartoonists, stuck in the Victorian public consciousness.

In *The Story of Ab*, we follow the youth and manhood of a young warrior of the same name. He has the full range of adventures that one would expect in a story of this kind; he wins his mate, earns the respect of his tribe, explores new lands, fights off rival tribes and lives to see a peaceful old age brought about by his own inventiveness. The story is set on the banks of the Thames, a tributary of the ancient Rhine which flows across the dry North Sea basin. Ab is one of the more advanced members of his group both physically and mentally. He is another example of the 'Great-Man' theory. Through inspiration and intuition he contributes to the development of the bow and arrow, the initial domestication of wolves and the concept of the village, while his partially-crippled mentor invents the polished arrowhead, and his young son invents parietal cave art (there is more than a hint of Percy Bysshe Shelly as a martyr to art in the character of Ab's young crippled son and this also features in *The Romance of the First Radical* below).

Ab's peaceful and settled community in the safety of its protected valley excites the jealousy of a neighbouring tribal chief. This is the origin of organised inter-community warfare, and even the dawn of military tactics (Ab's defence of the valley at the climax of the novel reads like the battle for Rorke's Drift). This was the same world as that of Red Lips and Yellow Hair. Life was relatively cheap and harsh, though individuals occasionally transcend this. The food quest was the primary motivation in life, followed by the need for shelter. Progress was slow and hard won. Under Ab's guidance the tribe evolves from self-sufficient family groups united by self-interest, to the beginnings of settled village life.

As a full length novel *The Story of Ab* gave Waterloo more space to develop his theme, so his story is more nuanced than Wells's *Story of the Stone Age*. Language is simple but effective, allowing for basic comprehension at the level of immediate necessity between individuals. It is clear that the limitations of language are, as with Wells, a hindrance to the development of more complicated cognition. Ab and his peers experience the same emotions we do, and just as intensely, but because they cannot reflect on their context, they are unable to comprehend them properly. The lack of linguistic subtlety inhibits not only cognitive development, but the development of society and ethics. Ab murders his boyhood friend Oak in a fit of jealous rage while competing for the attentions of Lightfoot who will become his mate. His conscience troubles him for the rest of his life, although the murder itself is not considered such a great event in his tribe. He dimly perceives a sense of a spiritual otherness in his reflections on his crime, but he cannot quite grasp the concept as a whole.

In this story, the roots of morality and emotion evidently date back at least to the transition between the Palaeolithic and the Neolithic, the time when Waterloo had set his novel. For Waterloo there was no hiatus between the two. Here the novel-length story comes into its own as the author presents a series of set pieces which develop human culture. Individuals whose superior qualities are self-evident invent new tools and practices. Their innovations become embedded within everyday usage and so pass on to the next generation. This is a clear usage of the 'Great-Man' concept, and of emulation/imitation as a mechanism to empower Lamarckian-like cultural inheritance. There is even a hint of Weismann on the last page of the novel. In describing a series of unnamed Victorian personalities, who would have been familiar to his readers (Ruddick, 2009 names them), Waterloo asserts that 'in the veins of each has danced the transmitted product of the identical corpuscles which coursed in the veins of those two (i.e. Ab and Lightfoot) who first found a home in the Fire Valley'.

Andrew Lang's *Romance of the First Radical* was a story woven around the discoveries of prehistoric skeletons in caves along the Côte d'Azur, near the French–Italian border. The first was discovered in 1872 at Mentone (Grimaldi). Many scholars such as John Lubbock accepted the skeleton as a genuine Palaeolithic burial. Pleistocene fauna occurred above and below the skeleton. Boyd Dawkins, as usual, claimed it was a Neolithic intrusion (Dawkins, 1874; Dawkins, 1880a). A number of other burials were discovered in the numerous caves at the locality in the early and middle 1870s (see Table 3.1). Lang's story suggests that he visited the area and knew the ground.

In the story, Why-Why is a social misfit, unable to accept the conventions, taboos and traditions of his day. He questions everything. This earns him the enmity of his tribe and costs the life of his sister. Details of her burial suggest Lang associated her with another skeleton discovered in the area. As Why-Why grows, he becomes a formidable warrior and so is able to stave off the retribution which the tribe believes he deserves for his temerity. He rescues a young orphan girl Verva who grows to love him. One day Why-Why's brother dies from overindulging at a cannibal feast. Tradition requires Why-Why should now murder somebody from the offending tribe by ambushing them. Instead he challenges that tribe's most fearsome warrior. He wins the contest but is left for dead. Verva, realising he is still alive, saves him. They declare their love for each other and live happily for a few years in a valley some thirty miles away from Mentone called Vallauris. In doing this Why-Why commits his greatest act of sacrilege. Both he and Verva are members of the snake totem and traditional tribal law forbids members of the same totem to marry. Why-Why realises that exogamous practices, established as incest taboos, need only apply to members of the same local group, not to members from different geographical localities that have had no contact. However the tribe from Mentone hunt him down. He and Verva are killed but not before Why-Why delivers a final messiah-like prophecy at the same time as forgiving the tribe their actions because they 'know not what they do'. The tribe realise their mistake too late. They reform their ways and renounce shamans and medicine men (proxies for the modern priesthood and the medical profession). Why-Why and Verva are buried in a cave together, though it is not clear whether this is at Mentone or Vallauris. Lang implies that their bodies too have also been discovered and excavated.

Why-Why fits the 'Great-Man' pattern, but he is not an inventor as in the other stories described. He is a social reformer; something close to the Victorian heart. Lang's main aim (Sparks, 1999; Ruddick, 2009) was to point out the deleterious effects of outmoded rituals and taboos (religious and secular). In the terminology of E.B. Tylor these were 'survivals', social practices which were actively maintained by tradition but which no longer served any practical or useful purpose. In most cases their original value had been long forgotten. When he wrote *Romance of the First Radical*, Lang was a staunch follower of Tylor and evolutionary anthropology (Stocking, 1995). He prefaced his story by telling his readers that all the social practices he described were validated by archaeology and anthropology. He was informing his readers that what they were to encounter in his text could be true. At the same time he was emplacing in their minds the power of the core methodology of evolutionary anthropology – the comparative method (see Textbox 3.1). As usual, underpinning this method was the universality of the human mind. Survivals validated the methodology as they proved the reality of successive cultural stages and thus the evolutionary nature of cultural development. They were therefore a useful tool for anthropologists to track backwards into cultural prehistory.

Sparks (1999) detects a strong hint of irony in Lang's prose, as if he did not really believe in the veracity of the method or the plausibility of the story. Certainly in *The Great Gladstone Myth* (Textbox 3.1) Lang had cleverly used satire to show up the potential shortcomings of the comparative method. In the late 1880s and through the 1890s Lang drew away from the evolutionary anthropological approach, possibly re-engaging with the more spiritual side of his nature (Stocking, 1995). But the point I would like to make here is that it doesn't really matter whether or not he believed in his story. His audience would have. He was a recognised authority on anthropology and had written a number of important academic books contributing to methodological and theoretical developments. His writings on myth and folklore were widely read. He prefaced his story with an appeal to the comparative method. The Palaeolithic world he described would have been familiar to readers of

Stanley Waterloo and H.G. Wells. It was the brutal world of 'all against all'; violent encounters, marriage by capture, and an uncaring society. Even if the reading public did recognise the irony, they would still have responded to the picture of life presented. Sparks (1999) suggests that Lang's approach blurred the boundaries between scientific writing, reconstruction of the past, and fictional narrative. But for the public this wouldn't have mattered because an expert was telling them what they already knew to be true.

In concluding this section I would add that the accepted picture of Palaeolithic life disseminated by popular writers in science fiction and prehistoric fiction was not vastly dissimilar to the more nuanced and archaeologically informed image of it presented by Tylor, Lubbock, Huxley, Boyd Dawkins and others. These researchers were widely read, but within a limited market. This was primarily the scientifically-minded section of the middle and upper classes who could afford such volumes (unlike Benjamin Harrison in Ightham). But as I suggested earlier, with the exception of a few anthropologically-minded researchers like Tylor and Lubbock (and after the 1860s Boyd Dawkins, Worthington Smith and J.A. Brown), few Prehistoric researchers presented their readers with reconstructions of Palaeolithic everyday life. As noted with the images of *Pithecanthropus*, there were few enough reconstructions of what Palaeolithic people may have looked like, at least from the 'professionals'. That role had fallen to the fiction writers and those who illustrated their stories. H.G. Wells, Stanley Waterloo, George Griffith and their peers were reaching a much wider audience because they published in the popular magazine market. The public's perception of the Palaeolithic and evolution was cemented into its wider consciousness through these broader media, and I suspect, it is the same today. It is worth re-emphasising that through their prose, the fiction writers not only interpreted and influenced, but invited their readers to share in the adventures of their characters. This was a very powerful and emotive device for emplacing a very specific vision of the past.

*Table 13.1. A list (not comprehensive) of the scientific romances published across the last decade of the Victorian era. Stories reprinted in later published anthologies appear with a number in parenthesis to indicate source. Stories with no parenthesis were read from the original source. (1) Russell (ed) 1979; (2) Moskowitz (ed) 1974; (3) Evans, H. and Evans, D. (eds) 1976; (4) Moskowitz (ed) 1974.*

## Other Scientific Romances Relevant to Human Origins

Science fiction began with what can loosely be called the scientific romances, but as will be evident this broad umbrella term covered a very diverse group of storylines. Table 13.1 lists thirty-three stories from a variety of magazine sources across the 1890s and 1901. Considering the number of magazines being published this was a trifling figure, and of these only eight had any real connection with evolution or human origins. Like Griffith's stories, these are difficult to locate nowadays and reliance has to be placed on anthologies or the internet. The list in Table 13.1 is undoubtedly an underestimate, but the total number of scientific romances and/or stories with evolutionary themes will still be small. The frequency of scientific romances is perhaps best illustrated by a review of the contributions to *Pearson's Magazine*, which,

| Year | Title | Author | Magazine | Month | Evolutionary theme | Synopsis |
|---|---|---|---|---|---|---|
| 1891 | Old Doctor Rutherford | D.F. Hannigan | The Ludgate Monthly (2) | September | No | A man with the Elixir of Life dies when he discovers the spirit of his lost bride is reborn |
| 1892 | The Doom of London | R. Barr | The Idler (2) | November | No | A killer smog destroys London |
| 1893 | The Angel of the Revolution | G. Griffith | Pearson's Weekly | January-October | Yes | A secret organization precipitates a world war in order to restructure human society |
| 1893 | The Syren of the Skies | G. Griffith | Pearson's Weekly | December – August 1894 | Yes | Sequel to previous. A future global utopia is disbanded resulting in a world war ahead of Earth's destruction from space |
| 1894 | The True Fate of the 'Flying Dutchman' | G. Griffith | Pearson's Weekly (4) | July | No | Ghost Story |
| 1895 | An Express of the Future | Jules Verne | The Strand Magazine (2) | January | No | A transatlantic tube system powered by compressed air |
| 1895 | The Purple Death | W.L. Alden | Cassell's Magazine (1) | February | Yes | A mad chemist invents a super virus to rebalance overpopulation. |
| 1895 | The Gold Plant | G. Griffith (as L. Carnac) | Pearson's Weekly (4) | May | No | Treachery in the African jungle as ruthless adventurers seek a lost white race |
| 1895 | Golden Star | G. Griffith | Short Stories | September-December | No | The mummy of the last Aztec prince, and his bride, are awakened by Europeans and claim their birth right - Mexico |
| 1895 | Valdar the Oft Born | G. Griffith | Pearson's Weekly | February-August | No | One of the *Aesir* is cast down to earth and fated to be reborn many times |
| 1896 | Citizen 504 | C.H. Palmer | The Argosy (2) | December | No | Two young people defy the arranged marriage system in a future dystopia |
| 1897 | The Microbe of Death | R. De Cordova | Pearson's Magazine (1) | November | No | A medical doctor uses his knowledge of immunology to kill a rival |
| 1897 | The Aerial Brickfield | John Mills | Magazine unknown (3) | | No | A man invents a way of turning air into house bricks with disastrous consequences |
| 1897 | The Thames Valley Catastrophe | G. Allen | The Strand Magazine (2) | December | No | A fissure volcano erupts and engulfs London |
| 1898 | The Curse of Ham | G. Griffith | Pick Me Up | August | | No data available |
| 1898 | London's Danger | C.J. Cutcliffe Hyne | Pearson's Magazine (1) | February | No | A great fire ravages through London destroying the city. |
| 1898 | The Lizard | C.J. Cutcliffe Hyne | The Strand Magazine (1) | February | No | Prehistoric reptiles awake from suspended animation in a cave. |
| 1898 | A Corner in Lightning | G. Griffith | Pearson's Magazine | March | No | Disaster results when the Earths magnetic and electrical fields are tampered with |
| 1898 | Where the Air Quivered | L.T. Meade & R. Eustace | The Strand Magazine (2) | December | No | Two young men enter Mecca in disguise and are pursued by Moslems in revenge |
| 1899 | The Monster of Lake La Metrie | W.A. Curtis | Pearson's Magazine (1) | August | No | A human brain is transplanted into a prehistoric reptile from an inner world |
| 1899 | The Master of the Octopus | E.O. Weeks | Pearson's Magazine (1) | September | No | An inventor offers a new form of lamp to a company which dominates the world's markets |
| 1899 | The Purple Terror | F.M. White | The Strand Magazine (1) | September | No | A story about a carnivorous plant |
| 1899 | The Wheels of Dr Ginochio Gyves | E. Douglass and E. Pallander | Cassell's Magazine (1) | November | No | A scientist builds a machine for to allow him to visit other worlds |
| 1899 | The Lost Continent | C.J. Cutciffe Hyne | Pearson's Magazine (1) | July-December | Slight | An adventure story in the last years of Atlantis |
| 1900 | The Abduction of Alexandra Seine | F.C. Smale | Magazine unknown (3) | | No | A chase story from early days of flight |
| 1900 | Stories of Other Worlds | G. Griffith | Pearson's Magazine | January-June | Yes | A newly wed couple honeymoon through the solar system |
| 1901 | The Man Who Meddled with Eternity | E. Tickner-Edwards | The Harmsworth Magazine (1) | February | No | A man possess a rivals body in order to win the love of his life, but it ends in disaster. |

| | | | | | | |
|---|---|---|---|---|---|---|
| 1901 | The Raid of Le Vengeur | G. Griffith | Pearson's Magazine | February | Slight | France invents a submarine, but a pre-emptive strike on Britain is thwarted |
| 1901 | The Fate of the Firefly | J.M. Bacon | Magazine unknown (3) | | No | An early experiment in flight goes wrong |
| 1901 | The Lady Automaton | E.E. Kellett | Pearson's Magazine (1) | June | Slight | A gifted inventor builds a mechanical woman. Human evolution is said to have taken 'thousands' of years by Darwin |
| 1901 | The Last Days of Earth | G.C. Wallis | The Harmsworth Magazine (1) | July | No | The last humans leave a dead Earth as the Sun dies |
| 1901 | The Last Stand of the Decapods | F.T. Bullen | The Strand Magazine (1) | December | Yes | Whales compete with giant cuttlefish for mastery of the ocean. |
| 1901 | Lord Beden's Motor | J.B. Harris-Burland | The Strand Magazine (2) | December | No | A ghost story from the early days of motoring |

along with the *Idler*, probably printed more scientific romances than any other magazine. These data are shown in Table 13.2.

In those stories in Table 13.1 which do comment on evolution (and some like the *Lady Automaton* only make a passing reference to it) there is a strong trend towards social Darwinism. Some of these stories have already been discussed. Only one, *The Last Stand of the Decapods* alludes to biological evolution. It presents a fanciful evolutionary history of the war between sperm whales and the giant cuttlefish of the deep oceans – the Kraken of ancient legend – today's giant squid. Reports of giant squid were not uncommon by the late Victorian period though the author of the story, F.T. Bullen, greatly exaggerates their size. Victorian readers would recall the memorable attack of the giant cuttlefish in Verne's *Twenty Thousand Leagues Under the Sea*, first published in 1870. Bullen's evolutionary scenario is made up, but the background of ruthless competition for survival between species would have been only too familiar. This would have given an authentic ring to even a fictional history.

I must be honest and state that the degree of social Darwinism present in the remainder of the stories is variable. It is usually more of a framework on which to hang the narrative rather than an integral element of the storyline. Nevertheless it is present and authors have clearly drawn ideas from common understandings of social evolution to construct their tales. In *The Purple Death* W.L. Alden plays on the terrorist theme by having a brilliant scientist develop an array of new and highly contagious diseases. Population increase and its control are the central theme here. His mad scientist intends to destroy the working classes thus creating more work, space and housing for the survivors, while at the same time freeing the world from 'Malthus's nightmare'.

C.J. Cutcliffe Hyne's *The Lost Continent* is a brilliant tale of the end of Atlantis and remains one of the best lost-continent stories ever written. I include it here, not only for its social commentary, but also because it shows the capacity of this type of literature to comment on contemporary society. Deucalion is an Atlantean provincial governor. He is recalled from Yucatan to the mother city after twenty years absence. Much has changed. New inventions and technology, new fashions, attitudes and standards of behaviour have appeared under the new 'self-made' empress Phorenice, a brilliantly realised and thoroughly despicable character. She has even abandoned the old gods. Deucalion is a very conservative and old-fashioned man, and a priest of the old religion. He is appalled at this new Atlantis. He falls out with the empress, falls in love with Naïs, a young woman from the rebel army besieging the city, and after many adventures escapes with her as Atlantis sinks beneath the waves. The story is a thinly disguised comment on society in the last Victorian decade – a classic *fin de siècle* tale with London easily substituted for Atlantis (see Alder, 2008 on this as well). There has been too much prosperity and progress, and all too quickly. The price is decadence, hedonism, spiralling poverty and a loss of moral fibre. Atlantis is a worn-out society. These themes would have resonated sharply with many readers, but the alternative is even worse; the Leveller philosophy of the besieging rebel army – no priests or chiefs, no authority at all – the chaos of democracy.

The sub-text of the story is that society needs to be ordered and controlled by an elite. Remove a strong moral authority and the result is mob rule and the bestiality of the rebel horde outside Atlantis' walls. The parallels with Griffith's *Angel of the Revolution* and *Syren of the* Skies, and a number of other stories are compelling. By virtue of his conservative nature and isolation in Yucatan, Deucalion is now the last moral man left in Atlantis. He is the equivalent of the innately superior individual, the 'Great-Man'. Also, there is an element of Tylorian survivals in the religion of the old gods. Though Deucalion's faith in the old gods remains, he begins to suspect, at the end, that

*Table 13.2. Table showing frequency of scientific romances in* Pearson's Magazine, *and some of the authors who wrote them, between 1896 and 1901.*

| Volume | Year | Total number of entries in volume (including fiction, non-fiction and poetry etc.) | Scientific Romances | George Griffith (also writing as Levin Carnac, George Chetwynd, and G.G.) | H. G. Wells | Others writing scientific romances | Articles on heredity and inheritance |
|---|---|---|---|---|---|---|---|
| 1 | 1896 January - June | 100 | 1 | 8 | 0 | 1 | 0 |
| 2 | 1896 July - December | 111 | 1 | 3 | 2 | 0 | 1 |
| 3 | 1897 January-June | 98 | 1 | 2 | 1 | 0 | 3 |
| 4 | 1897 July – December | 91 | 1 | 4 | 1 | 0 | 1 |
| 5 | 1898 January – June | 225 | 1 | 4 | 0 | 0 | 0 |
| 6 | 1898 July – December | 238 | 0 | 4 | 0 | 0 | 1 |
| 7 | 1899 January – June | 189 | 0 | 2 | 0 | 0 | 1 |
| 8 | 1899 July – December | 117 | 3 | 6 | 0 | 3 | 3 |
| 9 | 1900 January – June | 99 | 1 | 1 | 0 | 0 | 0 |
| 10 | 1900 July – December | 110 | 1 | 2 | 0 | 1 | 0 |
| 11 | 1901 January – June | 106 | 2 | 1 | 0 | 1 | 0 |
| 12 | 1901 July – December | 121 | 1 | 0 | 1 | 1 | 1 |

the institutions of the old ways were as worn out as Atlantis itself, and just as responsible for its impending doom. The continent's destruction is almost inevitable, but will Victorian society go the same way? Perhaps as much as any other story of its kind *The Lost Continent* invited the Victorian reader to reflect on progress and what actually stands for social evolution. The same theme is developed in the next story.

There is a strong body of scholarship and literary criticism contextualising the work of Grant Allen (Morton, 2005; Morton, n.d.). Like Wells he wrote from a scientifically-informed point of view and sought to educate and inform, as well as entertain. In *The British Barbarians* (1895), Allen has an anthropologist from the twenty-fifth century travel back in time to the Victorian era. He travels the world researching customs and taboos, finally ending up in England. Just as the English can never consider themselves foreigners when abroad, so they are unable to accept that many of their customs and social institutions are fetishes and taboos, in no way different to those of tribal people in Africa or the Pacific islands. This is a point his friends in polite society are unable to accept. Like Wells, Allen was targeting social arrogance and complacency. Racial arrogance is also evident as the traveller compares English and Muslim society.

The anthropological time traveller falls in love with a young married woman. On realising the truth of his criticisms of her world, she returns the sentiment. They elope and he plans to take her and her children away with him, although she is as yet unaware of his true origin. But her husband discovers them and shoots the traveller whose body dies while his spirit returns home. She is left contemplating suicide as the only way of joining him.

Perhaps not surprisingly the book was not received well, even H.G. Wells criticised it for being in bad taste (Morton, 2005). Allen is merciless in his unpicking of the polite social fabric of Victorian England. He takes it much further than Andrew Lang's attack on survivals in *Romance of the First Radical*. Allen gives the impression that the very fabric of English society is corrupted by these nonsensical traditions. The taboos surrounding marriage and male–female relations come in for particular attention, and are depicted as being no more relevant or civilised than the Palaeolithic capture-marriage of McLennan. Although social taboos are presented as Tylorian survivals, there is no acknowledgement of the significance of their historicity. They are never presented as an anthropological tool for revealing social evolution. This time traveller is more of an ethnologist than an anthropologist. In fact, taboos are a mirror in which the downside of cultural development can be revealed.

Allen was a follower of Herbert Spencer at this time and the book is more about the future than the past. In the preface he quotes Spencer's views that individuals with strong opinions could effect social change (mapping on to themes of the 'Great-Man' and emulation). As with *The Lost Continent* there are distinct similarities to Griffith's elitist views on the society of the Aerians and the evident nature of innate superiority (though interestingly in

Cutcliffe Hyne and Allen, that superiority does not automatically guarantee subservience in others, as it does for Griffith). Allen makes a powerful point: European civilisation bespeaks the ability to self-organise on a grand scale, but this is not the same thing as being civilised. Many complex urban societies practiced terrible barbarities in the past (e.g. Aztec ritual sacrifice). The qualities that single out a truly civilised society only come with a subsequent stage of social evolution as humans become more rational and ethically sensitive to each other.

For me, the structural underpinning of social Darwinism in these stories highlights a Victorian outlook that contrasts sharply with modern views. Today, very little in twenty-first century life leads us to believe humanity is special or somehow separate from nature. Archaeology, human origins research, DNA etc., all combine to give us a very clear perspective of our place in the natural world. But for the Victorians it was the opposite. Even after the Darwinian revolution, British/European society still instilled in its members a belief that they were different and that racial and biological distinctiveness was real; as was the innate and self-evident superiority of white European civilization. Like Griffith's these authors were telling the public what they wanted to hear as opposed to Wells who told them what they ought to hear.

## Lost Worlds/Lost Races Fiction

Lost worlds and lost races fiction was popular in the late Victorian period. It offered opportunities for authors to vent their spleen on the shortcomings of contemporary society and invent hypothetical ones to illustrate their preferred social remedies. The works of Henry Rider Haggard (1856–1925) were immensely popular He located lost races in the African or South American jungles, in the mountain ranges of the East, or in deserts at the margins of the civilised world. Haggard however did not engage with human evolution. His novels epitomised a school of writing that connected with the inner spiritual nature of humans. He was more in tune with Lang and the rediscovery of romanticism. He and his peers had to set their tales in impenetrable jungles or distant mountain ranges because by the late Victorian period most of the world had been visited by someone or other, even if hadn't been mapped. The world's empty spaces were rapidly shrinking. As usual, archaeology had followed in the baggage wagon of the explorers. Discoveries like Troy, Nineveh, Angkor Wat in ancient Cambodia, or the lost cities of the Yucatan and Peru, all faithfully reported in the *London Illustrated News* (Bacon, 1976), empowered lost race stories, granting them authenticity in an age when geographical discoveries had given way to archaeological ones (Pringle, 1981).

It is possible to dedicate a whole book to this class of fiction, but I have chosen to describe only the works of Jules Verne (1828–1905), not just because of his status as a founding figure in science fiction, but because his engagement with human evolution covers most of the topics discussed in earlier chapters. (I am aware this omits some very important contributions to this field of writing; Samuel Butler's (1835–1902) *Erewhon or Over the Range* (1872 – the title is an anagram of nowhere) and its sequel *Erewhon Revisited* (1901), and E. Bulwer Lytton's (1803–1873) perennial favourite *The Coming Race* (1871) which very effectively combined commentary on both biological and social evolution.)

Only two of Jules Verne's works contain any substantial engagement with evolution, and both are lost worlds/lost races novels.

*Voyage au centre de la Terre* was first published in 1864. The story is well known. Led by Professor Lidenbrock, a small party of scientific explorers follow the clues left by a medieval Icelandic adventurer. Through the volcano Snaefells they journey downwards towards the centre of the earth having many adventures along the way. They never reach the earth's core. In blowing open one of the blocked entrances to the next stage of their journey they release a maelstrom and are eventually carried out of the subterranean realms by a torrent of water and returned to the surface of the world. The first edition contained very little on human evolution, and was written prior to the discovery of any actual fossil human remains in the drift.

However, on the basis of the Moulin Quignon discoveries and finds elsewhere, a revised edition was published in 1867 (Butcher, 1992; Ruddick, 2009). Verne was a staunch Catholic and the implications of Darwinian evolution presented challenges to his faith. The revised edition contained new chapters which included the party's discovery of a huge graveyard carpeted with the bones of extinct animals. Among them they find the remains of ancient humans. There is flesh and hair on the bodies, indicating they died comparatively recently. Were they inhabitants of the subterranean realm? Or, had their remains, along with the bones of other extinct creatures, been washed down from the surface in some ancient cataclysm and been miraculously preserved? The explorers cannot answer this.

Not long after, they glimpse a twelve foot humanoid apparently herding mastodons. This creature is found in a Tertiary forest. The young narrator would rather believe the giant was an ape and not a man. Ruddick (*ibid*) cleverly suggests that what Verne may have been doing was resolving the conflict between his own religious convictions and the mounting evidence for Palaeolithic humans. By suggesting that under the earth (as perhaps a metaphor for the unexplored places of the world) other/primitive forms of humanity still persisted, he could maintain the belief that modern humans could not be descended from them (assuming a strict unilinear evolutionary sequence that is).

However, it is equally possible that Verne was simply acknowledging the contribution of France, via de Perthes and Moulin Quignon, to science, as well as his own acceptance of the discovery. Lidenbrock's speech on discovering the human bodies acknowledges French concern with the dating of the Moulin Quignon drift, but makes no mention of Falconer, Prestwich and Evans's doubts as to the authenticity of the finds. As Textbox 2.2 indicates, the affair was over by late 1863, though many French prehistorians continued to accept the jaw's authenticity. As Lidenbrock discusses the finds he bemoans the absence of Milne-Edwards and de Quatrefages (Chapter 2), both of whom played a significant part in the Moulin Quignon affair. Mention of the latter is particularly significant. He was a monogenist who believed that all humanity originated from the table lands of central Asia. The three races of humanity, black-, white- and yellow-skinned could all successfully breed with each other. Humanity had spread out from Asia in a series of migrations, during the course of which further differentiation in humanity had occurred. As noted in earlier chapters, migrations were an important mechanism for monogenists. For de Quatrefages acclimatization (as opposed to natural selection in A.R. Wallace's case) during migration explained human variability in different climates. The preserved body was a Caucasian, brachycephalic, and was not prognathic; in other words a European and related to the modern white race. There were stone axes, as well. At the very least Verne was accepting the deep roots of racial distinctivness, stretching back into the Quaternary. Invoking de Quatrefages is significant here for another reason. He expressly denied that the Abbeville jaw, as an example of the earliest human race then known, implied any derivation from the apes (de Quatrefages, 1863). Butcher (1992) notes that Verne relied heavily on L. Figuier's *La terre avant le déluge* (1863) for much of his information on prehistoric life. Figuier denied the existence of 'fossil man', believeing that all skeletal remains could be accommodated within the range of variation seen in the modern and historic races (Figuier, 1870). Verne may also have been aware of Carl Vogt's polygenist views published in the mid 1860s. I suspect that Verne's refusal to allow his heroes to commit themselves to any particular interpretation of human origins was more about scientific caution as much as anything else.

The revised edition was published in English in 1872 as *A Journey to the Centre of the Earth*. But the English translator took it upon himself to omit sections of Verne's original prose, as well as make changes to other parts of the text. Incredibly, the editor added sentences and at least one new incident not present in the original. In a dream sequence the young narrator of the story is attacked by a 'shark-crocodile' and flees into a cave where his pursuer is attacked by the 'Ape Gigans' (seen as the evolutionary precursor of the African gorilla). This is a very English insertion. Throughout the 1860s the travels and discoveries of Paul du Chaillu (1835–1903) in West Africa had excited the Victorian public as he described encounters with gorillas, many of whose skins he sold for display to the South Kensington Museum. Huxley had used the gorilla to demonstrate the morphological unity of humans and higher primates (Huxley, 1863; see especially Bray, 1973 on this). This 1872 translation of *Journey to the Centre of the Earth* forms the basis for most of the subsequent publications of the novel in English.

Verne returned to human evolution in *Le Village Aérien* in 1901. It was translated into English in 1964 as *The Village in the Treetops*. A party of hunters are returning home through the forests of the Congo Basin. They are attacked by a herd of elephants and have to take refuge deep in the forest. They continue on through this unexplored vastness said to be twice the size of France. They rescue what appears to be a young boy from drowning, but he is unlike anyone they have ever seen before. He is part of a new species of hominin, previously unknown to science. They in turn are rescued by the boy's tribe when their raft is wrecked. They are taken to a large treetop village where their rescuers, the Waggdi, live. Here, they are kept under semi-house arrest until they eventually escape with the help of the young Waggdi boy and his father.

Much of central and western equatorial Africa was unexplored, even in 1901. The Congo had been trail-blazed by the inimitable Welsh explorer Henry Morton Stanley (1841–1904) during his third great African Journey described in *Across the Dark Continent* (1878). He returned to the dark continent once more between 1887 and 1890 when he undertook his fourth and last African traverse. In this final journey he opened up a new Congo–Nile route while attempting to rescue General Gordon's last acting regional governor, Emin Pasha. This was published to great acclaim as *In Darkest Africa* in 1890. Both books contain large amounts of anthropological observation as well as new geographical discoveries. The great forests of Africa, Asia and South America were among the last places to be explored, and Verne assured his readers that they yet harboured many surprises for science. *The Village in the Treetops* was a story about one such surprise. Verne had in mind the discovery of a 'new species' of human by Stanley in the Congo and described in his book *In Darkest Africa*. These were the indigenous peoples of today's Ituri Forest, then known as Pygmies. They presented a puzzle to science. They were not atavars, nor were they microcephalic. They were small but normally-formed humans. De Quatrefages studied them in detail, and there was a suggestion that they represented 'one of the earliest families of mankind' (Daly, 1892). However there was much criticism of Stanley who had patently ignored earlier reports of them (Daly *ibid*). Du Chaillu in the 1860s had been the first to authenticate much earlier reports of them. Stanley was accused of a similar thing in relation to the discovery of the Ruwenzori mountain range, the fabled Mountains of the Moon, on the same expedition.

Ruddick (2009) suggests that in *The Village in the Treetops* Verne came to terms with evolutionary discoveries, particularly Dubois' *Pithecanthropus*, by shifting the emphasis of evolution from biology to race. Verne quoted the theories of Carl Vogt (Chapters 3 and 4) who stated that whites were the descendants of gorillas, while African blacks were descended from brachycephalic chimpanzees and negritos were descended from the brachycephalic Orangutan. Ruddick suggests that Verne intended the Waggdis to represent the missing link between the chimps and the African black. In so doing Verne was distancing the white race from the black race. By not providing a similar trajectory for whites, Verne made their origins seem less a part of nature.

Verne's citing of Vogt's polygenist views from the 1860s is unusual as the origins question had moved on significantly by the turn of the century. By the late 1890s, the likes of Vogt and Figuier were out of date. But then if his aim was to explore racial evolution, old-fashioned polygenist views may have been more appropriate. Although *Pithecanthropus* was mentioned in the novel, it was the characteristic of bipedality that was suggested as the reason for believing the fossil was a missing link. Verne did not appear to accept that *Pithecanthropus* was genuinely transitional; in two places in the text a relevant missing link is flatly denied to have been found – once in relation to a creature physically midway between apes and humans, and once in relation to intelligence. As in *Journey to the Centre of the Earth*, the reality of unilinear biological evolution remained difficult for Verne, so he resorted to scientific caution.

By the 1890s most biologists accepted a broad monogenist interpretation for the origin of the human species, with the racial question hingeing on whether the races had begun to differentiate relatively recently or not. In shifting to a discourse on racial origins (Ruddick, 2009), I suspect much of Verne's aim was to reassure his readers. It might not have been possible to determine whether mankind evolved from apes, but it was certainly possible to continue to demonstrate the moral and ethical superiority of the white race. So Verne's agenda here was to reassure his readers that they need not feel threatened by new discoveries. As with other books mentioned in this chapter and the last, this would have been a comfortable stance to adopt for many of his readers.

| The Waggdis don't have the following | The Waggdis do have the following |
|---|---|
| A sense of God or a sense of religion in general | They appear to have a primitive sense of ethics – stealing is known to be wrong |
| Curiosity | Communal living in large villages |
| A sense of the abstract or the ability to conceive generalities | Rituals and music |
| Any aptitude in the arts, sciences, or letters (music is considered a primitive art as is the sense it appeals to) | Tools, weapons, fire, etc. – also complicated items like canoes made from tree trunks hollowed out using fire. So they are material culture dependent |
| Cultivate cereals | Hierarchical society, also able to maintain a military/police force |
| | Family life |
| | Emotions as fully developed as their human guests |
| | Language and speech |

*Table 13.3. The attributes of Jules Verne's Waggdis from* A Village in the Tree Tops*, translated into English in 1964.*

In Table 13.3 I have listed all the attributes that the Waggdis have compared with all those they do not. The minus column could be read as a list of those characteristics they should have had if they were to be considered human. There is a noticeable difference between the more materialistic 'have' list and the more spiritual and morally influenced 'have not' list – echoing Grant Allen's assertion in *The British Barbarians* that civilization and acting like a civilized person are two different things.

There is little doubt from the novel that Verne had a very clear appreciation of the hierarchical worth of the different races. The black tribes of the Congo came in for biting criticism. They were slavers and cannibals, and bloodthirsty traders in children. Children were a must-have item on any cannibal menu. One of the Europeans in the book opines that there is little difference between the native black people of the Congo and its apes. The text contrasts these tribes with the selfless white missionaries for whom Verne had unbounded admiration.

Having said all this, it should be noted that Verne blurred the racial boundaries at the end of the novel as the party escaped and bade farewell to their rescuers. The young Waggdi boy and his father experienced deep emotional regret and sadness at the parting. Their tears finally force the white men to accept that there was some link between the Waggdis and humanity, despite their not having a sense of the divine. I think Verne was once more leaving a question open. Being human was not just a checklist of materialist characters. Many Victorians would have empathised with this too.

I will close this section with one last topic from Verne's *The Village in the Treetops*. The issue of speech has cropped up a number of times in discussing the popular visualisation of human origins, and was introduced in the last chapter. Verne's Waggdis clearly possessed language and were able to communicate relatively complicated thoughts. But Verne appeared to reverse the normal pattern of suggested development. Through the text he argued that thought preceded the spoken word; after all parrots talk beautifully, but they don't understand what they are saying. So a simple language based on the use of general terms would evolve to express pre-existing generalised thoughts. Normally authors expressed the reverse in their stories. Dr Moreau thought that the subtlety and delicacy of language would have to develop before any complicated cognition could evolve, a sentiment Wells repeated in *Human Evolution*. Ab, in Stanley Waterloo's tale had difficulty in grasping a burgeoning religious emotion because his vocabulary lacked conceptual sophistication.

Verne's Waggdis appeared to possess their simple language through an atavistic race memory rather than by social learning. This linked to Verne's discussion of an American amateur naturalist Professor R.L. Garner. In the early 1890s Garner claimed to have discovered a common language for monkeys based on eight or nine simple word-to-object associations, and which could be varied with context and when uttered at a different pitch. He had visited a number of American zoological gardens, and made phonograph recordings of monkeys and apes 'talking'. He then played these back to other monkeys, while learning to imitate the sounds himself. The language focused on words such as food or anything associated with it, danger, shelter etc. Different species of monkey could communicate at different levels since some species were more intelligent and articulate than others (Garner, 1891a; Garner, 1891b; Garner, 1892). Some species shared the same sounds, but they meant different things to different species. He believed the reasoning powers of monkeys were the same as humans, just less developed. The Capuchin monkey was the Caucasian of the simian world.

> To reason, they…[monkeys]… *must think*, and if it be true that *man cannot think without words*, it must be true of monkeys; hence they must formulate those thoughts into words, and words are the normal exponents of thoughts. (Garner, 1891a, 561; italics original, brackets mine)

So Garner followed the accepted sequence of speech then thought. His work was widely discussed, but not widely believed. An attempt to practice his linguistic skills in Africa on wild monkeys backfired and Garner returned to America and obscurity (Ruddick, 2007). In the novel Verne reported his failure in some detail – it clearly maintained the gulf between man and ape. But Garner did suggest that his simian language was at the root of all later human language arguing for a monogenist/unilinear evolution from apes to humans. He also suggested that modern indigenous peoples had simple languages, largely resulting from a lack of environmental stimulus. This would have been the same with the first human languages (he claimed blacks in America, despite several hundred years of habituation to whites, still had difficulty fully articulating European speech – Garner was the son of a slave owner).

## Conclusion

If it is accepted that early science fiction stories both reflected and possibly even shaped the public's views of evolution, what can we learn from them about the public's understanding of human origins?

- The past was brutal and inhumane. Relentless and merciless competition characterised our origins.
- The stories helped to visualise the remote past, make it more familiar and accessible, even if some found the images unpalatable. This was important because the time depth involved was often difficult for the public to grasp. It could almost be argued that the fiction writers took over the interpretation of the deep past after the 1860s when the evolutionary anthropologists moved on to other interests.
- The visualisation was aided by conscious or unconscious allusions to ethnology. The hunting and gathering tribal society was, by the

1890s, very familiar to the Victorians. Other fiction such as that by the popular writer James Fennimore Cooper will have reinforced expected stereotyping (Bowler, 2003).

- Depending on the views of the author, these stories could reassure, educate and heighten awareness; or they could warn and highlight dangers.
- These stories reinforced prejudices about progressive time which gave an added feel-good factor. Victorians could commend themselves – look how far humanity had travelled since the Palaeolithic, and the white English race (especially those who lived in the English Home Counties) were always portrayed as the pinnacle of evolution.
- In relation to the previous point, science fiction and pf often provided links and continuity between the past and the present/future (although this rather depended on the story). Pamboukian (2008) notes this relational element for a number of Victorian writers who engaged with evolutionary themes.
- I suspect that the imagery of these tales quickly stereotyped and then fossilised the basic image of a Palaeolithic society for a public already predisposed to accept certain perspectives on human origins.

# BIBLIOGRAPHY


A.C.H. (1912) Obituary Notices of Fellows Deceased. John Beddoe, 1826–1911. *Proceedings of the Royal Society of London. Series B,* 84, xxv–xxvii.

A.E.S. (1914) John Lubbock, Baron Avebury, 1834–1913. *Obituary Notices of Fellows Deceased. Proceedings of the Royal Society of London. Series B,* 87, i–iii.

A.G. (1908) Sir John Evans K.C.B. 1823–1903. *Obituary Notices of Fellows Deceased. Proceedings of the Royal Society of London. Series B,* 80, l–lvi.

Abbott, J. L. (1894a) The Ightham Bone Fissure. *Science Gossip,* 1 (New Series), 169–172.

Abbott, J. L. (1894b) The Ossiferous Fissures in the Valley of the Shode near Ightham, Kent *Quarterly Journal of the Geological Society of London,* 50, 171–187.

Abbott, J. L. (1894c) Plateau Man in Kent. *Natural Science,* 4, 257–266.

Abbott, J. L. (1897) Worked Flints from the Cromer Forest Bed. *Natural Science,* 10, 89–96.

Abbott, J. L. (1898) II. *Natural Science,* 12, 111–116.

Alder, E. (2008) 'Buildings of the New Age': Dwellings and the Natural Environment in the Futuristic Fiction of H.G. Wells and William Hope Hodgson. In McLean, S. (Ed.), *H.G. Wells: Interdisciplinary Essays.* pp. 115–129. Newcastle: Cambridge Scholars Publishing.

Allen, G. (1882) Who was Primitive Man. *Fortnightly Review* 32, 308–322.

Allen, G. (1885) *Charles Darwin.* London: Longmans, Green and Company

Alkon, P. (1994) *Science Fiction Before 1900. Imagination Discovers technology.* Oxford: Maxwell Macmillan International.

Anon (1856) Bibliographical Notices. Crannia Britannica. *The Asylum Journal of Mental Science,* 2.

Anon (1863a) The Abbeville Jaw Bone. *The Reader,* 24 (June 13), 580.

Anon (1863b) The Human Jaw of Abbeville. *Gentleman's Magazine,* June 1863, 713–714.

Anon (1863c) Notes on the Antiquity of Man. *Anthropological Review,* 1, 60–106.

Anon (1864a) Anthropology at the British Association, A.D. 1864. *Anthropological Review,* 2, 294–335.

Anon (1864b) The Fossil Man of Abbeville Again. *Anthropological Review,* 2, 220–222.

Anon (1866) Anthropology at the British Association. *Anthropological Review,* 4, 386–408.

Anon (1866–1867) Hugh Falconer. Obituary Notices of Fellows Deceased. *Proceedings of the Royal Society of London,* 15, i–xlvii.

Anon (1867) Review. Pike on the Origin of the English. *Anthropological Review,* 5, 49–55.

Anon (1868) Proceedings of the Special General Meeting of the Society Regarding the Proposed Expulsion of Hyde Clarke. *Journal of the Anthropological Society of London,* 6, clxxxii–clxxxix.

Anon (1869a) Anthropology at the British Association, 1869. *Anthropological Review,* 7, 414–432.

Anon (1869b) The Origin of the English. Pike v. Nicholas. *Anthropological Review,* 7, 279–306.

Anon (1869c) Le Hons Everett *et al.* 1879–1880. *Anthropological Review,* 7, 310–324.

Anon (1869–1870) Settle Cave Exploration. *Journal of the Ethnological Society of London (1869–1870),* 1, 388.

Anon (1870) The Origin of the English. Pike v. Nicholas (Continued). *Anthropological Review,* 8, 69–85.

Anon (1876) Untitled Report of a Lecture by T. McKenny Hughes to the Cambridge Philosophical Society in November 1876. *Nature,* 15, 87–88.

Anon (1880) Report on Lecture by F.C.J. Spurrell to the Geological Society Entitled 'On the Discovery of the Place Where Palaeolithic Implements were Made at Crayford'. *Geological Magazine,* 7, 378.

Anon (1881) Review of Boyd Dawkins' 'Early Man in Britain and his Place in the Tertiary Period'. *Journal of the Anthropological Institute of Great Britain and Ireland,* 10, 232–234.

Anon (1882) Report on Lecture by E.J. Jones to the Geological Society Entitled 'On the Exploration of Two Caves in the Neighbourhood of Tenby'. *Geological Magazine,* 9, 329.

Anon (1883a) Palaeolithic Implements from the Bed of the Thames. *Journal of the Anthropological Institute of Great Britain and Ireland,* 12, 436–437.

Anon (1883b) Report on 'Recent Opinions on Interglacial and Pre–Glacial Man in Britain'. *Geological Magazine,* 20, 37–39.

Anon (1886a) Anthropology in 1885. Extract from the Times Newspaper January 8th 1886. *Journal of the Anthropological Institute of Great Britain and Ireland,* 15, 388.

Anon (1886b) Report on Lecture by Henry Hicks to the Geological Society Entitled 'Results of Recent Researches in Some Bone Caves in North Wales (Ffynnon Beuno and Cae Gwyn)' by Henry Hicks, M.D., F.R.S., F.G.S., with Notes on the Animal Remains by W. Davies, Esq. F.G.S., of the British Museum (Natural History)'. *Geological Magazine,* 23, 39–41.

Anon (1888a) Abstract on Lecture by Henry Hicks to the British Association Meeting at Manchester, 1888, Entitled 'On the Migrations of Pre–Glacial Man'. *Geological Magazine,* 25, 29–30.

Anon (1888b) Lectures on Anthropology. *Journal of the Anthropological Institute of Great Britain and Ireland,* 17, 79.

Anon (1888c) Review of 'Prestwich, J. 1888. Geology, Chemical, Physical, and Stratigraphical, in 2 volumes. Volume 2, Stratigraphical and Physical. Oxford. Clarendon Press.'. *Geological Magazine,* 25, 158–164.

Anon (1889) Report on Lecture by Joseph Prestwich to Geological Society Entitled 'On the Relation of the Westleton Beds or Pebbly Sands of Suffolk to those of Norfolk, and on their Extension Inland; with some Observations on the Period of the Final Elevation and Denudation of the Weald and of the Thames Valley. Part I'. *Geological Magazine,* 26, 377–379.

Anon (1890a) Popular Anthropological Lectures. *Journal of the Anthropological Institute of Great Britain and Ireland,* 19, 441–442.

Anon (1890b) Report on Lecture by Joseph Prestwich to Geological Society Entitled 'On the Relation of the Westleton Beds or 'Pebbly Sands' of Suffolk to those of Norfolk, and on their Extension Inland, with some Observations on the Period of the Final Elevation and Denudation of the Weald and of the Thames Valley. Part II.'. *Geological Magazine,* 27, 92–93.

Anon (1890c) Report on Lecture by Joseph Prestwich to Geological Society Entitled 'On the Relation of the Westleton Beds or 'Pebbly Sands' of Suffolk to those of Norfolk, and on their Extension Inland, with some Observations on the Period of the Final Elevation and Denudation of the Weald and of the Thames Valley. Part III. On a Southern Drift in the Valley of the Thames with Observations on the Final Elevation and Initial Subaerial Denudation of the Weald, and on the Genesis of the Thames.'. *Geological Magazine,* 27, 183–184.

Anon (1891) Review of Articles by Lankester, Sollas, Hubrecht, von Graff, Bourne, and Herdman contributed to *Encyclopedia Britannica. Science,* 17, 236.

Anon (1894a) Obituary 'William Pengelly, F.R.S., F.G.S.'. *Geological Magazine,* 31, 238–239.

Anon (1894b) Report on Lecture by E.T. Newton to the Geological Society Entitled 'The Vertibrate Fauna Collected by Mr Lewis Abbott from the Fissure near Ightham, Kent.'. *Geological Magazine,* 31, 143.

Anon (1894c) Report on Lecture by Lewis Abbott to the Geological Society Entitled 'The Ossiferous Fissures in the Valley of the Shode, near Ightham, Kent. *Geological Magazine,* 31, 142–143.

Anon (1895a) Dr Dubois' Missing Link. *Nature,* 53, 115–116.

Anon (1895b) Dr Dubois' So–Called Missing Link. *Nature,* 51, 428–429.

Anon (1897) Abstract of Paper in Proceedings of the Royal Society, Volume lxi, pp 40–49 by Clement Reid Entitled "The Palaeolithic Deposits at Hitchin and their Relation to the Glacial Epoch". *Geological Magazine,* 34, 229–233.

Anon (1901) Untitled Report of an Exhibition of Eoliths Made by M R.D. Darbishire to the Manchester Literary and Philosophical Society. *Geological Magazine,* 38, 574–575.

Anon (1921) Benjamin Harrison. *Nature, 108, 251.*

Ashley, M. (1977) *The History of the Science Fiction Magazine. Part One 1926–1935.* London: Revised and Updated. New English Library.

Ashley, M. (2011) *Out of this World. Science Fiction but not as You Know it.* London: British Library.

Ashton, N. and Lewis, S. (2002) Deserted Britain: Declining Populations in the British Late Middle Pleistocene. *Antiquity,* 76, 388–396.

Bacon, E. (Ed.) (1976) *The Great Archaeologists.* London: Book Club Associates.

Ball, R. (1892) *The Cause of an Ice Age.* London: Kegan Paul, Trench, Trübner and Co. Ltd.

Barron, G. B. (1883) On a Human Skull Found Near Southport. *Report of the Fifty–Third Meeting of the British Association for the Advancement of Science Held at Southport in September 1883.* pp. 695–696. London: John Murray.

Beddoe, J. (1866a) Comments on. 'On the Evidence of Phenomena in the West of England to the Permanence of Anthropological Types'. *Journal of the Anthropological Society of London,* 4, xviii–xxii.

Beddoe, J. (1866b) On the Physical Characteristics of the English People. *Journal of the Anthropological Society of London,* 4, cxiv–cxix.

Beddoe, J. (1870) On the Anthropology of Devon and Cornwall. *Anthropological Review,* 8, 85–88.

Beddoe, J. (1870–1871) The Presidents' Address (Containing Obituary for James Hunt). *Journal of the Anthropological Society of London,* 8, lxxviii–lxxxiv.

Beddoe, J. (1880) On Anthropological Colour Phenomena in Belgium and Elsewhere. *Report of the Fiftieth Meeting of the British Association for the Advancement of Science held at Swansea in August and September 1880.* p. 629. London: John Murray.

Beddoe, J. (1885) *The Races of Britain: a Contribution to the Anthropology of Western Europe.* London: Trübner.

Bell, A. M. (1892) Exhibition of Pre–Palaeolithic Flints. *Report of the Sixty–Second Meeting of the British Association for the Advancement of Science held at Edinburgh in August 1892.* p. 900. London: John Murray.

Bell, A. M. (1894) Remarks on the Flint Implements from the Chalk Plateau of Kent. *Journal of the Anthropological Institute of Great Britain and Ireland,* 23, 266–284.

Bell, A. M. (1900) 80. On the Occurrence of Flint Implements of Palaeolithic Type on an Old Land–Surface in Oxfordshire, Near Wolvercote and Pear–Tree Hill, Together with a Few Implements of Plateau Types. *Journal of the Anthropological Institute of Great Britain and Ireland,* 30, 81.

Bennett, F. D. (1901) 'The Earliest Traces of Man'. *Geological Magazine,* 38, 427–428.

Bergonzi, B. (1961) *The Early H.G. Wells.* Manchester: Manchester University Press.

Blackmore, H. P. (1864–1865) On the Discovery of Flint Implements in the Drift of Milford Hill, Salisbury. *Quarterly Journal of the Geological Society of London,* 21, 250–251.

Blanckaert, C. (1988) On the Origins of French Ethnology; William Edwards and the Doctrine of Race. In Stocking, G. W. (Ed.), *Bones, Bodies, Behaviour.* pp. 18–55. London: University of Wisconsin Press.

Bolton, J. and Roberts, G. E. (1864) On the Kirkhead Cave near Ulverstone. *Journal of the Anthropological Society of London,* 2, ccli–ccliv.

Bowden, M. (1991) *Pitt Rivers.* Cambridge: Cambridge University Press.

Bowler, P. J. (2003) *Evolution the History of an Idea.* London: University of California Press Limited.

Bray, W. (1973) A Page of 'Punch'. In Strong, D.E. (Ed.), *Archaeological Theory and Practice.* pp. 45–60. London: Seminar Press

Brent, J. (1875) Notes on the Exhibition of Incised Flints. *Journal of the Anthropological Institute of Great Britain and Ireland,* 4, 88–90.

Brinton, D. G. (1897) On the Oldest Stone Implements in the Eastern United States. *Journal of the Anthropological Institute of Great Britain and Ireland,* 26, 59–66.

Broca, P. (1868) On the Crania and Bones of Les Eyzies: or the Ancient Cave Men of the Perigord. *Anthropological Review,* 6, 408–411.

Broca, P. (1869) On the Crania and Bones of Les Eyzies, Dordogne. *International Congress of Prehistoric Archaeology: Transactions of the Third Session (Norwich and London 1868).* London: Macmillan and Company, 168–175.

Brown, J. A. (1887) *Palaeolithic Man in North–West Middlesex.* London: Macmillan and Co.

Brown, J. A. (1889) On Some Small Highly Specialized Forms of Stone Implements, Found in Asia, North Africa, and Europe. *Journal of the Anthropological Institute of Great Britain and Ireland,* 18, 134–139.

Brown, J. A. (1893) On the Continuity of the Palaeolithic and Neolithic Periods. *Journal of the Anthropological Institute of Great Britain and Ireland,* 22, 65–98.

Bruce Foote, R. (1868) On the Distribution of Stone Implements in Southern India. *Quarterly Journal of the Geological Society of London,* 24, 484–495.

Bruce Foote, R. (1869) On Quartzite Implements of Palaeolithic Types from the Laterite Formation of the East Coast of Southern India. *International Congress of Prehistoric Archaeology: Transactions of the Third Session (Norwich and London 1868).* London: Macmillan and Company, 224–238.

Bryant, S. (1886) Experiments in Testing the Character of School Children. *Journal of the Anthropological Institute of Great Britain and Ireland,* 15, 338–350.

Büchner, L. (1894) The Origin of Mankind. *Fortnightly Review,* 55, 74–82.

Bullen, R. A. (1898) The Authenticity of Plateau Implements. *Natural Science,* 12.

Bullen, R. A. (1901) Eolithic Implements. *Geological Magazine,* 38, 426–427.

Busk, G. (1866) An Account of the Discovery of a Human Skeleton Beneath a Bed of Peat on the Coast of Cheshire. *Transactions of the Ethnological Society of London,* 6, 101–104.

Busk, G. (1874) Notice of a Human Fibula of Unusual Form, Discovered in the Victoria Cave, Near Settle, in Yorkshire. *Journal of the Anthropological Institute of Great Britain and Ireland,* 3, 392–395.

Busk, G. (1875) President's Address. *Journal of the Anthropological Institute of Great Britain and Ireland,* 4, 476–502.

Butcher, W. (1992) Introduction to World's Classics Edition of Jules Verne's Journey to the Centre of the Earth. In Butcher, W. (Ed.), *Journey to the Centre of the Earth by Jules Verne.* Oxford: Oxford University Press.

C.W.A. (1895) Reported Discovery of an Animal Intermediate Between Man and the Anthropoid Apes. Review of, *Pithecanthropus erectus.* Eine Menschenaehnliche Uber–Gangsform aus Java' by Eugene Dubois. *Geological Magazine,* 32, 1895.

Calvert, F. (1874) On the Probable Existence of Man During the Miocene Period. *Journal of the Anthropological Institute of Great Britain and Ireland,* 3, 127–129.

Carter Blake, C. (1863a) Note on Stone Celts, from Chiriqui. *Transactions of the*

*Ethnological Society of London,* 2, 166–170.

Carter Blake, C. (1863b) On Recent Evidences of Extreme Antiquity of the Human Race. *Transactions of the Anthropological Society of London,* 1, xxvi–xxxiv.

Carter Blake, C. (1864a) On Human Remains from Kent's Hole, near Torquay. *Journal of the Anthropological Society of London,* 2, cclxiii–cclxv.

Carter Blake, C. (1864b) On the Alleged Peculiar Characters, and Assumed Antiquity of the Human Cranium from the Neanderthal. *Journal of the Anthropological Society of London,* 2, cxxxix–clvii.

Carter Blake, C. (1867a) On a Human Jaw from the Cave of La Naulette, near Dinant, Belgium. *Anthropological Review,* 5.

Carter Blake, C. (1867b) Report on the Recent Investigations of Dr Édouard Dupont on the Bone Caves on the Banks of the Lesse River, Belgium. *Journal of the Anthropological Society of London,* 5, x–xiii.

Carter Blake, C., Pruner–Bey, D. and Bendyshe, T. (1864) The Neanderthal Skull. *Anthropological Review,* 2, 145–147.

Charlesworth, E. (1873) Note on Objects Found in the Red Crag Formation of Suffolk. *Journal of the Anthropological Institute of Great Britain and Ireland,* 2, 90–94.

Clodd, E. (1895) *The Story of Primitive Man.* London: George Newnes, Ltd.

Coffey, G. (1901) Naturally Chipped Flints for Comparison with Certain Forms of Alleged Artificial Chipping. *Report of the Seventy First Meeting of the British Association for the Advancement of Science held at Glasgow in September 1901.* p. 795. London: John Murray.

Conway, B. (1996) A History of Quarrying in the Swanscombe Area. In Conway, B., McNabb, J. and Ashton, N. (Eds.), *Excavations at Barnfield Pit, 1968–1972.* pp. 3–7. London: British Museum Occassional Papers 94, British Museum.

Crawfurd, J. (1866) On the So–called Celtic Languages in Reference to the Question of Race. *Transactions of the Ethnological Society of London,* 4, 71–100.

Crawfurd, J. (1868) On the Antiquity of Man. *Transactions of the Ethnological Society of London,* 6, 233–245.

Crawfurd, J. (1869) On the Theory of the Origin of Species by Natural Selection in the Struggle for Life. *Transactions of the Ethnological Society of London,* 7, 27–38.

Crawshay, d. B. (1892) Notes (Notes Added to Prestwich's Paper 'On the Primitive Characters of the Flint Implements of the Chalk Plateau of Kent with Reference to the Question of their Glacial or Pre–Glacial Age'). *Journal of the Royal Anthropological Institute of Great Britain and Ireland,* 21, 267–270.

Cunnington, W. (1897) The Authenticity of Plateau Man. *Natural Science,* 11, 327–333.

Cunnington, W. (1898) On Some Palaeolithic Implements from the Plateau Gravels and their Evidence Concerning 'Eolithic' Man. *Quarterly Journal of the Geological Society of London,* 54, 291–300.

Dallas, J. (1886) Primary Divisions and Geographical Distribution of Mankind. *Journal of the Anthropological Institute of Great Britain and Ireland,* 15, 304–330.

Daly, C. P. (1892) Who Discovered the Pygmies? *American Geographical Society,* 24, 18–22.

Darwin, C. (1859a) Letter to Charles Lyell December 2 1859. *Darwin Correspondence*

*Project Database http://www.darwinproject.ac.uk/entry–2565/(letter no. 2565 accessed 22 August 2010).*

Darwin, C. (1859b) Letter to Charles Lyell December 10 1859. *Darwin Correspondence Project Database http://www.darwinproject.ac.uk/entry–2575/(letter no. 2575 accessed 8 September 2010).*

Darwin, C. (1859c) Letter to Joseph Hooker June 22 1859. *Darwin Correspondence Project Database http://www.darwinproject.ac.uk/entry–2471/(letter no. 2471 accessed 10 August 2010).*

Darwin, C. (1863a) Letter to Charles Lyell March 6 1863. *Darwin Correspondence Project Database http://www.darwinproject.ac.uk/entry–4028/(letter no. 4028 accessed 28 September 2010).*

Darwin, C. (1863b) Letter to J.D. Hooker March 17 1863. *Darwin Correspondence Project Database http://www.darwinproject.ac.uk/entry–4048/(letter no. 4048 accessed 28 September 2010).*

Darwin, C. (1863c) Letter to T.H. Huxley February 26 1863. *Darwin Correspondence Project Database http://www.darwinproject.ac.uk/entry–4013/(letter no. 4013 accessed 28 September 2010).*

Darwin, C. (1864a) Letter to A.R. Wallace May 28 1864. *Darwin Correspondence Project Database http://www.darwinproject.ac.uk/entry–4510/(letter no. 4510 accessed 15 March 2012).*

Darwin, C. (1864b) Letter to J.D. Hooker May 22 1864. Darwin Correspondence Project Database http://www.darwinproject.ac.uk/entry–2357/(letter no. 4506 accessed 15 March 2012).

Darwin, C. (1871) *The Descent of Man, and Selection in Relation to Sex*. London: Murray.

Darwin, F. (Ed.) (1958) *The Autobiography of Charles Darwin and Selected Letters*. New York: Dover Publications. Reprint of 1892 edition by D. Appleton and Co.

Davis, B. (1865) The Neanderthal Skull; its Formation Considered Anatomically. *Journal of the Anthropological Society of London,* 3, xv–xix.

Davis, J. B. and Thurnam, J. (1856–1865) *Crania Britannica: Delineations and Descriptions of the Skulls of the Aboriginal and Early Inhabitants of the British Isles*. London: Published by Private Subscription.

Davison, C. (1900) Eminent Living Geologists: Rev. Osmond Fisher. *Geological Magazine,* 37, 49–54.

Dawkins, W. B. (1862) On a Hyaena–Den at Wookey Hole near Wells. *Quarterly Journal of the Geological Society of London,* 18, 115–125.

Dawkins, W. B. (1863) On a Hyaena–Den at Wookey Hole near Wells. *Quarterly Journal of the Geological Society of London,* 19, 260–274.

Dawkins, W. B. (1869–1870) On the Discovery of Platycnemic Man in Denbighshire. *Journal of the Ethnological Society of London (1869–1870),* 2, 440–468.

Dawkins, W. B. (1874) *Cave Hunting. Researches on the Evidence of Caves Respecting the Early Inhabitants of Europe*. London: Macmillan and Company.

Dawkins, W. B. (1878) On the Evidence Afforded by the Caves of Great Britain as to the Antiquity of Man. *Journal of the Anthropological Institute of Great Britain and Ireland,* 7, 151–162.

Dawkins, W. B. (1880a) *Early Man in Britain and His Place in the Tertiary Period.* London: Macmillan and Co.

Dawkins, W. B. (1880b) The Man of the Caves. *Science,* 1, 286–287.

Dawkins, W. B. (1882) On the Present Phase of the Antiquity of Man. *Report of the Fifty-Second Meeting of the British Association for the Advancement of Science Held at Southampton in August 1882.* pp. 597–604. London: John Murray.

Dawkins, W. B. (1888) Address to the Geological Section of the British Association, Bath 1888. *Geological Magazine,* 25, 459–467.

Dawkins, W. B. (1894a) Notes on Exhibits. *Journal of the Anthropological Institute of Great Britain and Ireland,* 23, 251–257.

Dawkins, W. B. (1894b) On the Relation of the Palaeolithic to the the Neolithic Period. *Journal of the Anthropological Institute of Great Britain and Ireland,* 23, 242–251.

Dawkins, W. B. (1896) William Pengelley. *Obituary Notices of Fellows Deceased. Proceedings of the Royal Society of London. Series B,* 59, xxxix–xli.

Dawkins, W. B. (1923) Sir Henry Hoyle Howorth K.C.I.E., D.C.L., F.R.S. *Man,* 23, 138–139.

De Quatrefages, A. (1872) *The Prussian Race Ethnologically Considered. To Which is Appended Some Account of the Bombardment of the Museum of Natural History etc., by the Prussians in 1871. Translated by Isabella Innes.* London: Virtue and Co.

De Quatrefages, A. (1875) *The Natural History of Man. A Course of Elementary Lectures.* New York: D. Appleton and Company.

De Quatrefages, A. (1879) *The Human Species.* New York: D. Appleton and Company.

De Quatrefages, A. and Rolph, G.F. (1863) On the Abbeville Jaw. *Anthropological Review,* 1, 312–335.

Deeley, R. M. (1893) The Glacial Succession. *Geological Magazine,* 30, 31–35.

Desmond, A. (1982) *Archetypes and Ancestors. Palaeontology in Victorian London.* London: Blond and Briggs.

Desmond, A. (1998) *Huxley: From Devil's Disciple to Evolution's High Priest.* London: Penguin Books.

Desmond, A. and Moore, J. (1991) *Darwin.* London: Michael Joseph.

Desmond, A. and Moore, J. (2009) *Darwin's Sacred Cause.* London: Allen Lane, Penguin Books.

Draper, M. (1987) *H.G. Wells.* London: Macmillan.

Dubois, E. (1896) On Pithecanthropus Erectus: a Transitional Form Between Man and the Apes (Abstract). *Journal of the Anthropological Institute of Great Britain and Ireland,* 25, 240–248.

Dubois, E. (1900) Données Justificatives sur l'Essai de Reconstruction Plastique du Pithécanthropus Erectus par Eugène Dubois. *Printed Pamphlet to Accompany Reconstruction of Pithecanthropus erectus at the World Exhibition in Paris in 1900.*

Duff, A. G. (Ed.) (1924) *The Life-Work of Lord Avebury (Sir John Lubbock) 1834–1913.* London: Watts and Company.

Duncan, P. M. (1869) Human Remains in the Cave of Cro-Magnon in the valley of the Vezère. *Anthropological Review,* 7, 422.

Duncan, W. S. (1882) Evidence as to the Scene of Man's Evolution and the Prospects of Proving the Same by Palaeontological Discovery *Report of the Fifty-Second Meeting of the British Association for the Advancement of Science held at Southampton in August 1882*. pp. 605–606. London: John Murray.

Duncan, W. S. (1883) On the Probable Region of Man's Evolution. *Journal of the Anthropological Institute of Great Britain and Ireland,* 12, 513–525.

Dunn, H. P. (1894) Is Our Race Degenerating. *The Nineteenth Century,* 34, 301–314.

Dupont, É. (1867) Discovery of an Habitation of Man in the Belgian Lehm. *Journal of the Anthropological Society of London,* 5, clxxvii–clxxix.

Dupont, É. (1868) Letter from Dupont Regarding Excavation in the Cavern of Le Trou Magrite. *Journal of the Anthropological Society of London,* 6, lx–lxii.

Dyer, J. (1959) Middling for Wrecks: Extracts from the Story of Worthington and Henrietta Smith. *Bedfordshire Archaeologist,* 2, 1–15.

Dyer, J. (1978) Worthington George Smith. *Bedfordshire Historical Records Society Proceedings,* 57, 141–179.

Ellen, R. (2011) The Place of the Eolithic Controversy in the Anthropology of Alfred Russel Wallace. *The Linnean,* 27, 22–33.

Ellen, R. and Muthana, A. (2010) Classifying 'Eoliths': How Cultural Cognition Featured in Arguments Surrounding Claims for the Earliest Human Artefacts as these Developed Between 1880 and 1900. *Journal of Cognition and Culture,* 10 341–375.

Evans, E. P. (1894) Pithecoid Man. *Popular Science Monthly,* 46, 183–185.

Evans, H. and Evans, D. (Eds.) (1976) *Beyond the Gaslight. Science in Popular Fiction 1895–1905.* London: Frederick Muller Limited.

Evans, J. (1860) On the Occurrence of Flint Implements in Undisturbed Beds of Gravel, Sand, and Clay. *Archaeologia,* 38, 280–307.

Evans, J. (1861) Account of Some Further Discoveries of Flint Implements in the Drift on the Continent and in England. *Archaeologia,* 39, 57–84.

Evans, J. (1863a) The Human Remains at Abbeville. *Athenaeum,* July 4 (number 1862), 19–20.

Evans, J. (1863b) The Abbeville Human Jaw. *Athenaeum,* June 6 (number 1858), 747–748.

Evans, J. (1863–1864) On Some Recent Discoveries of Flint Implements in Drift Deposits in Hants and Wilts. *Quarterly Journal of the Geological Society of London,* 20, 188–194.

Evans, J. (1864) On Some Bone and Cave Deposits of the Reindeer Period in the South of France. *Quarterly Journal of the Geological Society of London,* 20, 444.

Evans, J. (1865) On Some Discoveries of Worked Flints near Jubbulpore in Central India. *Proceedings of the Society of Antiquaries,* 3, 39–44.

Evans, J. (1869) On Some Antiquities of Stone and Bronze from Portugal. *Transactions of the Ethnological Society of London,* 7, 45–52.

Evans, J. (1872) *The Ancient Stone Implements, Weapons, and Ornaments, of Great Britain and Ireland.* London: Longmans, Green, Reader, and Dyer.

Evans, J. (1878a) On a Discovery of Palaeolithic Implements in the Valley of the Axe. *Journal of the Anthropological Institute of Great Britain and Ireland,* 7, 499–501.

Evans, J. (1878b) On the Present State of the Question of the Antiquity of Man. *Journal of the Anthropological Institute of Great Britain and Ireland,* 7, 149–151.

Evans, J. (1878c) President's Address. *Journal of the Anthropological Institute of Great Britain and Ireland,* 7, 515–534.

Evans, J. (1883) Chairman's Address. *Journal of the Anthropological Institute of Great Britain and Ireland,* 12, 563–566.

Evans, J. (1890) Untitled Address to Section H – Anthropology. *Report of the Sixtieth Meeting of the British Association for the Advancement of Science held at Leeds in September 1890.* pp. 963–969. London: John Murray.

Evans, J. (1897a) *The Ancient Stone Implements Weapons and Ornaments of Great Britain. Second Edition Revised.* London: Longmans, Green, and Company.

Evans, J. (1897b) Joseph Prestwich. *Obituary Notices of Fellows Deceased. Proceedings of the Royal Society of London. Series B,* 60, xii–xvi.

Evans, J. (1897c) On Some Palaeolithic Implements Found in Somaliland by Mr H.W. Seton–Karr. *Proceedings of the Royal Society,* 60, 19–21.

Evans, J. (1897d) Presidential Address to the British Association for the Advancement of Science held in Toronto, August, 1897. *Geological Magazine,* 34, 457–465.

Evans, J. (1900) Palaeolithic Man in Africa. *Proceedings of the Royal Society,* 66, 486–488.

Everett, A.M., Evans, J. and Busk, G. (1879–1880). Report on the Exploration of the Caves of Borneo. And Introductory Remarks. And Note on the Bones Collected. *Proceedings of the Royal Society,* 30, 210–324.

Fairbank, F. R. (1864) On Some Flint Arrow–Heads from Canada. *Journal of the Anthropological Society of London,* 2, lxiv–lxv.

Falconer, H. (1863) Falconer on the Reputed Fossil Man of Abbeville (Reprint of Letter to the Times of 25/04/1863). *Anthropological Review,* 1, 177–179.

Figuier, L. (1870) *Primitive Man.* (First published in French in 1870). London: Chapman and Hall.

Fisher, O. (1887) Interglacial Land and Man. *Geological Magazine,* 24, 238–239.

Fisher, O. (1892) Review of Robert Ball's 'The Cause of an Ice Age. Kegan Paul, Trench, Trübner and Co. 1891'. *Geological Magazine,* 29, 231–233.

Fleming, J. R. (2006) James Croll in Context. The Encounter Between Climate Dynamics and Geology in the Second Half of the Nineteenth Century. *History of Meteorology (http://www.meteohistory.org/2006historyofmeteorology3/),* 3, 43–53.

Flower, J. W. (1860) On a Flint Implement Recently Discovered at the Base of Some Beds of Drift–gravel and Brick–earth at St Acheul, near Amiens. *Quarterly Journal of the Geological Society of London,* 15, 190–192.

Flower, J. W. (1866–1867) On Some Flint Implements Lately Found in the Valley of the Little Ouse River at Thetford, Norfolk. *Quarterly Journal of the Geological Society of London,* 23, 45–56.

Flower, J. W. (1868–1869) On Some Recent Discoveries of Flint Implements of the Drift in Norfolk and Suffolk, with Observations on the Theories Accounting for their Distribution. *Quarterly Journal of the Geological Society of London,* 25, 449–460.

Flower, J. W. (1872) On the Relative Ages of the Stone Implement Periods in England. *Journal of the Anthropological Institute of Great Britain and Ireland,* 1, 274–295.

Flower, W. H. (1881) Untitled Chairman's Address to Section D, Department of Anthropology. *Report of the Fifty–First Meeting of the British Association for the Advancement of Science held at York in August and September 1881.* pp. 682–689. London: John Murray.

Flower, W. H. (1885) On the Size of Teeth as a Character of Race. *Journal of the Anthropological Institute of Great Britain and Ireland,* 14, 183–187.

Flower, W. H. (1894) Untitled Presidential Address to Section H – Anthropology. *Report of the Sixty–Fourth Meeting of the British Association for the Advancement of Science held at Oxford in August 1894.* pp. 762–774. London: John Murray.

Foot, M. (1995) *The History of Mr Wells.* London: Transworld Publishers.

Frere, J. (1800) Flint Weapons Discovered at Hoxne in Suffolk. *Archaeologia,* 13, 204–205.

G.H.D. (1912) Obituary Notices of Fellows Deceased. Sir Francis Galton, 1822–1911. *Proceedings of the Royal Society of London. Series B,* 84, x–xvii.

G.J.H. (1895) Untitled Review of 'The Great Ice Age and its Relation to the Antiquity of Man' by James Geikie. *Geological Magazine,* 32, 29–38.

Galton, F. (1873) Hereditary Improvement. *Fraser's Magazine,* 7, 116–130.

Galton, F. (1882) The Anthropometric Laboratory. *Fortnightly Review,* 31, 332–338.

Galton, F. (1886) Regression Towards Mediocrity in Heriditary Stature. *Journal of the Anthropological Institute of Great Britain and Ireland,* 15, 246–263.

Galton, F. (1892) Retrospect of Work Done at My Anthropometric Laboratory at South Kensington. *Journal of the Anthropological Institute of Great Britain and Ireland,* 21, 32–35.

Gamble, C. and Kruszynski, R. (2009) John Evans, Joseph Prestwich, and the Stone that Shattered the Time Barrier. *Antiquity,* 83, 461–475.

Gamble, C. and Moutsiou, T. (2011) The Time Revolution of 1859 and the Stratification of the Primeval Mind. *Notes and Records of the Royal Society,* 65, 43–63.

Garner, R. L. (1891a) The Simian Tongue. *The New Review,* 4 (June), 555–562.

Garner, R. L. (1891b) The Simian Tongue II. *The New Review,* 5 (November), 424–430.

Garner, R. L. (1892) The Simian Tongue. *The New Review,* 6 (February), 181–186.

Geikie, A. (1899) Summary of the Scientific Work of Sir Joseph Prestwich. In Prestwich, G. A. (Ed.), *Life and Letters of Sir Joseph Prestwich.* pp. 402–421. London: William Blackwood and Sons.

Geikie, J. (1877a) The Antiquity of Man. *Nature,* 15, 141–142.

Geikie, J. (1877b) *The Great Ice Age and its Relation to the Antiquity of Man (Second Edition, Revised).* London: Edward Stanford.

Geikie, J. (1889) Fifty–Ninth Annual Meeting of the British Assocciation for the Advancement of Science. Newcastle–Upon–Tyne, 1889. Address to the Geological Section of the British Association *Geological Magazine,* 26, 461–477.

Gillham, N.W. (2001) *Sir Francis Galton. From African Exploration to the Birth of Eugenics.* Oxford: Oxford University Press.

Gould, S. J. (1977) *Ontogeny and Phylogeny.* London: Harvard University Press.

Gould, S. J. (1987) *Time's Arrow, Time's Cycle.* Harvard: Harvard University Press.

Gould, S. J. (1989) *Wonderful Life.* London: W.W. Norton and Company.

Greenhill, J. E. (1884) The Implementiferous Gravels of North East London. *Proceedings of the Geologists' Association,* 8, 336–343.

Grey, G. (1870) On Quartzite Implements from the Cape of Good Hope. *Journal of the Ethnological Society of London (1869–1870),* 2, 39–43.

H.W.H. (1886) Review of John Beddoe's 'The Races of Britain; a Contribution to the Anthropology of Western Europe, 1885, Bristol, Arrowsmith; London, Trübner'. *Science,* 7, 84–86.

Haliburton, R. G. (1864) Extended Discussion Following the Reading of a Letter by R.G. Haliburton *Proceedings of the Society of Antiquaries,* 2, 330f.

Hammond, J. (Ed.) (1998) *The Complete Short Stories of H.G. Wells.* London: J.M. Dent.

Harrison, B. (1892) On Certain Rude Implements from the North Downs (Notes Added to Prestwich's Paper 'On the Primitive Characters of the Flint Implements of the Chalk Plateau of Kent with Reference to the Question of their Glacial or Pre–Glacial Age'). *Journal of the Royal Anthropological Institute of Great Britain and Ireland,* 21, 263–267.

Harrison, B. (1895) High–Level Flint Drift of the Chalk. Report of the Committee, Consisting of Sir John Evans (Chairman), Mr. B. Harrison (Secretary), Professor J. Prestwich and Professor H.H. Seeley. *Report of the Sixty–Fifth Meeting of the British Association for the Advancement of Science Held at Ipswich in September 1895.* pp. 349–351. London: John Murray.

Harrison, B. (1899) Plateau Implements (Eoliths) – Results of Recent Research. *Transactions of the South Eastern Union of Scientific Societies*, 12–17.

Harrison, B. (n.d.–a) Autobiography Volume 2. Draft Autobiography of Benjamin Harrison held in the Maidstone Museum as part of the Harrison Archive. Transcribed by Angela Muthana.

Harrison, B. (n.d.–b) Notebook in Harrison Archive British Museum

Harrison, B. (n.d.–c) Volume 8 of the Benjamin Harrison Archive held at the Maidstone Museum. Transcribed by Angela Muthana.

Harrison, B. (n.d.–d) Volume 12 of the Benjamin Harrison Archive held at the Maidstone Museum. Transcribed by Angela Muthana.

Harrison, B. (n.d.–e) Volume 13 of the Benjamin Harrison Archive held at the Maidstone Museum. Transcribed by Angela Muthana.

Harrison, B. (n.d.–f) Volume 14 of the Benjamin Harrison Archive held at the Maidstone Museum. Transcribed by Angela Muthana.

Harrison, B. (n.d.–g) Volume 15 of the Benjamin Harrison Archive held at the Maidtone Museum. Transcribed by Angela Muthana.

Harrison, E. (1928) *Harrison of Ightham.* London: Oxford University Press.

Harrison, E. (n.d.) Untitled 'Working' Copy of Harrison of Ightham.

Harrison, J. P. (1880) On the British Flint Workers at Brandon. *Report of the Fiftieth Meeting of the British Association for the Advancement of Science held at Swansea in August and September 1880.* pp. 626–627. London: John Murray.

Harrison, J. P. (1883a) Note to a Communication "On the Survival of Certain Racial Features.". *Journal of the Anthropological Institute of Great Britain and Ireland,* 12, 438–439.

Harrison, J. P. (1883b) On the Survival of Certain Racial Features in the Population of the British Isles. *Journal of the Anthropological Institute of Great Britain and Ireland,* 12, 243–256.

Harrison, M. (1963) *London by Gaslight 1861–1911.* London: Peter Davies.

Hicks, H. (1885) On the Fynnon Beuno and Cae Gwyn Bone–Caves. *Report of the Fifty–Fifth Meeting of the British Association for the Advancement of Science Held at Aberdeen in September 1885.* pp. 1021–1023. London: John Murray.

Hicks, H. (1886) Report of the Committee, Consisting of Professor T. McK. Hughes, Dr H. Hicks, and Messers. H. Woodward, E.B. Luxmoore, P.P. Pennant, and Edwin Morgan, Appointed for the Purpose of Exploring the Caves of North Wales. Drawn up by Dr. H. Hicks, Secretary. *Report of the Fifty–Sixth Meeting of the British Association for the Advancement of Science Held at Birmingham in September 1886.* pp. 219–223. London: John Murray.

Hicks, H. (1887) The Faunas of the Ffynnon Bueno Cave, and of the Norfolk Forest Bed. *Geological Magazine,* 24, 105–107.

Hicks, H. (1897) Obituary for Sir Joseph Prestwich. Delivered as Part of the Anniversary Address of the President. *Proceedings of the Geological Society (Included Within the Quarterly Journal of the Geological Society of London),* 53, xlix–lii.

Hodgkin, T. (1850) Obituary of Dr Prichard. *Journal of the Ethnological Society of London (1848–1856),* 2, 182–207.

Hooker, J. D. (1863) Letter to Charles Darwin February 23 1863. *Darwin Correspondence Project Database http://www.darwinproject.ac.uk/entry–4007/ (letter no. 4007 accessed 28 September 2010).*

Hooker, J. D. (1864) Letter to C. Darwin May 14 1864. *Darwin Correspondence Project Database http://www.darwinproject.ac.uk/entry–4494/ (letter no. 4494 accessed 24 October 2010).*

Hooker, J. D. (1865) Letter to C Darwin June 2 1865. *Darwin Correspondence Project Database http://www.darwinproject.ac.uk/entry–4849/ (letter no. 4849 accessed 24 October 2010).*

Howorth, H. H. (1901) The Earliest Traces of Man. *Geological Magazine,* 38, 337–344.

Hughes, T. M. (1878) On the Evidence Afforded by the Gravels and Brickearth. *Journal of the Anthropological Institute of Great Britain and Ireland,* 7, 162–165.

Hughes, T. M. (1886) On the Pleistocene Deposits of the Vale of Clywyd. *Report of the Fifty–Sixth Meeting of the British Association for the Advancement of Science Held at Birmingham in September 1886.* p. 632. London: John Murray.

Hughes, T. M. and Thomas, D. R. (1874) On the Occurrence of Felstone Implements of the Le Moustier Type in Pontnewydd Cave, near Cefn, St Asaph. *Journal of the Anthropological Institute of Great Britain and Ireland,* 3, 387–392.

Hughes, T. M. and Wynn, W. (1881) The Results of Recent Further Excavations in the Cave of Cefn, Near St Asaph, North Wales. *Report of the Fifty–First Meeting of the British Association for the Advancement of Science Held at York in August and September 1881.* p. 700. London: John Murray.

Hunt, A. R. (1901) The Late Rev. J. McEnery. *Geological Magazine,* 38, 428.

Hunt, J. (1866a) On the Application of the Principle of Natural Selection to Anthropology, in Reply to Views Advocated by Some of Mr Darwin's Disciples.

*Anthropological Review,* 4, 320–340.

Hunt, J. (1866b) On the Doctrine of Continuity Applied to Anthropology. *Anthropological Review,* 5, 110–120.

Huxley, T. H. (1863) *Evidence as to Man's Place in Nature.* New York: Appleton.

Huxley, T. H. (1869) On the Distribution of the Races of Mankind and its Bearing on the Antiquity of Man. *International Congress of Prehistoric Archaeology: Transactions of the Third Session (Norwich and London 1868).* London: Macmillan and Company, 92–97.

Huxley, T. H. (1870) On the Ethnology of Britain. *Journal of the Ethnological Society of London (1869–1870),* 2, 382–384.

Huxley, T. H. (1890) The Aryan Question and Pre–Historic Man *The Nineteenth Century,* 28 (November), 750–777.

Huxley, T. H. (1894) On the Methods and Results of Ethnology (First Published in the *Fortnightly Review* 1865). In Huxley, T. H. (Ed.), *Collected Essays by T.H. Huxley, Volume 7, Man's Place in Nature.* pp. 209–252. London: Macmillan.

Huxley, T. H. (1902) Emancipation – Black and White (first published in *The Reader* 1865). In Huxley, T. H. (Ed.), *Collected Essays by T.H. Huxley, Volume 3, Science and Education,* pp. 66–75. London: Macmillan.

J.B. (1868) Review. Davis and Thurnam's Crania Britannica. *Anthropological Review,* 6, 52–55.

Jacobs, J. (1886a) The Comparative Distribution of Jewish Ability. *Journal of the Anthropological Institute of Great Britain and Ireland,* 15, 351–379.

Jacobs, J. (1886b) On the Racial Characteristics of Modern Jews. *Journal of the Anthropological Institute of Great Britain and Ireland,* 15, 23–62.

Jones, E. J. (1882) On the Exploration of Two Caves in the Neighbourhood of Tenby. *Geological Magazine,* 9, 329.

Jones, R.T. (1873) On Some Implements Bearing Marks Referable to Ownership, Tallies, and Gambling, from the Caves of the Dordogne, France. *Journal of the Anthropological Institute of Great Britain and Ireland,* 2, 362–365.

Jones, R. T. (1884) On the Implementiferous Gravels Near London. *Proceedings of the Geologists' Association,* 8, 344–353.

Jones, R. T. (1894a) On the Geology of the Plateau Implements in Kent. *Report of the Sixty–Fourth Meeting of the British Association for the Advancement of Science held at Oxford in August 1894.* pp. 651–652. London: John Murray.

Jones, R. T. (1894b) 'On the Geology of the Plateau Implements in Kent'. Paper Read before Combined Sections for Geology (C) and Anthropology (H). *Geological Magazine,* 31, 416–417.

Jones, R. T. (1894c) Review of F. Noetling's 'On the Occurrence of Chipped (?) Flints in the Upper Miocene of Burma'. *Geological Magazine,* 31, 524–527.

Jones, R. T. (1899) Exhibition of Stone Implements from Swaziland, South Africa. *Journal of the Anthropological Institute of Great Britain and Ireland,* 28, 48–54.

Jones, R. T. (1901) Eolithic Man. *Geological Magazine,* 38, 425–426.

Jukes–Brown, A. J. (1887) Interglacial Land Surfaces in England and Wales. *Geological Magazine,* 24, 147–150.

Keith, A. (1917) How Can the Institute Best Serve the Needs of Anthropology. *Journal of the Anthropological Institute of Great Britain and Ireland,* 47, 12–30.

Keith, A. (1924) Anthropology. In Duff, A. G. (Ed.), *The Life–Work of Lord Avebury (Sir John Lubbock) 1834–1913.* pp. 67–104. London: Watts and Company.

Kennard, A. S. (1897) The Authenticity of Plateau Man: A Reply. *Natural Science,* 12, 27–34.

Kidd, B. (1895) Social Evolution. *The Nineteenth Century,* 37, 226–240.

Knowles, W. J. (1883) On the Antiquity of Man in Ireland. *Report of the Fifty–Third Meeting of the British Association for the Advancement of Science Held at Southport in September 1883.* pp. 562–563. London: John Murray.

Kropotkin, P. (1896) II. The Erect Ape–Man. *The Nineteenth Century,* 39, 425–432.

Lamdin–Whymark, H. (2009) Sir John Evans: Experimental Flint Knapping and the Origins of Lithic Research. *Lithics,* 30, 45–52.

Lane Fox, A. (1872) On a Discovery of Palaeolithic Implements in Association with Elephas Primigenius in the Gravels of the Thames Valley at Acton. *Quarterly Journal of the Geological Society of London,* 28, 449–465.

Lane Fox, A. (1873) Report on Anthropology, at the Meeting of the British Association for the Advancement of Science for 1872 at Brighton. *Journal of the Anthropological Institute of Great Britain and Ireland,* 2, 350–362.

Lane Fox, A. (1875) On the Principles of Classification Adopted in the Arrangement of His Anthropological Collection, now Exhibited in the Bethnal Green Museum. *Journal of the Anthropological Institute of Great Britain and Ireland,* 4, 293–308.

Lane Fox, A. (1878) Untitled Discussion. *Journal of the Anthropological Institute of Great Britain and Ireland,* 7, 178–179.

Lane Fox, A. (1977) Reprint of Lecture 'On the Evolution of Culture' Originally Published in the Proceedings of the Royal Institution 1875, vol. 7, p496–520. In Thompson, M.W. *General Pitt–Rivers.* Bradford–on–Avon: Moonraker Press.

Lartet, E. M. and Christy, H. (1875) *Reliquiae Aquitanicae; Being Contributions to the Archaeology and Palaeontology of Perigord and Adjoining Provinces of Southern France. 1865–1875.* Jones, T. R. (Ed.), London: Williams and Norgate.

Layard, N. (1901) Notes on a Human Skull Found in the Peat in the Bed of the River Orwell, Ipswich. *Man,* 1, 151.

Layard, N. F. (1890) On Reversion. *Report of the Sixtieth Meeting of the British Association for the Advancement of Science Held at Leeds in September 1890.* pp. 973–974. London: John Murray.

Leakey, R. E. and Slikkerveer, L. J. (Eds.) (1993) *Man–Ape Ape–Man. The Quest for Human's Place in Nature and Dubois' Missing Link.* Leiden: Netherlands Foundation for Kenya Wildlife Service.

Lewis, A. L. (1870–1871) The Peoples Inhabiting the British Isles. *Journal of the Anthropological Society of London,* 8, xxxiv–xl.

Livingstone, D.N. (2008) *Adam's Ancestors. Race, Religion and the Politics of Human Origins.* Baltimore: The Johns Hopkins University Press.

Longe, F. D. (1901) On a Piece of Yew from the Forest Bed on the East Coast of England, Apparently Cut by Man. *Report of the Seventy First Meeting of the British*

*Association for the Advancement of Science held at Glasgow in September 1901.* p. 798. London: John Murray.

Lowie, R. H. (1917) Edward B. Tylor. Obituary. *American Anthropologist,* 19, 262–269.

Lubbock, J. (1865a) On Mr Bateman's Researches in Ancient British Tumuli. *Transactions of the Ethnological Society of London,* 3, 307–321.

Lubbock, J. (1865b) *Pre–Historic Times as Illustrated by Ancient Remains and the Manners and Customs of Modern Savages.* London: Williams and Norgate.

Lubbock, J. (1868) The Early Condition of Man. *Anthropological Review,* 6, 1–21.

Lubbock, J. (1869) On Stone Implements from the Cape. *Journal of the Ethnological Society of London (1869–1870),* 1, 51–53.

Lubbock, J. (1870) *On the Origin of Civilization and Primitive Condition of Man.* London: Longmans Green and Company.

Lubbock, J. (1873) President's Address. *Journal of the Anthropological Institute of Great Britain and Ireland,* 2, 429–443.

Lubbock, J. (1875) Notes on the Discovery of Stone Implements in Egypt. *Journal of the Anthropological Institute of Great Britain and Ireland,* 4, 215–222.

Lubbock, J. (1887) Nationalities of the United Kingdom. *Journal of the Anthropological Institute of Great Britain and Ireland,* 16, 418–422.

Lyell, C. (1863a) *The Geological Evidences of the Antiquity of Man with Remarks on Theories of the Origin of Species by Variation.* London: John Murray.

Lyell, C. (1863b) Letter to Charles Darwin March 11 1859. *Darwin Correspondence Project Database http://www.darwinproject.ac.uk/entry–4035/(letter no. 4035 accessed 28 September 2010).*

Lyell, C. (1864) Untitled Address Reprinted in Anthropology at the British Association A.D. 1864. *Anthropological Review,* 2, 299–301.

MacGregor, A. (2008) Sir John Evans, Model Victorian, Polymath and Collector. In MacGregor, A. (Ed.), *Sir John Evans 1823–1908. Antiquity, Commerce and Natural Science in the Age of Darwin.* pp. 3–38. Oxford: The Ashmolean.

Mackenzie, K. R. H. (1867) Notes on a Stone Axe from the Rio Madera, Empire of Brazil. *Journal of the Anthropological Society of London,* 5, clxxxvi–clxxxviii.

Mackenzie, N. and Mackenzie, J. (1987) *H.G. Wells. The Time Traveller.* London: The Hogarth Press.

MacKie, S. J. (1863) On Some Human Remains from Muskham, in the Valley of the Trent, and from Heathery Burn Cave, near Stanhope, in Weardale, Durham. *Transactions of the Ethnological Society of London,* 2, 266–278.

MacKintosh, D. (1866) Comparative Anthropology of England and Wales. *Anthropological Review,* 4, 1–21.

March, H. C. (1898) The Twin Problems of Plateau Flint Implements and a Glaciation South of the Thames. *Proceedings of the Dorset Natural History and Antiquarian Field Club,* 19, 130–144.

Mather, J. D. and Campbell, I. (2007) Grace Anne Milne (Lady Prestwich): More than an Amanuensis? In Burek, C. V. and Higgs, B. (Eds.), *The Role of Women in the History of Geology.* pp. 251–264. London: The Geological Society of London.

McCartney, P. J. (1977) *Henry de la Beche.* Cardiff: Friends of the National Museum of Wales.

McKendrick, J.G. (1887–1888) On the Modern Cell Theory and the Phenomena of Fecundation. *Proceedings of the Philosophical Society of Glasgow*, 19, 71–125.

McLean, S. (2008) Animals, Language and Degeneration – In the Island of Doctor Moreau. In McLean, S. (Ed.), *H.G. Wells: Interdisciplinary Essays.* pp. 25–33. Newcastle: Cambridge Scholars Publishing.

McLean, S. (2009) *The Early Fiction of H.G. Wells.* Basingstoke: Pallgrave Macmillan.

McLennan, J. F. (1865) *Primitive Marriage; An Inquiry into the Origins of the Form of Capture in Marriage Ceremonies.* Edinburgh: Adam and Charles Black.

McNabb, J. (1996) Through the Looking Glass. An Historical Perspective on Archaeological Research at Barnfield Pit, Swanscombe, ca 1900–1964. In Conway, B., McNabb, J. and Ashton, N. (Eds.), *Excavations at Barnfield Pit, Swanscombe, 1968–1972.* pp. 31–51. London: British Museum Occassional Papers 94, British Museum.

McNabb, J. (2007) *The British Lower Palaeolithic. Stones in Contention.* Abingdon, Oxfordshire: Routledge.

McNabb, J. (2009) The Knight, The Grocer, and the Chocolate Brownies; Joseph Prestwich, Benjamin Harrison, and the Second 'Antiquity of Man Debate'. *Lithics,* 30, 97–115.

Miller, S. H. and Skertchly, S. B. J. (1878) *The Fenland Past and Present.* London: Longmans, Green, and Company.

Monnier, G. F. (2006) The Lower/Middle Palaeolithic Periodization in Western Europe. *Current Anthropology,* 47, 709–744.

Moore, J. and Desmond, A. (2004) *Introduction to Penguin Edition of Charles Darwin's The Descent of Man, and Selection in Relation to Sex (Second Edition with Additions 1879).* London: Penguin Books.

Morris, J. P. (1866) Comments on: Report of Explorations in the Kirkhead Cave at Ulverstone. *Journal of the Anthropological Society of London,* 4, cci–cciii.

Morse, E. S. (1884) Man in the Tertiaries. Vice Preseidential Address to the Section of Anthropology (H) of the American Association for the Advancement of Science at Philadelphia, September 4, 1884. *Proceedings of the American Association for the Advancement of Science,* 33, 3–15.

Morton, P. (2005) *The Busiest Man in England: Grant Allen and the Writing Trade, 1875–1900.* Basingstoke, England: Palgrave Macmillan.

Morton, P. (n.d.) Grant Allen Home Page. https://docs.google.com/View?docid=dfj62zqb_20cfsvg4&revision=_latest. Accessed March 2012.

Moser, S. (1998) *Ancestral Images. The Iconography of Human Origins.* London: Sutton Publishing.

Moskowitz, S. (1974a) Introduction: A History of Science Fiction in the Popular Magazines, 1891–1911. *Science Fiction by Gaslight. A History and Anthology of Science Fiction in the Popular Magazines, 1891–1911. (Reprint Edition).* pp. 15–50. Westport, Connecticut: Hyperion Press, Inc.

Moskowitz, S. (1974b) George Griffith – Warrior of If. In Moskowitz, S. (Ed.), *The Raid of 'Le Vengeur'.* pp. 6–47. London: Ferret Fantasy Ltd.

Müller, F. M. (1892) Untitled Presidential Address to the Anthropological Section of the British Association at the Meeting held at Cardiff in 1891. *Journal of the Anthropological Institute of Great Britain and Ireland,* 21, 172–192.

Neubauer, A. (1886) Notes on the Race–Types of the Jews. *Journal of the Anthropological Institute of Great Britain and Ireland,* 1886, 16–23.

Newton, E. T. (1895) On a Human Skull and Limb–Bones found in the Palaeolithic Terrace–Gravel at Galley Hill, Kent. *Quarterly Journal of the Geological Society of London,* 51, 505–527.

Newton, W. M. (1901) On the Occurrence in a Very Limited Area of the Rudest with the Finer Forms of Worked Stone. *Man,* 1, 81–82.

Nicholas, M. (n.d.) *Alexander James Montgomerie Bell.* England: The Other Within. Analysing the English Collections at the Pitt Rivers Museum. http://england.prm.ox.ac.uk/englishness–Bell–collection.html.

O'Connor, A. (2007) *Finding Time for the Old Stone Age.* Oxford: Oxford University Press.

Oakley, K. P. (1959) The Life and Work of Samuel Hazzledine Warren, F.G.S. *Essex Naturalist,* 30, 1–5.

Oldroyd, D. R. (1983) *Darwinian Impacts.* Milton Keynes: Open University Press. Second Edition.

Owen, J. (2008) A Significant Friendship: Evans, Lubbock and a Darwinian World Order. In MacGregor, A. (Ed.), *Sir John Evans 1823–1908; Antiquity, Commerce and Natural Science in the Age of Darwin.* pp. 206–230. Oxford: The Ashmolean.

Owen, C.M., Howard, A. and Binder, D.K. (2009) Hippocampus Minor, Calcar Avis and the Huxley–Owen Debate. *Neurosurgery* 65, 1098–1105.

Pamboukian, S. A. (2008) What the Traveller Saw. In McLean, S. (Ed.), *H.G. Wells: Interdisciplinary Essays.* pp. 9–24. Newcastle: Cambridge Scholars Publishing.

Partington, J. S. (Ed.) (2008) *H.G. Wells in Nature, 1893–1946.* London: Nature Publishing Group and Peter Lang GmbH Internationaler Verlag der Wissenschaften.

Pearson, R. 2007. Primitive Modernity: H.G. Wells and the Prehistoric Man of the 1890s. *The Yearbook of English Studies* 37, 58–74.

Perry Coste, F. H. (1897a) Human Evolution I. According to Mr H.G. Wells. *Natural Science,* 10, 184–187.

Perry Coste, F. H. (1897b) Untitled Note. *Natural Science,* 10, 244.

Philmus, R. M. and Hughes, D. Y. (Eds.) (1975) *H.G. Wells. Early Writings in Science and Science Fiction.* London: University of California Press.

Pike, L. O. (1866a) On the Physical Characteristics of the English People. *Journal of the Anthropological Society of London,* 4, cxiv–cxix.

Pike, L. O. (1866b) *The English and their Origin; a Prologue to Authentic English History.* London: Longmans.

Pope, M. and Roberts, M. (2009) "Clenching Authority": Joseph Prestwich and the Proof of the Antiquity of Man. *Lithics,* 30, 35–44.

Poulton, E. B. (1917) Obituary Notices of Fellows Deceased. August Friedrich Leopold Weismann, 1834–1914. *Proceedings of the Royal Society of London. Series B,* 89, xvii–xxxiv.

Prestwich, G. A. (1895) Recollections of M. Boucher de Perthes; Being Some Account of the History of the Discovery of Flint Implements. *Blackwood's Magazine,* 157, 939–948.

Prestwich, G. A. (1899) *Life and Letters of Sir Joseph Prestwich.* London: William Blackwood and Sons.

Prestwich, J. (1859a) Flint Implements in the Drift. *Athenaeum,* December 3, 740–741.

Prestwich, J. (1859b) Flint Implements in the Drift. *Athenaeum,* December 10, 775–776.

Prestwich, J. (1859–1860) On the Occurrence of Flint Implements Associated with the Remains of Extinct Mammalia, in Undisturbed Beds of a Late Geological Period (Abstract). *Proceedings of the Royal Society,* 10, 50–59.

Prestwich, J. (1860) On the Occurrence of Flint Implements, Associated with the Remains of Animals of Extinct Species in Beds of a Late Geological Period, in France, at Amiens and Abbeville, and in England at Hoxne. *Philosophical Transactions of the Royal Society,* 150, 277–318.

Prestwich, J. (1861) Notes on Some Further Discoveries of Flint Implements in Beds of Post-Pliocene Growel and Clay with a Few Suggestions for Search Elsewhere. *Quarterly Journal of the Geological Society of London,* 17, 362–368.

Prestwich, J. (1862–1863) Theoretical Considerations on the Conditions under which the Drift Deposits Containing the Remains of Extinct Mammalia and Flint–Implements were Accumulated; and on their Geological Age. [Abstract]. *Proceedings of the Royal Society of London,* 12, 38–52.

Prestwich, J. (1863a) The Human Jaw of Abbeville. *Athenaeum,* June 13th (number 1859), 779–780.

Prestwich, J. (1863b) *Letter to Edouard Lartet 05/05/1863 in G.A. Prestwich's Life and Letters of Sir Joseph Prestwich.* London: William Blackwood and Sons.

Prestwich, J. (1863c) On the Section at Moulin Quignon, Abbeville, and on the Peculiar Character of Some of the Flint Implements Recently Discovered There. *Quarterly Journal of the Geological Society of London,* 19, 497–505.

Prestwich, J. (1863–1864) On Some Further Evidence Bearing on the Excavation of the Valley of the Somme by River–Action, as Exhibited in a Section at Drucat near Abbeville. *Proceedings of the Royal Society of London,* 13, 135–137.

Prestwich, J. (1864) Theoretical Considerations on the Conditions Under Which the (Drift) Deposits Containing the Remains of Extinct Mammalia and Flint Implements were Accumulated, and on their Geological Age. *Phiosophical Transactions of the Royal Society,* 154, 247–310.

Prestwich, J. (1874) Report on the Explorations of Brixham Cave, Conducted by a Committee of the Geological Society, and Under the Superintendence of Wm. Pengelly, Esq., F.R.S., Aided by a Local Committee; with Descriptions of the Animal Remains by George Busk, Esq., F.R.S., and of the Flint Implements by John Evans Esq., F.R.S. By Joseph Prestwich F.R.S., F.G.S., &c., Reporter. *Philosophical Transactions of the Royal Society,* 163, 471–572.

Prestwich, J. (1880) On the Geological Evidence of the Temporary Submergence of the South West of Europe During the Early Human Period. *Report of the Fiftieth*

*Meeting of the British Association for the Advancement of Science Held at Swansea in August and September 1880.* pp. 581–582. London: John Murray.

Prestwich, J. (1881) On the Strata Between the Chillesford Beds and the Lower Boulder Clay, 'the Mundesley and Westleton Beds'. *Report of the Fifty–First Meeting of the British Association for the Advancement of Science Held at York in August and September 1881.* p. 620. London: John Murray.

Prestwich, J. (1886) *Geology, Chemical, Physical, and Stratigraphical. Volume I. Chemical and Physical.* Oxford: Clarendon Press.

Prestwich, J. (1887) Considerations on the Date, Duration, and Conditions of the Glacial Period, with Reference to the Antiquity of Man. *Quarterly Journal of the Geological Society of London,* 43, 303–410.

Prestwich, J. (1888) *Geology, Chemical, Physical, and Stratigraphical. Volume II. Stratigraphical and Physical.* Oxford: Clarendon Press.

Prestwich, J. (1889) On the Occurrence of Palaeolithic Flint Implements in the Neighbourhood of Ightham, Kent, their Distribution and Probable Age. *Quarterly Journal of the Geological Society of London,* 45, 270–297.

Prestwich, J. (1890a) On the Relation of the Westleton Beds or Pebbly Sands of Suffolk to those of Norfolk and on their Extension Inland; with some Observations on the Period of the Final Elevation and Denudation of the Weald and of the Thames Valley, &c, Part 2. *Quarterly Journal of the Geological Society of London,* 46, 120–154.

Prestwich, J. (1890b) On the Relation of the Westleton Beds or Pebbly Sands of Suffolk to those of Norfolk and on their Extension Inland; with some Observations on the Period of the Final Elevation and Denudation of the Weald and of the Thames Valley. Part 1. *Quarterly Journal of the Geological Society of London,* 46, 84–119.

Prestwich, J. (1890c) On the Relation of the Westleton Shingle to other Pre–Glacial Deposits in the Thames Basin, and on a Southern Drift, with Observations on the Final Elevation and Initial Subaerial Denudation of the Weald; and on the Genesis of the Thames, Part 3. *Quarterly Journal of the Geological Society of London,* 46, 155–181.

Prestwich, J. (1891) On the Age, Formation, and Successive Drift–Stages of the Valley of the Darent; with Remarks on the Palaeolithic Implements of the District, and on the Origin of its Chalk Escarpment. *Quarterly Journal of the Geological Society of London,* 47, 126–163.

Prestwich, J. (1892) On the Primitive Characters of the Flint Implements of the Chalk Plateau of Kent, with Reference to the Question of their Glacial or Pre–Glacial Age with Notes by Messrs B. Harrison and De Barri Crawshay. *Journal of the Anthropological Institute of Great Britain and Ireland,* 21, 246–276.

Prestwich, J. (1895a) Nature and Art. *Geological Magazine,* 32, 375–377.

Prestwich, J. (1895b) The Greater Antiquity of Man. *The Nineteenth Century,* 37 (April), 617–628.

Prestwich, J. (1895c) On the Primitive Characters of the Flint Implements of the Chalk Plateau of Kent, with Reference to the Question of Age and Make. In

Prestwich, J. (Ed.) *Collected Papers on Some Controverted Questions of Geology.* pp. 49–80. London: Macmillan and Company.

Prestwich, J. (Ed.) (1895d) *Collected Papers on Some Controverted Questions of Geology.* London: Macmillan and Company.

Prigg, H. (1885) On a Portion of a Human Skull of Supposed Palaeolithic Age from Near Bury St. Edmunds. *Journal of the Anthropological Institute of Great Britain and Ireland,* 14, 51–55.

Pringle, D. (1981) Lost Worlds. In Nicholls, P. (Ed.), *The Encyclopedia of Science Fiction.* p. 364. St Albans: Granada Publishing.

R.L. (1890) 1. Professor Gaudry on Dryopithecus. *Geological Magazine,* 27, 374.

Radick, G. (2007) *The Simian Tongue; the Long Debate About Animal Language.* Chicago: University of Chicago Press.

Reade, T. M. (1883) The Human Skull Found Near Southport. *Geological Magazine,* 20, 547–548.

Roberts, A. (2005a) *Science Fiction: the New Critical Idiom.* 2nd Edition. Abingdon: Routledge.

Roberts, A. (2005b) *The History of Science Fiction.* London: Palgrave Macmillan.

Roberts, A. and Barton, N. (2008) Reading the Unwritten History: Evans and Ancient Stone Implements. In MacGregor, A. (Ed.), *Sir John Evans 1823–1908. Antiquity, Commerce and Natural Science in the Age of Darwin.* pp. 95–114. Oxford: The Ashmolean.

Roe, D. A. (2009) Worthington George Smith (1835–1917). *Lithics,* 30, 85–95.

Ruddick, N. (2009) *The Fire in the Stone. Prehistoric Fiction from Charles Darwin to Jean M. Auel.* Middletown (CT. USA): Wesleyan University Press.

Rudler, F. W. (1874) Report on the Department of Anthropology, at the Bradford Meeting of the British Association for the Advancement of Science, 1873. *Journal of the Anthropological Institute of Great Britain and Ireland,* 3, 330–341.

Rudler, F. W. (1888) Fifty Years Progress in British Geology; Being an Address on the Opening of the Session 1887–1888. *Proceedings of the Geologists' Association,* 10, 234–272.

Rudler, F. W. (1903) 104. John Allen Brown, F.G.S., Born September 3rd, 1831; Died September 24th, 1903. *Man,* 3, 184–185.

Rudwick, M. J. S. (1992) *Scenes From Deep Time.* London: University of Chicago Press.

Rudwick, M. J. S. (2007) *Bursting the Limits of Time. The Reconstruction of Geohistory in the Age of Revolution.* London: University of Chicago Press.

Rudwick, M. J. S. (2008) *Worlds before Adam. The Reconstruction of Geohistory in the Age of Reform.* London: University of Chicago Press.

Russell, A.K. (1979) (Ed.) *Science Fiction by the Rivals of H.G. Wells.* Secaucus New Jersey: Castle Books.

Sayce, A. H. (1888) Presidential Address to the Anthropological Section of the British Association at Manchester. *Journal of the Anthropological Institute of Great Britain and Ireland,* 17, 166–181.

Schaaffhausen, H. (1868) On the Primitive Form of the Human Skull. *Anthropological Review,* 6, 412–431.

Schaaffhausen, H. (1880) On the Origins of the Neanderthal Skull. *Report of the Fiftieth Meeting of the British Association for the Advancement of Science held at Swansea in August and September 1880.* p. 624. London: John Murray.

Scott, B. and Shaw, A. (2009) The Quiet Man of Kent: the Contribution of F.C.J. Spurrell to the Early Years of Palaeolithic Archaeology. *Lithics,* 30, 53–64.

Seton–Karr, W. H. (1895) Some Implements in Somaliland. *Report of the Sixty–Fifth Meeting of the British Association for the Advancement of Science held at Ipswich in September 1895.* pp. 824–825. London: John Murray.

Shipman, P. (2001) *The Man Who Found the Missing Link. The Extraordinary Life of Eugene Dubois.* London: Weidenfeld and Nicolson.

Shone, W. (1894) Post–Glacial Man in Britain. *Geological Magazine,* 31, 78–80.

Shrubsole, O. A. (1885) On Certain Less Familiar Forms of Palaeolithic Flint Implements from the Gravel at Reading. *Journal of the Anthropological Institute of Great Britain and Ireland,* 14, 192–200.

Shrubsole, O. A. (1890) On the Valley Gravels about Reading, with Especial Reference to Palaeolithic Inplements Found in them. *Quarterly Journal of the Geological Society of London,* 46, 582–594.

Shrubsole, O. A. (1893) On the Plateau Gravel South of Reading. *Quarterly Journal of the Geological Society of London,* 49, 320–322.

Shrubsole, O. A. (1895) On Flint Implements of a Primitive Type from Old (Pre–Glacial) Hill–Gravels in Berkshire. *Journal of the Anthropological Institute of Great Britain and Ireland,* 24, 44–49.

Skertchly, S. B. J. (1876a) On the Discovery of Palaeolithic Implements of Interglacial Age. *Nature,* 14, 448–440.

Skertchly, S. B. J. (1876b) Untitled Letter to the Editor. *Nature,* 15, 142.

Slotten, R. A. (2004) *The Heretic in Darwin's Court; the Life of Alfred Russel Wallace.* New York: Columbia University Press.

Smith, F. (1892) Discovery of the Common Occurrence of Palaeolithic Weapons in Scotland. *Report of the Sixty–Second Meeting of the British Association for the Advancement of Science held at Edinburgh in August 1892.* pp. 896–897. London: John Murray.

Smith, W. G. (1879) On Palaeolithic Implements from the Valley of the Lea. *Journal of the Anthropological Institute of Great Britain and Ireland,* 8, 275–279.

Smith, W. G. (1880) Palaeolithic Implements from the Valley of the Brent. *Journal of the Anthropological Institute of Great Britain and Ireland,* 9, 316–319.

Smith, W. G. (1884a) Excursion to Homerton. *Proceedings of the Geologists' Association,* 8, 124–131.

Smith, W. G. (1884b) On a Palaeolithic Floor at North–East London. *Journal of the Anthropological Institute of Great Britain and Ireland,* 13, 357–384.

Smith, W. G. (1889) Palaeolithic Implements from the Hills near Dunstable. *Nature,* 40, 151.

Smith, W. G. (1894) *Man the Primeval Savage.* London: Edward Stanford.

Sommer, M. (2007) *Bones and Ochre. The Curious Afterlife of the Red Lady of Paviland.* London: Harvard University Press.

Sparks, J. (1999) At the Intersection of Victorian Science and Fiction: Andrew Lang's

Romance of the First Radical. *English Literature in Translation,* 42, 125–142.

Spencer, F. (1990) *Piltdown: a Scientific Forgery.* London: Natural History Museum Publications.

Spurrell, F. C. J. (1880) On the Site of a Palaeolithic Implement Manufactory, at Crayford, Kent. *Report of the Fiftieth Meeting of the British Association for the Advancement of Science Held at Swansea in August and September 1880.* p. 574. London: John Murray.

Spurrell, F. C. J. (1883) Palaeolithic Implements found in West Kent. *Archaeologia Cantiana,* 15, 89–103.

Stepan, N. (1982) *The Idea of Race in Science: Great Britain 1800–1960.* Houndmills, Hampshire: MacMillan Press.

Stevens, E. T. (1870) *Flint Chips. A Guide to Prehistoric Archaeology as Illustrated by the Collection in the Blackmore Museum, Salisbury.* London: Bell and Daldy, Covent Garden.

Stirling, J. (1869) On Some Flint Arrow–Heads and North–American Indian Pipes, Found in Kelby's Island on Lake Erie. *Journal of the Anthropological Society of London,* 7, cxi–cxii.

Stirrup, M. (1885) On the So–Called Worked Flints from the Miocene Beds of Thenay, in France. *Journal of the Anthropological Institute of Great Britain and Ireland,* 14, 289–290.

Stocking, G. W. (1971) What's in a Name? The Origin of the Royal Anthropological Institute(1837–1971). *Man,* 6, 369–390.

Stocking, G. W. (1973) From Chronology to Ethnology. James Cowles Prichard and British Anthropology 1800–1850. In Stocking, G. W. (Ed.), *Researches into the Physical History of Man. James Cowles Prichard. Edited and with an Introductory Essay by George W. Stocking Jr.* pp. ix–cx. London: University of Chicago Press.

Stocking, G. W. (1982) *Race, Culture, and Evolution. (Phoenix Edition).* London: University of Chicago Press.

Stocking, G. W. (1987) *Victorian Anthropology.* New York: Free Press.

Stocking, G. W. (1988) Bones, Bodies, Behaviour. In Stocking, G. W. (Ed.), *Bones, Bodies, Behaviour.* pp. 3–18. London: The University of Wisconsin Press.

Stocking, G. W. (1995) *After Tylor. British Social Anthropology 1888–1951.* Madison, Wisconsin: The University of Wisconsin Press.

Stopes, H. (1881) Traces of Man in the Crag. *Report of the Fifty–First Meeting of the British Association for the Advancement of Science held at York in August and September 1881.* p. 700. London: John Murray.

Stopes, H. (1890) Indications of Retrogression in Prehistoric Civilization in the Thames Valley. *Report of the Sixtieth Meeting of the British Association for the Advancement of Science Held at Leeds in August 1890.* p. 979. London: John Murray.

Stopes, H. (1893) On Palaeolithic Anchors, Anvils, Hammers, and Drills. *Report of the Sixty–Third Meeting of the British Association for the Advancement of Science Held at Nottingham in September 1893.* p. 904. London: John Murray.

Stopes, H. (1894) On the Evolution of Stone Implements. *Report of the Sixty–Fourth Meeting of the British Association for the Advancement of Science held at Oxford in August*

*1894*. p. 776. London: John Murray.

Stopes, H. (1895) On Graving Tools from the Terrace Gravels of the Thames Valley. *Report of the Sixty–Fifth Meeting of the British Association for the Advancement of Science held at Ipswich in September 1895*. p. 826. London: John Murray.

Stopes, H. (1899) On the Discovery of *Neritina Fluviatilis* with a Pleistocene Fauna and Worked Flints in High Terrace Gravels of the Thames Valley. *Journal of the Anthropological Institute of Great Britain and Ireland,* 29, 302–303.

Stopes, H. (1900) Unclassified Worked Flints. *Journal of the Anthropological Institute of Great Britain and Ireland,* 30, 299–304.

T.G.B. (1904–1905) Henry Hicks. Obituary Notices of Fellows Deceased. *Proceedings of the Royal Society of London,* 75, 106–109.

Thompson, M. W. (1977) *General Pitt–Rivers.* Bradford–on–Avon: Moonraker Press.

Thurnam, J. (1864) Comments on 'The Two Principal Forms of Crania Amongst the Early Britons'. *Journal of the Anthropological Society of London,* 2, ccxxxi–ccxxxiii.

Tiddeman, R. H. (1873) The Relation of Man to the Ice–Sheet in the North of England. *Nature,* 9, 14–15.

Tiddeman, R. H. (1876) The Age of Palaeolithic Man. *Nature,* 14, 505–506.

Tiddeman, R. H. (1878) On the Age of the Hyaena–Bed at the Victoria Cave, Settle, and its Bearing on the Antiquity of Man. *Journal of the Anthropological Institute of Great Britain and Ireland,* 7, 165–173.

Topinard, P. (1881) Observations Upon the Methods and Processes of Anthropometry. *Journal of the Anthropological Institute of Great Britain and Ireland,* 10, 212–224.

Trinkaus, E. and Shipman, P. (1993) *The Neanderthals. Changing the Image of Mankind.* London: Pimlico.

Turner, W. (1889) On Heredity. . *Report of the Fifty–Nineth Meeting of the British Association for the Advancement of Science Held at Newcastle–Upon–Tyne in September 1889.* pp. 756–771. London: John Murray.

Tylor, E. B. (1865) *Researches into the Early History of Mankind and the Development of Civilization.* London: John Murray.

Tylor, E. B. (1878) *Researches Into the Early History of Mankind and the Development of Civilization (Third Edition, Revised).* London: John Murray.

Tylor, E. B. (1880) Untitled Address to the Department of Anthropology of the British Association, Sheffield, August 21, 1879, by Edward B. Tylor, D.C.L., F.R.S., President of the Anthropological Institute. *Journal of the Anthropological Institute of Great Britain and Ireland,* 9, 235–246.

Tylor, E. B. (1894a) On Some Stone Implements of the Australian Type from Tasmania. *Report of the Sixty–Fourth Meeting of the British Association for the Advancement of Science held at Oxford in August 1894.* p. 782. London: John Murray.

Tylor, E. B. (1894b) On the Tasmanians as Representatives of Palaeolithic Man. *Journal of the Anthropological Institute of Great Britain and Ireland,* 23, 141–152.

Tylor, E. B. (1895) On the Occurrence of Ground Stone Implements of Australian Type in Tasmania. *Journal of the Anthropological Institute of Great Britain and Ireland,* 24, 335–340.

Tylor, E. B. (1898) On the Survival of Palaeolithic Conditions in Tasmania and

Australia, with Especial Reference to the Modern Usage of Unground Stone Implements in West Australia. *Report of the Sixty-Eighth Meeting of the British Association for the Advancement of Science Held at Bristol in September 1898.* pp. 1014–1015. London: John Murrat.

Tylor, E. B. (1900) 37. On the Stone Age in Tasmania, as Related to the History of Civilization. *Journal of the Anthropological Institute of Great Britain and Ireland,* 30, 33–34.

van Ripper, A. B. (1993) *Men Among the Mammoths: Victorian Science and the Discovery of Human Prehistory.* Chicago and London: University of Chicago Press.

Vogt, C. (1864) *Lectures on Man: His Place in Creation, and in the History of the Earth.* London: Edited by James Hunt and Published for the Anthropological Society of London by Longman, Green, Longman and Roberts.

Vogt, C. (1867) The Primitive Period of the Human Species. *Anthropological Review,* 5, part 1 204–221, part 2 334–350.

W.B.K. (1870) Prehistoric Archaeology. *Journal of Anthropology,* 1, 164–170.

Waddington, S. (1900) The Cradle of the Human Race. *The Nineteenth Century,* 48, 801–806.

Wagar, W. W. (1964) *H.G. Wells Journalism and Prophecy 1893–1946.* London: The Bodley Head.

Wake, C. S. (1872) Report on Anthropology at the Meeting of the British Association for the Advancement of Science for 1871, at Edinburgh. *Journal of the Anthropological Institute of Great Britain and Ireland,* 1, 268–274.

Wallace, A. R. (1864a) The Origin of Human Races and the Antiquity of Man Deduced from the Theory of Natural Selection. *Journal of the Anthropological Society of London,* 2, clviii–clxxxvii.

Wallace, A. R. (1864b) Letter to C. Darwin May 29 1864. *Darwin Correspondence Project Database http://www.darwinproject.ac.uk/entry-4514/(letter no. 4514 accessed 28 September 2010).*

Wallace, A. R. (1866) Mr Wallace on Natural Selection Applied to Anthropology. *Anthropological Review,* 5, 103–105.

Wallace, A.R. (1889) *Darwinism.* London: Macmillan and Company.

Warren, S. H. (1900) Palaeolithic Implements from the Chalk Downs of the Isle of Wight and the Valleys of the Rivers Western Yar and Stour. *Geological Magazine,* 37, 406–412.

Watson, J. A. (1901) A Suggested Link in the 'Break' Between Palaeolithic and Neolithic Man. A Reply to Howorth. *Geological Magazine,* 38, 424–425.

Weismann, A. (1891–1892) *Essays Upon Heredity and Kindred Biological Problems. Edited by E.B. Poulton and A.E. Shipley. In Two Volumes.* Oxford: Clarendon Press.

Weismann, A. (1893) *The Germ-Plasm. A Theory of Heredity.* New York: Charles Scribner's Sons.

Wells, H. G. (1891) Zoological Retrogression. *Gentleman's Magazine,* 271, 246–253.

Wells, H. G. (1897) Human Evolution III. Mr Wells Replies. *Natural Science,* 10, 242–244.

Wells, H. G. (2000) The Moth. (*Story Published in Pall Mall Gazette March 1895*). In

Hammond, J. (Ed.), *The Complete Short Stories of H.G. Wells.* pp. 84–91. London: Phoenix Press.

Wells, H. G. (2005) *The Time Machine* London: Penguin Classics (First Published in 1895 by William Heinemann).

Wells, H. G. (2007) *Clasics Co. Omnibus. H.G. Wells 2 books in 1. Tme Machine & The War of the Worlds & The Invisible Man & The Island of Doctor Moreau.* Mumbai: Wilco Publishing House.

Wenban–Smith, F. F. (2009) Henry Stopes (1852–1902): Engineer, Brewer, and Anthropologist. *Lithics,* 65–84.

Westropp, H. M. (1866) On the Analogous Forms of Implements among Early and Primitive Races. *Journal of the Anthropological Society of London,* 4, clxxxiii–clxxxvi.

Westropp, H. M. (1867) On the Sequence of the Phases of Civilization, and Contemporaneous Implements. *Journal of the Anthropological Society of London,* 5, cxii–cc.

Whitaker, W. (1894) On the Age of the Plateau Beds. *Report of the Sixty–Fourth Meeting of the British Association for the Advancement of Science held at Oxford in August 1894.* p. 652. London: John Murray.

White, M. J. (2001) Out of Abbeville: Sir John Evans, Palaeolithic Patriach and Handaxe Pioneer. In Milliken, S. and Cook, J. (Eds.), *A Very Remote Period Indeed.* Oxford: Oxbow Books.

White, M. J. and Pettitt, P. (2009) The Demonstration of Human Antiquity; Three Rediscovered Illustrations from the 1825 and 1846 Excavations in Kent's Cavern (Torquay, England). *Antiquity,* 83, 758–768.

White, M. J. and Plunkett, T. (2005) *Miss Layard Excavates: the Palaeolithic Site at Foxhall Road, Ipswich, 1903–1905.* Liverpool: Western Academic and Specialist Press,

Williams, K. (2008) Alien Gaze: Postcolonial Vision in The War of the Worlds. In McLean, S. (Ed.), *H.G. Wells: Interdiscipinary Essays.* pp. 49–73. Newcastle: Cambridge Scholars Publishing.

Wilson, D. (1865) Inquiry into the Physical Characteristics of the Ancient and Modern Celt of Gaul and Britain. *Anthropological Review,* 3, 52–84.

Woodward, A. S. (1931) Sir William Boyd Dawkins 1837 – 1929. *Obituary Notices of Fellows Deceased. Proceedings of the Royal Society of London. Series B,* 107, xxiii–xxvi.

Woodward, H. B. (1880) Review of Boyd Dawkins' 'Early Man in Britain and His Place in the Tertiary Period'. *Geological Magazine,* 7, 371–374.

Woodward, H. B. (1884) Discoveries in the More Recent Deposits of the Bovey Basin, Devon. *Geological Magazine,* 21, 131–132.

Woodward, H. B. (1893) Eminent Living Geologists No. 8. Professor Joseph Prestwich, D.C.L., F.R.S., F.G.S., F.C.S., etc.'. *Geological Magazine,* 30, 241–246.

Wyatt, J. (1862) On Some Further Discoveries of Flint Implements in the Gravels near Bedford. *Quarterly Journal of the Geological Society of London,* 18, 113–114.

Wyatt, J. (1863–1864) Further Discoveries of Flint Implements and Fossil Mammals in the Valley of the Ouse. *Quarterly Journal of the Geological Society of London,* 20, 183–188.

Wynn, M. (n.d.) *http://grantallen.org. Site accessed 22 August 2011*

# APPENDIX

## *Review of the Main Themes in Terms of Anthropology and Ethnology, Human Origins and Palaeolithic Archaeology, as Reported in the Journals of the London Societies in the 1860s*

| Year | Volume | Journal title | Society to which Affiliated. Pres. = president | Summary of volume's content. HO = human origins |
|---|---|---|---|---|
| 1861 | 1 | *Transactions of the Ethnological Society of London* (continues from the *J. of the Ethnol. Soc. London*) | *Ethnological Society of London* | Strong emphasis on descriptive articles on contemporary races and global ethnological themes. Only c. 14% of content is anthropological. Only two HO articles, but topic mentioned in others; all framed by polygenist vs monogenist debate, and teleological explanations common. Great diversity of papers, both specialist and non-specialists equally represented. |
| 1862 | No volume | | | |
| 1863 | 2 | *Transactions of the Ethnological Society of London* | *Ethnological Society of London* (Pres. Lubbock) | Strong emphasis on descriptive articles on contemporary races and global ethnological themes. Similar format to previous volume. Little HO material, introduced as part of broader discussion on race and racial origins. HO mostly discussed by a few individuals (Crawfurd and Hunt – so HO is reflecting their teleological and/or polygenist views) |
| 1863 | 1 | *Transactions of the Anthropological Society of London* | *Anthropological Society of London* (Pres. Hunt) | Comparatively less on contemporary world ethnology emphasis on physical anthropology. World ethnography also dealt with from physical anthropological perspective. HO strongly represented as a theme, with monogensist vs polygenist type debates mostly absent. Moulin Quignon discussed. Important review of Pliocene human evidence by Carter Blake |
| 1863 | 1 | *Anthropological Review* | *Independent but affiliated to Anthropological Society of London.* ( Hunt editor) | Limited world ethnology. Strong anthropological/physical anthropological basis. HO well represented though many entries more philosophical than empirical. Excellent anonymous overview paper of European data. Influence of polygenist Hunt as editor is marked. Moulin Quignon and Neanderthal skull discussed; interest in and polemical reviews of Huxley and Lyell's books |
| 1864 | 2 | *Journal of the Anthropological Society of London* (continues Trans. Anth. Soc. Of London) | *Anthropological Society of London* (Pres. Hunt) | Absolute frequency of world ethnology articles increases by comparison with vol.1, but greater range of articles and more of them. Contemporary world ethnology accounts by no means dominate. Emphasis on prehistoric (not HO) ethnology anthropology; physical anthropology themes well represented. HO present mostly focused on physical anthropology aspects. Important papers on Neanderthal skull in European scholarship and Wallace on racial origins. |
| 1864 | 2 | *Anthropological Review* | *Independent but affiliated to Anthropological Society of London.* ( Hunt editor) | Detailed reporting of anthropology at British Association (BA) meeting (in all volumes of AR); HO is included but most of discussants subsume it into racial issues. Reports of Lubbock and Lyell at BA. Other HO entries in volume mostly philosophical in nature. Neanderthal skull and Moulin Quignon still being debated. Anthropological and prehistoric subjects discussed with some world ethnology. |
| 1865 | 3 | *Transactions of the Ethnological Society of London* | *Ethnological Society of London* (Pres. Lubbock) | Continuation of pattern in vol. 2 with strong focus on world ethnology. Crawfurd reviews Huxley and Lyell's books from a teleological/polygenist stance. Important HO article on SW French cave sites by Christy, using archaeology to reconstruct lifestyle (commits to monogenist view). HO debate still mostly embedded in monogenist vs polygenist debate and continued by a small number of discussants after papers were read. |
| 1865 | 3 | *Journal of the Anthropological Society of London* | *Anthropological Society of London* (Pres. Hunt) | Stronger emphasis on contemporary world ethnology than in previous volumes. Physical anthropological themes still present. Little HO in this volume. British prehistory is better represented |
| 1865 | 3 | *Anthropological Review* | *Independent but affiliated to Anthropological Society of London.* ( Hunt editor) | World ethnology poorly represented but higher incidence of UK and European ethnology, linked with racial questions. Strong emphasis on political machinations at BA. Few specific HO articles, but some included in anthropological and racial entries. When present HO tends to be broader philosophical treatments. This volume heavy on book reviews including Lubbock and Tylor. |
| 1866 | 4 | *Transactions of the Ethnological Society of London* | *Ethnological Society of London* (Pres. Crawfurd) | Continuing strong emphasis on contemporary world ethnology and race. A good number of general prehistoric articles, many with a racial or linguistic-racial slant to them. Little HO. |
| 1866 | 4 | *Journal of the Anthropological Society of London* | *Anthropological Society of London* (Pres. Hunt) | Focus on global ethnology falls again, with a stronger showing of anthropology and racial themes. British prehistory is still represented with contributions on British racial ethnography by Beddoe. |
| 1866 | 4 | *Anthropological Review* | *Independent but affiliated to Anthropological Society of London.* ( Hunt editor) | Mixed bag of papers, mostly reviews. Few overtly anthropological or descriptive world ethnology in character. Race incorporated in other aspects. Most of the HO discussion is from reports of the BA meeting, with emphasis on Naulette jaw. Good overview paper by MacKintosh on range of evidence types and underlying theoretical assumptions in persistence of racial types (UK in this case), as commonly discussed. |
| 1867 | 5 | *Transactions of the Ethnological Society of London* | *Ethnological Society of London* (Pres. Crawfurd) | As in previous volume emphasis maintained on descriptive world ethnology and racial matters. Little anthropology, and less prehistory, mostly tied to racial differences |

| | | | | |
|---|---|---|---|---|
| 1867 | 5 | *Journal of the Anthropological Society of London* | *Anthropological Society of London* (R. Burton elected – abroad - so Hunt stands in) | Stronger presence of world ethnology at expense of prehistoric and anthropological subjects, British contemporary ethnology drops in frequency too. Excellent long discussion to paper by Wake on HO – very wide cross-section of possible view points. Two communications on Belgian Palaeolithic caves, discussion afterwards covers many views. |
| 1867 | 5 | *Anthropological Review* | *Independent but affiliated to Anthropological Society of London.* ( Hunt editor) | World ethnology present but mostly included in articles/reviews on race and racial differences. More HO pieces. Good cross section of French and German scholarship emphasising that continental research into HO different from that in UK; in same vein Carter Blake attempts to link fossil Belgian jaws with modern continental peoples. A number of teleological submissions. Possible fossil jaw from Foxhall, Suffolk. |
| 1868 | 6 | *Transactions of the Ethnological Society of London* | *Ethnological Society of London* (Pres. Crawfurd) | Continuing the world ethnology focus and reflecting on racial issues. Lubbock defends the idea of progress in human societies. Little HO or prehistory, what present mostly allied to racial origins as in previous volumes. |
| 1868 | 6 | *Journal of the Anthropological Society of London* | *Anthropological Society of London* (Pres. Hunt) | Strong showing for world ethnology articles much lower frequency for prehistory and anthropology. Higher incidence of communications on society business and untitled letters. More on Belgian Palaeolithic caves. Excellent discussion to paper by Schaafhausen on Darwinism as explanation for HO and racial origins |
| 1868 | 6 | *Anthropological Review* | *Independent but affiliated to Anthropological Society of London.* ( Hunt editor) | Good spread of subjects presented and/or reviewed. HO distributed amongst various papers on race and racial origins (Schaafhausen), and on man's place in nature (Lubbock on degeneration). Good presence of anthropological themes, relatively little world ethnology. Broca attempts to relate skeletons from Les Eyzies. As much as any other this and preceding volume show how parochial English anthropology/HO is |
| 1869 | 7 | *Transactions of the Ethnological Society of London* | *Ethnological Society of London* (Pres. Huxley) | Dominated by world ethnology, little on prehistory. Teleological orientation has diminished through the decade, but not entirely vanished. |
| 1869 | 7 | *Journal of the Anthropological Society of London* | *Anthropological Society of London* (Pres. Beddoe. Hunt dies August) | Fewer world ethnology articles but a stronger emphasis on British ethnology and prehistory than in previous volume. Race well in evidence. No specific HO papers |
| 1869 | 7 | *Anthropological Review* | *Independent but affiliated to Anthropological Society of London.* ( Hunt editor) | A good mix of anthropology and HO, with a strong presence of HO themes especially in reports from the BA. World ethnology limited but racial issues well represented, and some HO included in these too. Lubbock extends debates with Duke of Argyll who takes teleological stance, and whose work separately reviewed; Argyll denies the reality of a global stone age period |
| 1869 | 1 | *Journal of the Ethnological Society of London* (continues *Trans. Anth. Soc. London*) | *Ethnological Society of London* (Pres. Huxley) | Similar to preceding volume in series. Dominated by world ethnology, little on prehistory, slightly more on race than previous. Interesting discussion to paper on South African quartzite implements. Boyd Dawkins, McKenny Hughes on council, Lane Fox general secretary |
| 1870 | 2 | *Journal of the Ethnological Society of London* | *Ethnological Society of London* (Pres. Huxley) | Still strong on world ethnology, but big increase in number of papers overall and a higher percentage of prehistoric and British prehistoric papers as well. Possibly reflecting interest of new council members. JES continues the tradition of TESL in having almost no British ethnological papers. Comparatively few anthropological papers, some included in the submissions on race and racial differences. Important piece by Huxley on global race identification |
| 1870 | 8 | *Anthropological Review* | *Independent but affiliated to Anthropological Society of London.* | Little on HO in this volume. BAAS not reported. More on ethnology and on British racial ethnology Slim volume by comparison with others. Possibly reflecting death of Hunt and a new editor. |
| 1870-1871 | 8 | *Journal of the Anthropological Society of London* | *Anthropological Society of London* (Pres. Beddoe) | Very strong presence of world ethnology and racially related issues (possibly reflecting Beddoes as new president?). Anthropological articles fewer, with more European and world prehistoric submissions than British. HO only present occasionally |
| 1870 | 1 | *Journal of Anthropology* (continues *Anth. Rev.*) | | Ethnological focus maintained from last volume but high proportion of subject overviews and reports on other societies and reports about publications. HO virtually absent |

# INDEX

## F



## G



## H

**M**

## N



## O



## P

**Q**



**R**



**S**

**W**



**X**

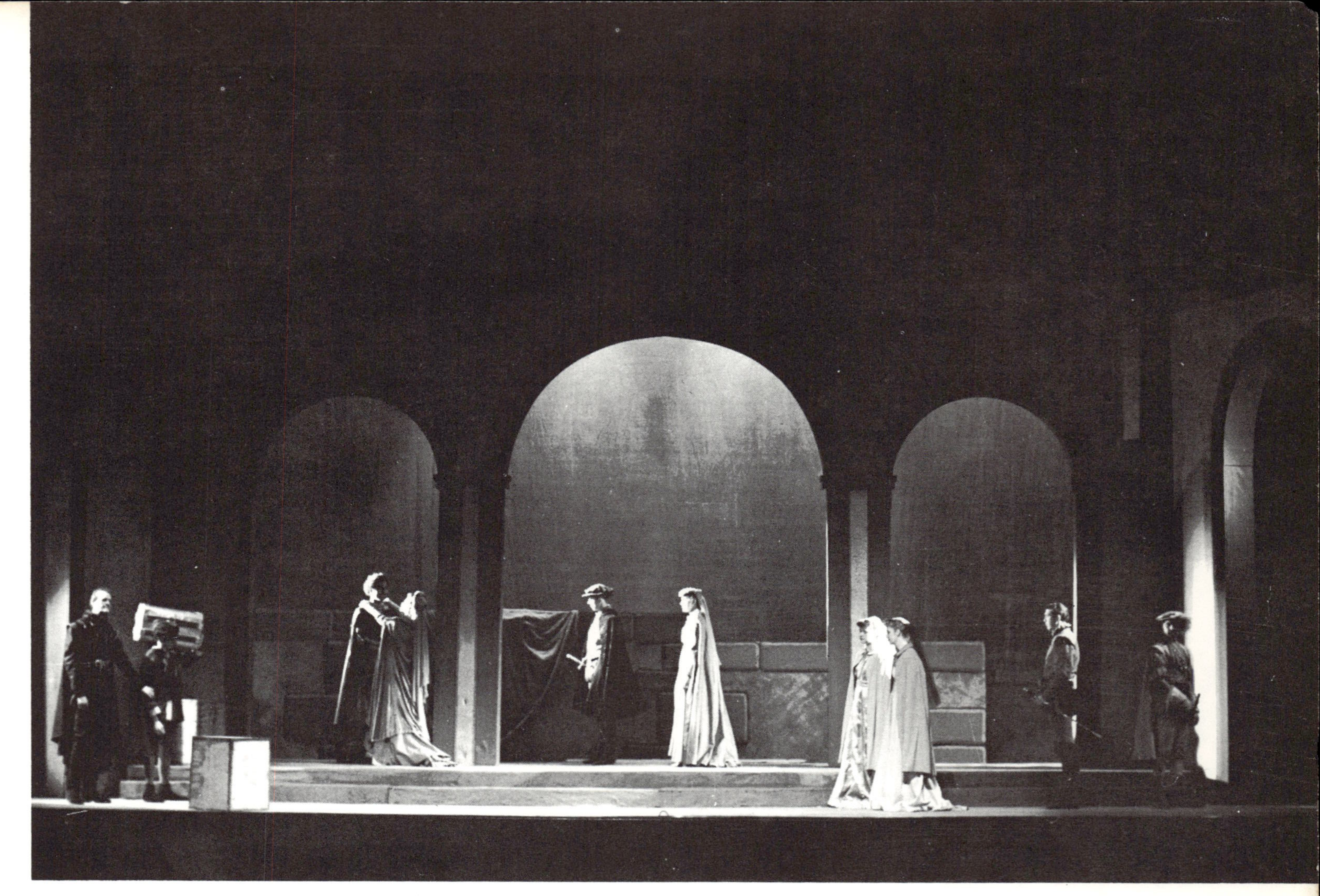

*Othello.* Basic setting consisting of three archways in the back, one on either side, and platforms stretching across the stage.

# *Theatre Scenecraft*

For the backstage technician and artist

*by*

*Vern Adix*

Professor of Theatre
University of Utah



Revised Edition


Illustrated by the Author with Drawings and Photographs of University of Utah Theatre Productions. Additional Drawings by Robert W. Weideman.

cover layout by Ron Crosby
ISBN-0-87602-013-9

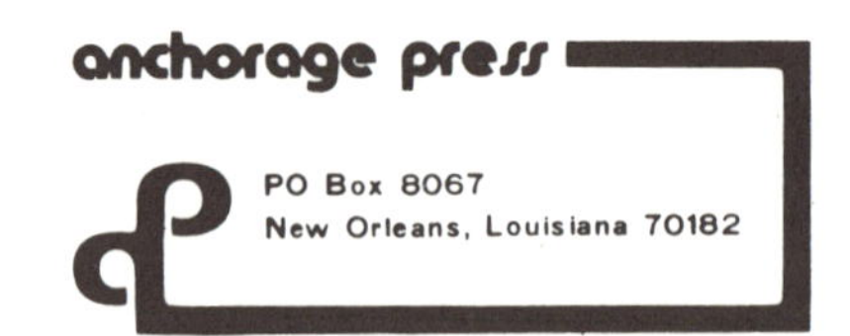




First Printing 1956
Second Printing 1961
Third Printing 1966
Fourth Printing 1970
Fifth Printing 1974
Revised Edition printed, 1981

*In memory of*

*my father*

*Jack-of-all-trades.*

# FOREWORD

Not many possess the ability to write in a concise manner when they must describe various technical processes; fewer still have the ability to illustrate their work, and only a very few indeed have the years of experience necessary to make what they have written about and illustrated of any great value to the reader. Yet this is the happy combination of facts found in Vern Adix's book *Theatre Scenecraft.* Some fifteen years of practical theatre work have been combined with a natural, easily read style of writing and with clean-cut illustrations, that have resulted in a text that should prove a boon to all those organizations whose business or pleasure it is to produce plays.

Of greatest value to the inexperienced designer-technician are the innumerable suggestions and methods described by Mr. Adix that, if followed, will result in the saving of many hours of time in construction, painting, rigging, and shifting of scenery. Of equal importance is the fact that these suggestions will not only lead to a saving of time and a reduction of probable cost, but to a higher standard of production as well.

—ARNOLD S. GILLETTE, *Managing Director*
The University Theatre
State University of Iowa

# *Acknowledgments*

This book is a compilation of some of the thoughts and ideas of one lonely non-professional theatre designer and technician. The loneliness has come from the pressures and strains of as many as 20 "opening nights" a year for a period of 15 years. Out of this "Sturm and Drang" experience have come some "short cuts", simplified processes and economical means of production that may be of value to others. Since continuously working against time that is always too short, trying to stretch budgets that are too rigid, and maintaining amiable relations with directors whose "opening night tempers" are too quick seem to be the lot of the non-professional theatre worker, it is hoped that the material in this book will provide a few helpful suggestions for overcoming these "opening nightmares."

A factual crafts book is either the result of years of thought, research and experience or the careful compilation of the ideas of others. The material of this book is, in a sense, a combination of both of these processes. The material has been accumulated piecemeal over the years as the productions have been planned, built, painted, lighted, staged and "struck." The facts, figures and information of this book have been collected, assimilated and now put together in a new form. The source of most of this material is so varied and nebulous that it would be all but impossible to state specifically where much of it came from.

In general I am indebted to groups of people: to everyone for whom I have ever staged a play, to everyone who has ever answered a question regarding the multitudinous list of supplies and materials that are used on the stage, to the authors of all of the books I have ever read, to the editors of the pictorial magazines that I have thumbed through, to the builders of the private and public buildings that I have seen, to the curators of the museums I have visited, to the authors of the plays I have staged, and to all of the students who have worked with me and given me an infinite collection of new ideas.

More specifically an author is indebted to those who have given freely of their time and energy to help collect and collate the material of his text. For this I am deeply indebted to:

PATRICIA CHRISTIANSEN, whose invaluable assistance and forbearance in reading, correcting, editing and retyping the material made it possible to complete this text.

ROBERT WEIDEMAN, whose drawings in the furniture section, and on the chapter headings contribute so much to the finished format.

JOHN C. CLEGG, who carefully checked many of the facts of the electrical sections of the book.

ROBERT HYDE WILSON, who read the text carefully and provided numerous helpful suggestions.

C. LOWELL LEES, who has tolerated me as a member of his theatre staff for so many years.

SARA SPENCER, who encouraged me to write this book, and her printers who have been so kind and understanding in preparing and printing the text.

NELLIE HARVEY, who first gave me a healthy interest in drama many years ago when I was a student in her junior high school literature class.

ARNOLD GILLETTE, who contributed the foreword. As a student in his classes I gained much of my technical theatre knowledge.

To all of these people and many others I extend my heartfelt thanks.

VERN ADIX —

# CONTENTS

# INDEX OF PHOTOGRAPHS

# INDEX OF LINE ILLUSTRATIONS

# Theatre Scenecraft

# *Introduction*

*"All the world's a stage,*
*And all the men and women merely players;*
*They have their exits and their entrances;*
*And one man in his time plays many parts,*
*His acts being seven ages."*

These words, spoken by Jacques in Shakespeare's *As You Like It,* have become an immortal description of the vastness of the stage. Shakespeare's concept of the stage may have differed from ours, but his words were never truer than they are today if we think of the vast utilization of dramatic production. The idea that all men and women, yes, and even children, are literally players is obvious when we look about. The stage includes professional and non-professional companies; civic and private theatrical organizations; adult, adolescent and child players.

This stage may be almost anywhere in the world. It may be in the corner of a backyard, in a musty basement, in New York's Radio City Music Hall, in the church around the corner, in a school house, in the center of a large room. The chosen place may become a stage the moment play activity begins, or it may take months to prepare it with elaborate machinery and equipment. The stage may be wide or narrow, deep or shallow, raised or on the ground, with or without proscenium. It may have revolving stages, traps and wagons; it may be a sylvan area consisting of a flat grassy space surrounded by trees; or it may be the end of a room with only the doors and windows normal to the living space of a home. But wherever it is, and whatever it is, the stage serves one major purpose: it is the realm of the actor, the place of the play's action.

The stage may be decorated or plain. It may be covered with acres of scenery or decorated only by human forms in motion. More often than not there is at least a minimum of decorative background for the action of the play. This may be a simple screen merely to suggest the locale of the action or it may be an entire enclosure describing the scene in detail.

Regardless of how simple or how complex the setting, the producer of a play should know how to design, build, paint, erect and light the scenery that is to provide the visual environment for the actors. He should know how to achieve satisfactory scenic effects with a minimum expenditure of materials and effort. To do this he must know the raw materials of the scene crafts. From these raw materials he must be able to form the scenic units which, when manipulated, produce the structure that is assembled and lighted to provide the environment for the actor and the stage picture for the audience.

Large or small, simple or complex, the stage setting is usually made up from the same types of basic units of scenery. And with a supply of standardized units it is possible to create any imaginable type of stage setting.

Designing the setting is the refined process of the stage decorator. Before it is possible to design a stage setting it is necessary to learn about the raw materials of the arts and crafts of the stage. It is necessary to know the construction of the basic units of scenery; to know something about the various painting techniques; to know the effects of colored light on colored pigment; to know basic electric wiring and the fundamentals of stage lighting; to know something about the operation of the physical theatre plant; and to know something about tools, lumber, wood joints, perspective drawing, stage properties and other paraphernalia used on the stage. In this book it seems logical to discuss all of these matters first, since they are basic to the planning and decoration of the stage setting, and then to discuss stage design as the culmination of all of the other assembled information.

A book of this nature is not, and should not be, a rule book. At best it can only be a helpful stimulant. It should point toward ideas and techniques that have been found to be helpful and practical in the past. There is no ONE way to stage a play and there is no ONE color for a stage setting just as there is no ONE way to build a stage tree. There may be procedures that have been tried and found to be workable, and there may be ideas that seem to be the best for a specific play in a specific situation. This does not mean that you or I should limit ourselves to this procedure. Our situation may allow us to try new ideas that will work better, or our budget may restrict us and force us to use simpler, more economical procedures, or our budget may be limitless so that there are no restrictions to imagination or cost and our wildest wishes stagewise can have full rein.

# 1

# *The Stage House*

Today there are so many sizes, shapes and varieties of theatres that it is difficult to describe one and state specifically that it is typical. However, for the sake of convenience in discussing the various parts of the backstage area and its operation, it will be assumed that the theatre in question is flexible enough that it may have any or all of the facilities that are used in normal productions. There actually are buildings having all of the equipment that is mentioned here. However, such theatre plants are in the minority. Most of them have only a meager amount of equipment. Fortunately, a building need not have elaborate equipment before plays can be presented. However, it is worthwhile to know what can be done to improve the existing theatre, and what is available if improvements are to be made.

STAGE—The term "stage" refers to that portion of the theatre used by the actors. In an arena type theatre this is the area not occupied by the members of the audience. Frequently this is interpreted so literally that the actors practically walk on the toes of the audience and sit in their laps. In the conventional theatre there is usually a narrow chasm separating the audience from the actors. This space just in front of the block of seats is the orchestra pit. It may be dropped down lower than the auditorium floor so that the name "pit" is logical, or it may be flush with the auditorium floor. This pit in a sense separates the audience from the performers.

The stage rises up just beyond the orchestra pit. The amount of its elevation above the audience level varies from one theatre to another. It may be one step or four feet. It is inadvisable for it to be higher than the eye level of the members of the audience who sit in the front row.

PROSCENIUM—The enlarged hole cut through the wall to allow the audience to view the stage is the Proscenium Arch, or simply the proscenium. This archway is, in a sense, the frame for the action of the stage. Conven-

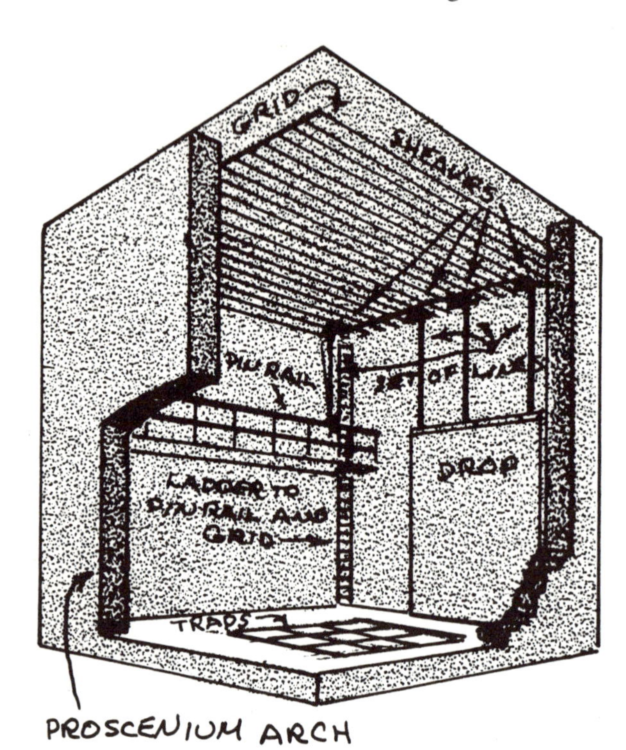


tional staging is, for this reason, frequently referred to as picture frame staging.

The arena type theatre has nothing to correspond to the proscenium unless it is the lighted area of the action into which the audience peers. The audience is separated only by its own lack of illumination. The separation of the actor and audience here is more hypothetical than real.

APRON—If the stage projects out beyond the curtain and proscenium toward the audience, the projection is called the Apron of the stage. Sometimes this area is found convenient to use with the curtains closed in order to cut off the rest of the stage for meetings, for lectures and for scenes that require a minimum of space. In complex stage productions it is sometimes difficult to move scenery rapidly enough to keep the play flowing along and if apron scenes can be contrived to cover while scenery is being shifted behind the act curtain this problem can be partially solved at least.

FIRE CURTAIN—Immediately next to the proscenium arch on the stage side is a fire curtain composed of a fabric woven from asbestos thread. This fireproof curtain raises and lowers in channel iron slots mounted to the sides of the proscenium. It is usually counterweighted and rigged so that the excess heat of a fire causes a low melting point metal link to melt, automatically dropping the asbestos curtain.

## Masking Devices

ACT CURTAIN—The proscenium arch is usually provided with a curtain large enough to fill its opening. This curtain can be opened and closed to reveal or conceal the stage beyond. This is the Act Curtain, or Front Curtain.

GRAND DRAPE—An overhead height adjustment curtain known as the Grand Drape is mounted either immediately in front of or

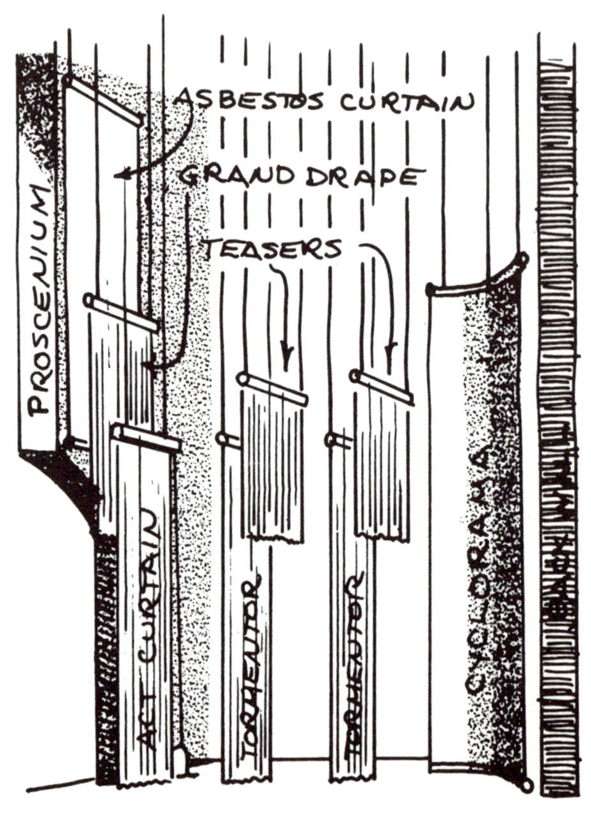

behind the Act Curtain. This curtain makes it possible to increase or decrease the height of the proscenium arch.

TEASERS—Overhead masking drapes upstage of the Grand Drape and also operated to increase or decrease the height of the stage revealed are referred to as Teasers.

TORMENTORS—Curtains installed at the sides of the stage and used to cut down the width of the stage are referred to as Tormentors.

### Stage Floor

ACTING AREA—The space where the action of the play takes place on the stage is logically named the Acting Area. This area usually extends across approximately the width of the proscenium arch and runs upstage approximately two-thirds the depth of the stage. Of course the acting area proportions will not only vary from stage to stage but from production to production on the same stage. When a stage setting is in place, the acting area is within its confines. Usually the acting area is the stage space that is visible to most of the members of the audience.

FLOORING—The stage, at least that portion of it known as the acting area, should be floored with relatively soft wood so that scenery can be anchored in place easily. Fir is frequently used. When this section becomes worn it can be refloored without having to cover the entire stage area.

TRAPS—The stage floor may be equipped with removable sections, or trap doors. There are times, in *Faust* for instance, when it is convenient for an actor to rise through the floor as if coming from the steam heated world. At other times people may descend into lower levels—into the sewers of Paris in Act II of *The Madwoman of Chaillot;* into the store below in Act I of *The Merchant of Yonkers;* or into the subway in *My Sister Eileen.* For such scenes it is convenient to have "traps" in the floor.

REVOLVING STAGES—Some of the more elaborately equipped theatres have revolving discs cut into the floor of the acting area. It is possible to set up a series of scenes on the disc and change from one to another by a simple turn of the disc.

TRACKS FOR HEAVY WAGONS—The stage of the University of Iowa theatre has tracks recessed in the floor that extend across the stage parallel to the curtain. Large wagons, the full width of the stage and half its depth, can be rolled onto the stage complete with setting, properties and furnishings. A revolving disc makes it possible to switch the wagon upstage or downstage onto another set of tracks where it can be rolled into a storage dock. This is an unusual innovation that may be used to quickly shift the settings of complex productions, otherwise difficult to move.

ELEVATOR STAGES—On another type of mechanized stage, portions of the floor may be raised or lowered by hydraulic lifts similar to those used for grease racks in service stations. Spectacular vertical movement, as well as simple floor level arrangements, can be achieved with equipment of this variety.

### Stage Directions

STAGE RIGHT—STAGE LEFT—To clarify directions on the stage, stage right and stage

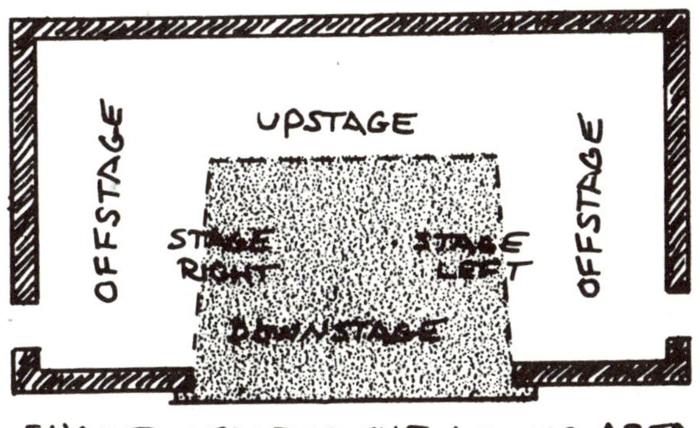


SHADED SPACE IS THE ACTING AREA

left have been set up as right and left of the actor as he stands facing the audience.

UPSTAGE—DOWNSTAGE—In like manner to the rear of the stage is referred to as upstage and to the front of the stage as downstage.

Anywhere on a line running through the center of the stage from front to rear is referred to as center stage. By using combinations of these directions it is easy to find one's way about the acting area of the stage.

OFFSTAGE—When one moves out of the acting area to the sides of the stage he is moving offstage.

WINGS—The offstage portions of the stage are the wings.

### Flying Systems

FLYWELL—The acting area of the stage is usually provided with an extremely high roof. The space between the floor and the strut work of wood, steel or wood and steel suspended about 5 or 6 feet below the roof is called the flywell.

GRID—The strut work referred to is the grid. The grid supports pulleys through which ropes or cables are threaded to support scenery, lights, drops, draperies and other paraphernalia of the stage that are suspended in the flywell.

PIN RAIL—If a rope flying system is used to raise, lower and suspend the scenery above the stage, then the ropes are tied off on a railing which is anchored near the wall on stage right or left, either on the floor level or part way up into the flywell. It should never be more than half way to the grid.

LOCKING RAIL—If a counterweight system is used in place of a rope flying system, a locking rail is anchored to the stage left or stage right wall on or above the floor level. And just under the grid is a loading platform where counter-balancing weights are added or removed from the carriages to balance the weight of the flown units of scenery.

SETS OF LINES—Flying systems, whether rope or counterweight, ordinarily have 3 or more lines evenly spaced across the stage in each set. The ropes are located so that battens as long, or slightly longer than, the width of the proscenium may be used to fasten backdrops, draperies and scenery preparatory to suspending them in the flywell. In the 3 line set the one nearest the operator on the rail is called the short line; the center one is logically the center line, and the one farthest away is the long line. If there are 5 lines in the set, these names hold but two more are added: the line between the short and center lines is the short center line and, of course, the one between the center and the long lines is the long center line.

ROPE SYSTEM—In a rope flying system Manila or hemp ropes may be tied directly to the objects being flown. On a unit of scenery they may be tied to ceiling plates or hanger

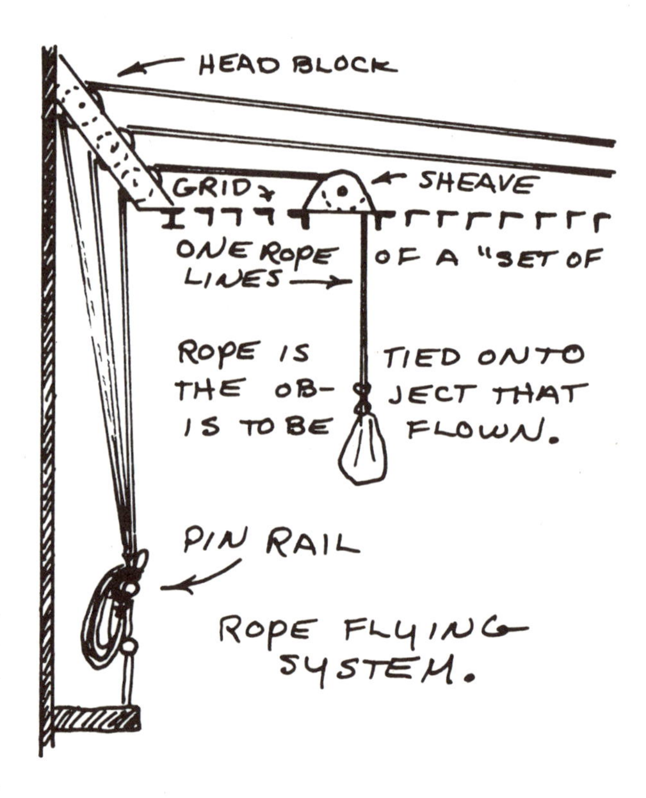


irons that have been mounted on the framework with screws and bolts. Or the ropes may be tied onto a long wooden or pipe batten and the scenery then tied onto this.

The ropes in the flying system extend up to the grid where they pass over loft blocks. The ropes of an entire set of lines cross the stage to a position above the pin rail and there pass through a multiple pulley, consisting of several sheaves (pulleys) mounted side by side called a head block, and from there the ropes drop down to the pin rail where they are tied off on pegs, or belaying pins, that are mounted in a pipe or rail that runs from the front to the back of the stage against or near one of the side walls of the flywell. To raise or lower scenery, one, two or more men grasp the set of 3 or 5 lines and "take up" or "let in" on the lines. Many times it helps if these men are on the "beefy" side so that their weight can be thrown onto the lines to balance the weight on the other end.

SANDBAGS—To make the job a little easier bags filled or partially filled with sand are fastened to the lines by means of a rope lock. Usually the weight on the stage end of the lines is lifted into the air so that the sandbag may be mounted at the pin rail level. Ordinarily it is advisable to put slightly less than enough weight to balance the load in the sandbag so that the stage load will tend to sink down by itself when the ropes are released. One man will then be able to raise the scenic unit without any trouble.

BLOCK AND TACKLE—If the load on a set of lines is extremely heavy it is possible to attach a block and tackle to the lines. The block and tackle is a means of achieving a mechanical advantage, at the sacrifice of speed, in lifting objects. The number of pulleys determines the amount of advantage gained. (See appendix for details.)

SELF LEVELLING SANDBAG—If the weight on a set of lines is evenly distributed across the stage a simple self levelling device can be made. If there are 3 lines in a set, the center one, at the pin rail end, is securely tied to a sandbag. A pulley or block is tied on at the connection of the line and sandbag. One of the other lines is led through the pulley and the other tied to it. The lines are then adjusted so that the object being flown is in "trim" (level) when the ropes are tied. Thereafter it is possible to level the load by simply slipping the rope that goes around the pulley one way or the other. Since the long and short lines are tied together, making one continuous line around the pulley, pulling one side down pushes the other side up and vice versa. This

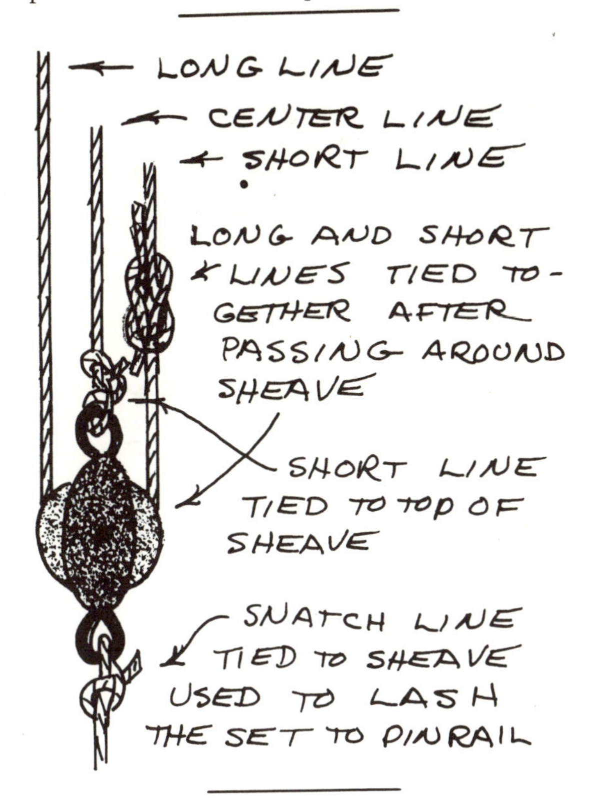


is an excellent device to use since rope tends to slip through rope locks sometimes and the entire set gets slightly out of "trim."

COUNTERWEIGHT SYSTEMS—The counterweight system uses cables in place of ropes and the cables are permanently fastened on the stage end to pipe battens. On the wall end the cables are fastened to weight carriers called cradles. The cradle is governed in its up and down movement by wire or angle iron guides, secured to or near one of the side walls of the flywell. The weights for the system are stored on a catwalk or platform just under the grid. Enough weights are loaded onto the cradle from there to balance the weight of the scenery. To raise and lower a set of lines together with the scenery and the counterweight, a line is fastened to the top of the cradle, run up to the grid over a pulley, extended back down to the floor under a spring tension pulley and extended back up to the bottom of the cradle where it is attached. This is rightly named the endless

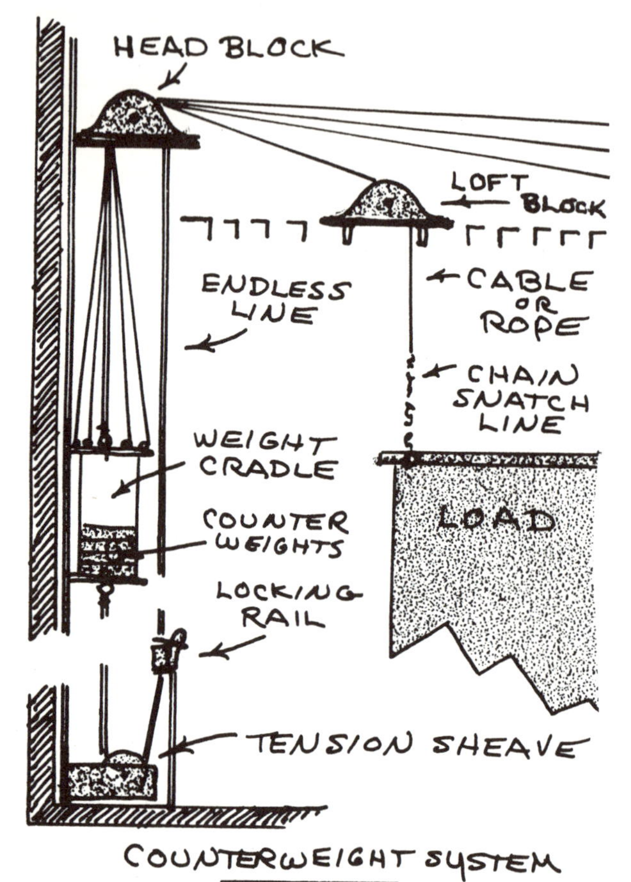


COUNTERWEIGHT SYSTEM

line. The load is locked in place with a simple concentric lever mounted on a rail. Pushing the lever in one direction wedges it against the rope, and locks it in place. Pulling it back in the other direction releases the tension so the rope can move freely.

COUNTERWEIGHT VS ROPE—The counterweight system, though more expensive, has many advantages over the rope flying system and should by all means be installed in new theatres or in theatres that are being remodelled. The counterweight system is built to carry heavier weights; the lines are permanently trimmed so that counterbalanced scenery units are not sagging at one end, off the floor in the center and trim on the other end; one operator can handle heavy loads when the carriage is correctly loaded with weights; and last, but far from least, there are no great coils of rope ends under the operator's feet.

DOUBLE PURCHASE COUNTERWEIGHTS—The regular counterweight sets mentioned above require the full height of the grid in order to operate correctly. The weights must move just as far up as the battens do down in coming to the stage floor. A special type of counterweight called the double purchase is designed so that the weights need travel only half the distance of the objects they are used to lift. The locking rail can be mounted up on the wall slightly less than half way to the grid and objects can still be lifted from the

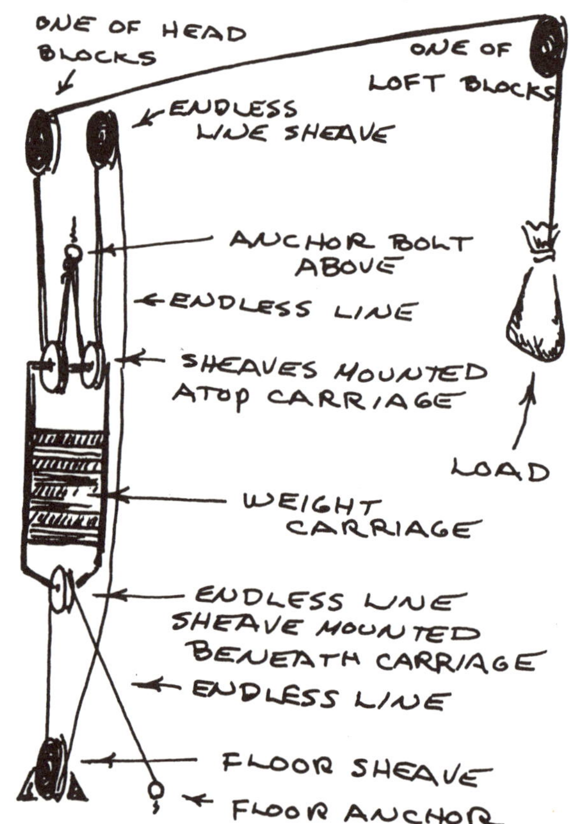


ONE LINE OF SYSTEM TO SHOW DOUBLE PURCHASE COUNTERWEIGHT

floor to the grid. This system has advantages and disadvantages over the conventional system that will make it useful as a part of the equipment on some stages. The individual units are more complex, hence more expensive, and it takes twice as much weight to counterbalance an object so the number of available weights must be doubled. However, since the pin rail is mounted halfway to the grid the wall beneath may be cleared of mechanical contrivances so that more wall stacking space is available for scenery and properties.

WINCHES—Another device for lifting scenery is the winch, which consists of a drum upon which the rope or cable is wound, a crank and a gear system. Turning the crank meshes the gears and winds the rope onto the drum or unwinds it off the drum. When properly geared tremendous weights can be lifted by one man with a winch. However, the drum is usually small in diameter and the gear system makes it necessary to turn the handle many times in order to lift the weight. Speed is sacrificed for mechanical advantage.

MOTORIZED COUNTERWEIGHTS AND WINCHES—Electric motors may be installed to operate both counterweight and winch systems. They save a great deal of human energy, but at the same time are so expensive that very few theatres can afford them.

### Scene Shifting Devices

Some of the permanent equipment of the stage has been mentioned. There are other scene shifting devices frequently used on the stage, and often considered to be a part of the stage house equipment, which may be stored away from the stage and brought in when needed. These devices fall into four categories: rolling, rotating, sliding and flying mechanisms.

### Rolling Mechanisms

The rolling type devices consist of such equipment as wagons, outriggers and tip jacks.

WAGONS—Stage wagons are platforms mounted on casters so that they may be easily rolled around on the stage. The frame work may be of wood, angle iron, or aluminum. Wood frames are the least expensive and easiest to build, though not the most durable. A welded angle iron frame, though heavier and more expensive, will outwear several plywood tops. Aluminum is ideal but expensive.

WAGON WOODEN FRAMEWORK—The size and weight load to be carried by the wagon will determine its structure. For normal use 4′-0″ x 8′-0″ platforms with 1″ x 6″ outside frame, two cross stringers, and ¾″ sheathing plywood top fastened together with 1½″ flathead wood screws or 6d or 8d cement coat box nails will make a good standard unit. The 4′-0″ x 8′-0″ size utilizes a full sheet of plywood. Several wagons may be fastened together with hinges or straps to form a scene carrying unit.

CASTERS—At least 6 casters, 2 on each side, and 2 halfway between, should be mounted on each wagon. Casters may be mounted on removable plywood blocks. 3″ or 4″ rigid or swivel ball bearing casters that are kept well-greased may be used. Larger casters roll easier under a load.

The casters should be mounted to lift the wagon off the floor ¾″, then danger of small obstacles wedging beneath the edge of the wagon will be minimized. Allow 360° turn space for swivel casters.

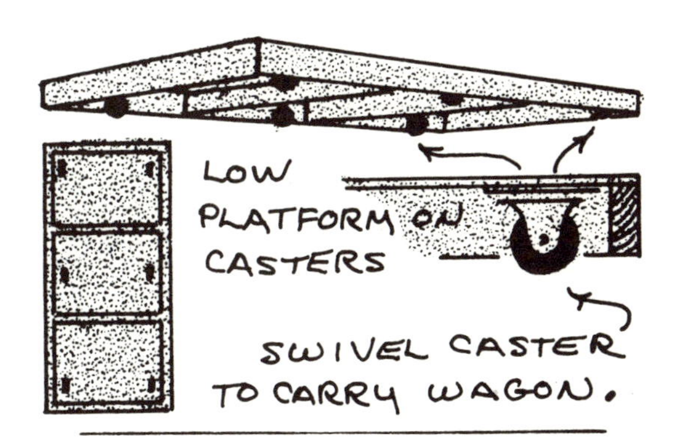


Tracks of wood strips, or angle iron may be mounted on the stage floor to facilitate moving wagon units in direct lines on and off stage. And for straight line movement use rigid rather than swivel casters.

MOVING WAGONS—It is difficult to overcome friction and start a wagon rolling. Stagehands usually push the scenery mounted atop the wagon rather than pushing on the edge of the wagon. This may loosen, or break, the scenery, so it is advisable to rig a device that can be grasped by the hand to pull or twist the wagon unit. Ceiling plates, large screw eyes, or eye bolts may be mounted on the tops or sides of the wagon. Then loops of pull rope may be tied to the wagon permanently. Or removable ropes with hooks tied to them, or iron pull rods, may be hooked on to pull the wagon.

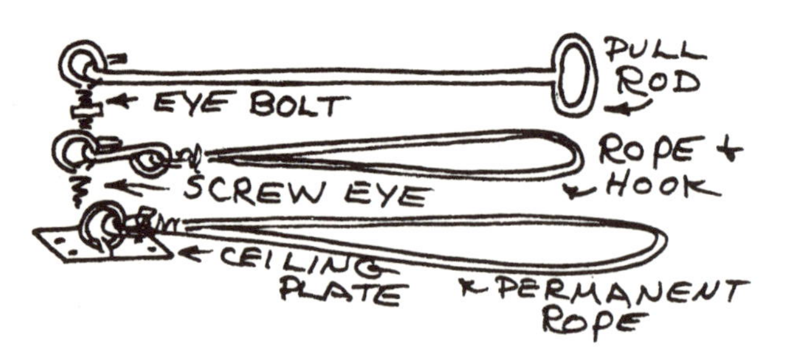


The stage wagon is a convenient scene shifting device if there is adequate offstage space in which to maneuver and store all of the units needed for a given production. Either partial or entire settings may be mounted atop one or more wagons and shifted complete with properties and stage dressing. Units such as thrones, fireplaces, judges' stands, podiums, speakers' stands, stairways, platforms, etc. may be handled easily if mounted on wagons. Hinges, iron straps, wooden cleats or plumbers' strap may be used to fasten wagons together.

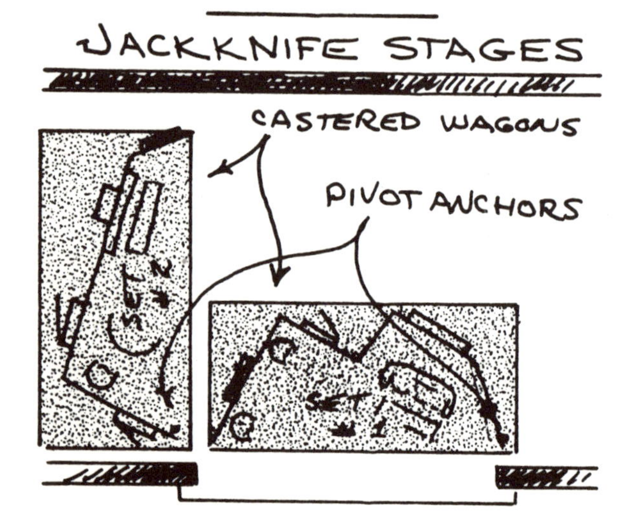


JACKKNIFE STAGES—Large wagon units are sometimes rigged so that they operate like blades on each end of a jackknife. The downstage off right corner of one wagon is fastened to a pivot and the downstage off left corner of the other wagon is also fastened to a pivot. Depth of the stage and offstage space must be checked to be certain that there is adequate storage for one wagon while the other is in use.

OUTRIGGERS—The outrigger, or outrigger wagon, is another caster device that can be used to help carry a good deal of the weight of heavy scenic units when they are shifted. Frequently heavy units need to be moved, but it is inconvenient to design a step or level into the setting so that the scenery can be mounted on a wagon. In this situation a wagon or an outrigger (which is a castered frame similar to a wagon frame but designed to fit into specific positions behind the setting) can be mounted to the back of the scenery. Before being mounted the wall section is lifted off the floor slightly so that the outrigger will carry it. This device makes it possible to roll heavy units on and off stage.

TIP JACKS—There are times when an outrigger is not practical. If the weight of the scenic unit tends to pull backwards slightly or straight down, the outrigger will work very well. However, if the weight tends to pull the scenery forward, a tip jack should be used. A tip jack is a right triangular shaped frame with one arm elongated almost to the height of the scenery and with a relatively short arm on the other right angle side of the triangle. The diagonal board connecting these two arms should be about the height of the flat when the frame is completed. Casters are mounted under the short arm and the diagonal is secured to the flat with nails, or loose pin hinges or is fastened in some other way.

Ordinarily, tip jacks are used in pairs. One is mounted near each end of the wall section and the two are then securely cross braced to one another and to the wall section. If they

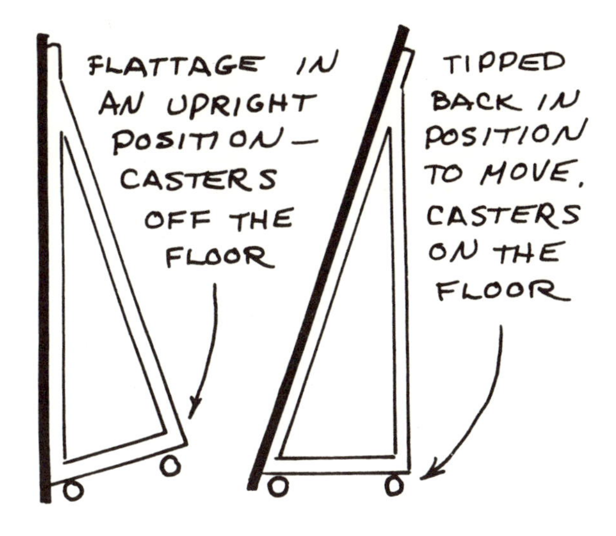


are mounted while the scenery is in place in the setting the front caster of the right triangle corner barely touches the floor. The other, because of the shape of the triangle, is then off the floor by several inches. When it is time to move the scenery the wall section is unfastened at each end and leaned back until both casters are on the floor. The construction of the jack causes the scenery to be lifted off the floor and the entire unit can be easily rolled into the wings. By means of tip jacks one man can move a unit that would require three or four if the unit were being dragged or carried.

Tip jack units can be turned around and mounted with the right angle side against the flat so that both casters are on the floor when the scenery is in position. The units then work in much the same manner as outrigger wagons.

### Rotating Devices

The revolving discs, previously mentioned, are installed in the floor as part of the regular equipment of very few theatres. It is possible to build units to fit on top of the stage floor that will revolve and serve basically the same purpose as the built-in device.

PERIAKTOI—Simple rotating scenic units can be built even without platforms under them that operate in somewhat the same manner as the disc. Three wide flats, or three sections each made up from two or more flats hinged together and stiffened, may be fastened together at the corners to make an oversized prism, similar to the periaktoi thought to have been used in the classical Greek theatre many centuries ago. Casters or large furniture gliders may be mounted under these units so that they may be rotated easily. The unit can then be turned to present any of its three faces to the audience. And of course it isn't necessary to be limited to three sides. Four or even more flats could be fastened together in this way.

ROTATING WAGON UNITS—Wagon units can be used singly or fastened together to form units that may be covered on all sides and then rotated to present their various sides to the audience.

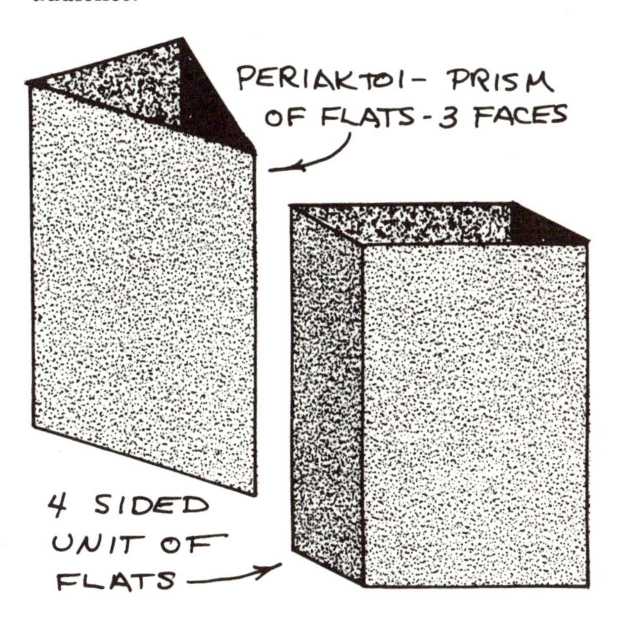


TEMPORARY REVOLVING STAGE—Stage wagons may be fastened together to form a square form and then curved sections added on each of the sides to make a disc. Three dimensional scenery units mounted on top may then be moved into the desired position by rotating the mechanism. Of course there is no reason why this device needs to be round except that the curved edge of a disc can be placed near the curtain and then the device rotated without the corners pushing into the curtain.

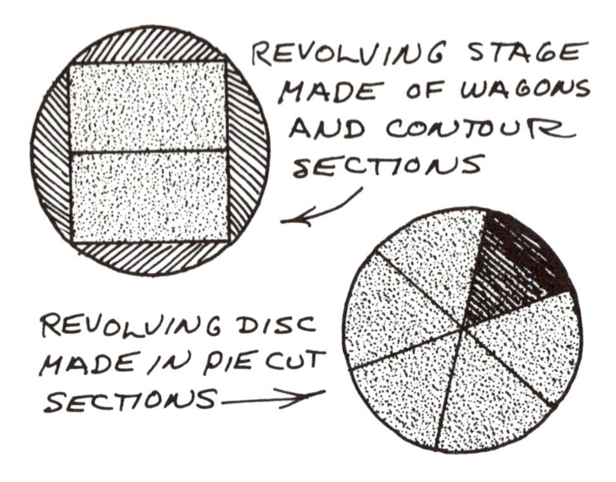


Discs can be used singly or in combination with other discs or other scenic units. A disc may be used on each side of the stage; a large disc can be placed in the center and a smaller one on each side; two discs can be used side by side; a disc can be mounted on a wagon, the wagon moved up and down and across stage and the disc revolved while it is stationary or moving. These are all variations that have been tried at various times.

Revolving discs may be made in wagon sections as indicated or they may be made in halves, quarters or pie sections. Usually they operate better if few sections and a minimum of casters are used. Non-swivel casters, mounted so they are in line with their path of movement, make it easier to move in either direction than when swivel casters are used because the swiveling of the casters requires a slight side movement of the wagon. Swivel casters are all right once they are set in motion in one direction, but it is difficult to change the direction.

A center pivot may or may not be used. If the unit is completely supported by the casters the pivot merely holds the disc in place so that it cannot shift its position on the stage. If a support is needed in the center the mounting of a swivel caster, with the wheel removed, can be used as a rotor. A specially designed free wheeling, ball bearing device, might work out even better. The wheel mounting and hub section of an automobile wheel work beautifully. Of course it is possible to manufacture a special pivot that will operate even better. This might be worth while if the disc is to be a permanent device that will be used frequently.

### Sliding Scenery

The simplest method of moving stage scenery is to drag it across the stage floor. This is usually referred to as "running" scenery. If the floor is fairly smooth some of the units may have large furniture gliders mounted under them so that they will slide more easily. However, gliders lift the scenery off the floor slightly and there might be light leak problems when they are in position for use. It shouldn't be difficult to remedy this, however, and certainly the advantage of easier sliding is well worth the small problem that arises because of it.

Tracks for Shutter Scenery—There are times when the designer might find it advisable to go back to his theatrical forebears and employ sliding scenery that operates in shutter fashion. Strips may be fastened to the floor and slider tracks mounted above so that

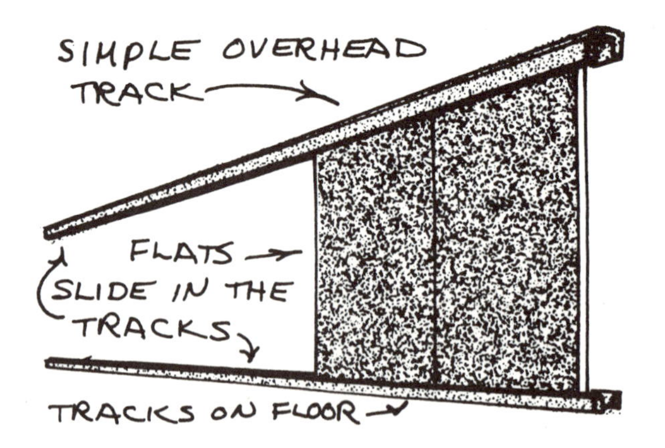


scenery can be slid on and off stage. If more elaborate rigging is necessary it is possible to devise a frame to provide tracks both above and below that are completely self sufficient. A truss may be built above and strips mounted to plywood below. Shutters can be pushed in from one or both sides to provide a solid straight wall across all or a portion of the stage. The track above should have just enough clearance so that the flats will move easily and readily. A smooth bottom track with a little grease, wax or tallow rubbed on it will allow the flats to slide smoothly. The tracks needn't be bothersome to the actors since they need not be high nor need the slots be very wide. The sides of the tracks can be made of strips of 1″ board and the slide space need be only about an inch wide.

When this sort of system is used it is necessary to have offstage space in which to push the shutters when they are open. If the flats are built new and of fairly light construction it is possible to hinge flats together and make wide sections that may be easily operated.

## Flying Scenery

If there is a flywell available it is probable that a portion of the scenery in the theatre is moved by means of the flying system. Flats, wall sections, ceilings, screens, cycloramas, actors, properties, backdrops and draperies can be moved up into and down from the flywell. With elaborate flying equipment it is possible to fly entire stage settings completely equipped with a ceiling on top. However, most theatres have fewer sets of lines and smaller available clear sections in the flywell than would be needed for this so the stagehands must be content to fly only portions of settings and other smaller units of scenery.

Units that are to be flown must have a support to which rope (or wires, chains or cables) may be attached. Ceiling plates or hanger irons are commonly used for this purpose. There are straps of iron with rings securely mounted to them. The straps are drilled for screws or bolts and are fastened to the flat, ceiling, wall section or drop that is to be flown. Ropes are then tied onto the rings. Sometimes it is advisable to have a ring at the top of the unit through which the rope passes and a fastener that hooks under the bottom of the unit to which the rope is fastened. Some hanger irons are designed with a foot that hooks under

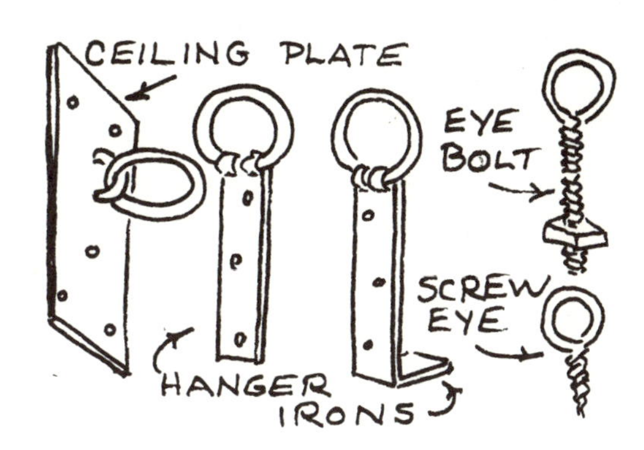


the batten. The rope then helps hold the unit upright by means of the upper support that it passes through and the strain is transferred to the bottom batten. This is a safer rigging method on heavy units of scenery. If necessary the rope can actually be tied at both places in order to release the strain on one part even more.

## Stage Draperies

The flying system is probably used more for suspending, lifting and lowering stage draperies than for any other purpose.

There are few stages without at least one enclosure type set of draperies and many of them have two or more. The term "enclosure draperies" refers to a set of draperies that surround the acting area to mask off the remainder of the stagehouse from the audience view. They may be used as scenery by themselves or in conjunction with other stage units.

Besides enclosure draperies the conventional stage has drapery units for its front curtain, grand drape, teasers, borders and travellers. These may be made of any one of a number of different fabric weaves and textures, but basically they are put together in the same way.

Drapery material is available in varying widths, depending partly upon the weave and partly upon the size of the panels desired. Ordinarily tall panels are sewn together with the seams running up and down while short ones, such as borders, may be sewn either up and down or lengthwise, depending a good deal upon the fabric, its width and probably its pattern if it has one.

FULLNESS—Draperies are much more attractive when they hang in heavy folds than when they are stretched almost taut. When the drapery is ordered (if it is purchased from a theatre supply outlet) or when the material is purchased (if it is to be made up locally) this must be taken into consideration. Add 100% or more fullness for attractive folds. Fullness of 100% means that the material is two times as wide as the space it is to cover. When the draperies are made the fullness may be gathered and sewn in permanently or the drapery may be sewn flat and the gathering done when the drapery is hung. Act curtains are almost always operated on a track and used specifically for one purpose so the fullness may be sewn in. Other sections may be used variously and it is usually a good idea to leave them flat so that their coverage, and consequently their percentage of fullness, may be varied. If a wide space is to be covered the fullness can be decreased and if the space is narrow more fullness can be allowed. This makes the draperies more flexible and practical for use on the ordinary stage.

CHAIN POCKET—Ordinarily the panels of material are sewed together with a simple flat seam. The bottoms of the sections have a 2″ or wider hem into which a chain is often slipped and stretched across the drapery. Each end of the chain should be fastened so that it cannot slip in the pocket. The chain provides enough weight to stretch the material down slightly when it is hung on a batten. This pulls out many of the wrinkles.

WEBBING—The top of the drapery is almost always reinforced with a strip of canvas or burlap webbing 2″ or more in width. It is securely sewed near both of its edges. If the material of the drapery is gathered into permanent folds it is gathered and stitched, then the webbing is sewn across to hold the folds securely in place. If the draperies are relatively light in weight and fairly short a substitute for webbing may be made from roll type window shades. This material is fairly tough and can be cut into strips that may be sewn onto the top of the drapery.

GROMMETS—The top of the drapery may have grommets inserted at intervals of from 1′ to 2′. Grommets may be described as hole reinforcers. They are metallic and consist of two parts: a washer and a washer with a short inner sleeve. The sleeve is slipped through a hole cut in the drapery material and webbing (in the center of the webbing). The washer slips over the sleeve and a special crimping device, called a grommet die, rolls the sleeve down tightly against the washer. When tie lines are put through these holes and the drapery tied to a batten the strain is transferred over a larger area of the material than it would be if the material were merely slit and the lines passed through.

TIE LINES—Although there are some special fasteners for draperies the usual practice is to loop pieces of shade cord or cotton tape through the grommets and tie the drapery onto the batten with these. The lines should be about

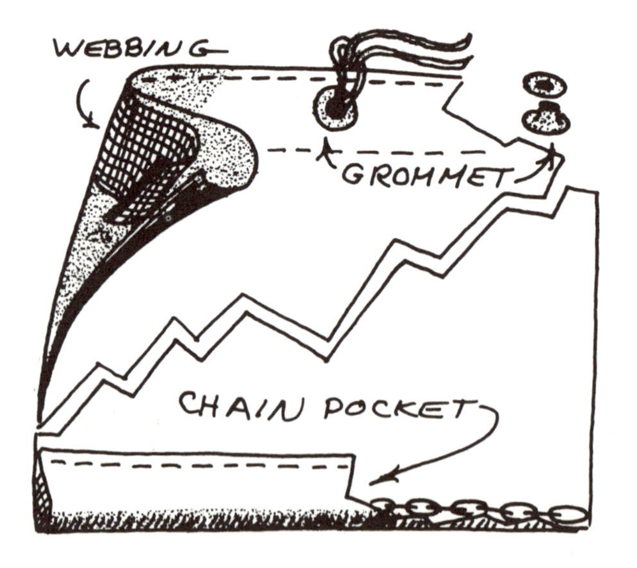

18″ long, and when they are tied a bow knot should always be used so they may be loosened easily and rapidly.

### Drapery Rigging

Various types of hanging and operating devices are used for stage draperies.

TAB, TABLEAU OR GUILLOTINE CURTAIN —Draperies may be tied onto pipe or wooden battens suspended from the flying system. They may then be raised or lowered from the pin rail as needed. The front curtain, when handled in this way, is called a tab or guillotine curtain. Actually most of the draperies on the stage are suspended in this way, but some of them have auxiliary devices so they may be operated differently on occasion.

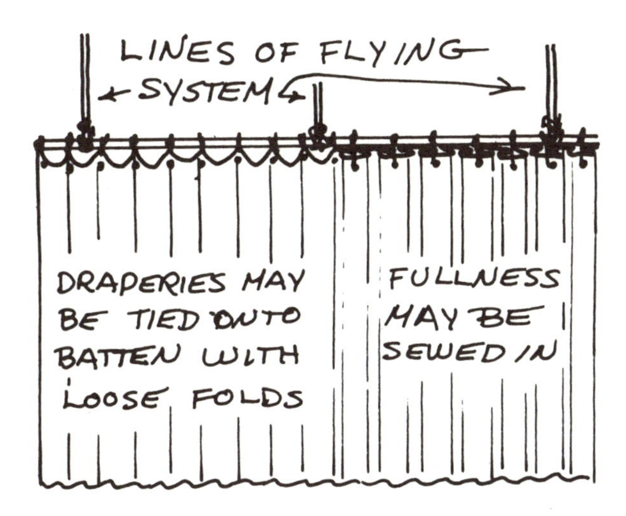


TRAVELLER—A traveller curtain is rigged in two sections that overlap in the center. In operation the sections part in the center and move off to the sides of the stage. Variations from this procedure may be devised so that sections of the drapery move in one direction or so that they move together into the center of the stage.

TRACKS—The drapery may be fastened at the top to rings that slide on wire or to sliders, balls, blocks or small wheel carriages that move in a track made of wood, steel or aluminum.

DRAPERY CENTER OVERLAP — Basically, traveller sections are operated in somewhat the same way. If wire is used there are two wires stretched as taut as possible across the space. Eye bolts may be used to fasten the ends and a turn buckle placed at one end may be turned to tighten the wires. Two wires are needed so the draperies may overlap one another at the center of the stage; otherwise, they will gap open. If a track is used it is built in two sections that overlap in the center. The overlap will vary but should be at least 18″.

DRAPE SIDE OVERLAP—The drapery sections should be made with sufficient extra width for the center overlap and for an overlap of at least 18″ behind the proscenium arch on the outside end of the drapery. This may have to be even more, so that members of the audience cannot see unmasked space between the proscenium and the drapery. The wires or tracks should be long enough to accommodate this extra length.

HANGING AND OPERATING A TRAVELLER —When the drapery sections are tied or hooked

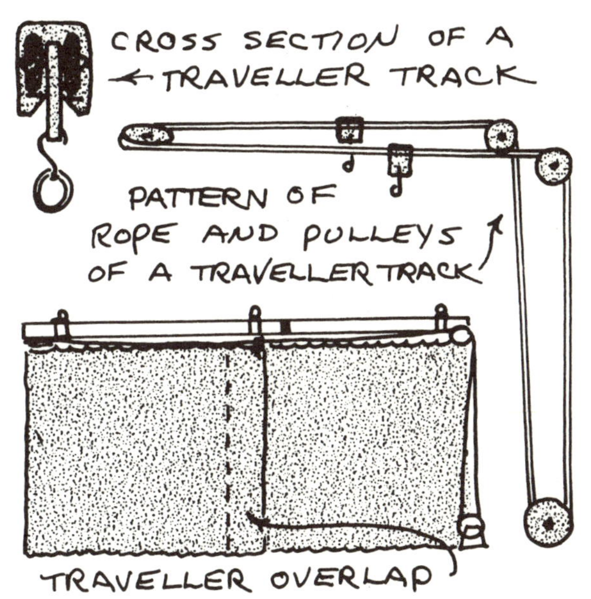


onto the rings on the wire or the carriers of the track, it is time to rig them for operation. The offstage ends are anchored to the wall or to the end of the track so they cannot slide onto the stage. The onstage ends are then tied to the operating ropes.

OPERATING LINE—On the floor, where the operator who opens and closes the curtains stands, a pulley is fastened. A rope is threaded through the pulley and the two ends are taken up to the end of the track above where each is led through a pulley. From there one end of the rope goes to the overlapping center end of each of the draperies and is securely tied after the rope has been cut. This is done so that when tied the draperies remain in their closed positions and the rope is snug around the pulleys. Another rope is then fastened to the tie position of one end of the rope just installed. The new rope extends from there to the far end of the track where it passes around another pulley and comes back to the other knot position of the first rope, where it is pulled snug and tied.

After this has been done it should be possible for the operator to pull one of the lines to open the draperies and the other to close them.

BUTTERFLY—Sometimes for variety it is desirable to drape the curtains open rather than travel or fly them. For this operation a line is

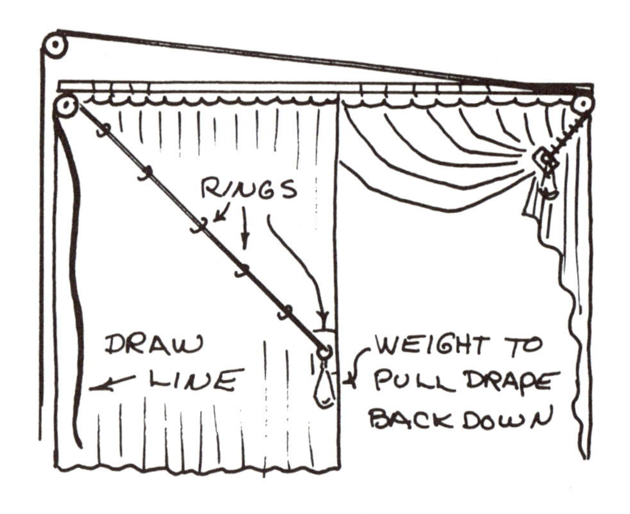


fastened to a ring securely fastened to a heavy reinforcing patch sewed down low near the onstage edge of one of the curtains. Other rings, approximately the same size, are fastened to patches sewed on in a direct line diagonally up to the outer corner of the drapery section. The rings should be spaced so they come at least on every seam and probably even closer. Others can be added after the original rigging is complete if needed to improve the appearance. The rope, after being tied to the lower ring, is passed up through the other rings, over a pulley mounted in line with the row of rings and then down to the floor. When the rope is pulled down it lifts the drapery diagonally and as each ring is reached by the one below the material is pulled into gathered folds.

For the butterfly to operate correctly, the distance from the lowest ring to the top of the drapery should be greater than the distance from the same ring diagonally to the pulley the rope passes over, unless it isn't necessary to pull the corner up that high. Chain in the bottom of the hem of the drapery (chain pocket) and a small sand bag or weight attached to the onstage ring will help in pulling the drapery back into its normal folds when the lifting rope is released.

A rope may be attached to the other drapery section in the same way. For convenience of operation, this rope is passed over a pulley, taken across the stage to another pulley so that the two sections can be operated together or separately in order that all or half of the stage may be opened.

CONTOUR—The contour, or scalloped opening curtain, is made in one large panel. Its rigging is similar to that of the butterfly. The ring rope is fastened near the bottom of the drapery. A series of rings is attached in a straight vertical line. The manipulating rope runs straight up through the rings and then over to one side of the stage. The rope runs through pulleys wherever it changes directions. Another set of rings and ropes attached and rigged in a similar manner is needed for the position of indentation of each scallop, including one for each end. The individual ropes may be led to the same place of operation where they may be operated together or individually. By operating them individually it is possible to secure a number of patterns in the opening, depending upon the height of the scallops. Here, as with the butterfly, it may be necessary to have chain in the hem and small weights on the bottom ring to pull the curtain all of the way down.

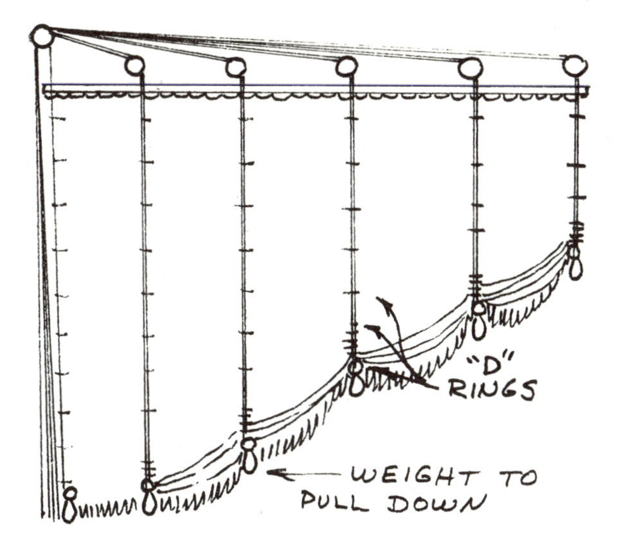


These various types of operation may be used for front curtain or for any other drapery that extends across the stage. They may also be used in false prosceniums and adapted for use with window draperies.

GRAND DRAPE — Frequently the grand drape is not tall enough. It should be possible to lower it and cut the stage down to the minimum height that may be desired for any production and still have it mask the top of the proscenium arch. It is also desirable to lift it up to clear almost the entire height of the proscenium. The grand drape is not only a device for altering the height of the proscenium, but also a masking for the traveller track, light bridge and any other scenic devices that need to be flown in downstage positions.

TEASERS—As in the case of the grand drape, teasers are frequently not long enough. Actually if they are tall fewer are needed to mask since it is necessary to mask the top of the one behind with the bottom of the one in front of it. Even on relatively small stages 6′ or more in height may be desirable. Teasers should have enough fullness so that they may be stretched to carry offstage a few feet.

TORMENTORS—Tormentors, or side maskings, can be fastened onto short pipes and suspended directly from the grid or they can be fastened to pipe battens and suspended from the flying system. They may be mounted on special 180° swivel arms so that they may be turned to extend across the stage or up and down stage. At times double faced draperies are used this way so that they can be reversed with the swivel. The swivel arms can also be mounted on a traveller track so that the legs (tormentors) can be moved on or offstage as well as turned.

Still another method of mounting tormentors is to mount them on traveller tracks so they can be pulled onstage a few feet or pulled offstage into a snug pack.

All of the methods of handling tormentors have advantages and disadvantages. The individual situation and budget will determine which method is desirable.

### Drapery Installations With No Grid

There are many stages being built with no grid space; consequently scenic units cannot be flown on these stages. In this type of situation it is possible to either fasten battens to the ceiling and tie the draperies onto them or to install tracks made of steel or extruded aluminum to travel the drapery sections. Some of these are equipped with elaborate switching devices so that the draperies may be switched from one track to another. This type of arrangement is sometimes designed with additional switches and track making it possible to pull all of the sections into a storage cabinet in one corner of the stage.

### Enclosure Draperies

Most stages are provided with enclosure draperies for general purpose use. They may be used for meetings, concerts, lectures, assemblies and plays. For these draperies to be flexible enough so they can be practical for all purposes, several factors should be borne in mind.

SIZE OF PANELS—The size of the drapery panels is important. In some installations the stage is surrounded by three large panels of material: one on each side and one across the back. With this design there are four entrances: down right and down left between the act curtain and the front edge of the side panels and up right and up left where the side panels and the rear panel meet. Although this may be excellent for some types of programs, it restricts the director too much when used for plays.

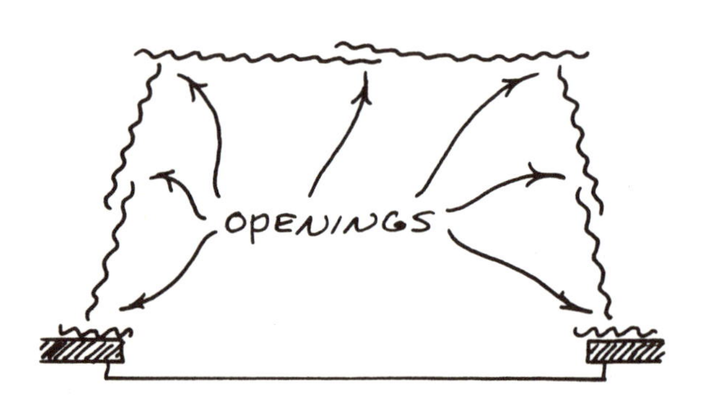


If possible, the sides and back should be broken into two or more sections, in order to add three or more entrances besides the four already mentioned. Each of the panels should overlap the one next to it to prevent gaps, so it is necessary to allow more material in width than would be required for single wide panels.

EXTRA NARROW SECTIONS—The addition of some narrow panels to fit above doors and windows, when they are inserted in openings provided by the parting and the retying of the regular drapery sections, will make simple staging much easier and more effective. These narrow sections can be full length and rolled or fastened up to allow the insertion of the door or window or they can be shorter panels that extend down only to the top of a normal door.

RIGGING OF DRAPERIES—There are various ways to hang and manipulate an enclosure type set of draperies. The rear panel can be hung on a traveller track so that it may be opened all or part way, or closed. The side panels can be on swivel arms or hung in from the grid. Or the entire set of draperies can be tied onto a large "U" shaped frame that is mounted permanently or flown.

DRAPERY HEIGHT—It is advisable to have the draperies as tall as possible to cut down the number of teasers needed to a minimum. Where there is a grid it might be advisable to have them no more than half the height of the grid so that they can be flown up out of sight lines. But height adds to the physical appearance of the stage. Low hanging teasers seem to give a feeling of claustrophobia whereas tall ones give a feeling of freedom and airiness.

Of course, here again, the budget will probably enter into the picture and many corners will have to be cut before the specifications can be set. Basically the idea of numerous openings and a few narrow panels to fit above openings is sound and should not be sacrificed.

DRAPERY FABRICS—Draperies may be made from materials that are both attractive and durable. Conventionally the materials used are in plain colors.

Front curtains are ordinarily made of velour or plush, both of which are pile fabrics. These materials have a beautiful surface and are fairly opaque, both desirable qualities for the front curtain.

The color of the front curtain is usually consistent with the decorating scheme in the auditorium.

Occasionally textured fabrics, nubby weaves, metallic weaves, textured patterns and colored patterns are chosen.

LINING THE ACT CURTAIN—Ordinarily it is advisable to sew a lining or backing to the rear of the act curtain to improve the opacity and to increase the wearing qualities of the fabric.

All of the other draperies may be made of the same rich quality materials as the front curtain, but usually less costly fabrics are used. One of the most durable of the other materials is cotton Repp. Repp may be obtained in a wide range of colors and with either a plain or patterned weave.

Almost any material could be used. Actually, there is no limit to the possibilities. However, the material should be chosen for durability as much as for any other quality since stage draperies do take somewhat of a beating. It may be well to bear in mind that the curtains are more attractive if the material is soft enough to hang in easy folds.

* * * * *

In the pages to follow there will be descriptions of some of the units of stage scenery, properties and lights that are used on the stage in conjunction with the stage machinery and draperies mentioned here.

# 2 Building Supplies

Throughout the discussion of flat type scenery, the supplies, equipment and tools needed in construction will be referred to with great frequency. This section is placed here so that it can be referred to easily when any of the materials mentioned are unfamiliar. Lumber, including plywood, fabrics used for flat type scenery, glue, tools, hardware and the various wood joints used in the construction of scenery are briefly discussed.

## Lumber

LUMBER FOR THE FLAT FRAME—1″ x 3″ white pine (abbreviated W. P.) is the size and type of lumber ordinarily used for the frame of the flat. White pine is plentiful, rather light in weight and relatively soft with slight difference in texture and hardness between the soft and harder portions of the grain (summer and winter growth). The better, or finish grades of lumber are ordinarily used since they have more of the necessary qualities for use when ripped into relatively narrow boards. These qualities include straightness, freedom from knots, checks, cup, and warp.

GRADES OF LUMBER—White Pine is graded as either "finish" or "common." "B" and better is the top grade, "C" and "D", almost free of knots, are best for flat frames but almost prohibitively expensive. #1 common has a few knots, #2 has more knots, and so on down the line. To reduce breakage, flat frames should be free of large knots. Poorer and knottier grades may be used for platforms, steps, thicknesses and contours.

In a period of price fluctuation it is impossible to quote exact figures but in spite of this fact it may be helpful to show how to compute the cost of wood for props and scenery. Adjustments must be made in terms of exact current prices.

COMPUTING THE COST OF LUMBER—Lumber usually will be priced at the lumber yard at a specified amount per thousand board feet. A board foot of lumber is a piece of wood 1″ thick, 1 foot long and 1 foot wide. At least this was the measurement when the lumber was first cut into boards, but the process of smoothing and drying causes the board to shrink and lose some of its ample proportions. Actually, lumber listed as one inch in thickness is only about ¾ of an inch thick. The width of a board is about ⅜ to ½ inch narrower than its specified width. For example, 1″ x 3″ is about ¾″ x 2½″; 1″ x 4″ is ¾″ x 3½″; 1″ x 6″ is ¾″ x 5½″; 1″ x 12″ is ¾″ x 11½″ and lumber listed as 2″ thick is about 1⅝″.

The lumber dealer may quote the price of white pine as $750.00 per thousand board feet ($750.00/1000 bd. ft.). Since a thousand board feet is $750.00, one board foot is 75¢.

To divide by 1000 move the decimal point 3 places to the left.

Cost of 1″ x 3″ Lumber—A board foot as previously described is 12″ wide. Since a 1″ x 3″ board is only 3″ wide it takes four pieces of 1″ x 3″ laid side by side to make the equivalent width of a board foot. These four pieces laid end to end make a piece of 1″ x 3″ four feet long. So it takes four lineal, or running feet of 1″ x 3″ to make one board foot of lumber; and conversely, one board foot will make four lineal feet of 1″ x 3″. Therefore, in computing the cost of 1″ x 3″ lumber it is necessary to divide the number of lineal feet by 4 and multiply by the cost per board foot.

Problem—A flat requires 4 pieces of 1″ x 3″ that are 6′ in length and 2 pieces of 1″ x 3″ that are 10′ in length. Four multiplied by 6 is 24. Two times 10 is 20. Twenty-four plus 20 is 44. So the total lineal feet of 1″ x 3″ is 44. To convert this to board feet, divide by 4. Forty four divided by 4 equals 11. If pine lumber is listed at $750.00/bd. ft. it is 75¢/bd. ft. 11 times 75¢ is $8.25.

Other Lumber—Although white pine is probably the easiest wood to work, there are other types that can be used. Here are a few of them and some of their characteristics.

*Ponderosa Pine*—Heavier than white pine, more resin content and splits more easily.

*Yellow Pine*— Heavier and harder than white pine and splits more easily.

*Spruce*—Many knots, may be firm, tends to twist and warp more than white pine.

*Redwood*—Excellent except that it is brittle and splinters rather easily.

*Cedar*—A little too brittle, splinters easily.

*White Wood*—Excellent, but expensive.

*Balsa*—Excellent, but expensive; extremely light in weight.

*Sugar Pine*—Good, soft, but has high resin content.

*Fir*—Heavier than white pine, strong, tough, coarse grained. Excellent for weight bearing structures and large platforms.

Lumber Sizes Used in the Scene Shop—The size lumber used more than any other in the scene shop is 1″ x 3″ white pine as already indicated, but other sizes are required from time to time. Following is a list of the more important widths and thicknesses that are used most often.

White Pine, Finish "C" or "D"

1″ x 2″ —for small frames and diagonal braces in flats.

1″ x 3″ —for flat frames.

1″ x 4″ —for large frames, battens for drops, framing for ceilings, for doors and door frames.

1″ x 6″ —for doors, frames, thicknesses, platforms.

1″ x 6″ —1″ x 3″ may be made by ripping 1″ x 6″ down the center.

White Pine, Common No. 1 or No. 2

1″ x 4″ —for stiffeners.

1″ x 6″ —for thicknesses.

1″ x 12″—for steps, platforms, contours, thicknesses, beams and arches.

White Pine, Shop

1¼″ —random widths for double covered frames and stage properties.

1½″ —random widths primarily for stage properties. Could be used for thicker units, such as frames for door shutters, double covered flat frames.

Fir

2″ x 4″ —used for units bearing weight; has considerable strength for size, but is splintery. Good for legs for platforms and frames for platforms.

2″ x 6″, 2″ x 8″, 2″ x 10″ —for long span platforms where considerable strength is needed.

## Plywood

Plywood is another lumber yard material used in constructing the flat. It has many uses in the shop since its structure provides maximum strength combined with minimum weight and thickness.

STRUCTURE OF PLYWOOD—Plywood is fabricated by gluing thin layers of wood together. When glued the layers are placed together so that the grains of alternate layers run at right angles to one another. In this way maximum strength with minimum thickness is obtained.

LAYERS OF PLYWOOD

Three layers or plies of wood are usually used to make ⅛″, ¼″, 5/16″ and ⅜″ thick plywoods. Plywood of greater thickness is made by using five or more layers. Ordinarily, there is an odd number of layers in a thickness of plywood so the grain in both exposed faces runs in the same direction. On a large sheet of plywood this grain runs the long direction. The bend test will easily show that the plywood is strongest parallel to the exposed grain, so when small pieces are cut to make corner blocks and keystones they are designed so that the grain runs across the joints of the pieces of wood that are to be fastened together.

THICKNESS OF PLYWOOD—Plywood as thin as ⅛″ may be used for irregular profile edges on flats or for covering curved surfaces of small diameter.

The ¼″ and 5/16″ plywoods are used for making corner blocks and keystones. These thicknesses may also be used for constructing profile edges on flats.

Plywood ⅜″ thick may be used for oversized corner blocks or for surfaces needing slightly more strength than that obtained with thinner wood.

Plywood ½″, ⅝″ and ¾″ thick may be used for making large contour cutouts of irregular form and for making platforms, parallels, and platform and parallel tops. The thicker the plywood the greater the strength is an axiom that may be used in working with this material.

GRADES OF PLYWOOD—*Sheathing plywood* is the most economical grade to use for most stage purposes. This grade is slightly rough on both sides, but not uneven enough to cause any concern or difficulty.

*Good One Side* is the next grade. This plywood has been patched wherever there were knots or knot holes on one surface and has been sanded.

*Good Both Sides* is the top grade plywood. It has been patched and sanded on both sides so there are no knots or holes and the surface is smooth on both sides.

*Boat Panel* is a special type of plywood that is bonded with waterproof glue and manufactured especially for the boat industry. It is available in pieces up to 16′ or 18′ in length. It is much more expensive than the other types of plywood, but might be needed for some special scenic units.

TYPES OF WOOD—The economical grades of plywood are made from the plentiful woods, fir and pine. Special surfaces of other woods may be obtained, but these will have little use around the scene shop.

PRICE OF PLYWOOD—Plywood is usually priced by the square foot or by the hundred square feet. A full sheet or piece of plywood is 4′ wide and 8′ long; thus it has a surface area of 32 square feet (4 x 8 = 32). So 32 times the cost per square foot is the cost per full sheet of plywood.

CORNER BLOCKS—Corner blocks are right triangular shaped pieces of ¼″ or 5/16″ plywood approximately 10″ on each of the right angle sides. If the flat frame is made with butt joints at the corners, that is, the boards after being cut off square are placed together so that the end of the one is butted up against the side of the other adjacent to its end, then the corner blocks are cut from the plywood so that the exposed grain runs parallel to one of the right angle sides. If mitre joints are used, that is, the end of each board cut with a 45° angle and the two pieces placed at right angles to one another so that the corner fits like the corner of an ordinary picture frame, then the exposed grain of the corner block should run parallel to the long side of the isosceles triangle shaped corner block. Since each corner block is half of a square with 10″ sides making an area of 100 square inches, each corner block has an area of 50 square inches.

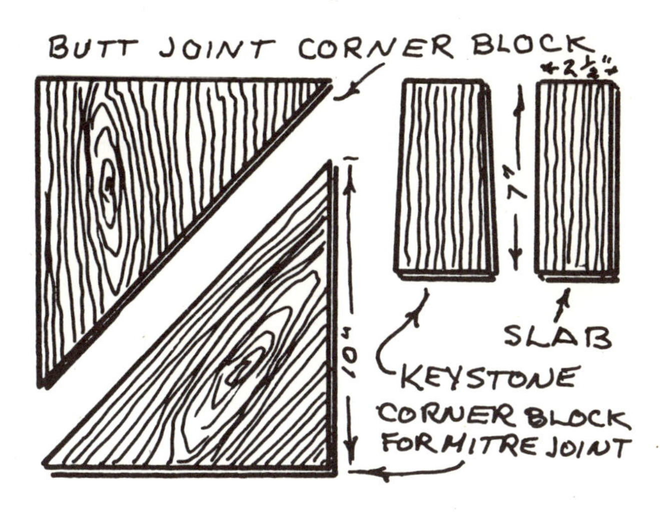


KEYSTONES OR SLABS—Keystones are used to secure joints other than the joints at the corner of the flat. If the keystone shape is employed the pieces of plywood are cut about 8″ long and tapered in width so that they are about 2½″ at one end and 3½″ at the other end. The exposed grain should run the long direction. If the slab shaped pieces are used they are cut so that they are about 8″ long and 2½″ wide with the grain running lengthwise. These may be cut from ¼″ or 5/16″ plywood. It is obvious that the slabs are easier to cut than the keystones and they will hold the joint about as securely. It will take 2½″ x 8″ or 20 square inches of plywood to make one slab.

### Flameproofing Compound

Flameproofing of everything flammable that is used on the stage is required in some places and should be standard procedure everywhere.

As soon as a wooden frame is built it should be coated either by spraying or brushing with a solution of one of the following or a regular standard mixture of flameproofing compound.

Mix No. 1:
- 2 pounds Ammonium Phosphate
- 4 pounds Ammonium Chloride
- 3 gallons water

Mix No. 2:
- 4 pounds Borax
- 4 pounds Ammonium Chloride
- 3 gallons water

### Fabric Covering For Flat Frames

A number of fabrics can be used for covering flat frames, such as muslin, pocket drill, scene duck and linen scene cloth. For the sake of economy, a good grade of muslin may be used to cover the frame. Type 128 is quite good moderately priced muslin. Muslin is sometimes known under other names such as sheeting, factory cloth and tobacco cloth.

FABRIC SPECIFICATIONS—Fabrics or textiles are frequently designated by their thread count (the number of threads per square inch woven in each direction) or by their weight per square yard. Fabrics are available in many widths. Some of the most common ones are 30″, 36″, 48″, 60″, 72″ and 81″. On special order it is possible to get material as wide as 30′ without seams. This material is excellent for large backdrops. Since most flats are not more than 6′ in width the most convenient widths for the scene shop are 72″ or 81″. For the sake of computing yardage it may be noted that running yards of the following widths are as indicated:

1 lineal yard of 30″ material is $\frac{5}{6}$ or .833 sq. yds.
1 lineal yard of 36″ material is 1 sq. yd.
1 lineal yard of 48″ material is $1\frac{1}{3}$ or 1.333 sq. yds.
1 lineal yard of 60″ material is $1\frac{2}{3}$ or 1.666 sq. yds.
1 lineal yard of 72″ material is 2 sq. yds.
1 lineal yard of 81″ material is $2\frac{1}{4}$ or 2.25 sq. yds.

PROBLEM—A flat is 10′-0″ high and 6′-0″ wide. 6 x 10 = 60. The area of the surface is 60 square feet. There are 9 square feet in a square yard so the surface is 60 divided by 9, which is $6\frac{2}{3}$ square yards. If the material is 36″ wide it will take $6\frac{2}{3}$ lineal yards to cover the flat. If the material is 72″ wide it will take only half as many lineal yards since each lineal yard is two square yards. This means $3\frac{1}{3}$ lineal yards of 72″ muslin will do the job.

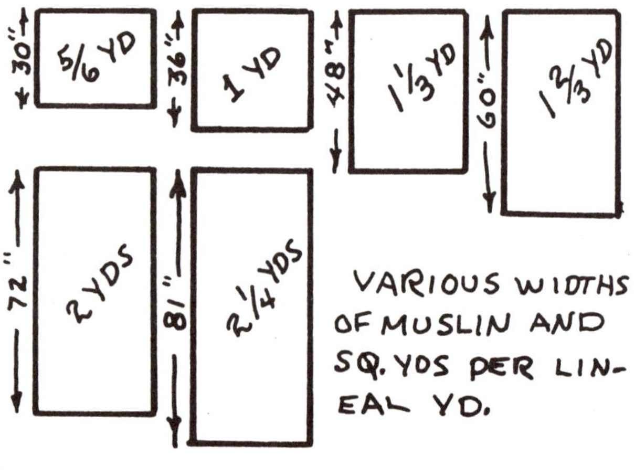

30″ — 5/6 YD; 36″ — 1 YD; 48″ — 1 1/3 YD; 60″ — 1 2/3 YD; 72″ — 2 YDS; 81″ — 2 1/4 YDS

VARIOUS WIDTHS OF MUSLIN AND SQ. YDS PER LINEAL YD.

### Glue

Fabric is secured to the flat frame by means of glue, glue and tacks, or glue and staples. White, all purpose, casein glue is an excellent adhesive for this. The D.A.P. product "Duratite" and Borden product "Elmer's Glue", are two glues commonly found in paint, lumber, and hardware stores. Although glue is obtainable in many sized containers the gallon size plastic container is probably the most economical. A hole drilled in the cap converts the container into a squeeze bottle applicator.

HOT GLUE GUN—This is a practical, useful backstage tool. Cartridges of glue are melted and squeezed out in spots or strips to adhere materials and to provide surface decoration.

## Hardware

A number of items that might be classed as hardware are needed when building flat type scenery.

Clout nails, staples and wood screws are used to fasten flat frames together and are probably first in importance.

CLOUT NAILS—Clout nails are made of relatively soft iron either tapered rectangular in shape with a round head or round in cross section with a round head and an elongated pinched or chisel shaped point. When driven into a piece of wood so that the point extends through the underside and strikes a steel clinch plate placed beneath the wood the point of the nail bends over, or "clinches." For regular flat construction 1¼" blued clout nails are used.

STAPLES—PNEUMATIC STAPLER—In a shop equipped with an air supply system, flats may be assembled with the aid of a pneumatic stapler. Wide crown staples may be used for "across the joint" temporary assembly and narrow staples for sturdier and more permanent corner block-keystone assembly of flats.

WOOD SCREWS—Wood screws are used extensively in the scene shop. Nails and screws may be used interchangeably for basically the same purposes. However, during scenery "strikes" screws can be removed more readily than nails with less damage to the salvageable materials. And flats fastened together with screws may be altered more easily.

Basically two types of flat head wood screws—slotted and phillips head—may be used. Phillips head screws have a + shaped slot and are easier to drive than the slot head. The screwdriver meshes into the screw more securely.

Wood screws are more expensive to use than clout nails but can be more easily removed and both wood and screws can be reused. When screws are used they should be long enough to extend almost through both pieces that are being fastened together. Since the wood is ¾" thick and the plywood is ¼" thick, the screws can be 1" in length. When ordering designate the diameter (# 9 will do for hardware also), the length (1" in this case) and type (flat head bright wood screws). So the total designation is 1" #9 flat head bright wood screws. Screws come packed a gross to the box usually and there are 144 screws in a gross. (A gross is 12 dozen.)

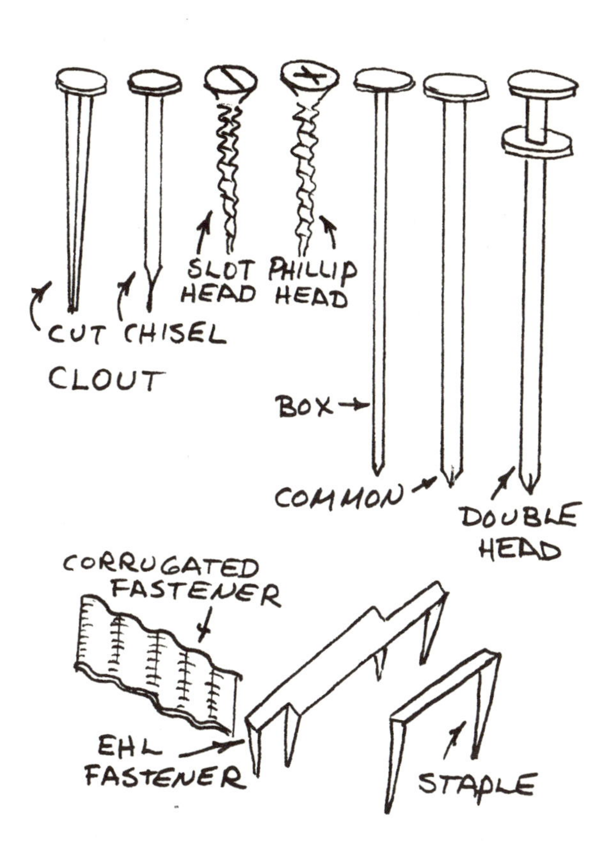


The 1" #9 wood screws mentioned may be used for most stage hardware. However, there are times when longer screws are needed. Almost all hardware is drilled to receive #9 so it will be just as well to use that diameter. It will also be wise to keep only one type, the flat head, since they will screw down flat on any hardware that has countersunk holes. (A countersunk hole is a hole drilled partially with a larger drill after the original hole has been completed. This allows the heads to pull part way into the hole.) Screws 1½" long will work very well for splicing two thicknesses of 1″ board together.

For most stage uses the stock of screws need include only:

1″#9 Flat Head Bright Wood Screws
1½″ #9 Flat Head Bright Wood Screws

FASTENERS—Corrugated fasteners and Ehl fasteners are sometimes used to secure or help secure butt and mitre joints. They are elongated metal holders that are driven in across a butt or mitre joint and into both pieces of wood that make the joint.

NAILS—A number of varieties and sizes of nails may be used around the shop from time to time.

Nail lengths are referred to in the United States in terms of "penny", the symbol for which is "d". Here are the penny sizes and length equivalents of the nails that will be used most often in the shop.

4d (four penny) nail is 1½″ long
6d nail is 2″ long
8d nail is 2½″ long
10d nail is 3″ long
12d nail is 3¼″ long
16d nail is 3¾″ long

The thickness of the nail is indicated as common or box. The slender box nail splits the wood less and is a little better for scenery use.

The type of head is still another nail classification. The box and common nails have regular round heads. Finish nails and casing nails have small heads just slightly larger than the nail. The double head, scaffolding, duplex or whatever other name they are known by, nails have two heads, one spaced beneath the other.

The surface of the nail is another classification. Regular nails are made of steel wire with no coating. Resin coated and zinc coated nails resist rust — and resist removal so should be used for somewhat permanent construction.

The nails used most often in the scene shop are:

Clout nails—1¼″
6d box nails (resin coated optional)
8d box nails (resin coated optional)
16d box nails
8d double head nails
16d double head nails
Small lath and Celotex nails (not previously mentioned)

The importance of double head nails might be mentioned before moving on to other hardware. The double head nail has many and varied uses. Since it can be driven in to the first head so that it holds securely, yet can be pulled out easily because the neck and top head stick above the surface, it is useful for fastening flats together in a stage set. It can be used in place of lash cleats, and can be used to fasten stiffeners in place. The 8d size will be used most frequently. The 16d size can be used for fastening temporary platforms together.

BACK FLAP HINGES—The most usable hinge is the back flap hinge. This hinge in either the 1½″ or 2″ size (the measurement indicates the size of each half of the hinge) is used to hinge flats together and to hinge doors into frames and for many other purposes. It is obtainable in loose pin (the pin may be pulled out and the hinge halves pulled apart) and tight pin. The pins may easily be removed from the tight pin variety by grinding or cutting the head from one end of the pin.

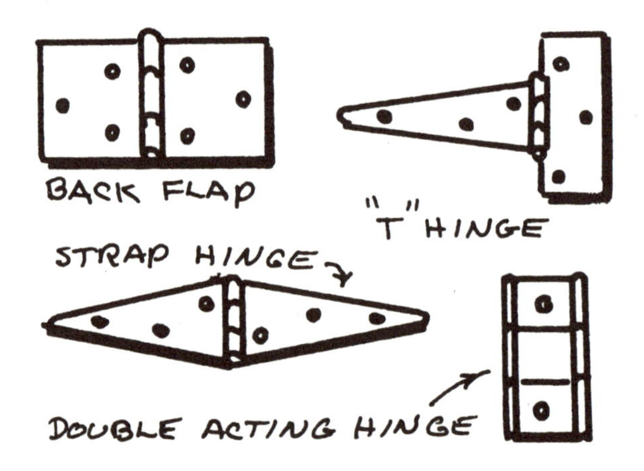


STRAP HINGES AND "T" HINGES—Strap hinges, each half of which is an elongated triangle, and "T" hinges, which are shaped something like the letter "T" with the joint between the top and leg of the "T", have limited use. They are used where larger or at least longer hinges are needed. Strap hinges are used on hinge-lock removable doors and windows.

DOUBLE ACTING SCREEN HINGES—When folding screens are built it is necessary to use hinges that will fold in both directions if the screens are to be completely flexible. The double acting screen hinge is used for this reason.

SPRING SCREEN DOOR HINGES — Stage doors may sometimes be kept closed by using screen door hinges that have a built in spring closer.

DOOR LOCKS, DOOR LATCHES, MAGNETIC DOOR HOLDERS AND FRICTION DOOR HOLDERS—There are many items of this nature that may be used for stage doors. The simpler varieties of regular door knob and latch sets really work the best for stage doors and since the doors are used quite frequently they may be the most satisfactory in the long run. There are many varieties. The ones that require the least work to install are the best to use.

TIE LINES—Two types of cotton tie lines are used a good deal on flats and drops.

¼" woven sash cord, which is a soft cotton cord, is used for lash line.

⅛" shade cord, which is also soft woven cotton cord, is used to make tie lines for back drops.

STRAP IRON—Strap iron is used for making sill irons and various pieces of rigging hardware. Strap iron may also be used for decoration on stage settings. It can be bent, twisted, curved and can be used to simulate wrought iron work on furniture, stair rails, and window grills. It is sold by the pound and is available in hardware stores and lumber yards. Width, length, and thickness should be specified when it is ordered. For most stage use the width can vary from ¾" to 1¼" and the thickness from 3/16" to ¼".

ALUMINUM TUBING AND THIN WALL STEEL TUBING—Aluminum tubing and thin walled steel tubing commonly used for electrical conduit can be used for decorative and abstract scenery forms. Tubing can be bent into interesting shapes with a simple "bender" tool that is regular equipment in most electrical shops. Tubing can also be bent around wooden templates. Connectors may be used to fasten sections of tubing together end to end.

WIRE—Various types of wire will be needed from time to time in the scene shop. Here are a few of the varieties that will be used most often.

*Stove Pipe Wire and Baling Wire*—Soft iron wire such as stove pipe wire or baling wire has many uses around the shop and on the stage. They are in the main utility wires and should not be used where great strength is desired.

*Piano Wire*—Piano, or music wire has great strength. A thin wire will support a weight of several hundred pounds. It may be used for flying actors or scenery. In a play such as *Peter Pan,* piano wire is hooked onto a harness worn by the actor and extends from there up just beyond the upper sight lines behind an overhead border where it is tied onto a line of the flying system. It is used in a similar manner to suspend scenery from the stage flying system.

*Aluminum Wire*—Large gauge aluminum wire, 3/16" or ¼" frequently used for clothes lines, is malleable and easy to work into intricate shapes for decorative and abstract forms. It may be used in conjunction with tubing, on furniture, or around screen frames to produce additional "lines" and forms.

CARPET TACKS—Carpet tacks are sold by the box or by the pound. The #6 size tacks are ½" in length.

BOLTS—Three types of bolts are commonly used on the stage. They are as follows:

*Stove Bolts*—Stove bolts have a round or flat screw type head. They are specified by thickness, length and length of threaded space.

Flat head #9 bolts may be used in most stage hardware.

*Carriage Bolts*—Carriage bolts have a round head and a square throat or shank. They are often used in wood because the square shank sinks into the wood preventing the bolt from turning when in place.

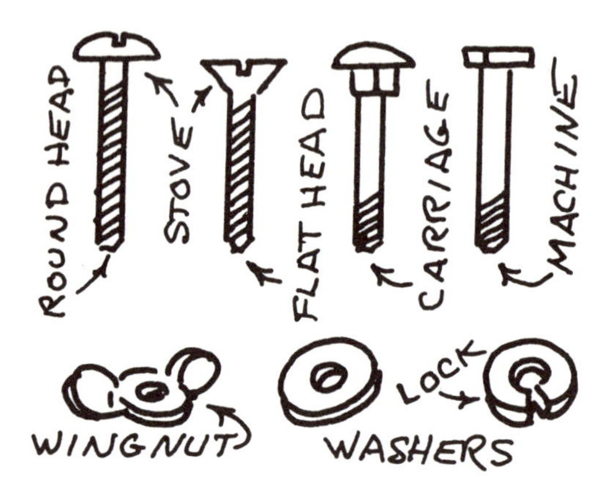


*Machine Bolts*—Machine bolts have a square or hexagon shaped head and are used primarily in metal work.

WASHERS—Various sizes of washers are used to enlarge the support area of a bolt head or nut. When washers are used under the nut it is easier to tighten the nut.

LOCK WASHERS AND SPLIT STEEL WASHERS—Lock and Split Steel Washers are designed so that they press against the nut when it is tightened. The tension of the washer against the nut prevents it from loosening.

WING NUTS—The broad flattened wings make it possible to turn the nut with the fingers. They are useful where nuts have to be put on and removed frequently and rapidly since no wrench is needed.

### Tools

A knowledge of the tools employed in the scene building process is important. Much time and effort may be saved if the correct tool is used and if tools are kept in good working condition. It is important to keep tools clean, sharp and in good mechanical condition. Cutting tools should be sharpened often; mechanical tools should be cleaned and oiled occasionally, and all tools should be stored in some kind of tool box or cabinet.

Tools may be roughly divided into two categories: manual and power. Manual tools depend upon the muscles of the operator while power tools have not only the muscles of the operator but also the power of another motive force such as an electric motor to make the job easier and faster. Many scene shops depend entirely upon the muscles of the building crew for their operation while the more elaborate ones have one or more pieces of power equipment. The hand tools are the most important and should be discussed first.

### Cutting Tools

There are a number of different types of cutting tools. The saw family is of primary importance in the scene building trade.

CROSS CUT SAW—The cross cut saw is used for sawing across the grain of the wood. Each triangular tooth is a double-edged blade. Two specifications are needed when ordering a saw: the length of the blade and the number of teeth per inch. Carpenters usually use saws 26″ long. Shorter ones may be used, but a longer blade makes possible a longer stroke and more rapid cutting. For rough cutting 8 point (8 teeth per inch) saws are used. For finer work 10 or even 12 point saws may be needed. Rarely would one need to do finer work than can be done with a 10 point saw.

RIP SAWS—Occasionally it is necessary to split boards and to cut with the grain. This is referred to as ripping a board. The rip saw used for this purpose has teeth that are filed straight across and designed to chip out wood. An 8 point rip saw would be the most useful and fastest cutting.

KEYHOLE SAWS—It is frequently necessary to cut curves and irregular forms. The keyhole saw is designed for this purpose. It has a slender blade sharpened on the end. The blade of the saw may be inserted into a small hole and the cut started in that way, or the cut may be started on the edge of the board. The narrowness of the blade makes it possible to cut where sharp turns are desired. Some keyhole saws are supplied with several blades, which increases the versatility of the saw.

SCROLL SAWS—Scroll saws have rather delicate blades and are used for irregular cuts in small work. They must be used with care or the blades break easily and readily.

HACK SAWS—The hack saw is a metal cutting saw with detachable blades. It is used for cutting strap iron, nails, bolts and other metal objects.

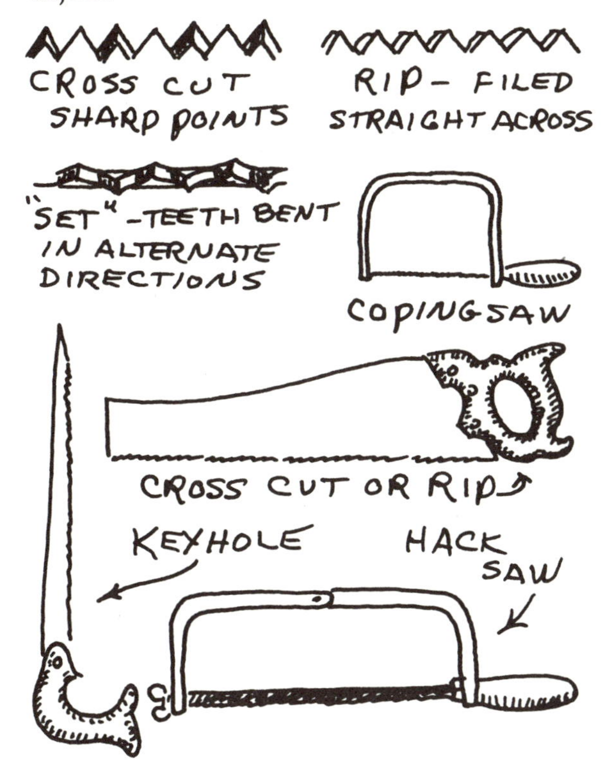


WOOD CHISELS—Wood chisels come in varying widths. Most practical around the scene shop are the ¼″, ½″ and 1″ chisels. Actually the chisel has infrequent use for rough construction, but it comes in handy quite often. Chisels should *never* be used as a substitute for screw drivers.

COLD CHISELS—The cold chisel is a round piece of steel 4″ or longer with a double taper on one end. It is used to cut metal objects.

WOOD PLANES—There are times when it is necessary to smooth the edge of a board, to trim off a door that fits too snugly, or to round a

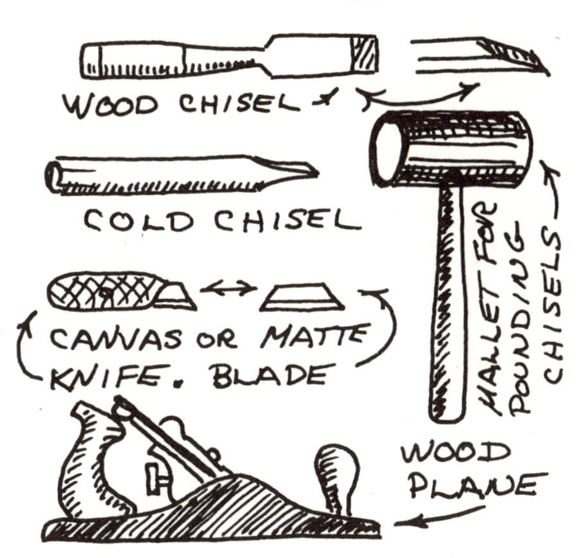


sharp edge. The wood plane is the correct tool to use for these jobs. It is a shaving tool. It cuts like a chisel, but the block in which the blade is mounted makes it easier to operate than a chisel. If one is to do a lot of straightening of long boards a plane with a long frame is the best. If the plane is to be used for small and short boards then a short one will do. The longer the plane the more expensive it is.

KNIVES—Once in a while a paring knife is useful in the shop. A canvas knife is a necessity. It is used to trim away the extra muslin when flat frames are covered. Many types of knives are available. Knives that are made for trimming wall board or cutting picture mattes are usable as are the various razor blade knife holders. Wall board and matte knives are superior since they use a heavier and more durable blade.

DRAW KNIVES—There are times when the draw knife is useful. If many props, such as guns, chair legs, rough poles, are built in the shop the draw knife is almost indispensable. A carpenter or shop man should be consulted about the proper way to handle this instrument. It is a dangerous implement!

SCISSORS — Muslin and paper frequently

need to be cut and for this oversize scissors are all but indispensable.

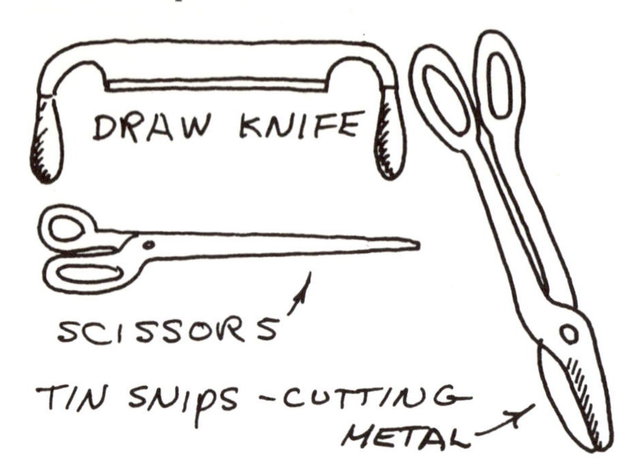


TIN SNIPS—Tin snips and circle cutter snips are made for cutting thin metal like aluminum, tin, sheet metal and sheet copper.

### Pounding Tools

There are several pounding or driving tools used in the scene shop. The most important of these are as follows:

CLAW HAMMERS—The claw hammer has a set of curved claws. This is the correct hammer to use for pulling out as well as driving nails. Hammers with wooden handles are more often broken than steel hammers. The hammer should be chosen for weight and balance. A large, strong individual can handle a heavy hammer; others should use lighter hammers.

RIP HAMMERS — The rip hammer has straight claws and although also used for driving nails it is not as good as the claw hammer for removing them because of the shape of the claws. This hammer is useful for ripping things apart; for instance, for ripping off corner blocks and keystones.

TACK HAMMERS — There are occasions when a smaller hammer is desired for driving in small nails or tacks. Tack hammers are designed for this purpose. Some tack hammers have magnetized heads and the tack will be held by the magnet until started in the wood.

MALLETS—Wooden or plastic headed hammers are made to use for pounding chisels. Chisels, especially those with wooden handles, should not be pounded with other types of hammers.

STAPLE GUNS—The staple gun is an invaluable tool for the scenery shop. Actually staples may be used wherever tacks are normally used. The staple gun is easier to operate than the tack hammer. A substantial gun with a steel frame probably will hold up better than one made of aluminum.

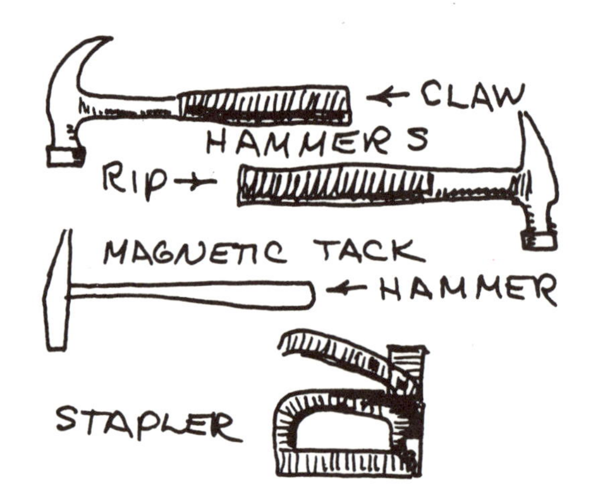


### Measuring Tools

Measuring, marking and squaring are important in the process of building scenery. A minimum number of tools of this nature should be included in the tool cabinet.

RULERS—Self encased steel tapes are probably the easiest to use and the most convenient type of measuring instrument. They may be obtained in many lengths: 6′, 8′, 10′, 12′, 25′, 50′ and 100′. The 10′ and 12′ tapes are probably the most convenient lengths for measuring boards for scenery. It will pay to buy tapes of better quality rather than inexpensive ones. They are not so easily broken and when broken it is possible to replace the tape portion.

YARDSTICKS—Yardsticks are usable, but the steel tape makes them somewhat unessential. Free yardsticks may usually be obtained from paint and hardware stores. Heavy yardsticks with metallic edges are preferred to plain wooden ones.

TRY SQUARES—The try square is a small right angle tool used as a guide for marking 90° angles across boards preparatory to sawing them.

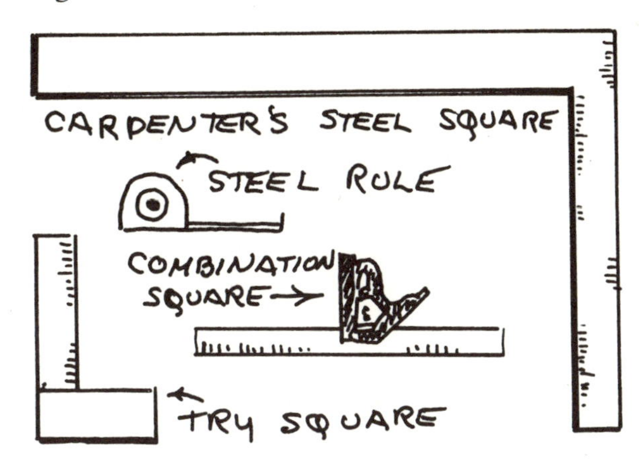


STEEL SQUARE OR CARPENTER'S SQUARE OR FRAMING SQUARE—The framing square is an indispensable tool. It is used for checking the squareness of corners on all types of scenery and for computing angles that are needed for steps and tops of angular flats. This square is "L" shaped with the short arm of the "L" measuring 16″ and the long arm measuring 24″.

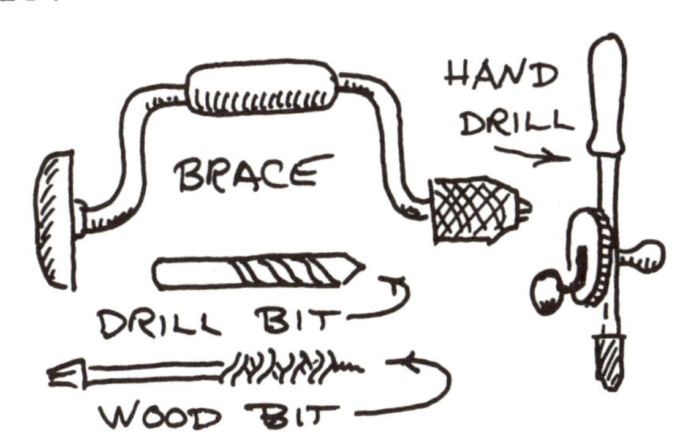


## Drilling Tools

Several tools are needed for drilling purposes.

BRACE—The brace is an instrument to hold bits or drills. The brace fitted with a bit is used to drill holes in wood or metal.

BITS AND DRILLS—Bits and drills are round spiral cutting devices which, when fitted into a brace or power press, are used to cut holes in wood or metal. Metal bits can be used on wood, but wood bits should *never* be used on metal.

## Miscellaneous Manual Tools

Additional tools used in the scene shop and operated by hand include the following items.

CLINCH PLATE—The clinch plate is a steel plate about 10″ square and ¼″ or less in thickness. Actually larger or smaller pieces may be used, but the 10″ square is a good size. The only requirement for the clinch plate is that it be flat and hard enough that nails will clinch as they are driven through a board and onto its steel surface.

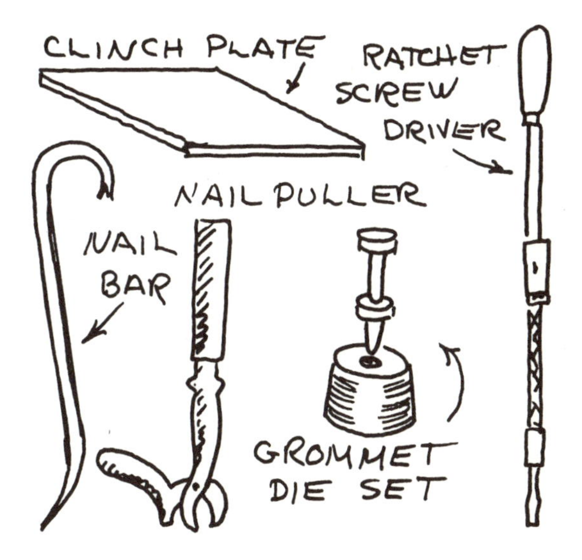


SCREW DRIVERS—The straight screw driver, or rigid screw driver, is used in electrical work and quite frequently on scenery.

MECHANICAL SCREW DRIVER—The mechanical, ratchet or automatic screw driver is a time saver when many screws must be driven; for instance, it is a great help when several flats are being hinged together. This type screw driver may be obtained in several lengths. The longer models are more expensive and more difficult to manipulate. However, screws can be driven more rapidly with them.

PLIERS—There are a number of varieties of pliers and each variety has its use. The ordinary pliers are used for miscellaneous grasping and holding jobs and for tightening nuts. Pointed nose pliers are used for bending a loop in the end of an electrical wire, for reaching into small holes to grasp and for many other jobs. This variety of pliers can also be purchased with a cutting blade far back in the jaws. Side cutter and diagonal or "dike" cutters are useful in electrical work.

CRESCENT WRENCHES—Crescent wrenches are often useful in the shop. There are a number of lengths. If relatively small bolts are to be used an 8″ wrench is probably large enough. (The size indicates the over-all length of the wrench.)

PIPE WRENCHES—If pipe battens are in the theatre and the pipe has connectors of one type or another and if pipe-type light stands are used, it may be wise to have a pipe wrench — 10″ or 12″ or even larger.

ALLEN WRENCHES — Various types of equipment require special wrenches such as the hexagonal "Allen" wrenches that are used a great deal on set screws.

GROMMET DIE — If many backdrops or sets of draperies are to be made in the scene shop, it will be a good idea to get a grommet die set. The grommet die must fit the grommets so sizes should be checked carefully.

FILES—Metal files are used to smooth and to hone off sharp edges of metal.

WOOD RASPS—Wood rasps are really wood files. They are used for cutting down and smoothing wood. The wood rasp is an extremely useful tool around the property making department.

NAIL PULLERS—The nail puller is a special tool used for pulling nails out after they have been driven completely in.

OIL STONES—Oil stones are used for sharpening many tools, including chisels and knives. An oil stone is indispensable to keep the fabric cutting knife sharp when covering flats.

## Power Tools

The operators of the scene shop that builds considerable scenery, that has adequate available space and that has money on hand will find it convenient to add some power equipment from time to time. It is a little difficult to decide which piece of equipment should be purchased first, but there are two items that are extremely useful, the power circle saw and the band saw.

POWER CIRCLE SAW—Either one of two basic varieties is available in the power circle saw.

*Table Saw*—This saw is designed with either a tilting arbor (motor and saw mount move but the table remains stationary) or a tilting table (only the table tilts). The tilting arbor variety is the most convenient to use.

*Pull Over Saw*—With this variety of saw, the motor has the blade mounted directly to it. This unit is mounted on an overhead arm. The saw is easily maneuverable for complex sawing. For cross cut sawing it is safer to use than the table saw. Since the saw is pulled over the work there is less danger of getting the hands into the blade.

The power circle saw is useful where a great deal of ripping, straight cut-off and angular cut off work is done. The pull over type of saw seems to be a little safer to use than the table type. An 8″ (diameter of the saw blade) saw is large enough for the moderate shop. Where extremely large amounts of work are done a 9″ or 10″ saw will work better.

BAND SAW—The band saw is so convenient in a shop where many irregular cuts are made on such items as contours and sweeps that it could be considered to be the most important piece of power equipment. The band saw has a single narrow continuous band blade 7′-0″ or more in length that runs around two large wheels spaced a foot or more apart. The blade is covered except for a small space where the cutting is done. It is important to get a band saw that will cut wide material. (This means the wheels the blade runs on must be large — the diameter of the wheel determines the spread of the blade in the machine.)

JIG SAW—The jig saw is an alternate for the band saw. The jig saw has narrow, short blades that may be either fastened at one or both ends. When the blade is fastened only at one end it is possible to start sawing from a small hole and make an inside cut. This cannot be done with the band saw since it has one continuous looping blade. The jig saw does not cut as rapidly as the band saw, but it has many practical uses. It is used for lighter cutting work than the band saw.

SABRE SAW—The sabre saw is a portable hand held jig saw. A heavy duty model should be purchased since it will probably have extensive use in cutting complex contours. This is a useful tool for inside cuts that cannot be made on a bandsaw.

PISTOL DRILL—The hand held pistol drill will have extensive use for drilling bolt holes when assembling scenery in the shop and on the stage.

POWER SCREW DRIVER—The power screw driver is a useful tool where screws are used in quantity in building and assembling scenery.

DRILL PRESS—In a shop where a great deal of metal work is done a drill press is an indispensable tool. It is difficult to drill holes in iron, steel, and pipe, with a hand held drill.

PNEUMATIC TOOLS—It is possible to equip a shop with an air system and a wide variety of pneumatic tools including: Paint spray guns, staple guns, drills, etc. This might be worth investigating.

SEWING MACHINE—A sewing machine should not be forgotten when the scene shop is being equipped. A small shop may need only a portable while a large shop should have a floor model machine. This is a necessary tool considering the amount of sewing that is done. Back drops and the coverings for large flats require seamed pieces of muslin.

Other power tools, such as wood turning lathes, power screw drivers and wood planers or jointers, have use only in large shops and shops where finished pieces of furniture are manufactured for the stage.

### Wood Joints

Woodworkers have developed many methods of fastening pieces of wood together. In special circumstances one type of fastening is more convenient, easier to make or provides a stronger bond. A basic knowledge of some of the joints that are usable in building scenery is essential.

BUTT JOINT—The most commonly used wood joint is the butt joint. When two pieces of wood are placed together so that the end of one is snugly fitted at right angles against the side of the other, or the end of the other, they are "butted" up against one another. This type of joint may be fastened by placing a piece of wood over the top of the joint and by fastening it to both pieces; or by driving nails in on an angle so that they enter both pieces of wood (toe-nailing). Or it may be fastened by using special fasteners — corrugated fasteners or Ehl fasteners — which are driven in such a way that they cross the joint. Or it may be fastened by using metal splicing plates that are placed over the joint and fastened to both pieces of wood.

MITRE JOINT—For a mitre joint the ends of both pieces of board are cut on a 45 angle and the boards fitted together to form a right

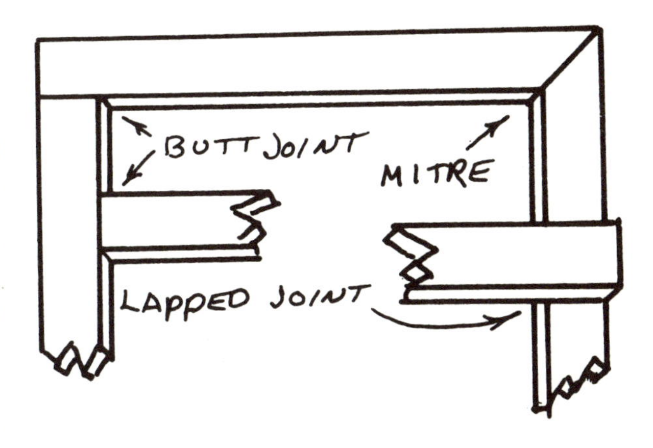

angle. The joint is fastened in a manner similar to the butt joint.

LAP JOINT—This is one of the easiest ways to fasten two boards together. The two boards are simply overlapped either running together in the same direction or running at right angles and then nails or screws are driven through both boards.

LAPPED SPLICE—When a board isn't long enough it may be lengthened by placing another board beyond its end and then placing a shorter piece over the top of the two and nailing the top one into the other two.

THE SCARF SPLICE—A more complicated, but at the same time satisfactory, method of splicing boards end to end is called the scarf

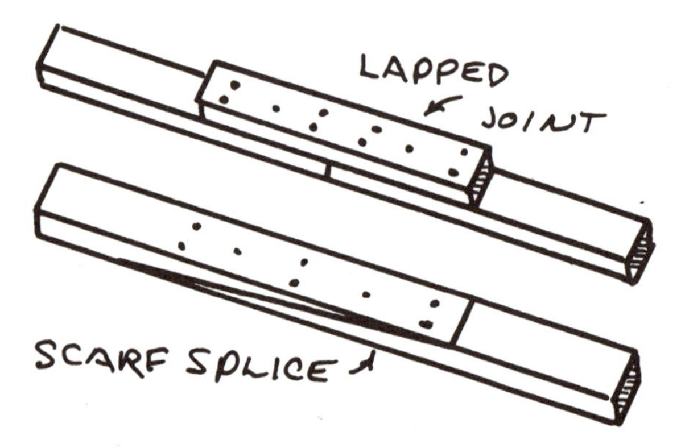


splice. The ends of the two boards to be spliced are cut with long tapers flat wise and then the pieces over lapped until they have the same thickness along the lap as along the board. The joint may be both glued and nailed. Clout nails will work very well on regular ¾" boards.

THE HALVED JOINT—If it is desirable to build frames without corner blocks so that they can easily be covered on both sides, the halved joint should be used. Half of the thickness of each board and the length of the width of the other board is sliced off. The two boards are then fitted together and the joint secured by gluing and nailing through the two.

THE NOTCHED JOINT—The notched joint is similar to the halved joint except that it takes place somewhere along each board rather than at the end of the boards.

THE MORTISE AND TENON JOINT—Mortise and tenon joints are used by professional scenic shops and by carpenters and cabinet

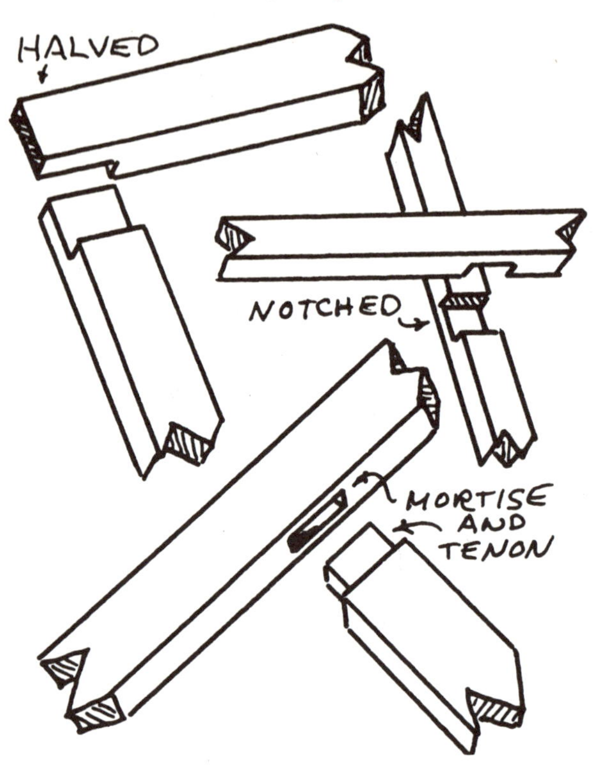


makers, but making these joints is somewhat difficult and time consuming for most theatre jobs. The mortise is a slot cut in one board and the tenon is a tongue cut on the other. The tongue fits into the slot to make the joint.

* * * * *

The small scenery shop will have only a few of the tools and supplies discussed, while the large, completely equipped shop will have all of them and probably many more. Some of the items will be used infrequently while others will be almost in constant use. By adding an item at a time, and carefully husbanding that item after it has been acquired, a good stock may eventually be built up whether on a large or small scale. As indicated previously, the project can be more easily and rapidly built if the correct equipment is available. But at the same time, some shortages stimulate one to try new ideas and procedures.

# 3

# The Flat and Flat Scenery

Practically every stage setting is made up primarily of flat surface units of scenery. Walls of rooms, surfaces of buildings, landscapes, tree forms, mountain forms, and even skies are made up of painted fabric, either free hanging and suspended from battens in the flying system or stretched across wooden frames. This chapter is devoted to a discussion of this important part of stage scenery.

## The Flat

The basic unit of framed scenery is the flat. This is a simple, light weight, rigid, wooden frame covered with fabric. Although it may be almost any size and shape there is a tendency in most theatre plants to use units that are small enough in size to be conveniently manipulated on the stage and stored away when not in use. The majority of a theatre's flats will be rectangular in shape, and have standard proportions for use on a specific stage. They will be built so that they can be fastened together in almost any conceivable combination to produce stage settings of great variety.

CONSTRUCTION OF A FLAT—Assume that a 6′-0″ x 10′-0″ flat is needed. The first step in its construction is the selection, measuring, squaring, and cutting of the various boards needed for the frame. Select 1″ x 3″ boards that are straight; then measure two pieces 6′-0″ in length. Square across them with a try square at the 6′-0″ mark and then saw them off at this point. It is essential to saw carefully so that the ends will be true when cut.

Place these two boards side by side and then take another board that is at least 10′-0″ long. Place this board at right angles to the other boards and butt it tightly up against them. Now measure across the width of the two 6′-0″ boards and down the longer board to the 10′-0″ point. Square across the board and cut it off. Another board is cut the same length. If the boards are measured in this way it is possible to automatically deduct the width of the two 6′-0″ boards from the 10′ board. The illustration of the flat frame will show that the 6′-0″ boards are for the top and bottom members of the frame which are called rails. The 10′-0″ boards, minus the width of the two 1″ x 3″ pieces, are for the side members or stiles of the flat.

Two more boards are needed for the cross pieces or toggle rails of the flat. In order to

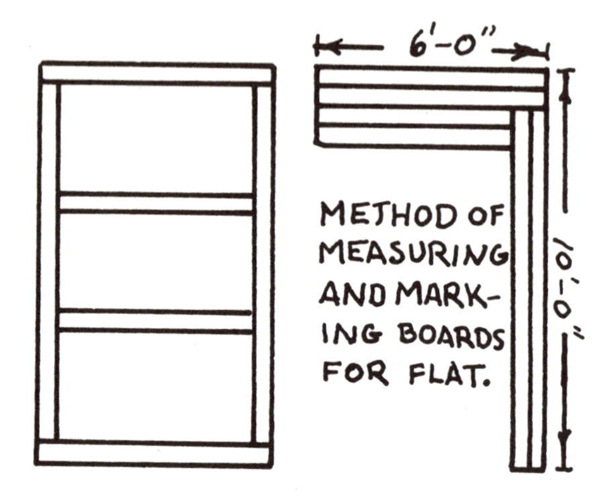


measure these correctly, place the two longer boards side by side and place one of the 6′-0″ boards atop and even with the outside board. Next measure the new board from a position adjacent to the inside of the longer boards to the end of the 6′-0″ board. In this way the width of the stiles is automatically deducted. It is important to measure and mark the toggle rails at this time so that they are the exact length needed.

The diagram will clarify the entire process.

Cut four corner blocks and four slabs (or keystones). This flat is to have butt joints so the grain of the corner blocks should run parallel to one of the right angle sides.

Butt one of the stiles against one of the rails so that the side of the stile is flush with the end

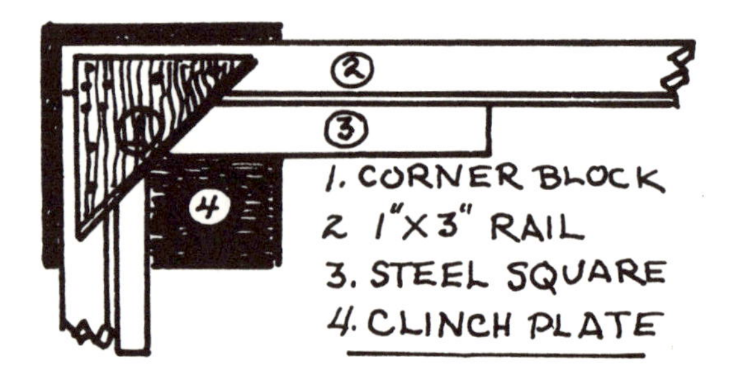


of the rail. The two boards should now be in the position of an elongated "L" or right angle. Place the clinch plate under the boards at the corner. Next place a corner block over the joint made by the ends of the stile and rail. Be certain that the exposed grain of the corner block runs across the joint so that maximum strength will be achieved. Move the corner block so that it is spaced in from the outside edge ¾″. This may be easily spaced by using a short scrap of 1″ x 3″ and placing it on edge next to the corner block.

A minimum number of nails should be used in securing the corner. A pattern with one

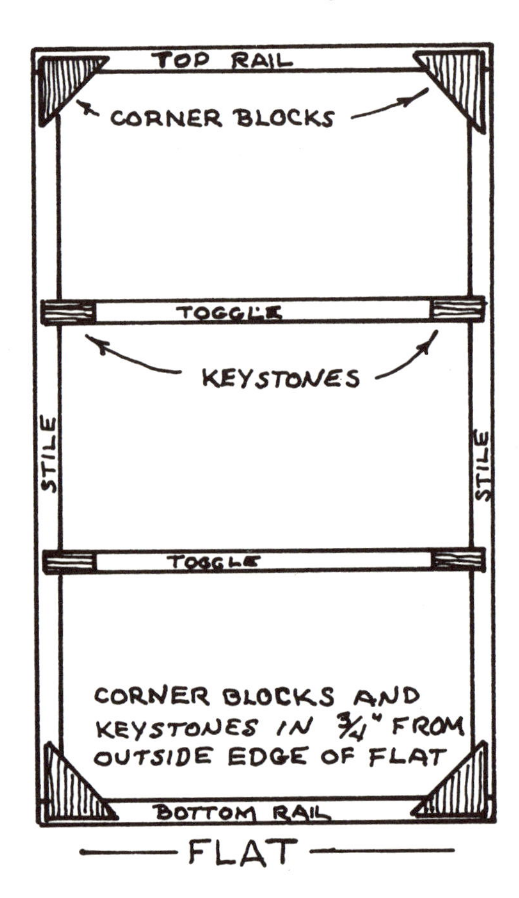


FLAT

nail in each corner, two on either side of the joint and one on each side halfway between the acute angles and the right angle is basic. If this pattern is followed nine nails will be used.

Nailing the corner with clout nails may be

done easily and efficiently as follows. Lay the boards in their approximate positions. Fasten the corner block to the rail using all five nails. Next drive one of the nails in the stile next to the joint. After this has been done the stile can still be easily twisted to the right or left to square the joint.

Place a steel square in the corner and adjust the boards so that the corner is square or true and drive a nail that has not been nailed as yet. The corner will now be held in place and the other nails can easily be driven in. The nails will be driven through onto the clinch plate and automatically bent over or clinched. It is important to note that clout nails, especially the tapered ones, will not hold unless clinched.

This process is repeated for the other three corners of the flat. A final check on the squareness and trueness of the flat may be made by measuring diagonally across the flat. If the two diagonal measurements are the same the corners are in their true positions.

When the corners have been secured and the frame checked for squareness, it is time to nail in the toggle rails.

Toggle rails should be spaced not more than 4′-0″ apart. The toggles are essential if the flat is not to be pulled out of shape by the shrinkage of the muslin when the surface is painted.

Special toggles are sometimes put in to support heavy wall hangings, such as pictures, mantels, animal heads, shadow boxes, wall phones, bracket lamps and other stage dressing units.

A minimum of nails may be used to fasten

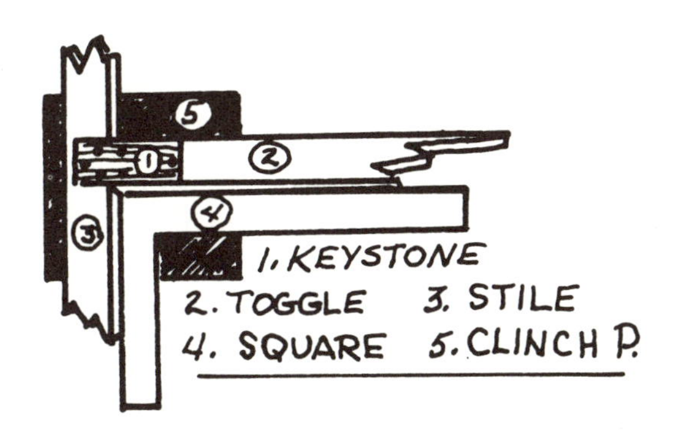


the toggles. Six or eight nails may be used in the keystones, depending upon the pattern. Two on each side of the joint are essential, but either one or two may be placed on each end of the slab or keystone. It is essential that the joint be close fitting or the flat may be sprung out, making it untrue. It is also important that the keystones be held in ¾″ from the outside edge of the flat.

The frame may be strengthened by the addition of diagonal braces. Braces should be placed in the upper and lower right corners as seen from the rear.

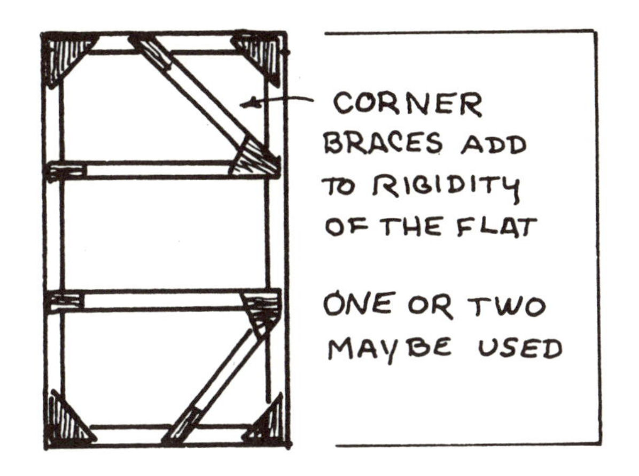

CORNER BRACES ADD TO RIGIDITY OF THE FLAT
ONE OR TWO MAY BE USED


Screws or staples driven by a pneumatic stapler may be substituted for nails in assembling the flat frames.

Covering the Flat—After the frame has been completed, it is time to cover it with muslin or some other type of fabric. Turn the frame over so that the corner blocks and keystones are underneath. Take a piece of fabric that is slightly larger than the frame (half inch to an inch all around). If 36″ material is being used for a frame that is 36″ wide or 72″ material for a 72″ wide frame the material can be used but the covering must be done carefully so that the material covers adequately.

Spread the material over the frame. Temporary tacks or staples may be placed in the corners to hold the material in place but these are not needed if the covering is done carefully.

Fold one side of the material back just far

enough to expose the wood of the stile. With a paddle or brush carefully spread glue on this

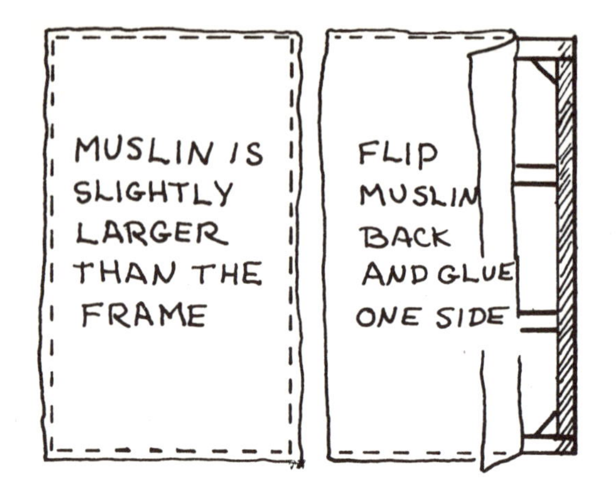


board. Work rapidly and use plenty of glue so that it will not soak into the board or "set-up" before the muslin has been pressed into place adjacent to the edge. Work the muslin down carefully and then rub it briskly. Next glue the opposite side. Then do one end and finally the other end. After the muslin has been glued all around it is a good idea to check and see that there are no loose places. The process completed, the muslin will be glued to the boards all the way around the outside edge of the flat. Do *not* glue to the toggle bars.

After the fabric has been glued to the frame,

the edges should be trimmed. A sharp knife or cutter should be used for this purpose. In order to prevent the muslin from fraying around the edge of the flat, it should be trimmed in 1/8" to 1/4" on the boards. After the knife has been run along cutting near the edge and the surplus material has been removed, the edge of the muslin should be carefully rubbed down. This edge will be sealed by paint so that it cannot be seen even upon close inspection.

NOTES ON CONSTRUCTION OF A FLAT—Corner blocks and keystones should always be held back 3/4" from the outside of the flat frame all of the way around. It is also advisable to hold the plywood pieces back 3/4" from the edges of openings in special flats. This facilitates in getting close joints when flats are put together to make corners in settings and when thicknesses are fitted behind openings in flats. This will be discussed more completely farther on in the chapter.

Check the corners carefully for squareness and trueness so that flats will fit together correctly to make stage walls.

Whenever possible use straight boards. There are times when boards that are slightly bowed must be used for stiles. If so, put them together in the frame so that one compensates for the other — so that both boards bow out or in — and then the toggle bars are put in to help straighten the boards and to force them into the correct position.

Cut all of the boards for the flat before assembling them and measure and cut them as indicated so that there is a relationship among them. If this isn't done the toggle bars especially may be cut too long or too short and the flat will bow in or out in the center.

Tighten the muslin only so that it is reasonably snug and wrinkle free and glue it only around the outside edges and/or around openings. Be certain to use unshrunk muslin, which is usually listed as brown or unbleached. The muslin must shrink during the sizing and painting process or it will not have a taut, smooth surface.

After the muslin has been glued onto the

frame, it should become thoroughly dry before any paint is applied. The glue must be set up until it is hard or the muslin may slip loose.

If mistakes have been made or if a flat is to be altered it is possible to remove clout nails rather easily with a carpenter's nail puller.

If no tacks or staples are used when covering the flat the process of removing muslin and recovering a frame is simplified a great deal. After all, a well built flat frame will probably outlive several coverings of muslin.

There are about 400 clout nails in a pound and the average flat takes about 70 clout nails.

Most materials are less expensive if purchased in quantity, and of course, if ordered in standard packaged amounts. For this reason it may be advisable to purchase glue in the economical gallon container, providing the budget will stand it. If the gallon container with a hole punctured or drilled through the lid is too unwieldy, pour part of the glue into a smaller empty clean plastic detergent container.

Substitute Materials—There are a number of substitute materials that can be used in place of materials listed in the process of building a flat. There are times and places where some types of materials are not easily obtainable and substitutions must be made.

Some other kind of lumber can be used. Certain requirements should be met if the lumber is to be entirely practical. The wood used should be relatively straight grained. Wood should be straight and free from cup or warp, light in weight, soft and easily workable and relatively inexpensive. Woods that can be used rather easily are: redwood, sugar pine, spruce, cedar and some soft native lumbers. White pine is probably superior to any of the others and should be used if possible.

There are some substitutes that can be used for plywood in making corner blocks and keystones, but plywood is probably the easiest and best to use. Prestwood, Masonite and other tempered hardboards can be used, but they get brittle with age. Tin corner blocks have been somewhat successful. Corner blocks and keystones made from heavy sheet aluminum would probably be excellent but expensive. The professional scene building shops use mortise and tenon joints and then make corner blocks and keystones from slabs of thin wood covered on both sides with cloth which is glued onto the wood.

Muslin is by no means the only material that can be used to cover flats, but it is probably the least expensive material that is durable enough to stand a moderate amount of abuse. Pocket drill, ticking, scene duck, linen denim, burlap, velour and almost any other textile could be used. Special textures and surfaces might be used for special flats to achieve results other than that achieved by the ordinary flat surface.

Stock of Flats—Building up a stock of scenery is a cumulative process that takes place over the years. Flats can be reused many times if they are built with a view toward standardization. One or two heights should be decided upon and then all of the regular flats built to conform. Many different widths can be used, but there is convenience in some unity in this respect, too.

Height—The standard flat height differs from theatre to theatre. This may be determined by the physical facilities of the stage and the scene storage space. Ceiling height, sight line height, proscenium height and shop space may be factors worth considering carefully before deciding that the scenery should be 24′-0″ or 8′-0″ in height. It is more practical to start with the minimum height than the maximum. It is possible to fasten extensions onto flats and then to return them to their normal size later, but tall flats will have to be cut off in order to make them shorter. In all probability flats of two standardized heights will allow the designer a certain amount of flexibility in his design and yet make it possible to work with a limited number of flats. Practical heights are 10′-0″ and 12′-0″, especially for modern productions.

Width—The maximum width of the flats

used on a stage may be determined by the doors through which the flats must pass on the way from the building area to the stage and from the stage to the storage docks. Usually 6′-0″ (commercially this is reduced to 5′-9″) is considered to be the maximum and is the widest flat that can be easily handled by one person. If, however, there are large spaces to fill it is possible to use few flats if they are wider. Sometimes 7′-0″ and even 8′-0″ flats have a definite place in the stockpile.

Many different widths are practical. However, anything narrower than a foot is a little impractical since a board can be used to fill such a narrow space. 1′-0″, 2′-0″, 3′-0″, 4′-0″, 5′-0″ and 6′-0″ widths are all needed at times. Occasionally beyond the 1′-0″ width it is desirable to progress at only 6″ intervals. The amount of storage space available may again determine at least in part the number of widths that are stocked.

STORAGE FOR FLATS — Storage space is needed for scenery. Less floor space is required and the flats are easier to store and remove from storage if they are stored on end rather than on their sides. Either of these positions, however, is better than stacking them flat like pancakes. Partitioning the scene dock into stalls that hold not more than 15 or 20 flats will make it easier to find the desirable pieces when needed and will keep them in better condition. The storage room should be dry. And if there is enough space there should be some orderliness to the stacking of the various widths and heights.

### Variations on the Flat

DOOR FLAT—The door flat unit is framed with an opening that may be used in one of three ways. A door may be hinged into it directly. A door unit may be fastened to the back of it. Or a door unit may be slipped into it from the front and be held in place by strap hinges that bind against the flat. These units will be described later.

The size of the opening in the door flat will depend on which of the three types of doors is used. The opening for the first two will be the regular door measurement while the one for

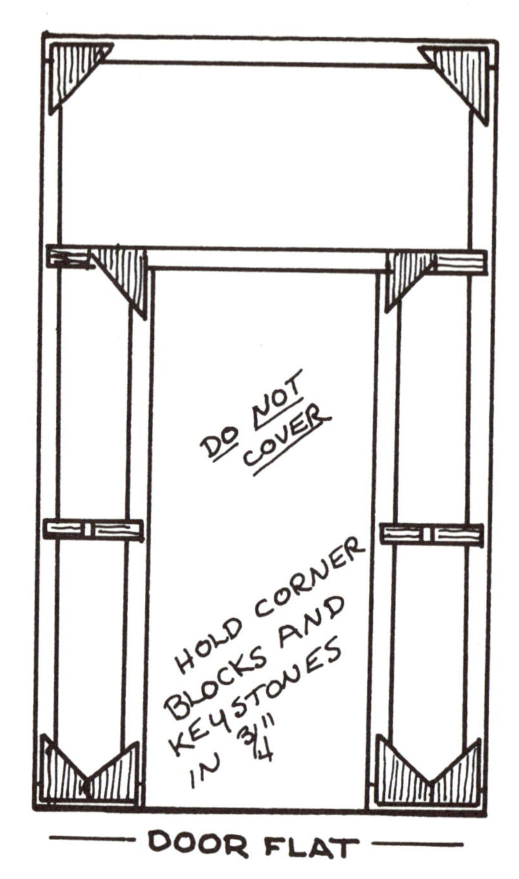


DOOR FLAT

the door that is slipped into the opening will be 2″ narrower and 1″ shorter than the door shutter.

The ordinary door in a home is 2′-0″, 2′-6″, 2′-8″ or 3′-0″ in width and either 6′-6″ or 6′-8″ in height. For convenience, on the stage either 3′-0″ x 7′-0″ or 2′-6″ x 6′-6″ doors may be used. Here again it is a good idea to set one standard size so that the units can be used interchangeably. Standardization shouldn't deter one from using other sizes for special effects, but in the main it is better to have one basic size. The door opening can be built into any flat that is wide and high enough to accommodate it. Although the opening need not be centered in the flat, actually for normal use this might be set up as regular procedure. If the door flat is to receive a 3′-0″ x 7′-0″ door a toggle bar should be placed so that the bottom of it measures 7′-0″ from the bottom of the flat. Stiles for the sides of the opening are cut 7′-0″ long minus the width of the bottom rail of the flat. These stiles are nailed in place so that the opening is correctly placed in relation to the sides of the frame. All of the joints should be checked for squareness as they are fastened. A set of short toggle bars is fastened in on either side at a height of 3′-6″ between the door stile and the outside stile of the flat.

To gain additional strength and rigidity, corner blocks may be used for the door stile fastenings at the top and bottom. Here again it is important to hold all of the corner blocks and keystone ¾″ from the outside edge of the flat and from the inside edge of the opening so that the thickness will fit into place without extra trimming.

To make it easier to square the corners, the bottom rail can be put in just as in an ordinary flat and then after the whole thing is assembled the section across the door opening can be cut out and a sill iron fastened to the bottom of the flat to keep the narrow legs on the sides of the opening in their correct positions and to make the flat more rigid.

SILL IRONS—Sill irons are made from strap iron that is anywhere from ⅛″ to ¼″ thick and from ¾″ to 1″ wide. Sill irons may be made in a number of ways. They may extend just beyond each side of the door about 6″ or 8″; they may consist of a piece of strap iron that is in the shape of a broad "U" with the short upright arms placed so that they fit within the opening of the flat; they may extend across the bottom and turn up about 6″ on the outside edges of the flat; or they may have extra pieces riveted or welded on so that they extend across the bottom, up on each side of the opening and up on each outside edge of the flat.

Sill irons are usually drilled for #9 wood screws. The holes may be drilled with a 3/16″ or ¼″ drill and then the sides where the screw heads will be should be countersunk (drilled in part way with a countersink or a large drill) so that the heads of the screws will pull up flush with the iron. They should be drilled so that at least two 1½″ or longer #9 flat-

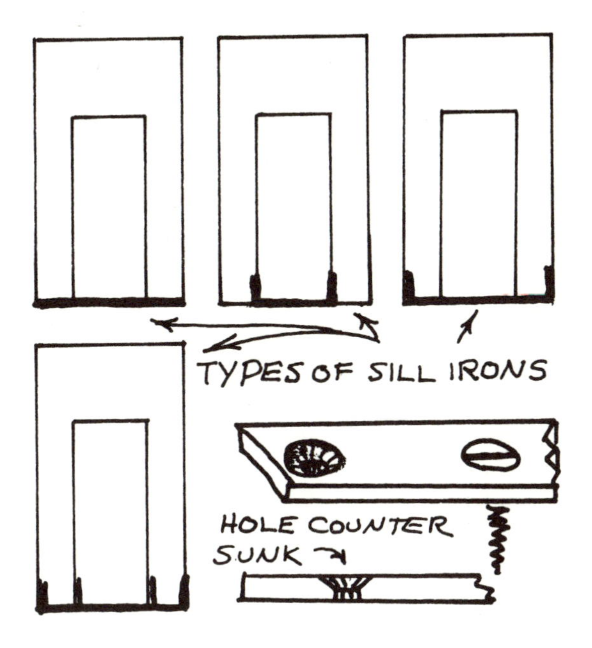


head bright wood screws can be placed in the section on each side of the opening. Before the sill iron is put on, file off any sharp edges that might cut into the floor as the flats are run across it.

WINDOW FLAT—The opening in a flat for a window may take many sizes and shapes. An ordinary window in a flat will probably be from 2′-0″ to 3′-0″ in width, from 3′-0″ to 6′-0″ in height and be placed from 1′-0″ to 3′-0″ above the floor. The outside of the window flat is exactly the same as that of a regular flat. The toggle bars are spaced so that the underside of the top one is the height of the top of the window and the upperside of the bottom one is the height of the bottom of the window. Extra stiles are then placed appropriately for the sides of the window and at the halfway point in the spaces at each side of the window short toggle bars are placed.

JOGS—Narrow flats, those of less than 2′-0″ width, are usually referred to as jogs.

PLUGS—Small flats are usually referred to

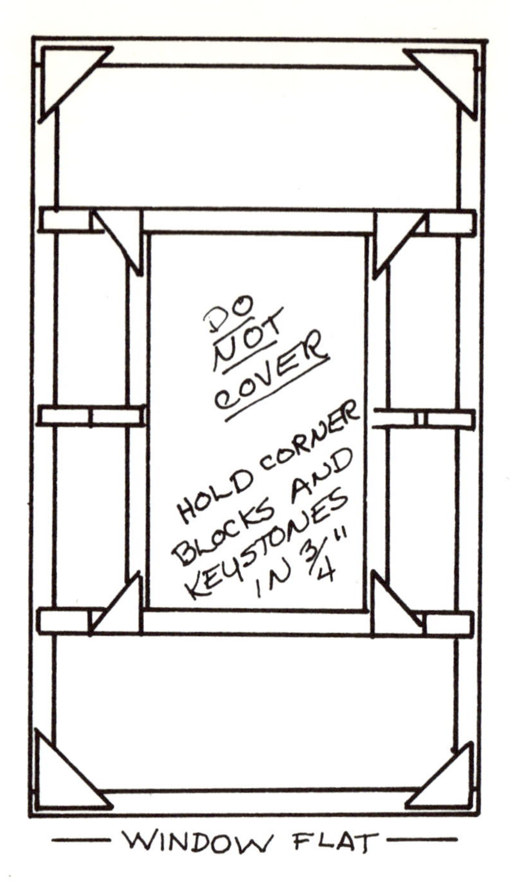


as plugs, especially those flats that are used to fill up, or plug up, openings in flats.

CONVERTING DOOR FLATS TO WINDOW FLATS—Sometimes it is easier to hinge or batten a plug into the bottom part of a door flat than to make a new window flat or cut into a regular flat to make an opening for a window. The hinges may be placed on the back and the cracks dutchmaned on the front. A dutchman

is a strip of muslin bonded to the surface of the flat with paint.

UNCOVERED FLAT SCENERY—Unusual scenic effects may be achieved by using combinations of uncovered flat frames which may be built in the regular manner. In some instances it may be desirable to use keystones in place of corner blocks in order that they will not be visible from the front.

### Wall Sections

It may be that no wall sections in a stage setting consist of more than one flat. This would be an exceptional setting. Usually two or more flats are combined to get the desired width for a single wall. When this is done it is almost always desirable to have the flats fastened together so that they may be folded for easier moving and stacking. Ordinarily, back-flap hinges are used for this purpose. Either 1½″ or 2″ hinges may be used. Each half of the back-flap hinge is the size of the measurement used to designate it.

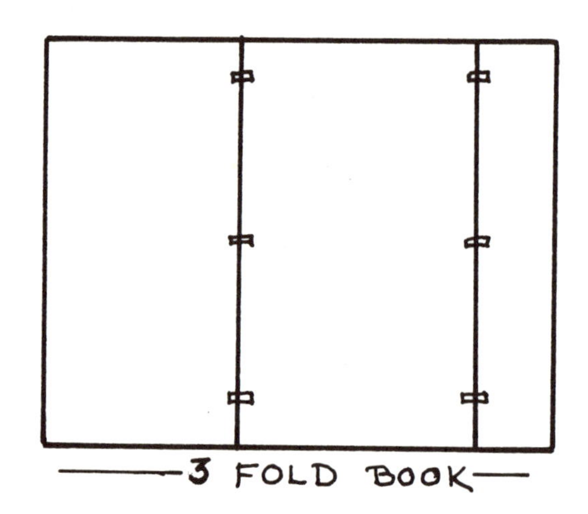


HINGING FLATS TOGETHER — Flats are hinged together on the face so that the joint may be covered with a piece of fabric, called a dutchman, giving the appearance of a single solid wall section when painted. Unless the flats are quite short, at least three hinges should be used. Hinges should be placed not more than 4′ or 5′ apart. Hinges are placed on the flat with the thickened roll up and directly in line with and centered over the crack between the flats.

Books—Books are wall sections consisting of two or more flats hinged together. A two flat section is referred to as a twofold, three flats as a threefold and so on.

Tumblers—Two flats of any width may be hinged together and they will present no problem in folding. As soon as three or more are hinged there may be some difficulty in folding unless tumblers are put in. A tumbler is a board, usually a 1″ x 3″ the height of the flats, that is hinged between two flats. If three flats of the same width are hinged together two of them will fold together but the third will bend only up past the halfway mark before it begins to bind against the one that has been folded in from the other side. With a board hinged into this position the board reaches the binding stage but since it is hinged on both of its edges the hinges on the edge opposite the binding edge lift the other flat over far enough so that it will fold down against the other flats. The tumbler isn't needed on all threefold flats. If the center flat is wide and the two side ones narrow the unit will fold without a tumbler. If two of the flats are the same width and one is considerably narrower the narrow one can be folded in first and the wide one last and it will fold. But as soon as the flats approach the same width the tumbler is a necessity if the unit is to fold into a flat pack.

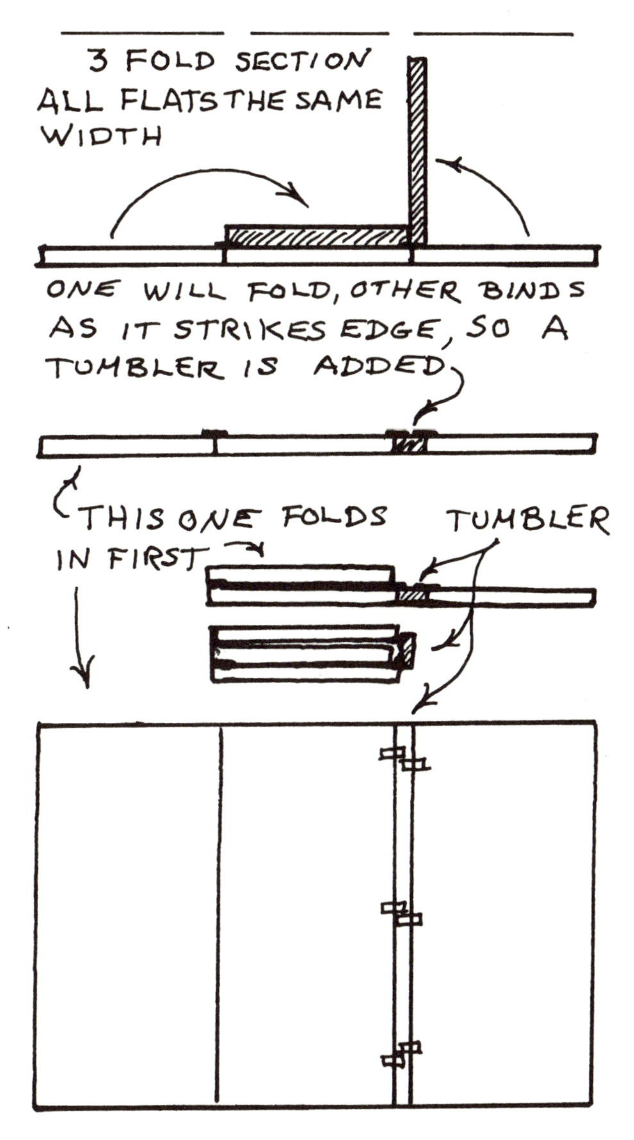


When more than three flats are hinged together two or more tumblers are frequently required.

Wall Sections Made Up of Flats—To keep the weight down to a minimum, it is wise to make a wall section out of as few flats as possible. Actually, there is no limit to the number of flats that can be hinged together to make a section, but there is a physical limit to the amount of weight that can be moved around the stage, from the shop to the stage and packed into a truck if the scenery has to be moved very far. Usually the widest flats practical should be used and the flats hinged together with appropriate tumblers hinged into place so the entire section will fold into a convenient pack.

Alternate Method for Making Door and Window Openings—Door and window openings may be created by using plugs to form the block above a door and the blocks above and below a window. When this method is employed it is possible to use regular flats throughout the structure and no window or door flats need be built. Units put together in this way are heavier and sometimes more awkward to handle than regular door or window flats, but the convenience of having flats with-

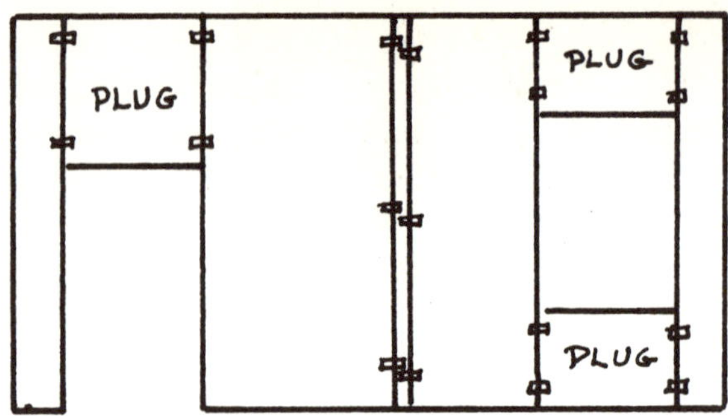


DOOR AND WINDOW OPENING MADE WITH PLUGS

out openings cut in them may be a compensating factor worth considering.

ALTERING THE SHAPE OF DOOR AND WINDOW OPENINGS — Archways, doorways and windows are not always rectangular in shape. In fact, they frequently have tops that are

RECTANGULAR OPENING TEMPORARILY ALTERED

curved, horseshoe shaped, scalloped, tapered or any of a dozen other shapes. Many times their shape is used as one of the major factors in suggesting the period or the locale of the setting to the audience.

### Archways — Sweeps

SWEEPS — Frequently regular doors, windows and archways can be reshaped by the additions of shaping corners or sweeps. Sweeps may be cut from plywood or inexpensive #2 or #3 common white pine lumber and then fastened into place with slabs of plywood and nails. After the scenery has been used and disassembled these may be removed and the flat returned to its normal shape.

Often half circle tops are desired. To lay these out it is necessary to have an oversize compass device of some kind. The most common device is the string and piece of chalk compass. The string is held at the center of radius and then the other end, equipped with chalk or pencil, is swept around to draw the curve. Frequently the results are not true since the string stretches rather readily on a large circle. A better method is to take a strip of board (thin and narrow) of about the cor-

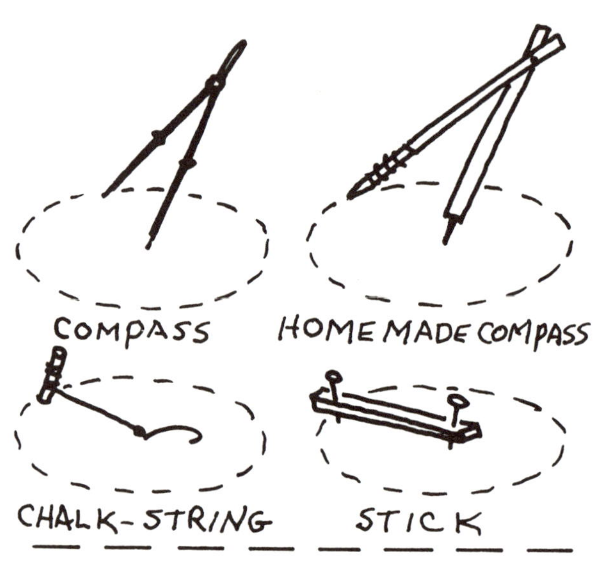


rect length and nail one end to the point which is the center of the circle to be inscribed. Then locate the position along the board that represents the outside of the circle (distance from center to outside is the radius) and either hold a pencil against it and sweep it around, or cut a notch to hold the pencil, or drill a hole in it, or if a piece of wood is to be marked drive a nail through and use it to scratch the line on the surface of the board.

Irregular curves may be either marked out free hand or by the aid of a simple jig that can be easily made from a scrap of lumber. A thin

THIN BOARD BENT TO CONTOUR

Egyptian motif with canted archway in a setting for *Antony and Cleopatra.*

strip of lumber, not more than 1/4″ thick, may be used. If the curve is small and tight an even thinner strip may be required. The strip is then simply bent to conform to the desired contour and while being held in that position a line is drawn along side of it with a pencil or piece of chalk.

NOTES ON MARKING AND CUTTING SWEEPS—To get a pleasing edge for the contoured opening it is necessary to mark and cut the sweeps carefully. Where a primitive feeling is desired a rough free hand quality is all right, but if any degree of sophistication is necessary the lines should be true and the curves gradual and regular in their contour.

Long gentle curves may actually be cut with a regular hand saw (rip or cross cut depending upon whether the cut is with or across the grain; usually the rip saw will do very well).

Relatively sharp contours should be cut with a keyhole saw, power band saw, stand type jig saw or a portable jig saw.

Usually the curve of the sweep where it departs from the straight line surface should seem to grow out or spring out of the surface. When completed it shouldn't have a tacked on look.

A paper template or pattern can be cut first and then the paper laid onto the wood in such a way that waste wood will be reduced to a minimum. This is essential when many pieces are to be cut.

ARCHWAY SHAPES—The imagination of the

Archways used in setting for *The Madwoman of Chaillot* (above) and *Le Bourgeoisie Gentilhomme* (below).

designer is the only limit to the shape, size and proportions of archways, windows and doorways. Some of the shapes are based upon architectural characteristics, some on structural materials, some on historical structures and some on the whim of the designer.

WIDE ARCHWAYS—In the alternate method for making door openings it was suggested that door units could be made by taking two

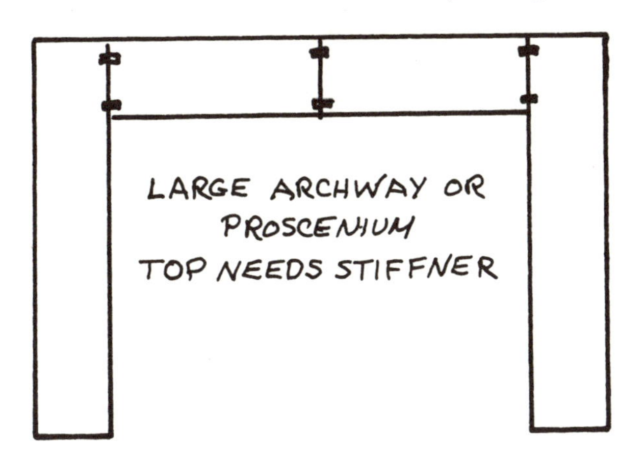


jogs and hinging a plug in between them to make the door flat. Wide archways may be made in the same way. If the opening is more than 6′-0″ or 7′-0″ in width two flats hinged together may be used for the overhead piece, or header as it is sometimes called. If constructed this way the unit will consist of four flats with three hinged joints. The widths of the flats should be checked carefully for folding and if a tumbler is needed it should be placed in the center. If the over-all width of the opening is critical, the extra width provided by the tumbler will have to be compensated for in the width of the flats. Building the archway and hinging it this way allows it to be folded into a compact book.

WIDE ARCHWAYS, ALTERNATE CONSTRUCTION—There are times when it is more convenient to have a long rigid spread across the top of the archway. In this case a single flat may be used for the header and legs hinged so that they extend down from one edge. Measurements should be checked carefully to be certain that when folded the unit will not exceed

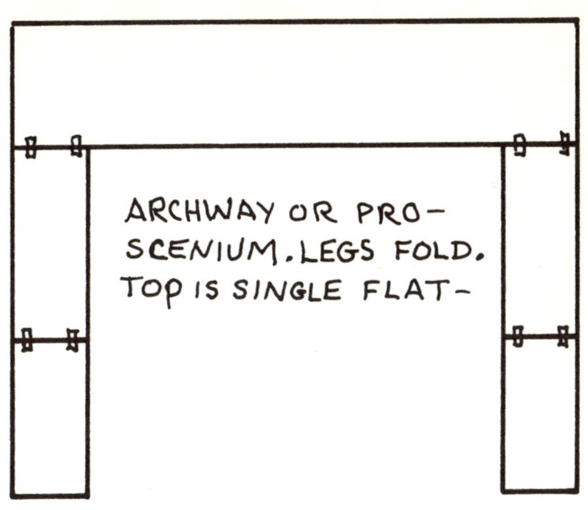


the measurement of doors between the place where it is constructed and where it is to be used. If necessary the legs might even be made with a joint in them in order to facilitate folding the unit.

### False Prosceniums

The false proscenium is basically an archway. It may be wider and higher than most other arches used on the stage, but it is still constructed in the same way. The unit may be so large that it is inconvenient to fold it into a single book. In this case it should be built and put together with loose pin hinges, bolts and wing nuts, or some other device that will allow it to be rapidly assembled and disassembled.

The false proscenium may not have thickness, but in all probability it will have an indication of depth, either painted or three dimensional. If the thickness is real and no more than 12″, lumber will be the best material to use; but if it is deeper, flats will be more convenient and lighter in weight. The thickness can be planned so that it serves as a stiffener for the archway. To do this the thickness should extend across the full width if the width is cut into segments and hinged, or the side pieces should run the full height if the legs are hinged on below the header.

If curtains or drops are to be rigged behind the proscenium it is best to plan a support that will carry the weight and strain of these. This support may be fastened to the archway, but if the stress and strain of pulling the curtains is thrown onto the flats the muslin covering them will probably show wrinkles.

Support for Curtains, Drapes or Drops—A relatively narrow, elongated box made of 1″ x 6″ lumber with sturdy legs will

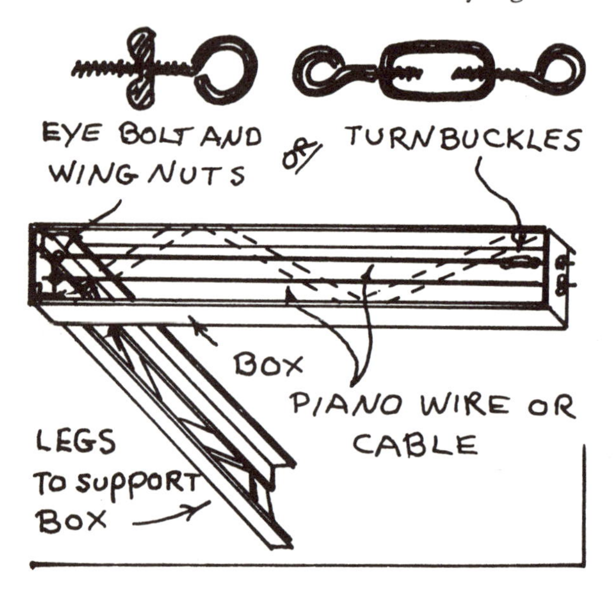


support any of the materials that are likely to be used with the false proscenium. The box should be designed specifically for the individual situation.

Roll drops will require a means of anchoring the top batten, pulleys for the operating lines to pass through (a single pulley above one end of the drop and a double pulley above the other end), and tie-off cleats to fasten the lines to when the drop is in the open position.

Traveller type drops or curtains that can be opened and closed like draw draperies will require a regular traveller track or parallel wires stretched across separated from one another by 3″ or so, depending upon the bulk of the material from which the curtains are made. If lightly painted or dyed muslin is used 3″ will probably be plenty of clearance. Two wires are needed so that the curtains when pulled together can overlap far enough to mask in the center. Some type of draw cord may be needed on this track to open and close the curtains, unless that can be done simply by pulling the ends of the material.

Full Sized Single Drops—It is possible to make full sized drops, drops large enough for a series of scenes to be painted in one continuous strip, or almost any other arrangement of drop combinations. One wire can be stretched across and if the wire and frame on which it is mounted extend offstage far enough, the muslin drop can be stored just behind one of the legs of the proscenium before it is used, pulled across the stage and stored behind the other proscenium leg after it has been used. If one long continuous strip scene is used the storage space on each side may have to be as much as 4′ or even more. Several wires can be rigged in the frame above the proscenium and individual drops rigged on each, or one stored behind each proscenium leg to be pulled on and off as needed. With the continuous strip method it is sometimes difficult to find the correct section in a hurry. If the other system is used any one may be easily found and pulled on. It is also possible with this system to have individual curtains behind one another so that the

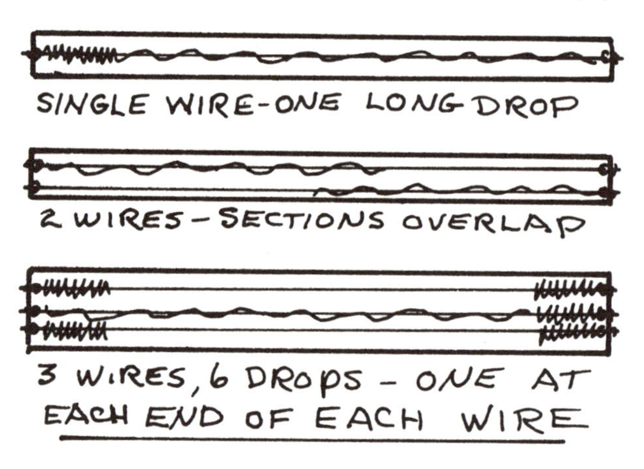


one in front can be pulled to reveal the one behind. Any number of other arrangements could be devised for this type of staging and the structure changed to fit the requirements.

### Curved Walls

Curved walls or curved corner sections present somewhat of a special problem on the stage. A curved effect can be achieved by hinging a series of straight flats together and setting them up so they deviate from the

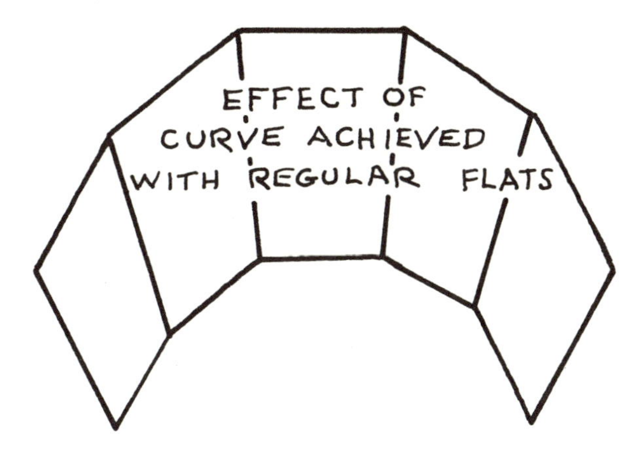


straight line enough to conform to approximately the curve desired. If a smooth curve is desired, special sweep sections must be built.

Curved surfaces should usually be covered with some type of firm surface material. Fabric on a curve will tend to shrink to a flat surface, which is usually undesirable. The ordinary wall board materials come in pieces that are 4′ x 8′. If the surface is designed to make use of this size, in as few pieces as possible, the surface will be more satisfactory since better sweep effects are achieved with large unbroken pieces. Plywood (with the grain running at right angles to the curve if possible), pulp boards, paper boards, sheet aluminum, plastic and corrugated box board are some of the materials that can be used.

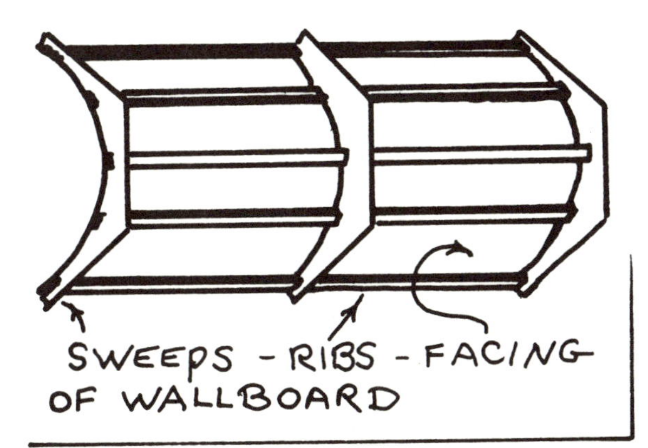


To build the curved section cut the desired contour shape from one or more boards. For this individual pieces of 1″ x 6″ or wider lumber that are cleated together, or single pieces of thick plywood (¾″ sheathing plywood) may be used. The plywood sections, although stronger, have a laminated edge. Wall board nailed into this edge isn't as firmly attached as when nailed into the edge of a solid board. The height of the section, as well as the type of covering material used, will determine how many sweep sections will be needed to hold the form together. It may be possible to make a section out of a single piece of wall board with one sweep at the top, one sweep at the bottom and one 1″ x 3″ stringer on each edge. The section can be strengthened by adding sweeps and other stringers. The stringer mentioned is an upright piece used to space the sweeps and fasten them together to make the basic form. When enlarging the sweep section in height or width it is necessary to increase the size and number of sweeps and uprights. The framework should be planned so that there are solid supports wherever two pieces of wall board are matched together.

### Ceilings

Stage settings are frequently covered, either completely or partially, with ceilings. A ceiling is an oversized flat section either made in one piece or in several pieces which are hinged together. The ceiling may be flown when not in use or built so that it can be taken apart and rolled up like a backdrop.

One Piece Ceilings—The one piece ceiling is built like a large flat. Since its size in at least one direction exceeds the length of normal lumber, scarf spliced pieces are used.

To achieve greater strength, the size of the corner blocks and keystones can be increased.

The ceiling should be wide enough to cover the width of any setting likely to be used on the stage. Under most circumstances it is not desirable to have the ceiling extend down as far as the curtain line since it then becomes much more difficult to light down under and inside the setting. How far upstage the ceiling is placed will depend upon the individual theatre, but it should probably be placed at least 4′ from the curtain line. To facilitate the overhead masking a flipper is usually hinged to the ceiling. A flipper is a narrow flat, as long as the ceiling, that is hinged to the ceiling so that it forms a facing and stiffener for the front or leading edge of the ceiling when it is in place on top of a setting. The flying lines for the

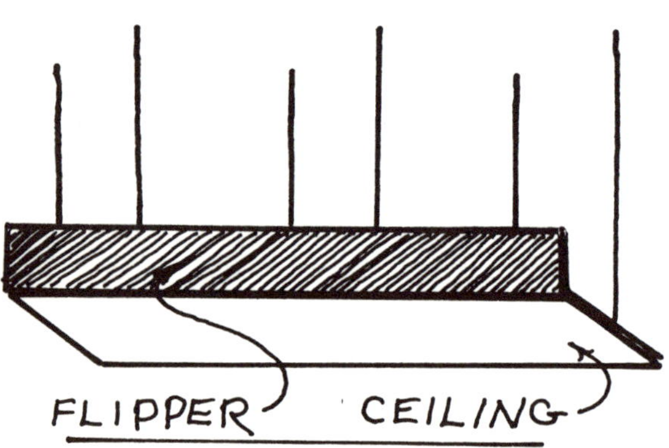


ceiling usually pass through rings attached to the top edge of the flipper and are tied to ceiling plates which are firmly attached with screws and bolts to the front batten of the ceiling. The ceiling is usually provided with two sets of lines, one in front and one upstage, so that it may be swung into position above the setting more easily.

Book Ceilings—A two piece ceiling is used where a single flat will be too large to handle. The two piece ceiling consists of two flats long enough for the width of the setting and half as wide as the depth. They are hinged together with backflap or strap hinges spaced not more than 3′ apart. A flipper is hinged to the leading edge as on the one piece ceiling and a stiffener is securely fastened to the edge of one of the flats at the hinged joint. This type of ceiling uses three sets of flying lines: one set across the front, one set across the center, and one set across the back. The center set is used to fly the ceiling so that it can fly folded and be lifted up higher into the fly well. The other

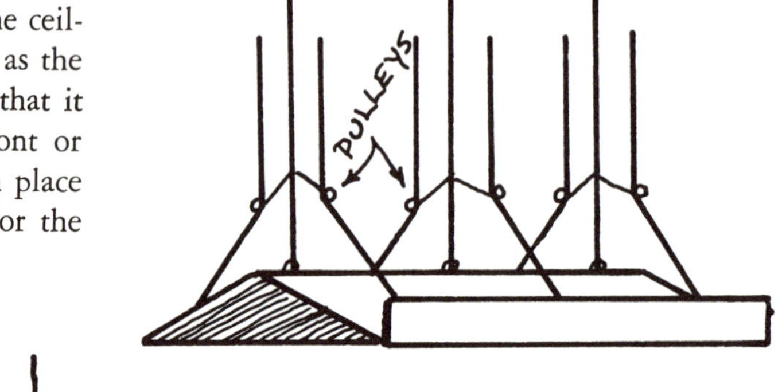


BOOK CEILING LINES CONCENTRATED-PERMANENT

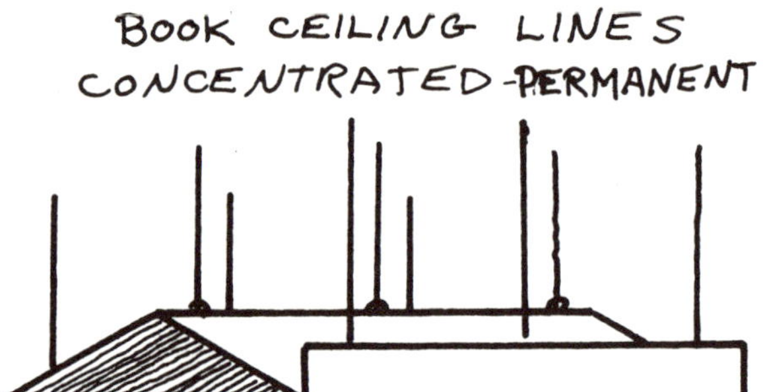

BOOK CEILING, LINES ABOVE FRONT, CENTER AND BACK

two sets may be removed when not in use or a complex system using three sets of lines closely spaced can be used to handle the ceiling and can be left on when it is in use and in storage.

Roll Ceilings—The roll ceiling is made with only the two long battens securely fastened to the fabric. The other battens, the ends, and all of the toggles are temporarily fastened in place with small steel plates (ceiling plates) that may be easily removed. The fabric is temporarily tacked to the end battens, or the battens are slipped through pockets sewn on for that purpose. When the ceiling is not in use it may be disassembled, the fabric rolled on the rails, the toggles tied up in a pack, and

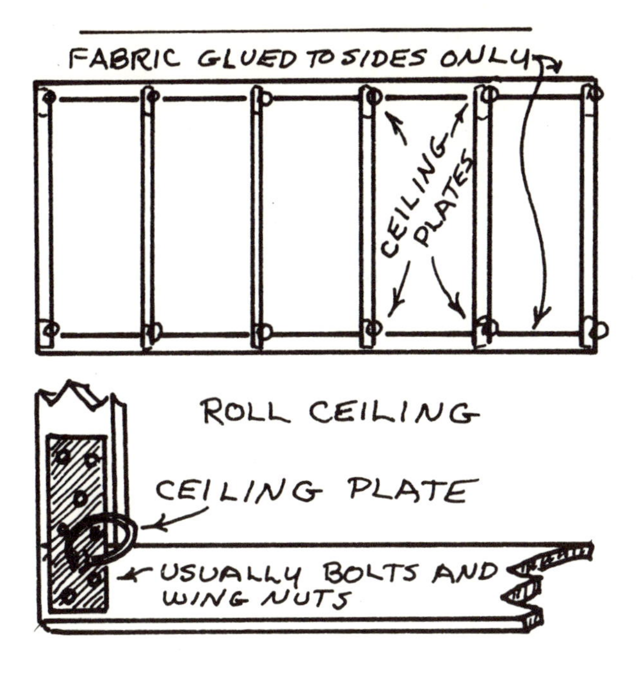


the whole thing stored on a rack or transported to the next stop.

### Screens

Double faced screen type scenery has many uses. The screens are much the same as decorative screens often used in the home, except that they are usually, though not always, a little larger. Since a smooth surface is desired on both sides of the screen frame, corner blocks and keystones are a little awkward to use for securing the joints. If they are used it is necessary to put strips around the back of the flat near the outer edge in order to make the entire flat the thickness of the 1″ x 3″ plus the plywood. Strips of plywood or thin, narrow strips of wood can be used for this purpose. It is really better to employ wood joints that will give a smooth surface on both sides. If the flats are made commercially the corners will probably be fastened with mortise and tenon joints. Unless there is special shop equipment available it is a complicated process to put the frames together this way. An easier and simpler type of joint is the halved joint. Although 1″ lumber can be used for screens, something a little heavier, like 1¼″ or even 2″ lumber, is better.

After the frames have been constructed they are covered on both sides. The covering material may be muslin that is painted after it is put on the frame or it may be dyed or patterned print material. The covering can be glued on the frame or sewed to form a tube and then slipped over the frame like a pillow slip is put on a pillow. If a tube is made the material will require gluing or tacking only along the top and bottom.

After the flats have been covered they can be hinged together. Two-way (double acting) screen hinges are used for this purpose. They are a little difficult to find, but most hardware stores of any size should have them somewhere in their stock. The hinges are placed on the edge of the flat rather than on the face or the back, and they are designed so that they will fold readily in either direction. In fact the flats will fold snugly together in either direction if the hinges are at least as thick as the wood of the frame.

### String

String may be used on uncovered frames to achieve interesting effects. "L" shaped frames, "U" shaped frames or complete rectangular

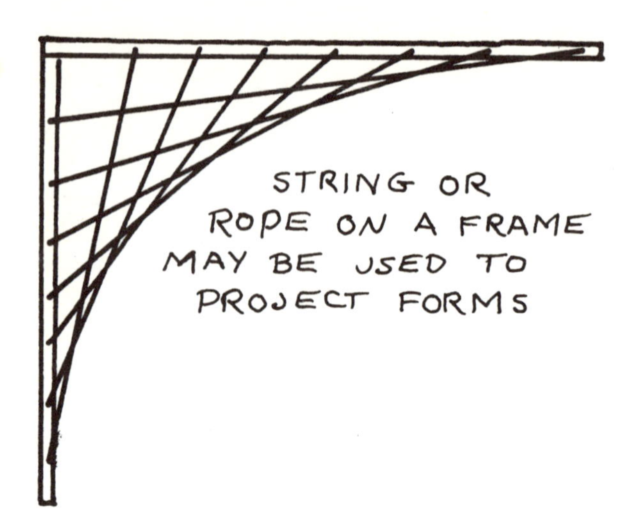


frames lend themselves to this treatment. Various patterns may be tried.

### Fireplaces

There are literally thousands of different types of fireplaces. The fireplace of each home differs in some respect from fireplaces in every other home. Basically they all have a place for the fuel and most of them have a mantel, but the shape and arrangement of both of these differ greatly.

UNIT FIREPLACE — The conventionalized stage fireplace unit that fits in front of the flats is complete in and of itself. It consists of a front and back, thickness on both ends, thickness around the fire box opening and mantel to cover the top. The entire unit, with the exception of the top, can be made of individual flats that are covered and then fastened together. The top should be of some type of firm material such as boards or plywood. The top can be put on so that it is flush all of the way around and then various moldings can be applied afterward to provide the conventional overhang on the front and ends.

FLAT AND MANTEL FIREPLACES—An alternate type of fireplace uses a flat with a hole cut in it for the firebox and a mantel that is designed to be hung on the flat above the firebox opening in a desirable position. This type of fireplace requires less storage space than the other and the flats are all available for other uses later. With this type of construction it is possible to use a flat the desirable width for the front of the fireplace, and fasten a jog on either side to provide depth for the chimney.

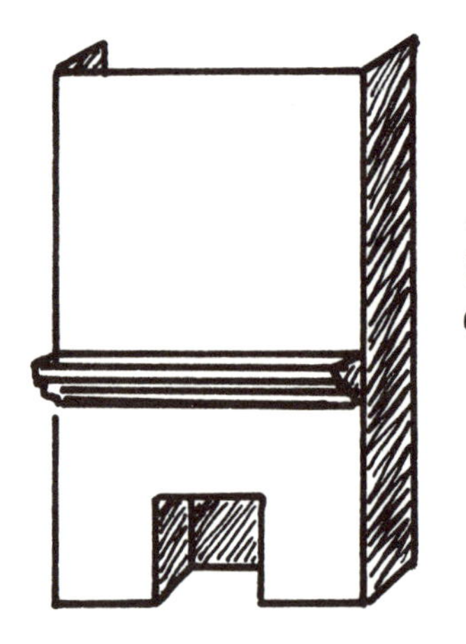

FIREPLACE
MADE OF A
FLAT WITH
CHIMNEY
THICKNESS
FLATS-MANTLE
AND HOLE
FOR FIREBOX


These two types of fireplaces have somewhat different appearances and it is very possible that both may be used at times. There may be other times when other designs will be needed. The setting for *Papa Is All* should be equipped with all of the paraphernalia of a fireplace in which the family cooking is done. The old fireplace in *Ladies In Retirement* must be large enough for a body to be stuffed in through the door. The fireplace in *Shop At Sly Corner*

must be rigged so that it can be opened to allow the shopkeeper to enter his secret kiln room. Other fireplaces have secret compartments, are large enough to hide inside, etc.

Usually it will be advisable to look at pictures until the correct type of fireplace is found and then build accordingly.

### Thickness

Walls of buildings are more substantial than the flat structures used on a stage. In order to project a third dimensional feeling of solidity it is necessary to suggest that walls have thickness. This is done around the openings in the walls such as doors and windows.

TEMPORARY THICKNESSES—Sometimes boards nailed at right angles around behind the edge of openings can be left in place for the duration of the production. This is possible on a one set show. It is also possible when the units are small enough to be handled without folding or when the thicknesses do not interfere with the folding. When permanent thicknesses are used it is also possible to allow some of them to extend across the entire wall section and serve as stiffeners. The header thickness above a door or window can be extended out this way. Boards used for thicknesses of this type may be nailed or screwed into place. Double head nails can be used to advantage as

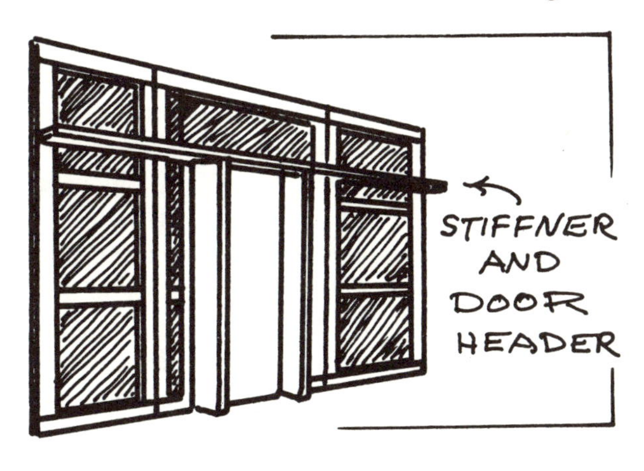


long as they are not in a position where they will catch on costumes.

CURVED THICKNESSES—One quarter inch or thinner plywood cut so the grain runs across the thickness direction can be nailed in place around contours to produce contour thickness. Small nails should be used so that the plywood

can be easily removed and kept for reuse.

PERMANENT THICKNESSES — The permanent thickness is carefully built to the theatre's own standard size and kept as part of the regular stock of equipment.

BEHIND THE FLAT TYPE THICKNESS — There is one type of thickness that fits on the back of the flat and its opening is the same size as the opening in the flat. Since this type of thickness is designed to be used many times it is desirable to set up size standards before the construction begins. For the present purpose assume that the doors are to be 3′-0″ x 7′-0″. It can also be assumed that the walls are about 6″ thick. (Most inside walls in homes are just about 6″ by the time they are plastered or covered with wall board.)

There are two basic methods of building this type of unit. The first style requires a little less time for labor and less lumber. However, the finished unit isn't as sturdy and will not stand as much abuse as the second type.

METHOD NO. 1—The opening in the flat is 3′-0″ x 7′-0″ so the finished door jamb, or thickness, should measure 3′-0″ on the inside from side to side and 7′-0″ from the outside of the bottom to the inside of the top. When in place the bottom rests on the floor and the underside of the top board will be flush with the top of the flat opening. This thickness resembles a low topless and bottomless box when completed.

The following lumber will be needed:

2 — 1″ x 6″ — 3′-0″ long (for top and bottom)

2 — 1″ x 6″ — 7′-¾″ long (for sides)

These boards are nailed together with the sides running through. 8d box coated nails, or 1½″ or 2″ #9 flat head wood screws should be used for fastening the frame together. The

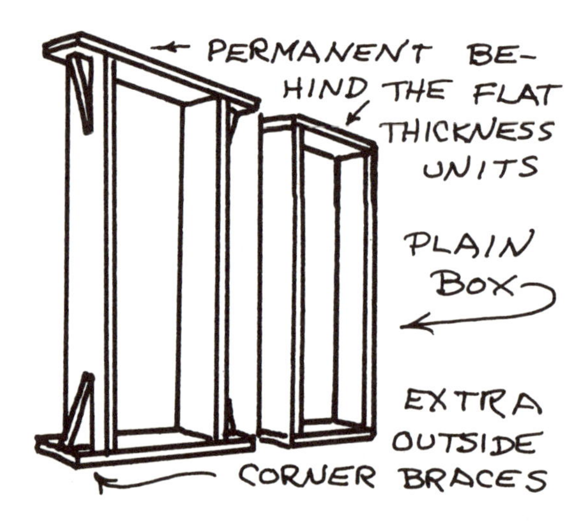


fasteners go through the sides and into the top and bottom. The nails or screws may gradually work loose and if they are put in from the sides they cannot cut into the floor.

Method No. 2—If the top and bottom members are cut to run through and are extended at least 6″ beyond the side members of the jamb on each side, it is possible to nail a short diagonal brace into place to stiffen the unit and help it retain its shape when in use and storage. A right triangular shaped piece of 2″ x 4″ may be cut for the brace. The braces are placed just beyond each corner of the frame on the sides where they fit into the corner made by the extended top and bottom and the sides.

Window Thicknesses — Window thickness units may be built similar to the door thicknesses just described. For normal use this thickness will work all right. If the window sill has to support considerable weight it should have supports running to the floor so that the strain isn't thrown onto the flat.

Window Faking—In many instances the window thickness does not show after the stage dressing has been completed. Window curtains, glass curtains, lace curtains, venetian blinds and other dressing make it unnecessary to have thickness. So unless there is window business or unless there are no trappings to cover the sides, the window thickness may be ignored.

Fastening Behind-the-Flat Thicknesses in Place—Double head nails, 1½″ or longer wood screws and loose pin hinges are the major fastening devices that can be easily used to fasten behind-the-flat thicknesses in place. The double head nails are especially easy and convenient to use.

Behind-the-flat thicknesses are inconvenient to remove during scene changes unless loose pin hinges are used. But of course it is possible to design in terms of using this type unit so that the thickness and wall may be moved all in one piece. However, there is a type of removable door and window unit that may be used advantageously. This unit will be called the Hinge Locking Unit.

The Hinge Locking Door Unit—The jamb of the hinge locking door unit slides into the flat from the front so it is necessary to construct the unit with its outside measurements at least an inch narrower and an inch shorter than the opening in the flat. When this thickness unit is built the trim or casing is nailed or screwed onto the face of the jamb across the top and both sides. The trim forms a stop so that the unit can be pushed snugly into place from the front of the flat. A strap hinge is mounted on the unit about half way up the jamb. It is mounted on an angle with the thin end of one strap toward the bottom of the jamb and about a quarter of an inch from the edge the trim is fastened onto. The center of the hinge is then positioned so that it is about 1″

or slightly more from the same trim edge. This half of the hinge is secured to the jamb with 1" #9 screws. The other half of the hinge remains free. Two hinges placed along the side of the jamb will allow for even more secure mounting in the flat but one on each side will do well enough in most instances.

This unit is mounted into the flat by pulling the unfastened hinge halves up and holding them in that position while the door is slipped

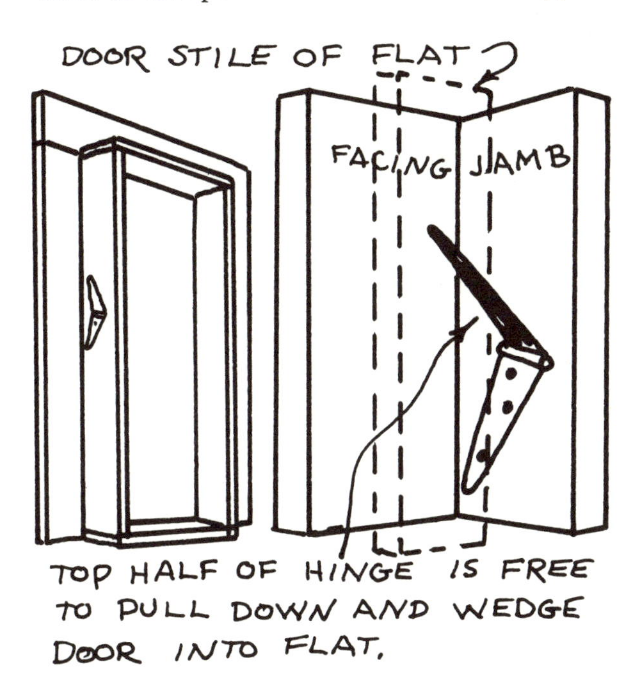


into place in the flat. The loose hinge halves are then forced down and wedged against the stiles of the flat, and the unit is locked into place.

Hinge Locking Window Unit — Window units can be made in the same way as the door unit described above. The window unit will have trim all of the way around it and may also have a sill strip mounted to the bottom member of the jamb projecting out in front an inch or so. This is to resemble the sill on an ordinary window. The locking hinges are mounted on the sides of this unit just as on the door unit.

Contour thicknesses to fit behind contour openings should conform to the shape of the opening. A contour piece may be attached flush with both edges of the thickness in the appropriate places. The space between may

be covered with muslin if the curves are complex; plywood or wall board may be used if the curves are simple.

Other methods of making and mounting door and window units might be devised that are easier to install and more convenient to shift and store.

### Door Shutters

A simple stage door, or shutter, may be made like a flat. It should be a quarter of an inch narrower and an inch shorter than the opening to allow for clearance and to prevent the door from binding when it is opened and closed. Muslin, canvas, or more substantial material may be used for covering the door. Heavy paper-type wall boards, wood pulp type material such as Masonite and Prestwood, ⅛" or ¼" plywood or any other firm material will do for a harder surface.

Although the harder surface materials will stand more wear and abuse than fabrics for the door covering yet at the same time they will increase the weight of the unit.

At times it is convenient to plan the hard-

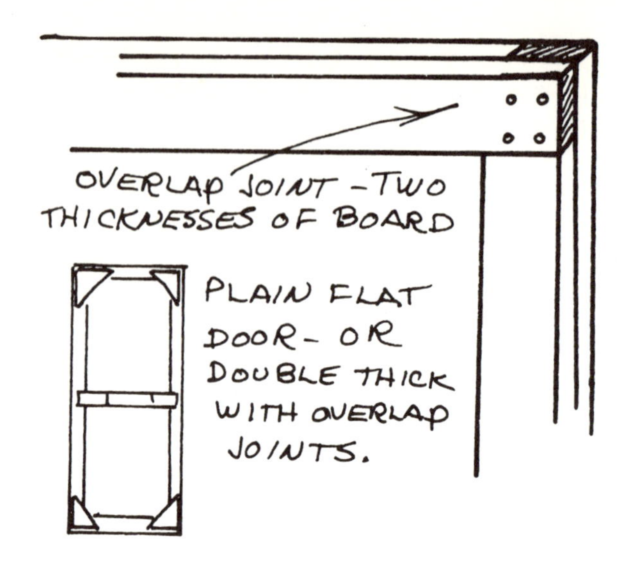


ware for the door at the time the frame is constructed so that door latches, locks or catches can be provided with toggle bars where needed.

Doors To Open Onstage—Even though most doors are planned so that they will open offstage there are times when the stage business makes it necessary to have one open onstage. This necessitates covering both sides of the frame since the audience can see the one side when the door is closed and the other when the door is open. Usually it is better to build the frame with more thickness so that the door seems to be more substantial.

A rectangular door may be mounted behind an irregular or curved top door opening. The portions above and beyond the opening may be painted black so that the door seems to conform to the shape of the opening.

Use of Regular House Doors—Regular house doors may be used onstage but frequently they are heavier in weight than doors built specifically for the stage.

Surface Finish on Doors—All of the detail on doors can be painted or applied to the surface. Moldings can be fastened onto the surface to simulate panelling or the doors can be built with actual recessed panelling. A plane surface door can be convincingly painted with panelling and changed easily and readily each time it is used.

Door Hardware—Regular door hardware, including bolts, locks, latches, mortised sets, rim lock sets and other types of catches, can be used. Other hardware that might be used includes spring type screen door hinges, hydraulic

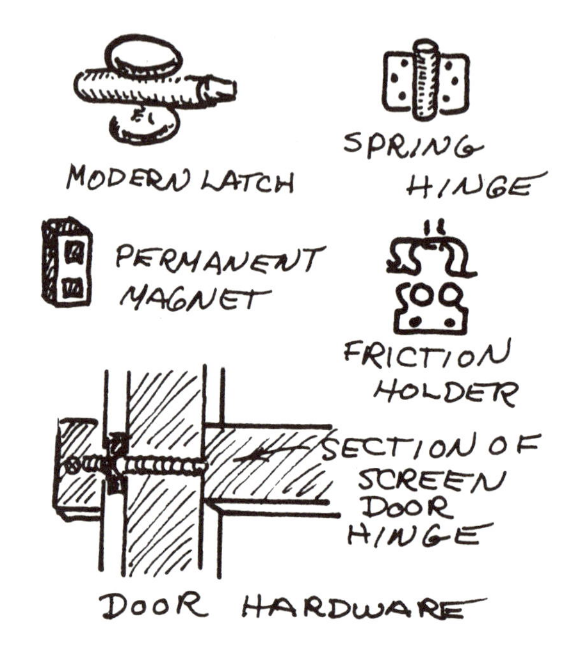


door closers, cabinet door closers, magnetic door holders and spring screen door closers.

Door Stop—If no thickness is being used a door stop may be easily made from a lash cleat, hinge half, piece of plumber's strap or a piece of wood. This may be placed on the back of the door extending out far enough on the swinging edge so that it will strike the rear of the flat when the door is closed.

When a door has a thickness the stop consists of a strip of wood not more than 1″ thick and not more than about 1½″ wide. With the door in its closed position the strip is permanently nailed around next to the shutter at the top and on the sides. This not only serves to stop the door but it also prevents light leaks.

It is possible to make double faced universal doors with the knob placed in the center near one edge. These doors may be placed behind the unit so they will swing offstage in either direction by turning them so that either end is the top or bottom, and turned around they will swing onstage in either direction.

### Window Sash

Window sash are needed occasionally. If they are plain sash that are never moved they may be built and nailed permanently onto the back of, or into, the thickness piece.

Standard home window sash show about 2″ of frame around the glass on the top and sides and about 3″ on the bottom. Double hung windows, with sash that can be raised and

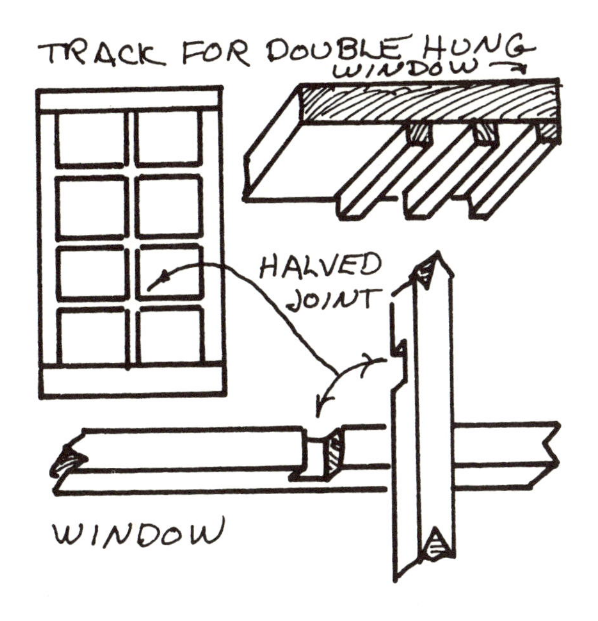


lowered, have a break across the center where the two sections overlap that is slightly narrower than the other framing members. These boards measure about 1½″ in width. Sash may be built with halved joints or butt joints with narrow strips of plywood to secure the corners. Cross bars (mullions) may be fastened together with notched joints where they cross one another and may be fastened with strips of plywood where they attach to the outer boards of the sash.

French doors and windows can be built in the same way. French doors will have slightly wider boards for their frame. 1″ x 4″ for the top and sides and 1″ x 8″ for the bottom will do very well and 1¼″ strips for the mullions that break the space into small panes.

Double French windows and doors are easier to handle if one side need not be practical. It can then be nailed permanently into place and the keeper for the latch of the other door can be mounted in it. A regular door latch or lock can be used to hold the practical side closed.

Double hung windows are easier to rig for operation if only the bottom sash need be opened. The top may then be fastened permanently in place in the thickness unit. The top sash is outside and the bottom sash inside in relation to one another.

Simple guides for the sash that move up and down are made by fastening strips of ½″ x ½″ lumber on the sides of the jamb so that the window can be slid up and down. A snug fitting sash will stay in place when lifted open.

### Glass

Glass is the poorest substitute for glass on the stage. There is too much danger of breakage. Frequently it is necessary to use nothing in the panes. But if it is necessary to simulate glass, a number of substitute materials may be used. If the glass is to be transparent try sheet plastic, aluminum or plastic screening, cheese cloth or marquisettes that have been dyed a light gray or blue gray. If translucent glass is desired muslin dyed gray or blue gray may be used. If fastened to the frame immediately after being dyed the muslin will tighten up as it dries and provide a smooth, taut surface.

Diamond Shaped Glass Patterns — When small diamond shaped panes are desired in a window unit, friction tape can be stretched diagonally across the back of the sash frame in both directions. Pressing the point of crossing of the tape will help fasten one strip to another.

### Miscellaneous Doors

There are times when doors of a slightly unconventional character are desired. Sliding doors and Dutch or half doors fall into such categories.

SLIDING DOORS—Sliding doors are rather easy to make and handle. As a matter of fact, the "pull" on the flattage isn't as great as it is for a regular hinged door. The jamb is built in such a way that the top and bottom members extend out beyond the side on which the door is to disappear far enough to provide a track in which to "store" the door when it is behind the setting. Narrow strips of wood may be used to make the sides of the track in which the door slides. A little heavy floor wax or paraffin (old candle) rubbed on the bottom member of the jamb where the door is to slide

will help make the door easier to open. Be careful not to allow too much vertical clearance or the door will twist and stick.

DUTCH DOORS — Dutch doors should be kept light in weight if possible and of course be securely hinged. This type of door usually swings onstage which necessitates double covering and added thickness. Usually additional bracing is needed on the door flat — or careful designing so that the wall section is narrow and braced by adjacent flats which are placed almost at right angles to the door flat.

ABUSED DOORS—Doors that receive great abuse — that is, doors that are slammed hard and slammed a great many times in a play — are always a problem since the hard slamming

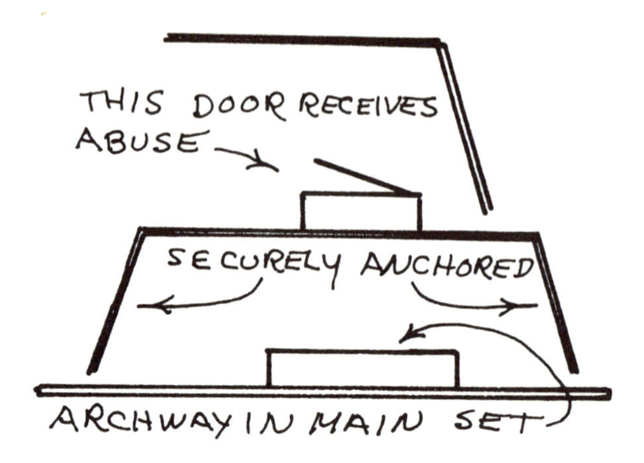


causes the entire setting to quake and quiver. The hallway door in *Junior Miss* is an example of this kind of door. If this type of door can be designed so that it is behind an archway, in an alcove or entry position, it is possible to separate the door's flattage from the rest of the setting. Through the archway only a small portion of the area around the door will show and the apparent shaking of the setting will be diminished a good deal.

SWINGING AND REVOLVING DOORS—Two other types of doors may be needed occasionally. These are the swinging door and the revolving door. Hardware for a swinging door is somewhat expensive and at times difficult to find. A simple substitute that works fairly well is a section of heavy screen door spring mounted to the jamb and then to the door. If the spring is mounted in such a way that the spring tends to straighten back into its normal position when the door is in the neutral position it will work quite well. However, it may be necessary to do a little trial and error work. A piece of spring mounted the same way on each side of the door will also work. Double acting screen hinges will work along with a spring also.

In order to build a revolving door that will work easily it is necessary to have a bottom swivel that operates easily and smoothly. The pin at the top merely keeps the door in position and is not as important. A good swivel may

be made by removing the tire or wheel from a caster. A caster with ball bearing type swivel will probably operate easier than the simpler variety. Grease the swivel well before using. The top of the door may be held in place by driving a nail down through the jamb into the top of the door or by drilling a hole and dropping a pin, bolt or nail into it. It is important that the fittings be carefully positioned so that the door will swing true.

## Ground Rows

Ground rows are usually low, relatively long, flat structures with contoured sides and top. They serve several purposes. They conceal the base of the cyclorama, mask base lights that are used to illuminate the lower portions of the cyclorama, and serve as a perspective device. When contrasted with relatively large foreground units and actors playing downstage the lowness of the ground row gives a feeling of greater distance than actually exists between forestage and backstage. Sometimes ground rows are placed in depth, one behind the other, to achieve still greater distance. And some-

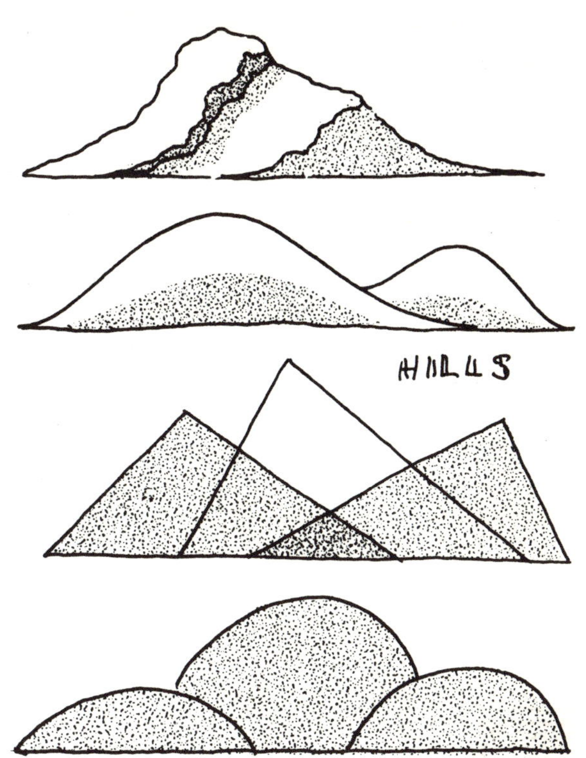


times, also to achieve depth, they are spattered with the cyclorama or sky color, so that they will fade into the background. At other times they are set in bold relief when the sky is lighted behind them.

Mountains, hills, bushes, trees, buildings, fences, flowers, city sky lines, rows of buildings,

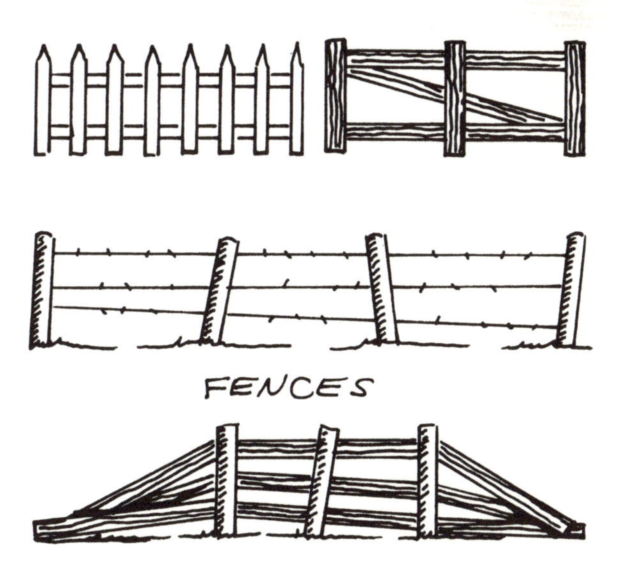

ships, seas and almost any other form desired may be represented by such cut out ground row forms. These may be made out of flat frames covered with fabric or may be made of solid surface materials cut to conform to the desired shape. They might be made out of cardboard, corrugated box board, plywood, Prestwood, Masonite or any of a number of different varieties of wall board. They might be built up out of functional three dimensional forms such as steps and platforms and shaped with chicken wire and papier-mache. Usually they should be designed so that they are self supporting if possible.

Construction of Ground Rows—Ground rows made from substantial materials may need no frames, or at most a strip of wood here and there to stiffen and straighten the surface. They may require a jack or some other contrivance to help them stand up. A jack is a triangular shaped wooden frame. The angle between the board resting on the floor and the one extending up the back of the unit is usually just slightly less than a right angle so that the unit will lean slightly backward against the jack. In this position there is less danger of the unit falling forward. Usually there is a diagonal board connecting the upright and the board on the floor. Sometimes space is left on the foot (board on the floor) of the jack to place a weight or sandbag and make the ground row even more substantial.

If the ground row is small it is possible to merely fasten a board to the back either flat or on edge and use that as a supporting foot for the unit.

Flat Type Ground Row Construction—If there is a stock of small flats in the shop it may be possible to start with one or more of these and then proceed to construct contours upon them until the desired shape is obtained. At other times it will be more desirable to start by building a framework that approximates the shape and then add contours to complete the structure.

1" LUMBER CONTOURS MAY
BE PARTIALLY CUT OUT TO RE-
DUCE WEIGHT.

Contours may be made in a number of ways. Plywood of ⅛″, ¼″ or 5⁄16″ can be cut and nailed to the frame. Plywood can be rabbeted into the frame (frame rabbetted or planed

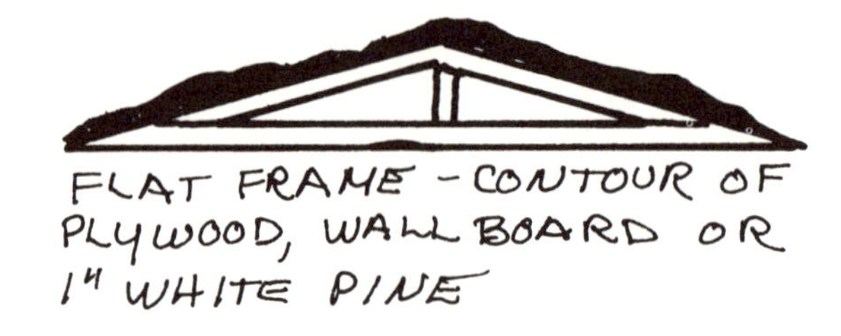

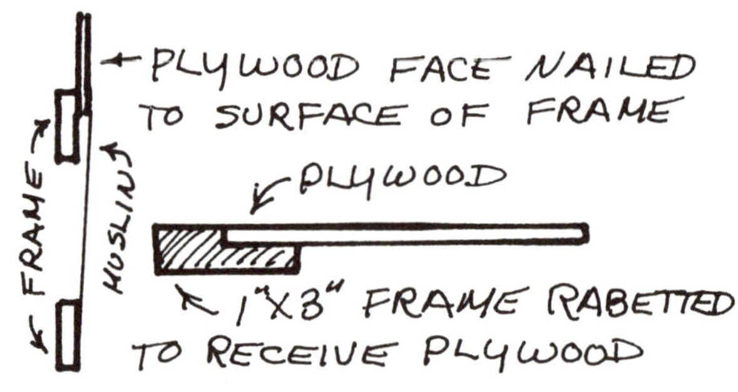


away enough so that the plywood fits into a sort of slot and is flush with the remainder of the frame). 1″ boards of # 2 or #3 common white pine can be used and cut to conform to the shape desired and then nailed onto the edge of the framework; or forms may be cut from thicker plywood and fastened onto the edge of the frame.

The use of thicker plywood (½″, ⅝″ or ¾″) enables one to build more elaborate forms that still have a great deal of strength. In fact, if the forms aren't too large it is possible to build an entire ground row from one piece of wood. The plywood can be cut to conform to the outside shape and then in order to reduce the weight, and also to save wood, the inside can be cut out leaving an outline or frame of wood about 3″ wide. To provide more rigidity in the finished product toggle bar shapes can be left in, wherever desired. The front is then covered with muslin and for all intents and purposes a flat has been built using one piece

of wood. The maximum size that can be made in this way with standard sheets of plywood is 4′ x 8′. For this purpose sheathing plywood can be used. This is the lowest priced grade of plywood. If ½″ plywood is joined to 1″ lumber (which is actually only ¾″ thick) keystone or corner block fasteners will need ¼″ plywood pads between them and the ½″ plywood. This makes the connection the same thickness as the 1″ board and provides a smooth joint on the face of the flat.

### Trees

Trees of small dimension can be cut out of single pieces of firm surface material such as cardboards or wall boards. Medium sized trees can be made as flats. Large trees can be built with separate trunks and foliage so that the trunk rests on the floor and the foliage is flown or supported from overhead.

Flat Type Trees—If the trees aren't too large the flat may be made in one piece. Of course, doors should be checked for size or it may not be possible to get the tree from the shop to the stage. The simplest way to lay out a form of this nature is to chalk it upon the floor and then cut boards to fit the contour as closely as possible. Some regard for solidity should be considered when cutting and assembling the frame. The outline should be made of as few boards as possible. If the frame is small 1″ x 2″ lumber will probably be heavy enough. If possible a fairly wide base should be built on the trunk so that the tree will have a firm foot to stand upon. After the frame has been completed the contour may be added. This contour may be made of thin plywood or wall board attached to the front or, if the contours are not too complex, pieces of 1″ board

that are nailed to the edge of the frame boards. When this method is used some of the surplus board can be cut away so that the piece is only about 3″ wide following the contour, but with a substantial edge where it is fastened to the frame.

To support this tree it will probably be necessary to have a jack to rest on the floor or hardware so that wire can be used as support from above — either connected to fly lines or to the ceiling above the stage.

Self supporting trees can be built if they are correctly designed to begin with. In general it is necessary to have a slightly enlarged base. This may represent rocks and earth, an enlarged

trunk or a clump of bushes. The remainder of the tree is designed so that it can be built in two pieces that split the center vertically. The two sections are covered, hinged together and the tree will stand folded slightly, like a screen. Since this tree needs no anchoring it can be easily and readily set up and struck.

Either of the types of trees discussed previ-

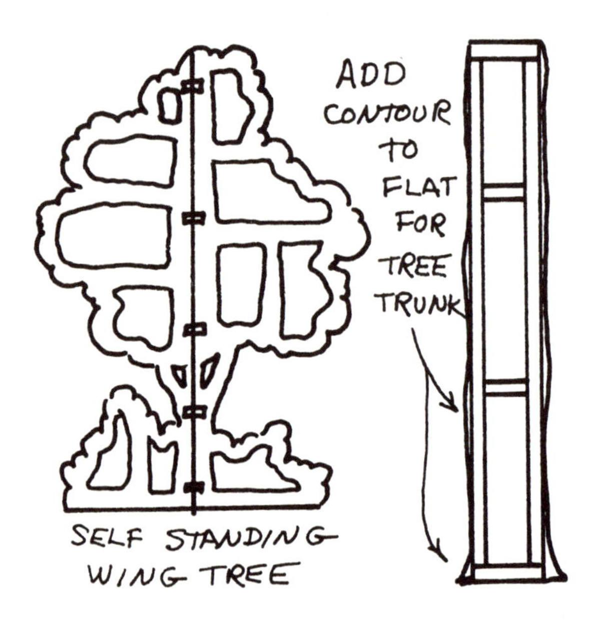


ously can be made using the heavier plywood one piece method described for smaller trees. One piece of plywood may not be enough to cut the entire foliage section so two or more sections will need to be fastened together. If a book or screen type tree is being made it is necessary to split the foliage section anyway so this can be done with two pieces of plywood. This method has much to recommend it. The cost may be slightly more, but the amount of work is no more and the rigidity of the finished product is probably greater and the weight less.

FOLIAGE-TREE-TRUNK TREE—The foliage-tree-trunk variety of tree is satisfactory and probably easier to handle where there is a flying system than the other types of trees. Tree trunks are made by taking jogs (narrow flats)

of the appropriate width and tacking contour strips along the edges to achieve the slight irregularities that a tree trunk might have. These can be cut on a band saw from scrap strips of lumber 1″ or 2″ in width. A piece of wood may be added to each side of the bottom of the flat to suggest the beginning of the root structure and with that the trunk is completed, unless a limb or two is desired, in which case a board cut to the correct conformation is attached so that it projects out and up toward the top of the flat. The foliage is then made in the form of a border which is suspended from the ceiling or fly loft. The foliage may be painted first and then attached to a long board or batten or attached to the batten first and then painted. While being painted it should be temporarily tacked to the floor or some other surface so that it will tighten flat and not shrink in an irregular fashion. The foliage may be designed in depth so that smaller sections are dropped down in front of and behind the tree trunk. The foliage must be planned so that it extends down far enough to mask the top of the tree trunk when this method is used.

Tree forms can be built in still another way which will be discussed with backdrops.

### Miscellaneous Standing Units

Other standing units may be built flat with contours to provide the silhouette shape and

then detail painted on to project their three dimensional form. Objects like houses, barns, sheds, columns, posts, and statues can be handled in this way. Sketch them first to determine their outline form and then build a unit to conform to that shape.

Railings for stairways, patios, theatre boxes and balustrades can be built to conform to the size and shape required. The face of posts and the balusters or spindles can be cut from flat

material such as plywood, fastened into place and painted to appear three dimensional. Thicknesses may need to be added to posts and rail-

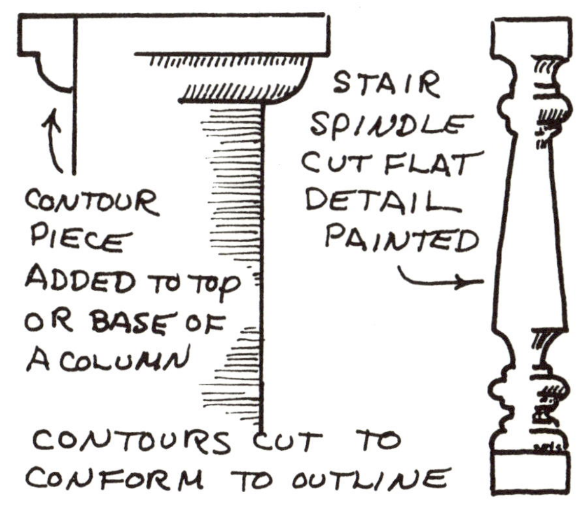


ings later to complete the unit, but the basic form can be built as a flat.

Three dimensional buildings can be built of separate flat sections for the walls and roof which are fastened together later with loose pin hinges, lash line, hooks and eyes or other types of fastening devices.

### Backdrops

The backdrop is a large sheet of muslin or other fabric. It may be permanently secured to a batten at the top, at the top and bottom, or it may be equipped with tie lines along the top so that it can be tied to the pipe or wooden battens of the theatre's flying system. If the top or top and bottom are fastened to battens permanently the drop is rolled up for transportation and storage. If the drop is equipped with tie lines it may be folded and packed in a box or trunk. Conventionally the drop is placed in a position on the stage where its top and sides can be masked by other scenery. However, drops can be designed and built with finished edges all of the way around so that they are complete units in themselves. There are several varieties of drops and they will be discussed individually.

Ordinary Backdrop—The ordinary backdrop consists of a large flat surface upon which a picture is painted that may represent a specific locale. The drop is usually suspended above the stage in such a way that it may be raised and lowered. There are times when a drop is permanently rigged in one position, usually as far upstage as possible, where it serves as the universal scene for use when outdoor settings are called for in any production.

This type of drop may be fastened to a batten by means of tie lines. If so, then the top of the

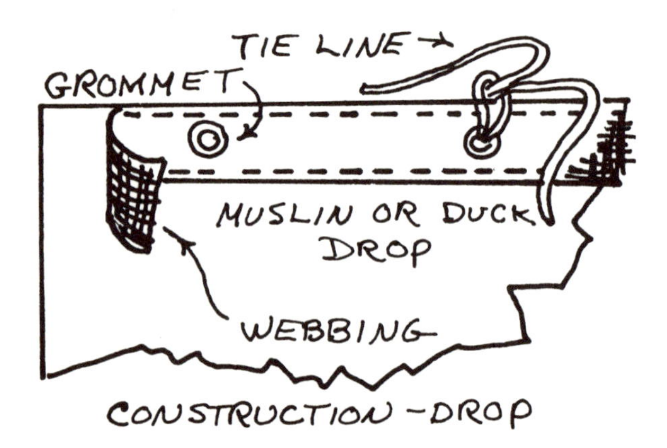


drop probably has a 2″ hem which may be reinforced with a strip of canvas or burlap webbing. Grommets are put through the drop and webbing and spaced about 1′-6″ apart. The lines, of shade cord, or ½″ heavy cotton tape, are tied through the grommets and used to fasten the drop to the batten of the flying system. If this construction is used the bottom of the drop may hang free or may have a pipe pocket sewn into it. The pipe pocket is a hem open at the ends and large enough to slide a pipe into. The pipe pulls the drop downward, tightening and stretching it.

The method just described would be used if the drop were to be moved frequently. A permanent drop will more likely be fastened securely and permanently along the top to a

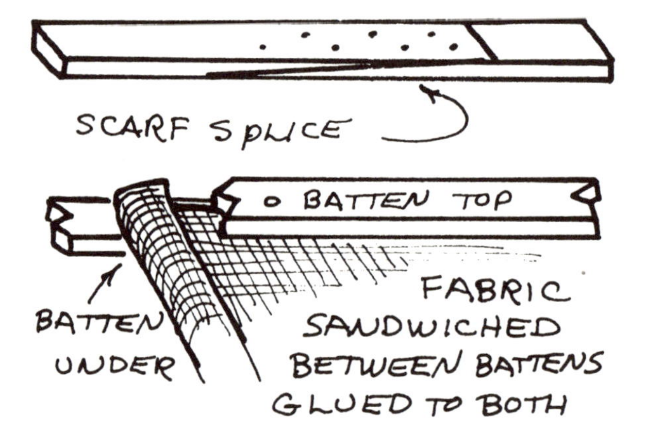


wooden batten. This batten is made of 1″ x 4″ white pine lumber. The boards can be extended to the full width of the drop by means of scarf splices. The batten will be stronger if it is doubled with the drop sandwiched in between the two layers of wood. Before fastening it together the top of the drop will be glued to one board and then glue will be applied to the muslin so that the other batten will also be glued in place. 6d box nails, which will go through two thicknesses of 1″ board far enough to clinch, or 1½″ wood screws will be used to fasten the two thicknesses of wood together. This batten may have holes drilled through it to pass rope or wire through to use for fastening it to the flying system or it may have screw eyes, ceiling plates, eye bolts, drop holders or other similar fastening devices for the same purpose. The bottom of the drop may have a batten, single or double, to which it is permanently secured or the batten may hang free.

FRAMED DROPS—A drop similar to the one just mentioned may be used with some type of stretchers on the sides. These are usually temporary and may be boards that are fastened

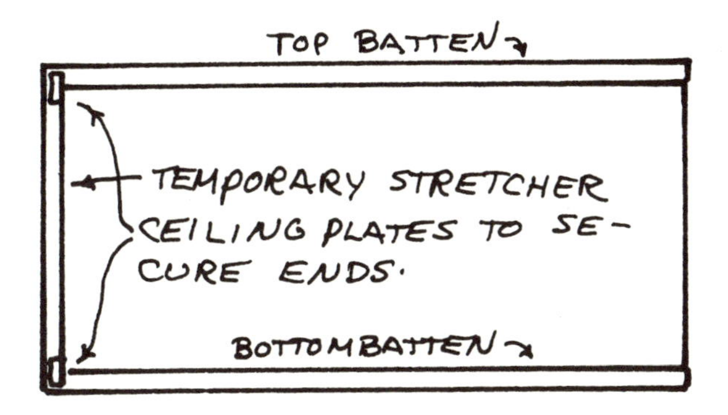


to both the top and bottom battens, probably with bolts and wing nuts, and then the sides of the drop may be temporarily stretched and secured with tacks or staples.

A completely different variety of drop has grommets on all sides and is laced into place on a framework. The framework is portable

LACED DROP OR SCREEN

and fastened together with bolts and wing nuts. The laces fasten to pegs or screw hooks that are appropriately placed around the frame. Motion picture screens and rear view projection screens of the translucent variety are rigged up in this fashion so that they can be stretched taut on the framework.

TWO FACED BACKDROP—A backdrop can be painted on both sides and then rigged so that an extra set of lines can be snapped into eyes on the bottom batten and the drop reversed by pulling up the bottom and lowering the top. The lines should be fitted with snap hooks so that they can be hooked and unhooked rapidly. In preparing this style drop the artist paints his designs so that one is upside down in relation to the other.

TRIPPING A DROP—The process used on the drop just described indicates a method of storing a backdrop where there is a relatively low grid. Attach a set of lines to both the top and bottom of the drop. The bottom lines can extend behind the drop and be in place all of

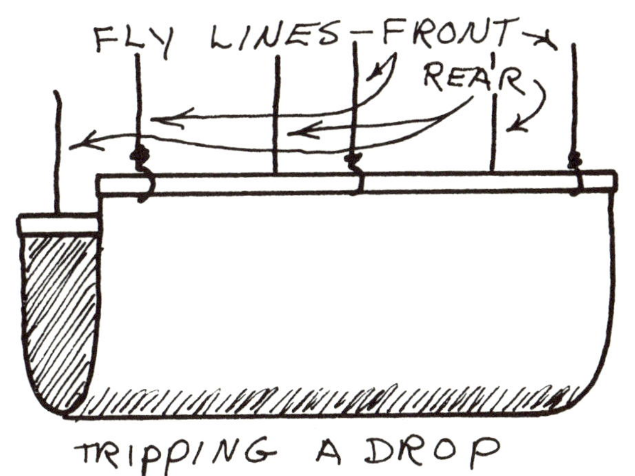


TRIPPING A DROP

the time. The drop is then raised as high as possible with the lines attached to the top and then the bottom is raised or "tripped" up far enough to take it out of sight lines.

TRANSLUCENT DROP — The design on a drop may be applied with thin paint or dye. When lighted from the rear this drop will have a pleasant glow that cannot be reproduced by front lighting, especially if the drop is painted with harsh opaque paint. It is also possible to vary this procedure and obtain effects that will alter as the light is changed from the front to the back of the drop. The face of the drop is painted with dyes and the rear of the drop is painted with heavy opaque colors so that sections will allow no light to pass through. The effect when the back lighting is used will be a combination of glowing sky and dark areas of mountains or trees or buildings in silhouette.

SCRIM OR TRANSPARENT DROP—Scrim is an open weave fabric like cheese cloth, marquisette, theatrical gauze, sharktooth gauze, camouflage netting or screening. It is possible to see through this material in the direction of the heaviest concentration of light. This operates like the lace curtains on windows in a home. During the day time when it is lighter outside than inside it is possible to easily see out, but no one can look into the house. At night it is possible to see in when the inside is lighted, but impossible to see out into the darkness. Sheets of material of this open weave nature used on the stage make it possible to conceal actors and objects behind the drop until the light is brought up on that side. The object can be made to fade out again by simply changing the concentration of light to the other side.

ROLL DROP—A regular drop rigged so that it can be rolled up like a window blind is called a roll drop. The top of the drop is securely anchored to an overhead support. The bottom of the drop is fastened to a round or cylindrical

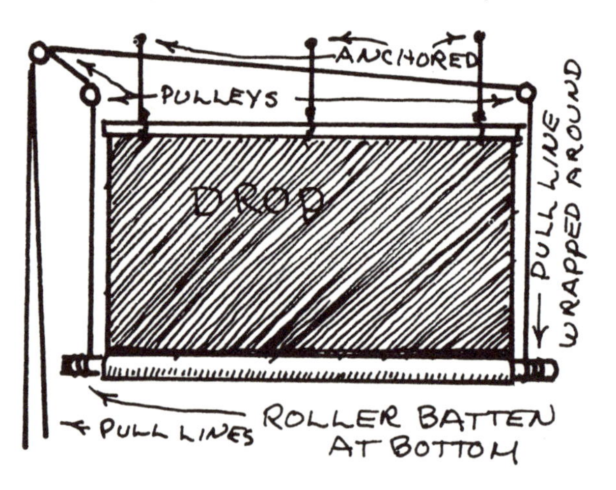


shaped batten. This batten might be made of heavy dowel, cardboard rug or linoleum rolls fastened together, a pole, a downspout from a roof drainage system or any other tubular shaped material that the drop can be glued or stapled onto. The bottom batten should extend out beyond each side of the drop from 1′-0″ to 1′-6″ depending upon the height of the drop. A rope is wound around this extension of the cylindrical batten enough times to use up a length equal to the height of the drop. This rope then extends up along the side of the drop at least as high as the top of it, is passed over a pulley and then allowed to drop back down to the floor. Another rope is fastened onto and wound around the batten on the other end in the same way — and wound around in the same direction — front under to back. After the same number of turns have been put on the roller the rope goes up and over a pulley at the top, extends across to a position beside the other rope and is brought down to the floor. Either another pulley can be placed beside the one for

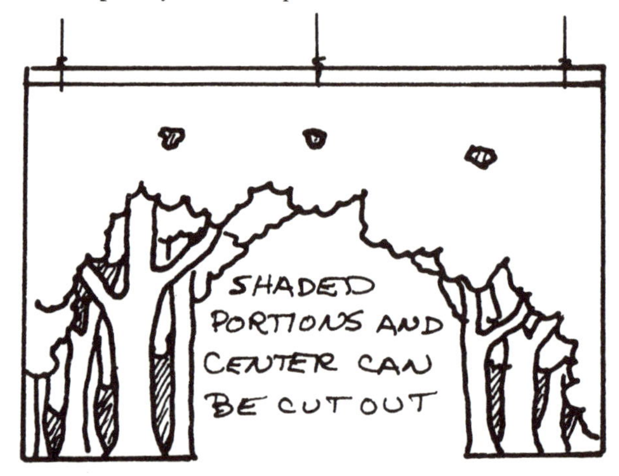

the first rope or a double one for both of them to come through. When the ends of the ropes are pulled downward the ropes begin to unwind from the batten and as they unwind the drop rolls up. It will be necessary to have some kind of anchor to fasten the ends of the lines to when the drop is rolled up or a long suffering curtain puller will have to stand and serve as ballast until the curtain is closed.

CUT DROP—The cut drop has sections cut out to represent openings in foliage, doors, windows, archways, cave entrances, spaces between trees or any other kind of opening imaginable. The holes are usually cut out after the drop has been painted so that the material is somewhat stiffened and stretched. Sometimes after the drop has been cut it is found advisable to turn it face down and glue netting across some of the openings to keep the cut edges from curling. If there are small projections extending out in such a way that they are likely to sag a thin strip of wood may be glued so that it extends horizontally across the drop and out into the opening far enough to support the projection.

CONTOUR DROP—The contour drop adds a new variation to the old drop. This type of drop can be suspended in space with all of its sides showing since it is designed with a finished edge on the top, sides and bottom. In place of a plain top batten this unit has a flat of the appropriate contour attached to it. The drop is then glued onto the bottom batten of this flat section and allowed to hang down from there. It may be a plain surface or a cut drop. When flying this unit it is advisable to use wire (piano wire will take the strain easily) for the support extending from the top of the drop up behind the overhead masking where it is attached to the flying lines or counterweight batten.

FABRIC FOR BACKDROP—Muslin, linen and scene duck are materials commonly used for backdrops. For convenience wide material is used. It may be that more than one width is needed for the height of the drop; if so, it may be sewed together with a plain flat seam. It is ordinarily advisable to sew the material so that the seams on the finished drop are horizontal rather than vertical.

## Cyclorama

One of the standard background pieces for the stage is the sky cyclorama. It has many uses as a sky, as a neutral background for scenery, as a screen for projected scenery, and as just a plain background for action.

The cyclorama is composed of a large sheet of fabric. The ideal cyclorama will be high enough and wide enough to mask the entire stage area with a minimum of assistance from borders or teasers above and from tormentors or other scenic units on the sides. It is physically impossible to mount a cyclorama on most stages that will accomplish this so usually something less is accepted. The ideal cyclorama begins from a position offstage and upstage from the

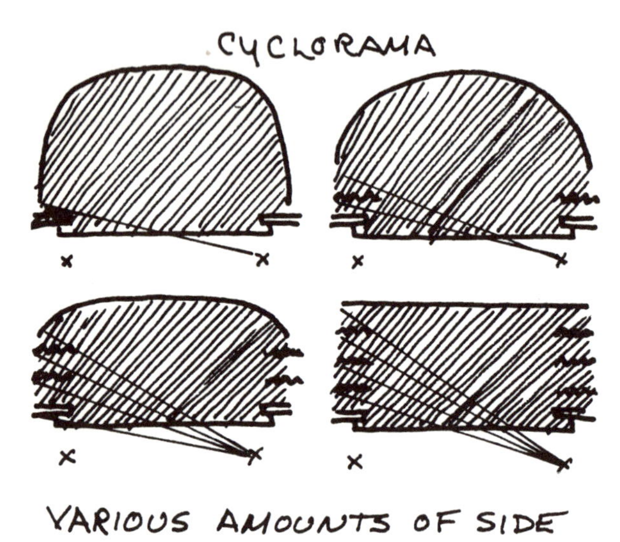


proscenium a few feet, moves toward the back of the stage, sweeps around the back of the stage and then moves forward to a position just behind and offstage from the proscenium on the other side of the stage. When this type of cyclorama is installed only a small amount of masking is needed on the sides for the extreme front corner seats; and a minor amount of masking is required above for the extreme front seats. If it isn't possible to have a cyclorama that provides this coverage then as much of the shape and proportions of the ideal cyclorama as possible should be kept. In order to determine the size and shape of the cyclorama gradually move back until the practical proportions for the stage are reached. In the extreme situation the cyclorama will consist of a sheet of muslin stretched straight across the back of the stage. This type of cyclorama should be as wide and high as possible.

Cycloramas may be purchased from theatrical supply houses or manufactured in the theatre. Material of maximum width should be used and the seams planned to run horizontally. The top should be reinforced with 3″ or 4″ canvas or burlap webbing and should be provided with grommets placed about every 1′-6″. The sides should be hemmed but need not be reinforced with webbing. A few grommets placed along the sides, maybe every 3′ or 4′, will be helpful if the material needs to be pulled taut when in place. The bottom may have a chain or pipe pocket hem or it may be reinforced and grommetted like the top. The bottom is treated this way to hold it down with weights or tie lines in order to get a wrinkle free sky.

Various fabrics can be used. The ideal material is pale blue or aqua colored velour since velour has such pleasant texture. However, this is also the most expensive material to use and for that reason impractical in many theatres. Material similar to light blue denim can be used. It should be checked carefully under lights to be certain that there is no dye variation within the width of a piece or from bolt to bolt. At best there will be slight variations that will cause the seams to be visible. Muslin, linen or scene duck may be used. These should be painted with dye after they have been sewn and stretched out. If possible they should be in place where they are to be used before any dying or painting is done. If dye is impractical the surface may be painted with scene paint. Although it may be possible to get a smoother paint job by using a spray gun yet at the same time the tiny spray that flies off into the air will probably be found all over the building upon completion of the job. At any rate some method should be used that will produce a smooth, even, blue surface.

There is probably no other single unit of scenery that will be used as much or be as pleasant with which to work as a smooth, beautifully painted cyclorama. On the other hand, nothing is quite so disheartening as to have to work with one that is dirty, stained and torn.

* * * * *

With a basic knowledge of the construction

of various types of flat and drop scenery it is possible for the theatre technician to plan and build other units that are similar. There should be no hard and fast set of rules concerning construction procedure or materials that can be used for constructing stage scenery. The scene builders of the past used many techniques that were sound and practical but their procedures need not be accepted as today's end standard. The easiest, fastest and most efficient method should be accepted as long as the end product will stand the abuse of stage handling.

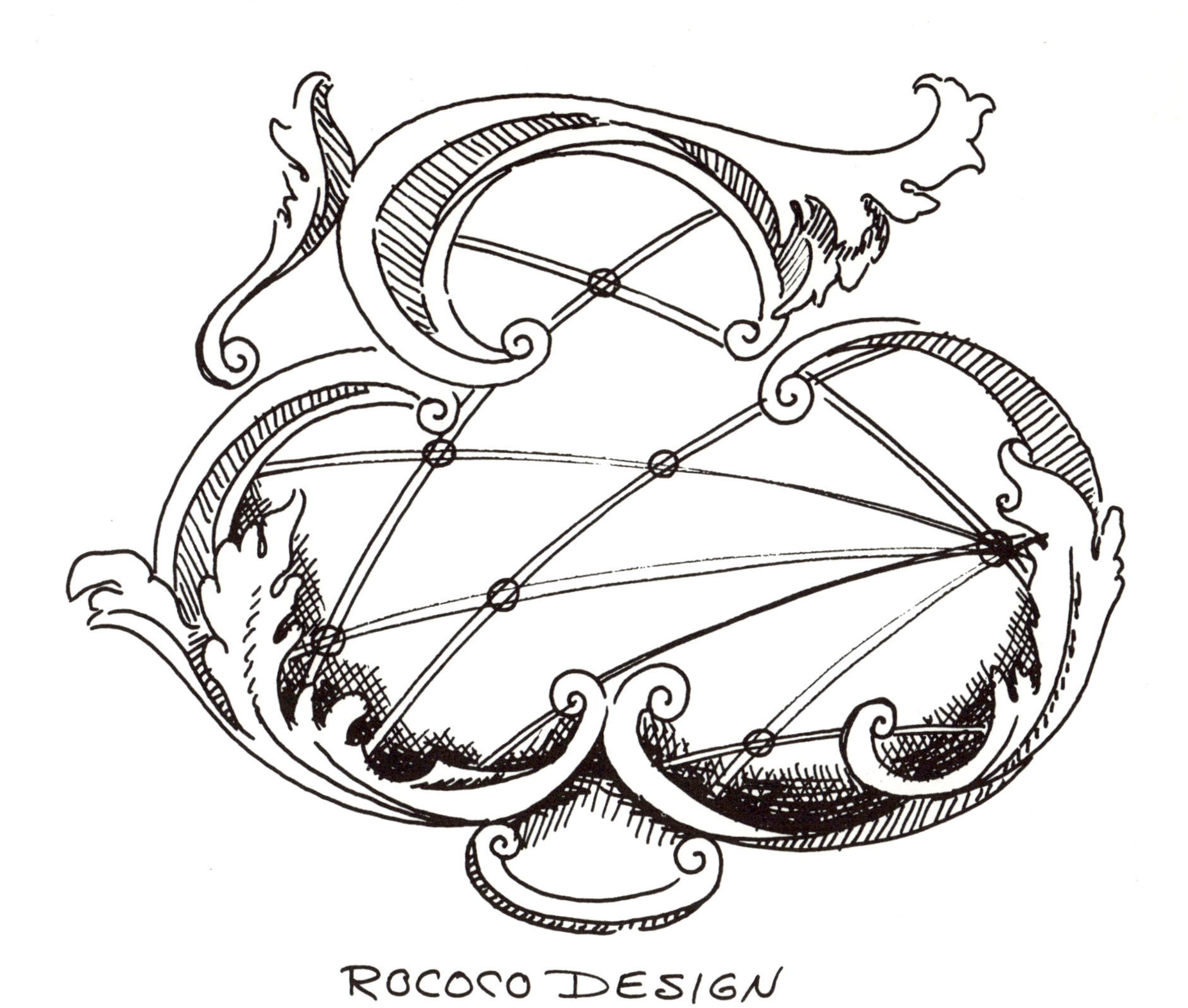

ROCOCO DESIGN

*The Ghost of Mr. Penny.* Setting consists of two rooms side by side.

*The Petrified Forest.* Numerous properties are needed to dress the setting for this Arizona desert eating joint.

*Another Part of the Forest.* Heavy cardboard forms are here used for columns.

# 4

# *Three Dimensional Forms*

Three dimensional and molded forms are used in conjunction with flat scenery in the theatre. Some of these forms are platforms, steps, built up rocks, columns, ramps, built up ground sections and mache units of various kinds.

If the theatre has a stock of stage wagons it is easy to convert them into platforms by equipping them with legs. The legs may be fastened on temporarily for a single production or may be made up in reusable sections that are bolted into place. Leg sections can be removed and stored away for future use. A 4′-0″ x 8′-0″ wagon should have 6 legs under it to make a substantial platform. If leg sections are used each section may consist of two legs tied and braced together and bolted onto the wagon. Then, on each side, a stringer is bolted across to space and fasten the sections to one another and at least one diagonal brace is fastened on to stabilize the unit. When this is done the wagon becomes an elevated platform. If it is desirable to cover the sides of the platform this may be done with flats, wall board, plywood or muslin. If muslin is used it will be necessary to fasten furring strips to the legs and braces to make them flush with the side of the wagon so that the face of the frame (which will be covered with muslin) will be smooth.

In case the platform is to be 1′-6″ or less in height it is possible to fabricate each leg from two pieces of 1″ x 6″ nailed edge to edge. This type of leg may be fastened in place on the wagon; thus without using ties or braces a fairly substantial platform is created. Bolts and wing nuts or double head nails can be used to make removal of the legs easier.

Short legs may also be made from pieces of heavy plywood which have been tapered so that they are wider at the top than the bottom. Two pieces of plywood should be fastened together edge to edge to form corner legs just as with the two pieces of 1″ x 6″. It may be advisable to nail both pieces to a corner strip of 2″ x 2″ in order to have a more secure corner since the

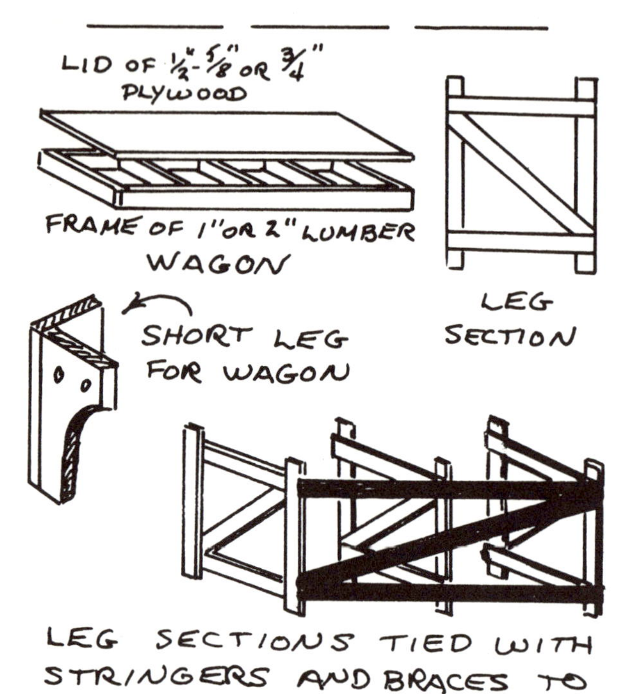


LEG SECTIONS TIED WITH STRINGERS AND BRACES TO MAKE PLATFORM OUT OF WAGON

edge of plywood does not hold nails satisfactorily.

### Parallels

Ordinarily stage platforms are made up in the form of parallels. The parallel is a collapsible platform consisting of a removable top and folding support. This type of unit can be stored away more conveniently than the stage wagon.

LIDS — Parallel tops may be made from single pieces of plywood ½″ to ¾″ thick and up to 4′-0″ x 8′-0″ in size. Sheathing plywood will work very well for this purpose and is much less expensive than any of the other grades. It may be necessary to fasten small stop blocks underneath near the corners so that the lids will fit snugly onto the frame, but this can be done very easily. The lid should then be covered with padding of some kind and muslin or canvas. Canvas is better than muslin because it will withstand much more abuse. The covering should extend over the top and around the edges and then it should be fastened underneath with glue and tacks or staples.

Parallel lids may also be made from square edged boards 1″ x 12″ or narrower or from tongue and groove flooring boards. If boards are used it will be necessary to cleat them together with 1″ x 4″ boards on the underneath sides. The cleats can be spaced to hold the top snugly in place on the frame. This type of lid should also be padded and covered in order to soften the sound of footsteps when the platform is used on the stage.

FRAMES—The support, or parallel, is made of frame sections that are similar to the frames for flats. A 4′-0″ x 8′-0″ parallel will be made of two sections the full length of the platform and the desired height minus the thickness of the lid. There should be two ends and at least

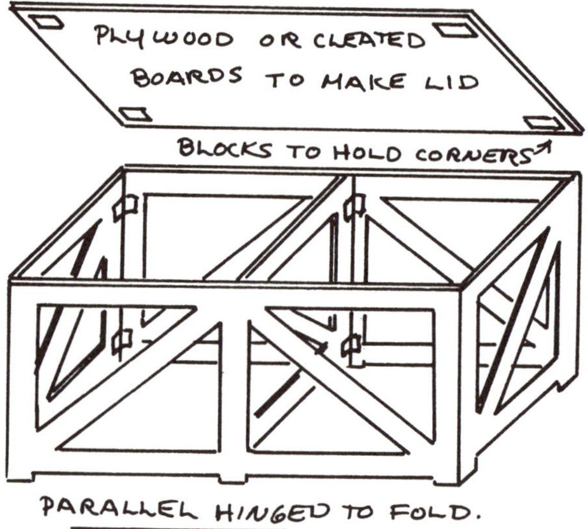


PARALLEL HINGED TO FOLD.

two other cross frames the width of the platform (minus the thickness of the sides where they are fitted in place) and the height of the platform (minus the thickness of the lid). It may be desirable to design the frames so that the legs extend down to the floor, but the bottom rail should be placed so that it is lifted off the floor by at least half an inch. This will insure a more stable support. There will be less chance of the platform rocking than if the rails are flat on the floor providing a support all of the way around the frame. All of these frames should be provided with diagonal braces to make the platform rigid in all directions.

An alternate method of building the parallel frame makes use of single pieces of plywood for each end, side and cross section. ½″ to ¾″ sheathing plywood may be used. The framework may then be cut out of the piece of plywood so that it has legs, braces and supports similar to the regular framework but made out of one piece of wood. This method works quite well especially on smaller parallels. It would be impractical to make this type of support for platforms that exceeded in any one direction the size of a regular 4′-0″ x 8′-0″ sheet of plywood.

HINGING—Hinging the parallel together is a tricky process. The position of the sections should be located rather carefully and then the

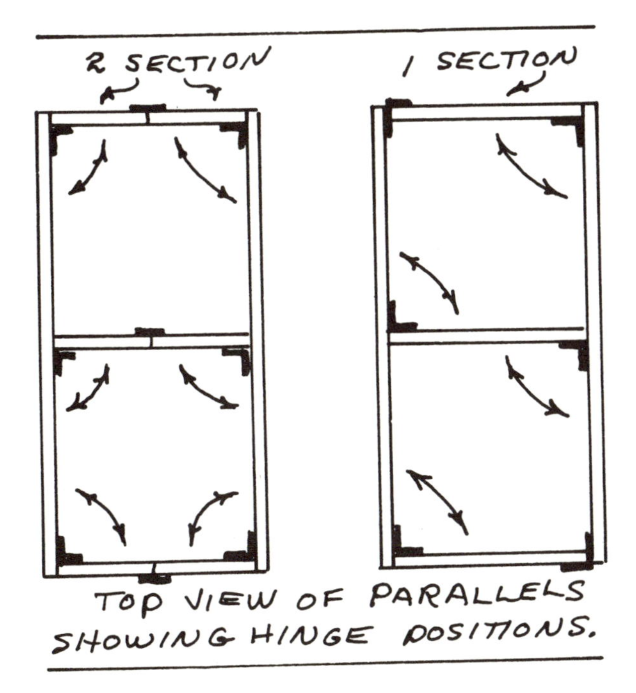


hinges placed so that they are lined up with the joint. If the end sections are spaced in about three inches it is possible to place the hinges on one side of the parallel so that they are on the right side of the ends and cross sections and on the opposite side of the frames on the other side of the parallel. If the ends are spaced flush it will be necessary to place one pair of the hinges so they are half inside the frame and half outside on each end of the framework.

Parallels such as the one just described fold into a pack that is equal to the length of the side plus the length of one end. A 4′-0″ x 8′-0″ parallel will be 12′-0″ long when folded and a 3′-0″ x 6′-0″ parallel will be 9′-0″ long. Cross members of parallels may also be folded in the center; then the folded length will be the same as the open length of the parallel. This type of parallel makes a better stack when stored but requires more time and material for its construction. The hinges should be placed so that the centers of all of the cross sections fold inward.

GENERAL FACTS — Parallels need not be 4′-0″ x 8′-0″. For a large stage it may be advisable to make them even larger and for small stages it may be advisable to make them smaller. Here as with most of the stage equipment it is a good idea to set up some standard sizes. If more than one size is decided upon it might also be advisable to choose proportions that will work with the other stage equipment. Usually it is necessary to have more parallels than lids since it is possible to use a lid with varying heights of parallels.

When the parallels are completed it may be practical to fasten gliders on the bottom of the legs so that they can be easily moved about the stage. Of course there are times when this can be a disadvantage. If the platform moves too easily it might scoot away when it is supposed to remain planted in one spot.

Parallel sections may be built so that they can be stacked on top of one another to make higher levels. It is also possible to heighten parallels by having extension legs that can be bolted into place when the parallel is assembled.

Frequently it is advisable to fasten the lid to

the parallel so that it will not be noisy when walked upon. Hooks and eyes, loose pin hinges, wood screws or nails may be used for this purpose.

### Ramps

Ramps, platforms of more than one elevation, work benches, stage risers for seating and other similar weight supporting stage levels may be built in the form of parallels. The unit may be in any parallelogram form and it may be hinged to fold. Odd forms with one raked

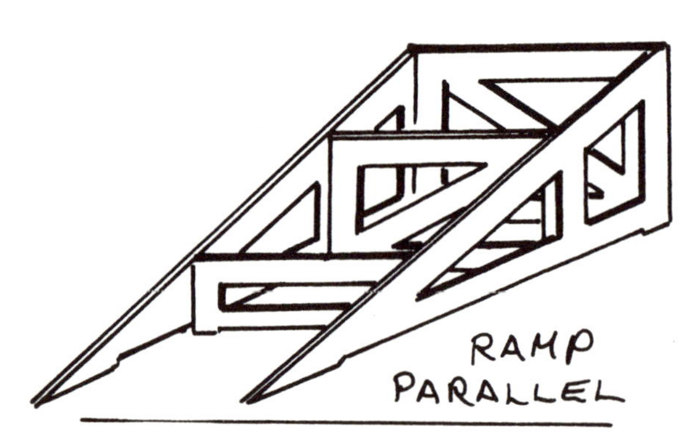


end may be rigged so that one corner is fitted with loose pin hinges. After the pins have been removed the rest of the support frame will fold into a compact unit.

To facilitate the handling of parallels, it is a good idea to fasten a tie line permanently onto the parallel so that it is always available to tie up the unit for transportation and storage.

### Boxes

Low rectangular boxes of varying sizes are extremely usable on the stage. These may be built one step high (6″ to 8″) and used in many combinations. They may be stacked and used as stairs, covered and used as rocks, or used together to make irregular platform units. ½″ plywood nailed on a 1″ x 6″ frame will make a box 6″ high and ½″ plywood on a 1″ x 8″ frame will make a box 8″ high. Square, triangular, round, ½ round, ¼ round and oblong boxes may all be useful on the stage.

A round, partially round or curved box may be made by cleating 1″ boards (1″ x 12″ # 2 Common W.P.) together after they have been cut to the desired shape. For a solid box the

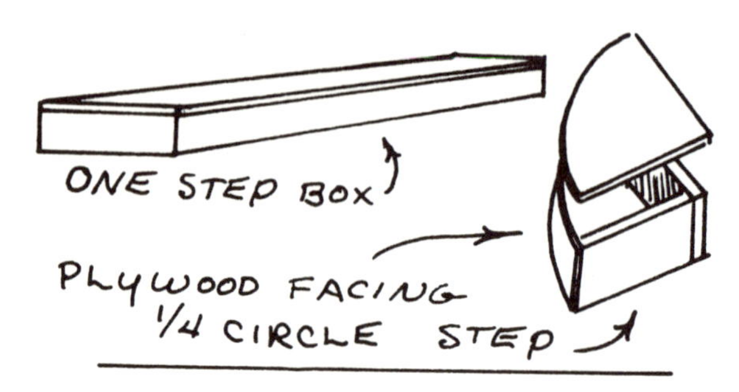


top and bottom forms should both be provided, although the bottom need only be the contour shape without being a solid support form. ¼″ or 5/16″ plywood facing, cut so that the grain runs across the narrow direction, can then be nailed around the contour to cover the riser. If the platform or box isn't too large, say not more than 3′-0″ in diameter, it is unnecessary to use inner support other than the plywood facing. If the curves are sharp it may be necessary to either use thinner plywood or paper-type wall board for the riser. Another method of facing curves is to use strips of board between the top and bottom contours for support and then cover the edge with muslin or canvas. If this is done a strip of muslin 4″ to 6″ wider than the thickness of the box is cut and placed so that the material can be glued to both the top and bottom of the box as it is applied.

### Step Units

As previously indicated it is possible to make step units by stacking several boxes together. Usually it is more convenient to build step units that can be added to the regular stock of scenery and be reused when needed.

The individual step has three measurements that must be considered when it is constructed. The part of the stair that supports weight is the tread. The tread has width, depth and height. The height is referred to as the "rise" of the step.

The proportions of a set of steps is determined

by the space in which it must fit. Ordinarily there is a landing that must be reached within

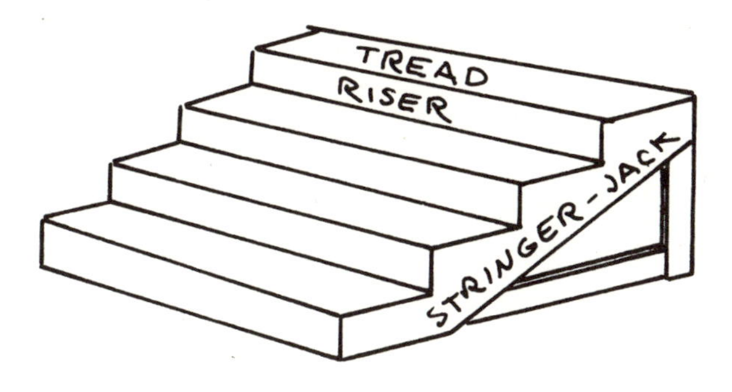


specific confines of width and length. The depth of the tread, height of rise and width of each step will be computed from this spacing.

Stage step units will differ in some details from the stairs in public and private buildings. On the stage the stairway is a device to be used for the actor. It is in a sense a functional acting area consisting of a series of levels one above the other and each tread should be large enough for easy, comfortable and graceful stage movement.

The tread of the step may be almost any width but it is rarely advisable on the stage to have less than 12″ depth to the tread.

The rise of the step unit should be determined partly in terms of other scenic units in stock. In order to work with parallels or platforms it might be advisable to choose a rise that will in multiples be 1′-0″, 1′-6″, 2′-0″, 2′-6″, 3′-0″, 4′-0″, 5′-0″ and 6′-0″ in height. 6″ rise steps are used in most public buildings. In homes rises of 7″ may be found. For the stage multiples of 6″ and multiples of 8″ will work out very well. An 8″ rise is actually rather high for comfortable day in and day out climbing but on the stage it has the advantage of working out with various levels and is a means of gaining height rapidly. A 6″ rise can be used on many stair units and this rise will also work with many platform levels. In general it might be said that 8″ rise steps should be used on large stages for visual proportions and where it is necessary to gain height rapidly and that 6″ steps should be used for graceful appearance and minimum elevation gain.

CONSTRUCTION OF STEPS—Several ways to build step units will be discussed.

The first method uses stringers or jacks for the side contour pieces. The stringer is marked off on a piece of 1″ x 12″ white pine lumber. This piece of lumber should be fairly clear of large knots in order to keep it as strong as possible — any finish grade or carefully selected #1 common grade lumber may be used. The carpenter's square is used for computing the various angles needed for the tread and riser positions. Turn the square on a diagonal across and near one end of the board. Maneuver it so that the 11¾″ mark of the long arm of the square is about 6″ from the end of the board and the 6″ mark on the other arm of the square is at the edge of the board. Mark along both edges of the square. When this is done a right

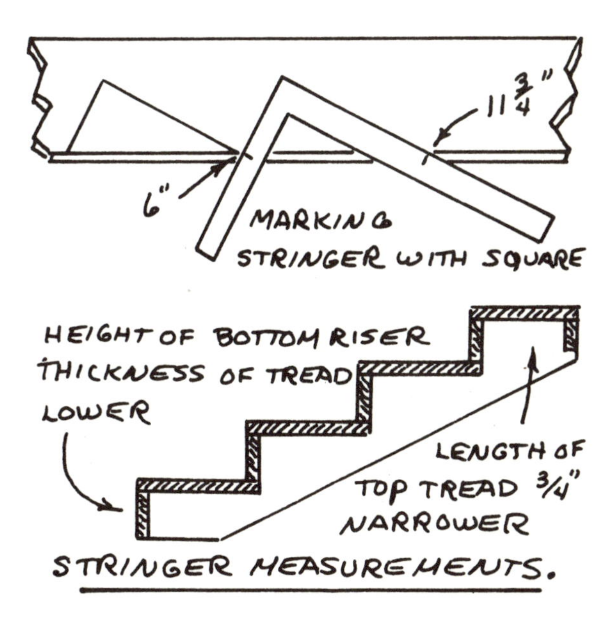


triangle is outlined on the board with the edge of the board serving as the hypotenuse of the triangle. The measurements suggested presuppose that the rise of the step is 6″ and the tread 11¾″ and those measurements on the triangle just drawn represent the positions of those elements on the stringer being marked off on the board. The square is now moved along the board and the process is repeated so that a series of triangles are drawn with their points just touching along the edge of the board. The number of triangles drawn represents the number of steps on the final unit. If two are drawn it will be a two step unit and consequently 12″ high; three will make it 1′-6″; four will make it 2′-0″ and so on. If 8″ risers are desired the measurements used on the square for marking out each step will be 8″ on the short arm and 11¾″ on the long arm. 11¾″ is used for the tread measurement in place of 12″ because the width of a 1″ x 12″ board will be about 11½″ to 11¾″ and these boards will be the most convenient to use for treads.

Before proceeding with the stringer it is necessary to determine what type of material is to be used for the risers. The unit may be made without facing for the risers, but the facing stiffens the step and makes the entire unit more substantial and durable. Plywood or wall board can be used but 1″ boards are superior since they stiffen the leading edge of the step better. For the present purpose the discussion will be based upon the use of 1″ x 6″ boards for the risers assuming that a set of steps with 6″ rise is being built.

Now take the square again. Place it with one edge along the first 12″ line that was drawn. This will be the first or bottom step. From the beginning point of the 12″ line draw a line at right angles toward the end of the board. This line should be measured to be 5¼″ long. This is for the riser of the first step. There will be a tread of ¾″ (the actual thickness of a 1″ x 12″ board) on top of this piece; hence the measurement of 5¼″ in place of 6″. Now draw a line parallel with the 11¾″ line diagonally across to the other side of the board. This will be the under side of the stringer and may be referred to as the bottom of the board. The line just drawn is the foot of the first step. This portion will rest on the floor.

The square is next placed on the 11¾″ line representing the top step of the unit. Since the riser is to be a 1″ x 6″ board and is to fit under the top tread, the support for the top need be only 11″ across. So with the square in position in line with the top tread measure 11″ on the long arm of the right triangle and at right angles in the opposite direction toward the bottom of the stringer mark across the rest of the board.

The board is now completely marked and the jack or stringer is ready to be cut. Carefully saw along all of the lines.

If this is to be a step unit with permanent legs a side frame may be completed by using 1″ x 4″ boards to extend back from the foot and down from the back of the stringer to complete a triangle. The finished side stringer section will then look like a right triangle with saw teeth on its hypotenuse side. The boards may be fastened together with slabs of plywood and clout nails. The full height of the side should be equal to the sum of the height of the risers and the length of the side should be equal to the sum of the tread depths.

If the step unit is to be narrow, say from 2′-0″ to 2′-6″ or even in some cases 3′-0″, a single riser on each side of the unit may make it strong enough. If, however, it is to be wider or if it is to carry considerable weight other stringers should be placed underneath and spaced not more than 2′-0″ apart.

Next it is necessary to determine the width of the completed step unit. The risers and treads are then carefully measured to that length, squared and cut. A riser and tread will be cut for each individual step. The only piece that will vary in width is the bottom riser which should be ripped so that it is ¾″ narrower than the height of the rise. When the tread is nailed on top of the riser the step will be the full planned height.

The risers should be nailed in place first. For a permanent unit it is wise to use screws or coated nails. 1½″ flat head screws or 8d box nails are good sizes to use. Nail all of the risers so they are flush on top of the triangle cuts and so their ends are even with the outside of the stringers. After the risers have been nailed in place it is time to fasten the treads onto the stringers and risers. The treads should be placed so they are flush with the front of the risers, and the stringers should be lined up to coincide with the end cut of the tread. If this is carefully done as the unit is being assembled the entire unit should be square and true as far as the steps themselves are concerned. The treads should be nailed into the top edge of the riser and into the stringer at each end. The step should be turned over and the bottom edge of each riser nailed into the side of the tread it is adjacent to.

In order to stabilize the rear edge of the top tread a 1″ x 4″ should be nailed across under this edge. The 1″ x 4″ can be notched into all of the stringers or it can be nailed between the end ones and notched into any others.

The entire unit is further strengthened by nailing a 1″ x 4″ across the back near the bottom to tie the right angle corners of the two side pieces together. A diagonal 1″ x 3″ brace is also advisable.

The step is completed as soon as the treads are padded and then the steps, risers and sides are covered with muslin or light weight canvas.

This type of unit can be built without permanent triangle sections below the stringer. These may be added when the step is being readied for a stage setting. Support sections may also be designed with hinges so that they will fold flat under the steps. Or, the step may be used without legs and supported on the legs of the platform or parallel. When this is done a cleat is nailed across the platform at the right height so that the supporting board under the top step will slide on top of it when at the correct height. Brace cleats or strap iron pieces may be screwed to the board on the step unit so they project down an inch or more. When the step is placed in position, the back of it is lifted enough to allow the cleats to slip over the board on the platform. This simple device will help hold the step in position.

Stringers can be made up in other ways. A straight piece of 1″ x 4″ or 1″ x 6″ can be fitted with triangular pieces fastened onto one edge to form stringers that are the same shape as the ones cut from a single piece of board. These pieces may be cut from scraps of 1″ x 6″. Actually the triangles cut out of the 1″ x 12″ when the regular stringer is made are just the right size. So these individual pieces are

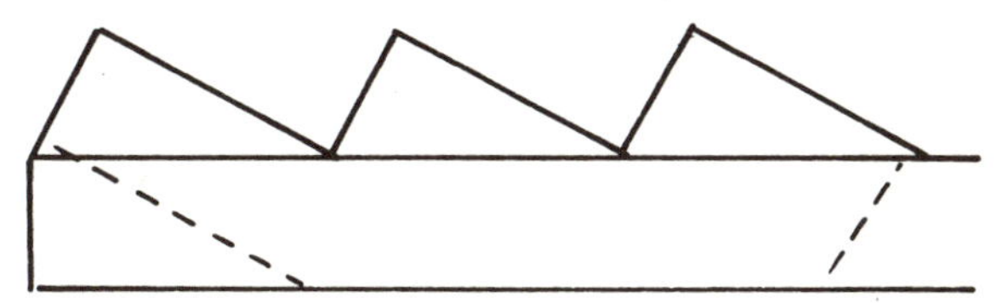

TRIANGULAR BLOCKS FASTENED TO EDGE TO MAKE STRINGER.

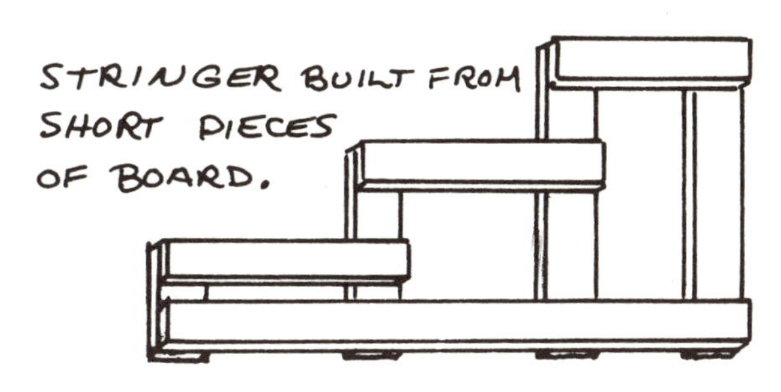


marked out in the same way using the carpenter's square. The simplest way to do this is to mark and cut one carefully and then use it as a pattern when marking out the others. These small triangles are then fastened to one edge of the stringer with slabs of plywood and clout nails. Be careful to get the angles all in the same direction.

Side frames for the step unit can be marked on a piece of plywood and cut out in one piece. It is more difficult to nail the risers and treads securely in place since they must be nailed into the edge of plywood and nails tend to split the plies apart.

The stringers may be solid pieces, marked as for the first type of stringer discussed. Instead of cutting the stringer it is possible to nail strips of 1″ x 2″ or 1″ x 3″ where the treads are to be placed and then nail the treads on top of these. Added strength is then gained by nailing into the tread from the outside of the stringer. Risers may be nailed in without extra strips since they will not carry as much weight as the treads. Risers may be eliminated in this type of step unit but actually they stiffen the treads and add stability to the entire unit. A step unit of this nature without risers may be used as an off stage return from a platform. If the space is restricted in which to reach the floor the steps may be steeper. When the riser is eliminated there is toe space to make climbing easier.

Still another type of stair side may be made as if it were to be a flat frame. Although this type of side support may take longer to build it uses many short pieces of lumber and may be more economical as far as supplies are concerned. This type needs to be carefully designed so that the strain of the treads is carried to the floor; otherwise, the framework will collapse.

CURVED STAIRWAYS—Curved stairways present special problems but as long as each step is supported down to the floor the problems are not too difficult. However, it will take considerable time and material to build one several steps in height.

The curve should first be laid out very carefully on paper and then chalked out on the floor. The floor plan of the curve is laid out and since it shows the treads as they need to be it is possible to easily cut out the pieces for them. Treads may be made of pieces of heavy plywood or pieces of 1″ lumber cleated together.

The major form supports may be built as riser sections that extend from the floor for each step. Except for the first riser these will probably be built in the same way as flat frames with corner blocks and keystones. The first one may be a piece of board the correct height minus the thickness of the tread. If the rise is to be 6″ this will be a 5¼″ wide board—that is, providing the tread is ¾″ thick, which it will be if regular lumber is used.

The front of the tread will be nailed onto the top of this board and the back of it, as well as the front of the second tread, will be supported by a board 11¼″ high or a frame that high with a strip of 1″ x 4″ spaced 5¼″ to its top fastened to the front of it. The third riser will be 17¼″ in height with a strip of 1″ x 4″ on its face 11¼″ above the floor to support the

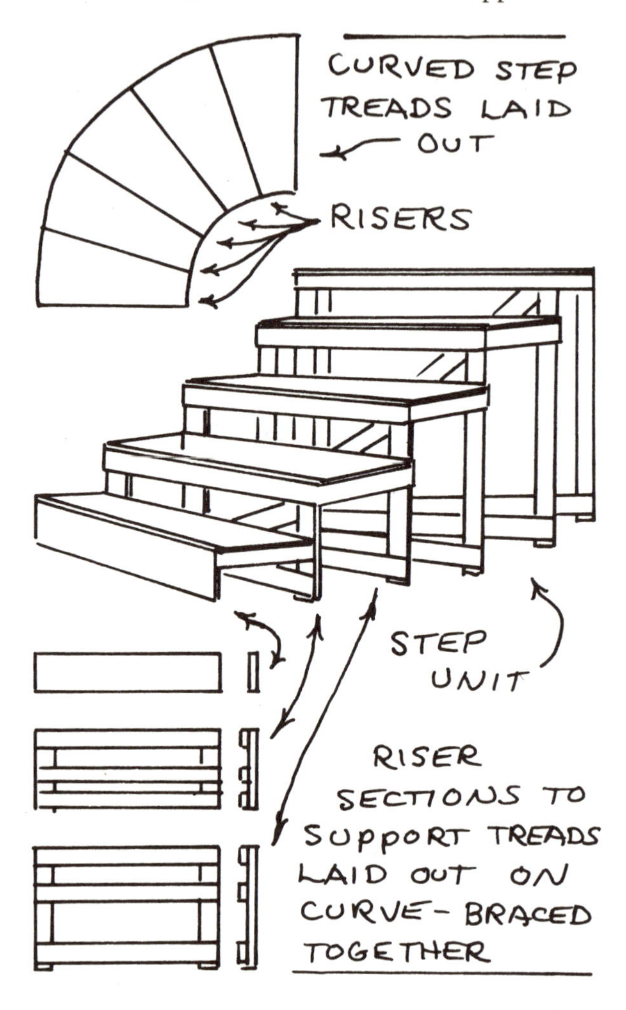


treads of the second and third steps. This process is continued until the top is reached. All the way along in the process the correct width must be maintained for both the risers and the treads. After all of the risers and treads have been made it will be necessary to nail the sections together with 1″ x 4″ strips to tie the riser sections together and to support the ends of the treads. Additional bracing should then be placed where needed.

Firm surface material, such as plywood or

wall board, makes the best covering for the sides of the unit. Muslin is not satisfactory for this purpose since it tends to pull in too much in shrinking and flattens the surface.

Curved stair units can also be made from individual step or box units. Each step may be provided with legs the proper height to elevate it into position and then the legs may be tied together with plumber's strap, strap iron or short pieces of wood to make a single stair of the various steps. After the step unit has been used the boxes can be salvaged and used for other purposes. These boxes need not be the exact shape of the desired treads. They may be oblong and overlapped to fit the curved contour shape.

### Tubular Forms

Another type of three dimensional form that is frequently needed for stage construction is the tubular form. When and where possible it will be easier to find tubular units that have been manufactured for some commercial purpose and then adapt these forms for use on the stage.

Mailing tubes, heavy cardboard tubes around which rugs, linoleum, plastic and yard goods are wrapped, tubes used in the manufacture of oatmeal boxes, cardboard casings for concrete column forms, cardboard barrels and packing cases — these and many other inexpensive tubular forms may be found in many communities. Round hollow wooden porch columns, dowel, closet poles, aluminum tubing, electrical conduit, iron pipe, asbestos chimney flue liners, aluminum and sheet iron stove pipe are some of the other tubular materials that may be adapted to stage use on occasion.

Smooth, regular, small diameter tubular forms are difficult to make since there are very few inexpensive materials that can be easily and readily bent to conform to the desired curve. A few of the materials that can be curved are: thin wall boards such as EZ Curve Upsom board, 1/8″ plywood cut so the exposed grain runs up and down rather than around the diameter of the curve, corrugated box board with the corrugations running up and down, corrugated paper with only one side faced, building papers such as Sisalcraft, roll roofing, cardboards of various kinds, sheet aluminum and sheet iron.

When building a tubular form it may be possible to use a simple process and simply wind tape or wire diagonally around the form to help it hold its shape. A strip of wood can be used to fasten the two edges at the seam. Tacks or staples can be driven through the material and into the wood.

It may be necessary to build a simple basic frame consisting of circles of the correct diameter and wooden strips to serve as spacers or ties to hold the circles in position and to fasten

BASIC FRAME FOR TUBE FORM

the seam edges of the material. If firm material such as plywood is used it will tend to spring back into its normal flat position so it must be fastened securely. If the surface of the material will stand to be soaked with water and can be softened slightly in this way it will sometimes facilitate the bending process and the material will maintain its new shape better.

In order to use muslin as a covering material for such shapes it is necessary to have a fairly substantial base. Any of the materials mentioned can be covered with muslin after they have been fabricated into the desired shape. A column can be made by nailing closely spaced narrow slats onto end and medial discs. The covered column will have vertical lines where the edges of the strips are located. Before covering the column some of these edges can be sliced down with a draw knife, shaved with a plane or sandpapered with a hand or electric sander in order to produce a smoother surface.

It is also possible to build up columns by laminating layers of wood together and then carving them with a draw knife, wood rasp and plane to produce the desired curve. This is a lengthy process and the resulting column is heavy, but it is certainly sturdy and will stand a great deal of abuse.

Another type of core may be made from one of the various types of wire mesh. Ordinary screen wire can be used if there is to be no strain on the surface. Chicken wire or hardware cloth can also be used. Hardware cloth of ¼″, ½″ or 1″ square weave will make the most substantial base of any of these materials. The wire from which it is manufactured is stiff, which makes it difficult to handle but it can be bent into shape and fastened with stove pipe wire. Chicken wire will also do, but the wire from which it is made is of lighter gauge usually so the form is less solid when completed.

The core, after it has been shaped, can be covered with papier-mache, muslin or possibly both.

Although fluting can be convincingly painted on a columnar surface there are times when it is desirable to have three dimensional columns with indented fluting. First build a core consisting of discs of the correct diameter and wooden slats. The slats should be fairly narrow and spaced an inch or more apart. The size of the slats and spaces will determine the width of the flutes and the space between them. The strips will represent the surface and the spaces the flutes. This is now ready to cover. Glue the muslin to each strip but allow enough material between the strips so that it will sag fairly deeply into the spaces between. If muslin that has been painted once is used there will be less shrinkage when the column is painted and better flutes will be produced than with new muslin.

### Miscellaneous Forms

Tubular or column shapes are not the only three dimensional forms that can be made with the wire core and mache. Stumps, logs, log walls, rocks, stone fireplaces and other pieces may be made with these materials.

PAPIER-MÂCHÉ — Basically the acceptable papier-mâché process uses a core or basic form that may be built solid and remain under the covering as a weight bearing portion of the scenic unit. It may be a screen wire, chicken wire or hardware cloth form that will remain permanently or be removed after the paper has hardened; or it may be a modelled or molded form into which the paper is cast and from which the paper is removed after it has hardened.

Weight bearing structures such as the tree and tree house in *The Grass Harp,* Peter's cave in *Peter Pan,* the river bank in *Ah Wilderness,* the rock levels of *High Tor* and other units of this nature should first be built in terms of support areas. These areas should have solid platform structures under them and substantial decks on top. They can be made from paral-

Step forms constructed with a wooden frame which has been covered with hardware cloth and newspapers.

lels, boxes, old chairs, benches, stools or from new material designed to have roughly the basic form needed. Chicken wire or hardware cloth can then be shaped to the correct contour for the finished form. The entire surface is then covered with papier-mâché and for additional strength a finish coating of cloth mâché is applied.

The paper is cut into strips, wide ones for flat surfaces and narrower ones for sharp contours. The strips are pasted in three or four layers over the wire. Any one of various binders may be used. Paper hanger's paste,

Toadstool with removable top. The stem has a disc top and bottom. The top is of hardware cloth bent to shape and covered with paper and cloth mâché.

wheat paste, flour and water (plain or cooked), a weak solution of glue — any of these will work all right. The paper may be dipped into the solution, the paste may be brushed onto it, or, after the first layer has been applied, the paste may be brushed onto the surface and the dry pieces of paper applied to the pasted surface.

The paper will tend to check and crack as it dries. It is a good idea to allow a layer to dry completely and then apply another. This will make a much more substantial surface. And in order to produce an even more durable covering strips of muslin may be applied over the paper. These may be pasted on or painted on like dutchmen. A mâché surface treated in this way with a coating of cloth over the paper will be much smoother and will withstand more shock than one merely covered with paper.

If the mâché is to be applied inside a cast or over a molded form and then removed when dry the first layer should be wet only with water rather than with paste so that it will not stick in the cast. Succeeding layers may be pasted on top of this. Additional precautions may be taken. If the form that is to be cast over is plain plaster of paris a thin coating of green soap (or any other soap) may be used first to fill the pores and seal the surface of the plaster. If the mâché work is done right after this has dried it should be easily removed. Aluminum foil is an excellent parting agent.

Many types of paper can be used for the mache work. Newsprint is all right but soft when wet. Wrapping paper may be used to produce a tough surface. Paper towels work well since they can be stretched a little and even toilet tissue can be used for comparatively fine finish work.

Sometimes mache surfaces are too fragile and harder finishes are desired. For this any one of a number of materials may be used. These materials are more expensive, but for some uses their permanence may make them more economical in the long run. Some of these materials are Celastic, fibre glass, plastic wood and medical plaster impregnated cast tape.

Plaster of paris or casting plaster and cloth such as net, burlap or cheesecloth may be used. The cloth is dipped into mixed plaster and then placed over the form of chicken wire or whatever it happens to be. Additional cloth and plaster may be added to build the surface to its desired form. The fabric supports the plaster and keeps it from cracking. If additional layers are added after the first have "set" the surface should be wet to reduce the suction and porosity; otherwise, the new layer may not bond firmly.

STYROFOAM and other factory produced flexible and rigid foam products that come in plank, bat, billet, or cast forms such as cubes, balls, rings, eggs, extrusions, ad infinitum, may be used extensively for surface bulk, bricks, stones, moldings, statues, etc. These materials, especially the rigid varieties may be cut and shaped with saws, knives, hot wire cutters, rasps, sanders and hand held sandpaper. Some thin sheet plastic pressed or vacuum formed objects for interior home decor such as: sectional panels of brick, stone, simulated wood, grillwork, fake carved wood detail, and on, and on, and on, may be found in do-it-yourself home improvement stores, variety stores, and paint shops. These materials may be nailed, glued or stapled onto flats, steps, and other scenic and prop units. (See also pp 143, 146 ff)

* * * * *

Many three dimensional forms other than those described in this chapter are needed for stage settings. Each play requires its own particular structures, objects and gadgets. Only a clairvoyant could foresee all of these needs. It is to be hoped that the discussion of some of the basic materials, structural problems and building procedures that are described here will aid the stage carpenter in contriving and manufacturing other three dimensional items that are needed for any one particular stage setting.

WROUGHT IRON PATTERNS

# 5

# Painting Equipment

A scene painting department can be set up with a minimum of equipment. And then, as the shop continues to operate, other items may be added from time to time. Many of these items need not be purchased, but can be made in the scene shop.

Paint and brushes are the most essential elements in the shop and should be discussed first.

### Paint

TYPES OF PAINT—There are four basic types of paint commonly used for painting stage scenery. These four, all water mix paints, are:

1. Dry Pigment—Size Water paint.
2. Casein base paint.
3. Acrylic, Vinyl—Rubber Latex Paint.
4. Aniline Dye.

(Although dry pigment—size water paint is no longer used extensively even by painters in the professional scenic studios, it is included here so that anyone interested may learn how to prepare the type of paint that was traditionally used in the theatre scene shops for literally hundreds of years.)

A brief discussion of the general types of paints listed above may prove of value to the novice not acquainted with materials in this field.

DRY PIGMENT—SIZE WATER PAINT—Size water, a mixture of melted glue and water, is the liquid carrier for water soluble dry pigments. For size water: Soak white flake or gelatine glue overnight in enough water to cover it completely. After being soaked, heat the glue in a *double boiler* until completely melted. Mix one part glue to 16 parts water (1 cup of glue to 1 gallon of water). If size water is to stand for any length of time, add a preservative to prevent spoiling. A teaspoon of formaldehyde or carbolic acid added to each gallon of size water is sufficient. If this precaution isn't taken the glue may decompose and give off an extremely unpleasant odor.

Pigment is added to the size water while it is being stirred. The paint may be tested by dipping a clean stick into it. When the paint covers the stick rather easily, without piling up, or running off and leaving streaks on the clear area of the stick, it is about the right consistency to use. By brushing some of the paint on a flat surface, the consistency can be determined more definitely. If the paint doesn't spread easily it is too thick; if it runs and leaves unpainted rivulet lines it is too thin. Obviously, if it is too thin more pigment is added and if it is too thick more size water is added.

Glue size paint has some definite advantages. There is a wide range of dry colors available; it is relatively inexpensive (especially in the less intense pigment mixes); it is easily cleaned from brushes and other equipment, and it may be scrubbed off surfaces (including flats), if necessary.

CASEIN BASE PAINT—Casein, once available in powder and thick paste form, is now primarily available in only a thick liquid form. Casein is a milk product with high binding qualities. Even thin solutions of the water-type paint have binding characteristics that allow its use for overwashes and glazes. There is minimal bleeding when new coats of paint are applied over old ones and although casein paint cannot be washed off after it has hardened it does not crack or peel. Coat after coat may be applied before flats require recovering. Brushes and other equipment, used for applying casein paint, should be thoroughly cleaned with soap and water immediately after being used.

Casein, mixed with water, is not completely stable and will spoil after prolonged storage.

POLYMER PAINT—Paint companies, almost without exception, now manufacture an extensive line of paints that carry the names "Vinyl", "Acrylic", and "Rubber Latex" in all varieties of combinations. These paint products are available in both oil and water mix bases. Since oil is flammable it is more convenient to use only the water base varieties, except for special applications.

Polymer paint is made of tiny globules of Acrylic or Vinyl Resin suspended in water thinned synthetic rubber latex emulsion. This is spread as a layer on a surface. The water then evaporates and leaves a continuous flexible waterproof film of plastic.

Vinyl and Acrylic paints are stable so they resist spoiling; and the pigment remains in suspension, so the color remains constant and little, if any, pigment settles to the bottom of the bucket.

Vinyl and Acrylic paints have a thick, creamy consistency and should not be thinned appreciably for good coverage. The paint dries with a solid, flexible surface so it works well on opaque backdrops that are rolled or folded for storage. Vinyl and Acrylic emulsion paints are excellent for stage use.

COLOR MEASURING AND MIXING SYSTEM—For a shop that uses a large volume of paint it is economical to obtain a color measuring and dispensing system similar to the ones used in retail paint stores. These machines are equipped with pumps or other devices to measure precise amounts of pigment. With a color swatch book, and a formula chart, it is possible to mix exact hues of color and later to match these colors by duplicating the formulas.

Modern paint systems use two or more basic vehicles into which pigment is mixed to create the vast number of color hues desired and needed by the scenic painter. A two vehicle system utilizes a white vehicle for tints and a gray vehicle for shades or deep colors. A second system uses three white vehicles while a third system uses four vehicles. The latter two systems use vehicles containing various concentrations of white pigment—more for tint mixes—less for shade or deep hue mixes. These systems use separate compatible vehicles for black and white paint.

A wide range of concentrated color pigments, that mix readily with the base vehicle of the same brand, will enable a painter to produce every desired hue. *CAUTION: Differing brands of paint should be intermixed with caution since they often differ chemically and may not mix compatibly.*

ANALINE DYE—Dye is an age old substance used since scene painters first began their trade early in the history of the theatre. Modern dye may be improved in color, quality and stability thanks to chemical technology, but the color use and concept remain the same as always.

Water soluble analine dye, available in many colors, dissolves in boiling water (App.

4 ounces to the gallon, or any proportion desired.) to create a thin solution, subtle or intense in color, depending upon the relative concentration of water and dye.

Analine dye is absorbed into and colors the threads of the fabric leaving the surface as translucent as it was before being dyed while other color media form a surface coating that covers threads and fills pores to form an opaque film over the entire surface. And since dye leaves the surface translucent both front and rear lighting may be used with pleasant results.

Analine dye, excellent for backdrops must be applied carefully and precisely for satisfactory results. Dye can be reworked to a certain extent but cannot be handled in the manner of pigment media that can be layered time and again to create complete changes in coloration.

To lighten or whiten dyed areas use bleach.

Portions of "flat scenery" may be treated with dye where translucency is desired. And portions of backdrops may be coated with translucent dye while other portions are coated with opaque paint to produce both translucency and opacity on various areas of the same surface.

Dye is a specialized color media for special select results that cannot be effectively achieved with any other color media.

*NOTE: Do not use natural bristle brushes to apply dye.*

METALLIC PAINT—Gold, silver, copper, brass, aluminum, and metallic colors such as red, blue and green are needed occasionally for painting stage properties and details on scenery. Gold is available in bright, dull and normal. These and other colors may be mixed with glue, bronzing liquid, or clear Vinyl.

GLITTER—Glistening chips of metal, glass, and plastic may be used for special effects. These may be sprinkled onto a fresh coating of glue, bronzing liquid, clear vinyl, or even paint.

FLUORESCENT PAINT—Paint containing minerals that glow when activated by ultraviolet light may be used occasionally to produce startling effects.

LUMINOUS PAINT—Sometimes it is convenient to paint onstage "spike marks" that can be located on a darkened stage. Luminous paint which stores energy while illuminated by stage or house lights will continue to glow after the lights are blacked out.

SHELLAC—Shellac will produce a slight gloss on a flat paint surface; orange shellac produces an amber glaze; white shellac alters the color only slightly. Resin spots in knotty boards may be sealed with shellac.

CLEAR VINYL—Clear Vinyl and Acrylic products will produce a glossy and/or protective coating on dull surfaces.

BRUSH CLEANERS—To clean brushes use: soap and water for water paint; alcohol for shellac; lacquer thinner for bronzing liquid; turpentine or paint thinner for oil paint.

FELT MARKERS—Felt tipped markers of various types, sizes and colors may be valuable for free hand drawing, or straight-edge lining on flats and drops. NOTE: When felt markers are used the marks will bleed through when the surfaces are repainted. The lines may be neutralized with shellac or clear Acrylic spray before repainting.

LONG HANDLED WINDOW WASHING BRUSHES that are designed to fit into a bucket may be used for basing, or laying-in work, when the scenery is flat on the floor, or for painting the acting area of the stage floor, or a floor cloth.

SMALL BRUSHES—Additional small bristle brushes may be obtained from theatrical supply companies to use for limited size detail. These brushes should have long carefully shaped bristles. They should be used carefully and should be thoroughly cleaned immediately after each use.

PAINT BRUSHES: One general way to judge a paint brush is by the length and quality of the bristles. A good brush has long, flexible bristles that hold their shape until they are practically worn down to the handle. Scenery brushes should be built to be used with water paint. Paying more for a good brush is worth while because the brush will last longer and give better results.

BASING BRUSHES—Brushes that are used for painting large areas should be relatively large and long bristled. It is obvious that surfaces can be painted more rapidly and efficiently with a large brush than with a small brush. Excellent for this purpose are 7″ or 8″ brushes,

frequently referred to as Dutch Primer or Calcimine brushes. The best brushes contain pure bristle but the new flagged-end-nylon are excellent and only about half the cost of the pure bristle brushes. Large brushes cost a good deal, so a small shop is not likely to have more than one or two. Although smaller brushes can be used, it is not advisable to use anything narrower than 4″ for basing in sizable areas.

DETAIL AND LINING BRUSHES — Brushes ranging in size from 1½″ down to ½″ in width with long pure bristles are used for striping and detail work. Brushes known to house painters as "sash tools" are excellent for most work. The 1½″ and 1″ sash tools may be used for almost all work except the finest details which require smaller brushes. Brushes in this category should have long bristles that are carefully "set" in the brush so that a chisel edge is obtained.

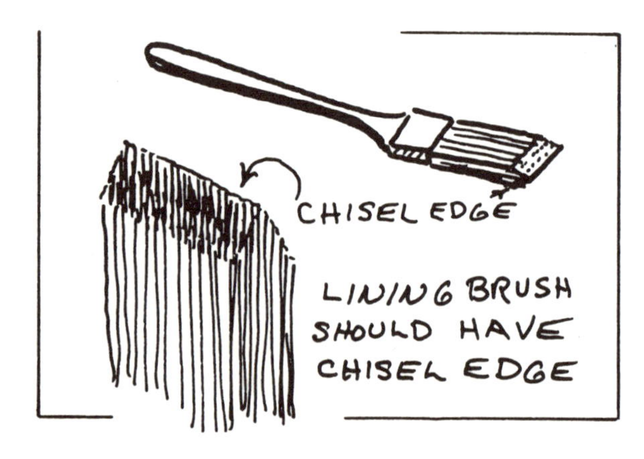


### Miscellaneous Supplies

There are a number of other supplies that should be added to the paint department's collection. A short description of some of these items follows.

SPONGES — Both animal and cellulose sponges have a use in the paint department. For most texture processes an animal sponge

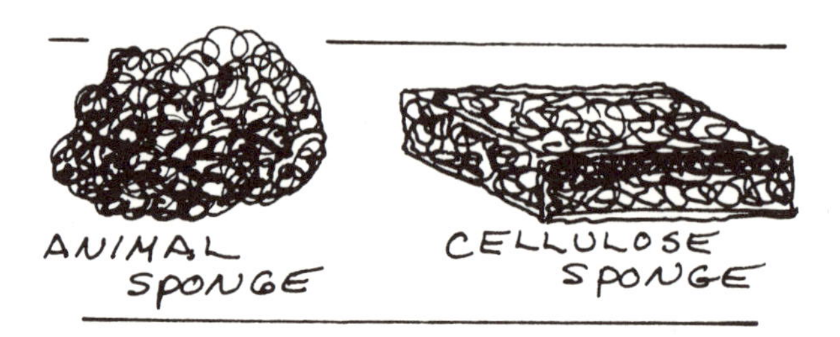


is superior to the cellulose since it has an irregular rather than a regular patterned surface.

CHALK LINE BOX—Frequently it is necessary to chalk a straight line of some length. A

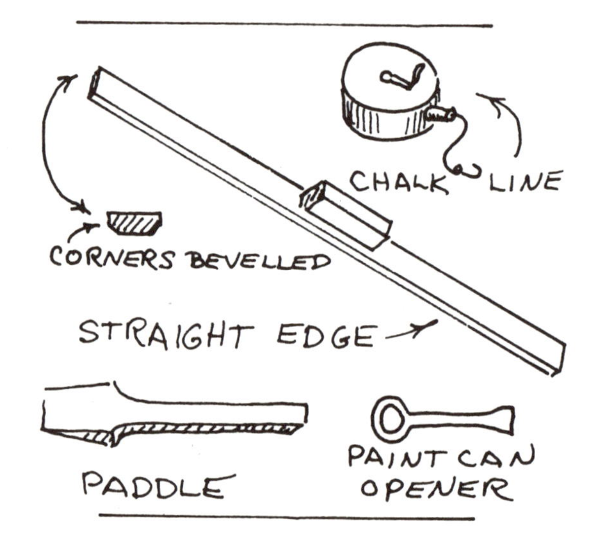


chalk line is excellent for this purpose and small boxes containing chalk, reel and string are available.

STRAIGHT EDGES — Straight edges can be made in the shop. The straight edge is a thin, narrow, straight piece of board with a small block in the center of one side for a handle and with the edges of the other side planed on a taper so that the paint will not run under and smudge as the brush is being drawn along the edge to paint a line.

PADDLES — Simple board paddles may be made in the shop. Usually the paint dealer will provide them free on request.

PAINT CAN OPENERS — The paint can opener is a small screw driver type instrument that can be obtained from the paint dealer on request.

PAINT BUCKETS AND CONTAINERS — Almost any type of container can be used to hold paint, including 2½ gallon pails, 5 gallon paint pails, gallon food cans, and large fruit juice cans.

LARGE COMPASSES—Compasses and dividers of large size are difficult to find and expensive to buy when found. Compass devices can be made in a number of ways. Chalk or a pencil tied to the end of a piece of string and then the string held on the center of the radius of the circle and the string or chalk end swept around is one of the simplest. If the string is very long there is danger of stretching and slackening. A strip of wood of the correct length held at the radius center by a nail and the chalk or pencil held in the correct position on the other end and swept around works very well.

TOOL FOR LONG GENTLE CURVES — A long thin strip of wood can be bent around to achieve a desired long gentle curve and while this is held in place a line can be drawn along its side.

STENCIL PAPER—A supply of stencil paper should be available so that stencils can be cut out when they are required.

HOMEMADE STENCIL PAPER — If regular stencil paper is not available a substitute can be made from almost any kind of paper, although heavy wrapping paper or butcher paper will probably hold up better than newsprint or other lightweight paper. Heat the paper with an electric iron and then while it is hot rub the surface with canning paraffin. The paraffin will melt into the paper giving it a waterproof surface that will make acceptable stencils.

MARKING SUPPLIES—Chalk, charcoal and large soft lead pencils can be used for marking out painting details.

PASTELS—Pastels can be used for coloring small areas and for drawing property pictures.

JIGS—Various crutches, aids or jigs should be made as needed and then kept for future use. Special devices for making forms such as curves, eggs and darts, dentils, molding ends, undulating lines, circles, etc., should be designed

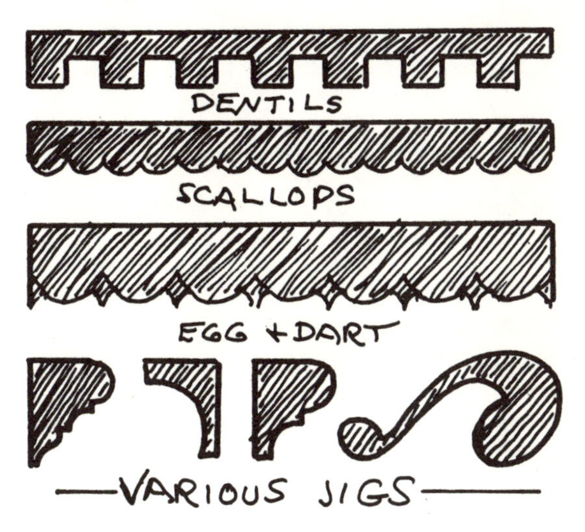


when repeat forms are needed. Spacing sticks, measuring sticks for standard spacings on a set of flats and many other types of devices will be contrived from time to time and should be retained for future use.

Some of the equipment already suggested as needed in the scene shop (such as steel tapes, carpenter's squares, hammers, staple guns, screw drivers and the various cutting equipment) is necessary, but may be assumed to be available in the normal shop.

* * * *

Although the above list is not complete, it at least includes the important items needed for painting. It is possible to start painting with a brush and a bucket of paint, but as more and more settings are painted many of the above items will be accumulated. Still other items are suggested in the discussion of *Scene Painting*.

*Cyrano de Bergerac.* Areas of the setting have been blacked out with paint to achieve the effect of a contour edge.

# 6

# Color

Since much of the mood and appeal of a stage setting comes from color, the designer needs to be aware of principles that govern its proper use and make it such a versatile element of set decoration. He should have a basic knowledge of the nature of color and the effect of colored light on colored pigment. Moreover, he must know something of the psychology of color.

### Colored Pigment

COLOR THEORY—There are a number of color theories, some simple and some relatively complex. For scenic purposes a fairly simple yet practical theory of colored pigments is advisable. The one described here will help the designer mix paints and help him determine how to get the color effects he wants.

COLOR VARIABLES—There are three variable factors in color:

1. Hue
2. Value
3. Chroma

The name usually associated with a color is its *hue.* For example, red, yellow, blue and green are all names of color hues. The *value* of a color is its apparent lightness or darkness or, in other words, its light reflective quotient. *Chroma* refers to the intensity of a color. A color of high chroma is brilliant or pure.

COLOR FORM—To show these qualities of color and their relationships to each other a three dimensional form is used. This form is

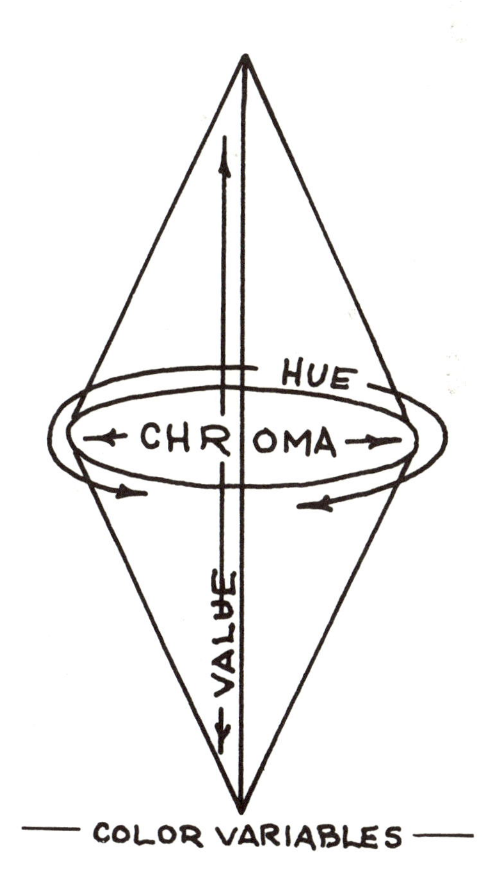

shaped like a double cone with the bases of the cones together and the points at opposite ends. The qualities of color are distributed on the form as follows: the hues of color are located around the circumference of the widened center; the values are located on an axis running from point to point through the form; chroma is the relative position of the color within the solid form extending inward or across the circular form toward the central axis.

HUE—Infinite variations of color are possible. Around the circumference of the center of the color form the hues of the rainbow are gradually blended with all of their countless variations. It is possible to start anywhere on the circle and pass around in almost any number of steps. The usual process is to begin with the primary triad of red, blue and yellow. The secondary triad, produced in pigments by mixing pairs of primaries, is composed of purple, green and orange. The tertiaries are produced by mixing adjacent primary and secondary hues: red purple, blue purple, blue green, yellow green, yellow orange, red orange. This process can continue endlessly until it is impossible to differentiate between one hue and its neighbor.

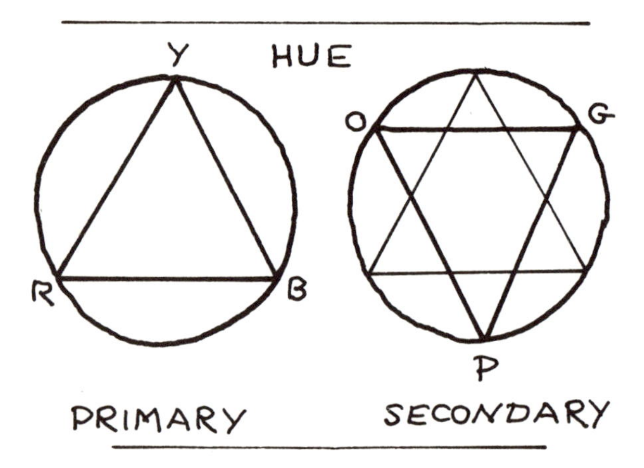


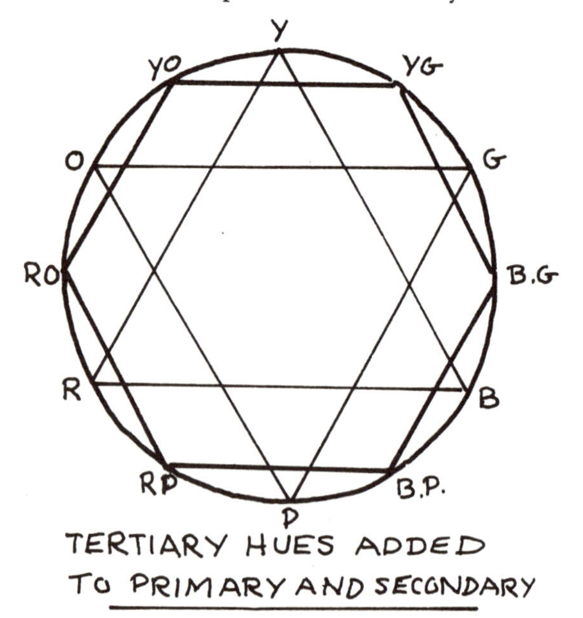


CHROMA—The intensity of a color, or its chroma, is altered by adding other hues. If the hue directly across the wheel is gradually added, there is a systematic graying until finally the center of the circle, or the neutral gray position, is reached. Adding one hue to another tends to move the final mixture into a position somewhere within the circle and away from the circumference. In other words, the color loses its brilliance and is reduced in chroma.

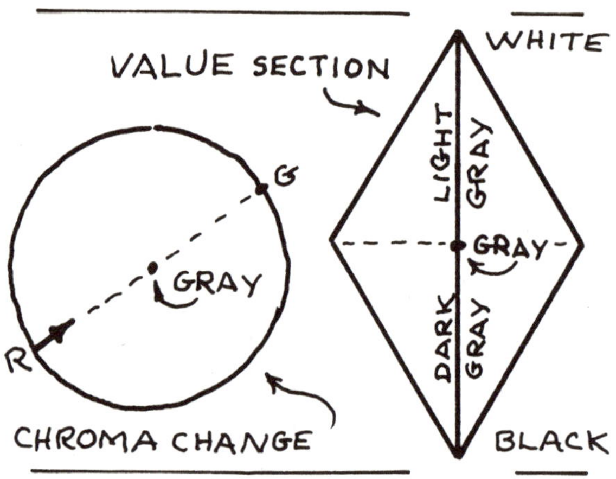


VALUE—The axis of the double cone represents the value or white-to-black scale. This is frequently referred to as the achromatic scale. If the axis is shown by itself it appears to be a pole, white at the top, progressively darkening to a neutral gray in the center and darkening to black at the bottom. Colors are changed in value by adding white or black. Adding white to a color produces a *tint,* which has a

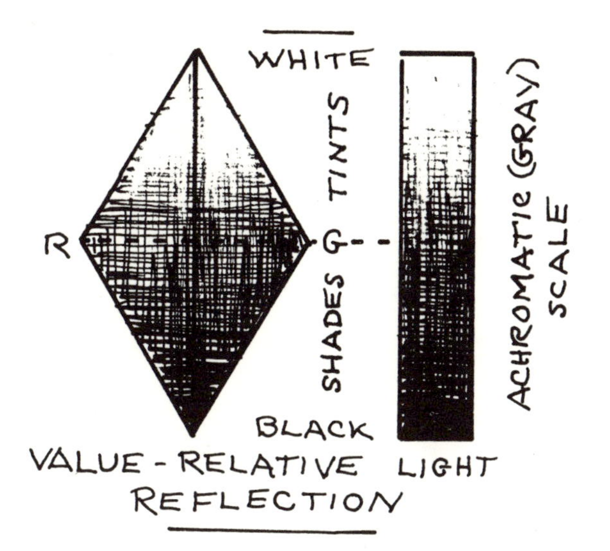


high light-reflective surface. Shades, which have low light-reflective surfaces, are produced by adding black to a color.

This theory, although brief, contains the basic information needed to mix and use colored pigments. However, the following additional facts about the color form and its use will help the designer in his work.

ACHROMATIC SCALE—An achromatic scale is one without color, composed completely of grays. Black and white photography, black and white printing, and black and white television all make use of this scale. The scale ranges all the way from pure white down to the deepest black. White reflects all of the light that falls upon it and black absorbs all of the light that strikes its surface.

In general light values tend to move forward and dark values tend to recede. In actual practice the opposite may appear true sometimes because of other factors. On the stage and in nature objects which tend to fade into the background appear to move away. In this type of situation the contrasting object gains dominance over the ones that fade into one another. Anyone who has lived in the mountains realizes the truth of this statement. When the sky is clear the mountains seem so near that one could reach out and touch them. They seem sharp, stark and rugged. At other times the sky is hazy and the mountains seem covered with a sky-colored glaze that makes them almost fade away. They seem distant, indefinite and mystical. Actually, this type of situation completely defies the idea of light value moving forward and deep value receding. The haze covered mountains are light in value but seem distant

because there is little color contrast between them and the sky. In the other situation there is a high contrast between the deeper values of the mountains and the soft blue of the sky.

Another demonstration that substantiates the theory of dominance and recessiveness may be graphically shown by painting patterns with tint colors on a flat black background. The patterns seem to be suspended in space because they pick up and reflect light, while the background of black absorbs light and seems to have infinite depth.

CHROMATIC SCALE—A chromatic scale of color is a scale of hues as they appear around the circumference of the widest part of the

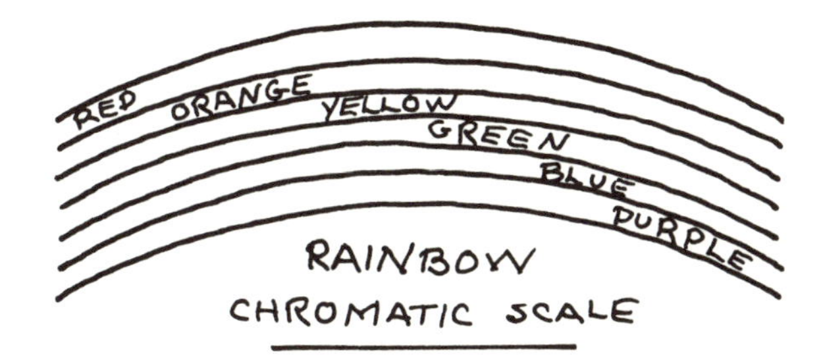

double cone form. The rainbow is a chromatic scale of color. In fact, the hues appear in what is considered to be their true order in the rainbow form.

MONOCHROMATIC SCALE—In a monochromatic scale of colors one hue and all its variations are used. If one hue is cut out of the color form its monochromatic scale will be in the

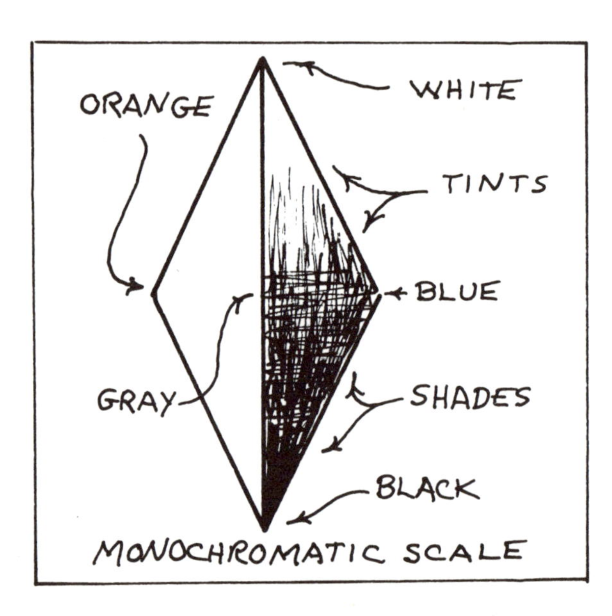


form of a flat isosceles triangle. Infinite variations are possible. The pure color, its complement, black and white are used to get all of these variations. Any amount of white or black—even to and including the pure pigment with none of the hue present—can be used. Only enough of the complement to achieve neutral gray when the two are mixed, which may be an equal volume, can be used. If more than an equal amount of the complement is used the resultant mixture moves to the opposite side of the axis.

Red can be taken as a specific example to show how this works out in actual practice. Saturated or high chroma red is decided upon as the color for a surface. After painting a small area it appears that this is going to be too garish, so a little of the complement is mixed with the red to reduce the chroma. The color directly across the wheel, the complement, is green. So green is used to gray the red.

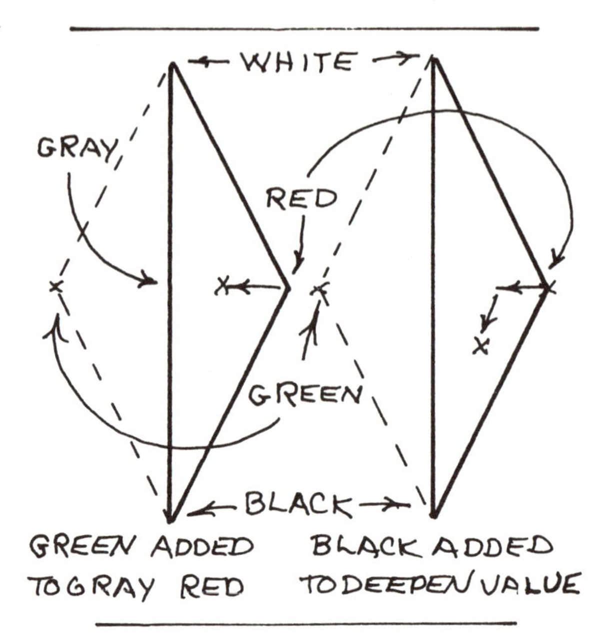


If the red happens to be too high in value and reflects too much light some black can be added. After the correct light reflective value is reached it might still be too red so a little green can be added to move it toward the gray scale.

Pink can be mixed in a similar manner. First, white is added to the red to get the correct value. The result might be a sickening pink,

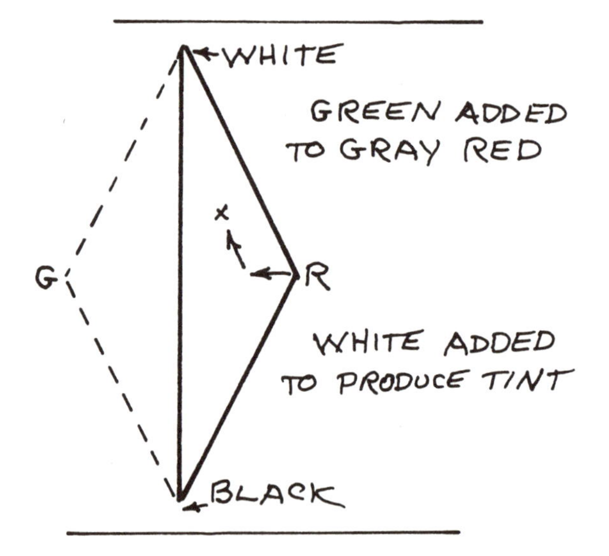

so next a little green is added to cut the effect of the red by reducing the chroma.

The addition of the complement will affect the light reflective value very little but will cut the brilliance or chroma of the color.

All of the red variants described are within the monochromatic scale of red—the lightest of pinks, the darkest of blackened red, the grayest of red and the most brilliant red. Any or all of them can be used with one another—at least in theory.

The above process can be used with any hue—blue, green, orange, purple, red purple or any other—with basically the same results.

COLOR RELATIONSHIPS—There are certain suggestions about using colors together that might be helpful to anyone who has worked very little in this realm. Actually the determination of color combinations that have appeal is a matter of individual feeling, but there are certain general facts to keep in mind.

Any single hue may be used with black, gray or white in almost any quantity.

Any monochromatic combination can be used effectively. A hue, its shades, its tints and its grayed chromas are attractive in combination.

Analagous colors—those immediately adjacent to one another on the wheel—may be used together. Many of them can be used effectively as long as they are not too far apart on the circumference of the wheel. And of course they may be used in varying values.

Analagous color combinations sometimes can be improved by the addition of a small area of the color that is complementary to the dominant one of the group.

Complementary colors of equal dominance and intensity should be avoided. When used together in this way, each tends to accent the other to the point of conflict of interest. So if complements are used together one must be dominant. The dominant one may be of a lower intensity and then the addition of a small area of the other will tend to accent each of them. Yet to say that complementary colors can never be used together would be denying nature some of its best shows. Is there anything as gorgeous as a red rose set off by its brilliant green foliage? In the fall we glory in the sight of leaves turned orange against the purple background of the mountains.

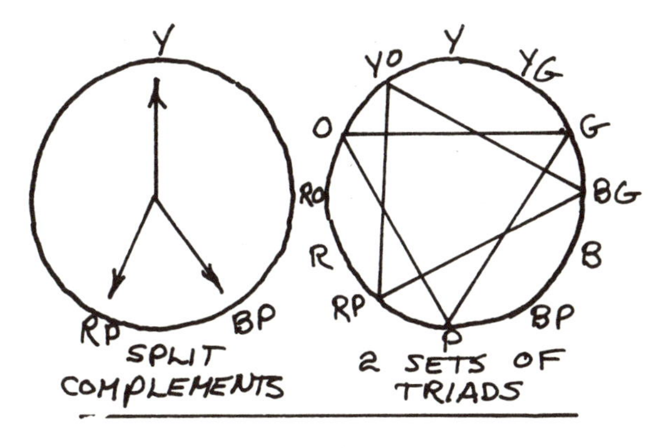


Split complements are frequently used. One hue is combined not with its complement but with the hues adjacent to the complement.

Triads—colors that are located at the points of an equilateral triangle placed upon the color wheel—may be used together.

COLOR DOMINANCE—In general when any color scheme is employed one hue is dominant over the rest. It is dominant in area, in intensity, in value or in all three. And in a scheme of this nature an important item of

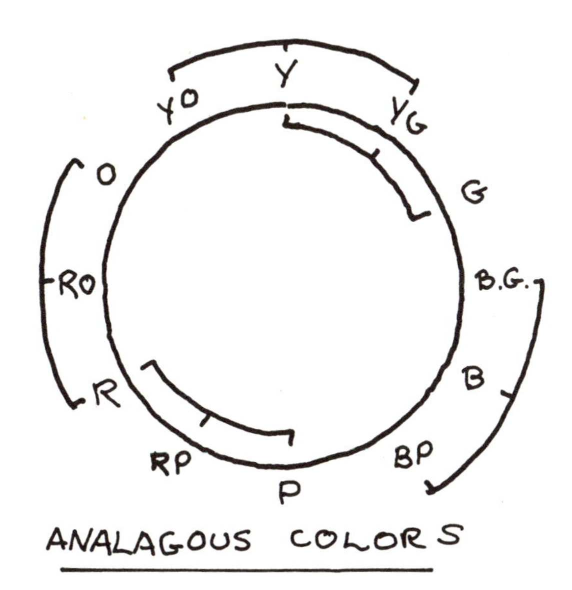

stage business such as a small property, a lamp, a box, a vase, a single flower, an item of wearing apparel or a picture on the wall may have a contrary color, an "accent" that either calls attention to itself deliberately or adds that "touch" which completes the picture.

COLOR FACTS—Unity and coherence can be achieved by using color carefully. If the stage is literally divided with one-half painted one hue and the other half another hue, the two sides seem not to belong to one another. By simply carrying a little of each color over to

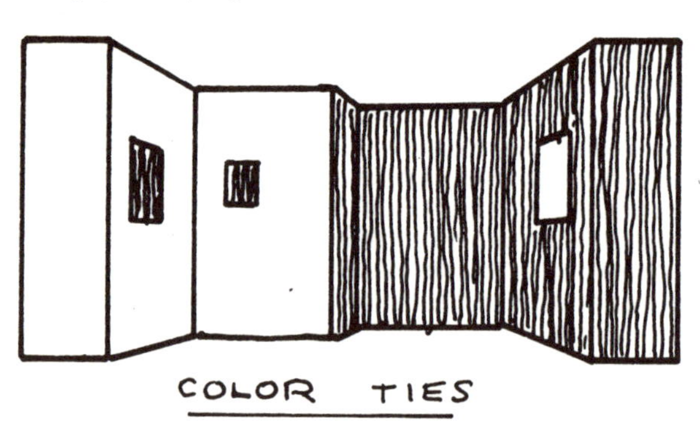


the other side of the room the sides will be at least loosely tied together. It is almost like tying the two sides together with pieces of string.

In general warm colors are better with black and cool colors with white. Warm colors appear to have higher reflective surfaces than cool colors, even though frequently they are the same. So it is logical that there is a higher contrast between black and warm colors and between white and cool colors.

Certain colors are associated with warmness while others are associated with coolness. In nature the sunlight seems to be associated with yellows and fire is associated with reds and yellows; these have become known as the warm colors. Coolness is associated with night and the blue of the sky is deeper at night. Foliage has a cooling effect so green is associated with coolness. Blues and greens are considered to be the cool colors. There are probably many other reasons why these associations have been made, but even this brief explanation may be sufficient to warrant the classification.

Other generalities regarding the use of color can be made, and although these needn't be observed, they are nonetheless true. Comedy usually has a feeling of warmth and a glow that is satisfied with the use of warm hues and tint values. The sombreness of tragedy frequently suggests cool colors and shade values. Mystery plays often carry an aura of shade values. And fantasy has an airiness that is interpreted best by colors that move at least somewhat into the tint area.

Too many tints tend to give an anemic feeling to a setting and too many shades will give a gray, musty feeling. And neither of these feelings can be overcome by the use of lights since the white of tints will reflect too much light and the black of shades will absorb too much and reflect too little light.

### Colored Light

Colored pigment cannot be discussed strictly by itself since in the theatre it is always illuminated and light can, and usually does, change the character of the color.

COLORED LIGHT—There are numerous colored light theories just as there are numerous theories about pigments. For the purpose of this study the theory that employs red, green and blue as the primaries will be used. This differs from the pigment color theory which employs red, yellow and blue as the primary colors.

In nature we receive direct and reflected light from the sun. The direct rays are obvious. What isn't so obvious is the fact that everything the sun shines upon reflects light and that there is an intermingling of all these sources of light. If this were not true our surroundings would take on a completely different character. We would live in a world with more blacks since every shadow receiving no reflected light would be perfectly black.

If a beam of white light from the sun is directed through a glass prism it is broken down into a wide range of colored rays of light varying all the way from red on one side of the

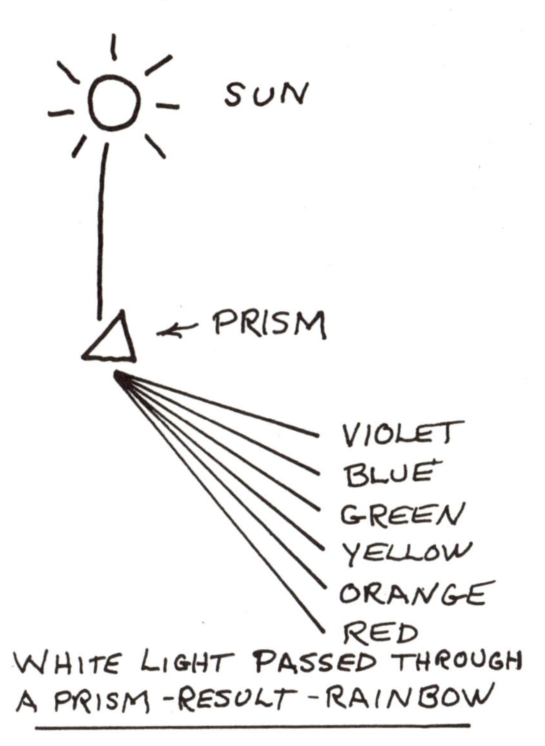


spectrum to purple on the other. When a rainbow appears in the sky basically the same thing occurs that happens when the beam of light is passed through the prism. In one case the light rays are bent as they pass through the glass, and in the other case they are bent as they pass through rain drops and moisture in the atmosphere. Hues fall into the natural order that we accept because they are bent progressively more until purple is reached. Ultra violet, used for special theatrical effects, appears below purple; and infra red, used in secret photography, appears above the red. However, neither ultra violet nor infra red are visible to the eye.

The foregoing discussion is designed to show that white light is a composite of the various individual colors. Experimentally it can be shown that not only is white light composed of many individual colors but that conversely white light may be produced by mixing the beams of various colors. Red, blue and green beams accomplish this effect and consequently can be accepted as workable primary colors for stage lighting purposes.

All objects have the ability to reflect or absorb certain wave lengths of light. The apparent color of an object is determined by a combination of the wave lengths of light ab-

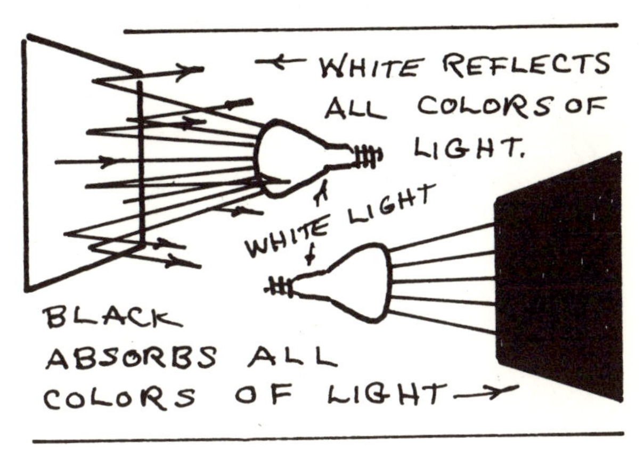


sorbed and those reflected by its surface. A white object reflects all wave lengths, hence it is white. A black object on the other hand absorbs all of the wave lengths and theoretically has no color at all. An object is red because it reflects red wave lengths and absorbs all of the others. A pink object reflects some of all colors but reflects more red than others and so becomes a light value of red or pink. A green surface reflects green, a blue surface reflects

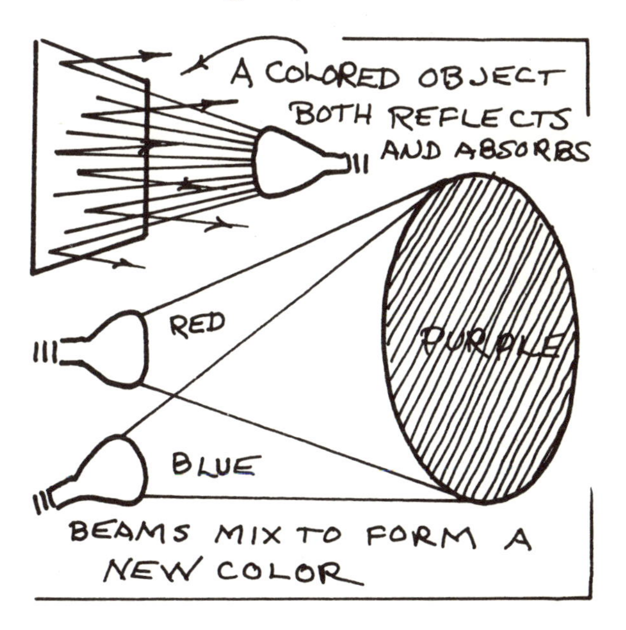

blue, a blue green surface reflects blue and green, a yellow surface reflects red and green, and a purple surface reflects red and blue.

PRIMARY COLORS OF LIGHT—The primary colors of light, as previously listed, are red, blue and green. By mixing additively, that is, by mixing beams of any two primary colors of light, a secondary color is produced. The sec-

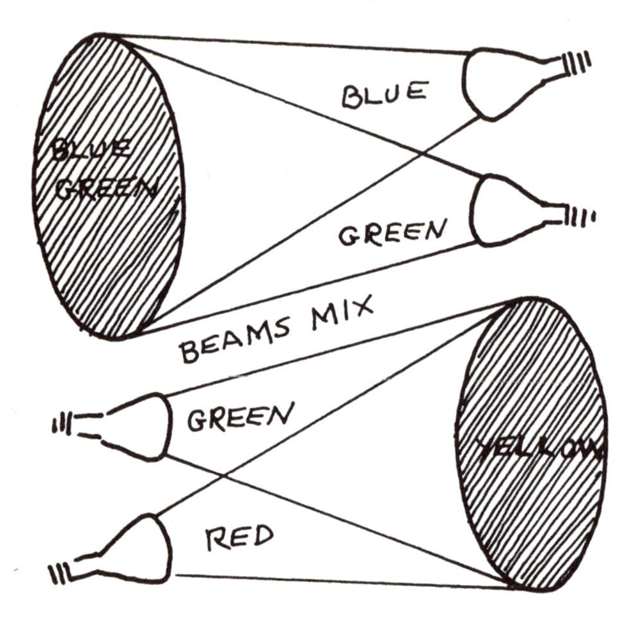


ondaries are produced with these combinations:

Red + Blue = Purple
Blue + Green = Blue Green
Green + Red = Yellow

Mixing can continue from this point combining primaries and secondaries to produce tertiaries and so on just as with pigments.

Now, return to the basic fact that objects are colored because they have the ability to reflect and to absorb certain wave lengths of light. This fact should be borne in mind by everyone who is working with colored light on colored pigment. Many startling things happen in the theatre lighting world that are traced back to this basic idea.

To point out the importance of this idea, start with a question. What happens when a beam of colored light falls upon an object that does not have the ability to reflect that color? The obvious answer, considering the previous

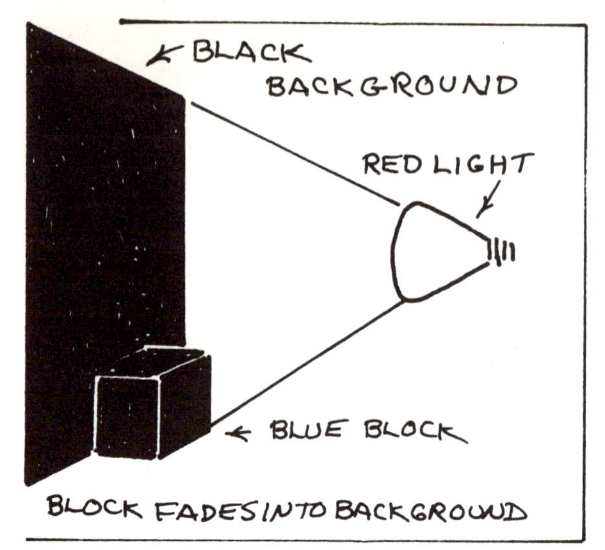


discussion, is that all of the light is absorbed and the object becomes black or at least extremely dark. A red light, for instance, is beamed upon a blue block. If the block is on a black background it will fade completely into that background. If the blue block is on a green background the background and block will both turn almost black. If the blue block is against a red background the block will appear to be black against red since red will reflect red.

To show another effect that operates in bas-

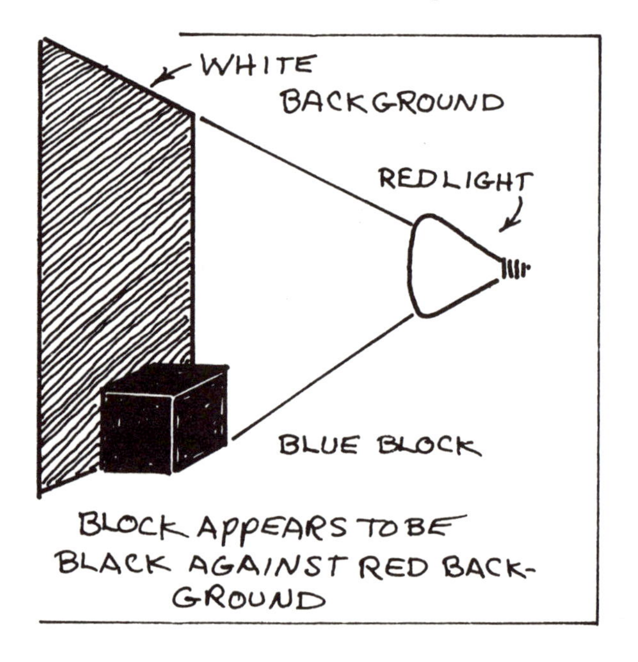

ically the same way, a colored light can be placed so that it shines on a white background. Since a white surface reflects all colors the background will assume the color of the light. If the light is red the surface will appear to be red. Now place a blue block on the surface. The blue block absorbs the light and appears to be black, while the surface reflects the light and appears to be red. Next, place a green block on the surface and the same thing happens — black against red. Finally, put a red block on the surface. The red block fades into the back-

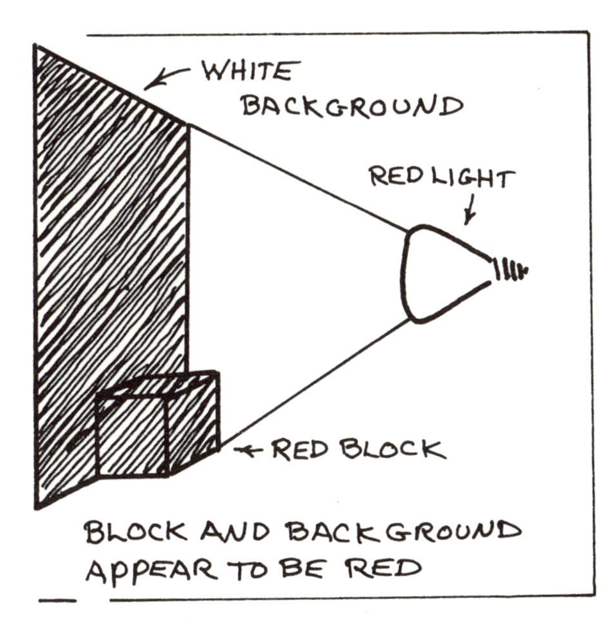


ground and cannot be seen at all because both the white and red surfaces have the ability to reflect red rays of light.

By using pure colored beams of light it is possible to make objects seem to appear and disappear. In actual practice on the stage exactly the same thing happens unintentionally at times. For example, a red color medium is placed on a spotlight and that light is used to illuminate actors. Immediately the makeup seems to disappear from the actors' faces. This is logical since the pink of the face and the red of rouge and lipstick all have the ability to reflect red light. They all turn red and the actors have completely red faces. Costumes in this

case might be made of blue and green materials, which under a red light appear to be black.

In another situation a midnight blue effect is desired. The lights are turned on and the makeup now changes in a different way. The skin takes on a sickening blue color except for the rouge and lipstick areas which become black holes in the actors' heads.

Color Media—Mixing the beams of light from various sources is referred to as additive mixing of light. There is also a subtractive method which is used when color media are employed to obtain colored beams. The color media filter out some wave lengths and allow others to pass through. Hence the light is "colored." The probable resulting color can be determined by looking through the color media. Various materials are used for color media, including gelatine, glass, cellophane and other plastics.

To demonstrate the subtractive color method,

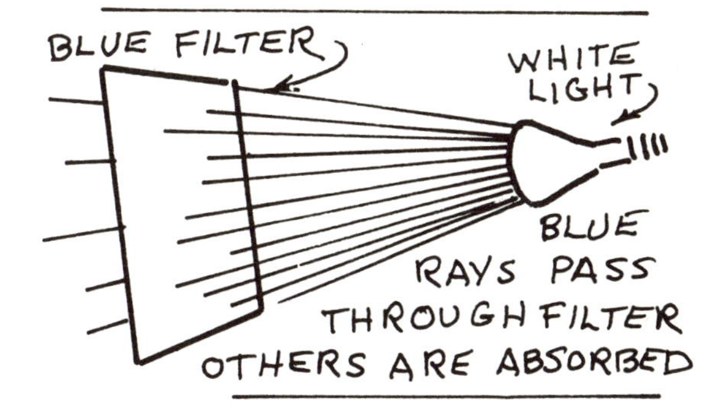

provide a spotlight with a blue color media. All of the beams except blue are absorbed or filtered out and only the blue rays pass through. Then place a green media in front of the blue media. For all intents and purposes all of the

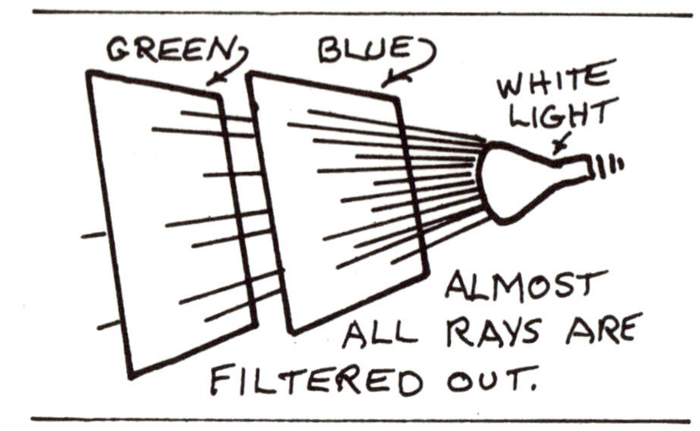


light is now cut out because the green will allow only green rays to pass through and there are no green rays passing through the blue. However, because of the impurity of the media a few rays will probably still come through. To cut these

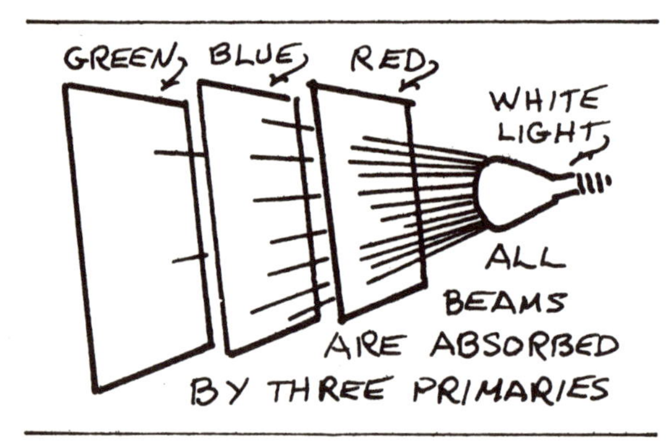


last rays out place a red filter in front of the blue and green. This should absorb all of the other rays.

### Colored Light on Colored Pigment

To discover which wave lengths of light will pass through any given filter it is necessary to analyze the color of the filter. Ordinarily in stage lighting pure color media are rarely used except for special effects. Usually tint colors—colors that allow a good deal of white to pass through—are used. They have some filtering effect and that effect is determined by the amount and colors of the dye in the media.

Here is a random list of colors of gelatine and an analysis of the colors that are likely to pass through.

| | |
|---|---|
| Light Blue | = White + Blue |
| Pink | = White + Red |
| Straw | = White + Red + Green |
| Lavender | = White + Red + Blue |
| Light Blue Green | = White + Blue + Green |
| Chocolate | = White + Dominance of Red + Blue + Green |

If two media are placed together in front of a light the resultant color can be approximated. If they are tint colors a good deal of white will pass through. If there are common colors in the dye mix of the color media they will be dominant. So one looks for the common denominators. Here are some likely results:

Light Blue = White + Blue
Straw = White + Red + Green

White is the only common denominator and much of it will be filtered out by the three primary colors.

Light Blue = White + Blue
Lavender = White + Blue + Red

Blue is a common denominator along with the white so there will be a dominance of blue.

Light Blue Green = White + Blue + Green
Straw = White + Green + Red

The green will likely be dominant.

A filter tends to lower the intensity of the beam since as the name implies it cuts out or filters out a portion of the light's rays.

To find the effect of colored light on colored pigment it is necessary to analyze the pigment colors in the mixes used. Here are just a few.

| | |
|---|---|
| Pink | = White + Red |
| Blue | = Blue |
| Orange | = Red + Yellow |
| Green | = Blue + Yellow |
| Lavender | = White + Blue + Red |
| Blue Green | = Blue + Green |

The only color that might prove to be confusing as regards its light reflective qualities is yellow. In colored light yellow is made up of red and green wave lengths, so the pigment yellow will reflect both red and green providing both are falling upon it. Yellow will also reflect either red or green alone and take on that color.

COLORED LIGHT AND PIGMENT COMBINATIONS—Here are some combinations of colored lights (L) shining on colored pigments (P) and the probable results. Of course there will be deviations because of color mixes and because of the varied gelatines available so that in actual practice some of the results will not be exactly as analyzed here. In general, though, here are the results.

L. Light Blue Green = White + Blue + Green
P. Lavender = White + Blue + Red

Since the blue will reflect from the blue surface and the red will tend to turn dark in blue or green light there probably will be a blueness and a darkening of the surface caused by this light.

L. Straw = White + Red + Green
P. Orange = Red + Green + Red (Yellow-Red)

Straw is basically yellow, although at times it has a touch more of red. This is dominantly yellow on yellow with maybe a slight dominance of red. (Pigment orange is a mixture of yellow and red; and since red and green produce yellow light the combination above is arrived at.)

L. Pink = White + Red
P. Green = Blue + Yellow (Red and Green will reflect from Yellow)

White light will show the true color of the green while the red portion of the light will tend to gray the green. Undoubtedly a little ruddiness will come through since some red will be reflected by the yellow in the green.

L. Light Blue = White + Blue
P. Orange = Red + Yellow

The white portion of the light will cause some of the orange to come through, but the blue will tend to neutralize it so the total effect will be a dusty grayed orange.

Although the probable effects can be predetermined when colored light shines on colored pigment, it is always a good idea to make tests under actual stage conditions before deciding definitely about the effect of light on pigment.

Certain light media produce skin tone effects that are flattering. These colors will likely be used. Fortunately, they are tint colors that allow a good deal of white light to pass through. Their general tone is in the pink range. Straws and light blues and lavenders are also frequently used. However, during the ordinary production there are light intensity and color changes not only from scene to scene but even within a scene. These changes tend to enhance the pigment color effects rather than destroy them. Interest is added through the changes in the color and intensity of the light.

## Psychology of Color

Certain psychological factors should be considered in any discussion of color. It has been proved experimentally in numerous tests that most people react to color in a similar manner. It is true that some are hypersensitive while others react only in a dull fashion to color. Ideas or experiences associated with specific colors in the past may warp the feelings of scattered individuals, but in general the reactions to particular colors are similar from person to person.

A few of the more important colors and their general feelings and meanings will be discussed. Since these are the basic colors and the colors used most often, a little knowledge about them may be helpful.

YELLOW—Bright, clear, cheerful yellow is associated with sunlight and is sacred to the Chinese and to the European Christian religions. Yellow was used freely in the 1890 period.

The darker yellows, neutralized or grayed yellows and the greenish yellows are frequently associated with sickness and disease; with indecency, cowardice and jealousy; and with deceit and treachery.

Yellow is sixth in a list of popular colors.

RED—Red is one of the most exciting of all

colors and universally the most popular. Defiance and violence are associated with red. The red cape of the bullfighter and the red flag of anarchism are symbols that reflect this spirit. The basic passions of rage, danger, courage, sex and strife are symbolized by red.

It should be noted that red used in large quantities is fatiguing.

GREEN—Green is relatively neutral in its emotional effect. It is the fourth most popular color. Green symbolizes youth and growth. It is restful and associated with faith and immortality.

PURPLE—Purple was one of the rare dyes of the Roman Empire and partly because of its rarity and great cost it was used to dye the Emperor's robes. It symbolizes nobility and also death and sadness and spirituality.

BLUE—Blue projects a feeling of cool serenity. Fidelity and aristocracy are associated with blue. Distance, boundlessness, nothingness are symbolized by blue.

BLACK — Black carries the sombreness of death and the depression of gloom. Yet sharp smartness is achieved by combining white and black.

WHITE — In China white is the color of mourning. In the Western World it is the color for the bride, symbolizing truth, chastity, innocence and purity.

NEUTRAL GRAY—Humility and passive resignation are symbolized by gray, as are sedateness and old age.

There are a few other general facts worth enumerating. In general the warm colors of yellow, orange and red are positive, aggressive and stimulating. The cool colors of blue, green and blue violet are retiring and serene. In general women prefer red and men blue.

There are certain preferred color combinations. Pure colors are preferred to shades and tints in small areas, while in large areas, logically enough, shades and tints are preferred. Contrasted color schemes are more popular than analagous or monochromatic schemes.

* * * *

Knowledge of many other factors concerning color would be of value to the theatre worker. This smattering of information about the nature of color, the effect of colored light on colored pigment and psychological factors concerning color may stimulate the investigation of books devoted to these subjects. There are many excellent books available, a few of which are listed in the bibliography.

GOTHIC ORNAMENT

# 7

# *Painting and Texturing*

When it's time to paint everybody wants to get into the act. And everybody can to a certain point.

Painting is an extremely important step in the preparation of scenery for the stage. Literally everybody can paint—under careful supervision. Painting must be carefully and painstakingly done if it is to be effective. Effectiveness does not depend wholly upon beauty or glamour but it does depend upon the desired effect being achieved. And to achieve effects it is necessary to know some of the basic painting techniques.

### Preparation of Surface

PATCHING FLATS — The first step in the painting process is preparation of the surface. If there are any holes in any of the flats they should be patched. Patches may be applied on the surface of the flats or on the back of them. Usually the patch shows no more on the surface than on the back if it is applied before any of the painting is done. Patches should be made from the same weave material as that used to cover the flats. Cut a piece of material large enough to cover the hole. (Patches should never be smaller than 5″ or 6″ square.) Apply paint to one side of the patch and to the surface to be patched. Place the patch over the hole, paint side down, and smooth it in place with the fingers and the paint brush. Be certain that the edges are carefully rubbed and smoothed down.

DUTCHMEN—After all of the patches have been put on it is time to apply "dutchmen." Dutchmen are strips of material—the same type material as that used to cover the flats—that are applied over the joints between the flats to cover the hinges and the crack. For a regular hinged joint a dutchman should be about 5″ wide, although it can be slightly wider or narrower. For a joint with a tumbler the dutchman should be cut 8″ in width.

When applying a dutchman, a coat of paint is put on one side of the dutchman and on the surface it is to cover. While both are wet the dutchman is carefully applied, paint side down, to the joint of the flat. With the fingers and the paint brush, carefully rub the edges outward on both sides in order to secure and stretch

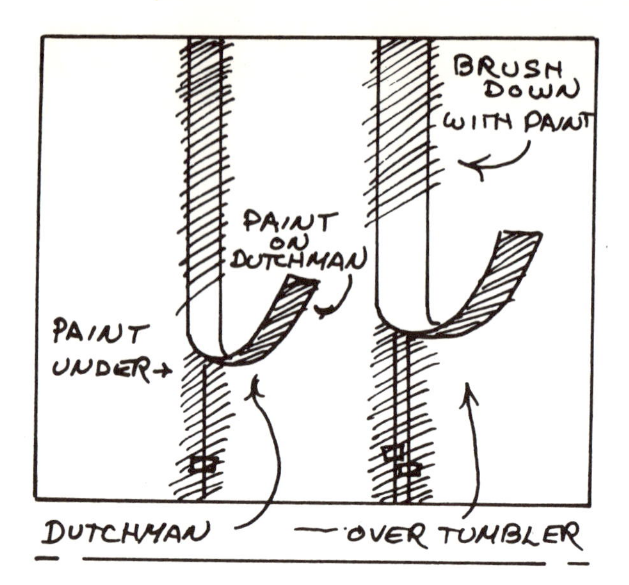


the dutchman so that it will not sink into the crack and so that any loose threads will be imbedded in the paint.

PAINTING SCENERY BEFORE—AFTER ASSEMBLY. There are advantages and disadvantages in painting scenery assembled in place on stage as opposed to painting scenery while it is lying flat on the floor, before final assembly. When assembled every corner of the setting can be fastened together and dutchmanned prior to painting and texturing, while certain corners of a setting to be painted while lying flat on the floor must of necessity be fastened and dutchmanned after the final assembly, then touched up to match the rest of the scenery.

Many texture patterns, including stencilling, may be more easily and rapidly applied with scenery on the floor than in an upright position. Spattering, especially, is easier to do with scenery flat on the floor. The spatter will usually run and "drool" when applied to an upright assembled setting unless it is done very sparingly.

Painters must do a lot of crawling around while painting scenery down, as opposed to climbing ladders and juggling paint buckets while painting scenery assembled. Make a choice!

When applying dutchmen it is a good idea to stand or lay the flats in a position where they may remain undisturbed until the dutchmen have dried. Otherwise, the dutchmen may pull loose.

## Base Painting

BASE COAT—More often than not a stage setting is put together with a conglomeration of flats that were previously used in a variety of productions. Consequently, their surfaces are of many colors and many patterns. To get a good final paint finish it is necessary to bring all these flats to somewhat the same color before starting any of the finish paint process. A base coat, which may consist of almost any neutral color, is therefore applied.

In general if the finish paint is to be a tint, the base should be relatively high in value; and conversely, if the finish coat is to be in a low key, the base should be low in value. If the shop is continuously active it may be standard procedure to mix all of the left overs from the previous production and use that mixture as the basis for the undercoat of paint.

PAINTING TECHNIQUE—When paint is evenly applied to a large surface, it may be referred to as a flat coat. When applying a flat coat use the largest brush that can be handled easily or that is available, and then use a rapid criss-cross or every-whichway stroke. This type of stroke leaves little or no distinguishable pattern. The surface may be painted while laid out flat on the floor, while leaned up against a wall or while secured to a paint frame.

RUNS, PUDDLES AND HOLIDAYS — If the flats are in an upright position the paint should be brushed out carefully and rapidly so there are no streaks or "runs" from the running of surplus paint. If the flats are on the floor the paint should be brushed out so that there are no "puddles" that cause dark and light patterns where superfluous paint was allowed to gather in pools. The puddles take on a different color than the rest of the surface when dry because of the longer drying time they require. The varying density of the pigments causes some to settle to the bottom while others rise to the top. In puddles there is no longer an even mixture of pigment as there is on the surface that dries more rapidly. Blank, unpainted spaces, or "holidays", should also be avoided.

### Ground Coat Mix

While the undercoat is drying, paint may be mixed for the ground coat and for the textures that are to be applied with it or over it. The ground coat should be basically the hue, value and chroma desired for the completed surface. The textures will tend to move the color in several directions from this, but the overall effect will return to the ground coat coloration.

### Texture Mixes

It is usually desirable to break the paint surface by using various texturing techniques. These are accomplished by employing variations of the ground coat color. Sometimes the textures do not work out as desired colorwise. Frequently they do not work out because the various paint mixes used are too far from the ground color in one direction or another. To correct this fault it is good procedure to begin with some of the ground paint and then move out in various directions by adding other colors to it. For instance, warm texture paint may be produced by taking a container of ground color and adding a little red to it; another container of ground color may have some yellow added to it. Cool textures may be produced by adding blue to one container of ground color and green to another. Lighter values are produced by adding white to some of the ground color and deeper values by adding black to some of the ground color. Any number of texture colors may be used and each time they are produced by adding a new color to a portion of the ground color. The closer the textures and ground color are in hue, value and chroma the more subtle the finished surface;

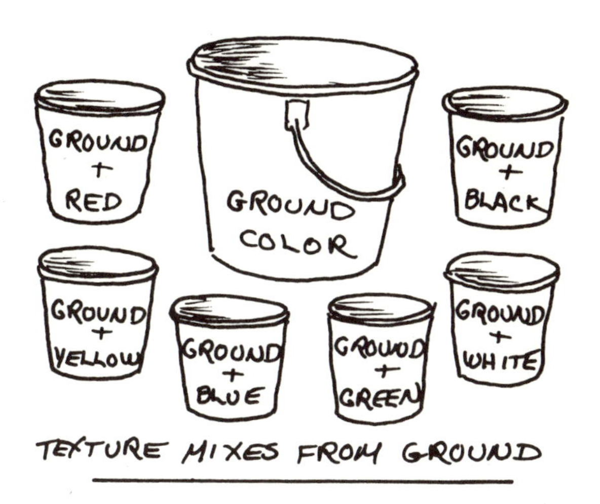

TEXTURE MIXES FROM GROUND

and conversely, the further they are apart the coarser the finished surface.

### Texturing

An almost endless variety of texture effects may be achieved through painting. But regardless of what effect is desired or of what devices are used to get the effect, the quality of the finished surface will depend in the main on how carefully and skilfully the undercoat and textures are applied.

Some textures are applied simultaneously with the ground coat while others are applied after it has dried. A few of the processes and how to achieve them are described below.

SCUMBLING—Scumbling is the process of applying texture mixes simultaneously with the ground coat. A brush is used for each color and the colors are alternately applied and partially blended together. The effect is lost with too much brushing since the colors are then completely mixed. This technique is used a

Flats for a large stage production of *Aida* spread on the ground for spattering.

good deal when there is a definite desire to move from one hue, value or chroma to another.

SPATTER—The process of spattering is literally throwing paint out of the brush. A large brush with long, somewhat flexible bristles is best for this process. The brush is swung downward or outward, depending on whether the flats are on the floor or standing up, and stopped when the forearm and brush are parallel with the surface being spattered. The brush may be stopped by snapping the wrist of the brush hand or by swinging the brush and striking it against the free hand. The wrist snapping technique wears out one hand and arm while the striking process wears out both hands and arms. Considerable practice is required before one becomes accomplished at spattering, but it is a process that is used more than any other both by itself and in conjunction with other processes. The spatterer will get better results if he keeps twisting and moving so that the swing is in constantly changing directions. The spatter will be applied in tiny dots if the brush is lightly charged with paint and the brush swung with great vigor. Large dots are made by loading the brush more heavily and swinging with less effort. The first swing, after dipping the brush, should always be an easy one to determine how heavy the spread is going to be. Then succeeding swings can be regulated accordingly.

When spatter is applied the dots can vary from the size of a pin head to the size of a nail head. They should be evenly distributed unless deliberately planned otherwise. And usually medium sized rather than large or small dots are desirable.

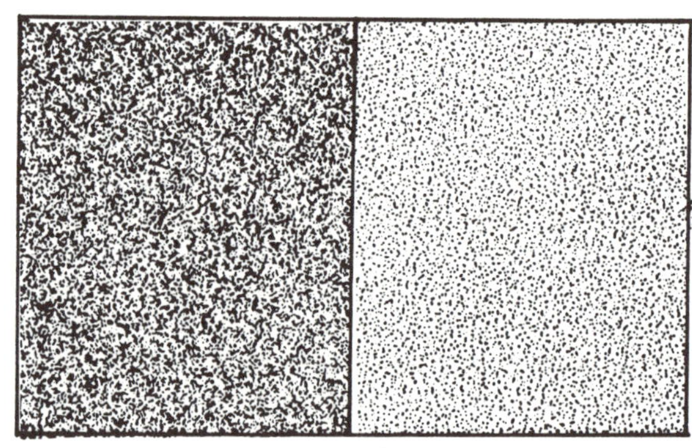


SPATTER WITH A GARDEN SPRAY—Most air pressure tank-type garden disinfectant sprays work well for applying a fine spatter. If used, they should be tried out first on something other than the finished paint surface. It is advisable to turn the valve on while the gun is pointed to a surface off the flats and then sweep into position for the spatter. And it is advisable to finish the run in the same way by sweeping off the surface as the valve is closed. As the valve is turned on and off there are likely to be spurts of paint coming out of the gun. Continuous movement in varying directions while spraying is necessary for successful use of this type of contrivance. It is an excellent device for final toning spatters.

SPATTER BLEND—Rapidly apply a section of ground coat and, while it is wet, spatter texture mixes heavily onto the surface so that there is a blending of the colors. This process works better, of course, with the flats on the floor than with them upright. The results are sometimes extremely satisfying. This process is also known as puddling.

DRY BRUSH—The dry brush technique is another that has varied uses and is one that should be mastered. After the ground coat has been applied, a streak type texture may be put on by pulling a partially charged brush lightly across the surface. The process is much easier if plenty of paint is used in the brush than if the brush is almost dry. The operator must then use a light touch, but he can spread paint for long distances and does not have to press so hard against the surface. This texture may be applied while the ground coat is wet or dry.

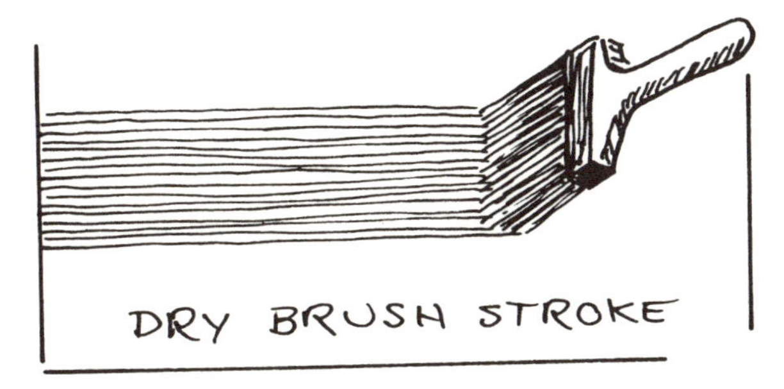


Usually when it is wet there is a great deal of blending, while on the dry surface definite streaks appear. This is a good texture for simulated wood grain, straw, thatch and many other effects.

CROSS HATCH — Cross hatching is a dry brush technique in which the brush is operated in a criss-cross fashion to streak the surface.

SWIRL STROKE — Another dry brush technique entails the use of a swirl stroke with the

brush rather than a type of criss-cross stroke. This gives somewhat the effect of roughly trowelled plaster walls.

WET BRUSH CROSS HATCH OR SWIRL — The cross hatch or swirl stroke may be used on a wet base as well as a dry base. With the wet base the stroke should be rather light or too much of the paint will be blended with the ground coat.

LONG STROKE DRY BRUSH—If it is desirable to get long, unbroken dry brush strokes a well loaded paint brush or a narrow push broom may be used. The broom must be handled carefully, but it does an amazingly good job. The broom enables a painter to cover large surfaces rapidly.

WET BRUSH BLEND—Definite movement from one color to another can be accomplished by gradually brushing from one color into another and blending between the two. If the blend isn't as smooth as desired it is possible to spatter the blend area with both of the colors to finish the process.

SPONGE TEXTURE—Texture colors may be applied over the ground coat with the aid of a sponge. The natural or animal sponge is usually better for this purpose than the artificial cellulose sponge since it has a less regular surface. The sponge is partially charged with paint, lightly squeezed and then the paint is applied by lightly pressing the sponge against the flat surface. The patterns should be close together and in order to avoid too much regularity of pattern the sponge is twisted slightly when raised in the air between applications.

Alternate sponge textures might be developed by twisting the sponge on the surface, by stroking it across the surface or by rolling it across the surface.

The sponge is a good paint applicator to use when stencilling a surface. The paint can be applied more lightly and with a less definite edge by means of a sponge than with a brush.

The ends of the bristles of the brush are used to produce the texture.

A push broom being used as a texturing tool.

STIPPLING—The process of stomping paint onto the surface with the ends of the paint brush bristles is called stippling. The term "stomp" should be interpreted lightly.

SLAP STROKE—Textures may be achieved by using the side of the brush and lightly slapping it onto the surface, leaving rectangular patterns. This stroke may be used effectively when painting foliage.

BURLAP STOMP—A texturing tool can be made by taking a piece of burlap about 6″ wide and 3′ long. Pull threads from one side lengthwise until about 2″ of the 6″ are without cross threads. Roll the burlap into a tight pack and tie a wire or cord around it. Dip the loose ends into texture paint and lightly press them against the flat surface.

WASHES—Thin solutions of color, referred to as glazes or washes, may be applied over a painted surface to cut the sharpness or brilliance of the pattern. Washes may be lightly stroked across or spattered onto the surface.

MISCELLANEOUS TEXTURES—Various other devices might be developed for applying textures. Flails, feather dusters, crumpled newspaper, wall texture rollers, rough textured pads, wrinkled burlap on a block, bunched up rags—all of these can be used. There are no rules regarding the tools that must be used to produce paint textures.

However, in spite of the fact that there are no rules, there are some general suggestions about textures that are helpful to the applier.

Use several colors. Do not be satisfied until the surface has a finished look.

To get a more interesting surface, "spot areas" of more intense color may be applied

The flat surface of a set piece for *The Madwoman of Chaillot* with shadows painted on to achieve a three dimensional effect.

here and there on the surface first, using the texture technique that will be used for the rest of the surface. The texture is then applied over the top and when it has been evenly applied the under colors will project slight alterations in the surface here and there.

When the overall texture has been completed the surface should be complex enough and the applications dense enough that the total effect from a short distance is one of a blend of color. It should be difficult to distinguish between the ground coat and the textures.

If the effect is too broad when the textures have been completed, the entire surface may be spattered with the ground coat to bring it back toward the original mix. The texture mixes may also be used as spatters for softening effects.

Spatters may be used over a texture to high light or to shadow or to soften the total effect.

If there is considerable doubt about the color of the light to be used for a setting, or if it is certain that the light color will change from time to time, it is good practice to use textures of many hues to insure pleasant and varied light reflective surfaces under all conditions.

### Specific Procedures

AERIAL PERSPECTIVE—If distance is to be achieved on objects, ground rows or backdrops and there is a sky effect in the background, aerial perspective may be used. To accomplish this the objects are spattered with the color of the background. The nearer they are to the background, the heavier they should be spattered. As they approach the color of the background they fade into it more and more. Near objects, or objects designed to appear to be in the foreground, should be painted in contrasting colors. Heavier textures, more intense colors or heavier values will all help to achieve this effect.

If the background is black, objects in the foreground will stand out more if painted in light values. And to achieve the effect of fading into the background, one uses dark spatters on surfaces. Other colors work in basically the same way. Dominance is gained by contrast and recession by similarity in color.

PROJECTING FORM WITH PAINT — With these painting techniques in mind it is easier to explain how specific effects can be achieved. It is frequently easier to paint a flat surface to give the impression of roundness than it is to build three dimensional forms. Tree trunks, columns, cannon barrels, posts and logs can all be simulated on a flat surface. The basic process of revealing the form is the same regardless of what the object happens to be. In this case it can truly be said that "It's a question of values."

Basically, it is possible to give any painted surface three dimensional form by the use of varying values of a color. To repeat a statement already made, light values tend to make a surface advance and deep values make it recede. The form of an object is revealed to the eye through the ability of the surfaces to reflect or to absorb light. Simple forms are made up of a series of simple planes; usually each plane has a different value depending upon its position in relation to the source of light. If one plane faces the light directly it has a high value. A receding surface might be less bright and have a slightly lower value since the light strikes it on an angle. A third area receives only reflected light and has a deeper value. Still another surface has hardly any light falling on it and appears to be the darkest value of all. In reproducing the form, the painter will use four values of paint to reproduce the ability of the four areas to reflect light. If he were using an achromatic scale the paints might be white, high value gray, low value gray and black. These four would give high contrast.

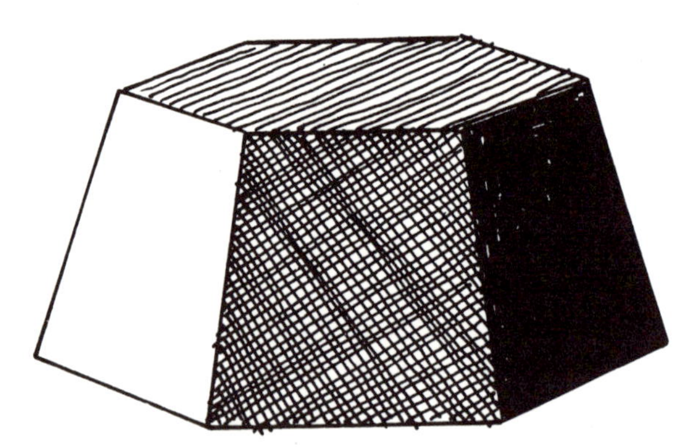
VALUES OF COLOR TO INDICATE VARIATIONS IN LIGHT REFLECTION

THE CURVED FORM—Now to return to the column form. This can be reproduced by employing three values of color. A ground color may be chosen. From one portion of this a high light value is made by mixing in some white. Another portion of the ground color can be mixed with black to make a shadow paint.

The form can be painted by using any of several texture processes. Assume that a 1′-6″ x 10′-0″ jog is to represent a column. Start at one end and paint the center 12″ the full length of the flat. With another brush rapidly brush some of the deep value along each side of the flat extending over to the 12″ strip of ground color. Then, where the ground color and deep value meet brush them into one another, blending as smoothly as possible. It may be necessary to use both brushes and work both colors together just a little. Be careful not to blend too far in either direction or all of the surface will become a mixture of the two colors. After this has been done, take a small brush and paint a streak of the light value down the center of the flat. Work this just a little, blending it slightly into the ground color. As the final step in the process, paint the remaining portions of the edge with the deep value. If there are areas that need just a little more blending,

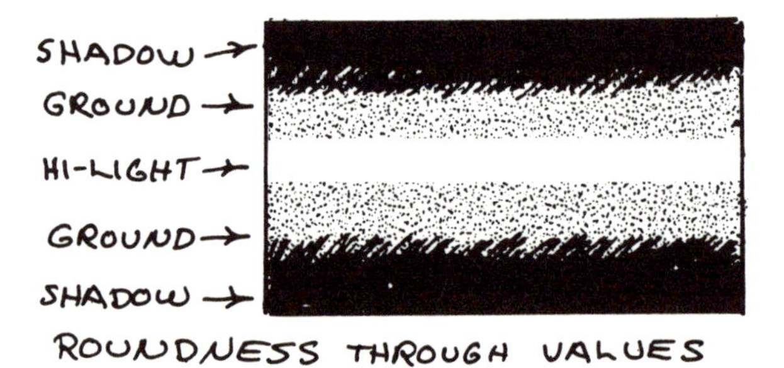

ROUNDNESS THROUGH VALUES

take a brush and some plain water and brush carefully over the surface. The water will tend to soften the paint and allow it to run together a little more. The water should be used only where and when necessary and used sparingly.

Now stand the flat up and move back a few steps and look at it. The first trial may be a bit ragged since this is a difficult process, but the form will begin to reveal itself. With practice, working more rapidly and carefully and painting relatively small areas at a time, more skill in this process can be achieved. The surface may be spattered to finish and soften the effect. It may be spattered with one or with all of the paints used.

The other methods of painting to achieve somewhat this same effect permit the complete painting of the surface with the ground color. After this has been done and the paint has been allowed to dry, it is possible to proceed in one of various ways.

If all of the surface except 2″ of the center is masked with boards, old flats or paper, it is possible to spatter a light value strip down the center. After this has been done, remove the masking and place a 10″ or 12″ board or strip of paper down the center so that only about 4″ of each side of the flat is exposed. These two side strips can now be spattered with the dark value. Remove the center masking. If there is too much contrast between the strips or if the result seems harsh, it is possible to spatter the entire surface with one or all of the colors. After one has become adept at spattering he can spatter a surface like this almost without using any masking. The maskings do provide a sharp edged demarcation between value areas that sometimes is needed to help sharpen the form.

The high light and shadows may be applied by using the dry brush technique or by using a sponge or by using a shadow wash of a little black in a lot of water and a high light wash of a little white in a lot of water. Whichever method is used, the basic plan is the same—darken the edges and lighten the center.

It is possible to move the apparent light source over to one side by painting the shadow area on one side wider than on the other side of the flat. The high light should then be moved in the same direction so that it will have the correct relationship with the shadows.

LIGHT SOURCE—When painting three dimensional forms on flat surfaces, the apparent light source must be considered in order to get the high lights and shadows in their correct relationship to one another and to the light. In general, where no definite light source such as a window, lamp or fire is indicated, it may be assumed that a natural direction and angle is motivated.

Sunlight shines down on an angle during a great portion of the day and forms reveal themselves best during those hours of the morning and afternoon when the angle is from 30 to 45 degrees. Usually the painting is not so exact that it requires precise angles for the light source, but general direction is a definite aid in determining the location of high lights and shadows.

MOLDINGS — Almost all moldings—wood, metal, plaster or plastic—are modeled from a few basic forms. These consist of the square, the oblong, the triangle, the circle and the oval.

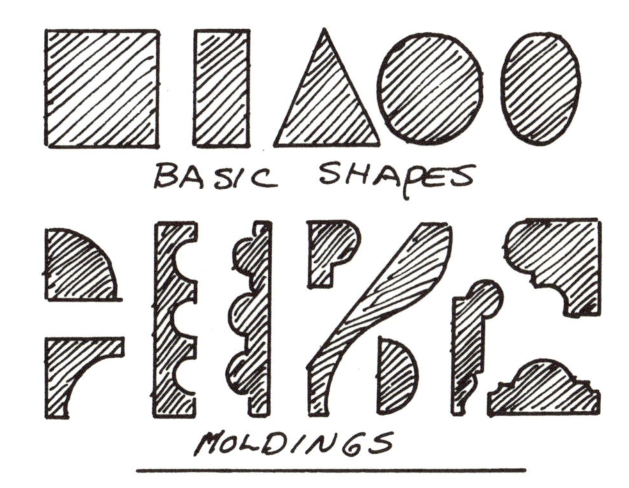


The inside or outside shape of the form and simple distortions may be used either alone or in combinations.

Moldings may be analyzed in a simple manner to determine how and where to use high light and shadows to give an impression of the form on a flat surface. In general, breaks in direction of the surface will show as sharp lines and may be painted that way. If one side of the break is especially sharp it may catch a glint of light. The light is indicated with a narrow stripe of high light paint. If the other side of the break recedes it casts a shadow so it will have a shadow value painted next to it.

Undercuts, straight or curved, will tend to be in shadow while overcuts, straight or curved, will pick up high light. Complex moldings are made up of many of these individual shapes placed one above the other and can be analyzed in terms of the individual shapes. It may be that the position of an undercut under a heavy downward curve will cast a shadow across the whole form. The painting should take such things into account, although at times it may be necessary to take some license to project the shape more clearly.

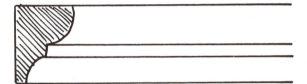

LINES REPRESENT CHANGES IN DIRECTION OF THE SURFACE

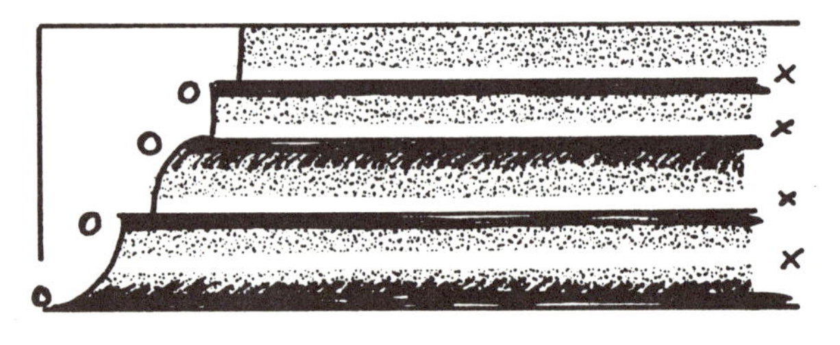

X = HIGH LIGHT
O = SHADOW

SHADOW WASHES—Shadow washes, mentioned before, are convenient to use on moldings and many other shapes since it is possible to lower the value of any color by using the same wash. The wash should be mixed carefully, adding just enough black to a container of water to get a wash that will lower the value of the surface it covers by one or two steps. It should be tested carefully before being used on a surface that is almost finished. The wash should be applied with careful, continuous, unbroken strokes of the brush.

Moldings may have complex carvings within their surface. Dentils (blocks), egg and dart forms, leaf forms and other forms may appear on a quarter circle curve or on a flat surface. These should be treated just like the rest of the molding. Outline the shape, use high light where the light strikes and use shadow wash where there are undercuts or where no light strikes the surface.

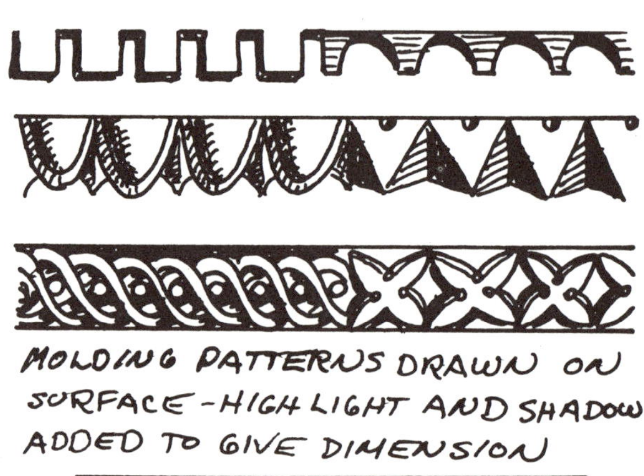

MOLDING PATTERNS DRAWN ON SURFACE—HIGH LIGHT AND SHADOW ADDED TO GIVE DIMENSION

WOOD GRAINING—Wood grain effects are achieved by means of the dry brush process. But before the dry brushing begins it is necessary to decide upon the desired general coloration so that the ground coat can be prepared accordingly. Finished wood surfaces range all the way from the honey color of pine to the black of ebony. After the ground coat has been mixed, use the texture mixing process and prepare three or four—or even more—colors from the ground coat.

After the ground coat has been applied, prepare for the dry brushing by drawing the board outlines with chalk or charcoal. Before the graining can start it is necessary to decide the directions in which the various boards run. Ordinarily, to gain strength, the grain of the boards will run in the longest direction of the

measured space. This is especially true in panelled rooms where some wood runs in all directions. The frame runs around a panel and so does the wood grain. Look at wood. See how it fits together, how the grain looks, what the malformations, grain patterns and texture look like. It is rarely necessary to be so exacting on the stage that specific kinds of wood have to be suggested, but the general effects are used frequently.

If pleasant color variations are desired in the wood area—especially if the area is large—dry brush a few areas of more intense texture mixes here and there at random. Be careful that it isn't too spotty. Make the strokes rather long but not too evenly distributed.

Now dry brush the other paints onto the surface in the correct directions with long, uninterrupted, but light strokes. Be certain that paint flows from the brush, but do not allow so much to flow that each brushing covers the ones underneath. When the project is finished it should be possible to see the strokes of all of the mixes even though, in general, there is a slight blending from one to the other.

Begin strokes carefully where a board ends, brush as far as possible and end with a light upsweep to produce a "feather-edge" rather than a hard edge. If it is necessary to start again and continue from the point where the brush was stopped in a feather-edge stroke, either sweep in lightly continuing in the same direction, or begin from the other end and finish by brushing into the feather-edge with another feather-edge. Heavy or hard edges should occur only at the ends of boards or at joints between two boards.

If the effect is too broad, or has moved too far from the original color, it is possible to dry brush some of the ground color over the top of all the other dry brush colors.

A room may have an entire wall panelled in wood, may have panel areas on each wall, or may have trim consisting of base board, casing around doors and windows, ceiling molding, picture molding, chair rail molding, or special wood areas. Whatever the wood areas they may be treated as indicated above. If it is wood, surface grain will be likely to show through even if it has been coated with paint.

GRAINING WITHOUT SPECIAL GROUND—In order to short circuit the painting process it is sometimes possible and practical to grain over the regular ground coat used for the walls of the set. If graining is carefully done with several mixes the results are perfectly satisfactory, especially if it is desirable to have the trim on the wood appear to be almost the same color, texture, and value as the walls.

LINING—After the graining has been completed the board edges are striped with lining color. The lining color may be black or a deep value of the ground color. Black has advantages in that it can be used as a standard outlining paint for all colors. However, it provides a high contrast against a light value color and should be checked for appearance. The painter will have to be the final judge as to which—dark value or black—should be used. While the striping is being done it is necessary to determine where edges will show and draw two lines to represent the thickness of the board. And where there are moldings, lines should be painted for all of the direction breaks in the surface.

DRAWING LINES—A brush about the size of a 1″ sash tool (window sash brush) can be used for painting lines. After dipping the brush in paint the bristles are carefully shaped on the edge of the paint container to get a trim chisel edge. A brush that shapes to a fine edge is important when painting lines. Place the straight edge in position where the line is to be drawn. Hold the brush at right angles to the surface so that the bristle tips all of the way across the brush touch the surface. Stroke the brush lightly along beside the straight edge, hardly bending the bristles, and an even, narrow line can be drawn. The pressure on the

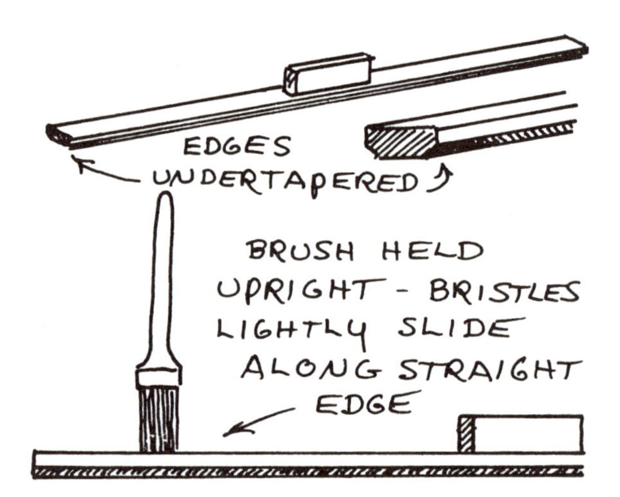


brush and on the straight edge should be so light that the straight edge need hardly be held in place. Wider lines may be drawn by placing slightly more pressure on the brush. By turning the brush sideways it is possible to get lines that are a full inch in width, and by applying pressure even wider lines can be painted. Use a larger brush and even the light touch lines are wider.

SHADOWS—Locate a light direction and determine the shape of the wood to find where shadows will be placed. A little freedom may be taken. If a strip of molding projects out from the wall a light shadow may be drawn above it on the wall and a wider one below it on the wall. The molding itself will require shadows for undercuts to reveal its shape. The side of a door casing will cast a slight shadow if it is away from the light. Here again a light shadow can be painted on the wall all of the way around to make the wood seem to project out from the wall.

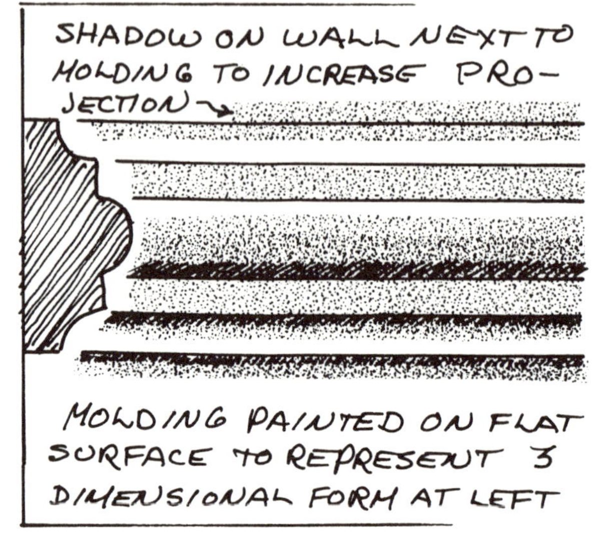


To reveal three dimensional form, shadows are generally painted under projecting surfaces and next to the side away from the light source.

HIGHLIGHTS—Highlights usually consist of narrow lines of light value or light wash paint. High light placed adjacent to a shadow will heighten the effect of the shadow. All sharp edges will usually catch high light.

To reveal three dimensional form, high lights should be placed on the surface edges of projecting forms and on the light source side.

PANELLING—Ordinarily panels are recessed areas with molding to contour the break from the edge of the framing board down into the panel. There are times when panels are built in reverse with the panel projecting above the framing members, but this is exceptional.

Draw the molding line breaks and the board edges with narrow lines of shadow value or black. If the panel is recessed there will be a shadow at the top where the molding cuts down to the panel and there will also be a shadow on the light direction side. High lights can be drawn between each pair of lines on the bottom and on the side away from the light. A high light may also be drawn next to the out-

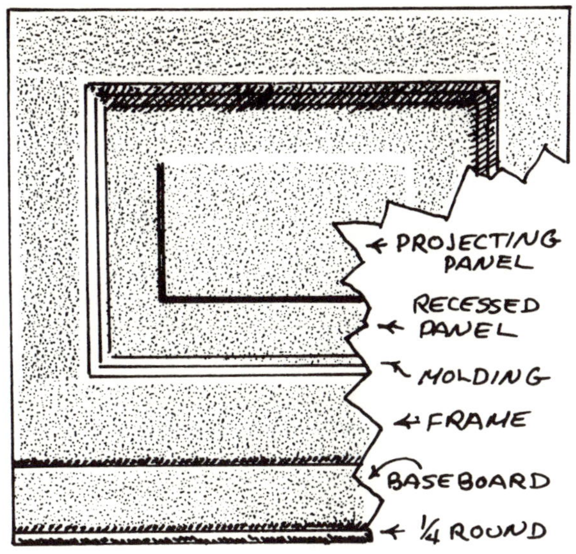

USE OF HIGH LIGHT AND SHADOW IN PAINTING PANELS.

side black line all of the way around. This high light accents the sharp edge of the framing board.

On a projecting panel the area pattern will be reversed.

Rough Boards — Boards are not always smooth. Sheds, barns, fences, shacks, piers and frontier houses may be built of rough timber. The texture is coarse, so the dry brushing should be broad and probably have some contrast. Boards do not fit together perfectly so there should be some double lines where the sides try to meet one another. The boards are random cut in width and probably have some knots and even knot holes in them. Knots may be made by painting rough circles of shadow wash

ROUGH BOARD FENCE

with grain lines that form concentric patterns moving around and out from them in a haphazard fashion. The boards may have checks and splits in their ends. And they may be fastened with oversized nails or pegs which may be indicated by drawing small circles or partial circles with the upper half light in value and the lower half dark in value. The upper half may even be drawn with chalk.

Fluted Columns — Various methods of painting to make a flat surface seem curved have already been described. Frequently when columns are painted in this way it is desirable to indicate fluting (indented half round carving) or reeding (half round applied to the surface). A round column appears to have flutes that diminish in width as they approach the sides. It may be difficult to draw the flutes free hand and get the correct proportional narrowing. Here is a mechanical method of computting the proportionate positions of the flutes.

Draw a horizontal cross section (through the column) having a diameter equal to the width of the column being painted. Determine the width and depth of the flutes and draw them in position on the circular cross section. Next

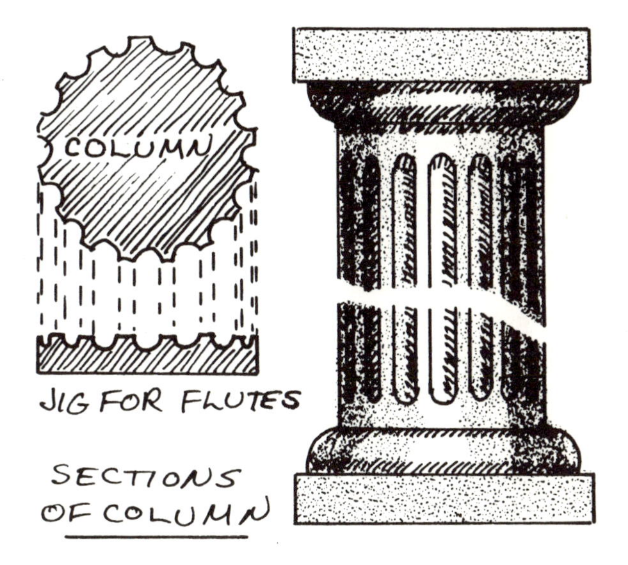

SECTIONS OF COLUMN

draw a line across the circle through the center of the radius. Then drop lines down at right angles to this from both edges of each flute on the half of the column below the line. This will provide the painter with the relationship between the flutes. A jig may be easily made from this by drawing the positions of these lines on a strip of board. The jig will be even more useful if the curves for the tops and bottoms of the flutes are cut into the edge of the strip of board. The center curve is a half circle. The rest of the curves will be gradually squashed in.

Use the jig to locate the positions for the vertical lines on the flute and draw them in place with chalk. The light source direction was determined when the surface was given its roundness treatment, so this direction should be used in drawing the flutes. Each flute has two lines to represent its edges. The one toward the light side will be striped with dark value or shadow paint. A little more than half the circle above will also be painted since it will have a shadow cast under it. Now the other line and most of the bottom will be painted with the high light and the flute is completed. Repeat the process for the other flutes, gradually diminishing the high light until none is used where the shadows will be the heaviest on the sides.

Although this isn't an exact duplication of the appearance of the column, it will project the shape. Under real conditions there will be a place where the high light and shadow will change from one side of the flute to the other, but that is for easel artists to paint.

CAPITALS—Capitals that top columns will have shapes that conform to the style of architecture used for the column. Some capitals are relatively simple while others are complex. The simple ones frequently consist of a combination of molding shapes and can be painted in much the same way as molding. If the capital moldings extend around the column it might be a good idea to extend the high lights only across the center half of the width of the flat to project the feeling of the light being less intense on the sides of the column.

If a more complex form, like the Corinthian capital, is to be painted it will be wise to cut a pattern out of paper first. Draw all of the detail lines on this. Then cut off a section and use the pattern to draw the line of the cut on the flat. Cut off another and trace the new con-

DRAW ENTIRE CAPITAL PATTERN – CUT TO PIECES – TRANSFER ONTO FLAT.

tour. This process may be continued until all of the pattern has been transferred to the flat. The lines may be drawn with a pencil, with chalk or with a brush and paint. If several columns are being made each set of lines should be put on all of the columns before another section of the pattern is cut.

After all of the lines have been striped in place determine the high light and shadow positions and paint them in. The process is really easier than it seems and the results are amazing.

STONE—Stonework has tremendous variation. There are coarse textured stones in rough and smooth finish, in regular cut stone shapes and field stone shapes. And there are smooth textured stones in broken and cut shapes with broken and smoothly eroded edges. It is necessary to determine the general color desired and the general feeling to be projected by the

A FEW VARIETIES OF STONES

stone before one can attempt to draw and paint it.

If a feeling of overpowering weight is desired huge blocks of rough textured granite in grays might be used for a wall. Heavy sponging textures with blues and greens and perhaps

some purple for warmth might be used as under textures. An over all texture of deep value gray sponged over the top might give the desired coarse texture. The outlines could be painted free hand with general rectangular form of slightly irregular lines. The high lights and shadows could be sponged carefully—sponged to keep the same rough feeling used on the rest of the surface.

If a setting using a paint texture like the stones just mentioned is fairly tall and enclosed or masked with black drapes, the upper portion of the flats can be heavily spattered with black, the spatter blending out as it comes downward and fading out by the time head height is reached. This treatment will give a feeling of great height, since the top cannot be seen, and at the same time a feeling of great oppression, since the weight of the black seems to pull downward. The large stones will give a feeling of vastness and strength.

Smooth, sharply defined stone might be used for the wall around the home of a well-to-do Victorian. The stones would probably require subtle texturing. The color might change slightly from stone to stone by either varying the texture colors used or varying the order in which they are applied. The stone will usually take on a little more of the hue of the last texture color applied. An over all spatter or a stone by stone sponging could be used. If the spatter texture is used, the change will be general with the areas of spatter varying, while with the sponging individual stones can be sponged. Of course a rather tedious process of masking all but one stone and spattering stone by stone could be used to change the spatter on individual stones. The outlines of these stones would be sharply drawn and the high light and shadow areas carefully defined. The edges of the stones might be tapered or flat; might have

A FEW TREATMENTS FOR SMOOTH CUT STONE EFFECTS.

a raised center panel or a recessed panel. These are but a few of the variations possible.

A fireplace in a cabin may be made of a variety of broken slab type field stones. The individual stones might have relatively smooth surfaces since they can be broken easily. They

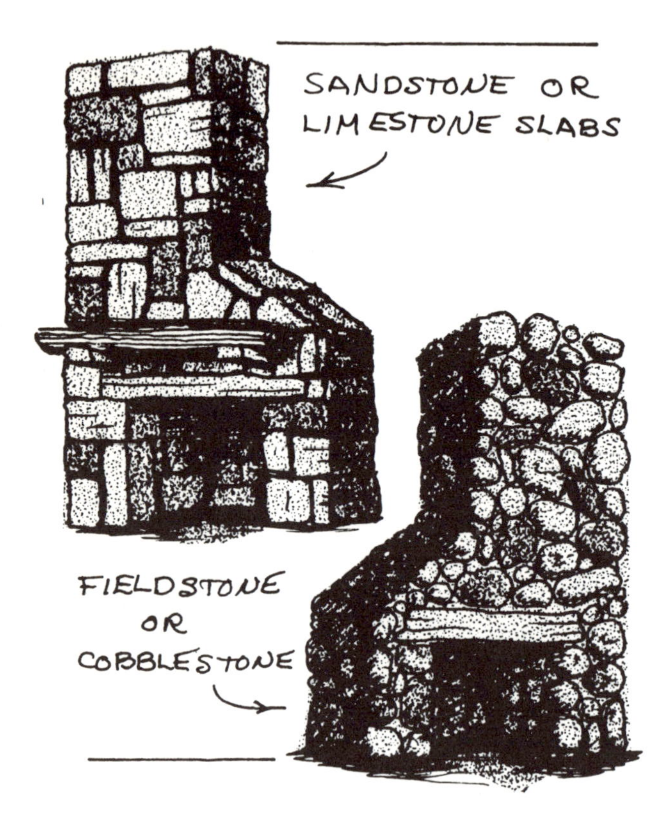


do not fit together too well, so irregular mortar joints fill the spaces between them. The smooth surfaces might be treated with subtle dry brushing, spatter or sponging. The textures will be near in value and color and evenly distributed to give a feeling of a relatively smooth surface.

The stones in the cabin may be rounded forms. They will be fairly smooth, but not all the same shape or texture. All may have a sponging texture, some smooth, some rough. There will be color variations. High lights and shadows should be carefully sponged in place to project the correct shape.

Stones are frequently used to top walls as sort of a cap. The stones may have rough texture, but are usually smoothed so the texture will be quite regular and the high light and shadow will be smoothly applied.

Some general facts about stones are worth considering. The textures of stones vary so the type of stone used should fit the general texture feeling of the play — rough, smooth, sharp, glazed. For rough texture stones use heavy sponging, heavy uneven spatter, stippling or the burlap stomp. For refined stone surfaces use light, evenly distributed spatter, sponge or stipple texturing. The outlining of the stone should project the same basic feeling as the texture of the stone unless there is a definite reason for anachronism. In general, straight even lines are used for refinement and uneven lines for coarseness. Uncut stones neither fit perfectly nor have even mortar joints. Usually mortar is used to fill the spaces between stones.

In walls there are times when it is advisable not to show all of the stones. Partially plastered areas with a few well placed stones show-

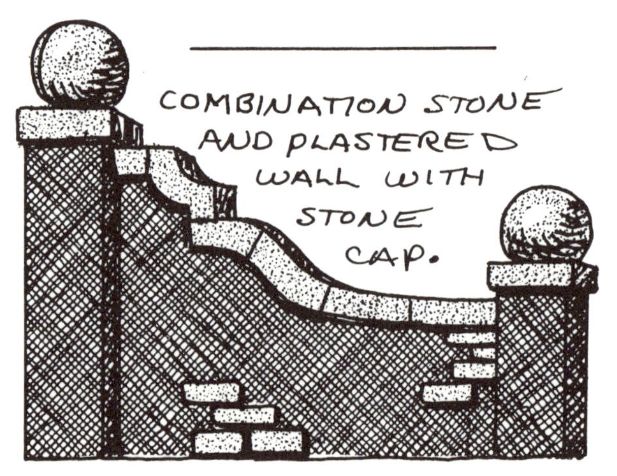


ing here and there make a pleasant surface. This treatment, of course, must fit the feeling of the play if it is used. The stones provide color to break up the wall section without there being so many stone forms that the recurring pattern projects a feeling of busyness.

Small set pieces, or cut-outs made to resemble stones and placed on the stage, depend a good deal on values to reveal their various surfaces. Flat painting of areas, somewhat softened by over all spattering, works out fairly well in projecting the shape from a flat surface. The

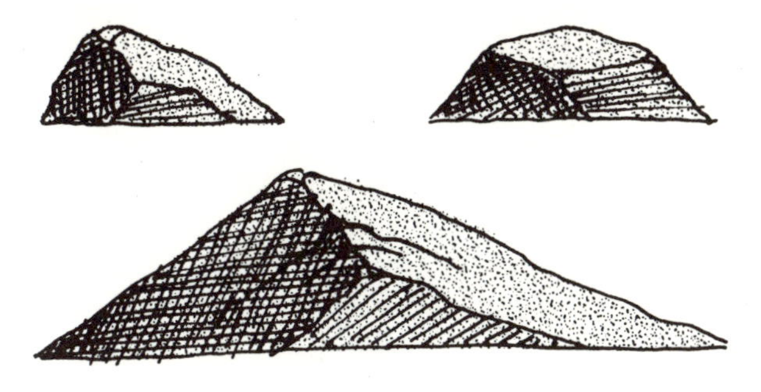

ROCK SET PIECES-VALUE CREATES FORM.

spattering can be done in all of the texture values used for the various facets. In applying them it is necessary to use some care and judgment or the surface will lose its three dimensional values.

Three dimensional built up stones (made of wood, chicken wire and paper or cloth mache) may need various values painted on the surfaces to help project their shape to the audience. An over all spatter on something like this is also helpful.

All of the discussion of stones so far has been based on how to achieve a somewhat naturalistic reproduction of stones. Stones need not be so real—as long as the feeling of the play does not preclude naturalism and as long as the rest of the design is consistent with the stones.

Stylistic stones could be simplified forms. Curved forms such as circles, ovals or distortions of these two are often used. Or the stones could be rectilinear forms—squares, triangles, hexagons. A monochromatic scale could be used; for instance, a tan surface with forms outlined in deeper brown. The areas could be patches of many colors with a solid color to fill each stone form. Color areas could be painted on with dry brush streaks having no definite form but placement to get the color where desired. And then stone outline forms

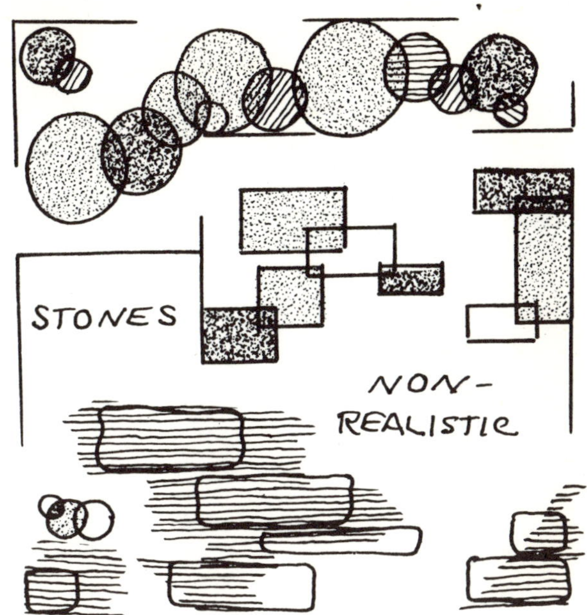


could be painted over them with no attempt at focusing the color patch and the stone outline.

But regardless of whether the stones are to be realistic or impressionistic, the best way to discover their forms is to go out and find real ones where they are scattered about the ground, where they are piled together by nature in hills and mountains or where they are piled together by man in walls, fireplaces and buildings. Observe their size, shape, texture and color and then try to either duplicate the effect exactly or draw an impression of the form and color.

BRICK—Brick surfaces are basically in the red and analagous color range—yellow, orange, red, red violet, purple. Solid colors or variations are possible on the surfaces of individual bricks. Although we usually think of one brick as being just like any other brick, there are many varieties. In texture bricks vary from old, soft, round edged sand brick to sharp, smooth edged, glazed brick to rough textured tapestry brick. Soft spatter, heavy spatter, dry brushing or sponging may be used to get the surface feeling for various types of brick. The mortar joints may be white or colored; may be flush with the surface, weeping or indented.

Bricks are about 2″ thick, 4″ wide and 8″

Simple painted brick around the openings in a setting for *Romeo and Juliet.*

long. Actually there are larger and smaller ones, but the standard conforms to this measurement. On the stage a little license may be taken and the size varied. Small brick could be used to give a busy effect. Large brick could be used to make an area seem smaller.

An entire wall with all of the brick carefully

BRICK AND MORTAR—

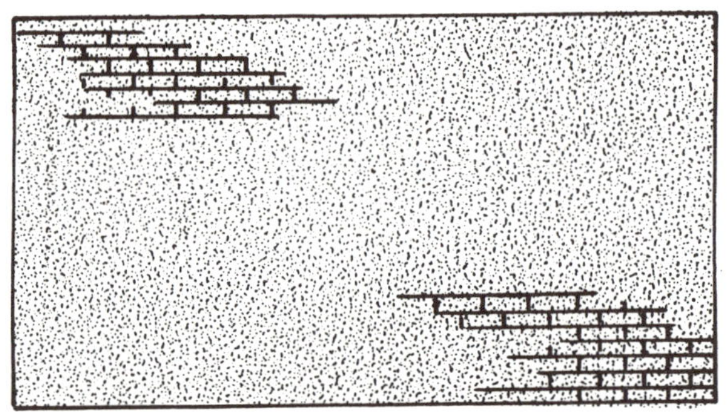

AREA LESS "BUSY" WITH ONLY A FEW BRICKS OUTLINED

painted frequently becomes too busy with all of its lines and regularity. Simplification of the brick wall will help allay this feeling and can be achieved by using a few areas of brick here and there and impressions of the texture on the remainder of the surface.

In small brick areas it is sometimes easy to paint individual brick on a mortar colored background by stroking individual bricks in place with a 2″ brush and brick colored paint.

This will work quite well for rough, irregular bricks. Various mixes can be used and the individual brick colored as painted.

Painting the surface brick color, texturing and then striping in the mortar joints is the usual process. Remember that brick courses overlap one another by half a brick when laid in the ordinary brick wall.

A little more form is given to the brick if it is outlined and if it has narrow lines of high light on the top and one end and narrow lines of shadow on the bottom and the other end to accent the edges. If the mortar joints are raked or indented the brick will cast a narrow shadow onto the mortar on the side and end away from the light.

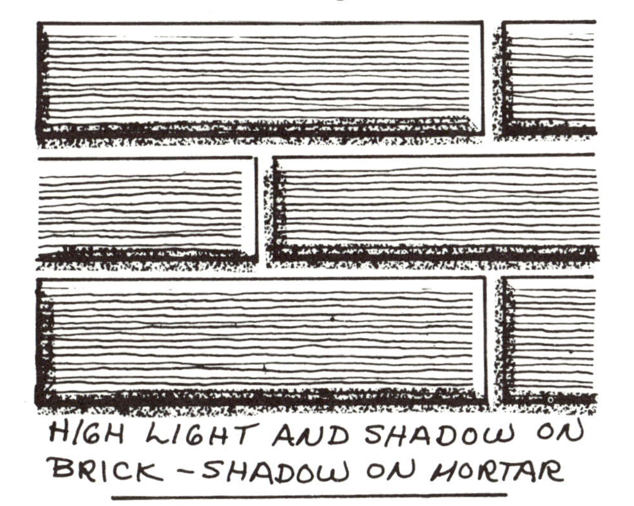

HIGH LIGHT AND SHADOW ON BRICK – SHADOW ON MORTAR


To get more information about brick look at brick walls, fireplaces, chimneys and buildings.

MARBLE—Although rough cut marble may have a relatively sandy textured surface it is usually smoothed and frequently polished to a glistening, shiny surface. There is a wide color range in the background and in the streaks. There are white, cream, pink, deep red, black, brown, green, gray and sienna marbles. The general color, texture and streak pattern should be decided upon before the reproduction begins. The streaks in some are thin-lined and sharp; in others the patterns formed make the surface look as if it were struck with a heavy hammer (shatter crack-like streaks run in all directions, intersecting one another and breaking the surface up into many fairly small patterns). And of course there are marble surfaces that seem to have hardly any pattern at all.

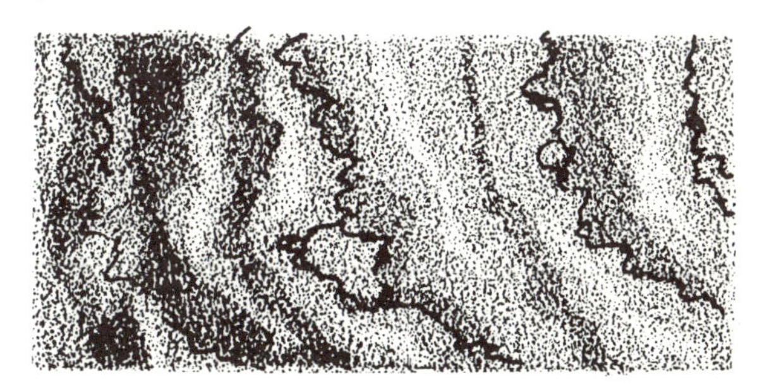

Here is a general process for reproducing marble. Employ the wet brush process. Apply the ground color and while it is wet streak in jagged, but somewhat parallel, lines of texture colors. The streaks can vary between high contrast and subtle variations in color and pattern. The surface will probably be broken into blocks or panels of marble and the streak patterns may vary from block to block. Some may run across, some on an angle, some straight up and down, but always the direction should be kept pretty much the same within a block. The streaks can begin thin, then widen out and then thin down again. Apply enough streaks—with enough variation — to make the surface seem complete, but be careful not to get it too busy and broken.

If the streaks have too much contrast after they have been applied, prepare a wash from the ground color paint. This is done by thinning the paint considerably. The wash when applied and dry should act as a glaze or screen placed over the surface, softening but not obliterating the streaks.

If marble columns are being painted, the wash process for creating the curve should be used. A light shadow wash should be employed. If necessary a second coat may be applied over the first to get the desired depth of shadow.

If the marble has a glossy finish, well placed dry brushed streaks of high light and a few high light windows may reveal the shiny surface. Here again a wash should be used rather than a solid coating of high light so that the base colors will show through.

Marble must be seen to be duplicated. There are so many colors and so many patterns that it defies description.

FOLIAGE—Bushes, trees and plants of every variety are called for in scenic design. Almost any texture device may be employed to produce desired surface effects. Except in the case of large leafed plants where individual leaves can be drawn easily and painted almost flat, after which veins can be painted in, it is usually advisable to attempt to project general impressions of foliage rather than to realistically duplicate individual leaves. Painting individual leaves is an endless job if the painted area is large and the leaf forms lose their identity in the extra busy surface.

FOLIAGE: SOLID COLOR PATTERNS — The foliage area can be broken up into basically rectilinear forms or curved forms. It can be broken so that there is a series of definite areas

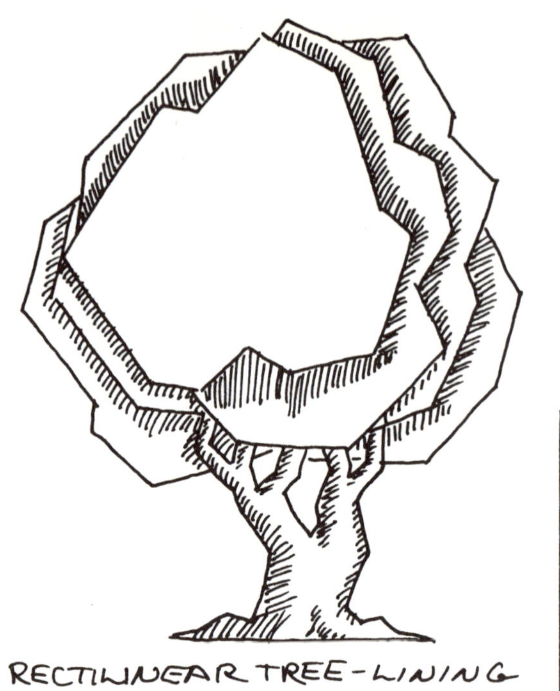

RECTILINEAR TREE-LINING AND SHADING TO GIVE FORM.

STYLIZED CURVILINEAR TREE PAINTED FLAT - LINE AND PLANE

and then the areas painted with various green mixes. The outlines can be left as they are or sharpened with black lines or with lines drawn using the mixes of the areas, but using them in positions where they have the most contrast. A yellow green line might be used to separate a blue green from a gray green; a blue green might be used to separate a yellow green area from a gray green.

FOLIAGE: SOLID COLOR PLUS TEXTURE—The interest value of the forms mentioned above might be increased by sponge texturing for variation within each area — or by using some other texture device or pattern.

FOLIAGE: COMPLEX PATTERN—Paint the entire foliage area using the scumble technique brushing various greens into the surface. If foliage pattern areas can be determined as the paint is being applied it will be possible to put in general high light areas by brushing in a

FOLIAGE PATTERNS

dominance of light texture colors. Shadow areas may be indicated by deeper texture values.

After the scumbled surface has dried, texture patterns to indicate leaf forms may be applied. A sponge, slapping paint on with the side of a brush, a burlap stomp, flogging with a feather duster or some other device might be used. Patterns are rapidly applied with a painting instrument that will produce patches of color. While applying these patterns it is possible to indicate areas where leaves are behind by using deep values; where they are in the foreground by dabbing on intense values of green; where they catch the sunlight by applying patterns of yellow green. Branch forms can be shown here and there passing through the foliage, but remember the structure of the tree so that the position and size of the branches are logical.

FLOWERS—The foliage patterns along the ground level that represent grass, bushes and hedges become somewhat monotonous. A few flowers here and there will add color and break the regularity of the surface. Simple flowers can be painted easily and quickly with little brush dab combinations.

TREE TRUNKS—The method of painting in the round has already been discussed. Tree trunks may have rough bark, like walnut, oak or maple, or they may have relatively smooth bark like quaking aspen, birch, apple, cherry. The bark may be light in color like aspen or birch; deep red brown like many fruit trees; or deep blackish brown like walnut, maple and oak. Determine the general feeling and then proceed. If the general coloration is used in painting to get the roundness of the trunk and

TREE TRUNKS

branches, then painting the surface structure of the bark is the main problem.

Smooth bark trees frequently have dark streaks in them that tend to go around the trees. On aspen they are not too close together, while on fruit trees they are spaced only a short distance apart. Rough-bark trees have patterns made up of a myriad of interlocking ridges. These may be painted by using deep value paint in the brush and drawing wavy diagonal lines in either direction across the surface. The angle of the diagonals should approach the vertical rather than the horizontal so that the shapes are squeezed diamonds rather than true ones. If the surface begins to lose its roundness, lighter values may be used in the center and deeper ones on the edges. Or a dark wash can be brushed over the side areas after the bark has been drawn.

SMALL TRUNKS—Simple small trunks and branches can be painted using one of the rounding processes—brown for the whole surface, dry brush deep value or wash along the edges and a streak of light value up the center.

These processes for painting trees are suggestions. Others might be developed that will work out even better. Here again it is advisable to observe, draw and paint sketches of trees. Become aware of their structure, form and color.

MIRRORS AND GLASS—Real glass cannot be used successfully on stage because of the danger of breakage and because of the ability of glass to reflect images of objects on, around and above the stage that should not be seen by the audience. A simulated glass surface can be painted with a middle gray background and then streaks of other gray values can be applied judiciously to the surface. Experiment to determine whether the streaks should be applied straight up and down or diagonally across the surface. A streak or two of high light in the same direction as the gray dry brushing will help project the feeling of a reflecting surface. A window here and there may also help. A window on a surface like this is a small rectangular form with square or distorted angles that represent the light coming from a real window, striking the surface and being reflected from it into the eyes of the observer. The window is painted with a high light value.

BOOKS ON SHELVES—The face outline of the bookcase (with lines for the shelves) should be painted first if a shelf of books is to be painted on a flat surface—and this is certainly the easiest way to provide the books for a bookcase. Various book cover colors are then mixed —red, blue, green, black, brown, tan. Recall

how books look in a bookcase. Remember the thickness, height and general shape of the bound end of the book. Here and there are shelves not quite filled so the books are slightly tilted. The arrangement of book colors may be carefully composed or mixed at random depending upon the desired impression to be conveyed to the audience concerning the owner.

Paint vertical, or nearly vertical, strips for the individual books. The bottoms of the books rest on the shelves, the tops are uneven; so paint the strips accordingly. Vary the heights and widths with here and there a set of several books the same size and color. The binding has a slight roundness at the edges so a subtle high light on one side and a shadow on the other or shadows on both sides (depending on the direction of the light) should be used. Tiny

A Traveller curtain for *Antony and Cleopatra* consisting of two large panels of camouflage netting, painted with a spray gun.

scribble lines drawn with a brush or a big soft lead pencil to represent the titles may be applied to a few of the books. The space above the books can be coated with shadow wash to give the bookcase depth.

WALLPAPER AND STENCILS — Patterned wall surfaces used to be more common than they are today. The patterns were—and still are—achieved with wallpaper or paint stencils. On a stage setting, unless real wallpaper is used, wallpaper and stencil patterns are applied with a stencil, by free hand brushing or by using cut patterns. Real wallpaper that is pasted on is almost impossible to remove from the flats.

CUT PATTERN—Interesting effects may be achieved by cutting out many pattern forms—the same or different—and laying them on the flat to form the wall pattern design. Each cutout may need a pin pushed through it to hold it in place. Then the surface is spattered. The cutouts prevent the spatter from striking the surface under them, so when they are removed the pattern appears. A spray gun can be used over them to produce somewhat the same effect.

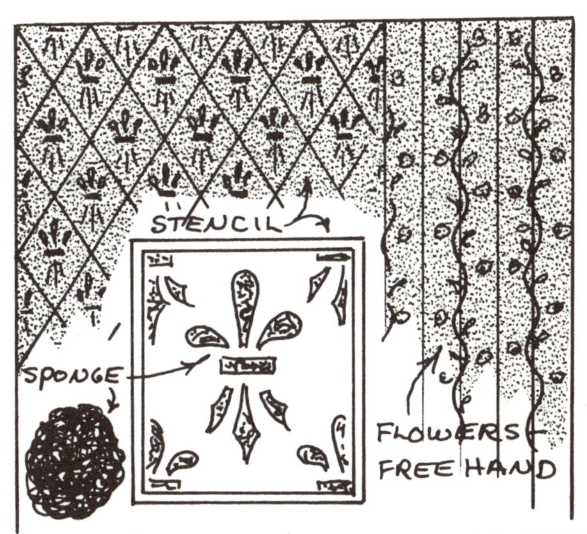


If this process is used the cutouts should be placed on the flats after the undercoat has been painted and before the texturing has been applied.

STENCILLING—It may be possible to buy some stencils from paint companies, but most of these will be small and inconvenient for most stage use. Usually it will be necessary to make them yourself. They may be cut from commercial stencil paper or home made stencil paper.

The pattern should be planned so that there are plenty of reinforcing strips between the cut-out portions of the pattern. Otherwise the stencil will become easily torn and corners will tend to curl up. In designing the stencil pattern it is also a good idea to have the beginning of a repeat of the pattern all of the way around so that matching is made easier. For instance, at the top of the sheet are two tiny cutouts that are placed in a position similar to two at the bottom of the stencil pattern; on the left edge are two that can be matched by two on the right of the stencil, and so on all of the way around. In this way the patterns can be lined up with one another as they are applied.

In some instances it may be necessary to square the surface off with chalk lines in order to space the stencil correctly each time.

A natural sponge works very well for applying paint through the stencil. It may be handled carefully so the stencil sections are not readily torn. When the sponge is used the pattern is less definite since the sponge surface's irregularities produce a combination open and solid pattern. Usually this is good on a stencil since it produces a more subtle pattern. Too often stencil patterns on wall are too sharp and attract undue attention with their harshness.

When planned the stencil pattern should be neither too simple nor too complex. It should give the impression of filling the space without clutter.

When the stencil is cut it is good practice to cut more than one or at least to transfer the pattern to another piece of paper. This may be done by simply stencilling it onto another sheet of stencil paper.

Surface areas may be panelled and a stencil used in the panel area. Picture molding may be spaced down from the ceiling and chair rails up from the bottom to restrict the area to be stencilled. Or the entire wall area may be stencilled. Stencil patterns may be used by themselves or in conjunction with vertical stripes. They may be placed in patterns set up by painting diagonal stripes in both directions forming diamond shapes. Or they may be used in any other conceivable manner either by themselves or in combination with other forms.

Straw — Straw in bales, straw stacks or straw on roofs may be painted by using the dry brush technique. Use various values of yellows and browns, and underneath use a little

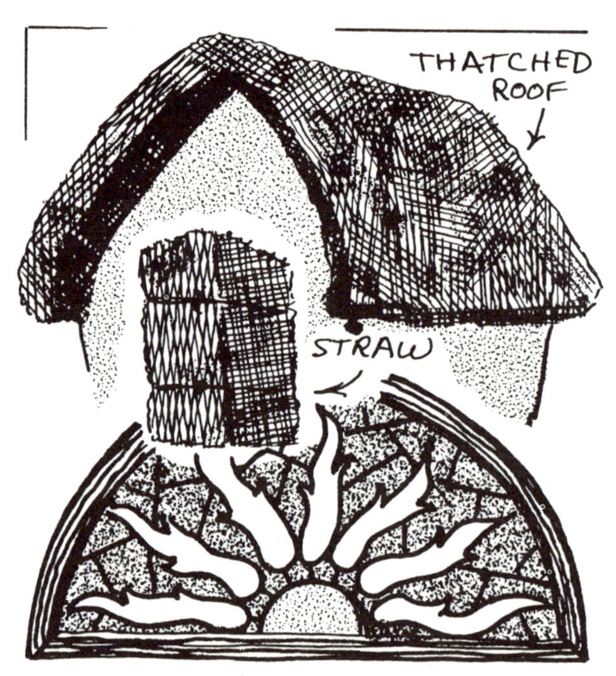


red and green to add interest and variety to the finished surface.

Stained Glass—Thin muslin may be used as the base for stained glass effects. Paint the color areas with thin washes, with dyes, with food coloring or with colored inks. The leading between the panes should then be painted on heavily with dark gray or black opaque paint. Regular scene paint will do. After the painting has been completed, coat the surface with shellac or varnish to soften the colors of the paint and to make the surface more translucent.

Celo glass, plastic glass, sheet plastic and rolled plastic can be painted with special translucent plastic paints. The materials are available in hardware and housegoods stores. The paint may be obtained from hobby shops. This is excellent for glass effects but might be somewhat expensive. Black adhesive tape can be applied for the leading.

Colored cellophane and sheet gelatine can be fastened to a transparent or partially transparent surface and used to simulate stained glass. The colors of these materials are excellent and more intense than painted muslin. Gelatine is water soluble and by moistening pieces slightly they may be softened enough to adhere to one another. An entire panel may be made of pieces stuck together in this way. Cellophane pieces can be glued to get a similar effect. Panels might be sandwiched between pieces of screen wire or hardware cloth. Black tape can be applied to represent the lead between the pieces of cellophane or gelatine in the simulated stained glass window.

Translucent glass effects can be obtained by dipping a piece of muslin in a wash and then stretching it onto the back of the window frame and fastening it before it dries. If a smooth effect is desired the dipping should be done carefully so that the wash soaks into the muslin evenly. If an uneven effect is desired the muslin can be thrown in casually and wrung out in the same manner. As a result the coloring will be uneven and the dried surface will be mottled. Any color may be chosen depending upon the desired results.

### Backdrops

In painting a backdrop it is possible to use all of the painting textures and all of the painting techniques one has learned and often it is even necessary to develop a few new processes.

The backdrop is mounted on a paint frame if one is available or on the floor if there is no frame. All of the edges should be securely fastened so that the surface will not shrink out of shape as it dries after being painted. (Information about construction of the backdrop may be found in the section on flat scenery.)

The process of painting, the desired end results and whether or not the drop has been previously painted will determine if the surface requires a ground coat of paint.

Interesting results can be obtained on new muslin by a simple process of dry brushing and outlining. This presents a finished product with much the character of pastel drawings. Areas of unpainted muslin remain so the entire surface has a light value and a pleasant translucent quality. A small amount of backlighting on a drop of this nature produces interesting results.

If new muslin is used and a ground coat is desired in order to obtain large solid color areas, the over-all painting process can successfully be done with a power spray gun. Spraying is more successful than brush painting on a new surface for several reasons: a thinner coat of paint is applied, less paint soaks through the muslin (consequently the muslin does not stick if it is painted on the floor), and the drop can be painted more rapidly. The spray process does raise a slight nap on the surface and unless a regular shop is used there is the problem of the spray that floats through the air and settles on everything in the room. Of course, if no spray gun is available the method of painting is easily decided without going into the other details.

If a previously painted drop is to be used, a ground coat is almost obligatory. It may be applied with a spray gun or a brush.

After the drop has been prepared for the painting of the scene it is time to refer to the design plate. This painting should be marked off in one foot squares in order to enable one to transfer and enlarge the design more easily. As a matter of fact, designs can be painted on squared paper so that the squares are already visible. The resulting picture isn't as nice as a piece of easel art, but it is far more practical

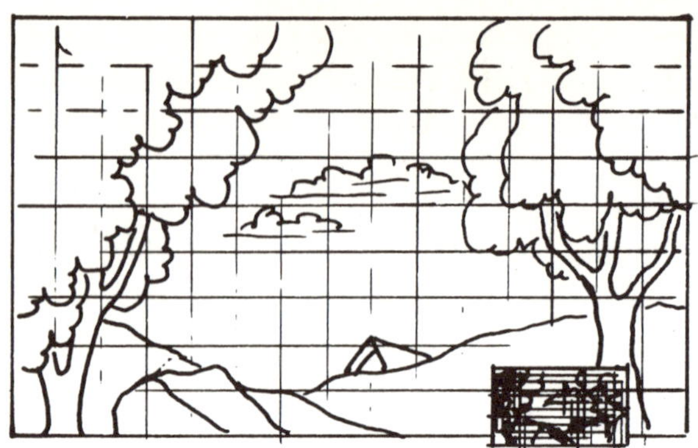

ENLARGE AND TRANSFER SQUARE BY SQUARE. FROM DESIGN

from the standpoint of the painter of the backdrop.

The design plate may be covered with a piece of refrigerator plastic or Saran wrap to keep dirty fingers and spattered paint from ruining it.

To transfer the design to the backdrop the drop is prepared by marking it off in squares to correspond to the grid plan on the design. This may be done with a chalk line.

With lines drawn vertically and horizontally across the drop corresponding to the ones on the design it is possible to transfer the design square by square, enlarging it to scale as it is drawn.

Chalk, charcoal, a heavy soft lead pencil or a brush may be used to do the drawing on the muslin. If a paint frame is being used it is relatively easy to do the drawing. If, however, the drop is flat on the floor the artist must stoop, squat or crawl. A stick attached to the drawing medium makes it possible to stand up. This is not only more comfortable, but makes it possible to see the results more easily. A long handled sketching brush is often used to draw patterns on backdrops for this reason (a handle may be easily wired onto a regular sash brush). If a brush is used, an extremely thin wash should be employed so that the mistakes can be covered as the paint is laid onto the various areas of the drop.

After the design has been enlarged and transferred to the drop, the areas are painted. As nearly as possible try to duplicate the effects indicated on the color plate. Everything including texture patterns should be enlarged. Aerial perspective may be used to push objects back into the distance. Aerial perspective is achieved by using a mix that is ever so slightly away from the background color and mixed from it. Clouds may be painted with a brush or a sponge or sprayed on. If the edges are too harsh, spatter them back out using the ground color of the sky for the spatter. Objects in the foreground should be sharp-edged and sharply defined color wise. Similar objects in the foreground are larger than those in the background and the distance between them determines the amount of difference in their proportions.

Whether painting with the drop on a paint frame or on the floor, be extra careful about spilling and dripping paint.

Occasionally step back and look at the work with half closed eyes to gain distance and perspective. If the drop is on the floor, climb as high as possible on a step ladder and look down on it.

Use high light, middle value and shadow to get dimension into the work.

SCRIM DROPS—Scrim drops are painted in very much the same manner as muslin backdrops except that the surface is much more difficult to work on. There is a tendency for the eyes to focus on and then through the material. If scrim is painted on a frame its resilience makes it difficult to paint. If it is painted on the floor a good deal of the floor surface will be painted, too, unless it has been covered first. Then, of course, there will be the problem of pulling the scrim loose from the material under it.

AGING SURFACES—Occasionally it is desirable to age a nice new shiny paint job. This is satisfactorily and easily done by spattering the surface with a thin black wash. The wash must be thin (very little black in the mix). If necessary two or three coats may be applied. Unless running and streaking are de-

Stencilled pattern applied to the walls of a setting for *The Damask Cheek.*

liberately desired this spattering should be done with the flats or drops laid flat on the floor. The spattering may be done with a brush, but an even better result can be obtained by using a tank type garden spray. This spray gives a fine drop spatter and it is possible to cover surfaces rapidly and easily with a minimum of physical effort.

PASTEL DRAWING — There may be times when it is easier to execute small details with pastels or colored chalks rather than with scene paint. Pastels offer an opportunity to work more leisurely when blending colors and it is easier to have a large pallette of colors available. Pastels can be used when it is advisable to paint pictures directly on a wall. The frame can be painted around them with scene paint afterwards. If they are used, the flats must be handled carefully to prevent smudging or the pastels must be sprayed with fixative or a plastic coating upon completion.

### General Comments About Painting

A few helpful facts about painting are worth mentioning before concluding this discussion.

If lighting units are to be mounted behind a section of scenery it is a good idea to "back paint" the rear of the flat so that light will not show through the muslin.

Once in a while a flat is bumped and the

muslin loosened. It may be tightened again by brushing or spattering water on the rear surface of the flat.

Texture mixes may be thinner than the paint for the ground coat.

A surface may be given a slight gloss by over painting with glue or shellac.

If a flat is damaged and a hole or tear results, place a patch on the back. Fasten it with paint and if possible use the ground color or a similar color of paint.

Remove chalk and charcoal guide lines by dusting with a soft brush, duster or moist sponge.

* * * *

The foregoing discussion of scene painting is by no means complete nor definitive. Each production, each setting, sometimes even each wall section, is a completely new problem and should be treated in that way. When painting scenery there should be more than "only one way to do it." There should be an infinite number of ways of getting desired results. For the individual painter there may be only one way at this particular time on this flat. But even for him there should be another way to do it next time. Painting is one place where creativity and freedom of expression have wonderful latitude. Even if the painter must tie himself down to someone else's designs he has great freedom of execution. If he executes his own designs his chances for expression are even greater.

# 8

# Rigging and Shifting

Erecting, rigging and shifting the scenery is usually planned, at least in part, when the setting is being designed and built; but the actual process cannot be completed successfully until the setting is on the stage.

ONE-SET PRODUCTION—A production having only one setting can be erected in a manner that differs from that used when two or more settings are required. The one set production can be erected and allowed to remain in place for the duration of the production. Even if it is necessary to remove all or part of it for another program to be held in the hall it is possible to make the change in a leisurely fashion. Consequently it is possible to use fasteners that are of a semi-permanent nature. Corners can be fastened with double head nails; stiffeners can be permanently nailed or screwed onto the wall sections; door units can be securely fastened into the flats with screws or nails (either double head or regular); and the entire setting can be anchored to the floor if necessary.

MULTI-SET PRODUCTIONS—If there are two or more settings in the particular production it will be necessary to rig all or part of the first setting so that it can be easily and rapidly moved off stage ("struck") so that other scenes can be moved on. If one setting is used more than once during the performance it might be a good idea to plan, where possible, to make portions of that setting permanent and shift the others around it. If all of the settings are about the same size, if some have odd contours, if some require a clear stage for outdoor effects or if others are larger than the recurring scene it will be necessary to make every scene completely "strikable."

If settings are to be erected and struck during the course of the production the rigging should be carefully planned so that all of the sections can be easily and rapidly assembled and disassembled. Shifting should always be accomplished in minimum time. Many theatre plants have special permanent stage devices to facilitate making rapid changes. Wagons, revolving stages, flying systems, jackknife stages and overhead track systems are all plant facilities designed for this purpose.

## Assembling the Setting

FLOOR PLAN—The first step in erecting a setting consists of chalking out a simple outline of the floor plan on the stage. This will save time later since there will be less juggling to locate the positions of the various units as they are erected.

STIFFENERS—Before setting up the scenery all of the wall sections consisting of two or more flats hinged together (twofold, threefold, fourfold books) should be provided with stiffeners. If the production has only one setting, or if the wall sections can be run on and off stage unfolded, the stiffeners may be fastened on with double head nails, hinges, bolts and wing nuts, or screws. The stiffeners may be made of almost any width of 1″ boards, but should probably not be less than 3″ wide. Sometimes even 2″ x 4″ lumber or small built up trusses may be needed. Usually minimum rather than maximum requirements will be met. Although it is desirable to have walls stiffened so they will not shake during the performance when a door is slammed or when someone accidentally bumps against a wall, still at the same time, economy of time and effort will usually specify that bracing be held to a minimum.

A board has greater stiffening quality when it is fastened onto the back of the scenery on edge than when it is flat.

If the stiffener is to be fastened on with double head nails the flat is "floated" (allowed to fall flatwise with the air cushioning the fall) to the floor face up, the stiffener is then slipped underneath and nailed in place. If the wall has little or no strain on it when in place the stiffener may be nailed near the upper edge against the top rails of the flats. This is the easiest place to fasten it since it is possible to see and feel where to nail. One stiffener may be sufficient but two, one at the top and the other at the bottom, will make the section more substantial. If there is a door or window in the wall section it may be advisable to place the stiffener immediately above the thickness and nail it down into the thickness as well as into the flat.

Stiffeners may be fastened with screws in a

similar manner. Usually it is advisable to drill holes through the flat and then sink 1½″ or 2″ #9 flathead bright wood screws through the flat and into the stiffener.

If bolts and wing nuts are to be used it will probably be easier to work on the flats when they are in an erect position. Ordinarily the stiffener is placed in a flat position when bolts are used. If the board is placed on edge it may be necessary to place blocks where the bolts come through in order to thicken the stiffener in these positions so that the bolts will not weaken the board. This can easily be done.

HINGED STIFFENERS—Hinged stiffeners require slightly more time and effort to install but are sometimes more satisfactory than those that are nailed in place. Loose pin hinges are used when the stiffener is to be removed. The wall section is "floated" face down for this installation. The stiffener is then laid in its desired position and hinges are put on in the proper places. Usually it is advisable to have at least one hinge near each end and one near each hinged joint of the wall. Sometimes even more are put on to make the wall more rigid. Hinges may be placed on either toggles or stiles. In order to keep the stiffner in its correct position against the rear of the flats the hinges are placed on alternate sides of the stiffner. If loose pin hinges are used it may be advisable to fasten the pins with string to either the stiffener or flats so that they will always be available when needed. Pins for this purpose should be slightly smaller than the pin holes of the hinges so that they can be easily put in and taken out.

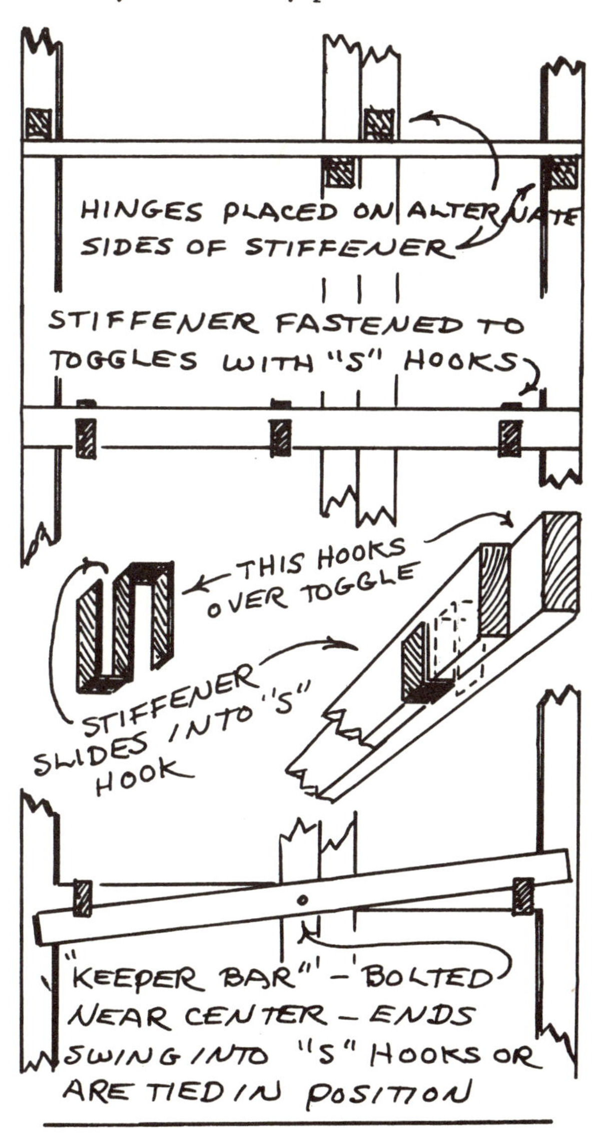


The wire of coat hangers is usually just about the right size for these pins.

S Hooks—The S hook is another device that can be used with stiffeners. An S hook is made from strap iron and bent into a modified angular S shape so that it can be slipped onto a toggle bar and then when it is in place a stiffener can be slipped into the bent up hook. Screw holes may be drilled into the hooks so that they can be fastened either onto the flats or stiffeners. The hooks may also be designed so that the stiffener can be placed in a flat or edge position. If the stiffener is to be placed in the edge position it will probably be advisable to fasten the hooks permanently to the stiffener.

Keeper Bars—Swinging keeper bar stiffeners may be rigged so that they are permanently bolted to the flat (near the center of the stiffener usually). When the flat is being shifted, stored or packed for shipping the stiffener is swung into a position parallel with the stiles of the flat. When the unit is set up on the stage the stiffener is swivelled around so that it crosses the flats and its ends are tied, fastened with hooks and eyes, fastened with bolts and wing nuts, screwed into place, nailed with double heads, or secured with keeper hooks or S hooks.

If more strength is needed in stiffening than can be provided with a 1″ board it may be necessary to use a 2″ x 4″ or even to make a simple truss. These are handled in basically the same manner as other stiffeners. However, if a truss is used, it may be necessary to provide some kind of support so that the weight of this frame extending behind the flats will not sag and pull loose from the wall section.

Assembling Walls—After the walls have been stiffened it is time to fasten the sections together. Provisions for fastening may have been made before the painting was done—this is especially so if all or part of the fasteners are on the face of the flats. If it is a one set show double head nails will work and can be put in and removed rapidly.

Double Head Nails—As few as three 8d

double head nails can be used to fasten two flats or wall sections together at right angles if they do not exceed 10′-0″ in height. If there is some strain on the joint caused by swinging doors it may be necessary to add a couple more nails. After the nail is driven in as far as its shoulder there is still enough head to grasp with the jaws of a hammer or crow bar for easy and rapid removal. If the corner made by two flats is on an angle greater than 90° it is easier to nail it at 90° and then twist the flats into position afterwards.

LOOSE PIN HINGES—If loose pin hinges are used to fasten walls together they should be mounted on the flats with the flats in position so the hinge will allow the flats to be placed on the correct angle. Pins can easily be removed from loose pin hinges but sometimes it is difficult to get the hinges aligned and the pins placed in a hurry.

LASHING—Lash lines are used almost exclusively on some stages for fastening wall sections together and they can certainly be used advantageously, especially where the flats form an outside corner at the back of the setting. It is more difficult to get snug joints that will remain tight on inside corners, but even these are perfectly satisfactory when handled correctly and when the flats have stop blocks and cleats to prevent them from slipping.

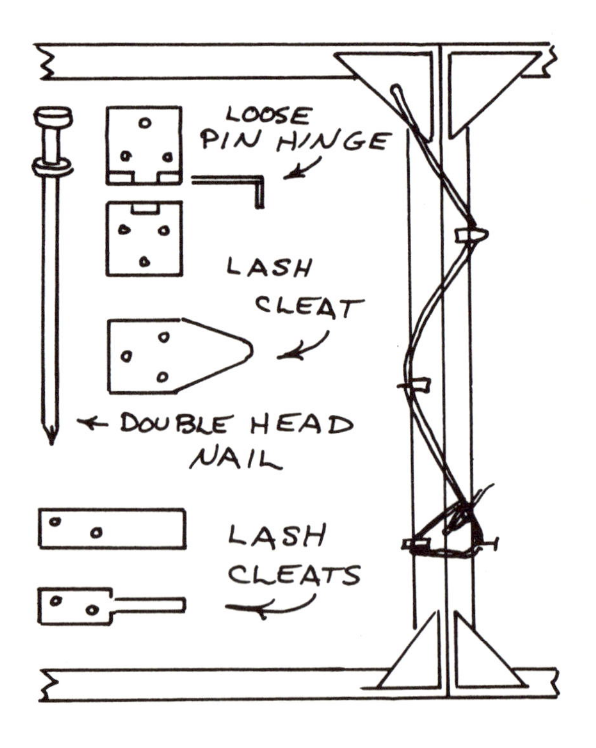


The lash lines should be uniformly fastened to the upper right corner of each unit as viewed from the rear (except in special cases—there are always exceptions to the rule). Lash cleats are then spaced so the line can be laced back and forth from one side of the joint to the other. A minimum number of cleats requires a pair of tie off cleats from 24″ to 30″ above the floor on the inside of each of the stiles of the joint and one cleat on the right side of the joint half way between there and the position where the line is tied to the corner of the flat to the left. As additional cleats are added the spacing is changed so that the lacing space is fairly even.

Lashing and tie off cleats may consist of regular commercial cleats, strips of strap iron drilled to receive screws for fastening to the stiles, long screws, double head nails or strips of plywood.

When lashing flats together a little experience will make it possible for the operator to snap the line around the cleats rather easily. If it is desirable to have the rope slip off the cleats easily they should slope slightly upward. If it is desirable to have the rope cling the cleats should slope slightly downward. In general all but the tie off cleats will slope slightly upward; the tie off cleats will slope downward.

LASHLINE KNOT—If the lashline knot is tied correctly it will be possible to untie it by jerking on the end of the line. If it is incorrectly tied much time will be wasted in fumbling with the line. To correctly tie the knot the line passes under one tie off pin, across and under the other one. A loop or bight is made in the line and passed over and behind the line where it goes up toward the lashing pins. The loop is pulled tightly downward without pulling the end through. A loop is now made in the line toward its end and this loop is pulled through the first loop. This loop is pulled until the other loop tightens down and binds against it. Notice that the end of the line is never passed through either of the loops. To untie

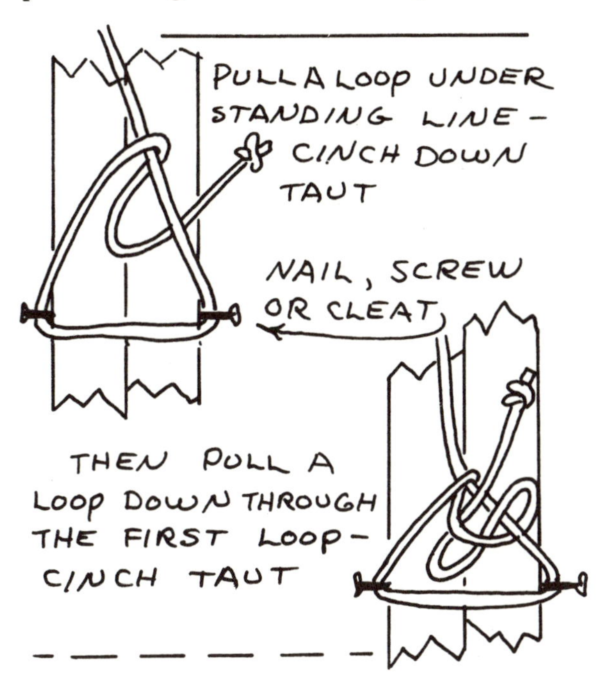


the knot the end of the line is pulled or jerked pulling out both loops and releasing the line.

Other devices might be used for fastening flat sections together. Hooks and eyes, picture hangers, short lengths of rope and screw eyes, ropes and small boat cleats, angle irons and removable bolts, and clamps are all contrivances that might be used.

In some cases where the entire setting is small it may be possible to hinge the setting into a single piece so that it may be moved as one unit. Loose or tight pin hinges can be used.

GLIDERS—If the scenery is being dragged across the floor (or "run") it may be advisable to consider mounting large "gliders" under the flats in places. The wall sections can then be slid across the floor quietly and easily. They also reduce the danger of marring the floor.

Many times it may be possible to leave doors, windows, pictures and all of the dressing on a section of scenery in place while it is being shifted. If this can be done the entire job will take less time and the process of re-setting the scene will be much easier. If there are many heavy units in and on the wall it would be wise to consider using tip jacks, wagons or outriggers on which the scenery can be mounted. These are all described in the stage machinery section.

Large units such as steps, platforms, fireplaces, oversized doors, and windows should be rigidly braced and provided with casters, gliders or other contrivances to make shifting them as effortless as possible.

BACKINGS—After the main setting is placed in position the backings are placed where they belong. If the backing fits snugly against the flat of a wall or against the side thickness of a door or archway it may be possible to securely fasten it to that unit by lashline, hinges, nails, or other means and use it as a bracing device.

BRACING—If walls, and especially doors, are not rigid enough it will be necessary to use stage braces, jacks or similar contrivances to provide more rigidity. The stage brace is an adjustable length device with a hook type fastener at one end designed for fastening into a brace cleat secured to a flat, and an angled foot iron on the other end provided with a hole through which a stage screw can be fastened by hand into the floor. Hardwood floors should be provided with some other type of fastener at the floor end if stage screws are to be used. Iron plates

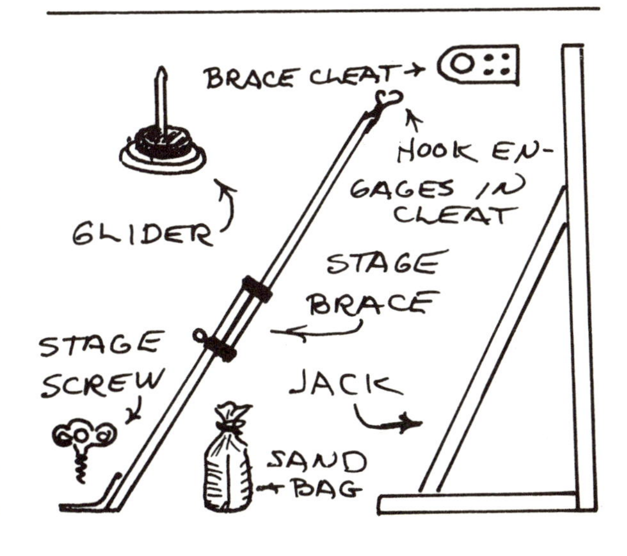

with threaded holes or bolt head holes may be used. If a jack is used in place of a stage brace the jack is fastened to the flat with hinges, nails, screws or bolts and its foot end on the floor is fastened down with a stage screw or weighted with a sand bag counterweight or any other heavy object. In the absence of these devices it is possible to fasten other flats or parallel lids to the rear of the set and also to the floor. If the base of the setting is secured in place it is also possible to lash the setting to the wall of the building with line. Sometimes a combination of a board that is the correct length to make a spacer and reach a nearby wall of the theatre building and a line that can be tied to a pipe or cleat fastened in the wall will provide stiffening support. The board will prevent the flats from moving one direction and the rope will prevent them from moving in the other.

When scenery is used on an outdoor stage lashing to a permanent or semi-permanent framework or to a solid deck is a good solution. The lashing lines in this case might be just short snatch cords rather than regular long lash lines. The outdoor scenery requires extreme precautions in anchoring since even the slightest gust of wind can twist and break a flat.

FLYING—Portions of scenery, back walls, drop sections, property pieces or even entire settings may sometimes be flown into the flies rather than moved on the floor level. When this is done it is necessary to carefully mount some type of hook or eye device to the flats. If the flats are not heavy it may be possible to use large screw eyes. Usually some piece of hardware similar to the commercial ceiling plate or hanger iron is used. These must be rigidly fastened to the flat where the rope or cable is to be tied. The rope may be fastened to the top of the object or passed down through rings or eyes near the top and extended down toward the bottom of the object where the rope is tied to other supports. The ropes should, if possible, be fastened so that the weight is somewhat evenly distributed and so that the weight will also level the scenic unit as it is raised into the flies; otherwise, it may get "fouled" in other units that are stored in the fly loft.

The flying of scenery can be more successfully handled in a theatre having a counterweight system than in one having only a rope and sand bag system. The permanent levelling of the counterweight is much more satisfactory than the constant struggle with ropes. But even with the rope system it is frequently advisable to lift some things into the flies in order to retain more available floor space for a complex production.

SHIFTING SCHEDULE — When the production being staged is complex and has more than one setting — as well as much furniture and many props — it is necessary to carefully work out a shifting schedule. The procedure should be plotted on a floor plan with the movement and storage of live scenery (that which is to be used), dead scenery (that which has already been used and will not be needed again during the current performance), and the setting that is being played. Storage area, path of movement and individual crew member assignments should all be worked out carefully.

CREW—Except for shifting small items it is advisable to work the crew in pairs. If the same two work together on all shifts they will help one another and one will remember what the other forgets.

Efficiency of packing, movement and storage are necessary. In order to achieve this it is advisable to do paper and trial work first and then have technical "run throughs" for just the crew. The entire effect of a play can be ruined by careless or lengthy manipulation of scenery.

SPIKING—In order to enable exact placement of scenery, furniture and properties it is necessary to "spike" or mark the floor with small dabs of paint for each object for each scene. If the work lights are to be turned on for the shift it is possible to use a different color of paint for each setting. If the setting is to be changed in almost complete darkness, white, light tints or even luminous paint should be used for spike marks. If paint cannot be used on the floor it may be advisable to use white adhesive tape for spike marks.

MARKING EDGES OF STEPS AND PLATFORMS—The actors and stagehands frequently have difficulty seeing the edges of stairsteps and platforms in the dark. A narrow stripe of paint wherever there is a change in elevation will help in alleviating this difficulty. White paint should be used if the stage is bright to dim and luminous paint should be used where it is dim to dark. Marking passageways and positions for actors and scenic units in theatre-in-the-round stages in this way is extremely important if the shifting is to be done in the dark and if the actors are to assume their positions on the stage during a "blackout".

NUMBERING SYSTEM—A numbering system will be helpful if there are more than two scenes. An acceptable system is to use the act number as a prefix and the scene number as a suffix. Thus 11 would indicate Act 1, Scene 1; 12 would be Act 1, Scene 2; 24 would be Act 2, Scene 4; and 33 would be Act 3, Scene 3. If a system of this nature is used consistently there will be little or no confusion about markings and identification.

SHIFTING PLAN — When the shifting and storage plan is worked out it is necessary to work out all of the various elements at the same time. There must be space for storage and movement of: 1. Scenery, 2. Furniture, 3. Properties and 4. Actors. In the case of the actors there must be space for them to move to all of the entrances and to wait at the entrances for their cues. If the production has a large cast this is often the most difficult problem of all.

HAND PROPERTIES — Stage properties carried by the actors and known as hand properties should be available as close to entrances as possible. It is good procedure to have a property table on each side of the stage and to regulate the flow of these items. If the properties are picked up just before the actor makes an entrance and checked back in to the table as soon as he is through using them there will be less trouble in keeping track of these items — and sometimes there are literally hundreds of them.

FURNITURE—Furniture may be stacked but it is necessary to be careful so that the various pieces do not get scarred or broken. Careful planning of the stacking may make it possible for one crew member to carry a chair and items that can easily be stacked on the chair.

FLOWN UNITS—If a number of backdrops or flown pieces of scenery are being used in a production it is advisable to check their order of appearance. Placing them in the correct sequence may make it possible to have one set behind the other ready to go as soon as the front one is flown away.

Correctly masking the stage should always be considered in the shifting plan. There are times when more space is planned than is available if this isn't taken into consideration. This is especially true when there are outdoor scenes consisting of nothing but a clear stage backed by a cyclorama. All of the scenery for the other scenes must be cleaned completely out of sight lines. When enclosure type settings are used, on the other hand, it may even be possible to have parts of other settings in place behind the one being played.

MANPOWER—The planning process should also take the available manpower into consideration. If units of scenery are flown it will be necessary to have one or more crew members at the pin rail. Frequently, especially if the pin rail is above the floor, the pin rail operator cannot be counted on to climb to the floor and help with the shift and then shinny back up to his perch. The strong should be assigned to move large objects and the weaker to handle the smaller objects. That is an obvious factor but sometimes when under pressure a stage manager fails to remember such simple things.

### Miscellaneous Facts

There are a few random facts that should be known by the stage crew. They should know the various methods of handling flats, common knots that are useful on stage and some miscellaneous facts about handling various rigging problems.

The lashline knot, already described, is used wherever joints are laced with rope.

CLOVE HITCH—The clove hitch is used where the rope is to be secured around an object for pulling or lifting where there is danger of the rope sliding. This is the knot usually used in tying a pipe or wooden batten to a set of lines in the flying system. The rope extending down from the grid is referred to as the "standing line." Assume that a line is being tied onto a pipe. The standing line passes down behind the pipe, and the end of the rope is led around the pipe once on one side of the standing line and started around a second time on the other side of the standing line. As soon as it passes just beyond the standing line it is slipped through under the first turn around the pipe. The end of the rope and the standing line are then pulled so that the rope tightens on the pipe. The end of the rope is then looped in two half hitches above the knot.

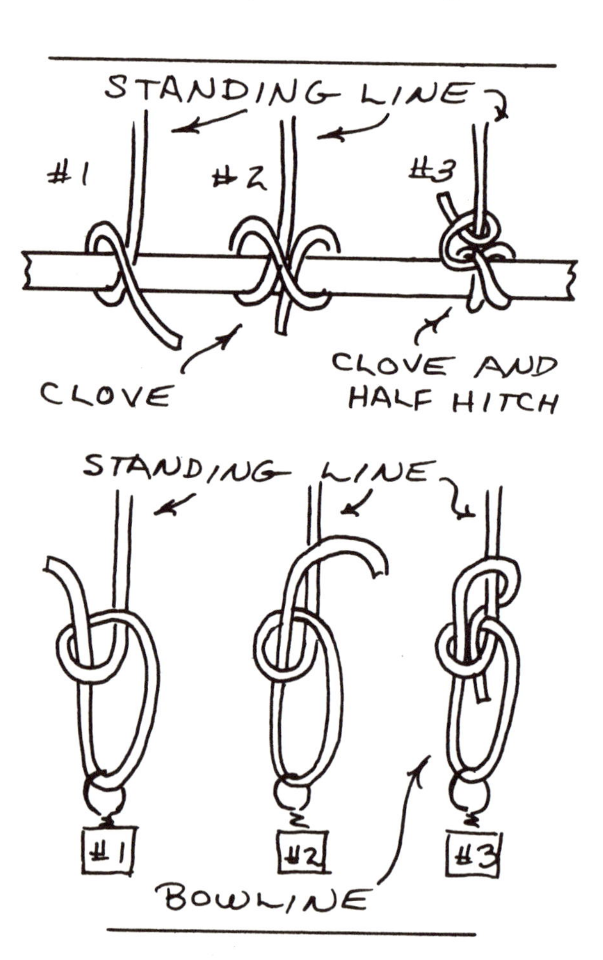


BOWLINE—If the rope is to be tied into a ring a bowline knot can be tied. A bowline has the advantage of always remaining easy to untie. In order to describe this knot assume again that there is a standing line coming down from the grid. Pass the end of the rope through the ring to be tied. Hold the standing line in the right hand just above the ring. With the left hand place the end of the line across the front of the standing line. Hold the crossing position with the left hand and with the right cast a small loop over the end of the line. With the left hand pass the end of the line around the back of the standing line to the front and down through the loop it just passed up through.

PIN RAIL TIE OFF—If the ropes on a rope flying system are correctly tied off it is possible for one person to release a set of lines and lower a heavy object to the floor. To tie this all of the lines of a set (usually 3 or 5) come down on the far side of the rail and are passed under, then pulled up and around the pin. For ease of handling, the ropes should not cross one another in this operation. The friction of the rope rubbing against the rail is utilized to give it holding power. The rope is then pulled down to the lower end of the pin and wrapped around it. Then a loop is twisted into the line before it is placed on the top pin again with the end of the line (as distinguished from the standing line) passing underneath. This will lock the lines in place. This process may be repeated for added security. If there is surplus line it should be coiled and hung on the pin so that it will not be under foot. If all of the ends are thus coiled after being tied off there will be less danger of lines getting fouled up in one another.

To prevent the end of a line from fraying it can be wound with heavy thread, friction tape or braided back upon itself or tied in a simple knot.

BROKEN BATTENS—Accidents are bound to happen on the stage. If a batten is broken—whether it's a stile, rail or flying batten—it can be temporarily patched by fastening a piece of wood along it to make a lapped splice. This splicing board should be long enough to secure the break without danger of the accident happening again right away.

TIGHTENING MUSLIN—Sometimes the fabric will be lightly bumped, loosening it. As long as the material is not broken it will be possible to tighten the area by spattering either the surface or back with plain water. As it dries out the material will usually shrink and tighten up again. Alum added to the water may increase the shrinkage.

PATCHING—If the surface is torn or slit it will be necessary to make a temporary patch. It may be possible to do this with Scotch tape, by stapling a small board to the back, or by adhering a patch to the rear with paint while holding a flat board against the face. When this is done carefully with paint that comes close to matching the paint coloration of the surface, the patch is almost unnoticeable from the audience. Glue should not be used for patches since it tends to pull the material unevenly and throw wrinkles into the surface.

CREW DEPORTMENT — The stage shifting crew should consider every performance and rehearsal as important. They should be briefed by the stage manager, who is the stage performance "boss", to work efficiently and quietly. The success of the performance depends on the dispatch with which their work is done. Absolutely no carelessness or horse play should be condoned back stage during the performance. There is a time for this after the show is over.

*Madame Butterfly.* Research into the architecture of a country and the history of an era often precede the design process.

# 9

# Stage Properties

The entire process of decorating the stage is important, but the final touches of the stage property department usually add the flavor that completes the setting and makes it appear to be lived in—whether it's a room, a garden, a patio or a barn. The term "stage property" is a general one that may refer to almost any conceivable object. As a matter of fact, the ordinary property storage room looks like a huge collection of castaway junk—and it usually is just that. Whenever someone wants to get rid of something it can profitably be added to the prop room stock. The stage technician looks upon every bottle, pillow, pin cushion, spear and empty whiskey bottle as an object that some day may be called for on stage. Some of the most impossible objects have been used in plays ranging all of the way from a mummy case in *The Man Who Came to Dinner,* to a horse in *Green Grow the Lilacs,* to a roll top desk large enough to hide an escaped convict in *Front Page,* to trees that grow on stage in a flash in *Sing Out Sweet Land,* to a pin ball machine that when defeated flashes lights, sends up an American flag and plays *America* in *The Time of Your Life.*

It is physically impossible to collect all of the items that may be used at one time or another in stage plays, but it is possible to gradually accumulate objects, such as dishes, pots, pans, furniture, pictures and vases, that are used most frequently. The number and variety of objects collected will depend a good deal upon the storage facilities available. Properties not available in the stock room may be begged, borrowed, rented or purchased. They may be found in old homes that are being torn down, in furniture stores, at auction sales or may be obtained by letting it be known that they are needed by the theatre. Sometimes the material is junk, but even this can be repaired.

If the properties are borrowed and are to be returned it is advisable to keep a close record of where they were obtained, check them carefully after each rehearsal and performance and return them as promptly as possible after the final performance. A written thank you note should either accompany the object when it is returned or be mailed out immediately thereafter. Actors should always be cautioned about the value of stage properties and urged to handle them carefully.

Since the term "stage properties" in itself refers to a vast conglomeration of objects this discussion will likewise consist of a conglomeration of information. Properties will be discussed in terms of where and how they may be used. Frequently the stage decorator is baffled about the objects that may at times be used on a desk, on a table or in a bar. For his convenience some of the possible objects will be listed. The unusual ones will be described and where objects can be faked or simplified or built this will be suggested and the method briefly described and illustrated. Stage furniture, extremely important in the stage dressing process, will be discussed in a separate chapter.

WALL PROPERTIES — Simple or complex shelf sections may be mounted on the wall. They may be fastened to toggle bars with double head nails, bolts and wing nuts, picture hangers or suspended from the top rail of the flat by wire. The shelves may be used to hold books, knick knacks, statuettes, statues, vases, toy animals, fish, stuffed fowl, planter boxes, dishes, clocks, bottles, medicine, photographs, lamps, lanterns, tools, decorated china or paper plates, pieces of drift wood and maps.

PICTURES — Pictures, including paintings, photographic enlargements, murals, pictures painted directly on the flat surface with moldings, painted or real, framing them, framed pieces of interesting wall paper patterns, beautiful magazine illustrations, cloth patterns or simple line drawing illustrations and cartoons, may be mounted on the walls. These may be mounted directly to toggle bars if they are correctly spaced; extra toggles can be installed or wires may be used to extend from the desired position of the picture up to the nearest toggle or rail of the flat above it. The wires may be concealed by piercing the flat with tiny holes behind the picture, pushing the wires through and extending them up to the nearest cross bar. String may be used in the same way. A picture will retain its balanced position better if two supports are used rather than one. One support will allow it to twist out of position easily, especially if the wall takes the strain of a slammed door or something of that nature. Water paint or pastels work quite well for painting directly onto the flat surface. If pastels are used they should be lightly sprayed with fixative in order to prevent smearing.

PICTURE TRANSFORMED INTO REAL PERSON—Sometimes it is desirable to have a picture "come to life." To do this the picture is painted on scrim and mounted on the wall. Cut the muslin out behind the picture and locate the actor in back of the portrait. When the real person is to appear illuminate him with high intensity light behind the picture and in place of the picture the actor will be seen.

PLAQUES—Plaques made of plaster of paris, wood, plastic, plastic wood, metal or cloth may be mounted on the wall. The method of casting some of these materials if they are to be created in the shop will be described later. The design on the plaques may be in bas relief or flat and may consist of masks, coats of arms, abstract designs, or figures. Plates, paper or china, trivets, antlers, and clocks are other objects that may be mounted on the wall like plaques.

SIGNS — The religious home may have framed decorative signs.

PLATE RAIL—A plate or dish rail may hang on a wall, especially in a dining room. It may have a china collection — demitasse, plates, cups, bowls — on it.

LAMPS — Lamps of many varieties may be mounted on the wall as decoration, light motivation or both. The lamps may be electric, kerosene, gas or of candelabra variety. They may be mounted on their own bases or on small bracket shelves. Usually for the sake of safety it is advisable to have flame variety lamps electrified.

MIRRORS—Mirrors of all sizes and shapes are used on the stage. Frequently they are mounted on the wall although there are times when free standing full length varieties or small mantel or desk varieties are used. When mirrors are used they should be checked carefully from the audience area to see that they do not reflect images of actors waiting offstage for entrances, images of untidy fly lofts, or images of spotlights. These images can be most annoying to members of the audience. If a mirror is on a sharp angle so the audience sees but little of the surface a false mirror painted with a middle gray base with streaks of tint and shade gray may provide a substitute. Placing net over a real mirror or spraying it with wax, scene paint or grease will diffuse the light from the reflective surface so that it is less annoying.

WINDOWS — Windows allow many and varied treatments. In fact a window effect may be achieved on a flat wall without even having a hole in the flat if the space is completely dressed with window paraphernalia. Pull down shades, Venetian blinds, split or toothpick bamboo blinds, accordion folded paper blinds, glass curtains, drapes, valances and cornices may all be used in decorating the window. The paper blinds are made by folding paper (butcher paper works very well) of the correct width into accordion pleats. This can be mounted to a molding at the top. A strip of wood is stapled to the bottom and the blind is completed. Strings may be run through holes drilled in the paper while the folds are pressed tightly together. These strings will make it possible to raise and lower the blinds.

GLASS CURTAINS—Patterned lace, marquisette, cheese cloth or any net material may be purchased in yardage. The top and bottom may be hemmed or left with the raw cut edge. The edges may be cut with regular or pinking shears. Glass curtains may be suspended in two sections with one section mounted to the top sash and the other to the bottom sash. They may be cut just long enough to extend from the top of the window down to the window sill or may extend all of the way to the floor. They may have fulness or not. They may hang straight to the floor; may be butterflied; or may be a combination of the two. Glass curtains may be used by themselves or in combination with draperies. They may have plain or ruffled edges. Glass curtain material may be obtained

in white, ecru, egg shell, dotted swiss and colors.

DRAPERIES — Window draperies may be mounted in a permanent position or may be rigged to open and close. They may open and close by having butterfly tie backs or by operating on a traverse rod or on wires. If they are on a traverse rod the pleats should be sewn in so that fulness will remain in place whether the draperies are opened or closed. Draperies may have a cornice at the top to conceal the rigging; they may have a valance; or they may have nothing at all. The cornice may be made of cardboard, wallboard, or wood and may be

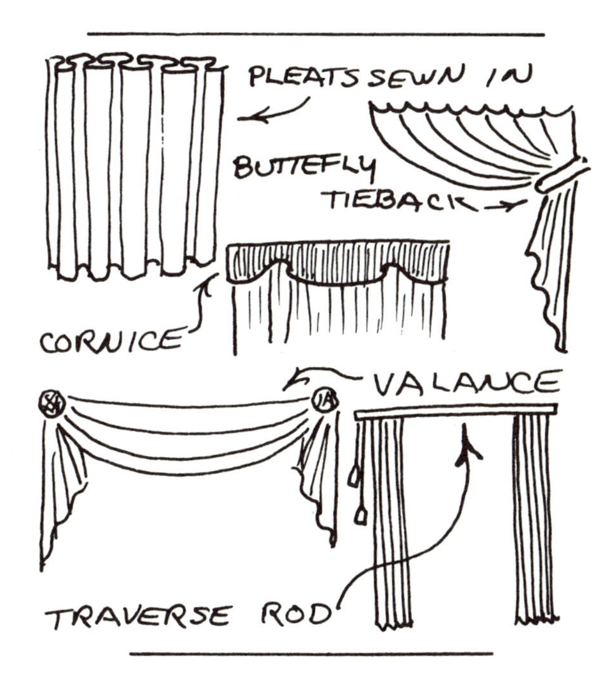

painted, unpainted or cloth covered. The draperies and valance may have a tassel trim, may have rope worked into the folds or may be plain. The draperies may be made of plain, striped or patterned cloth or may be made out of muslin and designs applied with scene paint and brush, textile paints, silk screen, or linoleum blocks. These processes will be described later. Draperies may be mounted directly to the flat on a rod, pipe or wire or to a board that is then mounted on the flat. If the draperies are to be removed rapidly for a scene change it is possible to mount them on a board and use hangers to hold the board, or rig up a pulley system to pull the board up into place and lower it for shifting.

The search for drapery materials should extend into the paper, synthetic and plastic, as well as cloth sections of the department store since these materials are frequently used to manufacture inexpensive draperies.

WINDOW SILL BOX—The window, inside or out, may have a sill planter box complete with real or synthetic leaves, flowers or both.

FIREPLACES—There are several decorative areas in connection with a fireplace. These are: the space above the mantel, the mantel and the hearth.

*Space Above Mantel* — The space above the mantel may be flat or recessed; it may be exposed stone or brick; it may be plain or panelled. Regardless what its finish there are times when essential props are required in this position. These may consist of masks, armament, pictures or plaques. At other times objects of the same nature are added to the decor to make the fireplace a more interesting or dominant unit in the total design.

*Mantel* — The mantel may be loaded with knick knacks or be tastefully decorated with one or two select pieces depending upon the total mood and feeling of the setting. Some of the objects that may be found in this position are: planter boxes, clocks, photographs, mirrors, ash trays, pottery, candelabra, vases, flowers, candy dishes, match boxes, books, figurines, boxes and candle snuffers.

*Hearth* — The hearth area may have the usual fire screen, andirons and fireplace tools consisting of shovel, poker, broom, tongs and stand to hold them. It may also have a coal scuttle, wood box, crane and iron kettle, cooking facilities, popcorn popper, log unit, logs, bellows and salt holder.

CHAIR DECOR—A chair may have a pillow tossed onto the seat, may have slip covers or robes casually or carefully thrown across them. In an immaculate household they may have antimacassers (a tidy or cover to protect a chair or sofa, usually placed where the arms and head are likely to rest and soil the upholstery finish).

SOFA DECOR—Antimacassars, robes, quilts, slip covers, pillows, dolls, teddy bears and many other objects may be placed on sofas to add color and atmosphere to the room.

BED DECOR — Regular bed linen, quilts, decorative bed spreads, rolled blankets, dolls, extra throw pillows, canopies, bed lamps, book holders are all items that might be used with a bed.

DESK PARAPHERNALIA—Desks have character that is determined by their topside clutter. They range all the way from completely cleared tops to the ones with animal collections, blotter, pen and ink, quill and ink, letter baskets, telephones, intercommunication speakers, typewriter, books, mirror, photographs, lamps, candles, ledgers, receipt books, stationery, figurines, vases, flowers, ash trays and general clutter.

TABLE PARAPHERNALIA — Table cloth (lace, white, checkered, oil cloth, tasseled, patterned), magazines, books, sewing basket, fruit basket or bowl, lamps, figurines, nut dish with nutcracker and picks, telephone, liquor services, decanter, tea set, coffee set, flower arrangement, newspapers, dinner service (plates, cups and saucers, crystal, napkins, silver, service dishes), playing cards, writing equipment and candy boxes are all items that may be found on a table.

A room may have much miscellaneous clutter—pillows on the floor, footstools, waste baskets, screens and grandfather clocks.

DISPLAY CABINET — Open display type shelves or cabinets may have select china (painted paper plates even), crystal or pottery, figurines or books.

BAR — A bar, whether it is a small home variety or a place of business, will have a variety of bottle types (wine, champagne, whiskey, Bourbon, beer, gin); it will have a variety of glasses for the various types of drinks; and it will have a variety of objects such as ice bucket,

tongs, stirring rods, lemon and lime squeezer, seltzer water bottle, decanter, ice pick, corkscrew, jigger, cocktail shaker, beer can opener, lemons, limes, soda water mixes, cherries, olives, onions, pretzels and napkins—and might even have beer kegs.

RESTAURANT—The restaurant setting may have a thousand and one items including: cash register, stools, counter, tables and chairs, piano, coffee maker, vendor (cigarettes, candy), juke box, stove, gum machine, dishes, silver, napkin holders, doughnut and roll jars, signs, cigars, cigarettes, toothpicks, china of all kinds, food, hat and coat racks (wall hangers or standing tree type).

DRESSING TABLE DECOR — Powder boxes, lamps, mirror, comb and brush set, atomizer, perfumes, powder puff, photographs, miscellaneous boxes and bottles and jewelry are all items that might be found on a dressing table.

MISCELLANEOUS SEATS—An odd collection of items can be used for seating in various places. Stumps, logs, kegs, barrels, boxes, crates, packing cases, stools, chairs, benches and dozens of other objects can all be used.

### Manufacturing Stage Properties

The property man with imagination and a modicum of mechanical ability will be able to manufacture many of the property items needed. Basic tools and a few materials can be used to make any number of items.

GLUE GUN—An invaluable tool for the prop person is a hot glue gun available in almost all hardware, hobby, lumber, and variety stores. This tool may be used to bond any number of materials together.

ROUND EXTRUSION FLEXIBLE PLASTIC—Flexible, round cross section flexible plastic (Sonofoam Backerod) available in sizes from about ¼″ to 2″ bends to contours. Can be used whole or split for edging or furniture decor.

STYROFOAM—Polystyrene, or Styrofoam, the feather-weight firm, white, sponge-textured material used a great deal by florists, has countless uses in making stage properties. It can be worked with regular wood working equipment including saws, knives, rasps, and sandpaper. It is carried in stock by some lumber yards and some florists supply houses. It is used as an insulation as well as decorative material. It may be obtained in various thicknesses, comes in fairly long pieces and is purchased by the board foot. Several thicknesses can be glued together or fastened together with wire, string or thread. It can be painted readily. Artificial fruit, cakes, roasts, hams, books, inedible rolls, doughnuts, steaks, mashed potatoes, cut out forms, letters, signs and a host of other things can be made from it. Wherever lightness of weight is desired it is an ideal substance to use.

DOWEL — Dowel is round cross section wood that comes in many sizes. It is obtainable in either soft or hard wood and can be used for making fake candles, gun barrels, stair rails, spears, flag poles, pennant poles, and tent poles. Dowel is obtained from lumber yards, cabinetmaker shops and some hardware stores.

PIPE AND TUBING—Iron pipe, steel pipe, aluminum and steel electrical conduit, aluminum tubing, rubber tubing, plastic and glass tubing can all be used for stage properties. The metallic tubings may be used for most of the same purposes as dowel. Rubber tubing can be used to conduct fluid, smoke or air onto or off the stage and can be used for making objects that need flexible tube structure.

LUMBER—White pine of varying widths and of thickness up to 2″ is useful for carving wooden properties. If more thickness is desired two or more thicknesses may be glued together. White pine may be used for gun stocks, musical instruments, stools, benches, tables, beds, etc. Plywood of varying thicknesses is also useful for table tops, bed tops, cabinet doors, drawer bottoms, base forms for papier-mache and boxes.

SPONGE RUBBER—Sponge rubber is another material that should not be overlooked. Although it is relatively expensive in large quantities, it has many uses and small pieces may be used to good advantage. Scraps of sponge rubber rug padding are useful for many property items. It curls, bends, twists and stretches for forming odd shaped items. Pieces of it can be used to simulate food, flowers, large leaves, etc. Thicker sponge rubber might be used for padding, for pillows, for various irregular objects. Department stores, upholstery shops and rug departments handle sponge rubber.

FELT — Felt, especially the heavy weight varieties, may be used for making some small decorative pieces. Felt may be steamed and worked into almost any shape. After it has been shaped it can be loaded with size, glue or paint to stiffen it and layers can be glued together. It can be used for brooches, for flowers, armour pieces, all kinds of head gear, epaulets and many, many more items. It may be gilded to look like gold and used for some jewelry items.

### Breakaways

UNFIRED POTTERY—There are times when it is necessary to have a vase or similar object that can be broken over an actor's head without damage to the head. Unfired pottery, especially if it is a thin "slip" casting in the green state, can be used for this purpose. In fact, it is usually so fragile that it must be handled with great care or it will be broken too soon.

BREAKAWAY FURNITURE — Old pieces of furniture should be used when a piece is to be broken when sat upon, thrown or used to strike an actor. The pieces can be pulled or broken apart and then loosely fastened together in such a way that they will easily crumple when pressure is applied. Scotch tape, loosely fitting dowel, joints that will easily slip apart, joints held together by easily broken thread, pieces cut or sawed through and then fastened with gummed tape or some similar material should be used. Sometimes the parts may be held together with loose pin hinges and the pins pulled at the opportune moment. Properties to be broken may also be made from materials that look substantial but which may be easily broken, such as sections of Styrofoam, hollow papier mâché, or sponge rubber.

### Cast Forms

There are many stage properties that can best be made as cast forms. Lightweight reproductions of bulky objects, many reproductions of the same piece, the outer shell of a heavy article, and rigid but durable forms may be made in this way. A master form consisting of the object itself or a clay model of the object may be used to build the negative cast. Puppet heads, bottles, book backs, artificial foods, masks, armour plate, helmets and statues are a few of the props that may be made in this way.

If the surface of the object to be cast is porous it will be necessary to coat it with a sealing material or build a facsimile form from sculptor's plaster or Plasticene (clay ground in oil so that it will not dry out). Since Plasticene has an oil surface it is possible to make shells of some objects over the top of a sculptured form Mâché, cornstarch-cheesecloth, plaster of paris and some other processes will work without further surface processing.

CASTS — Plaster of paris or casting plaster can be used to make the negative cast so that models can be made inside the cast. Formed this way, the object will have a surface conforming to the shape of the object over which the cast is made. If the object to be cast is merely a form raised from a flat surface in bas-relief it is possible to make a one-piece cast but if it is in the round it will be necessary to make the cast of two or more pieces. If the form is modelled in clay undercuts should be avoided so that the cast will pull straight off without hooking on portions of the clay.

A one-piece cast is made somewhat as follows. The object from which the cast is to be made is laid on a flat, nonporous surface such as a slab of glass or marble. If there seems to be danger of the cast sticking to it the surface should be lightly coated with vaseline, liquid soap, aluminum foil or similar material. Plaster is then mixed by sifting the dry plaster slowly into water until dry plaster begins to build slightly above the water. With the bare hands, or with a spoon or spatula, mix the plaster and water. The less the plaster is mixed the less danger there is of getting bubbles in the cast. Coat the surface with this plaster mixture. The cast should be built so that it is from ½″ to 1″ in thickness. After the surface of the object has been thoroughly coated the plaster should be allowed to set and dry. During the setting process the plaster will become warm to the touch. The cast should not be removed from the model until it has cooled and is thoroughly hardened. There are stages in the process when the plaster is crumbly and easily broken. It may take a couple of hours for the plaster to harden sufficiently so that the clay or object being cast over can be removed. If possible the cast should be left undisturbed even longer.

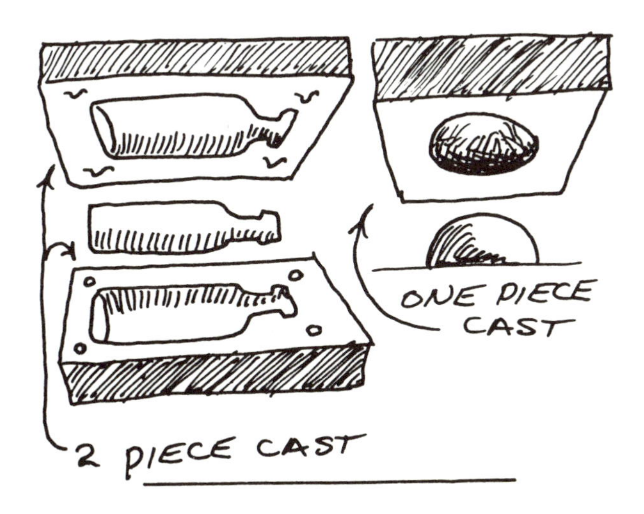


CASTS OF TWO OR MORE PIECES—If a three dimensional object is to be cast it will probably be necessary to make a two-piece cast (sometimes three or even more pieces may be needed). The object should be studied to be certain that the cast can be pulled off without destroying the object being cast—especially if more than one copy is desired. If the original form is Plasticene it will be possible to build separating walls from pieces of tin or aluminum. Strips can be sunk into the surface far enough to hold them in place. These will form a collar or fence around one area. Plaster can be poured into this area. After this plaster has set another area can be blocked off. Where plaster is to be poured against plaster notching indentations should be carved into the edge of the dry section and then this edge should be lubricated with soap, vaseline or oil so the new plaster will not stick to the old. This area is then coated with plaster. The process is repeated until all of the form is coated with plaster. After the sections have set they may be pulled apart and away from the original form. It may be that the form from which the casting is being made cannot be split into segments with metal strips as indicated above. Perhaps similar "dams" can be fastened on with Scotch tape, string, wire or some other material.

Another method may be used for a two-piece mold. A cardboard or wooden box can be made that is large enough to half submerge the object being cast. Soft plaster is poured into the box and the object half submerged. After this plaster has set the edges can be lubricated and plaster poured on the top half.

After the cast has been removed it may be prepared for any of a number of different kinds of castings.

MOLDING FOAM—Various molding foams are available under numerous trade names. Most fall into the category of Urethane Foam. Some are designed to be mixed in, and dispensed from, special machines while others are designed more for shop-mix by the "do it yourselfers".

In general the shop-mix consists of two ingredients which mixed together chemically react to expand and form a foam that congeals into either a flexible, or a rigid, mass depending upon the chemical formula.

The density and hardness of the rigid foam depends upon the chemicals used and can be as light and fluffy as one pound to the cubic foot or as dense as sixty pounds to the cubic foot.

Obviously the rigid foam will be most useful where pressure or weight is applied to the foam and, the flexible foam will be used where resiliency is needed. A rock or tree stump, upon which an actor is to sit, or stand, would be made of rigid foam, for instance, while flexible foam would be used for an object that was bent or compressed but expected to restore to its original shape after the pressure was removed.

Before using this material, experiment with a small amount of it, in order to discover the action of the foam, and in order to determine the approximate amount of expansion, as it turns into foam, and in order to determine the amount of mix needed to produce enough foam to fill the cast.

TO CAST—Fasten the pieces of the cast together securely. Mix the ingredients in the exact proportions listed in the directions, then pour the mixture into an opening left in the cast for this purpose. After it is thoroughly set, remove from the cast. Trimming and additional shaping may then be done with rasp, saw, knife, and a hot wire cutter. NOTE: Check parting agent to prevent sticking.

FREE FORM—Free form Urethane foam objects may be made by pouring foam on surface. Check directions to be certain about the need for a parting agent. And, of course, experiment with small amounts first.

PVC—Polyvinyl Chloride tubing used extensively for water, sprinkling, and sewer pipe may have many uses on the stage. This material available in white, black and gray may be purchased in garden centers, variety stores, and do-it-yourself building supply centers. It is light in weight, easy to cut, relatively inexpensive, is available in many diameters, can be fit together in a multitude of configurations with the help of available fittings and can be securely assembled with a special quick drying cement.

The tubing when heated to about 250° will bend and conform to complex configurations. Try a sample first if you plan to bend it. If it tends to collapse fill it with sand, then pour the sand out after it has been bent.

VACUUM FORMED OBJECTS—Prop, scenery, and costume objects such as brick and stone walls, balusters, bottles, books, shields, helmets, ad infinitum may be shaped from sheet vinyl by means of a thermal process that heats the plastic until malleable, then by means of a vacuum pump pulls the material into a cast to force it to conform to a pre-sculptured shape. If interested in this process check with a plastic products supply company.

*WARNING!—When working with plastic materials read the directions carefully and work with the materials as directed. Observe all of the cautions. Work must be done in a well-ventilated place and often workers should wear gloves and respirators. Some of the ingredients, fumes, solvents, and dust from plastic processes and/or fabrication are dangerously toxic.*

Plaster of Paris Cast in Plaster of Paris—If it is desirable to make a plaster casting in a plaster cast it is necessary to seal the pores of the cast before pouring plaster into it. One of two methods may be tried. The simplest, and possibly least safe, is to use soft soap. Liquid washing soap or tincture of green soap may be used. The surface is carefully coated with this material in thin layers until it feels almost like marble. The plaster should then immediately be poured in. The soap should seal the pores enough to prevent the new plaster from being absorbed. The second method, more complicated, is handled as follows. After the surface of the mold is thoroughly dry coat it with orange shellac. After the first coat of shellac is dry apply a second. When the second coat of shellac is hard coat the surface with stearine. Stearine is made by melting two tablespoons of stearic acid and adding to it eight tablespoons of kerosene. *Great care must be taken to extinguish the flame and add the kerosene where there is no danger of it catching fire.* This solution may then be brushed onto the shellac surface. This second process seals the pores much more completely than the first and the cast is ready to be used any number of times by merely recoating it with stearine. The plaster for the casting is mixed and poured into this form.

It is frequently difficult to remove a cast piece of plaster of paris and the mold must be broken before it can be removed, especially with a one-piece mold.

If it is necessary to patch either the mold or the final casting the pieces may be cemented together with a good grade of glue or plastic cement. If it is necessary to add more plaster of paris after a portion has dried it will bond only if this piece is soaked with water over the area where the new plaster is to be placed. If this isn't done the porous surface will cause the water to be absorbed from the new plaster before it goes through its setting process.

Papier-Mâché—Papier-mâché can be used to make a casting from the plaster form. A coating of liquid soap or vaseline will remove the porousness enough so that the paper will not stick. Three or four layers of paper in narrow overlapping strips should be used. If it is possible to place the whole works near a heat blower the paper will dry more rapidly. If too many layers of paper are put on at one

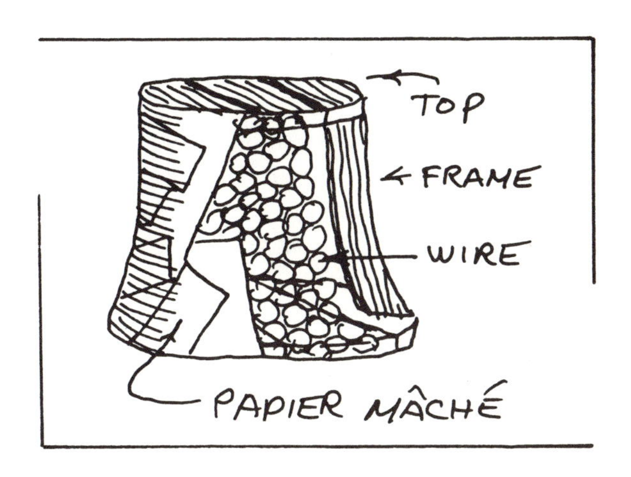


time and the drying process is too slow the mixture will become moldy before it dries. Wheat paste, paper hangers' paste, flour paste or a thin solution of glue may be used for the adhesive. Papier-mâché may also be made into a pulp by shredding the paper, soaking it in the adhesive solution and then pressing the pulp into the cast. For a fine finished surface tissue paper may be applied as a last coat. Brown wrapping paper may be used for sturdy objects.

Mâché may use paper (shredded, in strips or pulp), cheesecloth, muslin, asbestos (paper or pulp) and may be textured by scattering or pressing sawdust, sand, shavings, flock, flitter or similar material onto its surface. If the mache is cast in two or more pieces they may be attached to one another with strips of mache.

Cornstarch-Cheesecloth — Cheesecloth or other mesh cloth can be moistened to make it flexible enough to press into a mold and then cooked cornstarch about the consistency of thick cream can be brushed onto and into the cloth. Add a second layer and repeat the process. The layering can continue if a heavy "shell" is desired. This can be dried to a hard finish. Masks can be readily made from this material.

PLASTER CAST CLOTH—The cloth, impregnated with plaster, that is used by doctors to wrap broken bone areas for supporting casts is a material that can be easily used for making some items. It comes in a roll and is easily used. It needs only to be dipped in water and it is ready to wrap around or over an object for use. It dries rather rapidly and when dry has a hard surface. This with crepe paper mâché over the top for smoothing makes an excellent surface. The material is available at many drug stores and comes in a variety of widths. Strips of cheesecloth dipped in wet plaster may be used in the same way.

PLASTER OF PARIS AND BURLAP—Burlap impregnated with plaster of paris can be used to build up forms. A basic core of chicken wire or other material may be constructed first. Then pieces of burlap dipped in wet plaster are placed over this. Plaster can be molded over the top to get the desired modelling. It is necessary to work quickly since the plaster will harden rapidly when it begins to "set."

PAPER TAPE—Most dressmakers are familiar with body dress forms that are made from gummed paper tape. This same gummed paper may be used for making some property forms. It will work best if the object has rolling contours rather than sharp turns. However, the paper may be cut or torn into small pieces and used even for sharp curves and turns.

DIP AND DRAPE—A simple cloth substance impregnated with a water softening adhesive. Simply moisten and shape to create decor on a surface. Obtainable from Fabric and Hobby shops.

RUBBER LATEX—Rubber latex may be used for making either the cast or the casting. It is obtainable in a milky liquid form that is poured into a cast or over a form. The directions that come with it will help in determining how long it takes for it to set. Thin shells of objects can be easily made with this material. Special paints will be needed to color its surface. The finished form is tough and durable. Prop food forms such as loaves of bread, vegetables, fruits, fowls and roasts may be made in this way.

FIBRE GLASS—Fibre glass imbedded in resin can be used to make tough, durable properties. It requires special agents. Information may be obtained from the dealer who supplies it. Fibre glass may be purchased from some sporting goods stores where boats are sold since it is frequently used for building boats or for coating wooden boats. This is excellent for making permanent units like armour. It is rather expensive, which is its major drawback.

CELASTIC—Celastic is a sheet plastic molding material. It requires special softening and parting agents. The material and information about its use can be obtained from costume houses or from the manufacturers.

PLASTAB—Plastab is a colloid impregnated fabric similar to Celastic and handled as a molding plastic.

PAPER FLOWERS — Paper flowers may be purchased from ten cent stores, from window display houses and from specialty companies or they may be manufactured by the property crew. Since they are rather expensive to buy it is likely that the latter method of obtaining them will be the one most often used. There are some times in exterior settings when they can be painted directly on the scenery or set pieces. Large stylized flowers may be cut from cardboard, wall board or even plywood and painted. Flowers may be made from regular or double weight crepe paper, stiff painted muslin, plastic and cloth materials or combinations of all of these materials. Heavy textile material such as felt can be used beautifully for some flowers. The stems can be made of wire. Soft wire can be used to hold the petal sections together and to fasten leaf sections onto the steam and then the wire can be covered with florists green tape. Rubber flowers were used in the original Broadway production of *Knickerbocker Holiday.* The actors could step or sit on them and when they moved the flowers sprang back up into place.

A little experimenting will help in making flowers. Pamphlets on flower making can be obtained from the crepe paper manufacturers.

GRASS—The easiest thing to do about grass is to forget about it, but if the demands cannot be quieted there are a number of methods of supplying a substitute for grass. A colored section of floor cloth or rug might serve the purpose if it is painted in a grass green and then spattered with various other greens. The simulated grass mats sometimes used by funeral directors are excellent but expensive. Shredded green paper or cellophane can be glued to the surface of the grassy plot. Grass ground rows with painted cut out grass blades can be used and grass blades can be painted on the base of drops and some set pieces. Sometimes the suggestion of grass is better than an attempt at realistic duplication.

PLANTS—When plants are desired it is often easiest to buy, rent, or borrow real, or artificial plastic plants from a florist. Palm fronds can be purchased and rigged onto a tree limb which is then spirally covered with shredded strips of burlap and the result is a passable palm tree. Artificial large leafed plants may be purchased from window display houses. And of course it is possible to make stylized cut outs of plants if the technique fits in with the feeling of the rest of the production.

LEAVES—Leaves may be made of muslin, velour, felt, flocked paper, stiff paper, plastic, crepe paper or any of a host of other materials. The material can be stiffened with paint or glue size and after it has dried the leaves can be cut and mounted on wire stems to be fastened to the plant. Excellent durable plastic artificial leaves may be purchased from florist suppliers.

HEDGES—Potted real hedge bushes may be used for hedges or three dimensional built up flat units may be painted and used or realistic papier-mache on wire units may be shaped, covered and painted. Another effect to try here is to use chicken wire for the basic form and then stuff the holes with green crepe paper or paper napkin sections the way that parade floats are often decorated.

Grasses, weeds, grain, corn stalks and natural materials of this nature may sometimes be used. They can be sprayed with paint in order to color them and then be mounted on a board base to hold them in place.

VINES—Vines may be painted onto the scenery or simulated by stringing rope where desired. The rope can be painted and artificial leaves of muslin, paper or cloth wired on where needed.

SMALL TREE TRUNKS — Branches of real trees or shrubs will probably make the best tree trunks. At times extremely stylized tree trunks can be cut out of heavy plywood or be formed by twisting heavy wires together. They may be painted, wrapped with cloth or left plain.

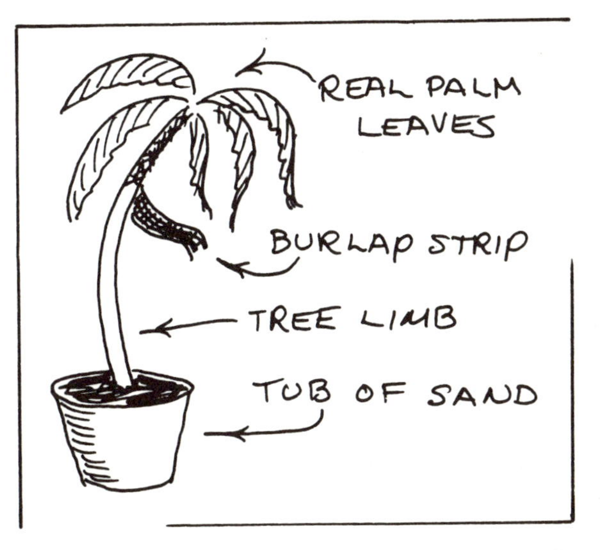


Leaves can be fastened on with wire and made as indicated above or can be completely nonconformist shapes made from twisted wire, pieces of sponge rubber, tufts of cotton or almost anything else.

ARMAMENT—A veritable arsenal of equipment will be used from time to time in plays. When guns are called for it may be possible to get real ones from private owners, the National Guard, police, sheriff or armed forces branches. If extreme naturalism is desired this should be done. They should be handled with great care and kept locked up when not in use. Some miscellaneous arms equipment of recent date may on occasion be purchased from War Surplus or Army and Navy stores. Costume houses also have equipment of this nature.

GUNS—Convincing wooden guns, both revolvers and rifles, can be manufactured if one has the time and material. Gunstocks can be cut out of 2″ shop pine and fashioned with a wood rasp, draw knife, plane and sandpaper. A piece of dowel, pipe, or electrical conduit can be used for the barrel. And the trigger guard can be made from a piece of plumber's strap or thin strap iron. Either the real thing or adequate pictures or illustrations should be used as models. Pictures may be found in a good dictionary or encyclopedia.

DAGGERS — Daggers may be made from wood or metal. Bayonets, often obtainable in surplus stores, can be ground down to dagger size. Obviously they should be dull so that no injuries are invited when they are in use on the stage. Sometimes wooden daggers are more practical for just this reason. Fake rubber daggers are obtainable in "trick" stores so that the dagger may actually be pushed against a person without injuring him.



SWORDS—Some swords can be manufactured of wood or metal but usually this type of equipment should be rented or borrowed. It is advisable to check rather carefully and determine the correct style and variety of weapon for the period and action of the play.

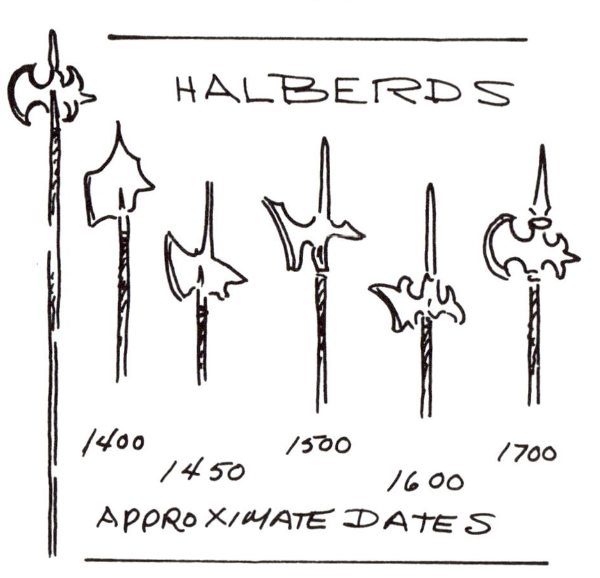


SPEARS — Convincing spears can be made from curtain rod, dowel, or aluminum conduit and a wooden head. Metallic heads could be made but the wooden ones will probably be easier to make and may be more convincing. These can be cut from ½″ or ⅝″ plywood. Here again it is advisable to check rather carefully in order to be certain that the head of the spear, halbred or whatever it is, has the correct shape. Spears may have a bit of cloth tied to them beneath the head to add color—again if it seems to fit the period and style.

HELMETS — The helmet liners used in the second world war are excellent for helmets. If they are not the correct style and shape for the period of the play the correct ones may be made from them by using them as a base and adding Celastic, wire and mache, or wire and plaster cast bandage.

PENNANTS—Pennants and flags add a good deal of color and smartness to a battle scene. This is especially true in productions of Shakespeare. At times they also help to keep the various factions straight in the minds of the audience. They may be made of muslin, rayon, silk, nylon or almost any material that will hang in nice folds. Correct designs may be applied free hand, silk screened, block printed or stencilled. The pennants can be mounted on poles of wood or metal and a fitting emblem or point mounted at the top of the pole. A bit of rope with tassels can hang down as added decoration. For the rope the twisted soft cotton rope used in upholstery for beading may be used. This can be painted or left white depending upon the color desired for contrast.

FLAGS — If real flags are desired they may usually be obtained from civic and service organizations.

POLICE EQUIPMENT—A policeman's night stick can easily be made from a piece of fat dowel or a straight piece of a tree limb. A badge may be cut from aluminum. It may be possible to borrow all of the equipment needed from the local police station.

MILITARY EQUIPMENT — When conventional and contemporary military equipment is needed it is usually advisable to get on the telephone and contact local or nearby installations of active or inactive military establishments before arranging to make or rent equipment. They are usually most cooperative and helpful in either providing equipment or furnishing information about where it may be obtained.

ARMOR—Armor, such as breast plates and other paraphernalia of that nature, can be rented or made up by the crew. Some sections might be made from aluminum, available in hobby shops and department stores. It may also be possible to mold armor on the body. If it is to be molded to fit the body a number of methods might be used. Obtain some plaster of paris medical cast tape. Use an old "T" shirt for chest covering, and the knit material that doctors use for leg or arm undercover for a cast; place them on the body and then over the top of these mold the shape desired. Several thicknesses of this material will dry to a hard shell that can be reused many times. Another method of making forms of this nature uses a screen wire base. Cut and shape the screen wire to the correct contour and then coat it with papier-mache, cloth mache, gummed tape or similar material. Still another process uses felt over a plaster of paris cast. In order to shape the felt correctly it is necessary to use steam or moisture. Over a cast the felt can be steamed into shape. A couple of thicknesses can be glued together if necessary and the whole thing heavily sized and painted with gilt or aluminum paint. Celastic, plastab, and fibre glass might also be used for armor.

SHIELDS — Shields can be made from plywood or metal. The plywood will probably be the easiest to use and most satisfactory in the long run. If desirable this may be partially or completely covered with metal for the sound effects. There is no reason for not using any of the other molding materials for the shield. A round, slightly dished shield could rather readily be made by cutting a circle of metal and then after slicing in to the center making the sections overlap slightly and fastening the joint with rivets. Wall board material such as EZ Curve Upsom board might be used in the same way. The EZ Curve shield wouldn't stand as much battle service as the metal one but for short time use might be perfectly satisfactory.

## Printing Processes

Patterns and designs may be applied to pennants, banners, draperies, costumes, signs and posters by one of at least four methods: 1. Free hand, 2. Linoleum Block Printing, 3. Stencils and 4. Silk Screen.

FREE HAND—The free hand method of designing and painting patterns is easy and takes less preparation than other methods. If an identical pattern is to be duplicated several times the other methods have certain advantages. Free hand patterns become tedious and boring to duplicate.

LINOLEUM BLOCKS—Heavy battleship linoleum may be used on a block, or without a block for some printing, but it is usually easier to handle when mounted on a board base. The linoleum may be purchased ready mounted or

LINOLEUM BLOCK IS NEGATIVE, UN CUT PORTIONS PRINT.

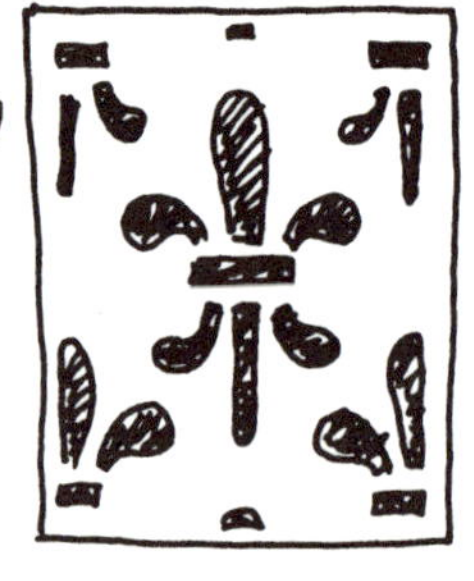

CUT-OUT PATTERN OF STENCIL – DARK AREAS CUT OUT LEAVING LINKS BETWEEN THEM.

may be cemented to a block of wood just before the pattern is cut. The pattern is drawn in reverse on the linoleum after having been planned on paper first. In fact it may be transferred with carbon paper if desirable. Linoleum knives or cutters are the most satisfactory tools to use for carving out the unwanted linoleum. Only the portions of the pattern that are to print are left. These raised portions are then inked and pressed onto the surface to be printed. In order to ink the block evenly a squeegee is used. The ink is placed on a flat surface, such as a piece of glass, and then the squeegee is rolled back and forth to coat the roller evenly. The roller is then pushed across the face of the block carefully to ink the printing surface. Any one of a number of devices may be used as a press. The material to be printed may be placed flat on the floor. The block may be inked, laid in place where it is to print, and then pressure applied to the block with one or both feet. The block may be placed in a regular press and printed on the surface. Or the material may be placed on top of the block and rubbed by hand to transfer the pattern. The pattern is applied wherever desired.

STENCIL — The stencil pattern may be cut out of the stencil paper with a sharp knife, razor blade, Exacto knife or matte knife. The portion of the paper representing the pattern to be colored is cut out. Great care must be taken to plan the pattern so that connecting links remain so that the paper will not tear out between stencil pattern sections. After the stencil has been cut the paper may be mounted on a wooden frame, similar to a picture frame, in order to make it easier to handle. A sponge, brush or cloth swab may be used to apply the paint. If a sponge is used it is possible to get open rather than solid coverage. Of course the paint may also be applied with a brush, but the sponge makes the process less regular and harsh. Almost any variety of paint can be used with a stencil which is one of its advantages. It is possible to use oil, water or textile paints.

SILK SCREEN — The silk screen requires a a little more equipment than the other processes but the finished products are often more satisfactory and most of the equipment will last for a long time and can be reused time after time. The first requisite is a sturdy wooden frame to fasten special silk onto. Although any tough silk will work it is best to get the silk that is specially prepared for silk screen work. The silk is stretched and tacked or stapled onto the frame. It is important to get the silk surface as taut as possible. After this has been done masking tape is placed over the tacks and edge of the silk extending at least half an inch toward the center of the frame. Next the frame is turned over and masking tape is placed around the inside of the frame to cover the tape on the bottom and extend up on the frame slightly in order to seal the edges where the paint might seep in between the silk and the frame. When this is done the frame is ready to use. A squeegee similar to a window washer's squeegee is used to pull the paint across the silk screen that has just been applied to the frame. A regular silk screen squeegee or a heavy floor squeegee or a heavy window washer's squeegee will do. The squeegee must have

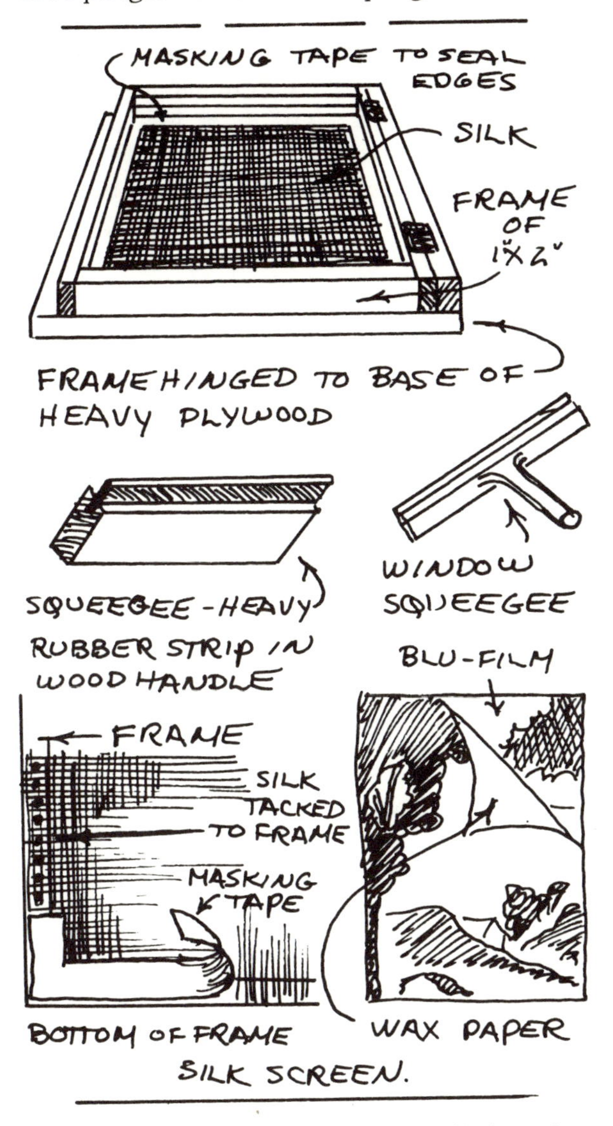


fairly heavy rubber or it will not pull the paint across the silk evenly.

Designs for the silk screen may be made in several ways. If the pattern isn't too complex it may be cut from paper. Paper that is not too absorbent will work the best. If a paper pattern is used it is placed under the screen on top of some old newspapers. The pattern is attached to the frame with masking tape. Some silk screen paint is then poured across at one end of the screen. The squeege is then used to pull the paint across the surface. Even pressure should be applied as the paint is pulled across. The paint will be squeezed through the screen transferring the pattern to the old newspapers and causing the paper pattern to adhere to the silk.

Areas of the silk screen can be opaqued (or sealed) with Le Pages liquid glue, with Block Out Lacquer or with Blu Film. Glue and lacquer are applied with a brush. The glue is really much easier to apply and at the same time is less expensive and easier to remove when the screen is cleaned.

Blu Film is a special thin film material mounted on a wax paper base that is manufactured specifically for use with silk screen equipment. It is rather expensive but for small detail work is superior to most of the other materials. With a sharp knife it is possible with extremely light strokes to cut portions of the film and remove them from the paper base. Fine detail can be thus removed. The film is removed wherever it is desirable to have the paint squeeze through the pattern. After the pattern has been completed the Blu Film is placed under the silk and adhered to the silk with a special liquid. This step in the process must be executed carefully but with a little practice and the expenditure of a good deal of energy it can be done successfully. Two soft cloths are used, one to apply the adhering liquid and a second to rub it vigorously to dry it. The liquid is applied rapidly to a small area and then with the dry rag the area is rubbed briskly with pressure. When this has been done inspection will reveal that the film has been pulled in to seal the spaces between the threads. After all of the film has been adhered and is dry the paper back is carefully peeled from the film thus opening the spaces through which the paint is to be squeezed.

All of the block-out substances may be used on the same job. Detail areas can be covered with Blu Film while the rest of the space is covered with paper. Dry brush effects can be accomplished by lightly brushing glue across the surface.

Multi-color silk screen printing is accomplished by repeating the process, cutting out other areas of the pattern and squeezing other colors of paint through the stencil thus created. Sometimes it is wise to hinge the frame onto a base of some sort so that the paper, cloth or material being printed can be kept in register for single or multi-color printing.

All of the paraphernalia will require cleaning. This is usually a messy process. The paper can be pulled off and thrown away. Glue can be washed off with soap and water, but special solvents are required for the lacquer, Blu Film and paint. Acetone will probably work very well for both Blu Film and paint if regular silk screen paint is being used (and this is the paint that is usually recommended).

* * * *

If there is space available it will usually prove advantageous to keep as many stage properties in a "morgue" as possible. When units are built this may be borne in mind. Genuine artificial simulated roasts, hams, Thanksgiving turkeys, fruits and vegetables as well as a myriad of other items will be needed for play after play and if they are available in the property stock room the job of "propping" the show will be made much easier. If the objects are durably made they will last for a long time.

Polonius' words from *Hamlet* may give a cue to the property manager:

"Neither a borrower nor a lender be;
For loan oft loses both itself and friend,
And borrowing dulls the edge of husbandry."

This will prove true if properties are borrowed, mishandled and returned without being carefully checked for damage. Broken, scarred and soiled objects should be repaired, cleaned or replaced. The kindness of the lenders must be respected if a repeat of the kindness is expected in the future.

LEAF PATTERNS

# 10

# *Special Effects*

It would be impossible to describe all of the special effects that could be encountered even in a dozen plays. However, it is possible to suggest a few of the more common devices and contrivances that might be called for from time to time.

### Manual Sound Effects

Many stage sound effects can be created with manual devices that can be manufactured in the shop. Among these are the following:

WIND—The conventional wind machine is composed of a slatted drum mounted so that it can be rotated. This is rigged with a sheet of canvas thrown across the drum. When the drum is turned the canvas drags against the slats and creates a wind-like sound. By placing pressure on the canvas an increase of volume may be obtained.

RAIN—The sound of rain may be manually created in a drum with hardware cloth or window screen sides. Dried peas, small pebbles or similar articles are placed inside. When the drum is rotated the particles roll around and produce a sound similar to that of rain.

THUNDER — The standard thunder sound machine is a large piece of sheet iron. It is usually rigged with a batten on the top and bottom. This should be suspended from above; then it is possible to grasp the bottom batten and give the thunder sheet a sudden shake.

HORSES' HOOVES — Coconut halves or plumber's helpers (plungers) may be used to reproduce the sound of horses' hooves. The coconut halves can be used on different surfaces

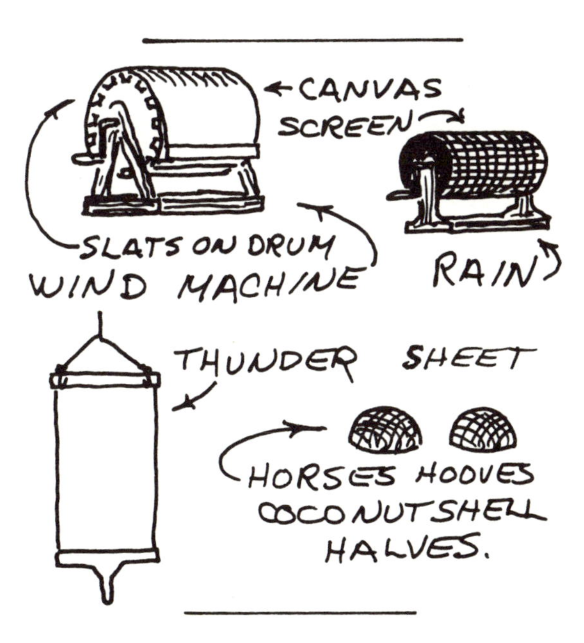

to create the sound of hooves under various conditions. The sound of hooves on soft ground is made by patting the chest with the coconut halves; a box of gravel is used for the gravel road effect and a piece of Masonite for the hard surface effect.

BREAKING GLASS — Pieces of broken glass can be placed in a box. A handful of the pieces dropped into the box create the sound of breaking glass. If a crash is to accompany the breaking sound the box holding the glass may be dropped onto the floor.

GUN SHOT—A blank gun may be used for gun shot sounds. On the stage even a .22 caliber gun may provide a loud enough sound. Shooting the gun into an empty barrel will give a reinforced sound. In an emergency such a simple thing as briskly slapping a stack of newspapers onto a table or holding a piece of board diagonally from the floor and then crashing it to the floor with the foot will provide a resounding bang!

SQUEAKY HINGE OR DOOR — Old rusty hinges may be used for squeaky sounds, but too often they lose their squeak at the crucial moment. Sometimes twisting a moistened cork stopper in a bottle will work if the sound need not be loud. New Year's eve rotating noise makers will work fairly well. Prying apart boards that have been nailed together with rusty nails may work. Effects of this nature are not always predictable and should be tried several times before being used in a production.

CANNON SHOT — The sound of a cannon shot may be simulated by striking a tympani or bass drum.

WHISTLES — Whistle sounds of varying pitch may be made with real whistles or organ pipes or they may be made by blowing across the tops of various sized bottles and jugs.

CHIMES AND GONGS — Real chimes and gongs, symphony chimes, old brake drums, sections of pipe or cymbals may be used for sound effects.

MARCHING SOLDIERS — A device to duplicate the sound of marching feet is easy to make. Short wooden pegs about three or four inches long are drilled near one end and wires strung through them. The wires with the pegs hanging on them are then mounted on a framework. By bouncing this so the pegs strike the desired type of surface the sound of many feet is created. This may be used with or without a microphone.

DOOR SOUNDS — Usually a real latch mounted on a small, heavily built door will provide a substantial door closing sound. However, there is usually a real door on the stage area that can be used for this sound.

### Electrically Operated Effects

BELLS, BUZZERS AND CHIMES—The sound effects department will havc need for various types of bells, buzzers and chimes. These may all be mounted onto a single board and operated off of a single power source. This will be convenient since most of these devices will require a 6 volt source of power and in order to get this power it will be necessary to use a transformer or batteries.

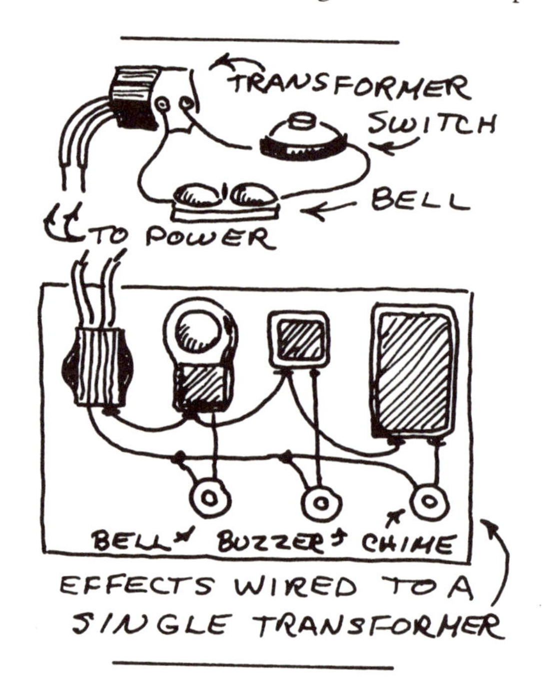


The transformer will have lead-in wires that

are to be connected to the ordinary 120 volt power source. These wires extend out from one side of the transformer. On the other side of it are 6-8 volt terminal screws. A simple series circuit is then arranged by extending a wire from one of the transformer contact screws to the bell, buzzers or whatever it is. A wire goes from the other contact on the transformer to a contact screw on the base of a push button. Another wire goes from the other contact on the base of the push button to the second contact on the bell. The circuit is then complete so the power passes from the transformer through the bell to the push button and back to the transformer.

If dry cell batteries are used as the source of power to operate the bell a similar series circuit can be wired. The power contacts in this

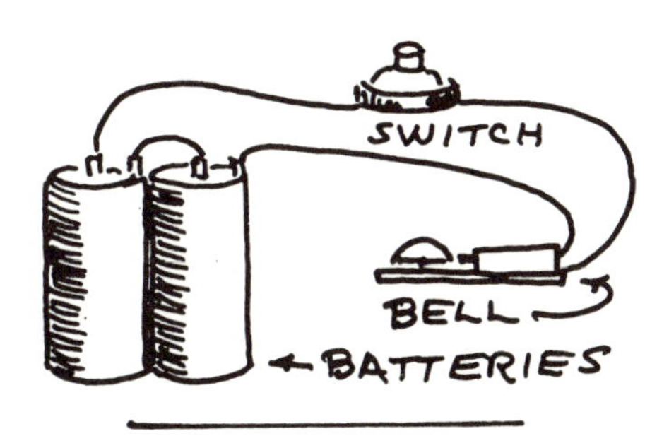


instance are the posts on top of the battery. Each battery is a 1½ volt source of power. If greater voltage is desired two or more batteries are connected in series. This is done by connecting the center contact of one battery to the side contact of the next and so on until they are all connected together.

SOUND AMPLIFYING SYSTEM—A sound amplifying system to be used with one or more backstage microphones, a record deck, and a reel to reel tape deck is an electronic device of value on a stage where there is need for background music and sound effects.

TAPE RECORDERS—A reel to reel sound recording and reproducing system should be part of the regular built-in equipment. Overture, incidental music, musical bridges, and sound effects may be recorded, edited and spliced with electronic or visible cuing tape in correct sequence to be cued in at appropriate places within the production.

SOUND EFFECTS RECORDS—Sound effects records are available from several record companies. Although the records never seem to have the exact sound desired they do contain a vast array of sounds, are extremely convenient, and are a must for every theatre sound department.

RADIOS AND RECORD PLAYERS—Scenes calling for the operation of record players and radios can be "faked" onstage and operated offstage with recorded sound. Record on tape with the other recorded sound, then cue and rehearse carefully and precisely.

MICROPHONE SOUNDS—If it is desirable to use a live microphone and produce sound effects in front of it or if it is desirable to produce the sounds for reproduction on tape a few of the more frequently desired sounds are created as follows:

*Chopping Wood*—Drive an ice pick lightly into a soft piece of wood.

*Motors*—Run a vacuum cleaner near the microphone or hold a piece of paper against the blades of an operating electric fan.

*Hail*—Drop dry rice onto glass, tin or wood.

*Rain*—Let free running salt run onto paper or glass; roll cellophane lightly in the hands.

INTERCOMMUNICATIONS SYSTEM—For the convenience of the operational crew of the theatre it is advisable to have some type of intercommunications system. The regular box type combination speaker and microphone will work all right and so will a regular powered telephone system. However, sound powered telephones perfected during the second World War are extremely practical for this type of theatre use. They require no power source except the human voice and maintenance is held to a minimum. One of the only disadvantages is that if a bell system is desired it must be wired in separately. Head sets for the men operating the show are probably more convenient to use during the performance and this minimizes the need for bells. Intercommunication between the stage manager, light control booth, pin rail and special follow spot operator may be all that will be necessary for the ordinary stage.

### Fire and Smoke Effects

FIRE—There are frequently occasions on the stage when it is essential to have fire effects. Log units can be rather readily built using pieces of real wood, papier-mache logs or screen wire and plaster logs. Sockets and small wattage lamps are mounted under them. A few pieces of colored gelatine covered over the lamps or placed in the spaces between the logs will make a fairly convincing fire. The whole affair is usually mounted on a board base so that it can be moved around more easily.

MOVING FLAME EFFECTS — In order to make the fire seem more realistic and give the appearance of moving flames, any one of a number of devices can be contrived. Here are a few of them.

1. Project motion pictures of actual flames on the surface where the flames are desired.

2. Allow flame colored beams of light to be broken by some moving object. Something as simple as hands moving through the beam of light will produce this effect. A spoked wheel placed in position can be turned to do the same work.

3. A mechanical device to break up the light pattern can be easily manufactured from an electric fan and a few strips of cloth, ribbon or paper. The strips are tied onto the face frame of an electric fan and the fan placed face up on the floor so that the air current set up by the fan will cause the ribbons to rise and flutter in front of the light source.

4. A color wheel device with irregular patterns of colored gelatine can be rotated in front of the light source to provide a changing light or flame effect. The wheel can be rigged up to an electric motor to facilitate in turning if desired.

5. The device used for fire insurance signs can be used also. This is a plastic tube with variable flames painted around its periphery. This is provided with a vented metal cap. The

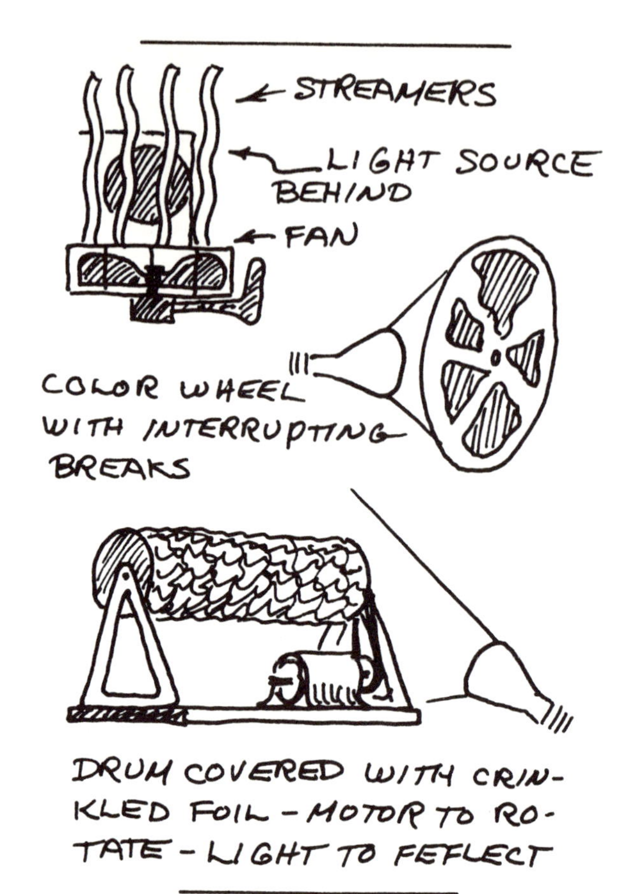


whole thing is then rigged to fit above and around a lamp. A needle shaft fastened around the lamp with a wire extends upward and supports the tube affair. The heat from the lamp moving upward sets up air currents that cause the tube to rotate slowly as the warm air passes up through the vents.

6. A rotating drum of crinkled aluminum foil with one or more lights rigged to shine upon it and be reflected from it onto the surface where the visible flames are desired is an extremely effective device. Devices of this nature are to be seen in many of the displays of fireplace equipment shops. Either the foil or source of light may be colored to produce colored flames and the light units may be rigged to be reflected wherever desired. They may be below and shine upward or above and shine downward. This type of unit, and all others,

too, for that matter, will probably be more effective if used in conjunction with a log unit.

SMOKE—A few of the methods of achieving smoke effects on the stage are the following:

1. *Sal Ammoniac*—Sal Ammoniac (or ammonium chloride which is its correct chemical name) will change from a solid to a gas form when heat is applied. Various devices can be used to furnish the heat. A hot plate, a cone shaped heating coil from an electric room heater, or even an incandescent lamp may be used as the heat source. The heater cone works beautifully since it is possible to place the powder on the inside of the cone. Sal ammoniac fumes are somewhat offensive and should not be breathed for any length of time.

2. *Dry Ice*—Dry ice (solid carbon dioxide) will change directly from a solid to a gaseous state when exposed to air. The change will take place more rapidly in water and if the water is hot the change is speeded up even more. Carbon dioxide is heavy and tends to sink rapidly to the floor. A device for getting dry ice "smoke" to rise may be easily contrived by using a sizable container with a narrow neck. If dry ice is placed in hot water within this container the rapidly escaping gas will be forced through the narrow aperture and shot into the air. (See Smoke Machine, page 160)

3. *Titanium Tetrachloride* — Titanium Tetrachloride should be used only on special occasions and where the fumes will not be breathed since they are extremely dangerous. This is very easy material to use since it changes from a liquid to gaseous state upon exposure to air. It is merely necessary to open the container and white smoke rolls out.

4. *Zinc Dust, Ammonium Nitrate and Ammonium Chloride*—Prepare a mixture of equal parts of zinc dust, ammonium chloride and ammonium nitrate. Squeeze a few drops of water onto this mixture and in a few seconds smoke will be created.

5. *Powder Blown into the Air* — A simple mechanical method of producing a smoke-like effect may be achieved by directing a puff of air onto or into a container of face powder, precipitated chalk or other lightweight material. The air used may be produced by mouth, released from a compressed air tank, emitted from an inflated tire or balloon, or blown from a vacuum cleaner or a compressor and it may be conducted to the container of dust through a small rubber hose or a piece of tubing. The powder material may be in any kind of container, including the air tube itself. A mechanism can be contrived by taking the dust bag from a cylinder type vacuum cleaner. The dust can be fed into one end and allowed to blow out through the hose at the other end.

6. *Steam* — There may be some times and places where it will be more convenient to use steam for a smoke effect than anything else. It will probably be wise to use only low pressure steam for the sake of safety and sound.

FLASH EFFECTS — Sudden flashes of light are frequently needed for spectacular effects. A number of devices can be used. Among them are the following:

1. *Flash Lamps* — The simplest and safest flash effects can be achieved by using photographers' flash lamps. These may be set off in a battery or A.C. circuit.

2. *Flash Powder* — *For safety and surety of fire use a commercial product, not homemade flash powder.* Flash powder is more spectacular than flash lamps but it is also more danger-

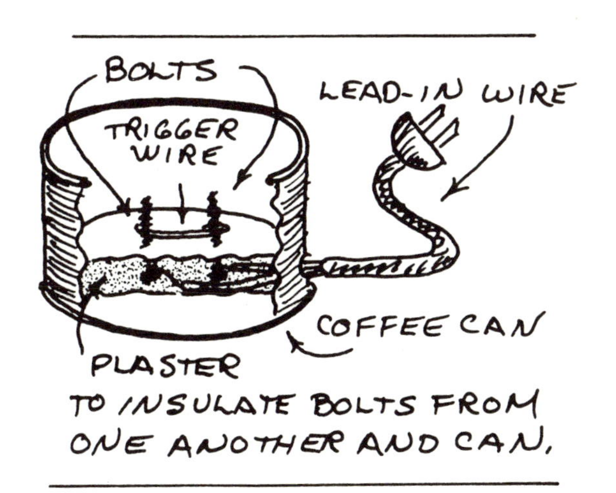

ous. The danger may be cut to a minimum if the material is handled carefully and set off in an insulated container. A "flash can" made from a one pound coffee container will do. Lead wires are passed through the side of the can and each is secured to a stove bolt about an inch long. Plaster of paris is then mixed and poured into the container. As the plaster hardens, the bolts are lifted so that they are clear of the bottom of the can and spaced 1½" to 2" apart.

To trigger the flash, stretch magnesium ribbon, 2 strands of rip cord wire, or a small wad of fine steel wool from bolt to bolt. Place ½ teaspoon of flash powder atop the trigger wire, plug the long cord from the pot to a "hot" circuit on cue. Flash!

ALTERNATE FLASHER. For an alternate flash device mount an electrical socket on a board. Cut the mica "window" from a small amp fuse (1 amp), screw fuse into socket, fill with ½ teaspoon of flash powder, cover with a thickness of tape, plug long cord from this into "hot" outlet on cue. Flash!

*WARNING—Flash devices should be handled with care. Materials should be prepared by reputable chemists or chemical firms. Be certain to have the flash pot disconnected while loading. Be certain to place the flashing device in a position where the flash will not start a fire in scenery or draperies and where it will not burn the actors.*

MINIATURE LIGHTS—Miniature lights, traditionally reserved for Christmas may be utilized in various ways. They may be strung on a tree branch to provide a handsome decor piece; they can be punched through holes in a proscenium to provide a colorful framework for a musical production; they can be suspended behind a translucent drop to serve as stars in a sky; or they can be mounted in the costume of a spectacular character (Trailing a cord, or provided with a power pack).

There are various sizes, styles, and varieties of lights. If they are mounted on a single continuous loop of wire they are series wired, and it may be that if one burns out they all go out. However these are sometimes wired to

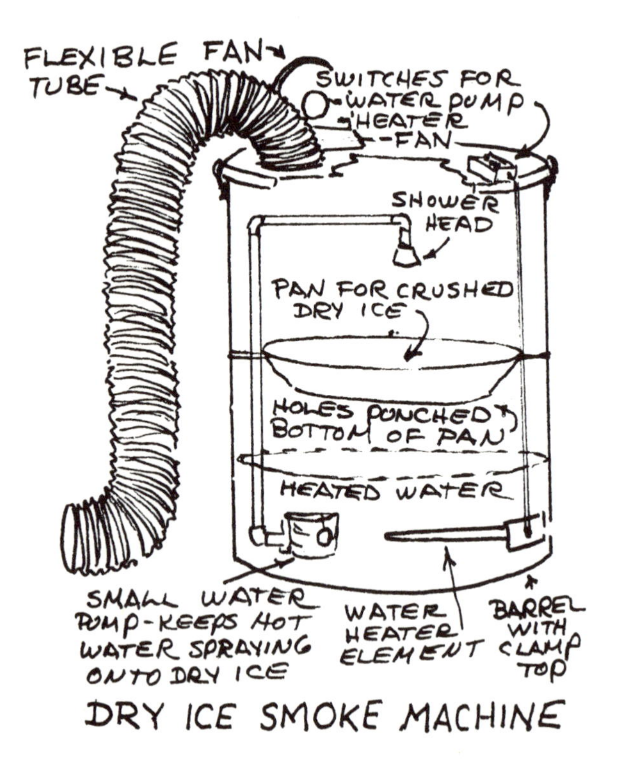

DRY ICE SMOKE MACHINE

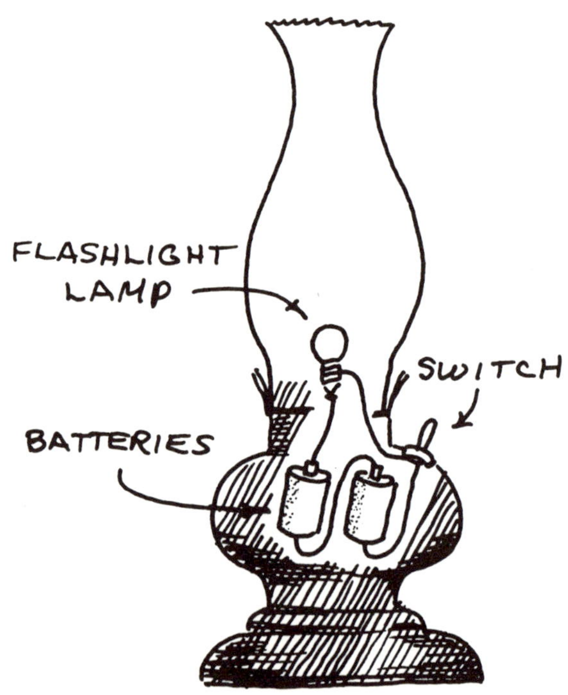

"short across" when one burns out (Check carefully). If two wires seem to go through each lamp base the wire system is parallel and one lamp burn out will not affect the others. This would be the preferred system if a choice exists.

### Electrified Lamps

Lamps, fake candles and flash bulb tripping devices are usually wired in series. They are simple to make providing there is space for batteries. Solder a piece of bell wire to the button contact of the lamp base and extend it to the base of the battery where it is soldered. Fasten a wire from the cap of the battery to a small switch or push button terminal. Then connect another wire from the other terminal of the button to the screw base of the lamp. If a socket is used the process starts with one terminal of the socket and ends with the other.

### Visual Effects

Many visual effects can be set up and operated rather easily. A few of these are as follows:

WAVES—Cut out waves moved up and down or sideways, using possibly two or three rows of waves in depth, will give an interesting effect. Painted or dyed scrim or cheesecloth stretched across the stage and manipulated from the sides so that air caught underneath causes the material to undulate produces an interesting wave-like effect.

TORCHES — The easiest and safest way to make a torch is to mount a flashlight in the end of a club and then cover the illuminated area with loose cellophane or gelatine.

Large candles can be enclosed in the torch, but the flame will be rather small and it will also probably be blown out.

Alcohol soaked cotton or "canned" heat may be placed in a container on top of or in the end of a torch. *There is danger of the alcohol dripping to the floor unless one is careful.*

Railroad Fusee torches can be used, but they are a fire hazard indoors.

BREEZE EFFECTS—A scrim or cyclorama may be shaken from one or both sides, setting up visual waves that move across the stage. A large fan can be set up on stage to blow lightweight material such as marquisette, silk, nylon or cheesecloth.

SPIRITS MOVING OBJECTS—A tiny hole can be cut through the flat and then wire or string can be threaded through the hole to move or manipulate objects. Vases can be knocked off the fireplace, pictures can be tilted, objects can be moved on a table—all by manipulating the wire or thread. If the object is to be lifted, fine piano wire or, if it is not too heavy, fish line can be used. Even a fishing pole—especially a long fly pole—can be used if there is a place for the operator to stand so that he can manipulate the pole out of sight lines.

PICTURES KNOCKED OFF THE WALL—Pictures may be suspended with thread that extends through the flat and up to a batten. The thread can be broken or released, dropping the picture.

LIGHTNING—Lighning may be simulated by setting off photo flash bulbs, striking a carbon arc or flashing a series of small wattage lamps (large wattage lamps take too long to heat and cool). A series of four 26 volt sealed beam spotlight lamps or signal lights will make beautiful lightning.

PRACTICAL RAIN — Rain effects may be achieved with an overhead pipe, pierced for water to come through, attached to either a hose leading to a tap or to a hose leading to an elevated container of water. A trough should be placed underneath the pipe to catch the water. This works well when the rain area is restricted, but utilizes much water and is messy if the rain area is wide. Designing the rainfall so the audience sees it through a window, across a doorway or across an arched opening is a way to restrict the rain area. A rain effects machine, such as the effects machine referred to in the section on electrical equipment, may work under the correct conditions. Actually, a darkened sky and the sound of rain will project the idea quite well. Audiences do have collective imaginations! It is also possible to "slosh" water over an actor just before he makes an entrance to show that he had sense enough to come in out of the rain.

SNOW—Devise an overhead slotted or punctured cradle of canvas or sheet metal that can be rocked. A sheet of canvas with battens top and bottom and suspended by two sets of lines to form a hammock will work. By juggling either set of lines the snow material used can be controlled so that it will be shaken out of the affair like salt out of a shaker. The snow

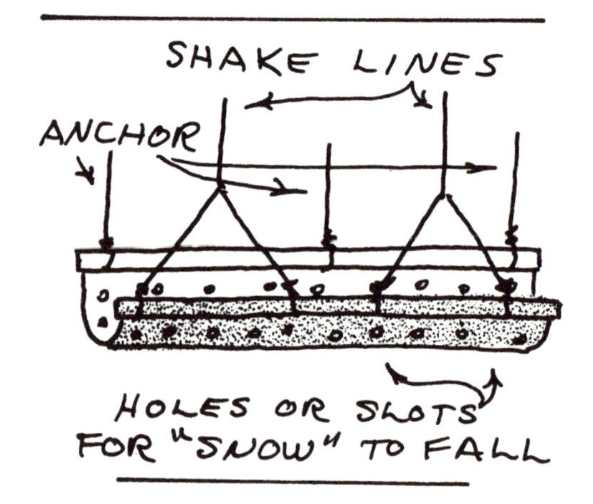


material may be mica flakes, corn flakes (unbaked if possible), Christmas tree snow, paper confetti, Lux Flakes or Styrofoam flakes.

LARGE MOVING OBJECTS—Cut outs can be pulled, pushed or dragged across the stage. Sometimes crew members can be inside or behind such objects.

COLLAPSING COLUMNS—Columns or similar objects can be built in sections and the sections can be bolted together at an overlapping point. The column can be suspended from a line or wire fastened to the top; the bottom of the column can be anchored to the floor. On cue the suspending line can be released or eased down, allowing the sections of the column to collapse. Flat forms will work very well this way. The tumbling columns of the opera *Samson and Delilah* can be made in this way.

"POP UP" OBJECTS—Objects that need to pop up or pop out on cue can have springs

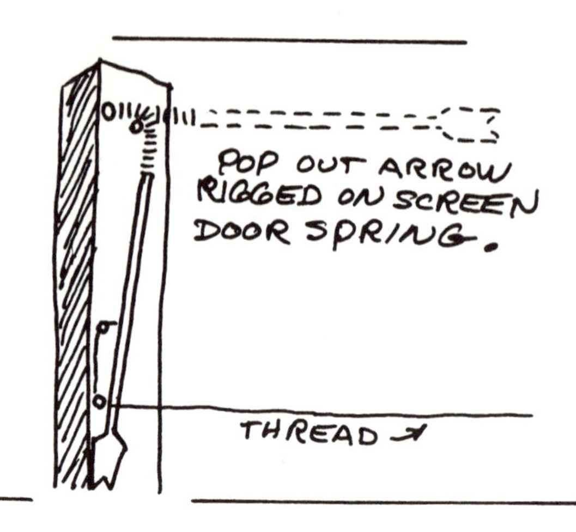


(screen door springs, for instance) rigged onto or within them. These objects can be "cocked" and held with thread or string. Then on cue the thread is broken or cut—and "boing"! The object springs up.

RUNNING WATER—A sink with taps is often needed on a stage setting. When this is desired it may be possible to make an artificial sink although a real one will probably prove more satisfactory in the long run. A hose may be run from the nearest backstage water source to the sink in the stage setting. If this method is used enough water should be run through the drain of the sink and into a bucket or tub below, or through another hose running off to the nearest drain, so that the air will all be removed from the hose. The tap at the source may be left on or the hose may be filled with water and the end raised up into the flies so that the hose itself becomes the water chamber. Another method may prove more satisfactory. A water container is hoisted up on a small platform or on a ladder behind the setting and a hose or tube is run from there to the tap above the sink.

PUMP—For a pump it is usually necessary to have a relatively flat tank of water under the pump platform, which is perfectly logical since that is the way most pumps are set up on farms. The pipe from the pump reaches down into the water in the tank. The pump will draw water out as long as the water level is

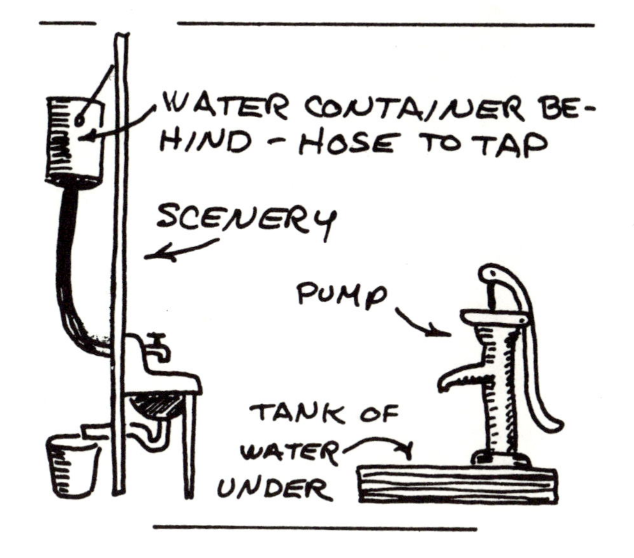


above the bottom of the pipe extending down into the tank.

FOUNTAINS — Small electric water pumps are available to pump water from a tank, spray it in the air and allow it to run back into the into the tank.

WATERPROOFING — When water effects such as those suggested above are used, the areas that the water will come in contact with should be coated with spar varnish or waterproof sealer so that all of the scene paint will not be washed off the surface.

* * * *

The effects listed above are the more common ones. A little ingenuity and imagination will help in contriving ways of solving other problems of this nature that may arise.

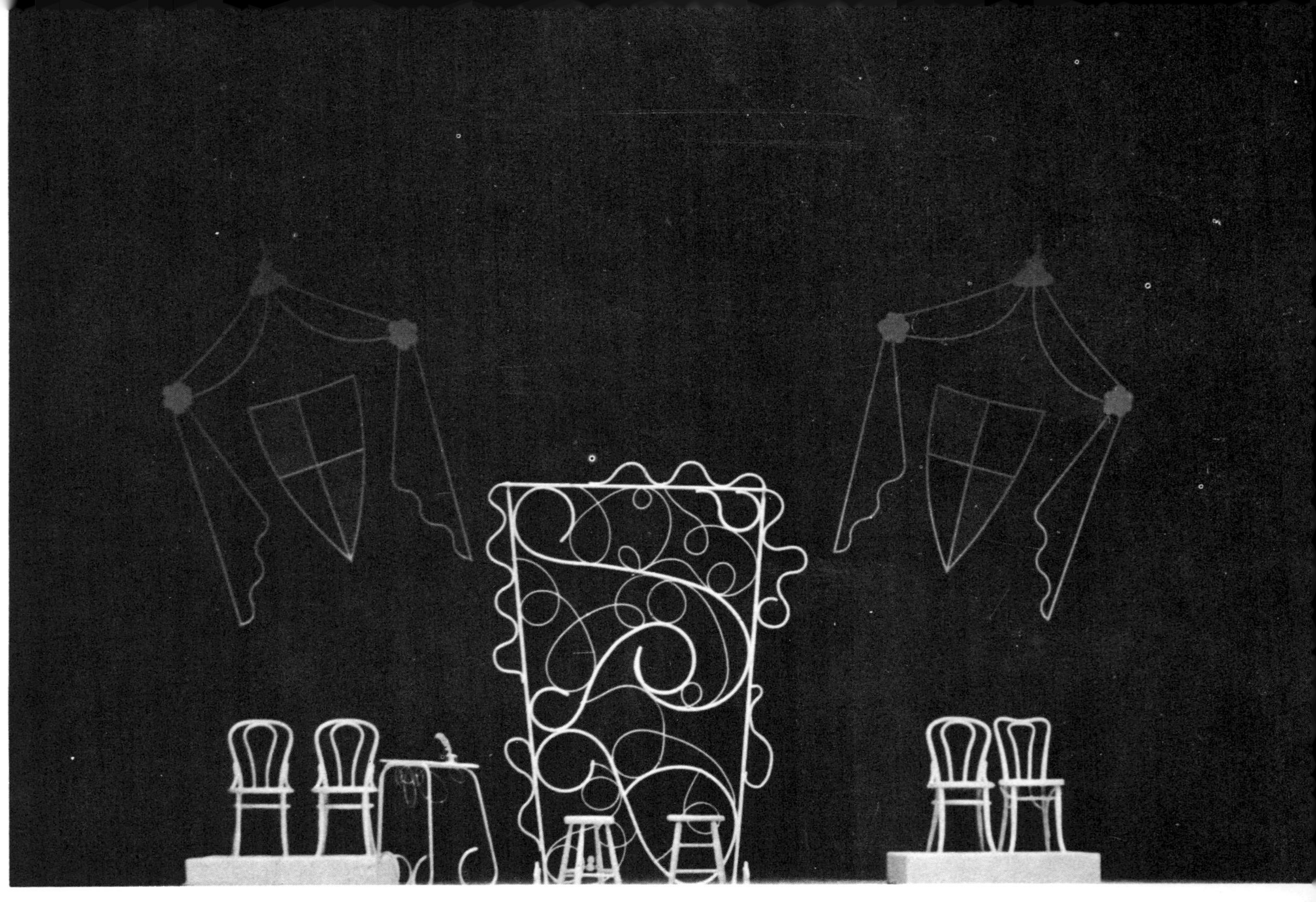

*The Marriage of Figaro.* Simplicity of setting achieved through the use of chairs, stools, aluminum tubing and wire.

# 11

# *Furniture*

The furniture and decoration of the stage setting frequently make the difference between a pedestrian and a finished production. Although it is not always possible to get exactly the correct chair or table for a period play it is helpful to know the salient features of the historical era so that an approximation can be made. Frequently it is possible to add color, surface decoration or slip covers to alter a piece of furniture and move it closer to the correct design stylewise.

The purpose of this chapter is to provide brief basic information rather than definitive historical facts about furniture. The text and illustrations are purposely brief and pointed. Included are basic facts about furniture and decorative motif that may be helpful to the stage decorator.

### Egyptian 1500 B.C.

The Egyptians used numerous pieces of furniture including chests, sarcophagi, tables, stands, stools, chairs and folding seats.

The folding seat, which persists down to the modern day, probably had its origin in Egypt. This seat had X-crossed legs and a leather seat. The Egyptians had low, rush covered seats, chairs with backs and throne chairs. The tables were plain but attractive. The legs on furniture frequently terminated in animal feet such as bull hooves and lion paws. Chairs were often painted white, while sarcophagi and chests were painted in brilliant colors and had bands of geometric designs. Stylized animal

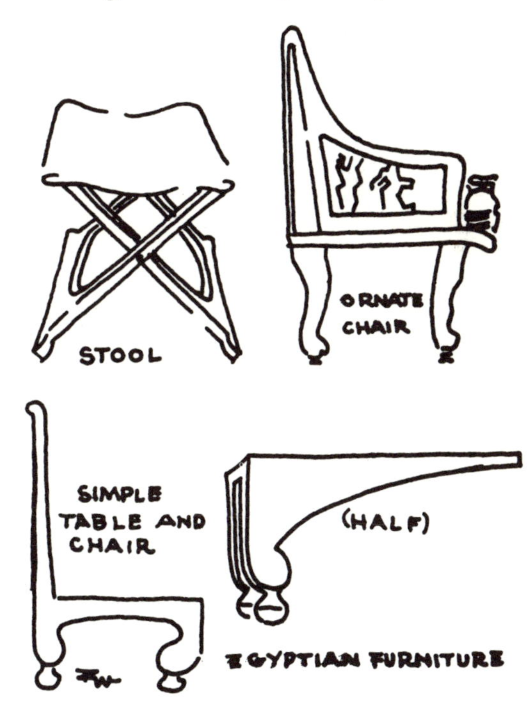

forms were used frequently in the design motif.

Wood was scarce so was frequently featured in the design. Various types of surface design such as inlaying and veneering were used. Inlays of mother-of-pearl, ivory, metal and semi-precious stones were used.

Color had symbolic meaning to the Egyptians and so was used rather carefully. Plant forms were used stylistically in design motif. These included forms of the lotus, papyrus, acanthus, palm, lily and reed. Geometric forms including the spiral, triangle, square and circle were used on the furniture and in wall decorations.

The throne, a prominent piece of furniture, may be seen in many of the Egyptian drawings. The thrones were wide and had low square backs and arms that curved down from the back on a diagonal to the front edge of the seat where they were held by straight posts. The sides of the thrones sometimes represented the entire figure of a striding lion.

Although some beds were merely piles of quilts on a simple frame others appear to be chairs with stretched out seats.

Furniture of the later Egyptian eras had cushioning and even some upholstery.

### Greek 12-300-B.C.

There is great variation between the furniture of the earlier and later periods in the Greek civilization. The early period shows influences from Egypt and other countries of the time. As the Greeks began to excel in architecture they also developed graceful and practical furniture. Well developed beds, chairs and tables appear in vase paintings. The chairs had sweeping curves on the backs and legs. The couches were elongated chairs similar to those of the Egyptians. The tables were low, portable and at times had bronze animal legs and feet. The chests had a pronounced architectural roof shaped top.

The Greeks used olive, cedar, yew and boxwood. The surfaces of the furniture were decorated with inlaying, painting, carving, gilding and encrustings of precious stones. Purple, scarlet and gold seem to have been popular colors and some of the pieces were upholstered with silken cushions.

The decorative motifs used included the egg and dart, bead and reel, honeysuckle, braid,

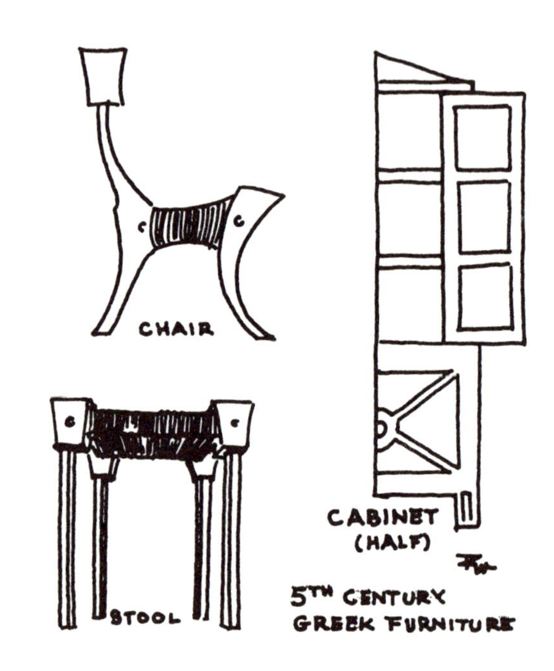


acanthus leaf, leaf and dart, guilloche and dentils.

The proportions and refinement of the Greek furniture influenced artisans of the eras that followed.

### Roman 1000 B.C.-500 A.D.

Rome borrowed heavily from the Greeks but the borrowings never had the gracefulness or beauty of the originals. Roman furniture made of wood or stone was rather heavy and ponderous.

There seemed to be several types of chairs, among which were: the Curule with a square seat and X-shaped legs, the Biseleum which was a double chair or settee made of turned wood and sometimes decorated with carved horses' heads, the Solium which was a throne like chair with a back and the Cathedra chair which was used solely by women. The legs on most of the chairs were turned.

The Roman beds and couches were similar to modern ones. The couch was an important piece of furniture since it was used for sleeping and semi-recumbent dining. For dining purposes it had arm rest cushions. This couch was composed of a platform supported by legs and piled with cushions.

There were chests for the storage of arms.

Many types of tables, tripods, pedestals and stools were used.

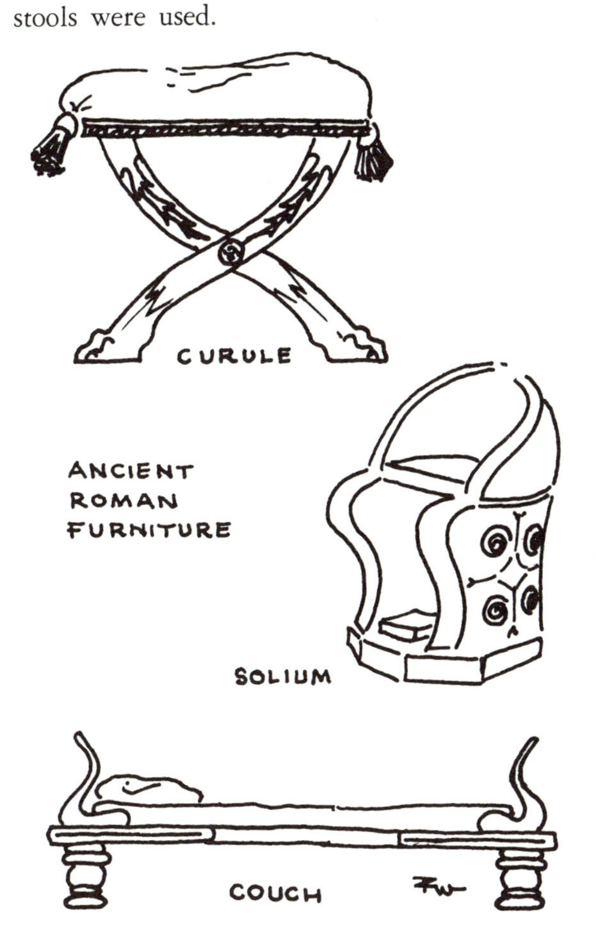


Painting, engraving, carving, inlaying, metal applique and varnishing were all used as finishes on the furniture.

## Gothic 1160-1530

The strong influence of the Gothic period was the Christian church. This was the era in which the great cathedrals were built. It was also the era of Byzantine showiness and of the barbaric Teutonic people. Furniture is usually heavy with carving that reflects the

structure of the Gothic arch and of various religious motifs. Chests were important items. They were portable and used to carry the family valuables. Chairs were often made from chests. They were simply chests with arms. There were a few portable chairs. These frequently had high backs and sometimes canopies. The beds were at first entirely enclosed and then later merely covered by canopies to provide warmth. The fireplace was moved from the center of the room to the side wall. Tables consisted of trestles and planks.

Architectural forms, including skeletal forms of Gothic buildings, linen fold tracery and finally painted designs were used on the furniture. The idea of skeleton framework for buildings was finally carried over into the furniture and the chests became lighter in weight.

With the skeleton construction of furniture came more ornamentation. The carving in oak was usually large in scale and made use of floral

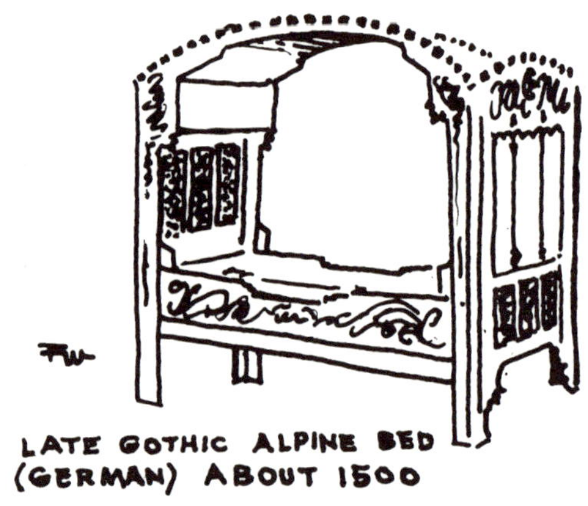


forms, vines and leaves, grotesque animal and human forms. Cusped arches, trefoil, quatrefoil, ogee curves and deep full moldings were freely used.

The greatest craftsmen built for their God and church. They built altars, screens, chests and great church doors.

### Italian 1100-1400

Although the Italian was similar to other Gothic furniture there were influences from the Near and Far East and Africa.

Heavy chests of planks and boards were used. The decoration varied with the provinces. Gothic details, including the pointed arch, pierced tracery carving and landscape patterns,

were painted, sometimes raised with gesso, sometimes inlaid with mosaic or marble and sometimes there were geometric patterns of bone or ivory in Moorish style.

Enclosed beds were unnecessary because of warmth, so simple frames were covered with Oriental fabrics and rugs.

### Italian Quattrocento 1400-1500

The Renaissance began in Italy. The growing wealth of families and cities, trade with the world of the 15th century and development of great art all had their influence on the making of furniture.

Chests and cupboards set on bases were decorated architecturally with pilasters and cornices. A form of settee appeared in the Cassapanca, a chest with sides and back. The credenza developed in Italy was a low sideboard with doors and drawers.

Chairs were large and rectangular with arms at right angles to the back posts. Some seats were padded. The X-chair was evident in variations, some showing the Moorish influence. The Dante chair, with a fabric seat and back, had four curved legs continuing into arms. Sgabelli wooden side chairs either had three legs dowelled into the seat with a flat board back or had two carved slab bases.

The trestle type table was further developed with baluster legs or shaped slabs. Four legged types with box stretchers appeared early.

The ornament was classical with pilasters, flutes, scrolled volutes, moldings enriched with

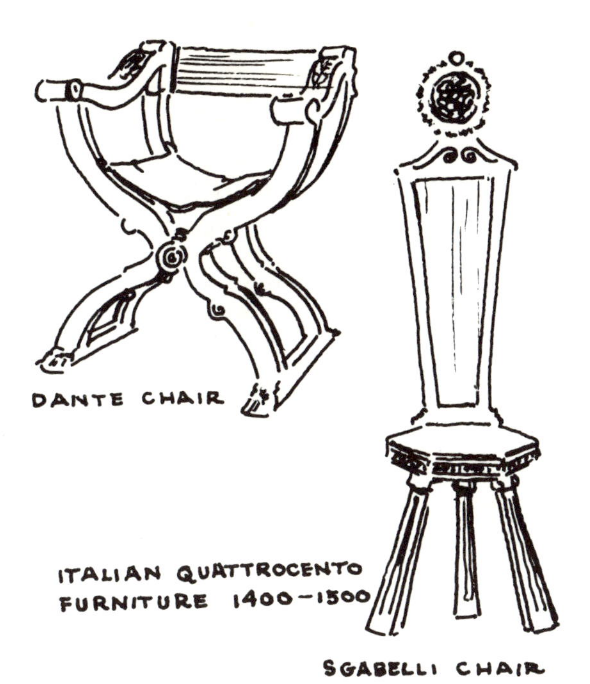


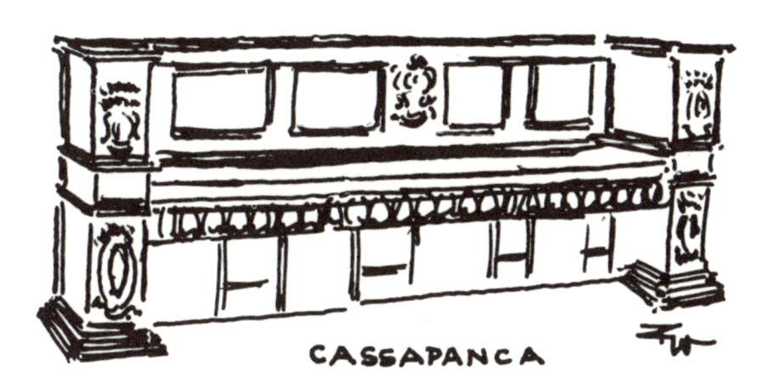

egg and dart and dentils. Decorated flat areas were painted or gilded.

### Italian Cinquecento 1500-1600

Further development in terms of classical motif appeared in the 16th century. Architectural cornices, guilloches, the acanthus leaf, rinceaux, pilasters, molded panels, gargoyles, scrolls and volute animal forms, caryatids, imbrications, gadroonings were used along with the newer forms like cartouches, strapwork, turned rosettes and broken pediments.

More chairs with cushions, more chests, more beds in four poster frames and more sideboards in many shapes appeared during this period.

### Italian Baroque 1660-1700

The Baroque in Italy is further development on a lavish scale of the decoration already begun during the earlier periods. The Baroque in general was a masculine, highly decorative style of furniture. The wealth and power of the owner was shown in his furniture. The designers employed broken pediments, deep moldings, huge scrolls and profuse sculpturing.

Furniture of this period was designed in terms of entire rooms so that individual pieces were designed to be units in an architectural whole.

Wall furniture was in evidence everywhere. Included were tall cabinets, console tables and wall seats. Dominating the room was one huge cabinet with a sculptured base and decorated with cherubim, mermaids, lions, eagles and negroes in combination with scrolls, shells and leaves. The top had great broken pediments gracing it while the center doors were richly carved with small veneered panels.

Table legs were elaborately carved and table tops were marble or painted imitations of marble.

Chairs were heavily carved and upholstered with rich silks and velvets. Nail heads were arranged in patterns.

The beds of the early period were light and graceful four posters. Later these, too, became

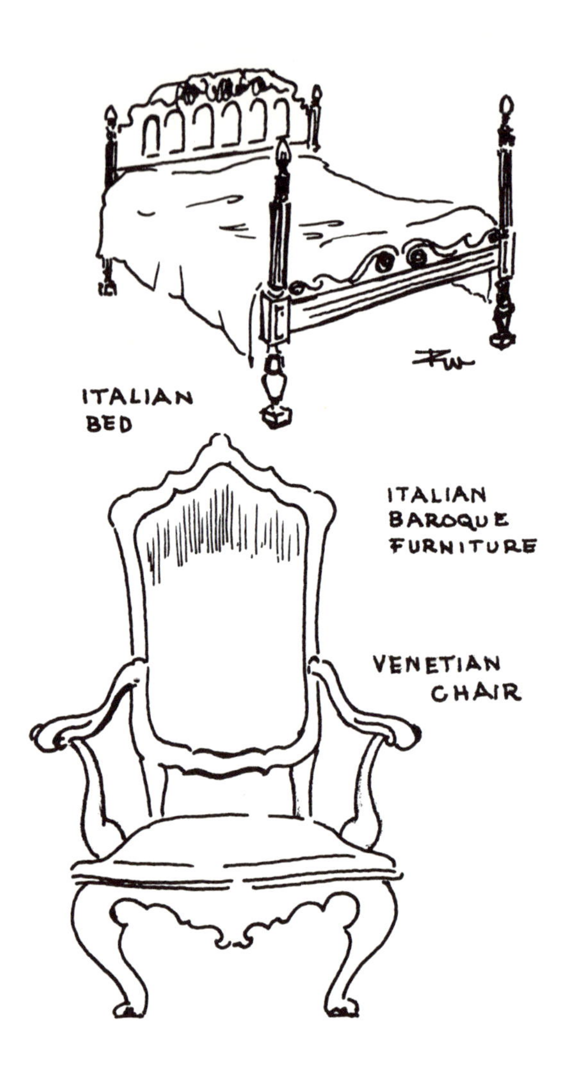

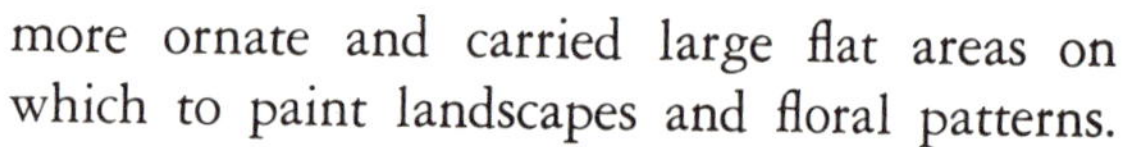

more ornate and carried large flat areas on which to paint landscapes and floral patterns.

### Italian Rococo 1700-1750

The furniture of the Rococo era was smaller and prettier than that of the Baroque period. Asymmetry and the curved line were the rule in Rococo design. The scale was smaller and more feminine. Decoration was lavish with the motif now toward foliage, ribbon, rocks and shell. Rococo did not reach the heights in Italy that it did in France.

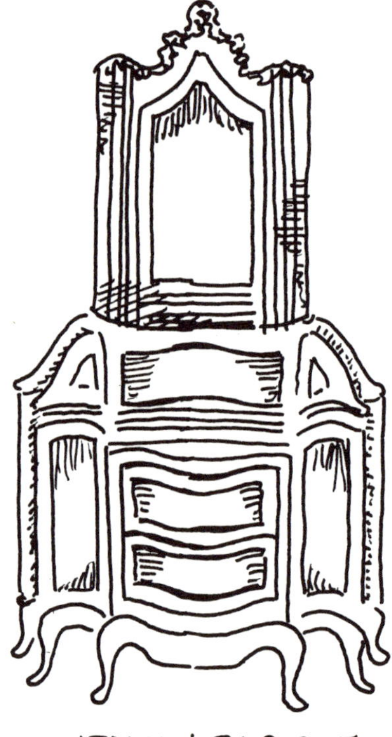


### Spanish Mudejar 1250-1600

Very little furniture was required by the Moors. Cushions were used for seating and tables were low. The Moorish influence in construction and decoration continued after Spain had once more become a Christian country. Moorish inlaying with ivory, bone, mother-of-pearl, metal and woods remained. The Arabesque designs came from this influence. This consisted of intricate and elaborate geometric interlacings. Ornamental bands using brilliant colors were popular design motif.

### Spanish Plateresque 1500-1556

Plateresque refers to the art of the silversmiths whose work became the dominant metal work of the period. Most of the wood craft was rather crude and joinery was poor. Inferior workmanship was frequently covered with polychrome painting. Turned work was flat or repetitious. The Portuguese influence showed in the wrought iron work.

The tables were trestle forms with turned or cut out members. The ties were of wrought iron. The tops were cleated and square cut.

A painted ladderback chair with rush seat appeared. Most of the chairs were rectangular in form with upholstery stretched across. The X-type chair was common, either in the heavy Italian or the light Moorish type.

Beds with iron posts and panels of decorative iron, turned spindles, arches, etc. were evident.

The Vargueno, a fall front desk mounted on a table support, was the major contribution of the period. The base had turned posts and the cabinet was fitted with many small drawers. The door had decorative lines and hasps of iron.

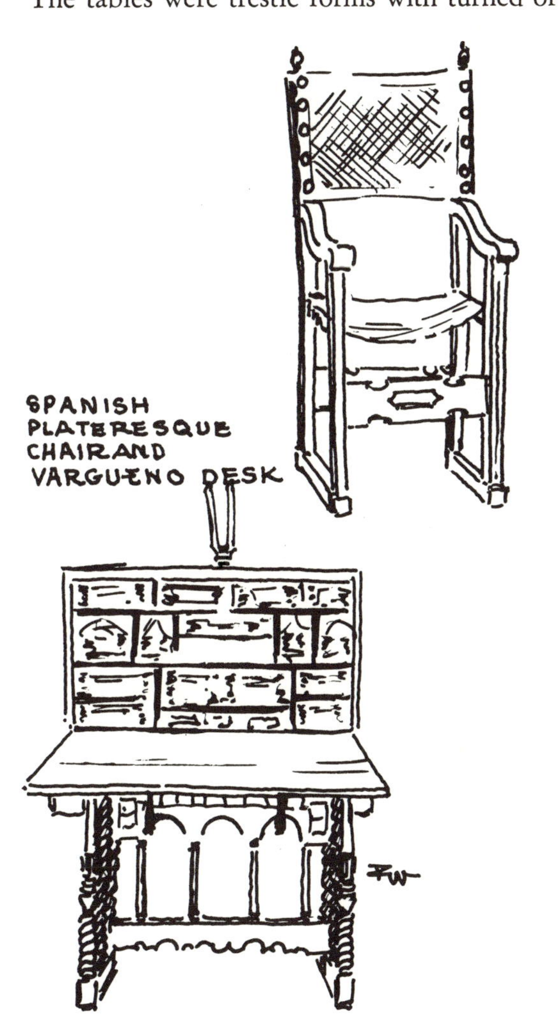

## Spanish Baroque-Rococo 1600-1700

The Baroque and Rococo influences from France and Italy mixed with the Moorish influence to produce furniture that never reached the decorative extremes of these movements in the other countries.

## French Renaissance 1515-1616

The early French Renaissance shows the influence of Italian, Spanish, German and Flemish artisans. Walnut became the most-used wood. Everything was covered with surface carving. The Gothic vegetable carving changed to the olive, laurel, acanthus and endive. Cabinets, tables, chairs and beds were the important elements of furniture during the early period.

Cabinets had irregular, jagged outlines and Baroque decorations. Pilasters were used often as motifs in conjunction with circle or lozenge panels. The diamond shape, stars and other geometric patterns were used as decoration on doors.

The table became an architectural form. The base was highly carved and began to show a strong Italian influence. Smaller tables also began to appear.

The bed with posts, canopy and draperies became a monumental affair with heavy posts and elaborate carving.

Chairs were scaled down and became more comfortable.

## Louis XIII 1610-1643 Baroque

The exaggerated Jesuit architecture of Italy was the basis for the French Baroque furniture. The furniture of this period had a masculine quality with straight lines predominating. Panels had curved transitions in the corners and inserted circle or oval curved tops. The carving was rich and abundant. Animal forms of nature, mythology and allegory were employed. Masks, lions' paws, satyrs, sphinxes, dolphins and griffins were used either in entirety or just their heads or feet. Various types of acanthus, water lilies, oak, laurel and olive leaves as well

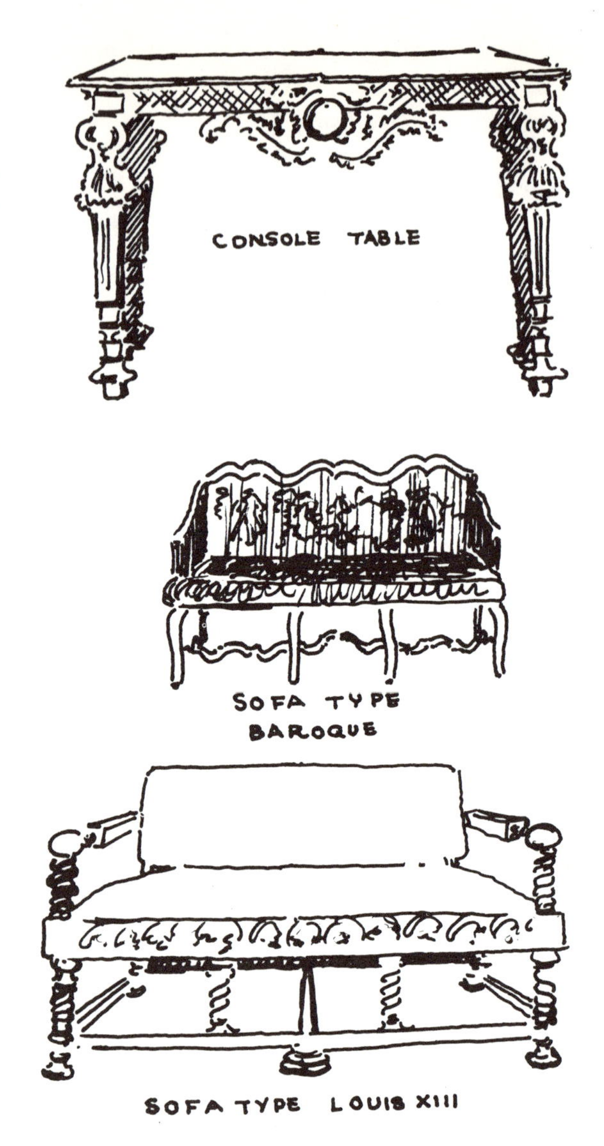


FRENCH
LOUIS XIII AND BAROQUE
SOFA TYPES AND TABLES

as ribbons, festoons, swags, agricultural implements, weapons and musical instruments were carved in the surface design.

Painting with strong colors such as red and green, inlaying of tortoise shell, ivory, bone, pewter, brass, mother-of-pearl and tin in complex detail was to be found in all kinds of wood. Real gold and silver were used where possible; otherwise, silver and gilding were substituted.

Bookcase cupboards and drawer cupboards with legs curved and ending in a doe's foot (destined to become the cabriole leg later) developed during this period.

Beds became more complex. Some canopies were detached from the frame and fastened to the ceiling. Partial canopies extended out above some beds. The bedstead, detached from the canopy, was also handsomely treated.

Heavy stone topped console tables with elaborate bases were immovable objects and were permanently installed against one wall. The wall side was undecorated. Table legs were turned or flattened balusters. Important tables were gilded. Smaller tables were natural or painted wood. There were coffee tables, candelabra tables, writing stands, toilet tables, night tables and all kinds of specialized game tables.

Seats were designed for protocol purposes.

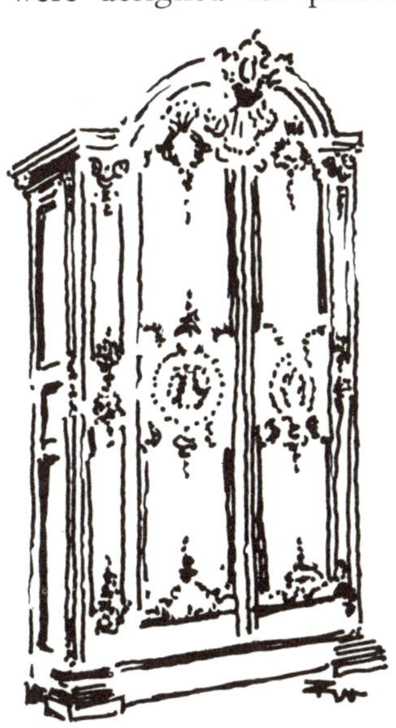

FRENCH REGENCE

In importance these ranged as follows: arm chairs, chairs with backs, joint stools, folding stools, hassocks with gold gimp and hassocks with silk edging. King Louis XIV's throne chair was, of course, the most elaborate in the kingdom. It was solid silver and was draped with crimson velvet. The back was 8′ high, draped with full gold embroidery carried by caryatids 15′ high.

Arm chairs always had stretchers, first H shaped and later X shaped. The legs were scrolled, flat or turned balusters. The arms were well molded and swung into the back with great curves. Carving in chairs was popular.

The Confessional was the first fully upholstered arm chair.

The sofa was the most important invention of the period. Almost bed in shape it was upholstered.

### French Regency 1715-1723

The Regency period is the short transitional span between the masculine Baroque of Louis XIV and the feminine Rococo of Louis XV. Straight lines became curved, ornament became freer, scale diminished and the Chinese influence as well as the rock and shell came in. The cabriole leg became the characteristic shape of legs.

### Louis XV Rococo 1700-1760

The Rococo of Louis XV really began to develop before he came to the throne. This feminine style furniture seemed to be constantly avoiding straight lines. It was florid and devoid of symmetry and right angles. One line always seemed to flow into another without any joint showing.

Ornaments were many and varied, including shells, flowers, musical instruments, pastoral objects like shepherd's crooks, and baskets.

Painting and lacquering of furniture became popular. Lacquering processes of great superiority were developed. The brightest reds, yellows and greens were used with black and these were emphasized with fillets of gold. Gilding was popular for console tables, mirrors, chairs and small tables.

Metal appliques were used functionally and decoratively. Carefully designed pieces were

used for locks, handles, escutcheon plates, keys, feet or fillets and covers for weak veneer edges. Many types of wood were used for marquetry, including rosewood, satinwood, amaranth, tulipwood, mahogany, cherry and plum. Marble of many colors, onyx and alabaster were used for tops of tables large and small. Imitation stone, mirrors and small china plaques were inserted in table tops.

The great variety of tables used in the Louis XIV period was increased considerably.

Chairs and chaise lounges were used in great numbers. They were designed to fit the human form and were fitted with loose cushions. Chair backs and seats were low. Armless chairs were designed for the voluminous costumes women wore.

Bedrooms were smaller and more efficiently heated than formerly. The canopy and draperies supported on four posts were retained but now became decorative. The Angel bed had similar head and foot boards and a half canopy that hung from the ceiling with two loop back draperies at the head. The Polish bed had the head and foot board and a curved dome canopy carried on four posts with four curtains looped up at the corners.

### Louis XVI 1774-1793

Renewed interest in the classics combined with weariness of the endless curves and elaborateness of the Rococo caused a movement toward simple forms and straight lines. Legs once more became vertical members jointed at right angles to the aprons. The vertical was emphasized with fluting and grooving and the capital and base were suggested in the moldings and feet. Curves were segments of circles or ellipses. Panels were flat and moldings simple. The ornament of the classics was employed, including laurel, acanthus, egg and dart, oak leaves, palm leaves, fret work, riceaux, ribbons, etc. Fluting was frequently partially filled in with the filling ending in small vase turnings or torch effects. Greco-Roman influence was seen in bound arrows, swans, urns, wreaths, festoons and fanciful animals.

Black and gold lacquer as well as gray and subdued grayish white and gray green paint were used. Geometric marquetry, in the forms of the diamond and lozenge, were employed in rosewood, tulipwood. The dominant wood was mahogany.

Most of the pieces of furniture used in previous periods were retained but the lines changed to conform to the new conservative classicism. Sofas with roll backs, Roman tripods for tables, smaller beds and the Angel bed were still popular.

BED WITH WALL MOUNTED CANOPY

SOFA TYPE

FURNITURE OF LOUISXVI TYPE

### Directoire 1795-1799 and Consulate 1799-1804

For the most part furniture was changed but little except that revolutionary symbols were added. These included the Phrygian cap, ar-

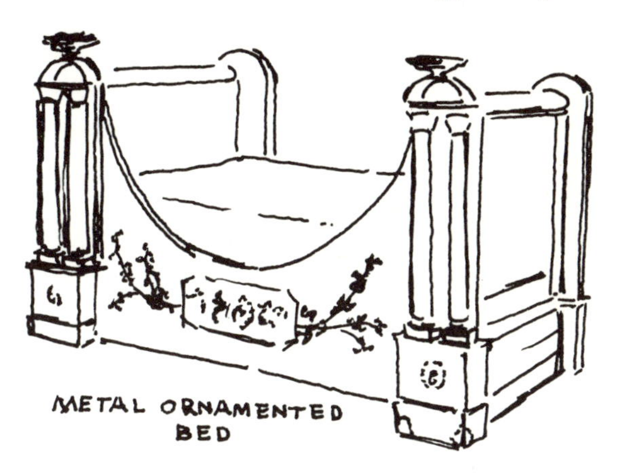


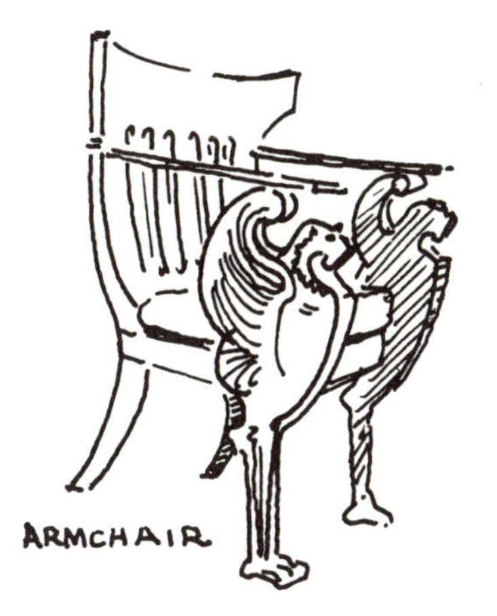


FRENCH EMPIRE FURNITURE

rows, pikes, triangles, wreaths, clasped hands, the fasces and lictor of Rome.

### Empire 1804-1815

Under Napoleon, furniture became classical. However, the specific needs of the day were ignored. Consequently much of the furniture was clumsy and uncomfortable.

Absolute symmetry, cubic, rectangular or geometric shapes and heavy solid proportions characterized all furniture. Large flat surfaces emphasized the quality of the polished wood. Bronze or flat gilt appliques were tacked on to provide ornamentation. Military symbols such as swords, shields, arrows, wreaths and winged figures were used plus a few of Napoleon's own inventions such as the letter "N", the bee and the Cornucopia. Carving was avoided except on the arms and posts of chairs and table legs. These were sometimes carved into lions, griffins or caryatides.

Fabrics rich in color (mostly reds, greens, yellows and deep browns) all in hard textures with large imperial patterns or diaper patterns with the usual stars were used. Mahogany was the wood most used. Rosewood, ebony and woods stained to simulate them were used often, also. Marble was classic and hence acceptable.

Tripod or pedestal tables with marble tops are typical of the era. Cabinet furniture seems to have been designed as if it were miniature architectural units. Desks with banks of drawers in temple facade form or large cabinets with flat fall fronts were used. The typical Empire bed was the boat style with richly scrolled ends of the same height.

### Provincial French

The furniture in the provinces outside Paris was more honestly designed for family life, and its restraint of proportion and style gave it distinct charm.

Furniture of the 17th century was composed of the bare essentials, including a closed bed, straw bottomed stools and a cupboard or hutch.

The buffet and cupboard and the chest were

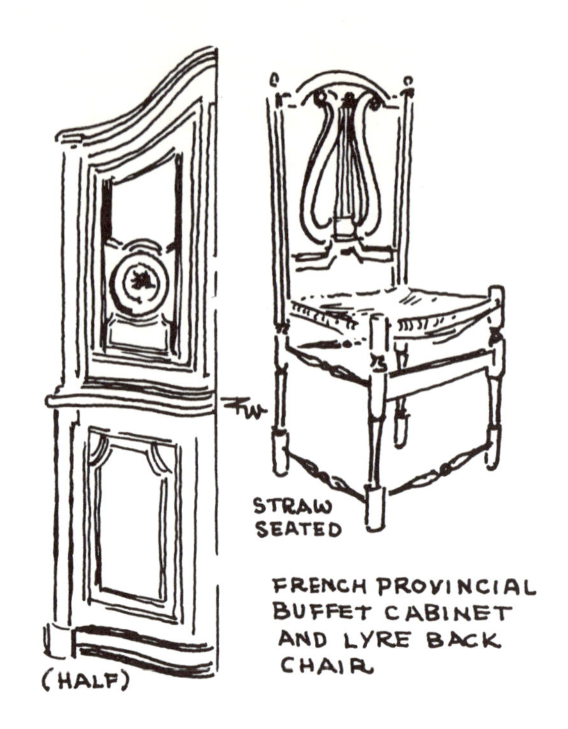


forms used everywhere. Beds depended upon the locale and climate. In the south they were open while in the north they were cabinets within a room (or a room within a room).

### German Gothic

North German furniture showed the Scandinavian influence and south German furniture showed the Italian influence. The chest was slightly different in Germany, having post and panel construction with the posts extending down to provide short legs. Coffers and cupboards with carved ornament painted red and green were found in south Germany. Turned chairs, trestle tables and beds with square posts, side pieces and a short wooden canopy were developed early.

### German Renaissance App. 1575

The Italian and classic influences were used during the German Renaissance. Pilasters appeared. Veneered panels of walnut and ash were to be found on a base of fir or pine. Chairs were four legged board types, folding chairs

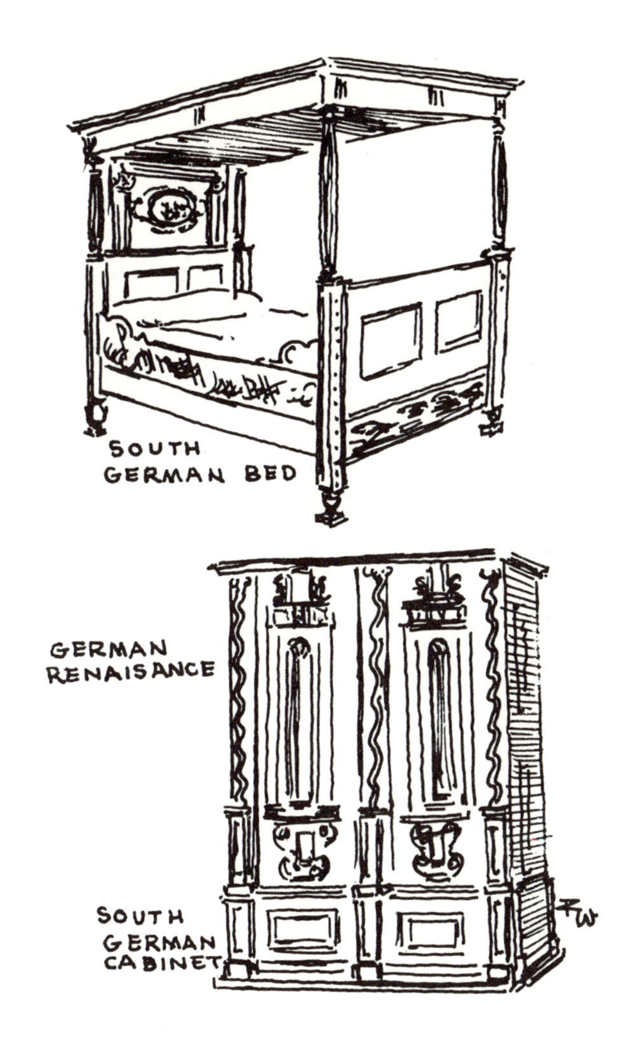


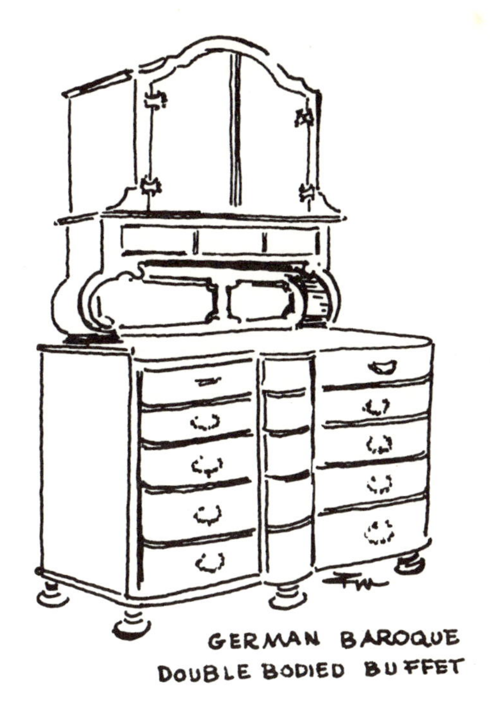


and later arm chairs with square legs. Turned legs appeared later.

Beds with carved posts and canopy frames appeared and state chairs were inlaid with ivory and silver. Cabinets were decorated with elaborate carvings and architectural features.

### German Baroque App. 1660

Heavy cornices, rich molded panelling, large bun feet with carved enclosed leaf, flower and fruit ornaments characterized the great walnut cupboards.

The cabriole leg gradually replaced the spiral turned leg and styles were borrowed directly from Paris. Bombe commodes and high chair backs with smooth wooden splats were in abundance.

### German Rococo 1730

The French influence became more pronounced and with it came consoles, mirrors, commodes, chairs decorated with fruit, flowers, garden tools, musical instruments. These decorations were carved, gilded and painted.

### German Classicism 1770

The classic lines of German furniture were more florid than in France. Much of the furniture was light and graceful but devoid of excessive ornament.

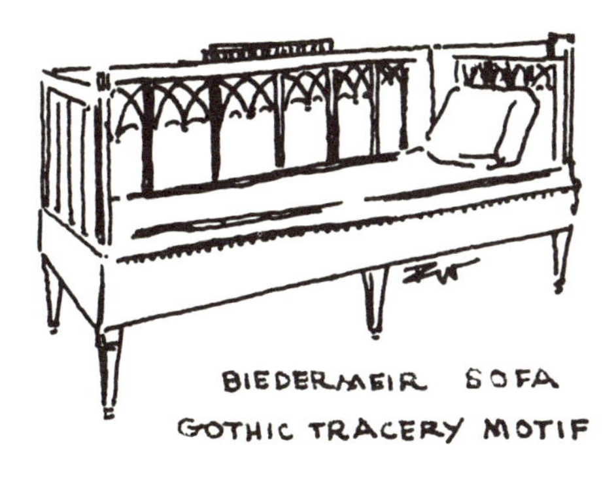

## German Biedermeier 1830

The Biedermeier is lightly based on the classic lines with a mixture of the Gothic. It is middle class and comfortable. Curved chairs and sofas were upholstered with horse hair, calico and rep, and were decorated with graceful ornaments of the swan, griffin, cornucopia, domestic flower and fruit variety.

## Scandinavian

Evidence of Romanesque, Celtic and even Far East influences are to be found mixed with the Viking system of intricately interlaced ornaments using birds, beasts and vines. Painting

and carving were used on flat surfaces. The furniture had a simple native feeling that was never completely lost in any of the other styles.

## English Renaissance (Elizabethan) 1558-1603

The Gothic style was well established in England and the Italian influence that came with Italian architects manifested itself only in details of ornamentation. Romayne work, scrolls and dolphins were added to the Tudor rose, palmetted band and zigzag that were already in use. The furniture was massive and large scaled. It was structurally simple with well braced right angle joints. The huge bulbous melon turnings appeared in almost all up-

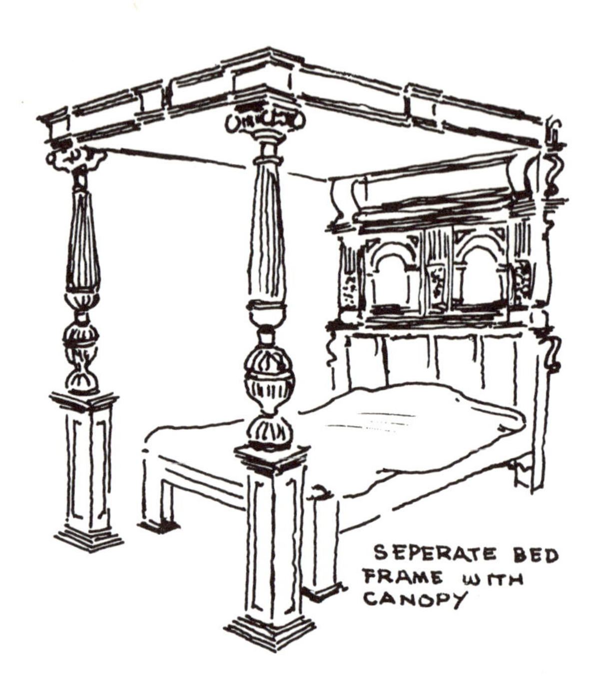


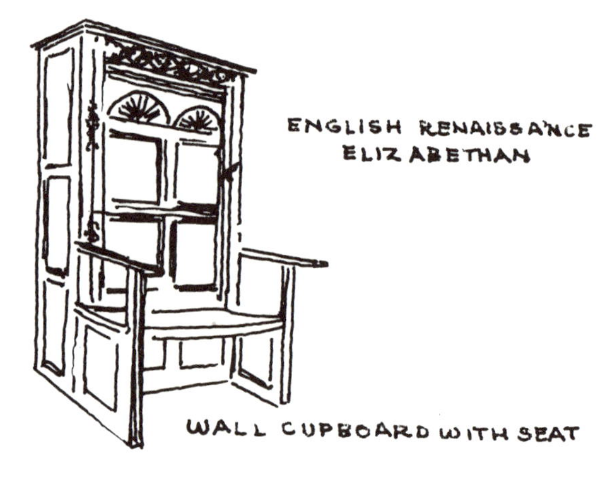

right members; stretchers were square and low.

The principal pieces of furniture used were chests, cupboards, wardrobes, desk-boxes, dressers for tableware, settles, chairs, stools, tables, beds, and cradles.

The wainscot and turned chairs were characteristic of the period. The wainscot had a nearly rectangular seat with turned or column legs, arms that were slightly shaped and a big solid back that frequently had low relief carving or inlay. The turned chairs were heavy in proportion with triangular wooden seats. This chair had heavy arms, back and legs composed of short, thick turnings. The X-shaped seat was in use.

The beds were composed of four heavily carved corner posts, often enriched with bulbous carvings, and an architectural capital that supported a heavy tester or canopy. Pull drapery hangings of rich velvet could be used to enclose this huge bed for warmth and privacy.

The permanent table began to replace the trestle and plank during the 16th century. Large refectory tables, some with extension tops, were built of solid oak. The host and guests sat on one side of the table while the serving was done from the other.

Chests were still popular and were used for storage, as seats and as beds. When used as seats and beds they were cushioned. The chests were richly carved.

Many decorative motifs were used including: Tudor rose, rosette, roundel, medallion, acanthus leaf, arabesque, dolphin, mask, quatrefoil, grotesques, vase, jar, lozenge, guilloche, rope molding, Gothic traceried arch, and coat of arms.

## Jacobean 1603-1688

Furniture in modern use dates from the Jacobean period in England, 1603 to 1688. The influence in the Jacobean period came primarily from the Elizabethans.

Jacobean furniture was masculine, vigorous, somber and austere. The wood used was primarily oak. The moldings followed geometric forms. And the strap work hardware had scrolls, acorn leaves and geometric forms carved into it. Slender baluster, Flemish scroll and

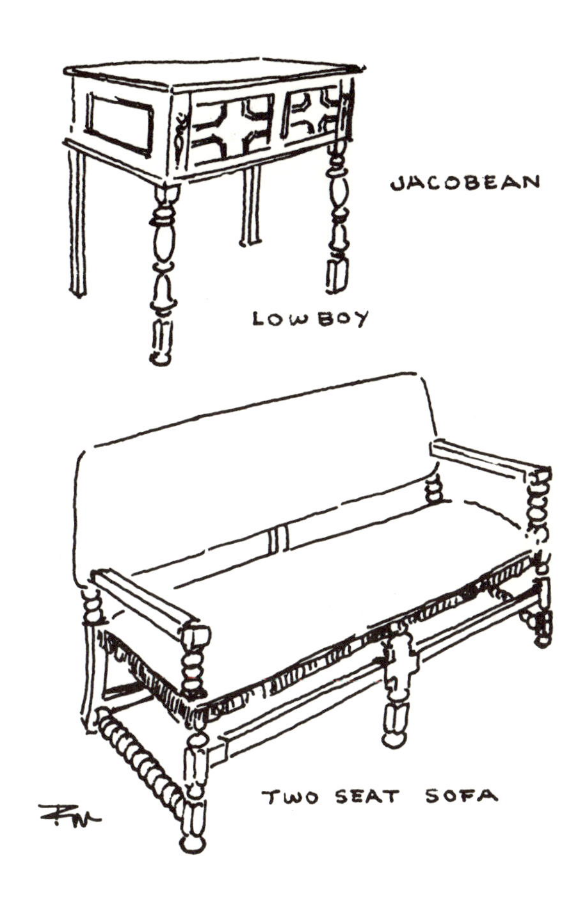


spiral legs appeared. Chairs became more comfortable and sometimes had padded seats and backs. These were upholstered with tapestry material and decorated with fringes. The seats and backs sometimes had caning. The wing back sleeping chair appeared during the reign of Charles II.

Jacobean furniture fits well into the English half timbered houses. It is highly masculine and massive and so will fit into men's apartments, large oak panelled libraries and dignified dining rooms.

## William and Mary 1688-1702

William and Mary ruled England from 1689 to 1702. During their rule life was simpler and more leisurely than it had previously been. Many cabinetmakers were brought in from the Lowlands. Display cupboards decorated with a canopy or hood were designed for porcelain. This same form was carried over to secretaries, chairs and settees. From this cupboard the first highboy was developed.

The table had a curved apron which carried down to the legs.

Chairs became more comfortable and graceful. Legs were cup turned, trumpet or octagonal tapered. A stretcher was used between the legs of the tables, chairs, settees and cupboards. The flattened bun foot was common as was the Spanish scroll and the ball and claw. Veneering and marquetry were used for surface design. Needle point and tapestries were used for upholstery. William and Mary furniture fits the Early English and Colonial architecture.

### Queen Anne 1702-1714

During the Queen Anne period, except for the characteristic shell carving, furniture depended upon its lines and the grain of the wood used for its beauty. Finishes included lacquering, marquetry and veneering. Curves were used rather than the straight lines of the William and Mary style and no stretchers were used between the legs. William and Mary and Queen Anne furniture became known as the furniture of the walnut age. Toward the end of Anne's reign mahogany became popular.

Chairs of the Queen Anne period were graceful and usually armless to accommodate the ladies' full skirts. The backs of chairs were slightly curved at shoulder height, and a splat down the center of the back was introduced. The fiddle back and the oval base back were characteristic shapes. Armchairs had continuous arms. The curved or cabriole leg was typical. The knee of the cabriole leg was decorated with shell carving. A club, or ball and claw, foot was used. The settee was a combination of two or more chairs. Needle point and tapestries were used for upholstery.

During the first quarter of the 18th century the Windsor chair appeared. This type of chair,

in many variations developed outside of court circles.

A variety of small tables were designed, including drop leaf tables, tilt top tables with plain tops or raised and carved pie crust edges.

Highboys, cabinets and china cupboards had carved, broken curve tops or broken pediments. Washstands, wigstands and tall clocks had characteristics similar to those of other cabinets with graceful drawer pulls and key plates to enhance their beauty. Drawer handles were frequently pear shaped.

### Georgian 1720-1810

Beginning with the Georgian period the furniture styles took the name of the designer. In this period from 1720 to 1810 there were four important designers: Chippendale, Adam, Hepplewhite and Sheraton. Chippendale was probably the most important of the four. This was the golden age of furniture and mahogany was the wood most used.

### Chippendale 1718-1779

Chippendale (1718 to 1779) understood and used all of the embellishments. He employed veneering, turning, carving, gilding, marquetry, inlay and metal mounting. Four influences are to be seen in Chippendale furniture.

1. *Queen Anne.* In the Queen Anne furniture Chippendale used the cabriole leg and the splat back for chairs. The splat back was pierced and carved.

2. *French influence.* The French influence was shown in the ribbon carving in the splats and in the cupid's bow top rail of the back.

3. *Revival of Gothic architecture.* Some of this, especially that with arched tops, pierced backs and straight squared legs, was good; but others in which he attempted to duplicate stone carving in wood was not successful.

4. *Chinese.* With the Chinese influence the backs of the chairs became nearly square and filled in with Chinese fret or latticework. The legs were square and straight and sometimes

resembled several bamboo rods bound together.

The most popular Chippendale chairs reproduced today are the pierced ladder back chair, the ribbon back chair and the wing back upholstered chairs called the "forty winks" chair. The forty-winks chair is more graceful with

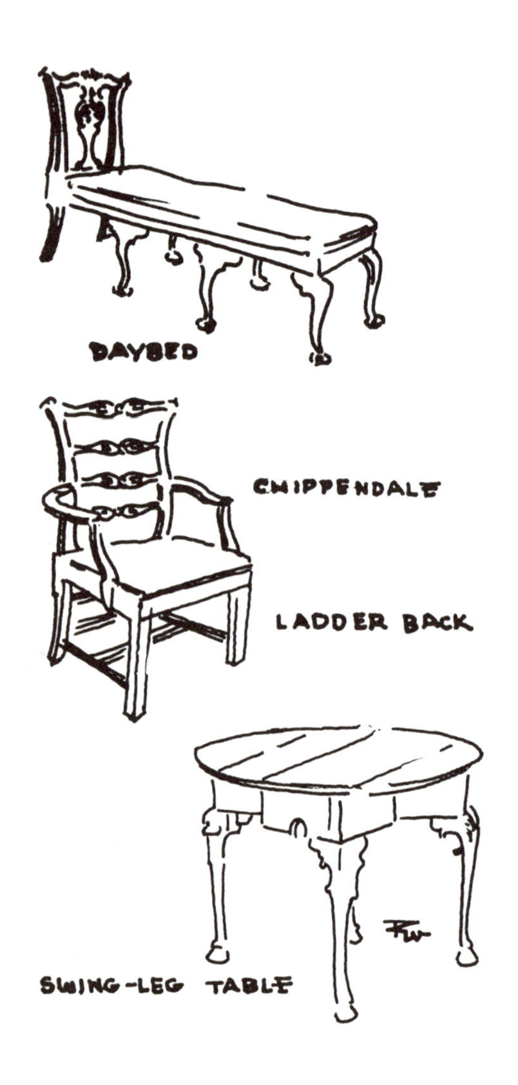


cabriole legs and ball and claw foot than it is with the plain straight leg. All of Chippendale's chairs had broad fronts and narrower backs.

Chippendale sofas and settees were of two styles: the all wood back combining two or three chair backs and the upholstered serpentine back with rolled arms which is popular today. Legs were cabriole or straight. Upholstery materials included fine leather, needle point, tapestry, embroidery and damask.

Chippendale tables were tilt top with a center pedestal and three legs, console or wall types supported by curved and carved supports, drop leaf types and tea and coffee styles. The intricate carving or interesting lacquer made them distinctly Chippendale.

Cupboards, straight or break fronted cabinets and desks had panelled and latticed glass doors with symmetrical swan neck decoration. The doors were usually topped with a carved broken-curve pediment and centered finial urn or eagle. The beautiful brass hardware added a distinctive finish to all of the cabinet pieces. Chippendale mirrors, either severely plain or elaborately carved, and grandfather clocks are among some of the best of his works.

### Adam 1762-1794

The Adam period, from 1762 to 1794, is frequently referred to as the neo-classic period. The four Adams brothers were primarily architects. They designed and decorated homes. They employed plaster on wall in place of wood panelling. They used raised plaster designs around ceilings, as borders and as panels. Wedgewood plaques were employed over doorways and fireplaces. Their fireplaces were especially attractive. They used festoons, frets, honeysuckle designs, swags, flat circular disc-like ornaments known as paterae, wheat ears, husks, urns, and rosettes. The Adam brothers were influenced by Louis XVI furniture. One of their most important pieces of furniture was the dining room sideboard with a commode at each end. On top of each commode was a large carved knife urn.

### Hepplewhite 17....-1786

Hepplewhite's furniture was delicate and beautiful. He designed some furniture for Adam interiors. The Pembroke table, a small four legged table with drop leaves, was one of beds with light delicate posts, and wardrobes.

Hepplewhite used mahogany and satinwood. Occasionally he employed artists to decorate his pieces with painting. Frequently the hardware was inconspicuous. He used veneers and inlays. His favorite upholstery fabrics were delicate brocades, horsehair, silks and satins with delicate flowers and stripes.

## Sheraton 1751-1806

Sheraton lived from 1751 to 1806. He was the first designer to introduce concealed

THREE PART SIDE BOARD

ADAM FURNITURE

Hepplewhite's favorites. He made many card, end, coffee, dining and side tables.

The Hepplewhite chairs are small, delicate and unusually graceful and usually have shield or camel backs. Within the backs he designed urns, Prince of Wales plumes and wheatears. Other backs included the oval, the interlacing heart and the wheel. The legs were straight, slender and tapering and might be square or round. The chairs often had upholstered seats. The sofas and settees had backs that repeated the chair backs.

Hepplewhite's other pieces included beautiful sideboards, dressing tables (with heart shaped mirrors), roll topped writing desks, washstands, secretaries, clock cases, four poster drawers, panels and compartments. Sheraton's pieces were characterized by rectangular forms. The fronts of his sideboards combined curved segments with straight lines in contrast to Hepplewhite's curved or serpentine fronts.

Sheraton chairs were delicate in line. Chair backs were rectangular in shape with vertical balusters, latticed bar work or ornamental splats between the upper and lower cross rails. Chair arms were delicate and had a continuous sweep out of the front legs and into the back. Chair legs were straight and tapering and terminated in a spoon or spade foot.

Sheraton used inlay, painting and multicolored marquetry. He often bleached and dyed woods for the marquetry. His favorite motifs were swags, festoons, urns, cameolike panels and latticework. He used mahogany, satinwood, sycamore and tulipwood. If pieces demanded upholstery he used delicate fabrics and colors.

Sheraton is well known for his Pembroke tables, his tripod base screens, his lattice door bookcases and his swan neck pediments.

### Victorian 1837-1901

The Victorian period from 1837 to 1901 was a hodge podge as far as design is concerned. It was a mixture of Gothic, Turkish, Venetian, Egyptian, Louis XV and Empire. Some of the pieces, such as chairs, tables and love seats, had a homey charm, but the styles in general were uncomfortable.

The principal woods used were black walnut, mahogany and rosewood. Carved decorations took the form of roses, buds and fruit. Wood pulls in the shape of fruits or pear drops replaced metal. Painting and mother-of-pearl were popular decorations also.

Chair backs were usually spoon shaped and the popular rose carving decorated chair and love seat frames. Many chairs were tufted and upholstered with plush or horsehair. Other popular pieces were drum shaped tables, terraced corner tables, whatnot stands, candle stands and drop leaf tables with center pedestals.

Most of the criticism is based on the over abundance of furniture and decoration during this period. Many of the individual pieces of furniture in correct surroundings are interesting and comfortable.

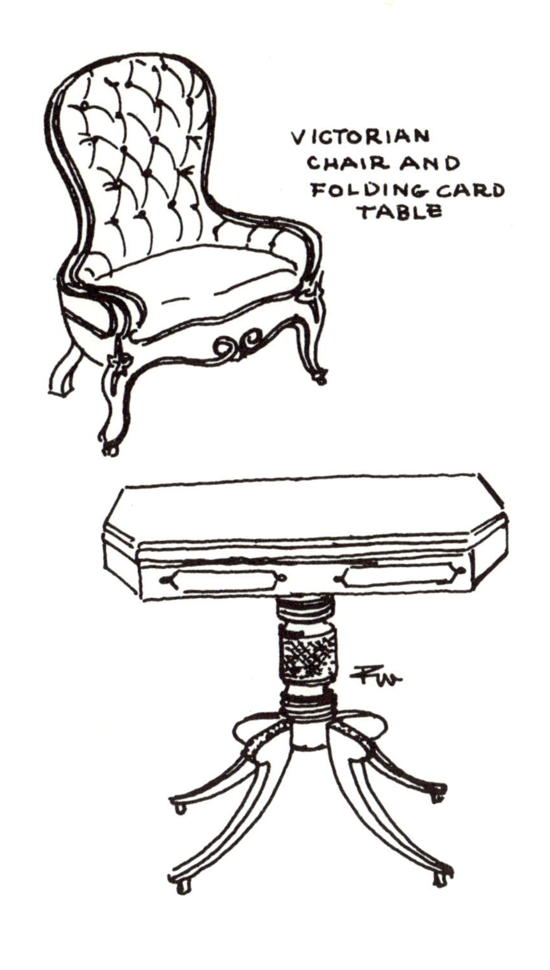


## Oriental

Chinese furniture is rare since the requirements are few. Surface decoration in lacquer and decorative painting is important. Simple coffers and chests stand on low bracket bases ornamented with intricate metal mounts. Tables and stands are low, usually with turned-in scroll-like feet. Ceremonial chairs have flat surfaces with elaborate carving or inlay on flat surfaces. Dragons, flowers, landscapes with figures, and geometric borders are employed.

Japanese life requires but little in the way of furniture even as the Chinese. Chests and cupboards are usually built in with sliding panels for doors. For sleeping and sitting, mats are rolled on the floor. The rare tables are low and portable. Furniture is lacquered and highly polished. The lacquer is frequently flecked with gold and decorated with fine scaled flowers, animals and landscapes.

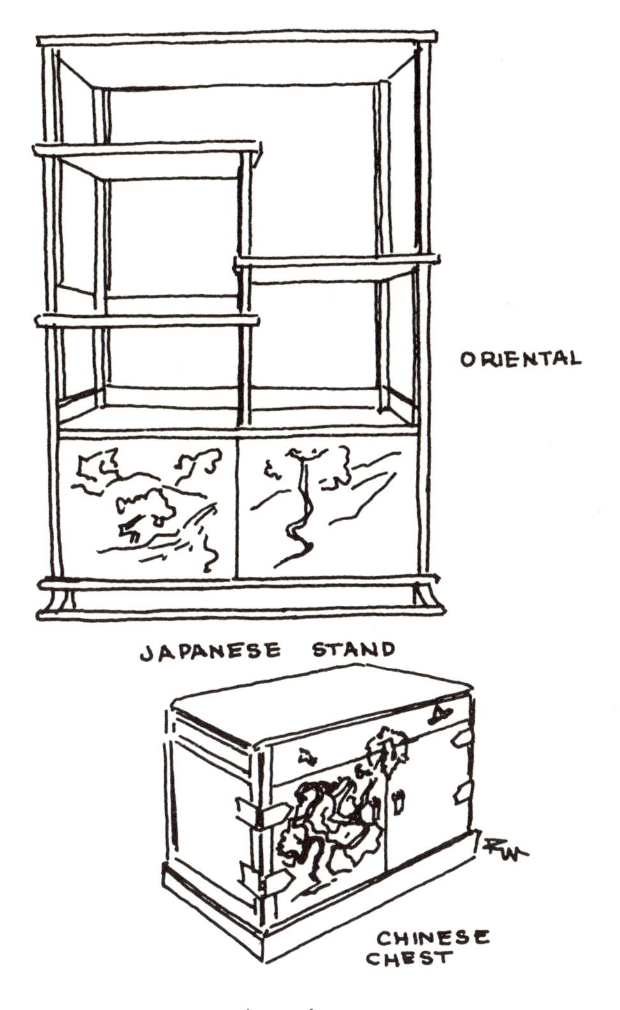


## American

In the main American furniture followed the designs current in Europe. The early furniture of the colonies shows the Gothic-Tudor influence.

Chests and cupboards had distinctly rectangular panels. Tables at first of the trestle type soon gave way to box styles and simple drop leaf types. Desk boxes, Bible boxes, forms or stools and a few crude beds made up the furniture. The woods at hand were used and these consisted of pine in great abundance, oak, birch and maple. The wood was left natural and aged to beautiful soft colors.

The Pennsylvania area shows the influence

of German, Swiss and Dutch peoples and the furniture has a distinct Medieval flavor. Chests, cupboards, tables and chairs are decorated with naive peasant type paintings. The wheel of fortune, tulip, heart and four leaf clover are typical ornaments.

Highboys, lowboys, chests, upholstered chairs with spiral turnings, elementary cabriole legs,

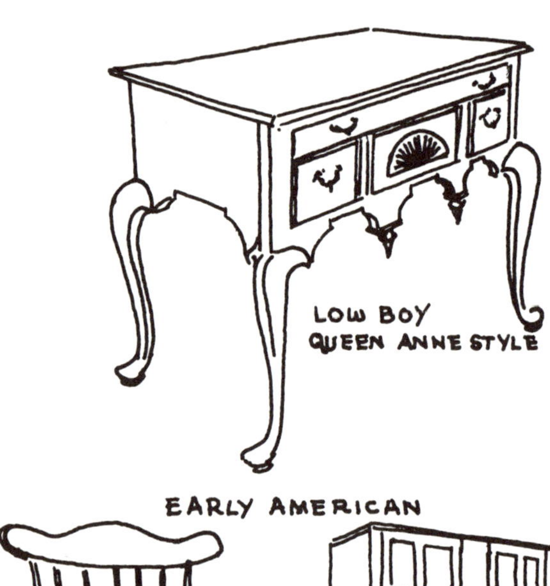

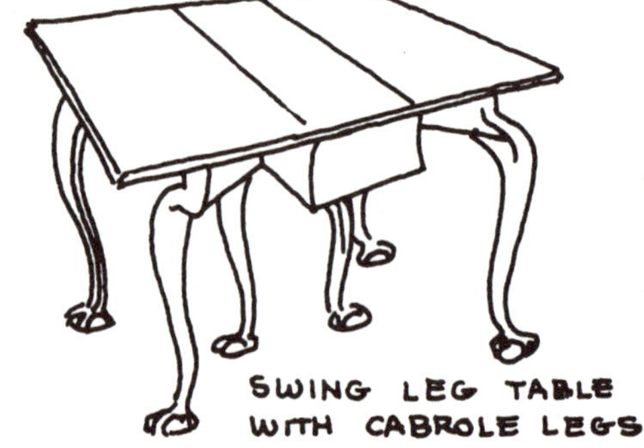


carved shells and pendants, and inverted cup shapes in walnut were typical of the 17th century.

The 18th century brought the Queen Anne and Rococo influences. Cabriole legs, shell carvings, pad or animal feet, Rococo curves and most of the other ornaments popular in England were used by the furniture makers.

The Windsor chair was a development of the Colonial furniture makers. Stools, chairs, benches, chests and cabinets made in pine, maple, hickory, oak, apple or cherry began to show the native talent. Beds with short posts, ladder back chairs, wagon seats, rocking chairs and writing chairs are all uniquely American.

### Federal Period 1780

The natural movement away from the English during and immediately after the war and the turn toward the French is evident in the furniture. However, it was not long until the designs of Adam, Hepplewhite, Shearer and Sheraton became popular.

### Duncan Phyfe 1790

Duncan Phyfe based his designs on Sheraton. His earlier work was in mahogany and his later work in rosewood. The lyre motif was characteristic of his work. This appeared in chair backs and table bases. His delicately carved lines in reedings and flutings combine with light carvings of leaves, plumes and animal motifs. He also used shells, bound arrows, pineapples, birds, lions' heads, eagles. Large areas on chests and drawers were veneered with beautifully grained mahogany in a V-shaped pattern. Hardware was used extensively in the form of metal or ormulu tips on legs, brass eagle finials on desks and mirrors, and metal lion masks and rings as drawer pulls.

### Southwest America

The Franciscan monks left their mark on the southwestern section of America. Hand hewn leather-backed chairs, oak refectory tables and finely carved painted beds show their influence.

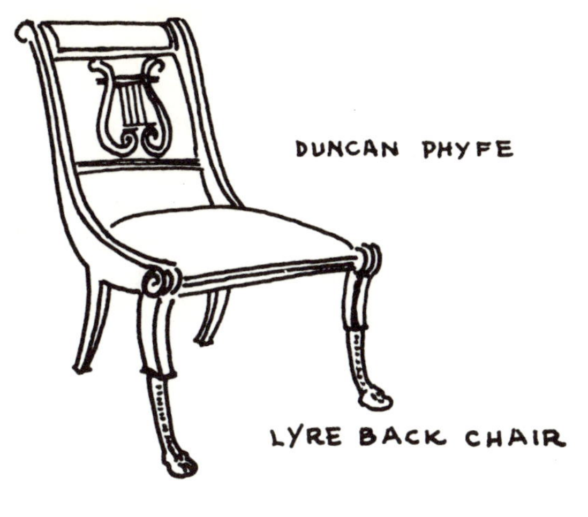


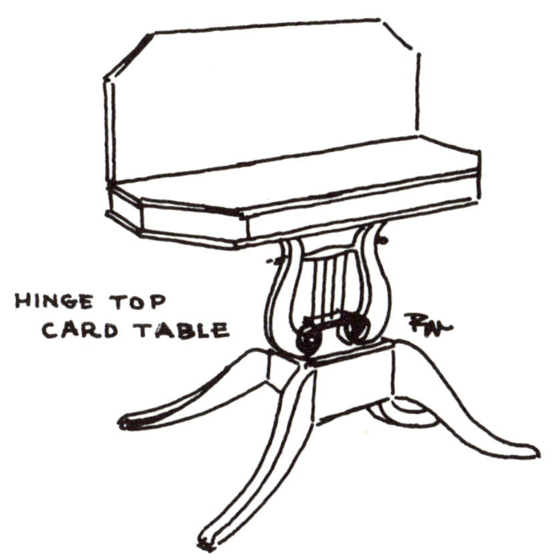


### Modern

Each age has its modern furniture. The current age is no exception. Today's modern moves toward extreme surface simplicity. It combines functionalism with relatively small scale and surface beauty of the natural material.

It is possible to combine elements of various periods. Careful matching in regard to scale, harmony of line and desirable textural feeling are essential.

### Furniture and the Stage Setting

The furniture used in a stage setting must be integrated not only with the other factors of design, including the general mood and atmosphere, scale of the various units, over-all texture pattern and color harmonies, but also with the general feeling of the play itself. The furniture is an important part of the over-all design and must be treated as such rather than as a completely divorced element. It must be chosen carefully. In some instances furniture more than any other element provides the audience with clues about the actors, their aesthetic taste, their conflicts, their emotional patterns.

The furniture may be chosen to harmonize

with everything else on the stage (setting, costumes, rugs, pictures and actors) or it may serve as an element of conflict. The stage may be deliberately divided into areas to show the difference between the varied characters by the furniture they choose and use. Single units of furniture may be used to attract attention deliberately, to serve as focal points. If this is done it should be done carefully and with motivation. The jarring note of an obviously mismatched piece of furniture can be extremely disturbing to the members of an audience and unless it is planned deliberately it should be avoided. Choose furniture carefully.

### Furniture — General Information

Certain information of a general nature may be helpful to the crew working on the stage furnishings. It may be necessary at times to make slip covers, upholster, repair broken furniture, remove stains or even manufacture simple pieces of stage furniture. The information here will be basic, general and rather sketchy. More detailed information may be obtained from the books listed in the bibliography.

### Cleaning Wood Finish

If furniture gets sticky wash it carefully with a solution of mild soap and water. Dry the area thoroughly and then polish.

Water stains may be removed by carefully rubbing the area with water to which has been added a couple of drops of ammonia. The area should then be dried carefully and polished.

Alcohol stains can be removed with alcohol or turpentine. Dry the surface thoroughly and polish.

Heat or burn stains will leave a white ring which can be removed with oil of camphor. After this has dried polish the surface.

### Cleaning Upholstery

In order to prevent the upholstery from becoming soiled it is advisable to keep the furniture covered with muslin, sheets, blankets or even paper when it is not in use. It is also advisable to instruct actors, stagehands and visitors that furniture is to be used only by the actors and only when it is on stage during the running of the scene for which it has been procured. This will minimize the danger of soiling and staining both the fabric and wood.

If upholstery is generally soiled it may be cleaned with any of the regular soapless furniture shampoos.

Grease may be removed with carbon tetrachloride, benzine or a commercial solvent.

Blood may be removed with several applications of raw starch paste.

Chewing gum may be loosened with benzine or carbon tetrachloride.

Mud may be brushed off if allowed to dry.

Candy may be washed off with soapy water and then the area should be wiped thoroughly with a cloth wrung out in clear water. Avoid letting the water penetrate the fabric too deeply.

Fresh oil base paint may be removed with turpentine. Water mix paint can be removed with soapy water. The soap should be removed with a rag moistened with clear water and then the surface should be rubbed with a dry cloth. If turpentine is used the last traces of turpentine should be removed with benzine.

Powdered cleaners that are on the market will clean some of the soil from upholstery.

### Upholstery and Slip Covers

If a piece of furniture is to be completely recovered it is easiest to remove the old upholstery and use the pieces as patterns for the new material. If this is done carefully the various pieces can be labelled as they are taken apart and the new pieces put together in the same order.

Slip covering or upholstering a piece of furniture which previously had no upholstery may seem to be a difficult job to tackle but for anyone who has simple sewing skill it is fairly easy. A little courage and faith in one's abilities will help, but no property crew member ever lacked these traits.

Inexpensive cloth or newspaper may be used for practice purposes. This should not be used as a pattern for cutting the finish upholstery cloth, however, since it will probably not retain its shape. The pieces will fit much better if the final cloth is cut to fit. Cut pieces to fit the various surfaces of the piece of furniture with about an inch to spare all of the way around for seams and strength. If the seam is too close to the edge it will tear or pull out easily. If the piece of furniture is symmetrical it is possible to turn the cloth upside down; if not it will be necessary to use a different procedure and mark the seam areas with pins. The pieces of cloth are cut to fit (with one inch to spare), pinned to the chair, and pinned together where there are to be seams. Cushions should be removed so that the material on the inside of the back and sides can be designed to fit down under them. The seams may then be sewn and the cover is finished. It will take considerable careful work if a professional looking job is to be done, but for stage use one need not be as careful as when doing a piece of furniture for display in the home.

## Welting

Welted edges, which usually give much better finish to the piece of furniture, can be put on the upholstery. Cotton cord of the desired circumference is placed in the center of a strip of the upholstery cloth about 2½″ wide. The cloth is then folded over the cord and the cord sewn into the fold. A zipper foot on the sewing machine will make it possible to sew a seam snugly against the cord. This is the welting strip. The two pieces of cloth that are to make the seam are then placed wrong side out so they are even with the loose edges of the welting strip. Using the zipper foot the two pieces of material with the cording strip sandwiched between are sewed snug against the cord. When this is turned right side out the seam will be covered by the bead of the welting.

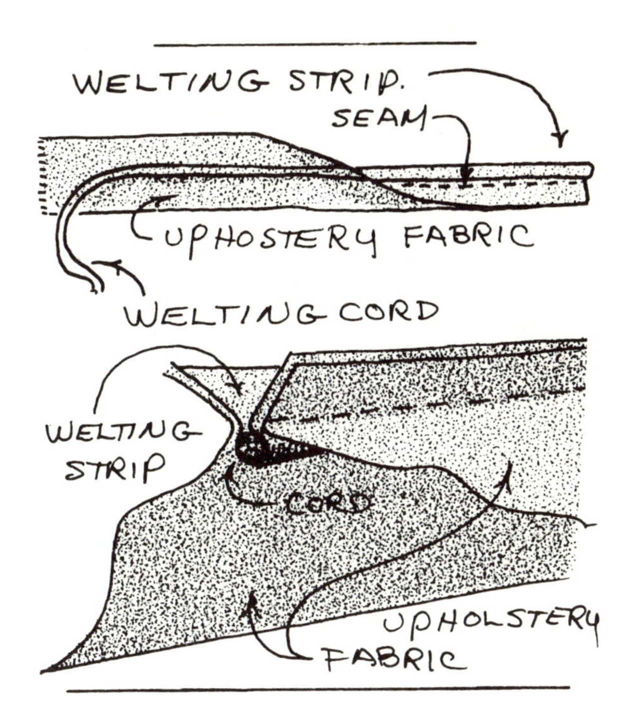


## Soft Seats

The springs are frequently too soft on stage furniture. An actor who sits in a chair or sofa sinks too low and is ungraceful while sitting and struggles when he tries to rise. A board or piece of plywood placed under the cushion will distribute the weight over the entire spring area and give the seat more firmness.

## Sagging Seats

Sagging seats on chairs, sofas or other pieces of furniture may be caused by loose springs, loosened webbing or lumpy padding. The chair may be turned upside down to check the webbing. A light dust cover may protect the bottom, but this can easily be pulled loose to inspect the webbing. It may be that the webbing can be tightened without disturbing the upholstery at all or it may be necessary to remove the muslin covering under the cushion and even some of the padding. The webbing may then be tightened and securely tacked.

If the springs are causing the trouble they can be retied. When this is done a strong hemp twine should be used and the springs tied to one another and to the frame. An inspection will show how this was done before and the same pattern may be followed in retying. The springs should be tied straight across in both directions and diagonally in both directions. The padding can then be replaced.

If the padding is lumpy it will be necessary to take the entire area apart and redistribute the horse hair and cotton or whatever the padding happens to be. When doing this, as when doing any reworking of a piece of furniture, try to replace the parts so that they are in their original positions.

### Broken Dowel

Sometimes furniture joints come apart or are broken. Most arms, legs, aprons, backs and ties on furniture are fastened with mortise and tenon or dowel joints. If the dowel or tenon is broken it will be necessary to replace or repair it. In the case of the dowel this is not too difficult. The broken pieces can be removed by drilling with a bit slightly smaller than the piece of dowel. This done the remainder of the dowel can be chipped out with a pocket knife. Pieces of dowel can then be cut to replace the broken ones. These should fit snugly into place. If the holes have been worn it might be advisable to redrill the holes a sixteenth of an inch larger in diameter. Be certain that the dowel is the correct size and length. With a pocket knife slice a thin strip off the dowel. Coat the piece of dowel with wood glue and push it into place in one side of the joint, then push the other part of the joint over the dowel. The sliver that was sliced off the dowel will allow the excess glue to seep out. Otherwise pressure of the glue at the ends of the dowel might crack the boards. Pressure should be applied while the glue is hardening. A large clamp, heavy cord tourniquet, or some other device can be used for this purpose.

If a tenon is broken it may be necessary to replace the entire board. Or it might be possible to mortise into the broken tenon side of the joint and then fashion a small block of wood that will slide into both the old and new mortises. The grain of the key should run across the joint, and the key should be glued into place in the manner suggested for the dowel.

### Building Simple Furniture

Simple pieces of furniture such as benches, stools, tables, ottomans, dressing tables, desks and footstools can be made in the shop if necessary. Such utility furniture can be made from white pine and various thicknesses of plywood.

For simple construction it is advisable to avoid complex joints and stick to butt, mitre

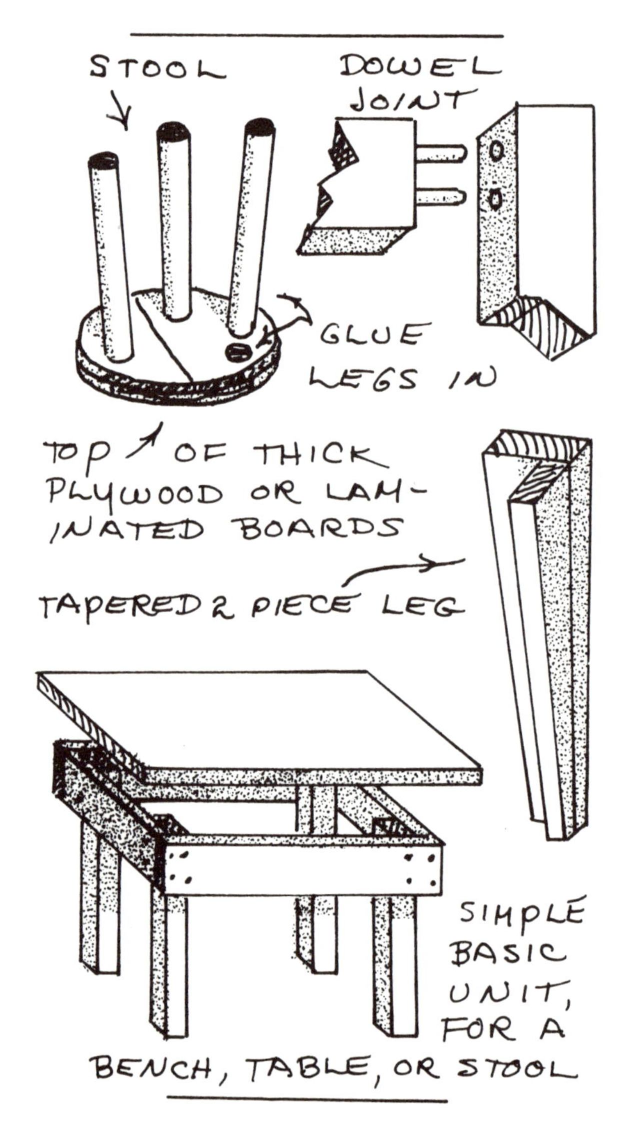

and plywood reinforced joints such as those used in making flat frames.

Coated nails or screws with or without glue will probably work best for fastening the joints. The glue will make the connection more substantial but is not always essential.

### Plywood in Construction

Sheathing plywood is relatively inexpensive and in thicknesses from 5⁄16″ to ¾″ can be used wherever it is possible to fasten through the thickness of the material. It is not practical to fasten into the plies of the edge grain since nails or screws will tend to pry the thicknesses apart. Plywood is strong for tops and panels especially. With plywood it is possible to have an unbroken area as large as 4′ x 8′. If regular 1″ boards are used it is necessary to cleat several boards together side by side to get this same area. The cleated section is usually heavier and not as smooth as the plywood surface.

### General Construction

Most of the utility pieces can be constructed in about the same way. Basically they will consist of legs that are fastened at the top inside of the corners of a rail frame. Over the rail frame, and usually projecting out beyond an inch or so all of the way around, is the top.

### Rail Frame

The rail frame is composed of boards on edge secured together at the corners with glue, nails, screws or all three. If the legs are to extend straight down, the ends of the boards will be cut square. If the legs are to cant, or slope out slightly, the ends of the boards will be cut on a slight angle. The legs are usually fastened snugly into the insides of the corners of the rail frame. Fastening into the leg from both sides of the corner, if possible, will make the corner and leg more substantial and durable. In order to fasten the leg more easily it is advisable to clamp the leg and board of the frame to be nailed with a "C" clamp. This will hold the two pieces together securely while the nailing or screwing is done. The size lumber used for the rail frame will depend upon the general design and style of the piece. Usually 1″ x 4″ will do.

### Post Legs

Legs for utility furniture may be made from rectangular stock (1″ x 3″, 1″ x 4″, 2″ x 2″, 2″ x 4″, 4″ x 4″), from round stock (dowel, closet poles, round stair rail stock), from pipe and flanges, or from two pieces of 1″ board fastened together edge to edge usually with the piece that butts up edge wise being ¾″ narrower to make the leg measure the same on both of the right angle sides. The legs may be straight or tapered. The tapered leg gives a more graceful finished appearance.

### Slab Legs

Slab type supports or legs are made either from one solid slab the width of the space between the front and back rails or from a hollow slab made up of two thin sheets of plywood or wall board with thickness spacers sandwiched between them around the exposed edges. Slabs may extend straight down, may cant outward

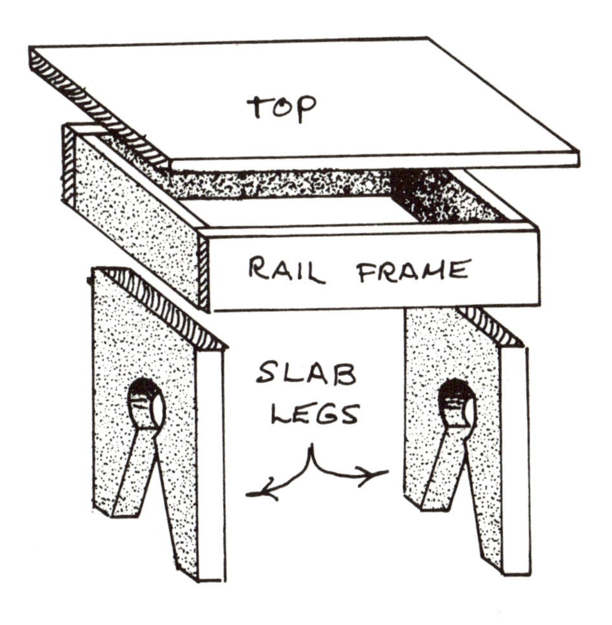

to provide a wider base, may have a split bottom or may have simple patterned open work cut out of them.

### Top

The top will usually project out beyond the rail frame by at least an inch all around and may project out even more. If the top is to support but little weight it may be composed of thin plywood. If it is to support great weight it may be composed of either heavy plywood or planks. A thickness edging around the top will aid in projecting greater thickness than the top actually has.

### Patterns for Furniture

Patterns or ideas for furniture may be obtained from magazines, books, photographs, sketches or pieces of real furniture. A little experience will aid one in simplifying the original design and duplicating only the essentials of the basic design.

### Surface Decoration on Furniture

Plain surface furniture may be altered and enhanced by adding commercial or shop-made decor. Plastic, pressed wood chip, and pressed saw dust decor may be found in paint and lumber stores. Materials for home made decor are: Plastic wood, gesso (a mixture of powdered glue and plaster), papier-mâché, and other plastic materials used for various crafts. Applique pieces can be cut from plywood, cardboard, Styro-Foam, aluminum, felt, linoleum, or Plastic Wood and can then be glued, nailed or stapled to the surface. Rope, string, twine, electrical cable, iron and aluminum wire are other materials that might be applied to the surface. The surface may also be painted with a flat coat, stippled, wood grained, marbleized, patterned or decorated. Surface textures can be obtained by using a coating of glue and then brushing, dusting or blowing flock, sawdust, shavings or any other material onto the sticky surface. The furniture may be coated with gilt paint, enamel, fluorescent or aluminum paint, glitter (on a glue, varnish or bronzing liquid base) or shellac.

* * * *

### Beg? Borrow? Steal? Buy?

Procuring furniture for use on stage may be a problem in almost any community, even for the wealthiest theatre. It is impossible to have stock furniture that will fit every production. It is possible for an organization to gather up a few standard pieces that can be used for rehearsal and for general all-around utility purposes. But there always comes the time when it is necessary to start out looking for items needed on a specific stage setting. Where does one look?

Furniture stores dealing in new furniture, old furniture, second hand stores, antique shops, junk shops, the homes of friends of the institution, clubs, lounges of the institution, offices, rest rooms — all of these may be places from which furniture may be obtained. It should be understood at the very beginning that there is great danger that the pieces of furniture will get damaged in transit, while being shifted on the stage or while being used by the actors. It is therefore always necessary to realize that it may be necessary to repair or replace any damaged goods. This should be understood by both the borrower or renter and the owner. With the assurance of care in handling the possibility of permission to use the furniture will be much greater.

While in the theatre the furniture should be kept clean. It should be picked up and taken to the stage only when absolutely needed and returned just as soon as possible after its use has been concluded. Complimentary tickets, if possible, a courtesy program note and a thank you note when it has been returned will all help make it possible for a repeat in the future.

Building a furniture "morgue" is valuable to any continuing theatre organization. Occasional purchases, acceptance of gifts (and gifts will be more likely to come if people know that the furniture is needed), and reclamation of discarded pieces will all be means of acquiring new pieces.

# 12

# *Electricity*

"What is electricity?" is always a startling question to the novice contemplating his first wiring job, and the answer is rarely completely satisfying or wholly revealing. Fortunately, in order to light a simple stage production it is not necessary to become an expert in the field of electricity. However, it is helpful to know a few of the basic general facts about electricity in order to understand the operation and maintenance of stage electrical equipment and to be able to handle simple wiring projects adequately and efficiently.

VOLTAGE—The source of power in most of the electrical equipment used every day is the dry cell battery of the flashlight or portable radio, the wet cell battery of the automobile, or the generator of the local power plant. These various sources of electrical energy build up a pressure or force which can cause tiny charged particles called electrons to flow along a wire. The unit of pressure is the volt and the number of volts between two terminals is the voltage of the generator.

The ordinary flashlight battery is rated at 1½ volts. When two flashlight batteries are stacked end to end the voltages are added together 3 volts are produced. Most flashlight bulbs are designed to be burned at this voltage. Automobile batteries are usually rated at 6 volts or 12 volts, depending on the model of the car, and the lights, horn, starter, radio and all other electrical equipment in the car are designed for this same voltage.

Most homes and stages are equipped with 120 volt power lines and so, as would be expected, most stage lighting fixtures and home appliances are rated at 120 volts. Vacuum cleaners, refrigerators, waffle irons, electric shavers and most light bulbs are designed for use on 120 volts. Certain appliances which use a large amount of power, such as electric stoves, water heaters and clothes dryers, are designed to use 240 volts. To get this voltage a third wire must be brought into the building from the power company's lines.

Every piece of electrical equipment must be used only with the voltage for which it is rated. If the applied voltage is below the rated value the equipment will not operate; lamps burn dimly and motors will not start. If the voltage is too high the equipment will probably go up in smoke. A small percentage of leeway is allowable in selecting equipment for use at a certain voltage. For instance, it is perfectly all right to connect a lamp rated at 110 volts to a 120 volt power line, but a 6 volt lamp would burn out instantly if connected to either 110 or 120 volts.

CURRENT—It was stated that voltage represents a pressure or force which tends to make electrons flow along a conductor. This flow of electrons is called an electric current. Current and voltage are not the same and should not be confused with one another. Currents are measured in amperes (amps), one ampere representing the flow of 62,250,000,000,000,000,000 electrons past any given point in a circuit each second. Don't be alarmed by this number. It will not be encountered in this book again.

POWER—Power is the rate of flow of energy in an electric circuit and it is measured in watts or in larger units called kilowatts. (One kilowatt is equal to 1,000 watts.) It is the power consumed which determines the amount of light given off by a lamp, the amount of heat given off by an electric stove or the horsepower output of an electric motor. Power is not the same as voltage or current, but is closely related to both by the following equation:

$$\text{Power} = \text{Voltage} \times \text{Current}$$

or

$$\text{Watts} = \text{Volts} \times \text{Amps}$$

In addition to a voltage rating most of the equipment used in stage lighting also has a power (or wattage) rating. These wattage ratings are very useful for they allow the technician to compare lamps which burn on different voltages. For instance, a 100 watt 6 volt lamp gives off just as much light as a 100 watt 120 volt lamp and the operating cost is the same for both even though the 6 volt lamp requires 20 times as many amperes.

Lamps are rated in watts and dimmers are usually rated in watts, but fuses, electric wires and cables, switches and connectors are rated in amperes. Therefore, it is often necessary to convert wattage ratings into current ratings, This is easily done since the power equation can be rearranged in the following way:

$$\text{Current} = \text{Power} \div \text{Voltage}$$

or

$$\text{Amps} = \text{Watts} \div \text{Volts}$$

With this form of the equation it is easy to find the current in any circuit if the wattage and voltage are known. For example, find the current in a 60 watt 120 volt lamp.

$$\text{Current} = 60 \div 120 = \tfrac{1}{2} \text{ ampere}$$

A 60 watt 6 volt lamp would have a current of $60 \div 6 = 10$ amperes

Other examples using the power equation will be given later in the chapter.

RESISTANCE—Every substance resists the flow of electrons to a certain extent, and this resistance is measured in Ohms. The resistance depends upon the character of the material and its cross sectional size. In general a thin wire, for instance, has a higher resistance to the flow of electricity than a larger one and consequently will carry less current.

The current which will flow through a resistance is given by Ohm's law which states in equation form:

$$\text{Current} = \text{Voltage} \div \text{Resistance}$$

or

$$\text{Amperes} = \text{Volts} \div \text{Ohms}$$

This means that in a circuit connected to a certain source of voltage the higher the resistance the lower will be the current. Some dimmers use this principle to reduce the current through a lamp, thereby reducing the light out put.

Whenever current flows through a resistance, heat is produced and this is why electrical equipment usually gets warm or hot when in use. Sometimes this heat is useful, as in the filament of an incandescent lamp or in an electric stove, but at other times the heat is not wanted. It wastes energy and can be a fire hazard if it gets out of control.

CONDUCTORS AND NONCONDUCTORS—It has already been stated that all materials resist the flow of electrons to a certain extent. In general the substances used in electric wiring may be classified as conductors or nonconductors. Silver, copper, aluminum, brass and most other metals resist the flow of current the least and are referred to as the conductors of electricity. These are the materials used for wire and contacts on electrical instruments. Porcelain, rubber, glass, wood, cloth and paper resist the flow of electrons almost completely and so are used for insulation, or covering, of the conductors.

WIRING SYMBOLS—Before proceeding further with this discussion it will be helpful to describe briefly and illustrate the symbols that are used in drawing electrical circuits.

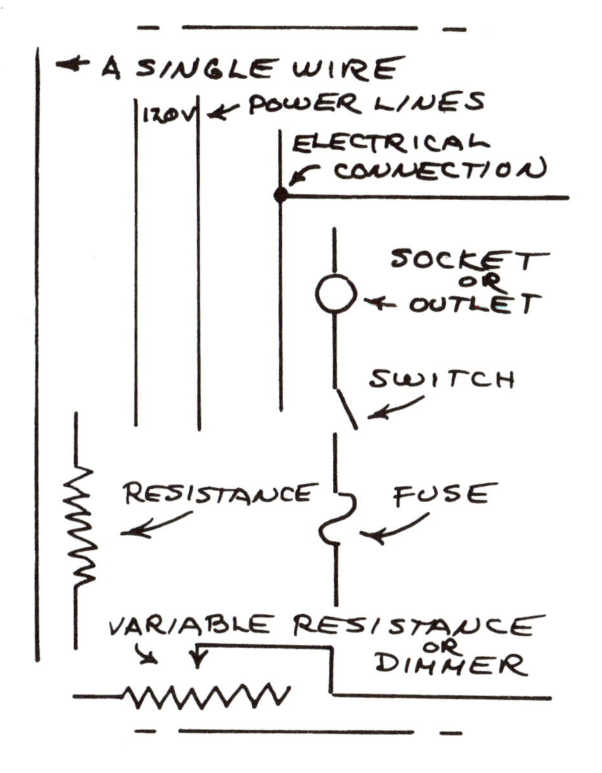


*Wire*—A single wire or electrical conductor is indicated by a single line.

*Power Lines*—Parallel lines with a numerical indication of the voltage between them will be used to designate the power lines.

*Electrical Connections* — Two wires connected together are indicated by one line meeting another and the point of meeting indicated by a heavy dot. If the dot is missing it will be assumed that the wires are not connected.

*Socket*—A socket, outlet or an instrument that need not be specifically designated may be shown by a circle with appropriate wire connections — a wire connected on each of two sides.

*Switch*—A switch is shown as a break in a wire. A little lever section of the wire represents the break.

*Fuse*—A fuse is shown by an "S" curve in the line.

*Resistance* — A resistance in the circuit is shown by a zig-zag pattern in the line.

*Variable Resistance* — A variable resistance or theatre dimmer of the resistance variety is shown by one line terminating in a zig-zag pattern. An arrow above this leads into a continuation of the line. The arrow shows that the length of the resistance can be varied.

These are the symbols that will be used most frequently in drawing simple wiring diagrams. Other symbols may be needed; if so, they will be described when they are used.

ELECTRIC CIRCUIT—In order to have a flow of current it is necessary to have a closed circuit; that is, it is necessary for the current to flow around in a loop or circuit that returns to the point of origin.

FLASHLIGHT CIRCUIT—A flashlight may be used to show how a simple electrical circuit operates. The battery represents the generator; it is the source of power. When the switch is pushed on, contact is made between the lamp socket and the side of the flashlight case and the circuit is closed. The voltage or electrical pressure of the battery forces electrons to flow from the base of the battery, through the case, up through the bulb and back down to the button on the top of the battery to complete the circuit. The conducting case of the flashlight in this instance is a part of the electrical circuit. When the switch is turned off the contact between the case and lamp is broken so that no current flows through the circuit.

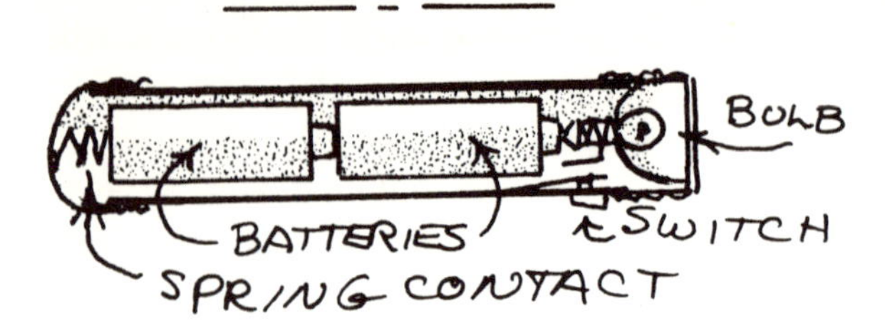


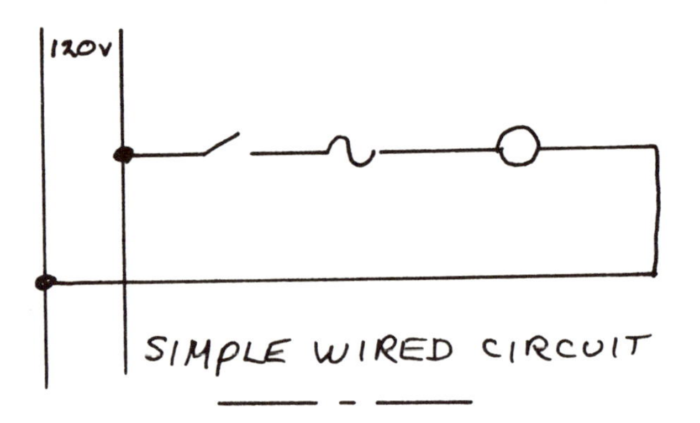


SIMPLE WIRED CIRCUIT—A simple electrical circuit is similar to the operation of the flashlight. Several of the symbols already shown are used together to form the completed loop. Included here is a source of power, a switch, fuse, socket and wires connecting all of these units together.

DIRECT AND ALTERNATING CURRENT—In the wiring system of the flashlight, or of an automobile, the flow of current in the closed circuit is constant in one direction, from negative to positive. This is referred to as a direct current. There are some commercial electrical power plants that produce direct current. However, since most power plants serve large areas and must send power to distant points, it has been found that a pulsating, or alternating, current is more practical and efficient. In the alternating current system the current alternately travels in one direction and then the other. Most systems employ 60 cycles, which means that there are 60 pulses of current in each direction every second. Direct current is designated by the letters D.C. and alternating current by the letters A.C. Practically all stage lighting equipment except arc lamps will work on A.C., but only resistance type dimmers can be used on D.C. Many arc lamps require direct current and rectifiers are used to convert A.C. to D.C. for this purpose.

TYPES OF CIRCUITS — Basically there are two types of circuits that are used frequently for simple stage wiring: parallel and series. Almost all wiring consists of one or both. For the present explanation it may be assumed that there are two wires representing the power source.

SERIES CIRCUIT — A series circuit is, in a sense, one continuous closed line of switches, wire and outlets. A single wire can be used and wherever an outlet is desired the wire can be cut and each of the ends fastened to one contact of the outlet. In order for the circuit to operate it is necessary to have something operating (a lamp or other electrical device) in each outlet. The circuit is not closed until contact is made through all of the sockets.

PARALLEL CIRCUIT — The parallel circuit consists of a continuation of each of the two lines of the power source. If the two wires are laid out in a line parallel with one another it may be explained that outlets are placed between the two wires so that the current flows

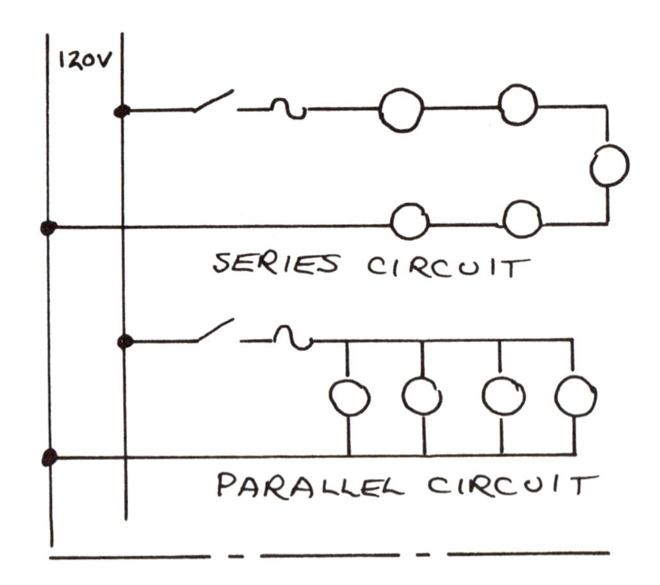

from one wire to another through each individual circuit. In the series circuit the current runs through one outlet before reaching the other and so on to the end of the series.

CHARACTERISTICS OF A SERIES CIRCUIT—Current flows in one straight path through several units as if flowing through a continuous wire. The voltage applied to the whole circuit is equal to the sum of the component parts. If there are two lamps of the same wattage and voltage each receives one half of the voltage. If there are three lamps each receives one third of the voltage. If there are four each receives one fourth and so on. As the voltage diminishes so also does the brightness of the lamps. The resistance of the circuit is equal to the sum of all of the resistances. And finally the amount of current flowing through the circuit is the same throughout.

CHARACTERISTICS OF A PARALLEL CIRCUIT—In a parallel circuit the current flows through several individual circuits or paths. Each unit is connected from one line voltage wire to the other. The voltage of each of the parts is the same as the voltage of the supply line. Adding more outlets does not diminish the voltage. The current flowing from the voltage source is equal to the sum of the currents flowing through the individual outlets.

USE OF PARALLEL WIRING — Almost all electrical outlets are wired in parallel so that the maximum voltage and maximum efficiency can be utilized from each individual circuit.

USE OF SERIES WIRING—Most switches are wired in series with the outlets since it is necessary to interrupt only one wire in order to stop the flow of current in a circuit. Turning off a switch is in principle the same as breaking or cutting one of the wires of the circuit.

FUSES—Fuses, or safety valves as they might be called, are also wired in series with the outlet. The fuse is designed so that it will allow only a specified amount of current to flow through a circuit before the fuse wire will heat enough to melt and break the circuit. Again, as with the switch, it is only necessary to interrupt the flow through one wire of the circuit so only one wire need be fused to shut off the flow through the entire circuit when the fuse "blows out."

Inexpensive sets of Christmas tree lamps are wired with eight lamps in series because it takes less wire to make a loop with a single wire than to make a loop of the same size with two wires. The lamps, or bulbs, are designed to burn with the lowered voltage ($120 \div 8 = 15$ volts). If the ordinary Christmas tree lamps designed to be burned at 120 volts are used in one of these circuits they will burn only very dimly. If, on the other hand, the 15 volt lamps are placed in a 120 volt circuit they will burn out immediately.

In rare cases where instruments having voltages lower than 120 are to be wired into a circuit it is necessary to have a transformer (see next paragraph) or to wire enough of the units in series to make the total of 120 volts. As an example, the use of 26 volt sealed beam lamps may be cited. In order to determine how many will be needed it is necessary to divide 26 into 120. The answer is about 4.6. It would be necessary to use either 4 or 5. If one wishes brilliance and is not disturbed about the lamps having a shortened life it is possible to use 4. If, however, the brilliance is to be sacrified for long life 5 will be wired in series.

TRANSFORMERS — It is often desirable to have more than one voltage available for special purposes on the stage. With alternating current it is possible to use a simple transformer for this purpose. Usually the transformer is built with fixed voltage input and output capacities so that it can be used only for the special purpose for which it was designed. As its name implies, the transformer is used to transform power from one voltage to a higher or lower voltage. A transformer has two sets of terminals—one set called the primary would normally be connected to a 120 volt A.C. source and the other set called the secondary would be connected to the equipment requiring a different voltage (probably 6 volts for bells, buzzers, low voltage lamps, etc.) If the secondary voltage is lower than the primary voltage, the secondary current rating will be greater than the primary current rating so that the power equation is satisfied.

Primary Power = Primary Voltage × Primary Current

Primary Power = Secondary Voltage × Secondary Current

Primary Power = Secondary Power

WIRING SYSTEM—The wiring system of a building starts where the service lines come in from the street. These lines are large enough to carry the total power that will be consumed

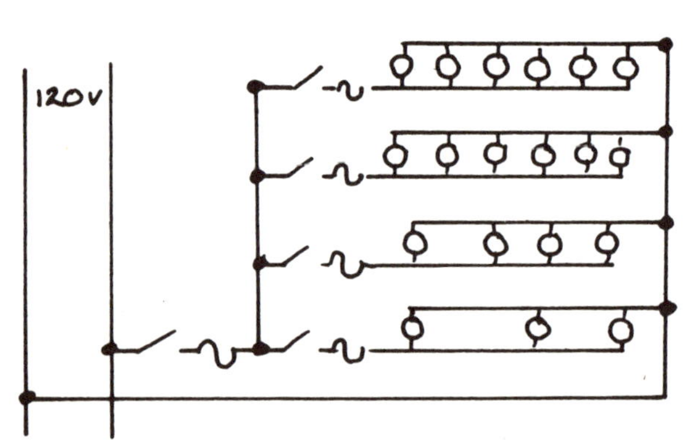


SIMPLE WIRING SYSTEM WITH MASTER AND 4 CIRCUIT SWITCHES

on the premises. Where they enter the fuse or control box they pass through a master switch and are then fused to carry that specified amount of electrical current. Wires of smaller current capacity are connected to the main fuses and then pass through smaller fuses having the same capacity as the wire. These become either individual circuits that extend out to the outlets or are branch circuits that are further broken down into smaller circuits. Usually the final break down is made as near to the place where the power is to be used as possible in order to economize on the amount of wire needed to complete the circuits. The capacity of each circuit is determined by the size of the wire that leads to it and the capacity of the fuse should not exceed the capacity of this wire.

120 VOLT WIRING SYSTEM — Where the 120 volt system is used throughout the building two wires enter the meter. One is covered

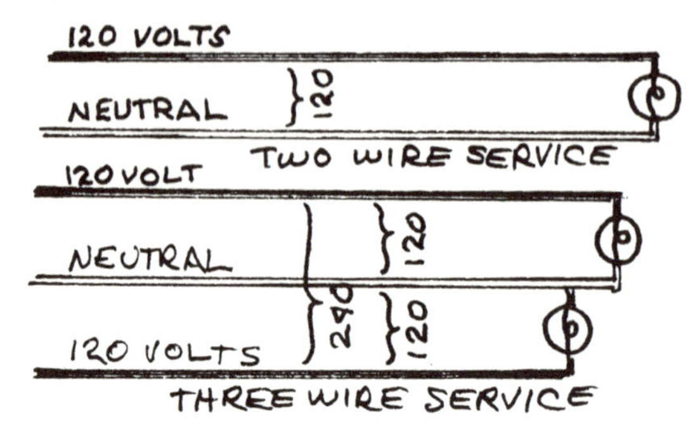


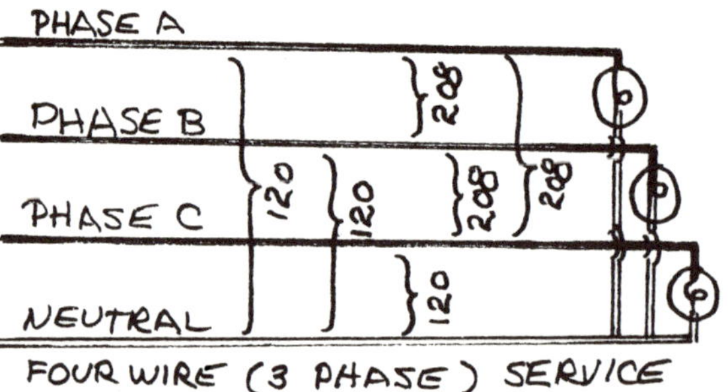


with white insulation, the other with black; white wire is neutral; black wire is "hot"; switches and fuses fastened into "black" wire throughout the system. Wires of various sizes connect to these and extend throughout the entire system.

240 VOLT SYSTEM—In this system 3 wires

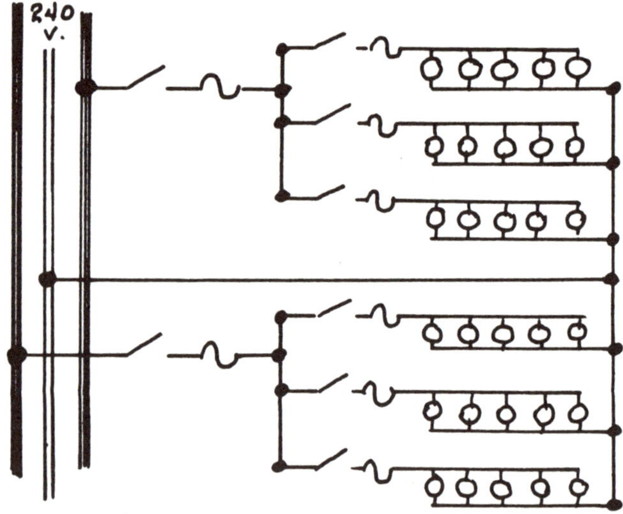


240 VOLT SYSTEM - TWO EQUAL BRANCHES.

come from the meter. The white wire is neutral; the two dark wires are "hot" with switches and fuses cut into them. If 240 volt power is required connections are made to the two dark wires. This system is usually divided into two sets of circuits with one dark wire and the neutral wire connected into each system. The total consumption of power should be balanced between the two systems.

4 WIRE (3 phase) SERVICE—In general use, this system utilizes 4 wire service; the white wire is neutral; while the three dark wires are "hot". The white wire and each one of the colored wires, individually, are utilized to create three different 120 volt power sources. So three circuits, or sets of circuits, may be established with this system. Connections between any pairings of dark wires will provide a 208 volt power source. This system is especially valuable since it can be used to wire in three phase motors.

SHORT—When an electric current takes the path of least resistance and jumps from one wire to another because of faulty insulation a short circuit, or "short" exists. Usually a "short" will cause a fuse to "burn out" or a circuit breaker to "kick out". Find the cause before replacing the fuse or resetting the breaker. Look for bare wires or faulty connections.

*NOTE: Fuses, switches, and other devices should never be cut into the white or neutral wire.*

PRACTICAL APPLICATION OF THE POWER EQUATION—Now that circuits and fuses have been explained here are a few examples of how the power equation is used.

1. The wire in a certain 120 volt circuit is heavy enough to carry 15 amperes safely but no more than 15 amperes. (A) What is the wattage capacity of the circuit? (B) What size fuse should be used?

*Ans.* (A) Power = Voltage $\times$ current
Power = $120 \times 15$
Power = 1800 watts

No more than 1800 watts can be used at one time on this circuit.

*Ans.* (B) Since the wiring can carry only 15 amperes, a 15 ampere or smaller fuse must be used. A larger fuse might allow the wiring to overheat and cause a fire.

2. It is desired to connect three 1000 watt 120 volt spotlights in parallel to a single circuit. What must be the current capacity of the circuit wiring and fuse?

*Ans.* $3 \times 1000 = 3000$ watts, the total load
Amps = Watts $\div$ Volts
Amps = $3000 \div 120 = 25$

The wiring and fuse must be rated to carry at least 25 amperes. The fuse must not have a higher rating than the wire.

3. A dimmer is rated at 550 watts, 110 volts. What should be the circuit current capacity and the fuse capacity?

*Ans.* Amps = Watts $\div$ Volts
Amps = $550 \div 110$
Amps = 5

The circuit should be fused at 5 amperes in order to protect the dimmer. Most stage circuits will carry 15 amperes without overheating, but a 15 amp fuse in this circuit would be too high.

* * * * *

Although this discussion of electricity and electric circuits is brief it will suffice as an introduction to stage electrical work. The chapters that follow will provide information concerning practical applications of the basic ideas expressed in these few short paragraphs.

BORDER PATTERN

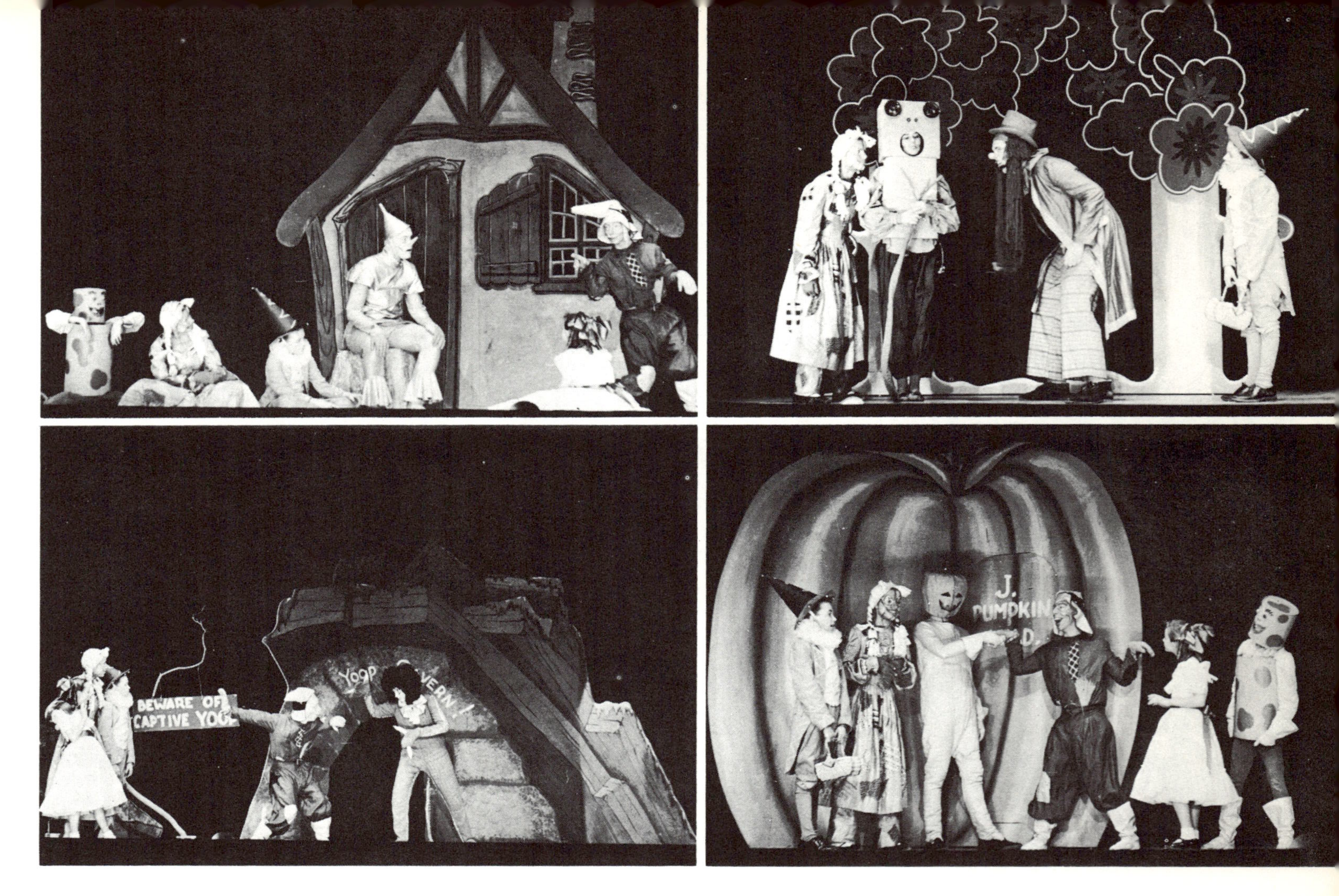


*The Patchwork Girl of Oz.* Small contoured book units that require a minimum of bracing are here space staged against black drapes.

# 13

# *Lighting Equipment*

Stage lighting has become one of the major factors of production in the theatre. Equipment and facilities for lighting vary tremendously from one theatre to another so any listing must be fairly general, yet inclusive enough so that one desiring to make purchases and improvements may know what is available.

### Tools

A limited number of tools are essential in the stage electrical department. Screw drivers, pliers and a knife are the major tools that will be used constantly.

SCREW DRIVERS—It will be convenient to have a set of varying width and length blade screw drivers. There are times when a long handled screw driver is convenient and times when only a short handled one can be used. Screw drivers for electrical work should have wooden or plastic handles and the width and thickness of the blades should fit the screws of the electrical connectors.

PLIERS—Although regular snub nosed general utility pliers may be used for stage lighting purposes there are a number of varieties that are more practical.

*Long nosed pliers,* with or without cutting edges in the jaws, are useful for shaping the ends of wires before making connections with screw contacts on outlets. These are also useful for reaching into narrow apertures. Long nose pliers are available in 5″, 6″ and 7″ lengths. The 6″ length is a good general purpose size.

*Side cutting pliers* have rugged jaws and cutting edges. They are extremely useful and practical for electrical work since they can be used to cut almost any size of wire and their jaws grip firmly for pulling, bending and twist-

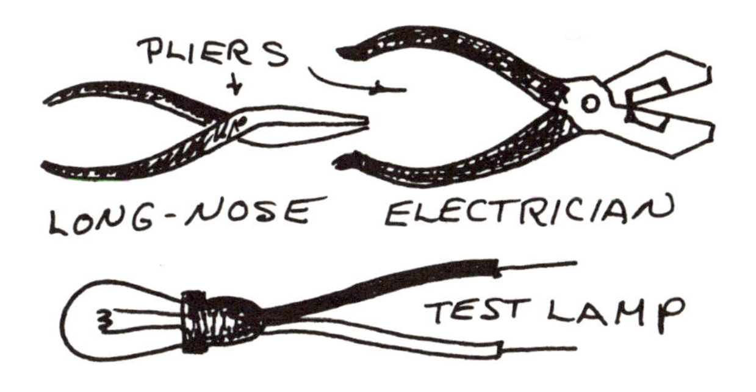

ing. They are available in lengths from 5″ to 9″. The 8″ size is large enough for most work.

*Oblique cutting pliers* are convenient to use if a great deal of wire cutting is done. Heavy side cutters can be used for most of the same work.

*Insulation of pliers*—When pliers are being used for electrical work it is a good idea to cover the handles with friction tape to insulate them and prevent shock to the operator.

CUTTING TOOLS—A knife is needed for removing the insulation from wires. Any type of sharp knife will do. If all of the work is being done on a work bench a paring knife will do. Usually a rugged pocket knife or a sheath knife with a belt case is more convenient to use if it must be carried around on the stage.

SCISSORS—A medium sized pair of scissors will work best for cutting color media.

WRENCHES—Most of the spotlights, floodlights and clamps have set screws that must be adjusted. Although pliers will work for loosening and tightening these, a crescent wrench will do the job better. An 8″ wrench is large enough for most of the nuts on lighting instruments.

PIPE WRENCHES—If there are "pipe stands," "trees" or "battens" on which lights are to be hung it may be necessary to have a pipe wrench. A pipe wrench has adjustable jaws with "teeth" so that it will grip the smooth surface of pipe and enable one to tighten or loosen a connection. A 12″ pipe wrench will be large enough for light jobs; 16″ or 18″ pipe wrenches will handle almost any pipe that will be used.

SOLDERING TOOLS—An electric soldering iron, small blow torch or cartridge type torch are convenient for heating wire and connectors for securing with solder. Often permanent connections need soldering in order to insure a sound contact.

SOLDER—It is advisable to use resin filled rather than acid filled solder for making electrical connections.

TEST LAMPS—A testing device consisting of a socket and lamp and pieces of insulated wire a few inches long extending from the socket is a convenient device for testing electrical circuits. The wires are touched to the contacts of a line. If the lamp glows, the line has voltage; if not, there is no voltage. By using the test lamp in various places along the circuit it is possible to trace the position of a faulty connection. Test lamps may be purchased or easily assembled in the lighting shop.

LABELS FOR EQUIPMENT—Electrical material, spotlights, cables, panel boards and swatch cords frequently need labels for rapid identification as to size, length, capacity, use, etc. Tie-on tags or regular bandage adhesive tape may be used for this purpose. It is easy to write on these and fasten them onto the object needing identification.

### Light Bulbs or Lamps

LAMPS—The electric lamp, frequently referred to as a light bulb, is one of the most important elements in illumination. It consists of a number of parts. The globe is a sealed glass container. It may be evacuated or contain any of a number of gases, usually inert. Most of the common lamps are gas filled. The glass may be clear, frosted inside or out, made of stained glass, dipped in color or may be partially silvered on the inside or outside to create

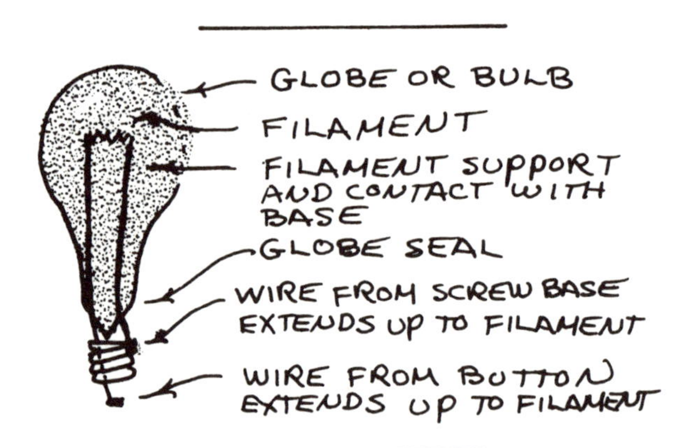

a reflective surface. Inside of the globe, glass supports or stems extend up from the base. Through these are wires which connect, one on each end, to the filament. The filament is a high resistance, high melting point wire that becomes white hot when the electric current is forced across it. The base of the bulb consists of two contacts separated by an insulation cement used to fasten the base to the globe. If it is a screw base lamp one of the wires fastened to the filament extends down and is fastened to the screw base while the other wire is fastened to a brass button at the tip of the base.

LAMP TYPES AND SHAPES — When lamps are ordered it is necessary to designate the shape of globe, type and size of base, voltage and wattage. The shape of the bulb is designated as follows: candelabra—C, flame—F, globular—G, ordinary lamps either pear shaped or modified pear shaped—A or PS, tubular—T,

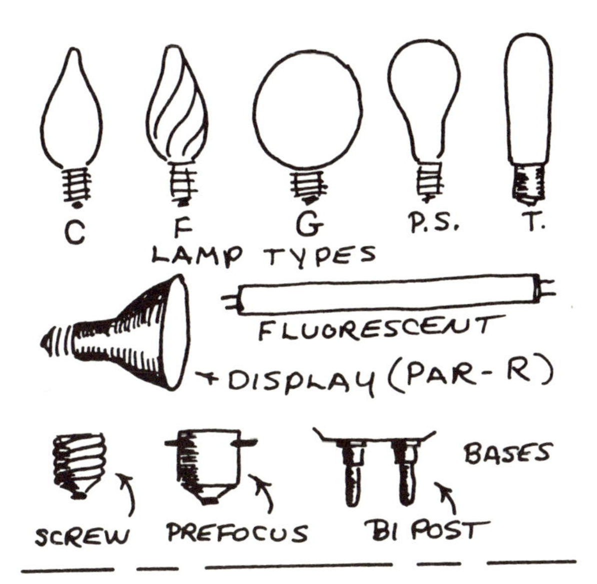


sealed beam display lights in the outdoor variety with heavy glass globes—PAR-38, display lights with indoor type globes—R-40, and long fluorescent tubes. The fluorescent lamp differs from the others and will be discussed later.

SIZE OF BASE — There are six commonly used sizes of lamp screw bases: candelabra, intermediate, medium, ad-medium, mogul and 3-lite (for lamp fixtures). Medium screw base is the type usually used in the home.

TYPE OF BASE—Although the majority of lamps have bases that simply screw into the socket there are other special varieties used extensively in the theatre. Prefocus is a variety of base designed to place the filament in a specific predetermined position. These are used in spotlights. The twist lock type of base is turned a quarter turn to lock it in position after being placed in the socket. The bipost base is another type of prefocus lamp that fits an instrument with the filament facing in a specified direction. The bipost has two connector prongs that are pushed into a special connector device.

LAMP SOCKETS—The receptacle portion of the lamp socket conforms to the size and type of lamp base to be used in it. The socket itself is made of insulation material and has contact screws for a pair of lead-in wires. There are numerous varieties of receptacles. Those used most often around the theatre will be the pendant variety that is supported either by hanging on the electrical cord or supported by a tubular conduit such as that of an ordinary floor lamp or table lamp; the porcelain socket that can be mounted on a flat surface; the sign socket that unscrews so that it may be mounted in a hole on a metal sign. The sign socket sandwiches the sheet metal with one half the porcelain on one side and the other half on the other side.

FILAMENTS—Filaments in lamps designed for stage use are shaped for specific lighting instruments. Lamps designed for floodlights and strip lights have open filaments that spread and disperse the rays of light for maximum area coverage. Lamps designed for spotlights have filaments concentrated into a small space to approximate a point source of light. (The carbon arc spotlight with its source a small spark gap between 2 carbons probably comes closest to the ideal point of light desired but it is not practical for most lighting instruments.)

LAMPS—The shape of globes or lamps and the position of the filament within them are designed for maximum ventilation and maximum burning life. They should be burned only in the position designated on the shipping carton and/or lamp.

FLUORESCENT LAMPS—Fluorescent lamps not used extensively in the theatre differ from incandescent lamps in that they contain a gas and interior coating that causes the tubular lamp to glow when current passes through the tube from an electrical contact at one end to a contact at the opposite end. Fluorescent lamps have long life, produce more lumens per watt than incandescent lamps, and produce a soft "line of light" that is good as a striplight source, but, unfortunately they cannot be dimmed effectively.

### Electric Wire

There are basically two types of wire: the solid composed of a single wire and the

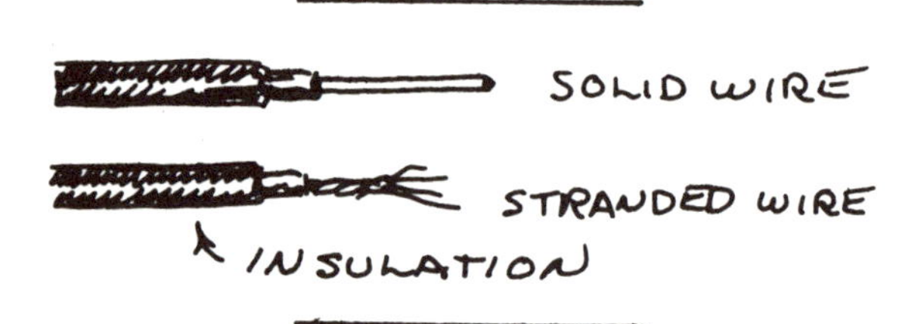


stranded composed of a number of small diameter wires twisted together to form one conductor. Single strand wire is used for permanent installations and is relatively inflexible while stranded wire is flexible and is used for cords that are moved about a good deal. Theatre cable is composed of stranded wire.

SIZE OF WIRE—As suggested previously the cross sectional size and substance from which the wire is manufactured will determine the load carrying capacity of a wire. Copper is a good conductor, relatively plentiful, and is used throughout most electrical systems. A standard has been set up so that the capacity of a wire can be determined from its size number. The smaller the number the greater the capacity of the wire. The "load" indicates the amount of current that can be safely carried continuously over the wire without danger of heating which might cause disintegration of the wire.

SIZE OF WIRE AND LOAD CAPACITY—Rubber covered cable is listed as follows:

| | | |
|---|---|---|
| #18 wire— | 7 amps— | 840 watts |
| #16 wire— | 10 amps— | 1200 watts |
| #14 wire— | 15 amps— | 1800 watts |
| #12 wire— | 20 amps— | 2400 watts |
| #10 wire— | 30 amps— | 3600 watts |
| # 8 wire— | 40 amps— | 4800 watts |
| # 6 wire— | 55 amps— | 6600 watts |
| # 4 wire— | 70 amps— | 8400 watts |
| # 2 wire— | 95 amps— | 11,400 watts |
| # 1 wire— | 110 amps— | 13,200 watts |
| # 0 wire— | 125 amps— | 15,000 watts |
| #00 wire— | 145 amps— | 17,400 watts |
| #000 wire— | 165 amps— | 18,800 watts |
| #0000 wire— | 195 amps— | 25,400 watts |

LOAD—All electrical equipment has a load capacity that should not be exceeded. This capacity is listed in amperes (amps) or watts. If it is listed in amperes it is necessary to multiply by 120 to get the wattage capacity (using the formula P=EI) assuming the circuit has 120 volts. The process is reversed to determine the amperage from the wattage. If the capacity rating for a circuit is not listed on the switches, door of the box, control board or elsewhere it will be a good idea to have an electrician check over and label the entire system. Overloading of circuits will at the least burn out fuses or "kick" off circuit breakers and at the most may start a serious fire.

INSULATION ON WIRE—Wire used for conducting electric current is covered with nonconducting material or insulation. The insulation varies according to the capacity and specific use of the wire. Wire leads for instruments that are likely to heat considerably such as heaters, waffle irons and spotlights have a special asbestos fiber insulation. Most other wires, including theatre cables, have a rubber covering on the outside and various impregnated thread and compositional materials between this and the individual rubber covered wire on the inside. Usually there are two conductors within a cable so that both wires needed to complete a circuit can be handled as one.

WIRE SPLICES—When wires are connected or spliced together the ends of the wires should be carefully bared of insulation without cutting the individual strands. The ends to be connected should then be carefully twisted together. If they can be twisted back upon one another the connection will probably be stronger and if a spot of solder is put on the electrical contact will be even better. The solder is easily applied. Heat the joint with a soldering iron or small torch. As soon as it is hot enough touch the end of a piece of resin core solder to the wires at the joint so that it will melt and run in between the strands. It will take but a moment for this connection to cool. Soldering is especially important if the connection is permanent or semi-permanent. After the wires have been appropriately fastened together the connection is covered with electrical tape that insulates the splice. The beginner might find it advisable to complete the connection of one wire of the pair through the taping stage before even beginning the other one. It is extremely important to insulate the wires from one another wherever there is a chance for a short circuit ("short").

"SHORT"—A "short" in a wiring system is a short circuit, which means that the electric current selects the easiest and shortest connection to complete a circuit. It is easier for it to flow across to its neighboring wire of the circuit if there is a bare section on both wires than it is for the current to flow through a lamp where it has to do work. Most short circuits are prevented by carefully insulating all connections.

INSULATING MATERIALS—Formerly it was necessary to use two different tapes, a rubber insulation tape, and a cloth friction tape, in order to insulate and cover an electrical splice. Now, however, it is sufficient to use a single vinyl plastic electrical tape that both insulates and covers the splice. This tape is "UL" approved.

"UL" APPROVAL—It is advisable to check all electrical equipment before purchase to be certain that it has the "U.L." stamp of approval. This stamp means that the electrical device has been approved by the "Underwriter's Laboratory". The lab checks for proper operating safety standards of electrical materials and equipment when used under normal operating conditions.

STAGE CABLE—Flexible 2 or 3 conductor electrical cord (3 conductor has an additional safety factor consisting of a "ground wire" that hopefully extends throughout the electrical system and is grounded somewhere on the premises, in addition to the regular "hot" and neutral wires) is cut into sections and used as extension cord or cable on the stage. A single cable, which may be of any capacity, and any length, is fitted with a female connector with no exposed contacts at one end, and a male connector with exposed prongs at the other end.

### Plugs and Connectors

Plugs or connectors are devices used to fasten two stage cables or a stage cable and an electrical device together to establish an electrical connection. The connector consists of two parts: the male, which is the portion with two or three prongs (3 if the circuit is grounded), and the female, into which the male is plugged and connected. In ordinary home use the male is connected to the end of a cord that is permanently wired to a light fixture or appliance, and the female is in a wall receptacle. The parts of the connectors have matching contacts and by plugging the male into the female one is in fact extending the length of the power line. The male portion of the connector is ALWAYS connected to the instrument or LOAD and the female connector is ALWAYS connected to the LINE or power source.

TYPES OF CONNECTORS—There are several types of connectors used on the stage. Among them are the following.

STANDARD CONNECTORS—The plugs ordinarily used around the home are standard connectors. These will carry only limited loads and they are connected by pushing into a standard receptacle. Some of these can be connected to lightweight utility cord without the use of a screw driver.

TWIST LOCK CONNECTORS—The twist lock has three prongs with a short extension on one side. After it has been pushed into place it is twisted about a quarter turn which locks

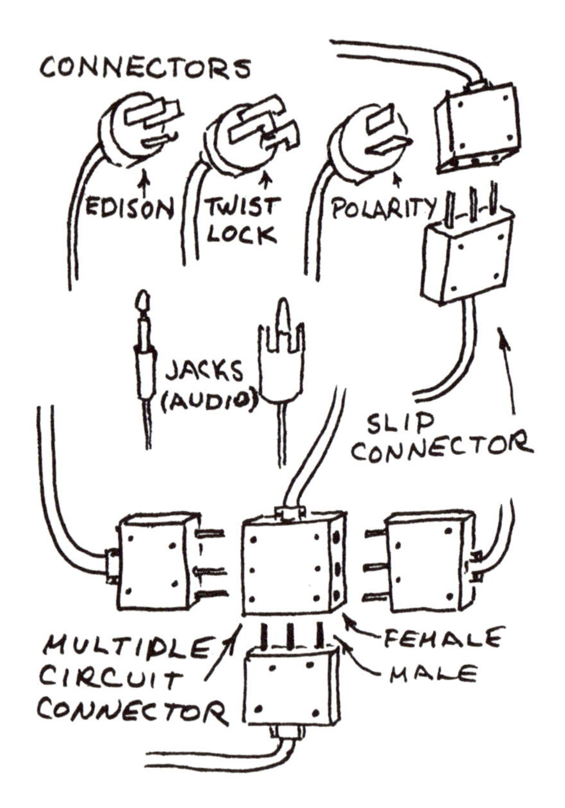


it into place. Twist locks are somewhat heavier than standard plugs and are excellent for stage use since they cannot be disconnected easily by accident.

POLARITY PLUGS—There are times when a plug must be used that can be inserted in only one specific position. Polarity plugs with mismatched prongs are used for this purpose. They either have one large and one small prong or they have one prong at right angles to the other prong.

MULTIPLE PRONGS—Sometimes it is necessary to have connectors with more than two prongs so a multi-prong connector is used.

SLIP CONNECTORS—Slip connectors are used on most stages. The round prongs are mounted in blocks of bakelite for the male half and metallic sleeves are mounted in similar blocks for the female half of the connector. These are designed for heavy duty use and will withstand a good deal of abuse.

FLOOR POCKET PLUGS—Floor pocket plugs may be of the slip connector variety or any of a number of other designs. Many of the heavier ones consist of a block of bakelite with a copper or brass contact plate on either side. These slide into pockets that have holes with similar copper plates on two sides to make the contact.

JACKS—Jacks are sometimes used as plugging devices. The jack may be used as a single or multi-wire connector. If it is a solid brass or copper rod an inch or more in length with an insulated handle to cover the wire connection it is probably a single conductor jack. If the rod is broken near the end with a segment of insulation it is undoubtedly a double pole conductor. Swatch panels in some interconnection systems use jacks.

## Fuses

The fuse is a safety device placed in the electrical circuit to prevent overloading which may result in damage to the wire or the instrument wired into the circuit. The ordinary fuse contains a link or wire of fairly low resistance, low melting point metal. As soon as the load on the line exceeds the capacity of the fuse the link will heat and melt, thus breaking the circuit.

SCREW BASE FUSE—The screw base fuse

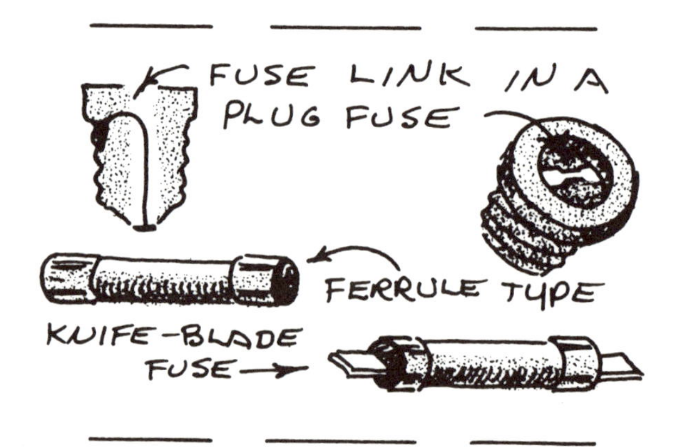

fits into an ordinary socket. This type fuse is available in sizes up to 30 amps.

FERRULE CONTACT FUSES—Ferrule contact type cartridge fuses are available up to about 60 ampere capacity. These are made in renewable link or one time fuses. With the renewable type it is possible to unscrew the ends and replace the fuse strip.

KNIFE BLADE FUSE—The knife blade fuse is a cartridge fuse with heavy copper plate contacts on each end which make the contact in the receptacle. Knife blade fuses range in size from about 70 to 1000 amps.

CIRCUIT BREAKER—The circuit breaker is an automatic switch type fuse. An overload or short causes it to trip off automatically. The trouble must be corrected before it can be permanently switched on again. The circuit breaker insures against over-fusing a circuit.

## Lighting Instruments

There are a few factors that should be considered when any type of lighting instrument is being selected for possible purchase.

1. *Size.* The physical measurements of the instrument should be known. Will it fit into the space where it is to be used?

2. *Shape.* Is it shaped so that it will be convenient to handle? Shape may be important if the instrument is to be in an exposed position where it will be readily seen by the audience.

3. *Weight.* Is it as light in weight as possible? Aluminum may be used by the manufacturer to cut down the weight of the instrument.

4. *Flexibility.* Can the instrument be used for varied purposes and in various places? The burning positions of the lamp may be checked to be certain that it will burn efficiently in the desired positions.

5. *Sturdiness.* Is it ruggedly built so that it will withstand rough handling?

6. *Convenience.* Can it be easily mounted and rigged? Can it be mounted in various ways — pipe, stand, tree, batten?

7. *Accessibility.* Is it easy to service? Can it be easily cleaned? Can the lamps be easily replaced?

8. *Ventilation.* Will it operate without undue heating? Is it ventilated without at the same time allowing undue light spill?

9. *Efficiency.* Is it as efficient or more efficient than similar instruments? Does it have a good reflector surface? Is the lamp positioned carefully in regard to the reflector and lens (if it is a spotlight)?

10. *Color holder.* Does it have a well designed color holder? Is it easy to mount the color media? Will the color frames stay in place when the instrument is mounted in the positions where it will be used?

Lighting instruments used for the illumination of the acting area of the stage may be classified under four headings for convenience: striplights, floodlights, sealed beams and spotlights.

## Striplights

Striplights, as the name implies, consist of a series of units mounted side by side. Striplights may be employed as footlights, overhead borderlights, cyclorama base and overhead lights, backing lights and sometimes, when mounted vertically, as side lights.

The simplest striplight consists of a series of sockets mounted side by side on the wall, floor or a movable board. Plain lamps, colored lamps or display lamps may be screwed into the sockets.

A more complex striplight may be built by adding a reflector trough built of wood, sheet metal or aluminum. Large size eave troughs may be used for this purpose. When the reflector is added the efficiency of the striplight is increased and the amount of spill light is minimized.

Commercial striplights composed of a series of socket-reflector-color media units placed side by side provide for more efficiency and flexibility of operation. These may be purchased in various length sections and with various sized reflectors for lamps of from 40 to 500 watts. If the strip is to be close to the surface to be lighted, for instance footlights on a small stage, it is more advantageous to have closely grouped reflectors and lamps, especially if red-blue-green combinations are used so that the color mix will be close to the strip. With large units, obviously the color mix is farther out from the individual lamps to the point where the beams from the lamps intermingle. If the striplights are to remain in place at all times

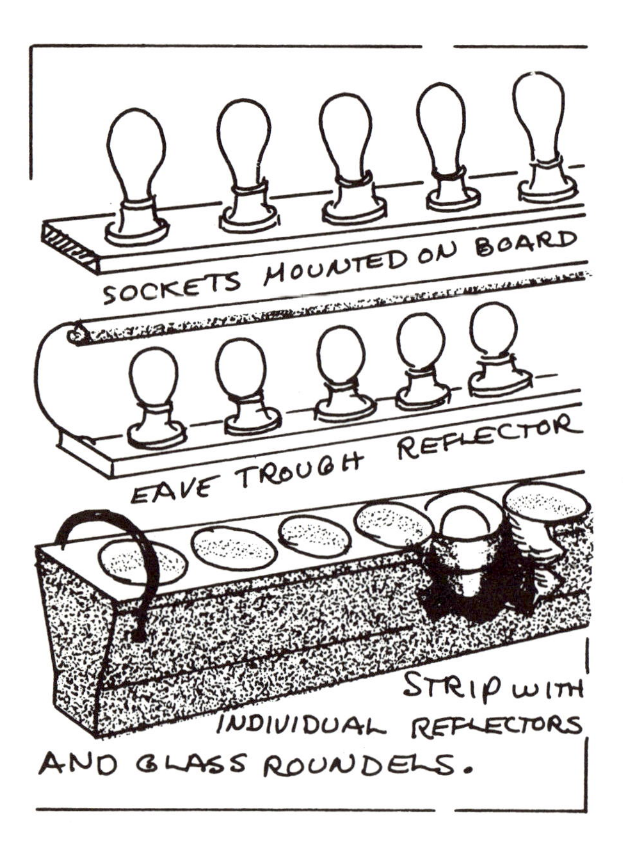


the length of the sections is immaterial. But if the strips are to be moved about on the stage and used for varied purposes and if they are to be stored while not in use, it will probably be advisable to get sections that can be handled rather easily. Some strip sections are available in 5′ or 6′ sections. The length of the sections depends to some extent upon the size of each reflector unit. If the three color system is to be used it is also advisable to get strips that are wired in three circuits so that the individual colors can be controlled separately.

FOOTLIGHTS—Footlights are strip lights permanently installed, removable, or designed to fold into the floor on the front lip of the stage. Rarely used today.

An indirect footlight in a floor slot has the lamps facing toward the audience, reflected toward the stage and actor by mirrors. The reflected beam of light is broadened and diffused.

HALOGEN LAMPS—A recent development in lamps is the quartz tube, iodine gas-filled, tungsten-filament lamp. In this lamp the tungsten filament is mounted in an atmosphere of iodine. In most lamps as the tungsten deteriorates the interior of the globe darkens but in the halogen lamp the tungsten filament is gradually replaced which tends to keep the globe clean, retain brightness, and lengthen the life of the lamp.

The extreme heat of this lamp necessitates having a small quartz crystal glass housing. The life expectancy of halogen lamps is reduced if the lamp is burned at less than full brightness for prolonged periods of time. Use it where excess dimming is not required, in housings designed for its use.

*NOTE: Observe warning about touching lamp with bare hands. Oil from hands will cause the quartz crystal tube to blister.*

### Floodlights

Although floodlights usually have rather large wattage lamps—500, 750 or 1000 watts—they may be any size. Basically a floodlight is a simple reflector housing equipped with a socket, lead in wires, color frame holder and frame, and a mounting yoke. For small wattage lamps the housing can be made of wood and shaped like a small box with one side removed. Or a 5 gallon cooking oil can (available at restaurants) can be used if one side is removed and a socket is mounted in the bottom. The inside of the reflector may be coated with white or aluminum paint. White will provide a high diffusion reflective surface which is large-

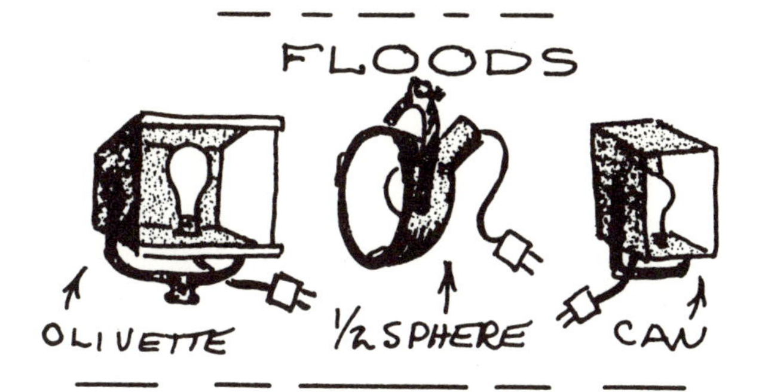


ly non-directional. Sheet aluminum, available at "do-it-yourself" counters, can be used to make a reflector.

COMMERCIAL FLOODLIGHTS—A variety of floodlight types are available from theatrical supply houses. These range all the way from box shaped Olivettes (a type that has been used for many years) to spherical reflector floods that are probably more efficient. Since the flood is used primarily for over-all lighting its requirements are simple. It should have a lamp housing with a high diffusion reflective surface which may be white or silvered; it should have a substantial color frame holder and frame; and it should have a yoke mounting that will allow it to be rigged on a pipe stand, tree, pipe or batten. Most floodlights are relatively inexpensive.

### Reflector Lamps (PAR38 and R40)

Electrical stores sell lamps for display purposes that are self contained spot or flood instruments. The PAR38 and R40 Reflector lamps are probably used more frequently than any other size. The PAR38 lamp is made of pyrex glass and will withstand considerable punishment. It is designed for indoor or outdoor use. The R40 series, on the other hand, is made of thinner glass and should be used indoors only. The number designates the diameter of the lens across the face of the lamp in eighths of an inch. R40's are therefore 5′ in diameter and PAR38's are 4¾″ in diameter. These are the numerical designations for the 150 watt Reflector lamps. Reflector lamps are available in other wattages (75, 200, 300 or 500 watts), some of which have other number designations.

A theatre with a scant budget and no equipment might start with Reflector lamps as a basis for a lighting system. The lamps are efficient and easy to use. Part of the inside of the glass is silvered to provide a highly reflective surface and the front of the glass is designed to serve as a lens to distribute the light. The flood type tends to spread to a fairly wide beam while the spot type tends to restrict the light to a narrower beam.

Reflector lamps may be obtained with clear or colored glass. The color is baked onto the glass and is quite permanent. Unfortunately, the range of colors is somewhat limited but the variety is sufficient for simple lighting.

A number of devices can be purchased or manufactured to use with Reflector lamps. Some of these follow.

SWIVEL SOCKETS—Home made swivelling devices can be made. Two short pieces of 1″ x 3″ boards about 10″ in length may be hinged together at one end. Fold the two boards together flat and drill a hole through the two about an inch and a half from the unhinged end so that an adjustable bolt can be passed through both of them. The bolt can

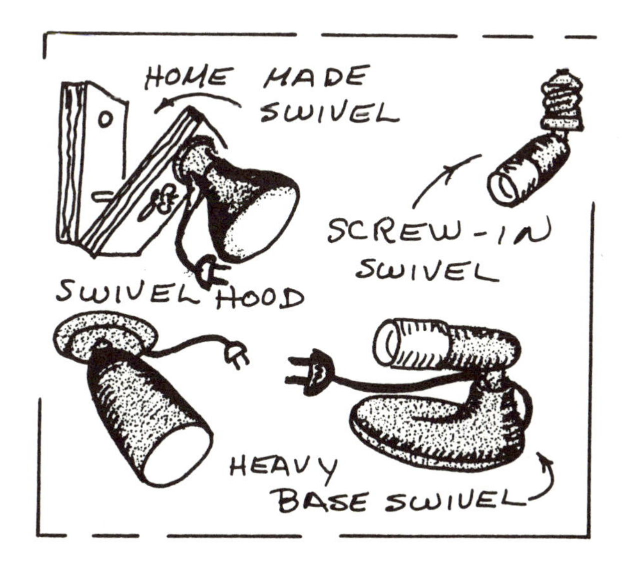

be fastened securely to one of the boards by threading the nut down tight. This is the board that will be fastened in place wherever the unit is to be mounted. Cut a slot down toward the hinge for about 3″ in the other board. The socket can be mounted below this slot. If the bolt is long enough it will be possible to adjust this light to shine in almost any position by tightening or loosening a wing nut that is placed on the bolt after it goes through the socket board. Although this is a makeshift affair it will work fairly well.

Various types of swivel sockets and swivel units can be purchased from electrical shops. Some are available attached to "bullet" shaped housings. The housings will cut out some of the spill light, especially if they are fairly long so that the lamp mounts down inside of them. There are permanent swivels mounted in strip sections, screw in swivels, clamp on swivels, bolt down swivels and swivels with bases that will stand on the floor. The type purchased will probably be determined by the position in which they will be used.

LOUVRES—Louvres are concentric rings that fit into the housing in front of a reflector unit. They are designed to cut some of the spill (or extraneous) beams that fall outside the area to be lighted. To be effective the rings should have an inch or more of depth and be painted flat black. If they are shiny their surface will reflect light; if they are flat black their surface will absorb the spill light.

HOME MADE HOUSINGS—Home made housings for reflector lamps can be rather easily made from gallon cans. Paint, syrup or any other type of can that has a pry off lid will work the best since the edge of this type can is reinforced. A two piece sign socket can be mounted in the bottom of the can and 2′-0″ of asbestos cord and a plug attached. "U" shaped hangers can be made from strap iron and fastened to the sides of the can. The lid might be adapted as a color holder if the center can be cut out and a fastening device contrived. Color frame holders can easily be fashioned from thin strap iron or the strap steel used for packing cases. Color frames to slip into these holders can then be made from cardboard, tin or aluminum.

FUNNELS—Unfortunately it is not possible to change the area coverage of reflector lights without devising a special mechanism. Funnels can be made for this purpose from 5″ stove pipe. It is possible to make adjustable funnels by having inner and outer sleeves so that one will slide inside the other. Spill light that falls outside the area that needs lighting can thus be partially restricted and the light area increased or decreased in size. A funnel 2′-0″ long can be used where it is desirable to have a pin spot.

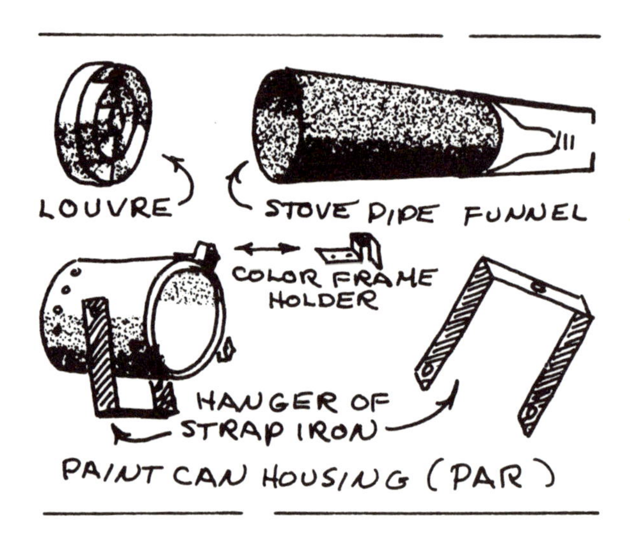


Reflector lamps have many uses on larger as well as smaller stages. They can be used for special effects, for footlights, striplights, backing lights and cyclorama base lights. They are inexpensive, efficient, easily obtainable and should not be overlooked as stage lighting equipment.

### Spotlights

The most important light units on any stage are the spotlights. Although reflector lights can be used in place of regular spotlights they are never quite as effective where complex lighting effects are desired.

Basically the spot is a directional light and consists of:

- Lamp housing with vent holes to cool the lamp but covered by baffles to prevent undue spill light.
- Focusing device to change the size of the area lighted.
- Reflector.
- Support or hanger.
- Lens or lenses.
- Shutter, cutoff and iris devices (sometimes).

Frequently spotlights are loosely classified as large (if they are 1000 watts or more), standard (500 to 1000 watts) and baby (100 to 500 watts).

Several factors should be borne in mind concerning the selection and use of spotlights.

A spotlight gathers light rays from the lamp and redirects them so that they travel almost parallel to one another. The rays that cannot be redirected are absorbed by the blackened inside of the housing.

A spotlight is used where concentrated light is desired in a restricted area.

A spotlight operates more efficiently if it is close to a subject than if it is at a great distance from the subject.

The beam of light is hottest when the lamp is fairly far back in the housing and the beam is narrow.

There is a correct burning position for the lamps of almost all spotlights and the instrument should be positioned so that the lamp will be in this position for greatest lamp efficiency.

Spotlights should not be operated over long periods of time unless they are absolutely required. Spotlight lamps have relatively short burning lives and are expensive.

The spotlight is designed to capture a maximum number of the light rays emitted from the lamp and by means of a reflector and usually one or more lenses to bend those rays so that they pass through the front of the instrument in lines that are almost parallel. The ideal spotlight would make it possible to have evenly distributed light or a hot center and fading toward the edges of the area being illuminated and would make it possible at the same time to light either the maximum or minimum sized area desired on any stage. Since the ideal light has not been—and probably will not be—developed it is necessary to have a variety of instruments to achieve all of these desired effects. These instruments differ in many respects, including the shape and size of the housing, reflectors, lenses and lamps. These elements and their variations will be discussed first and then some of the general types of spotlights will be considered briefly.

REFLECTORS—The lamp in the spotlight is usually backed by a mirror surfaced reflector to project as many of the beams of light forward as possible.

*Spherical Reflector*—The spotlights that have been used predominantly have a spherical reflector. It is a mirror surfaced portion of a sphere designed to be placed in a fixed position in relation to the lamp so that it will capture the rays of light that are projected to the rear

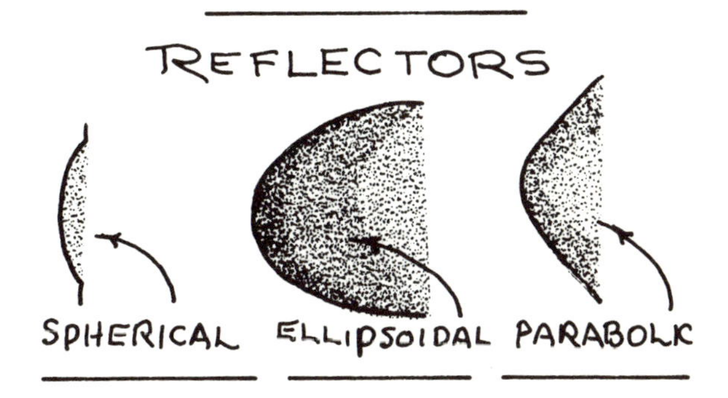


by the lamp. With this type of reflector the rays of light from the lamp that strike the reflector and the ones that pass directly toward the opening in the front of the instrument are projected to the stage. The other rays are absorbed in the interior of the spotlight housing.

*Ellipsoidal Reflector*—The spherical reflector just mentioned is relatively small, surrounding as it does just a portion of the lamp. The ellipsoidal reflector is much greater in area. The lamp projects into it through a hole. The reflector extends around the lamp and projects considerably forward. This reflector captures many of the light rays that would be absorbed by the blackened interior of a spherical reflector spotlight.

LENSES—Most spotlights have a lens or a lens system in front of the lamp to bend the rays of light into approximately parallel lines as they leave the spotlight.

*Plano-Convex Lens*—The plano-convex lens is used on the majority of spotlights. This type of lens in cross section is flat on one side and curves outward on the other. Lenses are available in varying diameters and focal lengths. A given spotlight is built for one diameter but it is possible to get varying focal length lenses of this diameter. The focal length may be determined by allowing sunlight to pass through the lens. The straight rays of light from the sun are bent as they pass through the lens. The point at which they converge is the focal point and the distance from this point to the lens is the focal length of the lens. Theoretically a point source of light placed at the focal point will, when passed through the lens, be sent out in straight lines. However, the filament of the lamp is larger than a point and irregular in shape, so when used with the lens at exactly

FOCAL POINT OF PLANO-CONVEX LENS

the focal point an image of the filament will be projected. In use the lamp socket on an adjustable slider is allowed to slide from slightly in front of the focal length to a position a fraction of an inch from the lens.

*Long Focal Length Lens*—The long focal length lens is thinner than the short focal length lens and can be used in a spotlight where it is desirable to cover only a small area.

*Short Focal Length Lens*—The short focal length lens is thicker than the long lens and

LENSES

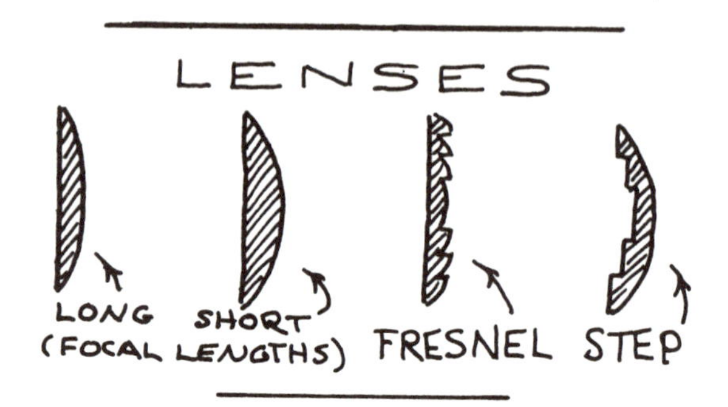


can be used where it is desirable to cover a wide area with a spotlight.

*Step Lens*—The step lens is a plano convex lens with a regular convex side. The plano, or flat side is partially carved out in "steps" to make the lens thinner which in turn reduces the weight, and cuts the light absorption of the lens.

*Fresnel Lens*—The Fresnel lens is a "step lens" with some of the characteristics of the plano-convex lens, but without its great thickness. The surface of this lens is composed of a series of concentric curved sections, each section having approximately the same curve as a plano-convex surface of the same diameter. Since the concentric breaks allow the curve to start again it is possible to reduce the maximum thickness of the lens appreciably. Reduction in the thickness of the glass reduces the danger of breakage from sudden heating and cooling, and reduces the amount of light lost by the absorption of the thick lens. The Fresnel lens is made of pyrex glass which will withstand greater heat than the optical glass of the plano-convex lens. Although the Fresnel lens is more efficient, at the same time, it usually has more diffusion qualities than the plano-convex lens.

### Spotlight Types

With these few facts in mind, it is possible to

describe briefly some of the general types of spotlights available. In many instances the types of spotlights are known, by those who possess them, as "Leko", "Kleiglite" or some other trade name. There are many companies that make instruments that are similar in nature. So here the general type will be named rather than the trade name of the instrument of one manufacturer.

PLANO-CONVEX LENS SPOTLIGHTS — The standard spotlights that have been used most universally have plano-convex lenses and spherical reflectors. They are a general purpose spot light and may be used for area lighting. The beam is sharp edged and tends to have a hot

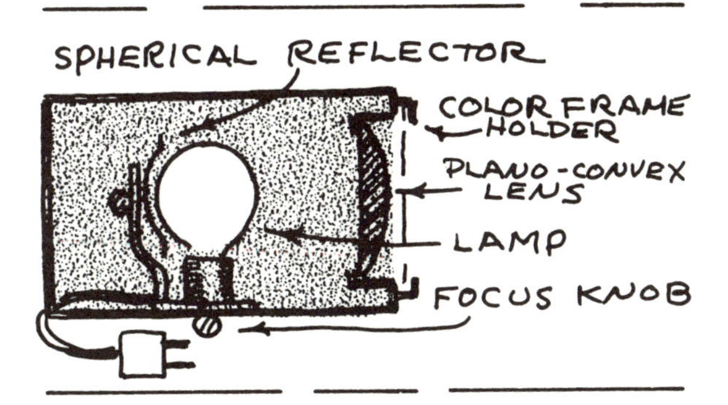


center and at times there is chromatic aberration around the edges of the lighted area. (Chromatic aberration is a colored "rainbow" effect at the periphery of the beam of light.) Usually they are less expensive than either the ellipsoidal or the Fresnel lens spotlights.

FRESNEL SPOTLIGHTS — The Fresnel spotlight gets its name from the Fresnel lens employed. Ordinarily it has a spherical reflector.

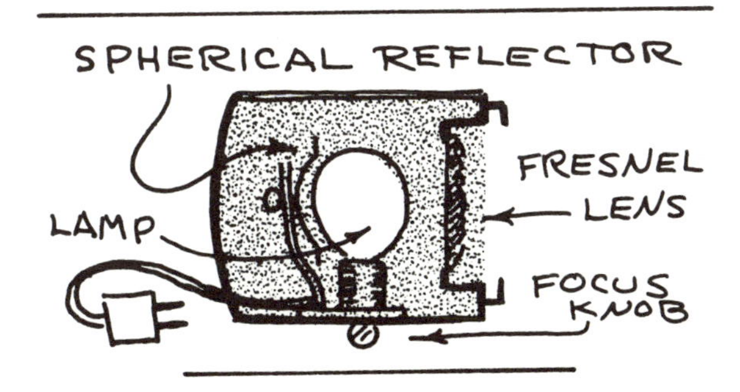


It is highly efficient, produces a soft, even beam of light with a fairly wide spread. Its tendency toward slight diffusion makes it better for use on stage than from the house positions since the diffusion tends to spread the beam.

ELLIPSOIDAL SPOTLIGHTS—Ellipsoidal spotlights have, as the name implies, ellipsoidal reflectors and they usually have one or more plano-convex lenses. This is a highly efficient light with a well defined beam. Many ellip-

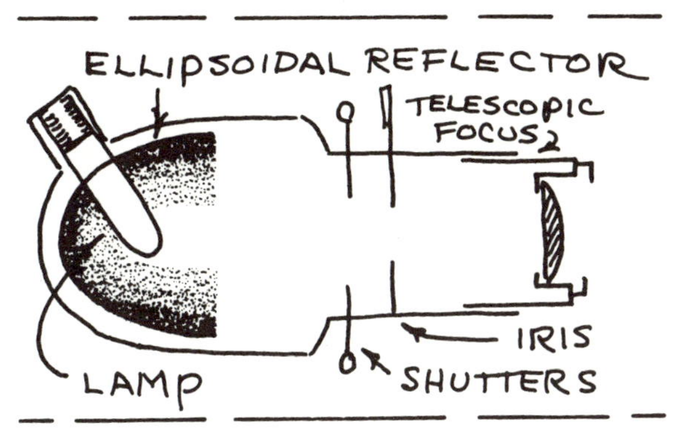


soidal spotlights are available with a telescoping lens mounting for increasing or decreasing the beam spread, with shutters to shape the beam and with an iris to diminish the size of the beam. Ellipsoidal spotlights are excellent for front of the house and other long throw use. Beam or balcony lights of this nature can be shaped by means of the shutters to fit the rectangular shape of the proscenium. Ellipsoidal spots give a sharp, well defined beam.

REFLECTOR SPOTLIGHTS—There are some spotlights that have no lens, or at least no lens that serves to bend the light rays. These may be referred to as reflector spotlights.

*Parabolic Spotlights* — The giant spotlights that sweep the sky to announce that a new supermarket is opening are of the parabolic variety. This is ordinarily a long focus spotlight with a shallow housing. A reflector is usually placed in front of the light source in order to eliminate the wide spill light. The light emitted from this instrument is all reflected light. These ordinarily have high watt-

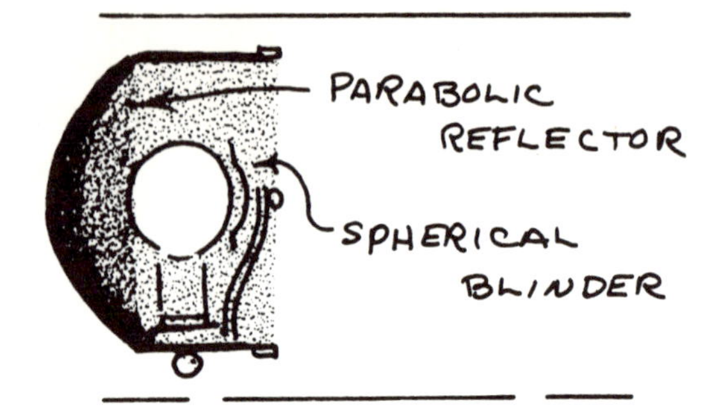


age lamps—or even carbon arcs—as the illuminant and are for long throw use.

*Reflector Floods*—The reflector floodlight is not as efficient as most spotlights, but when equipped with deep louvres a certain amount of concentration of the beam is achieved.

*PAR Lamped Spotlights*—Spotlights of a highly efficient nature with a lens or a lens system, and with cut-off shutters and an iris, are on the market. The Ariel Davis Company has a spotlight of this nature with various focal length lenses and with facilities for shadow pattern image projection.

THROW—When choosing a spotlight it is important to consider the distance the instrument will be from the object or area it is to illuminate. This is referred to as the "throw".

It is difficult to set specific standards, but the following suggestions might be helpful in selecting instruments to try out.

| | |
|---|---|
| 250 watt | 15′ throw |
| 500 watt | 25′ throw |
| 750 watt | 35′ throw |
| 1000 watt | 40′ throw |
| 1500 watt | 50′ throw |
| 2000 watt | 60′ throw |

In order to determine relative efficiency, throw, spread and adaptability to a specific situation, the lighting instruments should be mounted side by side, if possible, and compared with one another. The amount of light falling on an area can be measured with a photographic light meter.

Spotlights should be chosen for the features desired. If maximum illumination is desired they should be chosen for efficiency. The amount and evenness of the beam can be checked with a photographic light meter. If a small area or "pin spot" is desired the instrument should be checked for efficiency when the area coverage is small and it should also be equipped with an iris or shutters, or possibly with both, in order to facilitate in diminishing the area coverage. If a spot with a hot center is desired it should be checked for this quality. If smooth, even illumination across the area is desired—and this is the quality that will usually be the most desirable in regular area lights—this should be checked with the light in position where it will be used. Long throw, short throw, image projection, weight, ventilation, construction, servicing and mounting are other factors that should be taken into consideration. Although many salesmen are honest in evaluating their equipment, it is advisable to test the equipment side by side before making a purchase.

### Instrument Supports

In order that spotlights can be focused correctly and remain in focus it is necessary to have strong, well anchored supports. Some of the various types of devices of this nature are as follows:

1. PIPE BATTENS — Pipe battens may be suspended from the flies on a rope or counterweighted flying system. When lights are mounted it is usually advisable to have or contrive some type of counterweighting so the rigged batten can be raised and lowered easily for focusing the spotlights. Pipes may be mounted permanently on walls, ceilings or elsewhere for this same purpose.

2. LIGHT STANDS—There are several types of bases used on stands. These are crow foot, solid base and caster base. *Crow foot* bases, which have three or four feet extending out at the bottom, are light in weight and frequently the legs are mounted so that they can be folded up against the upright pipe for storage or transportation. The *solid base* is composed of a heavy cast iron disc threaded in the center to

receive a pipe upright. The *caster base* is equipped with casters so that it can be easily moved around on the stage. Sometimes the casters have locks so they will not roll when positioned.

All light stands should be substantial enough to support the weight of the lights that will likely be mounted upon them. They should have substantial bases and should be capable

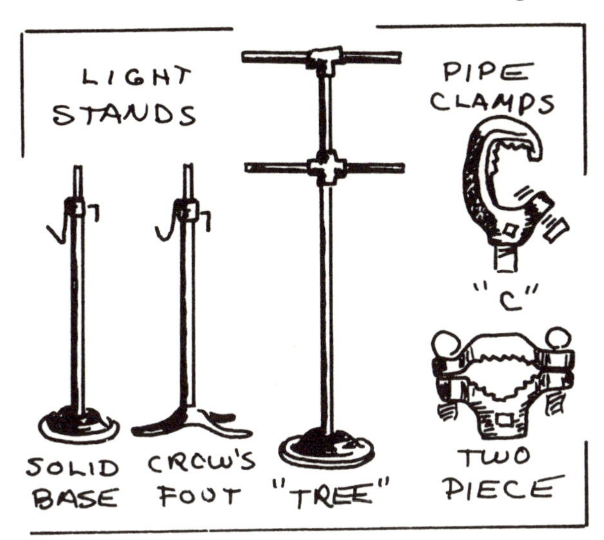


of storage and maneuverability on the stage where they are to be used. They should have telescoping sections that will allow them to be raised as high as desired for the specific stage.

3. TREES—Trees are light stands equipped with projecting arms on which lighting instruments may be mounted. These require a broad, heavy base that can be securely anchored.

4. TOWERS—A light tower may be made up with a wooden or pipe framework. It is frequently mounted on a caster base for maneuverability. Light towers should be designed for the individual stage so that they can be conveniently maneuvered and stored when not in use. For operation they should also—if tall and designed for many lights—have platforms for operators to stand on while focusing and repairing the lights.

### Mounting Clamps

Instruments are mounted to the supports by means of "C" clamps or two piece clamps. The "C" clamp easily hooks over a pipe after which a set screw is adjusted to lock it against the pipe. The two piece clamp is composed of one toothed piece that fits on one side of the pipe and another piece that fits on the other side. A bolt at either end is used to tighten the jaw sections together and hold them firmly against the pipe. The "C" clamp will work very well where the lighting instrument hangs straight down from the batten. Where it is necessary to cinch the clamp up tight to hold the instrument support out at an angle the two piece clamp is probably better.

### Image Cut-Off and Shaping Devices

Auxiliary devices may be mounted on some spotlights for cutting off or shaping part of the beam. These consist of two or four-way cut-offs with sliders that may be moved in across the beam, iris dissolvers that may be opened or closed like the iris of a camera, and "barn doors" (hinged doors) that may be partially or completely closed like a set of double doors. The funnel is another device of this nature.

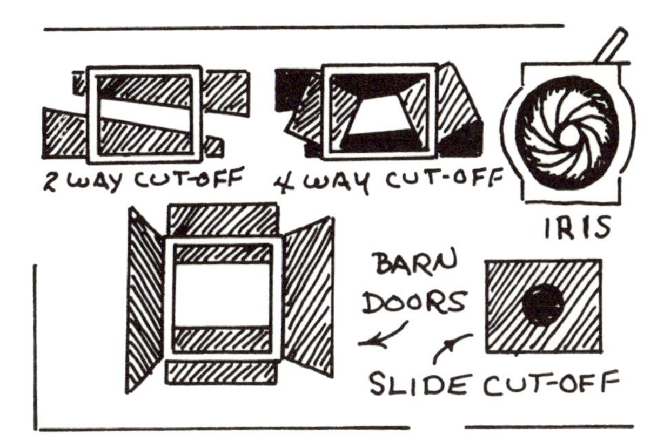


This is a long tubular contrivance, similar to a section of stove pipe. If the funnel is adjustable in length it is possible to cut out spill light and obtain a small pool of light that is almost without any halo spill area. The inside of the funnel should be painted flat black.

### Special Lighting Devices

There are special lighting devices that should

be discussed. These are: follow spots, and projection instruments.

FOLLOW SPOTS — Musicals, variety shows, operas and ballet productions frequently make use of follow spots to accent the action of the moment. Although there are times and places where this can be accomplished with ordinary spotlights usually a more intense spot is desired so special instruments are used.

The follow spot will ordinarily be placed in the front of the house somewhere. It may be at the back of the auditorium, back of the balcony, front of the balcony or at one side of the balcony. Its greater distance from the stage requires that it be a long focal length light with cut-off or iris controls, or maybe even both. It should be capable of being pinned down so that it covers no more than the upper half of the human body. Usually it will be used to illuminate an entire person or a small group, although there may be times when it is desirable to open it up and cover the entire stage. Sometimes it is used wide open and then as a number starts it is pinned down onto a single person.

There are many standard and special spotlights that can be used as follow spots. In fact, any spotlight that will provide a higher illumination than the ordinary stage lights will serve in a pinch. If the throw from the position where the spot must be operated is long it may be necessary to obtain a special incandescent spot of high intensity or a carbon arc follow spot. The carbon arc, if well designed, will probably be the most effective and most efficient to use.

Some lighting companies manufacture excellent carbon arc follow spots with ellipsoidal reflectors.

LINNEBACH PROJECTOR — The Linnebach projector is basically a shadow projector, hence it is most effective when the screen upon which it is projected is light in color value and receives very little illumination other than the illumination from the projector. A Linnebach projector can be built quite easily. It consists of a concentrated filament lamp, the filament being as close to a point source of light as possible; a housing painted black on the inside; and a cut out slide proportioned for the desired image to be projected. The entire instrument

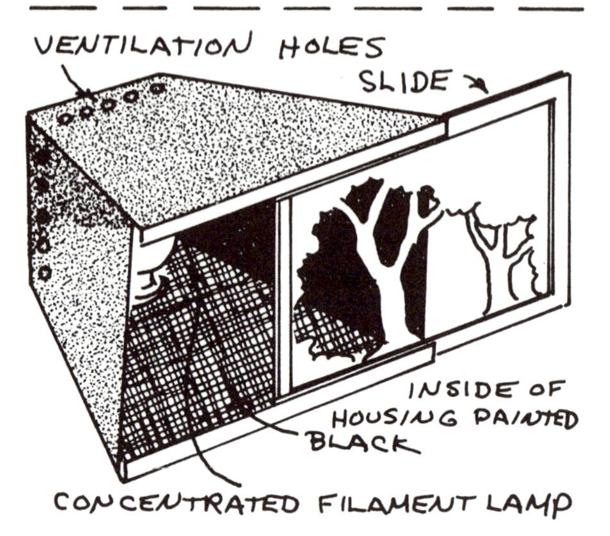


can be designed for the specific projection situation. Since the light shines out in a pyramid form with the lamp as the apex and the screen upon which the image falls as the base, it is easy to compute the desired width and height. The formula for computing this is as follows:

$$\frac{\text{Size of the slide}}{\text{Size of the image}} = \frac{\text{Distance of slide to the lamp}}{\text{Distance of the image to the the slide}}$$

The height and width are computed separately, using the same formula. When making the slide it may be necessary to distort the cut out in order to compensate for the distortion that will be caused by projection from a high, low or side angle. A little trial and error work will aid in making this compensation. The housing for the projector can be made from sheet metal or wood and the slide or cut out can be made from cardboard, metal or wood. If wood is used for the housing, asbestos paper can be pasted on the inside of it to reduce the fire danger. Slides can be made from sheet plastic and painted with special plastic paints if colored projections are desired. The flammability of all the materials used in the instrument should be carefully checked, especially if a large lamp is employed.

The Linnebach projector will give unrefined definition and will be most effective where relatively bold effects are desired. If more defined images are desired it will be necessary to use a projector with a lens system.

SLIDE PROJECTORS—Photographic or painted slides can be projected onto a backdrop or screen on the stage by means of a lens projection machine. If the area to be covered is relatively small a regular slide projector or stereoptican can be used. If, however, the image is to be much larger, it may be necessary to use a projection lens system that fits onto a regular spotlight. This may be rented or purchased from almost any stage lighting supply house.

GOBOS—Metal shadow-pattern projection slides, affectionately referred to as "Gobos" by lighting designers, can be used effectively with ellipsoidal spotlights that have been especially designed with slots built in immediately in front of the shutters. Fairly limited size but sharp patterns can be projected onto scenery, actors, or the stage floor by a spotlight equipped with a gobo. A number of patterns are commercially available and others can be readily made of thin sheet metal by a good lighting technician.

MOVING PROJECTIONS—Special effects machines are used where it is desirable to have special moving effects such as clouds, rain, snow, flames or water falls. This machine is similar to the slide projector in that it contains a lens system, but it also has a large disc that is rotated by means of an electric or clock motor. Since machines of this nature are expensive to purchase, only the completely equipped theatre is likely to have one as regular equipment. It is usually more advisable to rent the machines when needed and obtain the specific effect desired. Most of the large theatrical supply houses have projectors and discs available for rental.

At this point it should be stressed that projections are effective only if they are not washed out by the other lights on the stage. Acting area lights cannot be allowed to fall upon the surface and even high value spill light should be restricted from the surface upon which the image is projected. Snow and rain effects from a moving effects machine are worthless if the sky is brightly lighted with strips and floods. So if the stage is brightly lighted it will be just as well to forget about ever trying projections.

### Lighting Controls

Lighting control devices enable an electrician to cause all or part of the stage lights to increase or decrease in brightness.

Throughout theatre history various devices have been used to control the illumination of the stage. In the early days of burning embers, wicks in oil, and candles the number of units was increased or decreased as an actor or servant brought on, removed, lighted, or dowsed the flames. When gas was introduced the number and brightness of flames could be altered by adjusting the gas flow through the offstage gas table. Then, when theatre moved into the modern era a concealed offstage operator could flip switches on and off, and run dimmer handles up or down to regulate the flow of electricity to the lamps that brighten or darken the actors world. Some of the dimmers developed for this purpose have been simple; others have been complex and cumbersome.

SALT WATER DIMMERS—The salt water dimmer though primitive and crude is simple, inexpensive and works on a fixed electrical load.

This dimmer is composed of a jar of water and two electrodes (probably copper plates), one at the bottom of the jar, the other rigged to be raised and lowered in the jar. A circuit is set up with one well insulated wire connected to the electrode at the bottom of the jar. The other well insulated wire with fuse, switch, and socket (containing a moderate sized lamp of perhaps 75 watts) wired in series extends on, and is fastened to, the electrode that is rigged to be raised and lowered in the jar. The movable electrode is placed just under the surface of the water and salt is added to the water until the lamp barely begins to glow. It will then be found that the lamp may be brightened by lowering the electrode, and dimmed by raising it. When the electrodes touch, at the bottom of the jar, the lamp will be at full brightness and when it is out of the water at the top it will be blacked out completely. The resistance of the circuit will be determined by the height of the water column, and by the amount of electrolyte (salt) present in the water. The volume of the water and the strength of the electrolyte must be kept constant if the dimmer's operation is to be maintained.

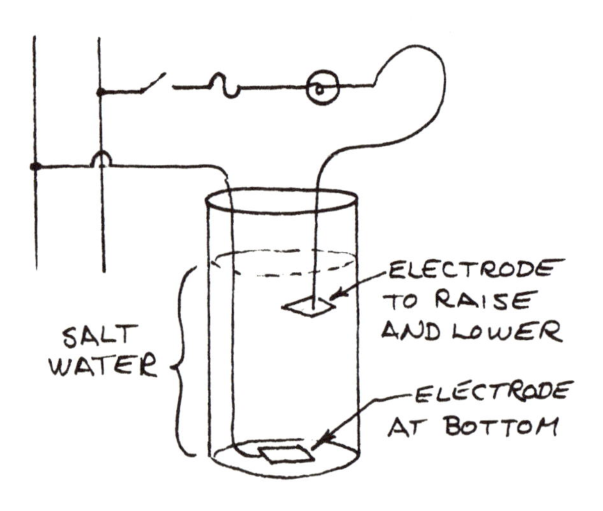


Resistance Dimmer—The principle of the resistance dimmer is similar to the salt water dimmer, except that wire of high resistance and high melting point is used in place of salt water. The electric current flows through the wire before coming to the load contact. As more wire is introduced into the circuit it becomes more difficult for the current to flow. In actual practice, a long wire of diminishing size and increasing resistance is mounted inside a vitreous enamel base. (The enamel is a non conductor.) Periodically along the wire, button contacts are placed. By moving from contact to contact along the resistance wire it is possible to tap off an increasing or decreasing amount of electric power. With a slider contact attached to the dimmer handle it is possible to brighten or dim the lights by moving from contact to contact. In resistance and salt water dimmers the energy lost by dimming is dissipated in heat.

Resistance and salt water dimmers each have a single capacity load and dissipate energy in heat.

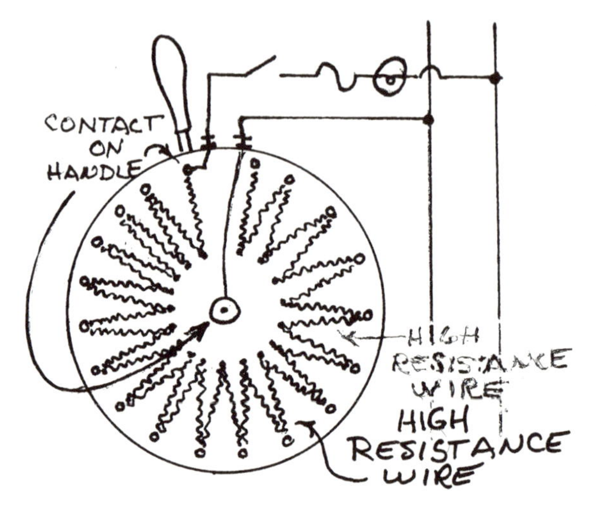


Auto Transformers—The auto transformer is a device for tapping off any desired voltage. By moving a contact (rotating knob, or sliding contact), manually or with motors, the operator taps off any desired voltage from zero to the top voltage of the line. This type dimmer will operate satisfactorily on any size load from zero to the total capacity of the dimmer.

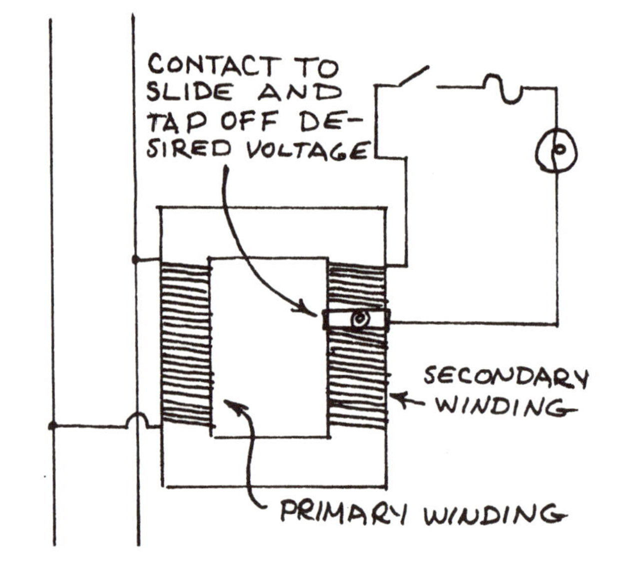

Auto transformers are available in many sizes. Some are available with heavy duty windings and several sliders to operate as if the single large dimmer was several smaller dimmers.

The variable load capacity makes this an extremely useful dimmer.

The various resistance type dimmers may be used on either A.C. or D.C. whereas the auto transformer is strictly an A.C. dimmer. However most lighting systems are A.C.

REACTANCE, SATURABLE CORE, THYRATRON TUBE, MAG-AMP, ETC. DIMMERS—The invention and development of the dimmers listed, as well as various others not listed, was part of a movement toward a lighting system that could be remotely controlled. These various systems, many still in operation, utilize low voltage-amperage controls to suppress or stimulate the flow of normal voltage-amperage in the lighting circuits to dim or brighten the stage lights.

Using minimal power for the controls reduces the installation cost since small, low amperage, wire can be strung from the dimmer bank to the control board and back. In some theatres this is a considerable distance since the control board is frequently located in the rear of the auditorium where the operator can see the production from the vantage point of a spectator.

SILICON CONTROLLED RECTIFIER (SCR) DIMMER—The invention and development of the Silicon Controlled Rectifier, frequently referred to as a solid state, or semi-conductor, dimmer, is partly a spin-off from the miniaturization required for the sophisticated electronics of space programs, satellites, and computers.

The simplest way to describe the operation of an SCR is to compare it with a switch. If you could operate a switching device at any desired speed and could then smooth out the flow of electricity it would be possible to control the amount of illumination of a lamp from "full up" to "full down" simply by operating the switch faster and slower. In the dimmer this all takes place electronically rather than mechanically.

The SCR is "packaged" small, can be remotely controlled, is highly efficient in its use of power, and is effective in its control of light. The character, design, and operation of the SCR makes it superior to all prior developments in theatrical lighting control.

### Stage Lighting Control Systems

PERMANENT WIRING—NO DIMMERS—In the most primitive control boards there are no dimmers. The circuits are merely turned on and off by means of switches. In this type of board each branch should be correctly fused so it cannot carry more than the capacity of the cable or wire leading to it. Nowhere should the total of branch circuits leading to a master switch be fused to carry more load than the fusing of that master. And the total of the entire board should be controlled by the appropriate size breaker or fuse to control all of the power of the board.

*SAFETY WARNING—Never under any circumstances should the capacity of a circuit be increased simply by increasing the size of a fuse or circuit breaker unless it has been determined that the wiring in the system leading up to, and continuing from, that point will safely carry the load of the fuse or breaker that is substituted. If this is done there is serious danger of an overload and a fire may result.*

PERMANENTLY WIRED DIMMERS—A simple board with dimmers may have the dimmers wired permanently into place or have some type of plug-in by-pass system so that a minimum number of dimmers may be used where they will be the most serviceable during a production.

The system having dimmers permanently wired to specific circuits lacks flexibility. This type of dimmer board might be improved by rewiring and adding a simple plugging system to allow the dimmers to be used on any number of different combinations of circuits.

INTERCONNECTING DEVICES—A variety of plugging systems are being used on control boards today. They are all designed to achieve the same thing in lighting-flexibility. A simple board may have only 3 or 4 dimmers that may be interconnected to 6 or 8 outlets. A complex board may have 100 or more dimmers that may be connected to at least twice as many outlets. The swatch cord system, push button selectors, slide contactors, and other interconnecting devices are all designed to accomplish the same thing; to enable a technician to connect any outlet to any dimmer, providing many lighting combinations, and making the operation of the board more convenient and practical. This is similar to a telephone system where it is possible for one person to call anyone else in the system. This makes it possible to connect lights that work together near one another and to master control groups that need to dim or brighten at the same time.

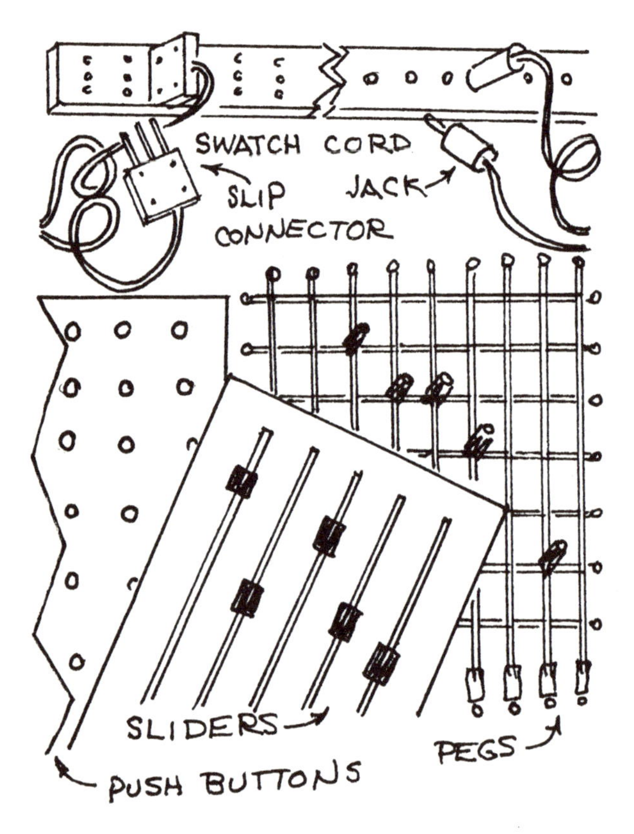


Simple Mechanically Controlled Dimmers With Mechanical Masters—A more complex board has a simple cross connect system, a limited number of dimmers that can be locked into or unlocked from a common axle that runs through all of the dimmers. A handle is rigged to the axle and a handle rigged to each of the individual dimmers. It is then possible to run all or part of the dimmers by means of the various handles.

Group Dimmers With Large Capacity Masters—This type of board consists of one or more groups of small capacity dimmers mastered by a large capacity dimmer or dimmers. When there are several masters, and a cross connect system, it is possible to achieve considerable flexibility of operation. Levels may be set on small dimmers that are connected in select combinations to master dimmers and in this way certain areas or portions of the stage may be dimmed and brightened independently of the remainder of the stage.

Remote Control Dimmer System—A remote control system may be used in a theatre having any variety of dimmers that utilize a small voltage-amperage electrical system to regulate the device that does the actual dimming of lights. This includes small motors on auto transformer dimmers, the suppressing devices on reactance, saturable core, thyratron tube, mag-amp, and SCR dimmers.

A remote system is desirable since the controls may be relatively small and the operator's console need not be adjacent to the lighting instruments nor the equipment that does the actual dimming of lights. Which means that the operator and console can be located somewhere in juxtaposition to the audience—in a floor rise—in a balcony facing—or in the rear of the auditorium. In these positions the board operator sees the acting areas as they are seen by members of the audience.

Preset System—The advent of miniaturized controls ("pots"—potentiometers or small resistance dimmers) has made it possible to design control boards that are provided with preset systems. In a preset system there is a duplicate set of controls for each preset. Usually these are set up in rows on a panel board and each row has a control for each individual dimmer in the system. The entire set of light levels for a single scene may be set up on the first row of controls, or "pre-set". The levels for the next scene may be set up on a second "pre-set". If there are three presets the third scene may be set up on the third preset and so on. And, of course after the presets have all been used up additional ones may be set as scene 1—2—3 and so on, are played.

In order to transfer from one preset to another, two dimmer controls are located in the control board that are referred to as "faders". Beside them are "punch-up" switches to switch in the desired presets. The two faders are handled as desired to cross fade by bringing one scene "up" while the other is being faded, or to fade one out to "black", then bring the other one up to "bright"—or to cross in any other desired manner.

MEMORY CONTROL BOARDS—Space age electronic development resulting in the miniaturization and refinement of electrical control components and in the creation of memory control computer systems has revolutionized both the art and craft of stage lighting. The ponderous cables, heavy lights, and laborious job of hanging instruments on pipe battens, brackets, and "trees" is no easier than it ever was but the aesthetics of lighting control, once the lights are set, is completely different than it was fifty years ago.

With a memory system settings are recorded in sequence. One of the first and simplest memory systems involves punch cards or cards with activating fasteners in them that electronically set the dimmer levels for a scene. In order to alter the setting for a succeeding scene a new card is inserted in the "read out" device and the levels are electronically set for a new scene.

Electronically recording dimmer readings on discs and tapes has followed as memory system stage lighting control boards have become more refined, more complex, and more automatic.

With a true memory system the lights and levels for each individual scene are recorded in sequence. These can then be called up as desired. In most systems any level can be changed as the scene is playing in order to alter the setting as recorded, scenes may be called out of sequence, or scenes may be played as recorded.

DIMMER/CIRCUIT—Lighting control equipment is becoming inexpensive enough so that, in some installations, the high cost of heavy wire and the cost of the labor of wiring, compared to the relatively low cost of dimmers and low voltage control systems to replace heavy duty cross connect systems, is beginning to make it economically feasible to equip each electrical circuit with its own dimmer.

DIMMING LAG—Regardless how complicated the system and how rapid the operation of controlling equipment one factor still determines the speed of "blackouts"—that is the cooling time of the filament of the instrument. There is always a time lag on a large wattage lamp—spotlight or floodlight. The lag diminishes as the size of the lamp filament diminishes so several smaller wattage spots or floods will "blackout" faster than one large wattage instrument.

## Color Media

Since the light from an instrument used on the stage is almost always colored and rarely white some type of media must be used as a light filter in order to achieve the desired color. The media most often used are: glass, gelatine, plastic, Mylar, and occasionally lamp dip.

GLASS—Glass color media is probably the media with the most permanent color but glass is not used often on the stage except for "roundels" on strip lights and occasionally in slat form on some flood lights. The danger of breakage, the cost, and the narrow range of color available in glass are the disadvantages.

PAR AND R LAMPS—These lamps with self contained reflectors are obtainable with color baked on the face, or lens, of the lamp. And although the color is pretty well permanent for the life of the lamp they are obtainable in only a limited number of colors.

GELATINE—Gelatine is used more than any other color media. Available in a wide range of colors, it is non-flammable, but the ordinary variety is water soluble, becomes limp when wet; consequently dehydrates and becomes brittle when dry. Colors fade with use, but this also happens to other color media since the beams of stage lights produce considerable heat.

PLASTIC COLOR MEDIA—Numerous nonflammable plastic color media are on the market that possess relatively stable dyes. The plastics are slightly heavier, more expensive, and more durable than gelatine. Colors will fade but not as rapidly as in gelatine. Punching rows of tiny holes in the plastic with a rolling leather punch, fabric pattern tracing wheel, or even a small nail, will help ventilate and prolong the life of the plastic color media.

MYLAR—Mylar is a still more permanent, if more expensive, color media worth investigating if durability is desirable.

LAMP DIP—Special dyes are available for coloring the outside of the lamp globe. These colors will peel or flake off after extensive use, but for special purposes they will probably be useful, especially for low wattage lamps.

* * * * * *

There is no substitute for an efficient, easily operable lighting control system. The recent proliferation in construction of theatres, TV studios, and other facilities that utilize electronic controls, within the past two decades, has led to mass production of the hardware and parts used in dimmer and control board fabrication. Consequently S.C.R. dimmers, preset equipment, memory systems, and similar equipment is probably no more expensive than the old fashioned, less efficient, and more cumbersome equipment installed a few years ago.

If a new control board is being installed, or an old board is being remodelled it is advisable to investigate obtaining the most modern equipment available. And when writing up the specifications be certain to advocate that the equipment be obtained from a reputable company, that it be properly installed, and that it be provided with continuous maintenance and repair.

# 14

# Lighting

Stage lighting in practice is the manipulation of color, light and shadow. Although the primary objective in any form of lighting is to achieve visibility other end results are also desirable when a play is being lighted. Either consciously or unconsciously the lighting artist attempts to: concentrate the audience attention, create mood and atmosphere, model forms and create dramatic design.

It has often been proved by experiment that hearing is dependent to a certain extent upon visibility. If members of an audience have difficulty seeing the action upon the stage they also have trouble hearing and understanding the dialogue. And obviously if they have trouble hearing and seeing they also have trouble following the action and story. Therefore, having sufficient light for visibility is of first importance in stage lighting.

By manipulating the quantity and distribution of the light on the stage it is possible to control the audience attention. In general, attention is drawn toward the point of higher illumination where there is a variation in light intensity. By subtly shifting the light during scenes it is possible to move the audience's attention from one area of the stage to another. In order to do this effectively the lighting operator must have complete control over every light on the stage.

The mood of a scene can be set by the amount and color of the illumination. There is usually a feeling of gayness when the illumination is in a high key and everything on the stage sparkles. On the other hand, low key lighting consisting of dim light and deep shadows will project a feeling of uncertainty, mysticism and fear. The mood of a scene as well as the mood of an audience can be controlled to a great extent by the proper handling of the color, intensity and distribution of the stage lighting.

The three dimensional qualities of form are revealed through the use of proper lighting. Subtleties of form are revealed by variations in the levels of intensity and variations in the color of the light. In nature the eye distinguishes the shape of objects by the slight variations in light values. Light strikes the object from all directions since it receives not only direct sunlight but also reflected light from every surface around it. Some areas will receive direct light, some high reflected light, others modified amounts of direct or reflected light. These variations in the quantity and quality of light reveal the form to the eye. The stage lighting man can make use of variations in light intensity to reveal the nature of forms on the stage.

Light may be used as a design element on the stage. Surfaces may be lighted in patterns that come from regular or special lighting instruments. These patterns may be abstract or projections of photographic reproductions of real objects; they may be shafts of white light or interminglings of colored light forms. Light may be used in many ways to create all or part of the stage design. Evening and morning, summer and winter can all be suggested by the color and visible distribution of the light on the acting area of the stage.

Many or all of these effects may be used in lighting the stage for a production. The total effect of stage lighting enlarges the emotional impact of the total production. Lighting is used to assist the visual design, the emotional content and the physical movement of the play.

It is impossible to list a set of rules and say specifically "This is the way to light a play." There are too many physical variations from one stage to another. The lighting system for each stage must be worked out in terms of the relationship between the auditorium and stage, number and capacity of circuits, the type and number of dimmers and the size and arrangement of the various elements of the stage itself.

Basically there are two types of stage lighting. These types are actor illumination and shadow softening illumination. Ordinarily they are both used in lighting a stage.

Actor Illumination—Although a successful dramatic production consists of many elements fused together into a closely integrated whole, the actor almost always remains dominant. Consequently the major portion of the stage lighting is designed for lighting the actor and his realm of the stage. Spotlights are usually employed for this purpose. With the spotlight it is possible to concentrate a major portion of the light on a restricted area.

Shadow Softening—Although the major portion of the illumination from the spotlight will fall on the acting area each of the surfaces upon which this light falls will serve as a reflector and cause diffused rays to spill off in many directions. This spill light will tend to soften the shadows around the rest of the stage. Usually this isn't quite enough to "kill" some of the heavy shadows so a limited amount of additional general illumination is needed. This shadow softening light comes from strip lights, such as footlights and borders and from floodlights.

Position of Spotlights—The spotlights used for actor illumination will probably give the best results if they can be located in front and above the area of the stage they are to light. The angle of the light beam with the stage floor should probably not exceed 45°. If these lights are located on a low level and shine almost straight forward toward the stage,

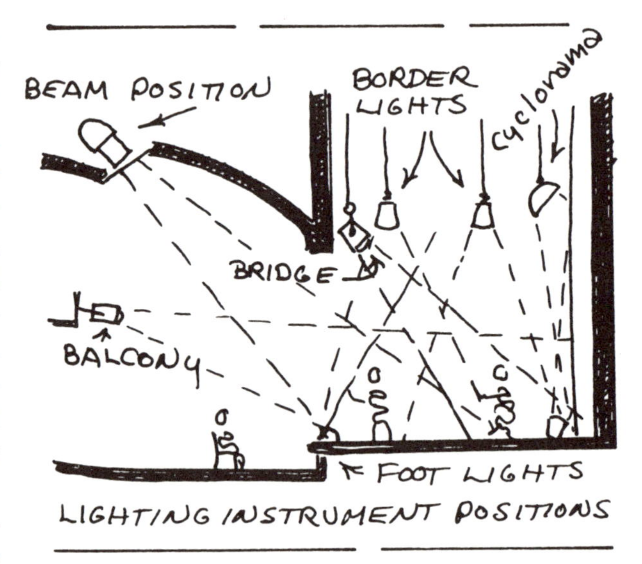


LIGHTING INSTRUMENT POSITIONS

large shadows of the actors will fall upon the scenery in the background. If, on the other hand, the angle is too steep the shadows under the eyes, nose and chin of the actors will be extremely heavy. An angle of from 30° to 45° will help modify both of these excesses.

Stage Areas—For convenience the stage is divided into areas for lighting. The size and number of areas will depend to a great extent upon the size of the stage and the size and number of spotlights available for lighting it. Ordinarily it is best to have a row of downstage areas and a row of upstage areas. If considerable depth is being used it may be necessary to have a third row or even a fourth row across the stage. The areas may be designated in terms of the stage directions accepted by theatre workers: upstage, downstage, right stage, left stage, right center, left center and so on. If there are spotlights in the auditorium they may have titles that refer to their relative positions. Lights in the ceiling are frequently referred to as "beam lights," those on the balcony as "balcony lights," those on the stage immediately behind the act curtain as "bridge lights." There are also "footlights," "border lights," "Cyclorama lights," "trees," and so on.

Area Lighting Plan—A small stage having a proscenium of 24′ width or less may be lighted by having as few as 6 areas, 3 downstage and 3 upstage. The downstage areas are lighted, if possible, by spotlights located in the auditorium ceiling or beams. Each area should have two units focused on it. This means that there will be at least 6 spotlights in this downstage position. These should be spaced so that there is approximately the same distance from the various lights to the stage area they are lighting. The two lights illuminating each area are placed so that their beams reach the area from different directions. If the 6 instruments are equally spaced across the beam the first and fourth may be used to light the stage right area, the second and fifth may be focused on the stage center area and the third and sixth may be on the stage left area.

Numbering the Acting Areas—In order to further clarify the relationship between

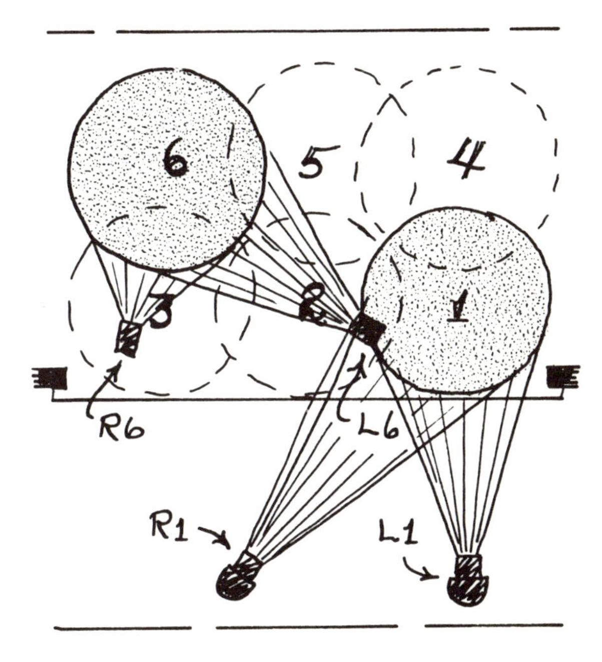


the acting areas and the lighting units it might be wise to number the areas. Looking at the stage from the actor's position, since the right (R) and left (L) directions are so located, number the down left area as "1", down center as "2" and down right as "3". The upstage areas will continue in the same order with up left as "4", up center as "5" and up right as "6".

Spotlight Designation—The spotlights may now be designated in terms of the area they light and the direction from which their beams reach the stage. Each area will have a right and left spot. Area "1" will have an "R1" and an "L1" and so on.

Alternate Positions for Front Area Lights—If there is no possibility for mounting beam lights it will be necessary to investigate other positions. It may be possible to mount them in high side windows, although the distance from the stage and the angle are not as good as for a beam position. Large chandeliers are other possible positions. If none of these possibilities will work it may be possible to rig ladder frames or tall flat frames in positions near the side walls and mount lights on these. If something like this is used it will be necessary to mask the supports so they are not too obvious. Sometimes towers of this nature can be rigged in the orchestra pit immediately in front of the stage, or even on the apron of the stage. However, the individual situation will have to be studied carefully to determine the best place for mounting such lighting units so they will not seriously affect sight lines

Balcony Rail—Under most circumstances the balcony rail is the last solution to the problem rather than the first one because, as suggested before, lights mounted in this position usually are so low in elevation that when the actors move about on the stage they cast heavy shadows on the back wall of the setting. However, this position is better than having no front lights. If spots are mounted here it might be advisable to use a larger number of smaller units rather than a few large ones so

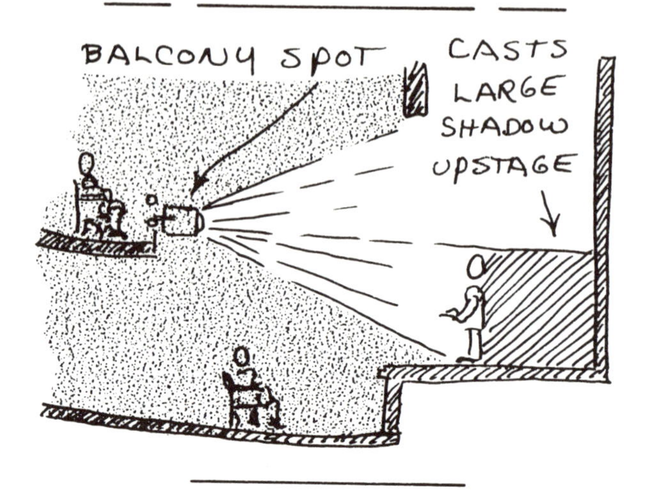


that three or four from varying angles can be focused on a single area. In this way the back wall shadows will be minimized.

Upstage Areas—The upstage areas of the stage can best be lighted by instruments mounted behind the top of the proscenium arch immediately upstage of the act curtain. These lights, frequently referred to as the bridge lights, may be mounted on a catwalk or a batten

suspended immediately upstage of the act curtain or on towers upstage of the proscenium side walls. The overhead batten position is probably superior since it lessens the distance from the lighting instrument to the stage and enables one to get greater intensity than from the more removed side positions.

BRIDGE—The bridge spotlights are mounted and arranged in positions somewhat similar to those in the beams. Areas 4, 5 and 6 are lighted, each with two spots, one coming from the right and one from the left. So here again there are 6 spotlights.

If the stage is medium sized or large it may be necessary to increase the number of areas and consequently the number of spotlights. If the stage is also deep it may be necessary to have another bridge position part way upstage.

SIDE LIGHTING — In some instances side lighting may be used exclusively or for some areas. When used it is most effective if the beams are focused to shine across and slightly up stage. If the actor moves ahead of the beam his facial features do not receive the front light needed and unnatural shadows appear. Side lighting is probably more effective for ballet and modern dance than for legitimate drama. The bolder movement projects all right but subtle facial expressions need front light for the best effect.

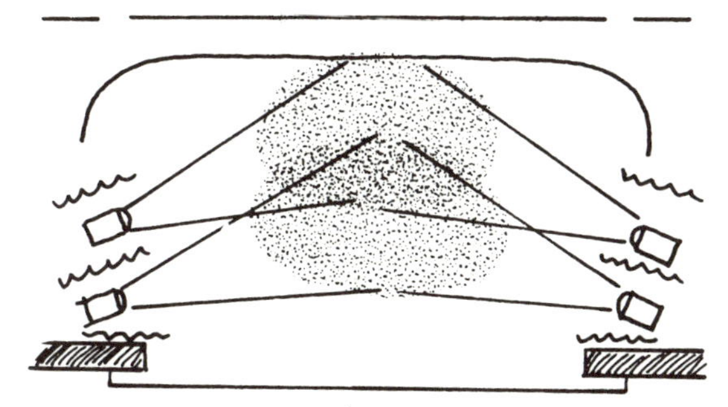

SIDE LIGHTING - SPOTS ARE DIRECTED SLIGHTLY UP STAGE

AUXILIARY LIGHTS — The 6 beam and 6 bridge spotlights may be sufficient to supply the actor illumination on a small stage. For a simple setting it is possible that the spots will provide enough illumination through their direct beams and reflected diffused beams from the floor and scenery to serve as both actor and shadow softening light. However, it will usually be necessary to have some auxiliary light from footlights, striplights and floodlights to serve the latter purpose. Using spotlights alone usually provides lighting that is harsh. The shadows are a little too heavy and the highlights a little too high. A touch of footlights to cut the eye socket, undernose and underchin shadows gives the face of the actor better definition. A touch of overhead borderlight helps cut any shadows cast by the footlights and softens the pool effect of spotlights.

Other lights are also necessary at times. There are various motivating or source lights used on the stage. Sunlight or moonlight shining through a window is affected by having a floodlight or flooded spotlight in a position offstage shine in through the stage window. Floor, table, gas or oil lamps may be placed on the wall, the mantel or on a table. These instruments, with extremely low wattage illumination, serve to indicate the source of the lights used on the stage.

BACKINGS — Areas off the setting (behind doors, windows, archways, and stairways), visible all or part of the time may also have to be lighted. These backing areas representing other rooms or the out of doors are frequently important to the action of the play and must be correctly lighted as a motivation for stage business or for the mood of a scene or to give credence to the lines of the play. These will be lighted in various ways depending upon their size and their position in relation to the setting.

Backings representing rooms are frequently lighted by placing a small striplight or floodlight immediately beyond the stage door so that it will shine upon the backing and also upon the actors as they enter. The position above the door provides face lighting for the actor and at the same time prevents awkward shadows from falling onto the backing. Of course the contrary is often desirable. The shadow falling into the room in advance of the entrance may be extremely important in a melodrama or mystery play, in which case a spotlight or floodlight should be positioned so that the body of the actor casts a shadow that precedes him into the room.

If the backing outside a door or window represents the sky it may be necessary to use two or more sources of light in order to get an even over-all illumination on the surface. Cross lighting from both sides, lighting from both top and bottom, or even lighting from

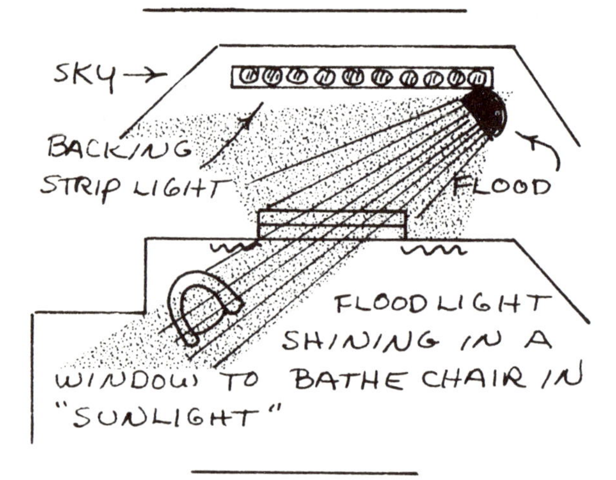


all four positions may be necessary. It may be desirable to represent dusk, morning, high noon or night on this surface. Using striplights from the bottom only and changing the color of the light may achieve some of these effects. Lighting from just the top may be a method of achieving other desired effects while combining the two will be desirable at other times.

FIRELIGHT—Indoor settings often have fireplaces. A fire on the grate indicates that it is cool or cold outside. The costumes of people entering from outside, the appearance of scenery outside a window, or the dialogue will usually be devices which with the condition and color of the fire will indicate the degree of coldness. The firelight may be the major motivating source of light. Obviously it will not under ordinary circumstances actually be the only light used for the action. It may be at high intensity while the other lights are dimmed but usually it will need some help from beam and bridge. This light may be used for highly dramatic scenes and will, of course, usually be in dominantly warm colors: red, straw, amber and violet. At this point it must be recalled that colored light can have dastardly effects on makeup, costumes and other pigment-color surfaces; so the lighting colors should be handled carefully.

The firelight may be comparatively calm or have moving effects. This movement may be achieved in any of the ways suggested previously. Effects of this nature should be studied rather carefully to determine whether they add to or detract from the scene. Sometimes it might be advisable to start the scene with the moving flame effect and then diminish the flame activity as the scene begins to move toward a climax. At another time the flickering flames might be used to accentuate the movement toward the climax, increasing in tempo with the scene. This sort of thing must be adapted to the individual scene, director and production.

SWITCHES—Light switches on stage are always a problem. It is difficult to know how to handle the turning on and off of lights. This is something that must be rehearsed carefully and both the actor and the light operator must know just how it is to be handled. And once it has been set changes should not be made unless they are rehearsed carefully. Frequently it is better to have the control board operator handle both the switch on the motivating source light, a lamp for instance, and the acting area lights around it. In this way he is able to turn both on at once. The actor in this situation must be directed to place his hand on the switch as a cue for the lighting man and hold the position until the lights go on, or off, as the case may be.

Lighting a lamp or a fire should be handled in a similar manner. Real matches may be used but the lighting process should be masked. When and wherever possible on the stage real flame lamps should of necessity be avoided to conform to fire ordinances. Torches, fireplaces, candles and oil lamps should be wired to either regular lighting circuits or flashlight batteries that are enclosed in the instrument itself.

SPECIAL LIGHT—Special lights are frequently needed on the stage besides the area lights. In fact some productions require that almost all of the lighting be done with "specials" rather than areas. In such cases the lights are focused so that they light only the table, the hat, the face of an actor, or a ghostly picture on a wall. Too many special lights may make the total job of lighting more difficult since the sheer number of instruments may suddenly "get out of hand." So it is usually advisable to start with area lights and then add a minimum number of specials. Specials may be mounted in strange and unorthodox places. Small units can be mounted about the setting, shining through windows, concealed in bookshelves, fireplaces, heavy picture moldings or in pieces of furniture, or they may be mounted in overhead positions above or at the side of the setting.

OUTDOOR SCENES — Outdoor settings on a stage have special problems of their own. To begin with there is usually more area to cover, including a large space of cyclorama sky. The acting area may be no larger than the area for a regular enclosure type of setting, but the open space around it may make it necessary to use more floodlights and striplights if there is supposed to be over-all sunlight or moonlight illumination.

The sky may be lighted from above and below with striplights or floodlights and striplights. The floor and overhead striplights will both usually need to be masked. This is more true of the floor lights unless the theatre is provided with a trough into which they are recessed. Ground rows as part of the scenic design will take care of this matter.

Usually the sky is lighted better with a large number of smaller units than with three or four large units. The large units tend to provide hot and cold areas whereas the smaller ones can be located in positions somewhat more

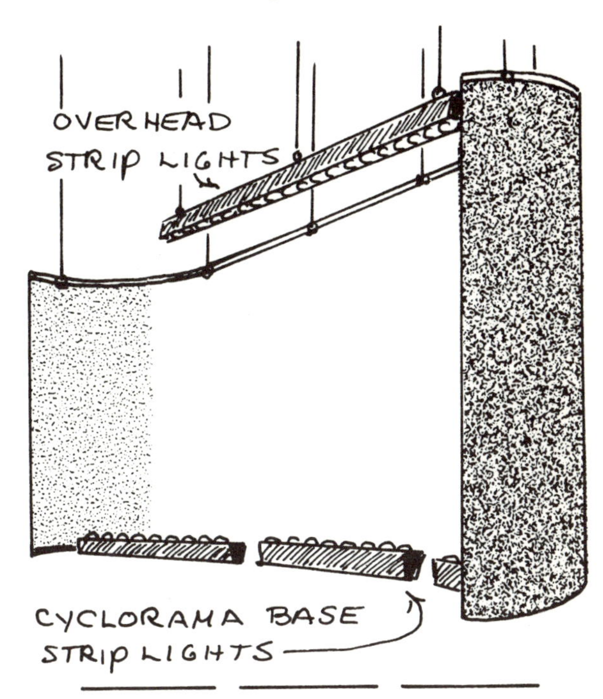


equidistant from the surface so that the entire surface is evenly illuminated. Striplights, floodlights or fluorescent tubes will probably serve best for cyclorama illumination.

SPACE STAGING — The process of staging productions with a series of levels as the only scenery and pools of light to locate an area as the space for the action of the moment is frequently referred to as space staging. The background for this type of production is frequently black drapes. The black velour will literally gobble up the light. If the costumes are light in value the problem of lighting is not too difficult. If, however, the costumes are dark in value it will be difficult to get enough light on the subjects to make them stand out from the background. Some side lighting or back lighting added to the front lighting will probably aid materially. Because of the absorbency of the black drapes, the areas for this type of staging will probably take more wattage than in productions where the background is light in value.

Cyclorama Space Staging — Interesting effects can be achieved when a sky or other light background of a neutral nature is used since it is then possible to make use of silhouettes as well as front lighted subjects. Whenever the source of light is behind the actor he will stand out as a dark figure against a light background. Front area lighting with no cyclorama lighting will place him against a dark background lighted only by the spill light from the front spotlights. If side lighting is used it may be possible to make the cyclorama almost black out since the spill to the rear will then be minimized. There will be some spill but not as much as from front lighting.

Cyclorama Color Effects—If the cyclorama is lighted with strips, floods or both of evenly distributed primary colors (red, blue and green) it will be possible to give the cyclorama, if it is light in value, almost any conceivable color by manipulation of the lights. Blues may be used for night. As dawn begins to break greens may be brought in and then gradually reds. The blue may be diminished allowing the red and green to mingle, producing yellow. The blue may be brought back in with the red to produce lavender. And all three may be brought up to produce white sunlight. Any combination may be tried. Sunrises and sunsets are rarely the same. It is worth experimenting to achieve the desired results. Color, especially on a cyclorama, can be used to great advantage.

Psychological Color in Light — Colored light may be used on a cyclorama to achieve psychological feelings. This must be studied rather carefully to be certain that the effect carries the correct mood. It may be that a blood red sunset in Macbeth will suggest the purge to come; that a deep blue or blue green will project the coldness of the Arctic; that intense yellow will project the heat of the Sahara; that surprise pink will suggest the glow of youth; that green will give the feeling of unworldliness. It is usually worth while to try an effect and see if it is going to work. In the case of color it will be necessary to consider more than just the color of the light beams. Remember that the light beams fall on surfaces and that the final color effect is determined by the color of the light *and* the color of the surface.

Designing With Light—In the chapter on Scene Design there is some discussion of using light as the media for both color and form. This might profitably be discussed here also. Simple shadows, either of moving or stationary forms, can be used to advantage. Actors moving between the light and the background, especially if it is a cyclorama, can create a dramatic effect. The performer is lighted from the front with a single source of light

(usually one large floodlight from below so that a fairly sizable area can be covered) and he casts one shadow. The shadow will be large as he approaches the light and will grow smaller as he approaches the background until it becomes his size and he and the shadow merge into one. White or colored light may be used. The performer may be lighted from the sides without blotting out the heavy projected shadow since the light on his figure and the spill light will be shining across parallel with the front of the stage. As soon as front light is used the shadow begins to lose its intensity unless the front lights are focused low so they strike close to the front of the stage.

Stationary objects may be used to project shadows onto the cyclorama in somewhat the same manner. They are placed in a position between the source of light and the background so that part of the light rays are intercepted. Plants, flowers, cut outs, objects with distinguishable or indistinguishable shapes can all be used. The object will be chosen for the shape desired whether it is of a real or an abstract form. Projection may be from the front as suggested or from the side. The front light from below will give an upward widening distortion; side lighting will give a lateral and possibly also an upward distortion.

MULTIPLE SHADOWS — Multiple shadows will be projected if more than one shadow casting lighting instrument is used. Each unit will cause a shadow to be cast since the figure or object will cut the light from that instrument. A single dancer may have three definite shadows if there are three light sources close together. If the sources are widely separated they probably won't work out quite as well if it is desirable to have multiple shadows of about the same size and intensity.

COLORED SHADOWS — When two or more colored sources of light are used it is possible to get colored shadows. The colors on the surface of the person or object will be a mix-

SIMPLE PROJECTION OF PLANT

SHADOW OF ABSTRACT FORM

OBJECT

RED BLUE

COLORS MIX ON BACKGROUND
ONE SHADOW IS BLUE - ONE IS RED

ture of the colors if the units are close together but the shadows will differ in color. The form will create a series of shadows; each instrument will have a shadow cut from its beam. The shadow of each instrument will take on the mixed color of the other light sources. If red, green and blue lights are used, for instance, the shadow from the red light will appear to be blue green since only the red will be cut out; from the blue the shadow will be yellow since only the blue will be cut out; and from the green the shadow will be purple since only the green will be cut out. Multi-colored foilage can be projected onto a cyclorama by shining various colored lights through the foliage of a house plant.

Color may be a significant factor in the projection of images. It should be chosen for the psychological effect desired and for its effect upon the colored pigment being illuminated or being used with the projection.

The type of projected scenery used and its effectiveness will be determined to a great extent by the physical structure and facilities of the building as well as the ingenuity, time and budget of the lighting technician. Effects such as those described above can be done with a minimum of equipment on a stage of almost any size and shape. A little experimenting will prove what can be done. The projecting light or lights can be located in the footlight position, off stage right or left from a low, medium or high level, part way up stage concealed in a scenic unit or under a platform, or almost immediately in front of the cyclorama. Of course if the stage has great depth, or if a cyclorama or backdrop is placed down stage leaving space at the rear, it is also possible to have rear projections. Rear projections are frequently far superior to front ones, but most theatre buildings lack the depth required for their use.

Shadow Projections Through a Backdrop — The rear projection may be achieved through a backdrop with portions painted with translucent paint and portions painted with opaque paint. The translucent sections will allow light to shine through while the opaque areas will hold it back. Other forms may be used in conjunction to cut off portions of the light. Even forms placed against the drop at the rear may be used to cast heavy shadows.

Scrim Lighting—Scrim is frequently used on the stage to effect rapid changes. A pattern can be painted on a gauze or scrim drop and the surface will appear opaque as long as there is no light behind. As soon as light is brought

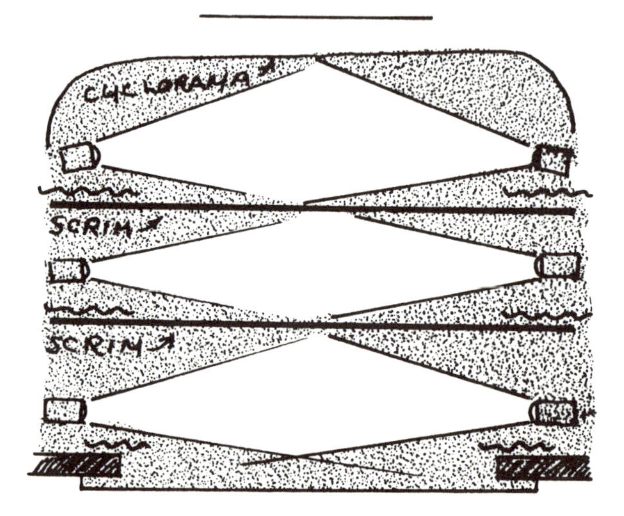


SCENES SET UP IN DEPTH BEHIND SCRIM—EACH IS REVEALED AS LIGHT COMES ON BEHIND IT.

up behind, the curtain will either disappear completely or fade into a dim mist. This is a place where fairly high angle front light and side light can be used to good advantage. Lights from straight front will usually spill through so much that objects behind the drop can be seen.

Scrim in depth can be used in a similar manner. Two or three drops may be painted and placed one ahead of the other with a playing space between them. As scenes are played in depth the lights are brought up on that area. It will then be possible to see back as far as the scrim behind the scene being lighted. When the scrim is to fade out it is better if the light is dimmed in front of it so that no halo or

Setting for *Aladdin.* The rock structure above is partially covered with scrim. When lighted from the front it appears to be a solid surface but as the lights are brought up behind, the scrim begins to disappear and the rear area can be seen as in the lower photo.

fuzziness will result. Obviously heavy thread weave materials, like camouflage netting, will become opaque in depth with more than two or at most three thicknesses between the actors and audience while material like theatrical gauze, sharktooth gauze, cheesecloth or marquisette might allow more scenes in depth. The main point to remember in using scrim is that visibility moves in the direction of the highest concentration of light.

FLUORESCENT EFFECTS — Objects painted with special fluorescent paints will glow when activated by ultraviolet light. The intensity of the glow is not as great as it first appears. It takes very little white light spill onto a fluorescent surface to destroy the effect of the ultraviolet light. If the fluorescent area is on a backdrop the acting area can be lighted from the sides with more success than from the front, that is, if the fluorescent patterns are not to be washed out by the actor light. Overhead beams used with such light should be aimed fairly straight down.

Ultraviolet lights usually provide low illumination since most of the rays are filtered out in order to get just the invisible ultraviolet wave lengths of light. Consequently it is important to have the light units as close to the surface activated by fluorescent paint as possible. The closer they are the brighter the glow of the surface. Ordinarily fluorescent paint-ultraviolet light effects are most useful for unusual or spectacular effects rather than for prolonged dramatic effects.

LIGHTING HIGH CONTRAST OBJECTS — If extremely light value objects are placed against a dark background almost the same effect as that of fluorescent paint may be obtained. The light values contain much white pigment so even the smallest light intensity is reflected from them while almost all of the light is absorbed by the dark background. This type of effect has advantages over the fluorescent effect since there is visible light so that actors can see their way around the stage and, of course, all objects that are not black will show to a certain extent. This last point may not be advantageous, however, since the desire is usually to have everything on the stage that is not coated with fluorescent paint disappear into nothingness.

* * * * *

The above discussion is not intended to be definitive nor all inclusive, but rather is intended to present enough basic ideas to start the prospective lighting artist working with his medium. Imagination, a certain artistic sense and a great deal of hard work and inventiveness will help him to achieve the effects he needs for any given production. It isn't necessary to have elaborate equipment to achieve beautiful, stimulating light effects. On the other hand it is never advisable to stand in one spot and admire the lighting effects achieved by a lamp on the end of a drop cord either. Constantly try to improve both the supply of the lighting equipment in the theatre and the degree of artistry achieved with a minimum of effort. A hundred thousand dollars worth of fine equipment is worth no more in the hands of a lout than one PAR in the hands of an artistic soul.

# 15

# *Mechanics of Design*

A number of basic mechanical drawing processes will prove of value to the designer of stage settings. Among these processes are the planning and drawing of sight line sections, scaled floor plans, working drawings, simple and complex perspectives; and the planning and building of models of stage settings.

## Sight Lines

The audience is one of the most important elements of the production. If possible the setting should be designed so that every member of the audience can see the action and scenic background of the entire play. This can be accomplished by checking the sight lines carefully throughout the entire process of designing the stage setting.

SIGHT LINE FORMS—A scaled floor plan of the theatre, and a scaled drawing of the lengthwise cross section of the theatre, are needed for checking the sight lines. It is highly advisable to make a number of "ditto" or mimeographed copies of each section so that floor plans and elevations of projected stage settings can be drawn to scale directly upon these forms.

In order to place the essential portions of the plan on a single sheet of paper it is necessary to use a fairly small scale. A scale of ⅛″ = 1′-0″ might be suggested, but the scale will have to be determined by the measurements of the specific building.

The drawing need not include the entire building plan, but only the portions required to check the sight lines. The floor plan should include the entire stage area and at least the extreme end seats of the front row in the auditorium. In rare cases seats may be positioned

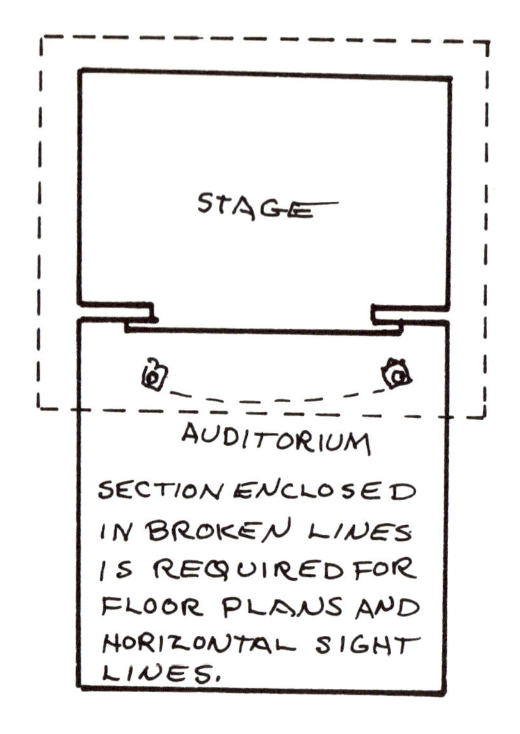


farther out to the sides of the auditorium; if so, then these positions should also be included.

The cross section drawing of the theatre, made to show the appearance from one side of the building split endwise, should include

the entire stage and enough of the auditorium to get the position of the lowest and highest seat levels.

HORIZONTAL SIGHT LINES—To check the horizontal sight lines the floor plan section of the theatre is used. Lines are drawn from the extreme right and left seat positions of the auditorium. The line from the left seat extends up past the proscenium arch on that side to the back wall of the stage. The line from the right extends in like manner to the back wall on that side. The space between these two lines is the area of the stage that can be seen by all of the members of the audience. If the setting fits within this area the sight lines will be ideal horizontally. Actually, if the auditorium is extremely wide and the proscenium arch proportionately narrow, the sight lines will cut in so sharply that only a pie cut area will be left when the lines are drawn in. In a case like this designing becomes a much more difficult problem than in an auditorium where the block

of seats extends out only as far as the width of the proscenium arch. But regardless of how wide or narrow the sight lines are, they must be checked on individual floor plans so that important portions of the setting aren't placed where they cannot be seen by sizable portions of the audience.

VERTICAL SIGHT LINES—Vertical sight lines may be checked by drawing a line from a seat in the front row across the leading edge of the stage apron and extending on to the back wall. This line will supply the critical upward angle. By drawing a line downward from the uppermost seat (in the balcony if there is one) in line with the lower edge of the grand drape and on to the back wall of the stage, the critical downward angle is determined. If the grand

drape can be raised and lowered obviously this angle could change.

The four sight lines now computed are used to determine how far offstage in each direction and above the stage (on platforms) the action can be moved and still be seen by all of the members of the audience. These sight lines are all needed in designing the setting.

SIGHT LINES FOR MASKING — Another set of sight lines is needed to plan the masking for the stage. If each of the patrons seated in the critical seats, except that fellow away up in the balcony, looks in the opposite direction from the ideal sight line direction, he will see more than he is supposed to. These patrons will see off stage into the wings and up into the fly well.

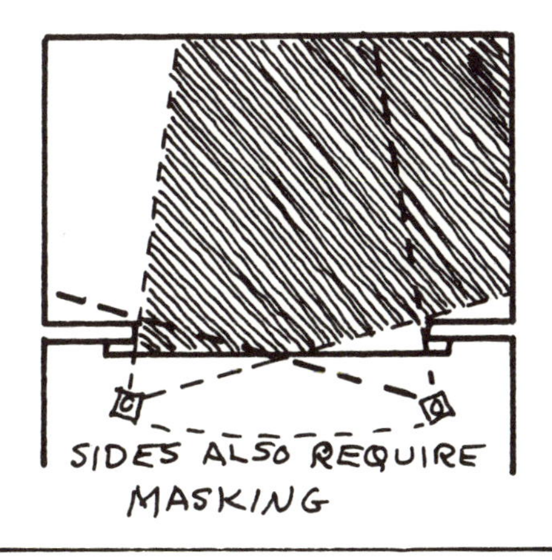


So another set of sight lines drawn from these same positions but in the opposite direction will indicate the areas that must be provided with some type of masking device to prevent the audience from seeing the backstage paraphernalia.

These sight lines should be checked carefully as the drawing of the floor plan and elevation proceeds. Even after the setting has been completed there probably still will be masking problems and some sight line problems, but

careful planning shows up most of these difficulties early enough that the director can foresee his problem.

SIGHT LINES AND THE SCRIPT—With the sight lines in mind, the script is studied carefully. Mechanical elements are noted: entrances and exits; the use of windows and fireplaces; business on stairways and platforms; references to elements of the stage or stage properties; manipulation of lights, draperies, windows, doors, fireplaces, stage properties; the use of furniture in stage business; movement of stage furniture; movement of the actors; important entrances and important actor stage picture arrangements. All of these factors and many others may affect the design of the setting.

After the script has been studied carefully, conferences should be arranged with all of the other people involved in the production, but especially with the director. In these conferences preliminary plans should be worked out. These plans will be fairly simple and consist of

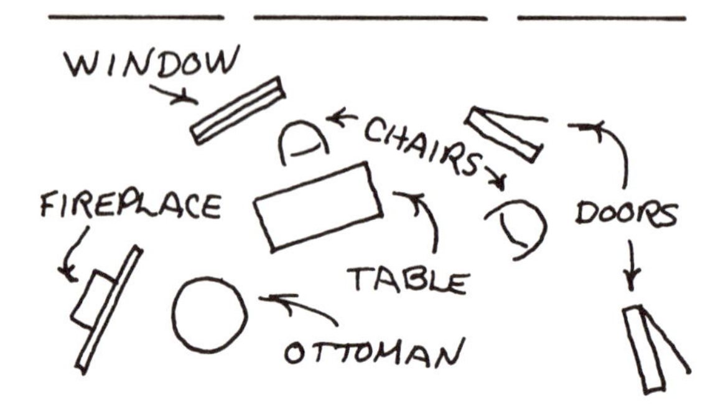


PRELIMINARY SKETCH WITH ESSENTIAL UNITS - APPROXIMATE POSITIONS.

determining the approximate positions and relationships between stage units, properties, furniture and major stage movements. After this conference it is time to begin devising possible stage floor plans.

### Floor Plans

The floor plans drawn to scale (on the stage plan, at least to check sight lines) represent a view of the setting as seen from directly above. On this floor plan accepted symbols are used to represent architectural features. Walls appear to be lines. Walls on a floor plan of a home rarely go off on odd angles since architectural forms are more economical to build when they meet at right angles. However, on the stage, in order to improve sight lines and add interest to the design, there may be many strange angles.

The walls will be broken here and there by architectural units. Archway openings are shown as boxes to the rear of the wall showing the proportional width and thickness of the opening and the wall it is cut into. Doorways are drawn like archways except that the door and the direction it swings must also be shown. So the shutter (door) is shown part way open and attached at one side. Usually, to simplify

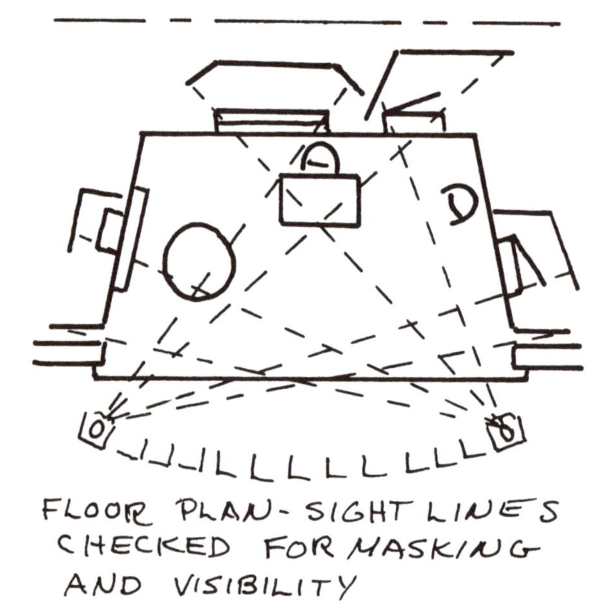

FLOOR PLAN - SIGHT LINES CHECKED FOR MASKING AND VISIBILITY

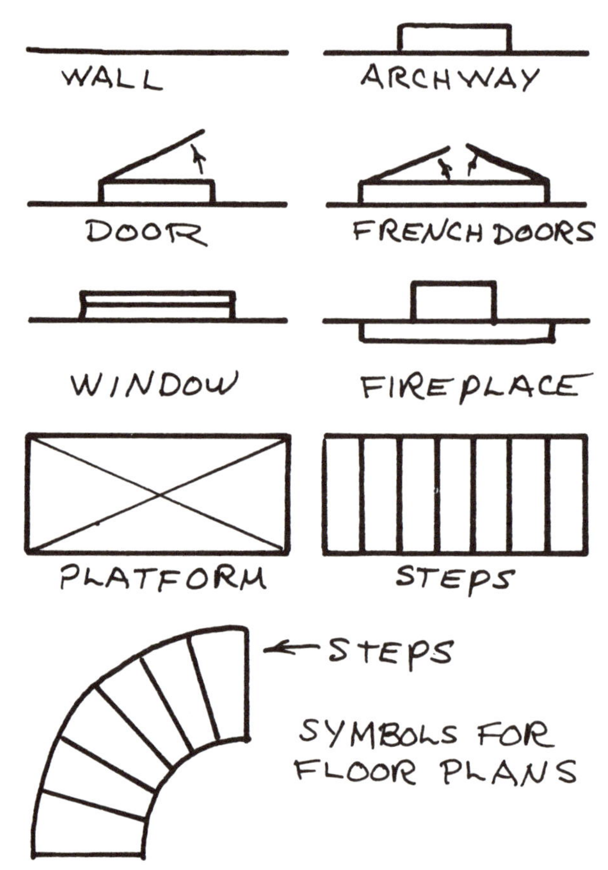


SYMBOLS FOR FLOOR PLANS

construction, the door opens offstage; and to help stage business and masking it is hinged on the upstage side. Windows are represented like archways with an extra line drawn through them lengthwise to indicate the sash in the frame. French windows or French doors will have one or two swinging sections like doors, depending upon whether they are single or double, and they probably will be labelled. A fireplace is shown with the position of the mantel indicated onstage and the firebox section indicated behind the flat. Platforms are shown by rectangles of the correct size and shape and each individual section will have diagonal lines drawn lightly across it to indicate the dimension lines of the unit.

The positions of steps in stairways are shown by a series of forms representing the width and depth of the individual treads. It is also wise to draw the major pieces of furniture in their individual positions. The form and approximate proportions of suggested units should be drawn so that stage business can be planned.

The various essential stage units can now be arranged in their approximate positions on the stage and the location of the walls of the setting drawn in place. Logical relationships as well as stage business relationships between the various doors, windows, stairways, and fireplaces should be considered. The flow of stage movement is important to consider also when arranging these units. Furniture placement should be included in the floor plan so that this flow can be more easily charted. The furniture should be drawn approximately to scale so that it is related to the rest of the stage equipment.

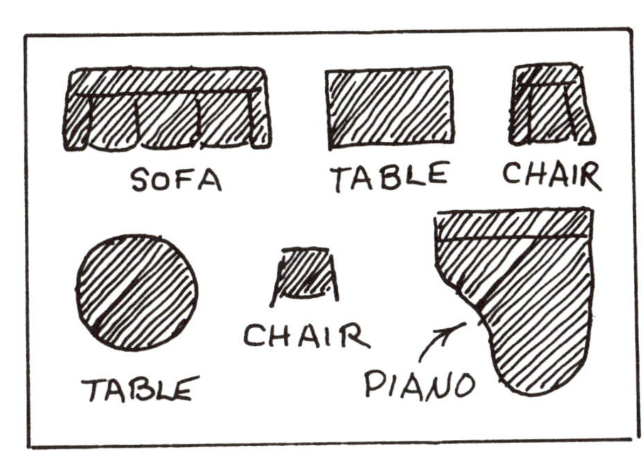


While all of this is being done the sight lines should be constantly checked to see that there is not too much blocking of entrances, action, playing area, furniture, and even important stage properties.

On the floor plan it will be necessary to indicate raised step and platform areas. These are functional devices. Sometimes only 2 or 3 steps are used to indicate an entire stairway. They are located so that the audience is led to believe that the rest of the stairs is beyond the wall or door that is conveniently next to them. In one play the step unit may be a place for stage business so it must be designed in dominance and in a good sight line position. An upstage area may be raised to set it apart from the rest of the stage, and to elevate the actors playing in that area so they may be seen above the furniture in the foreground. A jog is introduced into a wall in order to project the upstage portion of the setting onstage into sight lines, or it is designed into the floor plan to help brace and support a portion of a wall section, or a door, or a fireplace. Everything of this nature should be indicated on the floor plan so that it can be made useful to the director, the stage manager, the designer, and the builder of the setting.

After the floor plan has reached this stage, it is time to plan the backings, which are masking pieces placed beyond all of the openings to suggest other rooms, the out-of-doors, or just to block off the audience view beyond the setting. Sight line checks through the openings will help determine how much space must be covered by the backings. They can be drawn in position accordingly.

Labeling the Floor Plan — After the floor plan has been drawn to scale, complete with backings and approximate furniture, it is wise to label the wall sections. A system that works out fairly well is to start with the stage right return or tormentor (which is at the left of the floor plan) and letter it "A". The next wall section is "B" and so on around to the return on the other side of the stage. Usually it is advisable to skip the letter "I" since it looks so much like the number "1". The backings are lettered in a continuing sequence. This same system will be continued until the final setting reaches the stage and may even be used in setting up a stage shifting plan. It is a good idea to indicate on the floor plan areas that are elevated, such as platforms and steps. Each elevation can be shown by a number in a circle. The number indicates in each case the number of inches or feet off the stage floor. If the elevation from the stage floor rather than from another elevation is used fewer mistakes are made.

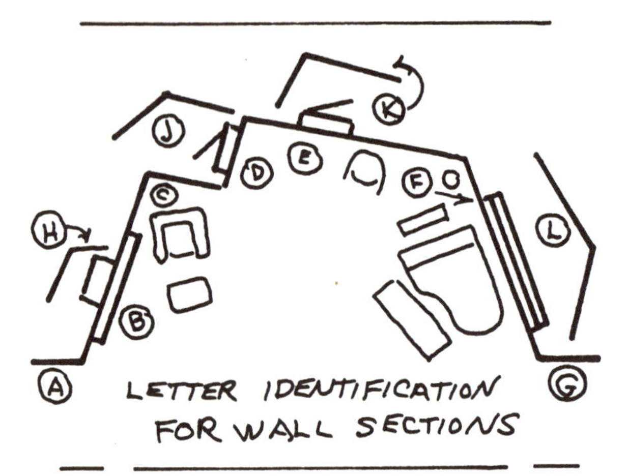

To clarify the floor plan for anyone who looks at it (especially for those who are not familiar with stage terminology) a legend frequently is included in a clear area on the paper on which the floor plan is drawn. The legend includes the symbols used, use of abbreviations and any further explanatory information needed. The scale used in drawing the floor plan and identification of this particular drawing (including the name of the production, the scene and the name of the designer) should also be shown on the floor plan.

## Perspective Drawings

The floor plan provides only one simple view of a stage setting. It is rather difficult to conceive of the appearance of the stage setting unless one has the vivid imagination of an artist, architect or stage designer or the powers of a clairvoyant. So to arrive at a more tangible visual concept of the finished stage setting it is necessary to carry the drawing process a few steps further and make perspective drawings or simple models of the proposed stage setting.

The perspective drawing is usually made even though a model is also prepared. The drawing can be reproduced in color and prepared in such a way that it gives a simple audience impression of the finished product.

PERSPECTIVE — Knowledge of a few facts about perspective will make drawing easier. A

cube seen from various positions takes on differing appearances. If looked at from dead center the front is all that will be seen. If the eye is moved straight upward the front and top can be seen, but the far edge of the top will appear to be shorter than the near edge. If the eye is moved to one side while still above, three sides will be seen—the front, top and one side. If the eye is moved downward, again while at one side, the top will begin to disappear and soon only the front and one side will be visible. The eye may be moved into many other positions and the cube will continue to present still different combinations of sides.

If a row of objects of the same size and height are viewed from various positions their apparent heights and widths will vary. Objects that are close to the viewer will appear to be larger than those located at a distance. When

BASIC PERSPECTIVE - OBJECTS GROW SMALLER AND APPROACH ONE ANOTHER WITH DISTANCE

a row of poles are placed along a railroad track, the poles, the tracks themselves and the ties under the rails all seem to diminish in size as the eye moves along the tracks toward the horizon. Finally they disappear as all of the lines converge at a point. This point is actually the vanishing point.

If a building is viewed from a position just beyond one end of the face, the side that can be seen extending away will appear to diminish

FORESHORTENING OF SIDE WALL GIVES PERCEPTION OF DEPTH HORIZONTAL LINES LEAD TO A POINT AT HORIZON - THE "V.P."

in height as it approaches the far end. If the top and bottom lines are continued beyond the building they will converge and again a vanishing point will be located—the vanishing point for all of the horizontal lines of that building. The face of the building will appear to diminish in size, too, but the angle is not as apparent since we are looking at it almost head on.

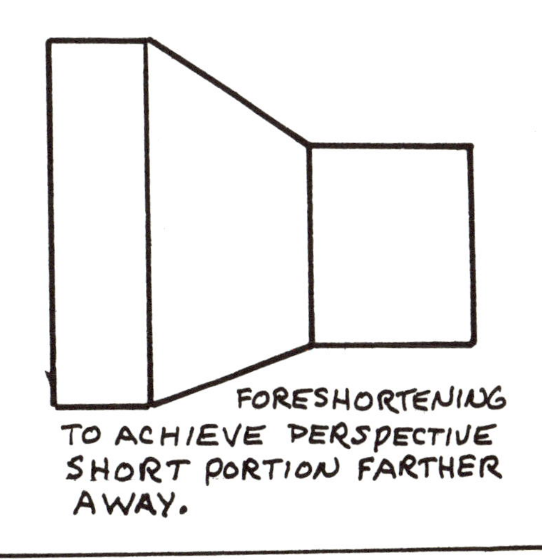

FORESHORTENING TO ACHIEVE PERSPECTIVE SHORT PORTION FARTHER AWAY.

Actually, the top and bottom lines if extended out, moving away from the viewer, will also converge, but the point of convergence will be at a much greater distance.

From this brief description it becomes apparent that objects at a distance seem to be smaller than near ones and if a rectangular flat plane is placed so that one edge is near and the other at a distance there will be apparent foreshortening of the surface as it moves away from the viewer. When reproducing this effect on paper, vanishing points (V.P.) are set up to aid in getting the correct angles.

Bird's Eye Perspective — The bird's eye perspective is easy to draw, but it is not the most satisfactory type of reproduction. To see the stage setting as it appears in this conception the

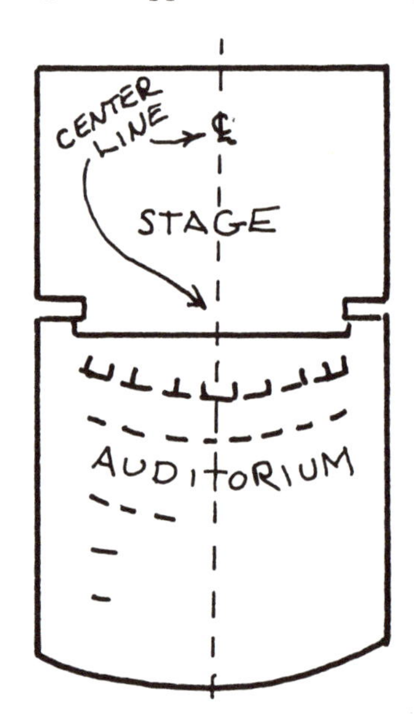


observer would have to perch himself in the thickness of the proscenium arch above the stage and look down. The distortion produced by this method makes the top of the setting much wider than the bottom.

In drawing the bird's eye perspective, as in all other perspective drawing, preparation of the floor plan is the first step. Then a vanishing point (hereafter referred to as a VP) is located a short distance below the front of the stage setting and on the Center Line (℄) of

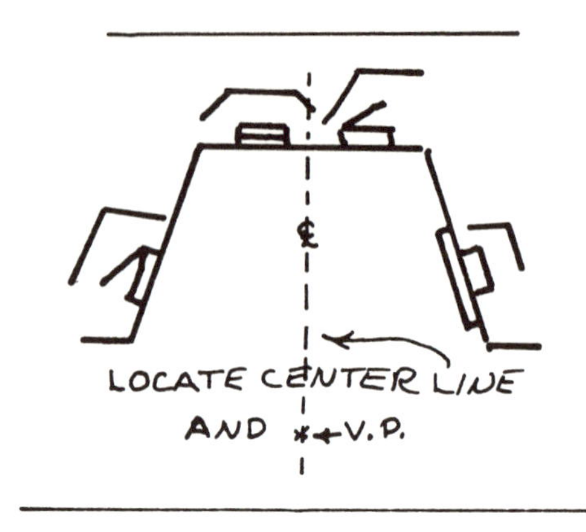


the stage. The center line (℄) is an imaginary line drawn through the center of the building extending from the back wall of the auditorium to the back wall of the stage.

From the VP lines are now drawn upward and outward through all of the corners of the

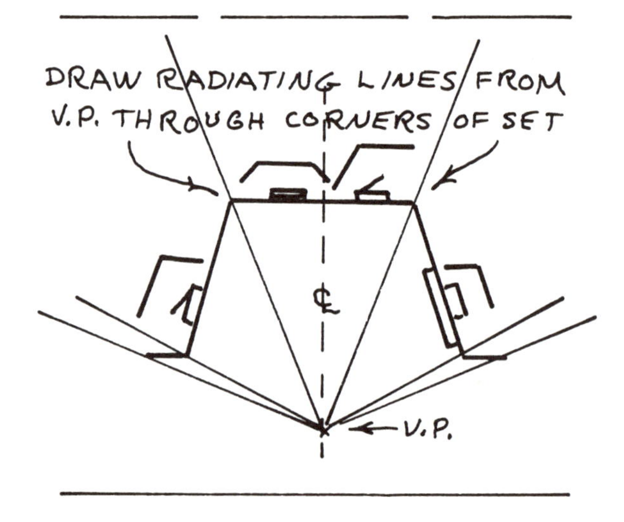


setting, that is, all of the places where walls begin, end or change direction. These lines are extended a couple of inches above the floor plan lines.

The lines just drawn represent the vertical lines of the setting. After they have been drawn, a convenient height is decided upon. For the first elevation try two inches. Beginning with this height, on the left side of the setting draw lines parallel with the floor plan lines following across all of the lines of the setting to the final line projected from the right side of the setting. This done, the setting in its bird's eye form appears. Doors and windows may now be drawn in a similar manner. Side lines are drawn up from the VP and the

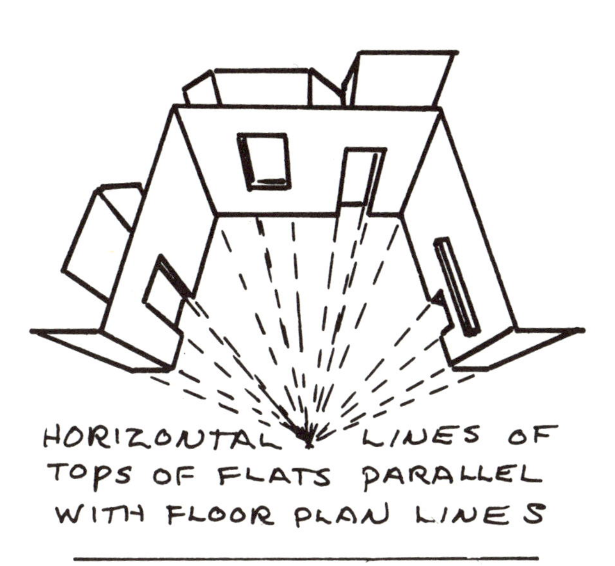

HORIZONTAL LINES OF TOPS OF FLATS PARALLEL WITH FLOOR PLAN LINES

tops and bottoms are drawn parallel to the base line on the floor plan. Furniture is drawn in a similar manner. This type of drawing may be completed with any detail that is to be included in the setting.

The finished bird's eye perspective is a means of rapidly drawing a rough impression of a stage setting. An angular type drawing that requires no VP may also be drawn. This reproduction has distortion, too, and is viewed from a position never assumed by any of the members of the audience. This might be called the pin rail projection.

PIN RAIL PROJECTION — For the pin rail projection, draw a floor plan. Now from each of the wall break points draw a line up and out on an angle of 30 to 45 degrees (choose any angle, but use the same one throughout). These lines will be parallel to one another. After the lines have been drawn, a height is determined for the flats; here actual scale can be employed and the top lines drawn across the flats parallel to the lines of the floor plan. Doors, windows and other units may be drawn in a similar manner.

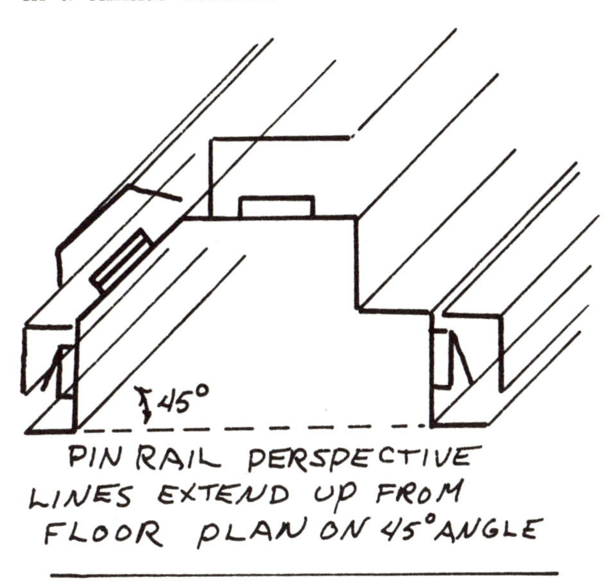


PIN RAIL PERSPECTIVE LINES EXTEND UP FROM FLOOR PLAN ON 45° ANGLE

The pin rail projection gives an angular view of the setting as it might appear from the pin rail above and to one side of the setting. This projection has an advantage over the other types

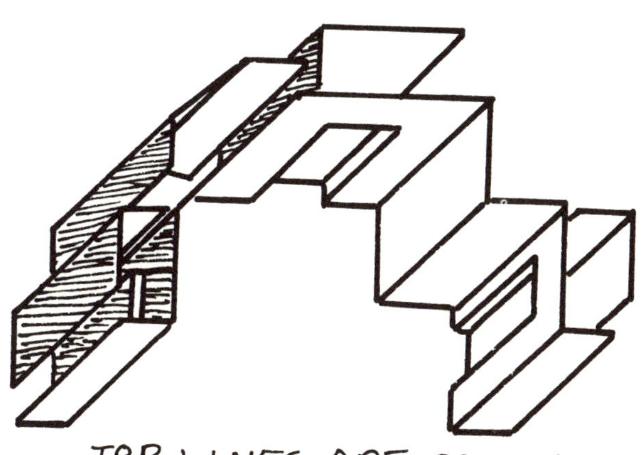

TOP LINES ARE DRAWN PARALLEL WITH LINES OF FLOOR PLAN.

of drawing since it is possible to draw everything to scale. Of course, one side of the setting is seen from the rear rather than from the front, but if it is desirable a view from the other side could also be drawn. This is almost as simple to draw as the bird's eye view and is probably more satisfactory.

One Point Perspective — The regular perspective system is more difficult to handle

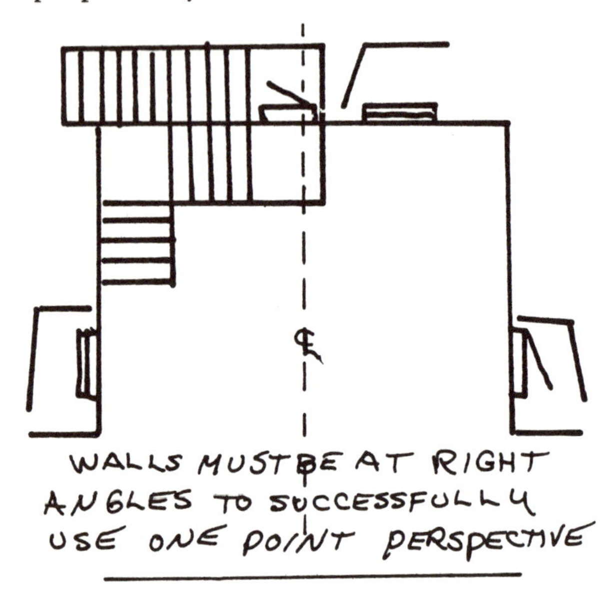


than either of the other methods of reproduction suggested, but the result is more in the nature of an audience view.

The one point perspective system may be

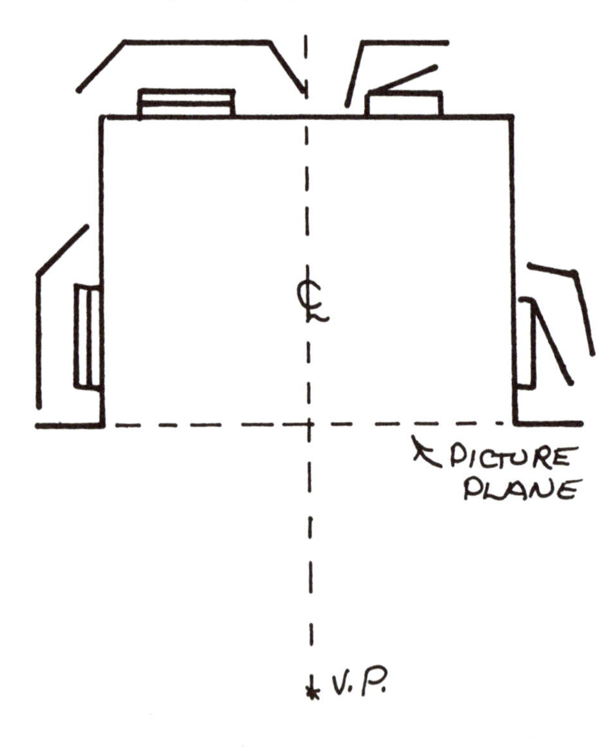


used as long as the walls of the stage setting are at right angles to one another and parallel with or at right angles to the curtain line.

The perspective drawing is started from the scaled plan. For the first step in the process of drawing, locate a vanishing point (VP) on the center line ( ℄ ) the equivalent of half way back in the auditorium. This point represents the position of the eyes of the patron seated in the center of the auditorium. Actually the position of any seat could be used, but this one will be the average view. The VP located will be used to determine the relative positions of the vertical lines and the foreshortening slope of the horizontal lines of the walls of the setting as they extend away from or come toward the audience.

For convenience, and to keep the scale exact, on some portion of the drawing the setting will

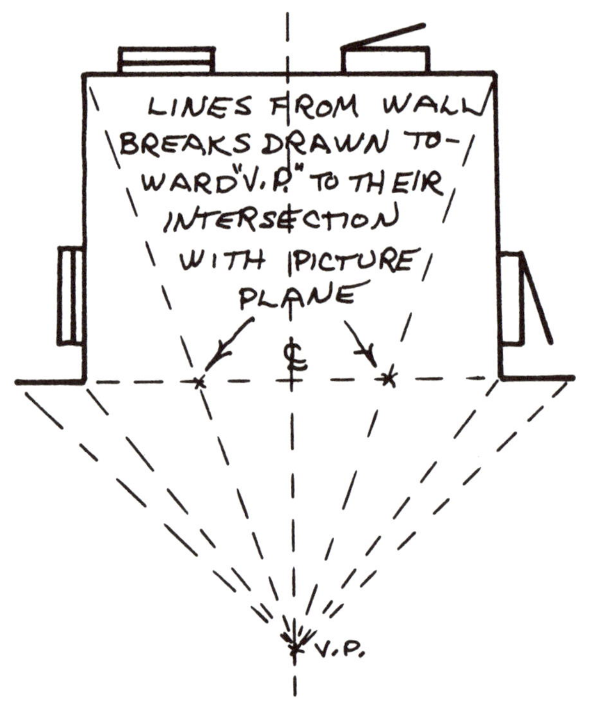


be drawn as if it were flattened out along the line of the tormentors of the setting parallel with the curtain line. For convenience this will be referred to in the descriptions to follow as the *picture plane.*

To get the relative positions of the vertical

lines of the stage setting, lines are drawn connecting the VP and the breaks in the floor plan that indicate inside or outside corners in the walls. These lines need only be drawn from the corners to their points of intersection with

LINES OF RETURNS DROP STRAIGHT DOWN

LINES FROM OTHER CORNERS DROPPED FROM POINT OF INTERSECTION WITH PICTURE PLANE

the picture plane. From the picture plane they are dropped straight down, at right angles to the curtain line. These lines are drawn so that they extend down below the VP.

The measurement to scale from tormentor to tormentor should be the same as it is on the floor plan.

Next in the drawing process, the height of the setting should be determined. And at this point in the process it is also necessary to determine the distance from the floor line to the VP. The members of an audience seated on the ground floor will view the stage from any of a number of eye levels ranging from even with the stage floor to a level about half way up on the stage scenery, depending upon whether they are seated in the orchestra or farther back in the auditorium. For the scene designer the best position is probably about 1/3 the height of the flats.

If the flats are to be 12′-0″ the VP for the plan shown will be about 4′-0″ above the floor line, or above the bottom of the setting. A line is therefore drawn parallel with the curtain line and 4′-0″ below the VP. This line need only extend across the lines that have been dropped down from the sides of the tormentors since they are the only portions of the setting that extend down to the picture plane. The tops of the tormentors may now be sketched by drawing a line parallel and 12′-0″ above the one just drawn. This done, the tormentors will be completed and should measure, by scale, 12′-0″ high and the width indicated by the floor plan.

The side walls adjacent to the tormentors may now be drawn. The vertical lines already have been dropped down into their positions. To draw the top and bottom lines, which will approach one another as they are drawn toward the center of the picture (because the wall section moves away from the viewer and appears to be shorter as it moves upstage), lines are drawn from the top of the tormentor toward the VP and from the bottom of the tormentor toward the VP until they reach the point of interception with the next vertical line or wall break. This done, the two side walls are drawn in their correct positions.

Since the back wall is parallel with the footlights, it may be drawn straight across from the interception of the corner vertical and horizontal lines just completed.

The walls are now completed and other structural units, such as doors and windows, may be

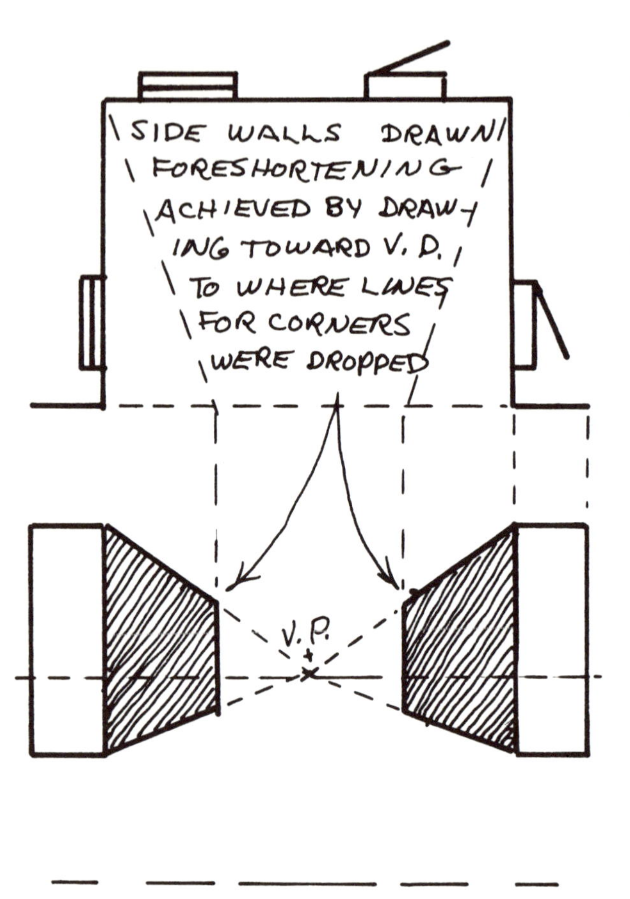

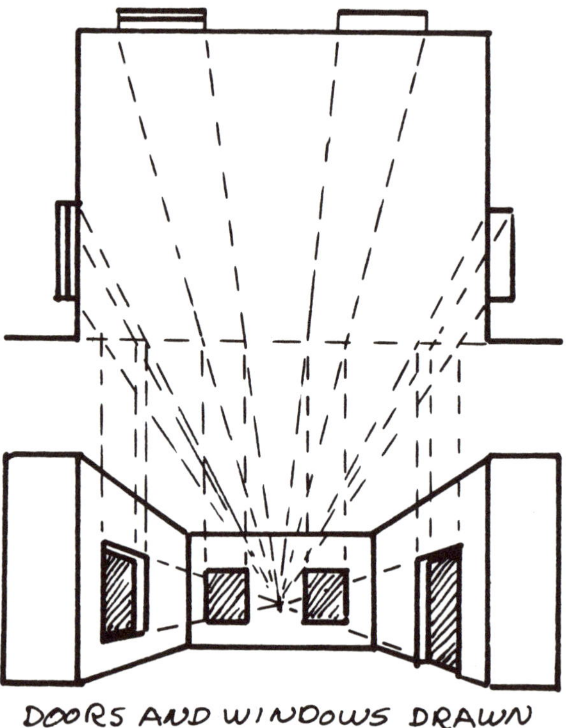

DOORS AND WINDOWS DRAWN AND LOCATED USING V. P.

drawn in a similar manner. First they are located on the floor plan. Then the relative positions of the vertical lines are located by drawing toward the VP to the picture plane, after which they are dropped straight down into position. The height of such units is measured on the tormentor since this is the only place where true scale may be measured. Then the measurement can be taken around to the correct position by following the process of using the VP.

PERSPECTIVE USING THREE OR MORE POINTS — If the side walls of the setting are raked, that is, if they move in on an angle, and if other wall sections are not parallel with or at right angles to the curtain line, it is necessary to add other vanishing points or the true foreshortenings of the walls will be lost. The process of drawing with three or more vanishing points is exactly the same as with the one point system through the step where the drawing of the tormentors is completed.

To minimize the distortion and foreshortening of wall sections a new vanishing point is introduced for each angle on the floor plan deviating from the square. If walls are parallel with one another they use the same VP and if they are parallel with the picture plane or curtain line they are drawn parallel with the top and bottom of the tormentors.

In perspective systems using more than one VP there is always a relationship between the original VP and any others. Each additional VP is located in terms of the original one.

To find the VP for a wall section that is

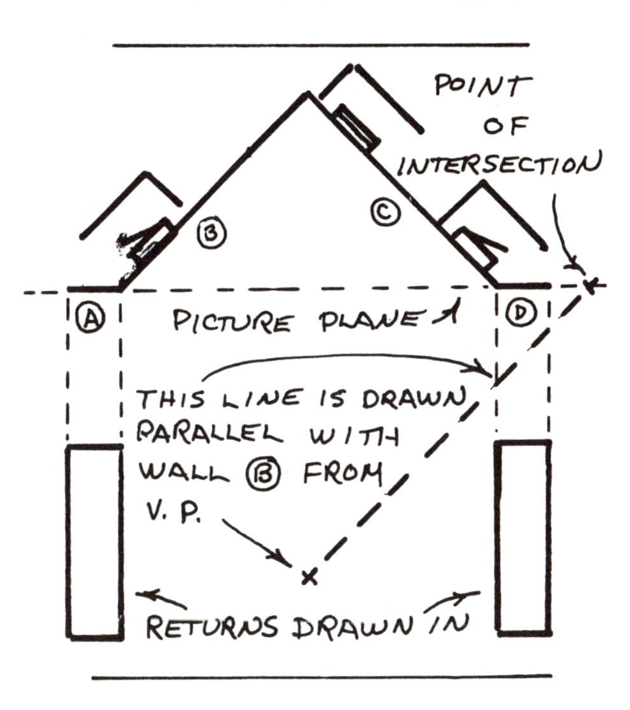


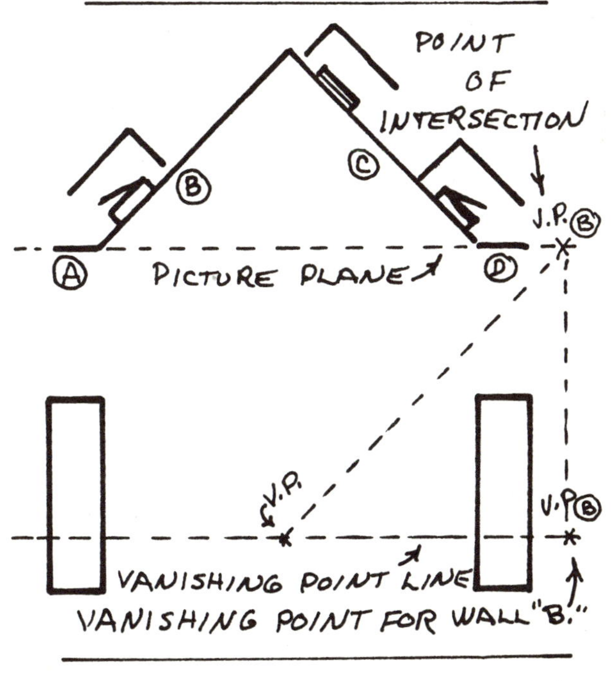

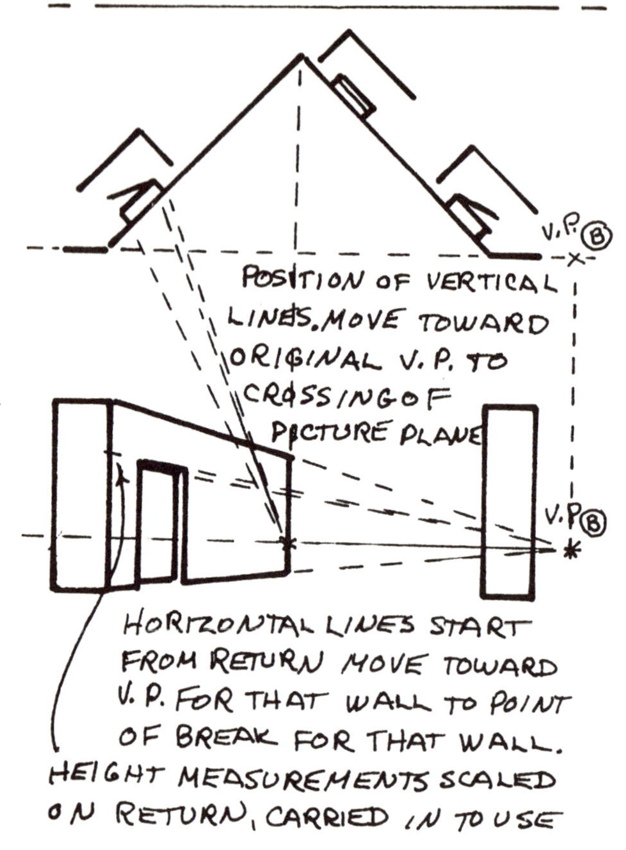


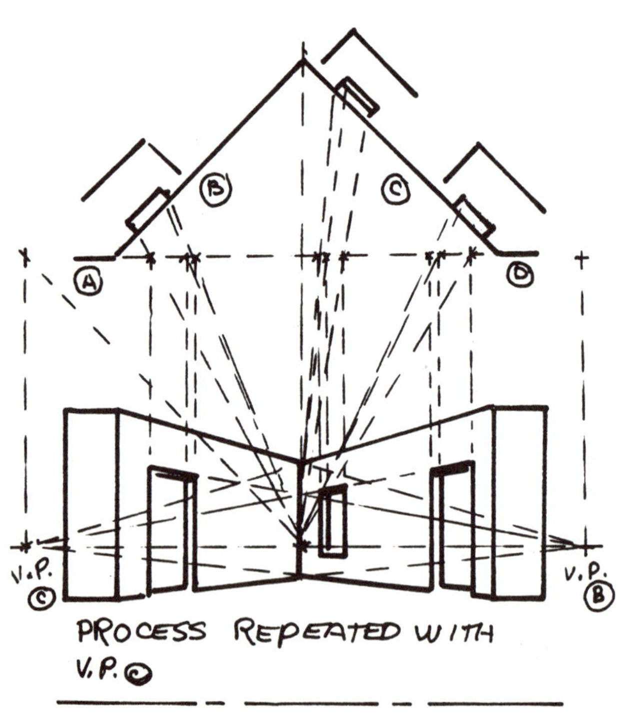


raked on an angle, draw a line from the original VP toward the picture plane and parallel to the line of the wall. Stop the line where it intercepts the picture plane and then drop it straight back down at right angles to the picture plane until it intercepts a line drawn through the original VP and parallel to the picture plane. In this way there is a horizontal and vertical relationship between the new VP and the original one. All VP's located in this way are related to one another and to the original. Each VP should be labelled with the letter identifying the wall or walls with which it is to be used. The VP located in this manner is subsequently used to draw any horizontal line on that wall, be it baseboard, chair rail, top and bottom of a picture, door header, cornice over a window, mantel, window sill or shelf.

Model of Stage Setting — A simple model of a stage setting can be made rapidly if squared paper is used. Paper with ¼″ squares is probably as convenient as any other since a ruler can be used to do auxiliary measuring after the model has been completed.

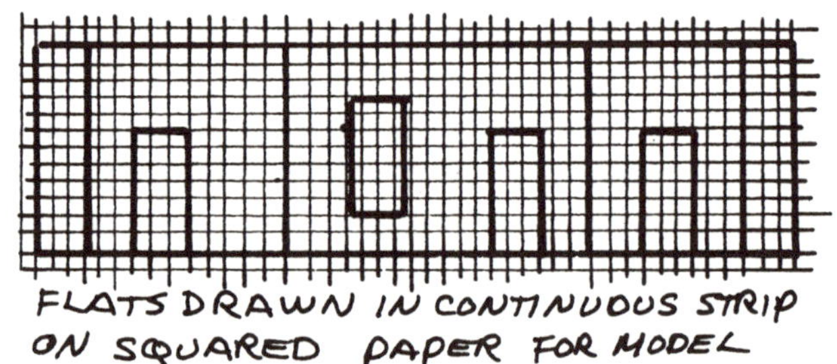


The first step in making the model is to draw a front view of the flats transferring the sizes indicated on the floor plan onto the squared paper. At this time the height of the setting must be decided upon. If a relatively small model is desired a scale of one square equals one foot can be used. Begin with the stage right tormentor (the left side of the floor plan as you look at it—so begin on the left side of the paper and draw in the logical direction toward the right) and draw all of the sections with each immediately adjacent to the other on the paper. In case one piece of paper does not give enough length for all of the flats (and it probably will not) overlap and glue two or more pieces together to get the desired length. After the flats have been drawn—with doors, windows, fireplaces, pictures, moldings, etc. drawn in place—the strip may be cut out. Fold the strip where corners are indicated and place the finished product on the floor plan, or in a model of the stage, and the setting will be seen in its true form, proportion and shape. More complicated and complete models can be made from cardboard, matte board or similar material, and stage properties can be indicated by little cut out forms of blocks of wood. Even little figures to represent and scale actors can be cut out of pieces of cardboard and placed at

various points on the stage. The model will make it possible to determine space relationships. Some of the directors and stage manager's problems might be solved before the setting reaches the stage if a model such as this is made and manipulated.

WORKING DRAWINGS—Selecting and processing the flats for an individual setting is made much easier if simple working drawings are prepared. If squared paper (preferably with quarter inch squares so that a ruler can be used with it to make more rapid calculations) is used for this purpose the drawing process is simple. Let one square equal 1′-0″. Beginning with the stage right tormentor or return, draw the wall sections in sequence. Drawn in this way the front of the surfaces is shown. If it is desirable to have both the front and back shown it is easy to turn the paper over, hold it up against a

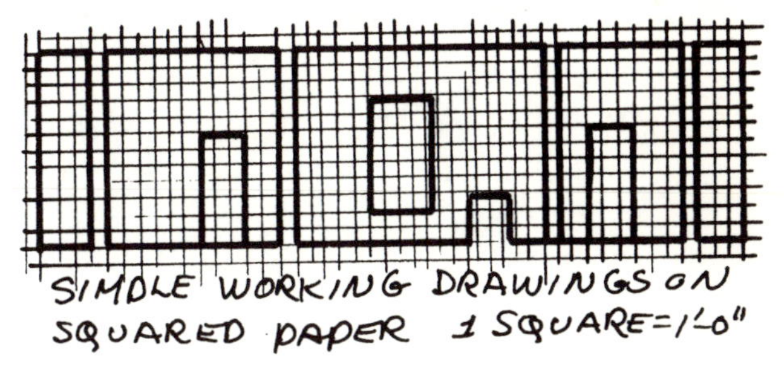


window and trace the image that shows through. The detail can then be drawn on the flats to help in building, rebuilding and assembling them. Be careful to label in large letters which is the front and which is the rear view of the flats. The breakdown into flats may be indicated after the wall sections have been drawn.

When drawing the wall sections it is a good idea to leave a space a square wide between sections to show which areas may be fastened together later and which are to be hinged to make unbroken walls. On the drawing the sections should be lettered to conform to the lettering on the floor plan and then as the flats are "dug out" they also should be lettered. If the setting is complex, some type of checking system should be set up. The University of Utah system is to place a check on the drawing when a flat is found, cross the arm of the check when the wall section is assembled and ready to paint, and place a circle next to it when the section has been given a base coat of paint.

* * * * *

Although this discussion of the mechanics of drawing perspectives, floor plans and working drawings is not extensive it may be enough to start the designer in the right direction. Actually, every production a designer works on is a completely new project and new challenge and he continues developing procedures that make the entire job easier for him.

*Jacobowsky and the Colonel* (above). The size of the stage reduced with a double false proscenium. *The Green Pastures* (below). Contour flats in depth are here used with a platform.

# 16

# Designing the Stage Setting

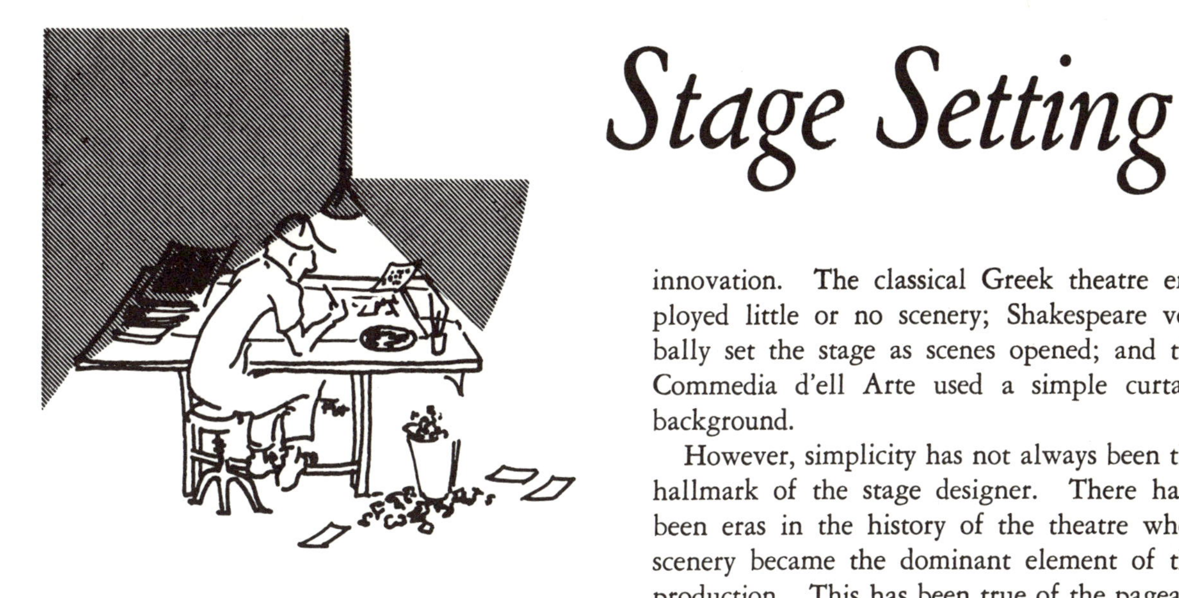

"This is our doctor's house—Doc Gibbs; this is the back door. (Two arched trellises are pushed out, one by each proscenium pillar.) There's some scenery for those who think they have to have scenery. There's a garden here. Corn . . . peas . . . beans . . . hollyhocks . . . heliotrope . . . and a lot of burdock."

Thornton Wilder, in this portion of the opening speech of the stage manager is brilliantly setting the stage for his play *Our Town.* He is setting it in such a way that every member of the audience must be pleased or rebuke himself for his lack of mental imaginative artistry. By employing Wilder's method, stage hands, stage machinery and act curtains are replaced by a twist of the head, flash of the imagination and a flick of the eyelid. Poof! The set is changed.

Wilder's idea of dispensing with elaborate stage decorations is by no means a new or novel innovation. The classical Greek theatre employed little or no scenery; Shakespeare verbally set the stage as scenes opened; and the Commedia d'ell Arte used a simple curtain background.

However, simplicity has not always been the hallmark of the stage designer. There have been eras in the history of the theatre when scenery became the dominant element of the production. This has been true of the pageant stage of all ages; of the elaborate court masque of the Renaissance; of the melodrama of the nineteenth century; of the naturalistic theatre of the twentieth century; of the Hollywood motion picture productions and of the great television spectacles. The technicians of these theatres have employed literally wagon loads of scenery and machinery on their stages. Their machines have been used to make actors and scenery appear, disappear, fly and sink into the floor. They have striven to outdo nature in depicting the great destructive forces of flood, fire, storm and war; they have striven to reach new and greater peaks of mechanical and spectacular beauty and grandeur.

In the contemporary theatre the designer usually tries to achieve a level of staging somewhere between the two extremes noted. He avoids both overabundance and poverty of scenery and stage effects.

FUNCTIONS OF STAGE SCENERY — Stage scenery may serve one or a number of functions. It may:

1. Serve as an enclosure for the action of the scene.
2. Serve as a motivating force for the action of the play.
3. Serve a purely decorative function.
4. Serve a symbolic function.
5. Serve as a reflective surface for light and sound.

The ordinary setting, especially if it falls into the category frequently referred to as a "box" setting, serves as an enclosure for the action of the play. It is in a sense a corral for the players and the very structure tends to define the space. It also serves to shut out distracting elements, elements that might be present just beyond the acting area such as actors waiting for cues, lighting units used to illuminate portions of the setting, and stray actors making cross overs behind the setting. When the *Our Town* style of production is employed it is impossible to accomplish these ends with the setting since there is no setting.

The various elements of the design may have specific use in the action of the play. This is more often true than not. In fact the designer usually begins his planning by placing doors, windows, fireplaces, stairways, platforms, trees, stumps and columns where they can be used to the best advantage for the stage business. At times the stage business may depend almost completely upon the manipulation of the actors by the setting. This practical and functional purpose of scenery is frequently the most important reason for its existence.

The elements of the stage design may be so arranged that they add beauty and interest to the production. Usually this visual beauty is enhanced by the actors and their costumes which also serve as a part of the scenery. Together the moving and stationary scenery present a pleasant picture. This is important even if the structure represented is a junk heap. There must be enough attraction to keep the audience's interest for the duration of the play.

Scenery may serve a purely symbolical function without being what one usually considers to be symbolism. The historical period, season of year, social level of the inhabitants and general mood may be symbolized by the elements of design and their treatment. Or abstract forms may be used to project specific moods and impressions.

And finally, scenery may serve as a surface from which sound and light may be reflected. An enclosure, setting, especially if it has a ceiling, will serve as a sounding board to help direct the sound into the auditorium, thus helping the acoustics of the hall — and many of them need all the help they can get. The painted surface of scenery also serves to reflect light, especially if it is painted in relatively high values of color. The light will be picked up and redistributed about the stage. If low light values are desired the setting can be painted to absorb light.

But whatever its function, the setting must be designed carefully and the design executed in such a way that it is functional, practical of manipulation upon the stage, and its composition pleasant and free from distracting elements.

### Composition

Certain factors of art composition are important to the stage designer. He should be conscious of line, plane, mass, rhythm, symmetry, proportion, contrast, unity and coherence, in costumes, properties and scenery.

LINE—The simplest element of composition is the line. In theory line has only one dimension—length. However, it is the direction and apparent movement of the line that is important to the scene designer. He must be aware of the lines formed by moldings, tops of doors and windows, ground row forms, platforms and the flats of his setting. Certain general factors may be of value to him as he plans his setting.

A sharp, smooth line has smartness; a wobbly line may suggest indefiniteness and even carelessness; a curved line may have beauty and conformity; an angular line may suggest con-

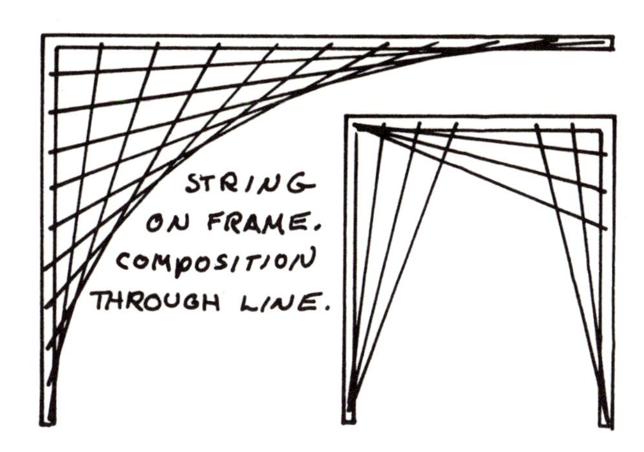


flict. Lines may lead into or out of the stage picture. They may lead the audience attention toward or away from the point of major conflict.

Simple, smart, abstract designs may be created by using lines exclusively. String, rope,

SPANISH STREET SCENE. LINE DRAWING EFFECT USING TUBING AND WIRE. SUSPEND FROM GRID ON WIRE.

pipe, rod, aluminum tubing or slats of wood can be arranged to make interesting patterns.

PLANES—Lines are combined to form planes or two dimensional forms such as squares, oblongs, triangles, circles, ovals, trapezoids, hexagons and other flat surfaces. The plane has two dimensions—height and width.

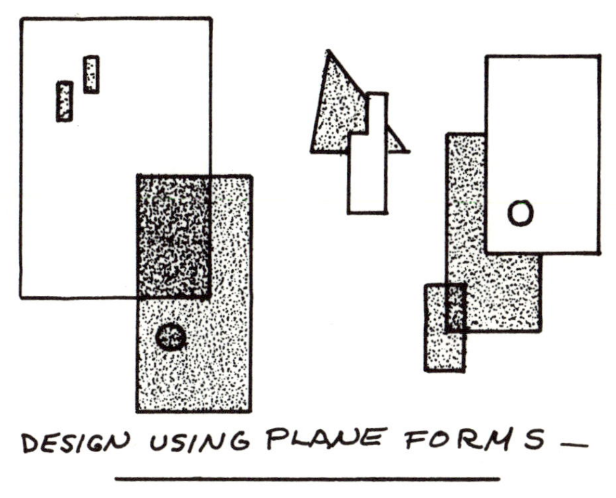


There are times when stage designs are composed of combinations of flat forms or planes with no attempt at three dimensional forms. Smart patterns can be created in this way for essentially abstract designs.

MASS—Usually planes are combined to form three dimensional forms known as mass. Mass forms have height, width and depth and seem to have weight which is somewhat absent in planes and lines. Practically every object in

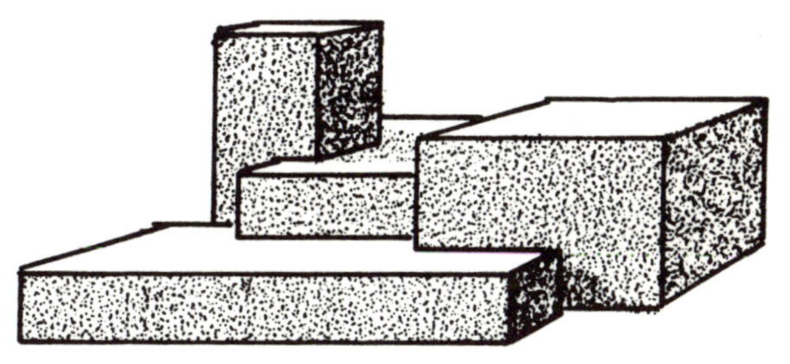

DESIGN THROUGH MASS. THREE DIMENSIONAL FORMS.

nature and every structure of man is in the form of mass. So when these objects are created on the stage the mass form is usually duplicated by using real or painted height, width and depth dimensions.

RHYTHM—Rhythm and movement are evident all about us. Life itself is composed of complex rhythms: the rhythm of breathing, of the heartbeat, of footsteps, of day and night, of seasons, of the rotation of the earth and of the growth of plants. In design the repetition of forms similar or like and the flow of lines are

RHYTHM THROUGH REPETITION

examples of rhythm. Two factors concerning rhythm are important in design: having it and not overdoing it. Interest is achieved by repeating color and form, by the orderly flowing of lines and forms, and by the recurring of patterns and textures. But too much repetition results in monotony. Repeating exactly the same color, the same form, the same mass, the same line will result in dullness while the correct use of rhythm and variety will result in pleasure and excitement.

BALANCE—The operation of a seesaw may be used to demonstrate in part the matter of balance. A single weight placed across the

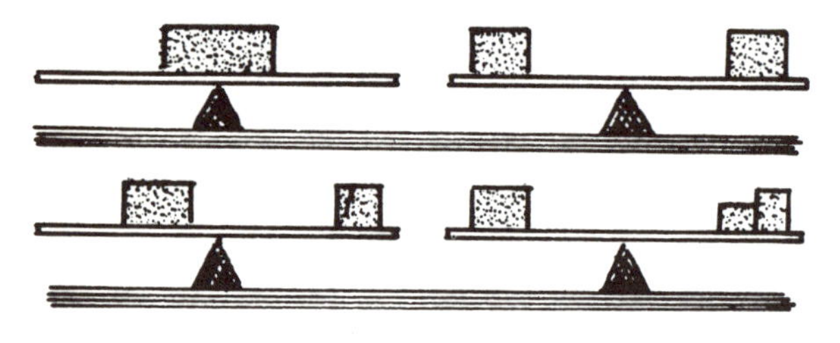

VARIETIES OF BALANCE

center of the seesaw will remain in balance. Two objects of equal weight will balance one another if they are placed on the board equidistant from the center. A large weight may be balanced by a smaller weight if the larger is nearer the center or fulcrum. Two smaller weights together may balance a larger. Balance in a picture is achieved in a similar way. Weight, size, color, area—all of these elements may be used in accomplishing balance in the picture. There are times when deliberate unbalance is desired to project a specific feeling. The designer begins working in terms of balance as he arranges the floor plan for the setting and continues through the processes of design, execution, dressing and lighting of the finished setting. He balances a large area of grayed red on one side of the stage with a small amount of brilliant red on the opposite side. He balances textures by using similar ones on both sides of his setting. He balances a heavy

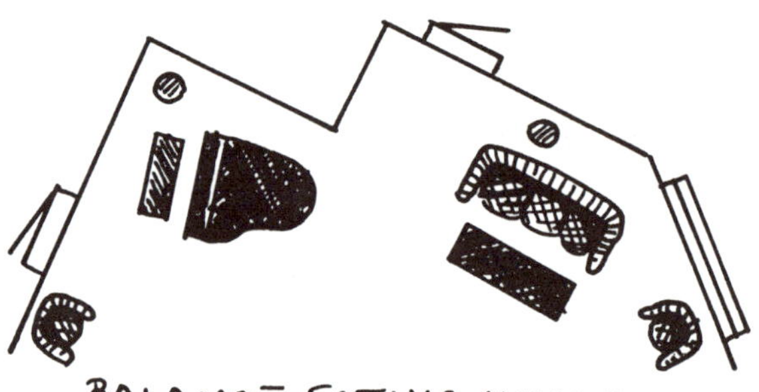

BALANCE SCENIC UNITS AND PROPERTIES.

weight on one side with an area of heavy color on the other. If he fails to balance his original design he can use pictures on the walls, pieces of furniture, boxes or whatever dressing materials are available to correct the imbalance. And, of course, if there is an apparent reason and motivation for imbalance, then that is what he strives to achieve in his finished setting.

SYMMETRY—Symmetry is in a sense a factor of balance. Perfect symmetry is achieved when the two sides of the picture are evenly and regularly balanced. When similar objects of the same weight or size are placed equidistant from the center of the stage the design is symmetrical. Informal symmetry is achieved by using dissimilar objects having the same feeling of weight, volume or color. Informal symmetry is usually more pleasant to the viewer than perfect symmetry. Perfect symmetry projects a feeling of orderliness in the extreme. Precision and mechanically developed aesthetic

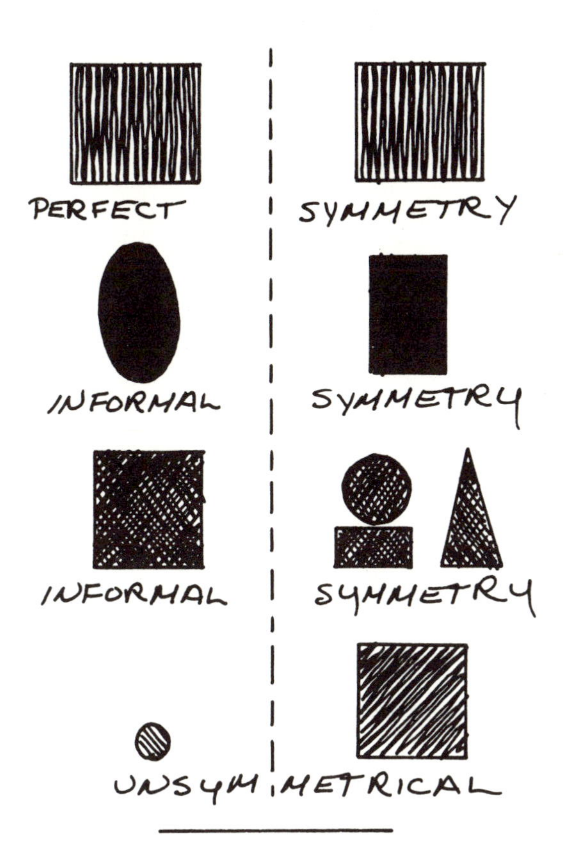


"taste" are suggested in this way. An asymmetrical design is unbalanced in one direction. This state of over and under balance may be deliberately used. When unintentionally employed it may be disturbing to the audience so it is a factor worth studying rather carefully before being employed.

PROPORTION — Proportion is important in theatrical design. This may also be referred to as scale. It is important to relate objects to one another in size, weight, color and form. A door 5′-0″ high looks silly if all of the actors who use it are 6′-6″ in height—unless there is a definite reason for this apparent abnormal relationship. A tall object will seem even taller if placed next to a short one and a short object looks shorter when placed next to an extremely tall one. Actors may be dwarfed when playing next to tall scenery but seem like giants when playing in front of a short setting. If everything in the setting is scaled in the same proportions this relationship will be heightened even more. If actors are playing the part of insects in *Under the Sycamore Tree* or *The Insect Comedy* they seem small when scaled against a king sized tree trunk, but as soon as the stage hands walk on dressed in regular "man type" clothes the illusion is destroyed. This feeling for proportion is well shown in marionette productions where everything is scaled to the size of the marionettes. After a few moments the members of the audience begin thinking the figures to be of human proportions. At the end of the performance the

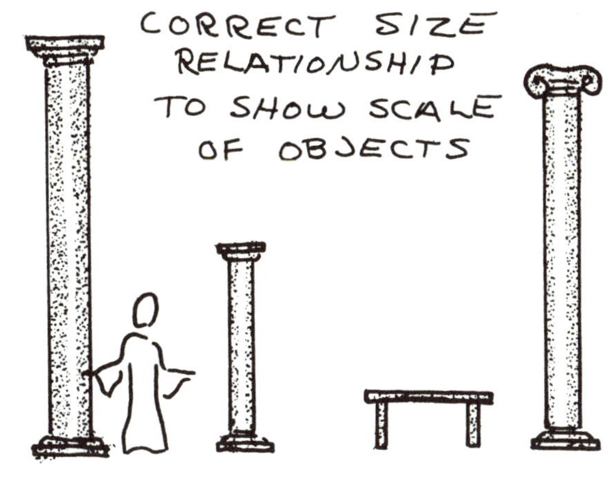


operator climbs onto the stage and immediately the mechanical figures become tiny. The effect is startling but indicates how easily the eye may be fooled. Proportional relationship or scale is extremely important in designing a stage setting. The designer should be constantly aware of this in planning his units of scenery and their relationship with one another, with the stage, and with the actors.

Openings for entrances must be proportioned for stage business at times. Hoop skirts, rapid entrances and exits of mobs, movement of fencers, dancers whirling onto the stage—all of these require amply proportioned doors or archways.

Units of scenery designed to be acted upon and designed to provide desired scale and proportion relationships between the actors and the setting. *Romeo and Juliet.*

CONTRAST—Contrast between the elements of the design may add to or detract from the total design. If deliberately planned, contrast between the sides of the stage or between the elements within a given stage unit may be of value. Frequently in a design where there are many rectangular forms one curved form is an interesting deviation. The same is true of a setting that has a dominance of curved forms with a spot or two of recilinear forms. The contrast may be subtle or broad depending upon the effect to be achieved. Contrast of color, texture, general mood, mass, form and line is possible. Clashing colors might be deliberately used to startle or "jar" the audience and actors. Extremely rough and shiny surfaces might be used to achieve a feeling of contrast or conflict. The name character of *The Hairy Ape,* fresh from the stokehold of the ship, thrust into a clean, smooth, shiny atmosphere is completely out of his element. Tweedy suits might be out of place in a world of mirror surfaces. However, contrasts are needed. If everything in the setting has exactly the same feeling there will be a weak, sterile feeling to the whole. Tweed may be just the thing to make the mirror surfaces more interesting. A touch of brilliant color may touch off a large drab surface. Great contrast is desirable where a startling effect is desired; subtle contrasts where normal effects are desired. In the main it is advisable to have one dominant element in the design.

CONTRAST THROUGH FORMS

This element may be dominant because of its size, proportions, shape, color, dissimilarity, texture or pattern.

UNITY AND COHERENCE — Unity and coherence should be achieved when and where possible. Even if portions of the stage setting are dissimilar a certain amount of unity is desired. The stage setting should seem to fit together to make a whole. Too much dissimilarity is disturbing and calls attention to itself. There may be times when even this is desirable. If it is correctly planned then it is perfectly acceptable; if not, then it is probably distracting. Basic similarity of forms, patterns, colors, general style, mood and atmosphere, architectural period and scale are all essential if the parts of the setting are to be cohesive.

COSTUMES — Costumes, an important element of stage design, provide form, color, mass, rhythm, repetition, comparison and contrast. They are moving elements within the static stage picture that provide sometimes subtle, sometimes broad variations upon the original plan or theme.

The general structure and design of some of the elements of the scenery are largely determined by the structure, bulk, color, historical period, texture, movement and "feeling" of the costumes. There are times when the scenery is designed to fit the costumes and other times when the costumes are made to fit the scenery. More often than not the plan is a mixture of these two so that the elements dovetail into one another. Deliberate conflicts and harmonies are planned to indicate the relationships that exist between character and character and between character and setting.

The brevity of costume discussion in these pages is not intended to minimize the importance of this element of design. A more thorough, complete and informative body of information on this subject is to be found in the costume books that are listed in the bibliography.

## Types of Settings

It has been suggested that a scene designer may use a poverty or a surfeit of scenery. It may now further be stated that there are many steps between these two extremes. Although there are no sharp lines of demarcation along the way, yet for convenience a number of types of settings ranging all of the way from simplicity to complexity of staging will be suggested. It is hardly necessary to say that these may be intermixed. The actual style or type of design used may be determined by one or all of the following factors: budget, physical facilities, style of stage direction, general atmosphere of the script, general audience character. In practice these will probably all be influential factors. However, in the finished design, regardless of style or form, there should be harmony, not conflict, among the playwright, the director, the actor, the audience and the scenic designer.

## Projected Scenery

A bare stage with human forms moving into and out of pools of light can be extremely dramatic. Shadows cast on a background by

COMPOSITION OF HUMAN FORMS

the same moving forms can also be effective. There are occasions when no other scenery could be more desirable.

SIMPLE PROJECTIONS—Simple forms projected onto a background can also be effective. It isn't necessary to have complex equipment to achieve results of this kind. Giant palms can be the result of shining a spotlight through the stems and leaves of a potted plant. If the spotlight is on the floor the image produced in this way will be distorted and greater in size as it shines on the surface at a distance, but even this distortion can be valuable in some instances. Almost any form—china figurines, branches from trees, forms cut from cardboard or wood, rope patterns—may be used. This type of design will require some trial and error,

SIMPLE ENLARGED SHADOW OF A SMALL OBJECT.

but in the theatre many of the best results come from this method of procedure. Every design is, in a sense, devised through trial and error. In some cases this process takes place on paper; in others it takes place in the scene shop or on the stage.

LINNEBACH PROJECTIONS — The Linnebach projector is essentially a shadow projector. The machine consists of a housing containing a large concentrated filament incandescent lamp. The front of the housing is equipped with a frame to hold a cut out pattern. When the lamp is turned on, portions of the beam of light are restricted or absorbed by the pattern on the front slide—the cut out. The resulting pattern which is allowed to fall on a surface at the back of the stage is basically a light and dark image. A certain amount of distortion may be eliminated by shaping the front of the box so that it conforms to the contour of the back wall. The pattern can easily be colored by introducing a color media to fill the open spaces of the cut out pattern. This colored effect may be enhanced even more by using a plastic slide which has a design painted on it with special plastic paint. Paint of this nature may be obtained from almost any hobby shop. Nonflammable plastic should be used if possible. The projector housing serves as a holder for the lamp and as a device to prevent spill light from spoiling the image. This piece of equipment, whether home made or commercially manufactured, should have adequate ventilation holes so that the heat may escape. The Linnebach projector is usually used for relatively broad effects. When used where other illumination is needed, that light should be directed so that it does not fall upon the projected image or it will be "washed out".

SLIDE PROJECTOR—Regular slide projectors with hand painted or photographic slides may be used on some stages and for some purposes. When this type of equipment is used it is possible to project extremely sharp focused images. However, unless it is possible to obtain a wide angle projector (with wide angle lens system) or possible to mount a projector in a position where it will have a long throw (at some distance from the screen surface) it is impossible to obtain a large image. If there is space at the back of the stage it is possible to use a rear projection on a translucent screen—a process much used in both motion pictures and television. When this type of projection is used the light falling on the other areas of the stage must be restricted. If too much direct or re-

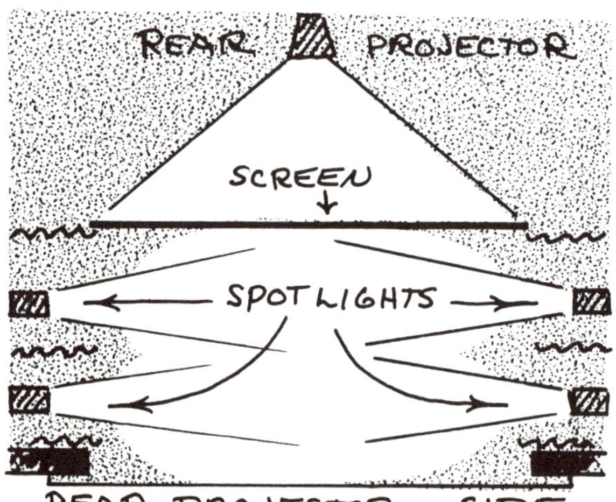


REAR PROJECTOR – SIDE LIGHTING TO PREVENT "WASHING-OUT" THE IMAGE

flected light falls on the screen the image of the projection will disappear. Although a regular picture screen surface will provide the most brilliant image when projections are used it is possible to get a fairly good image on almost any light colored surface. Scrim that is light in color value makes an interesting screen and introduces rather novel possibilities. As soon as the light is brought up behind, both the projected image and the scrim disappear.

Projected scenery when well designed and executed can be extremely effective. Although the process of building is minimal as far as the expenditure for supplies is concerned, yet the expenditure of time in working out an effective pattern or design and getting it projected onto the surface may be rather difficult and time consuming so this isn't something that should be left for the last moment. Experiments should be started early and continued until something satisfactory is achieved. Those who are easily discouraged probably should be dissuaded from trying this type of scenery—also those who are reluctant to spend time "fiddling around" and experimenting should keep away from it—but anyone who is anxious to try something different will have a wonderful time and come up with startling results.

In order to use projections one needs some type of lighting equipment, although nothing very elaborate, and some kind of background upon which to project the images.

### Economy Staging

Almost all stages have some type of draperies or curtains used as a masking or enclosure for all of the miscellaneous lectures, recitals, concerts, meetings, movies and programs that invariably take place in the hall while it is not being used as a theatre. Many times when the budget is small, or nonexistent, the designer will have nothing but these draperies with which to work. Interesting things can be done with drapes alone, with drapes and a few odds and ends, or with drapes and a few odd pieces of scenery.

There are many variations possible in rigging and handling stage draperies. They may be hung in folds straight down in rectangular panel sections; butterflied to form sweeping contours; rigged in combinations of one straight line and one butterfly; draped with one section high and the other section low; hung in one, two, three or even more swags or scallops; or hung on a half circle or full circle support so they form a column or tree trunk shape.

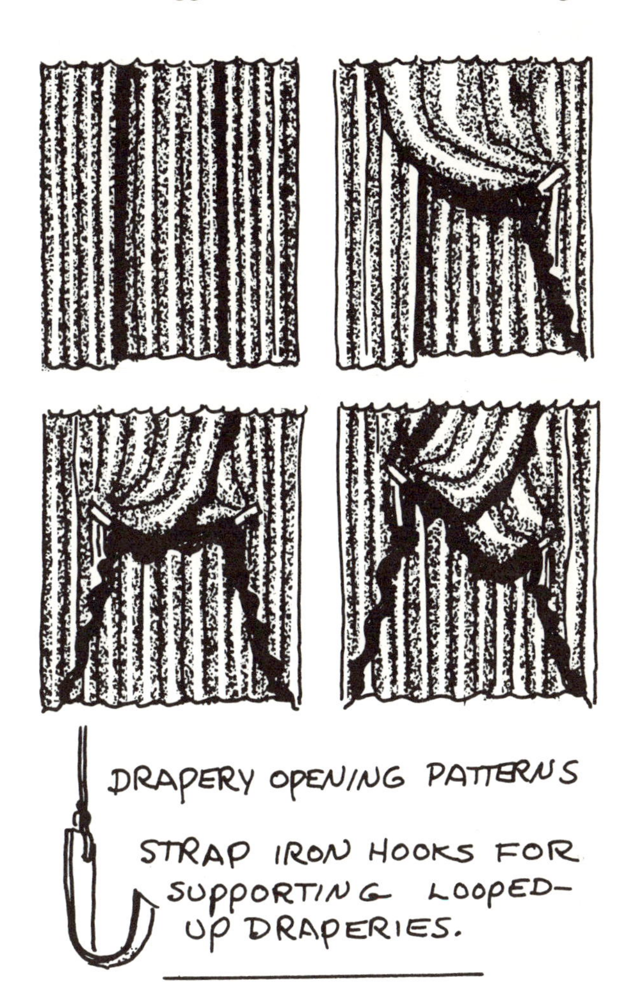

DRAPERY OPENING PATTERNS
STRAP IRON HOOKS FOR SUPPORTING LOOPED-UP DRAPERIES.


Scenery units such as doors, windows and fireplaces may be used in conjunction with draperies. If it is desirable to use pictures with this type of scenery the pictures may be hung

DRAPERY AND SCENERY UNITS

from the batten above rather than on the drapery itself.

Flats—plain, door, window, fireplace—may be placed behind draped-open curtains to add interest to the drapery setting.

Scenery units may be placed in front of the curtains. When possible these should probably be self supporting, self standing units rather than those which require bracing. In

SCENIC UNITS MAY BE USED FREELY WITH DRAPES.

fact, whenever possible scenery should support itself so that bracing may be held to a minimum. This is especially important when there are several settings in a production.

Screen type scenery units, faced on both sides and hinged with double acting decorative screen hinges, may be used to excellent advantage with this type of setting. The screen can have a scene painted upon either side and the scene can then be changed by merely turning the screens around. One or more screens may be used. Screens of this nature may be made with

A screen used to suggest a hospital room in *All of the King's Men.*

two, three, four or even more sections and still fold into a pack that can be handled. If the bottoms have heavy furniture gliders they will move around easily and may rarely, if ever, have to be carried. By using several screens the entire stage can be covered and the effect of a box setting achieved. Many patterns and designs can be used on the screens. Decorative motifs that suggest nations and historical eras might be used as well as general patterns or colors to suggest mood or atmosphere.

If three scenes are needed when screens are used a trick that was devised for a production of *The Mock Doctor* (Fielding version of Moliere's *The Doctor In Spite Of Himself*)

A simple screen with a Chinese design motif in a scene from *The Lute Song.*

might be used. Old over-sized window blinds that had been used in school rooms were obtained and cut so that they were as wide as the flats. Special holders were made from plumbers' strap and the blinds were then mounted atop the screens. When the time came for the scene change the blinds were pulled down over one of the sides of the double faced flats. The scene painted on them was executed with a light stroke dry brush and lining technique so that the paint did not load on heavily and make the blinds difficult to roll.

A PULL DOWN WINDOW SHADE
MAY BE MOUNTED ON TOP
SCREENS MAY HAVE
ANY DESIGN PAINTED
ON THEM - DOUBLE ACT-
ING HINGES ALLOW
THEM TO FOLD BOTH WAYS


Small drops to fit inside the curtained stage, cloth cut out patterns, pennants, spears, flags, colored rope, festoons of colored cloth and many other materials may be used with and on stage draperies. The limitations for this type of staging are set only by the stops on the designers imagination. If there is little with which to work then try to make that little work for you.

### "Line Drawing" Design

Aluminum and thin-walled steel electrical conduit and aluminum or iron wire can be used to produce interesting and graceful design patterns for the stage. The tubes can be bent into graceful curves with a regular conduit "bender" tool or bent around a wooden circle-shaped template that has been securely fastened to a work bench. Straight pieces, long gentle curves, "S" curves, "C" curves, wavy forms, and many other shapes can be combined to produce abstract trees, bushes, flowers, buildings, or hills. Aluminum or soft iron wire can be used in conjunction with the tubing where more delicate lines are desired. Tubing and wire used in this way produces design forms that are similar to "line drawings" in art work.

Pieces of tubing and wire can be fastened together and used by themselves or they can be used as added decoration on flat frames, doorways, windows, furniture, or decorative screens. The tubing can be drilled and fastened with bolts, nails, or wire, or pieces may be lashed to one another with stove pipe wire. The aluminum wire can be fastened to tubing or other wire with string or stove pipe wire. It can be fastened to wood with staples.

Tubing and wire adapt themselves to designs

The secret compartment behind the fireplace is an important design element in *Shop at Sly Corner.*

for imaginative and fantasy type plays especially. These materials have a light "feeling" that can be used to produce flamboyant, rococo, and baroque qualities in design. Used in their natural color or painted in pastel or tint colors and placed against dark gray or black drapes or flats they produce handsome and striking design patterns.

### Minimal Staging

Another type of staging might be referred to as minimal staging. This has many names, but in the main it is a type of staging in which everything is thrown out except the absolute essentials of the design. With a minimum of scenery quantitatively the designer suggests the flavor of the locale whatever it might be. If there are walls they are low except where large units like doors and windows are needed. As suggested before, there are many times when great height is unnecessary. Walls might be

MINIMAL SETTINGS WITH DRAPERY BACKGROUND

Simple set piece cut-outs for *Sing Out Sweet Land.*

only 6′-0″ high around an upstairs or attic room except on the gable end. The garret room occupied by the orphan in the children's play *Radio Rescue* could be designed in this way, as could the upstairs bedroom in *Mary Poppins.* With this type of design pleasant effects can be achieved by having a portion of a room set in space. This portion contains all of the important essentials for the stage business.

The minimal type setting might center around the important unit of scenery: The oven in *Ladies In Retirement,* the fireplace in *Shop At Sly Corner,* the bed in *The Four Poster.* Of course when the production requires several locales for its action it will be necessary to have several units to replace one another. The basic plan is to use only the barest essentials, but execute them well. The total effect should be one of scenic wealth rather than scenic poverty. Design, build, paint, mount and light the stage so that you project the feeling of completeness even though the entire setting consists of a single column.

Needless to say, this type of setting is ideal for variety shows where there are many short scenes or "black outs" since it allows scenery to be set up and struck rapidly. Many children's productions adapt themselves to this type of production. *Alice In Wonderland*—with scenes like the Duchess' kitchen in which the action centers around an outlandish stove, the Mad Hatter scene where an endless tea table is the main essential, the court room scene where the judge's stand is the essential element, the tree in which the Cheshire cat appears and disappears, and many other scenes where one small unit is essentially important—is an example.

*The Patchwork Girl of Oz* is another play in which there are many scenes but each requires only a small screen type unit to depict various places, such as the Voracious Tree, Jack Pumpkinhead's home, the Cave of Yoop, the Woozey's cage, the throne room. Many adult plays fall into a pattern much like this, too. *Of Thee I Sing* has many scenes but for the most part a single dominant idea can be used for each. The hotel room, which is a smoke filled campaign party headquarters, needs only a simple suggestion of a room with a symbol to suggest the party. The president's office requires a double desk and some symbol to suggest "Love", the party slogan, upon which the candidate was elected. The speakers' stand for the convention scene needs the names of the states on poles and a speakers' rostrum loaded with microphones. The Atlantic City Board Walk judges' stand for the bathing beauty scene needs nothing but a pennant decked platform with some semblance of a canopy for a background. The Senate chamber where Throttlebottom finally finds his job needs only an oversize desk so that he will seem as lost as he always is in the play.

*The Green Pastures* is another play requiring many settings, most of which can be small. The Lord's office, the dining room in Noah's

THE LORD'S OFFICE-GREEN PASTURES
SMALL BOX SET BEFORE DRAPES.

home, the tree in the Garden of Eden, the garden where the Lord talks to the flowers, the street lamp under which the young sinners are shooting craps, Noah's Ark and Moses' cave are a few of the scenes.

Small scenes of this nature may be made as booked flat units or built as screens to make them self standing. In order to make them easier to shift around on the stage the bottoms can be fitted with furniture gliders. Portions of the surfaces on these units can be blocked out by painting them the color of the background or they can be made with contoured irregular tops and sides, or they can be left basically rectangular and painted contour patterns can be applied to achieve the feeling of irregular forms. There are times when this process is more satisfactory since it requires less time and material in the building stages and the flats are immediately ready for reuse without having to remove the contour edges.

Units of this nature can also be made from drops. A drop need not fill the entire stage laterally or vertically. Piano wire may be used to extend from the top of the drop up to the overhead masking so that it seems to stand in space. Drops may be shaped on the top by building contoured batten sections and may be cut to shape on the sides to a certain extent if the shaping is planned to avoid flap over and curling.

### Functional Staging

There are times when the designer has no desire to describe the place of the action but does want to indicate it graphically. He desires to use units that have specific physical feeling of mass. He wants to differentiate between spaces on the stage while at the same time relating them to one another. The stage setting, consisting of platforms, steps, ramps and blocks, becomes strictly functional with no purposeless or decorative elements in the design. Careful lighting then allows the designer to select any portion of the stage he desires for a given scene. He bathes that area in light while the rest of the stage remains in darkness. This lighted section is the place of the action. Later in the play other areas are used, people move up and down the steps, group themselves upon the platform, sit on the blocks. This is above all a setting in which the actor is the dominant figure. He is moving in space, facilitated by the functional elements designed for his use. The units may remain in the same position throughout the play or be moved into new positions giving new mass relationships, giving the director a chance to arrange people in new compositional arrangements. Norman Bel Geddes projected many settings of this nature. His designs for *Lazarus Laughed* by Eugene O'Neill, for a production of Goethe's *Faust* and for Shakespeare's *Hamlet* are in this style. Lee Simonson's designs for *The Tidings Brought to Mary* by Claudel and Toller's *Man and the Masses* were also designed in this general style. A school of design known as Constructivism evolved in Germany and was extensively developed by the Russian theatre, especially the Kamerny Theatre during the period immediately following World War I. The Constructivists carried functionalism to the extreme. They believed not only in using the elements that had practical functions, such as platforms, steps, ramps and slippery slides, but they made no attempt to conceal the structures of these elements. The legs and braces of platforms were there for one and all to see and so were many of the spotlights, flying lines and any other mechanical devices utilized in the production. This frankness no doubt led to an increase in the number of distracting features of the production.

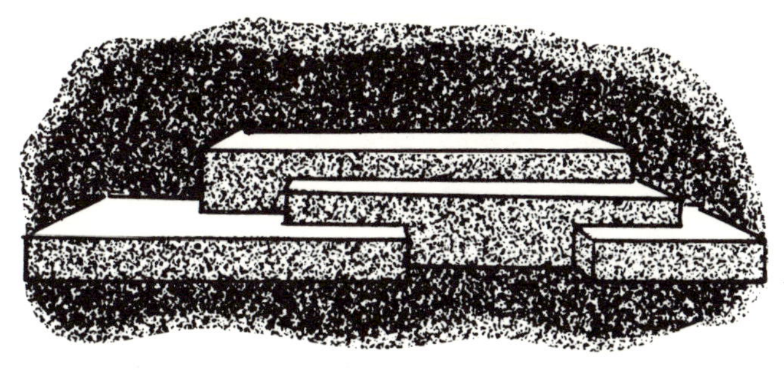

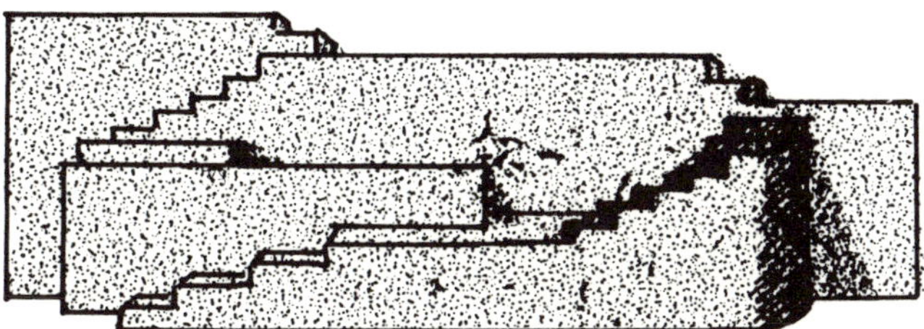

PLATFORMS - STEPS AND RAMPS.

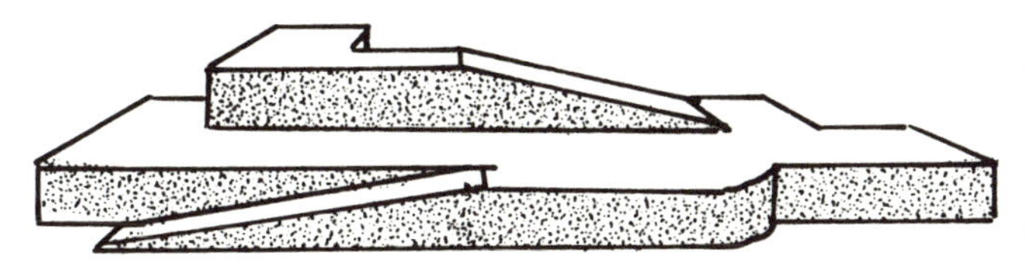

The formal units utilized in most productions today are covered with muslin or canvas so that they are seen as mass and form rather than specifically as portions of a scaffolding. Most of the scenic designers desire even more description than they can project with the mere

Numerous steps and platforms provide many acting areas for *Romeo and Juliet.*

Formal Balance. The platforms are covered and decorated. *The Taming of the Shrew.*

BASIC SETTING

SCENIC UNITS ADDED AND—

CHANGED!

forms they arrange on the stage for the actor's stage business. So after arranging the basic floor levels some background forms are designed to give the scene a little detail. A platform becomes a throne by placing an appropriate throne chair upon it; it becomes a judge's bench by erecting a facing across its front; it becomes a parapet by enclosing it with a battlement facing; it can be the edge of a cliff by facing it with stone forms; or it may be the deck of a ship by placing a portion of the ship's structure around it. In similar manner a step becomes a stairway or a rough hewn stone rise; a ramp becomes a hill, a stone slab or a river bank.

### Basic Setting — Variable Plugs

A still more elaborate variation of the functional type setting carries it another step toward the theatrical type setting. The designer adds basic, if somewhat nondescriptive, architectural forms to his mass units. He may use a set of formal openings which with his basic platform structure remain intact throughout the play. During the course of many scenes he may fill these openings with various "plugs". These plugs may be draperies, wall sections,

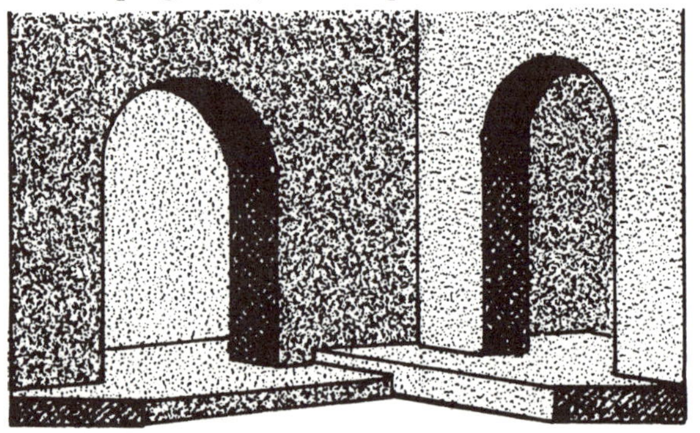

BASIC SETTING

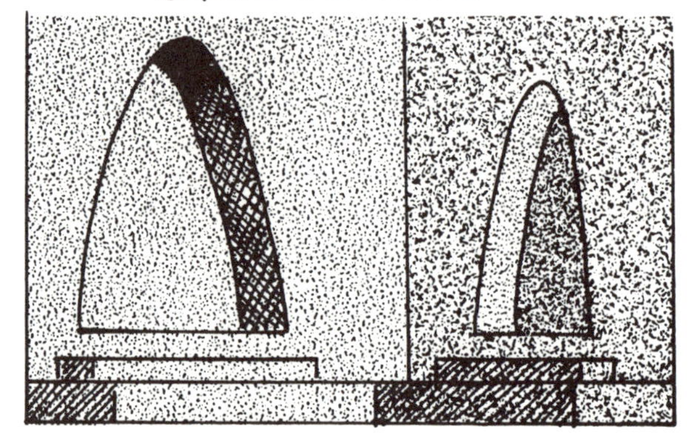

doors, windows, archways, battlements, portcullisses, hallways, fireplaces, thrones or any conceivable forms he desires. If the general flavor remains the same throughout the play, the permanent openings may have an oriental feeling for *Lute Song,* a Medieval quality for

The plugs behind the columns may be changed. *All of the King's Men.* The columns are flat surfaces with painted dimension.

An archway or proscenium rigged with plugs, drops, drapes, and scrim. *Lute Song.*

*Macbeth,* an Italian Renaissance character for *Romeo and Juliet,* an English Renaissance character for *Elizabeth the Queen,* a primitive design for *The Emperor Jones;* or the openings may have no specific character so that the general flavor may change with the inserts for *Marco Millions, Back to Methuselah, As You Like It,* or *Othello.* Lee Simonson's designs for *Marco Millions* graphically show the possibilities of this type of design.

Functional staging with formal openings, has possibilities for the complex production problem since it allows major changes through the use of smaller units. In order to change the entire picture appreciably it is not necessary to shift all of the scenery on the stage.

### False Proscenium Staging — Modified Functional

In a sense staging with a false proscenium falls into the same category as the Basic Setting With Variable Plugs since a portion of the stage scenery remains intact throughout the entire production. Scenery behind or within the proscenium may take any form as long as its design is consistent with the basic mood and feeling of the proscenium. It may consist of flattage, backdrops, set pieces, or movable curtains that may be drawn to one side or both.

The proscenium when used can be placed far down stage so that the action takes place within and upstage of its confines or it can be moved further upstage so that it becomes more of a box type setting with a large movable section of back wall. If the latter method is employed, the proscenium becomes more like a giant picture frame without a bottom. The picture is then changed in the frame.

FALSE PROSCENIUM AND DROPS

A combination of the two methods is also possible so that scenes can take place both in front of and within the archway stage. This of course makes it possible to change scenes behind while scenes are playing on the front stage.

If the proscenium is to be used so that much of the action takes place on the forestage area it will be advisable to design a wing or portal on either side of the proscenium with openings for entrances. These portals will also serve to help brace the proscenium.

Tracks — overhead and on the floor — for "sliding shutter" scenery on an outdoor stage. *The King and I.*

A framework above and behind the proscenium supported by its own legs will be equipped with rigging for roll drops, backdrops, curtains or whatever type device is to be used.

It is possible to use one long continuous drop with scenes painted continuously one beside the

Movement toward realism but retaining simplicity of material and detail.
*The Young Idea.*

other in which case there should be sufficient space at either end of the track for the entire curtain to be stored. Another method allows

DROP SUPPORT TRACK
PROSCENIUM

SMALL DROPS ON A SINGLE STRIP OF MUSLIN—ONE SCENE IN PLACE BEHIND PROSCENIUM—ONE IN "STORAGE" AT EACH END OF TRACK.

the curtains to be pulled on from one end and stored at the other with each wire carrying only one or two scene curtains. A third method would allow storage space for one curtain on each side of the stage. One curtain could then be stored on each end of a wire and pulled on stage for use, then pulled back into its own storage place and the other pulled in like manner from the other end of the wire and stored again at its own end of the proscenium.

A series of roll drops can be rigged on a support behind the proscenium and used to make the changes. The rigging mechanism and support for the drops will have to be more rigid, and probably heavier and more complex, in order to support the strain of lifting the drop.

The false proscenium type of staging will provide the designer with an interesting and flexible device for multi-scene productions. This method of staging may be a convenient way to handle the complex problems of touring productions.

### Theatrical Settings

Most of the plays written today, as well as many of those written yesterday and the day before, literally cry for more than mere functionalism in their settings. They demand treatment that is theatrical. They innately require decoration to project the correct mood, atmosphere and environmental feeling. In order to fulfill this seeming need the designer becomes an architect, builder and interior decorator. In the previous types of settings listed, the designer described the environment of the play's action in outline form only. In theatrical staging he begins to fill in the details. He describes it in one case with a duplication of reality—he uses the elements of the real structure or exactly duplicated facsimiles. Real moldings are applied to the walls, doors, cabinets and furnishings where needed; growing plants are installed in pots and planter boxes; expensive pictures are mounted in selective positions on the walls; furniture and furnishings to suit the exact period are provided; tables are set with the crystal, china and silverware required by the book of etiquette; correctly shaped and patterned rugs are placed on the stage floor; wall treatment as pictured in the history of decoration book is exactly duplicated on the walls. This is literally a real room, or a real exterior of a specific structure. Nothing is left to the imagination. This *is* a living room, circa 1910, complete to the last antimacasser and hair flower shadow box; it is the speakeasy of 1922; it is a Child's Restaurant in New York City in 1913; it is George Washington's study. This is the essence of decorative naturalism and it is theatrically effective. Extensive research must be done before the designer can do much beyond the floor plan of this type setting. He refers back to architectural magazines if it is in the 19th or 20th centuries. He may find his information in catalogues, history books, periodicals of the period, picture albums, home decoration books of the period, or any written work that will provide descriptions and illustrations of the style and period of the desired architecture and decoration. His desire is to duplicate as nearly as possible exactly what was used for this particular structure or portion of a structure. He must be a stickler for detail.

Equally effective theatrically is a variation on the above style of design in the direction of selectivity. The designer eliminates a few of the nonessential details but retains the basic important elements. In this way he retains the flavor of naturalism while at the same time moving a step back toward minimal staging. But by and large he is still remaining in the realm of theatrical staging. He retains the essentials without the clutter. Part or all of the moldings are painted rather than three dimensional; book cases are filled completely or partly with false book backs in place of the real thing; furniture suggests the period rather than duplicating it exactly; a table is set with a minimum rather than a maximum of silver, china and crystal; and rugs on the floor are there to deaden the actors' footsteps rather than to serve as a decorative part of the setting. This has the savor of naturalism without the frenzied braggadocio of exact, minute duplication. The majority of contemporary plays can be set in this style. They have this real, yet theatrical, quality.

The designer of theatrical type scenery begins with functionalism, the bare essentials for the stage business, and then he embellishes. He literally adds the seasoning, the sugar and spice, to the audience meal. His main cue for the design must come from the script, the style of action, the style of stage direction and the style of interpretation of the script. His spice may consist of salt, pepper and sugar, or he may move into the heady aroma of garlic, horseradish and ginger. He may remain in the realm of naturalism or may slip into the embellished world of romanticism and the artificial society of comedy of manners, or may enter into the biting but roisterous realm of satire. As he begins to move into this world he departs slightly from the real world. He becomes a bit tipsy, a bit flamboyant and a bit color happy. Elements of the Rococo and Baroque creep in and design becomes even more theatrical. Moods become gay and flippant. The scenery begins to smile with the audience. This is poking fun at realism without departing from it. Panels are enclosed by scroll work in place of stiff, formal moldings; furniture is gay and enchanting; lines are broken and free flowing; and colors are more dominant and talkative. Restoration comedies, comedies of manners and romantic comedies may well be staged in this way.

Staging of this nature must be handled with

Portion of a courtroom for *The Vigil.* Wood grain and molding details painted on a flat surface.

*The Skin of Our Teeth.* Slight deviations from realism. A smattering of many architectural types and periods (above). And decadence suggested by the cut outs, the impending storm suggested by the background (below).

great care lest the scenery become an overpowering element. It may become another actor but its voice must not be so loud that the other actors cannot be heard. And as with other types of design it must conform in general flavor and feeling to all of the other elements of the production. Its degree of theatrical flamboyance must be innate in the script and in its interpretation.

The deviation from naturalism can easily move beyond the realm just described and over into an outer world—a world of extra sensory projection. Psycho scenic staging may result in designs that move away from the natural in any direction and any degree. Various devices may be used to suggest the inner life of the introvert or the outer life of the extrovert; the imaginary sight of the blind or hearing of the deaf; the visionary world of the dreamer or the mental world of the demented; the unstable world of the neurotic or the blank world of the dead. Almost anything within the realm of the imagination is possible of recreation upon the stage. Angular walls, endless tunnels, ever spiraling staircases, bottomless pits, growing trees, shrinking tables — all of these can be created.

Form and color may be used to project emotions and ideas. Abstraction of one or both of these elements may be difficult but challenging. The designer must try to use his elements so they can be universally interpreted by his audience or his abstraction will become a confusing element that attracts more attention than it deserves.

Varying degrees of abstraction are possible. The forms may be distorted but still recognizable. They may be twisted and contorted to seem other than what they actually prove to be. Or the designer may move completely into the realm of sheer form and color.

A mass through its size, position and relationship with other masses may be used to project mood and atmosphere without representationally describing any known form. In some respects this is a step that the designer passes through in creating the representational types of design. He decides upon the correct juxtaposition of forms and the desirable colors to project his basic emotions. In abstraction he goes to the left and uses nonrepresentational form to achieve his final ends, whereas in representational design he moves to the right and creates familiar forms in familiar color patterns and arrangements.

There aren't too many occasions when the completely abstract design can be used. There are modern ballets that may be interpreted in this vein. And some of the plays of Eugene O'Neill, August Strindberg, George Kaiser, Karel Capek, Elmer Rice and Gertrude Stein move into this realm and can be treated wholly or partially in the abstract vein.

It is possible to intermix these styles whenever and wherever desirable. It is possible to use abstract elements of color or form in a setting that is basically representational or naturalistic and through the combination of elements to project moods and feelings that may not be achieved in any other way. It is possible also to add naturalistic clutter to a nonobjective setting in order to relate it to something familiar to the audience. However, when mixing the forms the designer must be careful that his ideas and units of scenery fit into a pleasingly coherent whole.

### Shifting and Storage Space

During the course of the designing process the scenic designer will have to be constantly conscious of all of the settings for the production so that he can devise a shifting and storage plan if there is more than one setting. He must plan space for the setting that is in the acting area position, the settings that have been used for previous scenes and are now dead storage, and the scenes to come that are in live storage. Obviously the dead storage scenes are no longer important for this individual performance so the storage can be somewhat disorderly, but the live storage is extremely important. Usually scene shifts should be accomplished efficiently, rapidly and quietly. In order to do this it is necessary to plan with great care. Everything must be arranged to be moved easily and readily. Sometimes it is advisable

4 SIDED WAGON UNITS. USE TOGETHER, WITH FLATS, WITH DROPS, OR CYCLORAMA.

to fasten as many units together as possible so that there are just a few units to move.

Where and when possible it may be advisable to plan using special shifting devices such

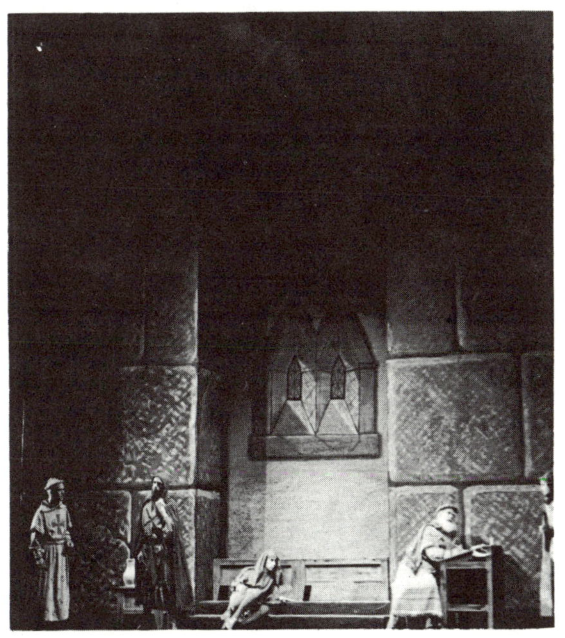

*King Lear* using four sided pylons mounted on wagons so they may be moved and rotated easily.

as wagons, revolvers, the flying system, jack-knife stages, self standing units, or any other device that may be convenient. Double faced scenery, small drops, screens, wings, three dimensional units with two, three or more faces may all be used. In fact, any contrivance that is convenient and helpful is usable if it is practical of manipulation and can be incorporated into the design without changing the basic plan.

Sets may on occasion be placed one inside the other. Then as one is used it is taken out and struck and the next one played.

It is advisable to be aware of all of the possible problems that are likely to arise when the scenery reaches the stage. It isn't possible to think of all of them, but careful planning will cut the last minute nightmares to a minimum at least.

### The Design

The stage design may be executed in any medium (pastels, colored pencils, water colors or even oils) and on any surface (water color paper, pebble matte board, tracing paper, canvas, squared paper or even wrapping paper). The design can be beautifully or sketchily done. If the designer is to follow through and supervise the execution of his design he may by-pass many of the steps that must of necessity be taken when others are to reproduce the design in scenery. The major fact to bear in mind is that the finished product and not the steps along the way is the important element. Beautiful designs are useless if they cannot be transferred to muslin or canvas, and if they are not functional and practical when the transference has been completed. The design is complete only when it is on the stage, lighted and has actors performing in and around it. It must be an integral part of the production.

When it is necessary to make drawings and plans of details, short cuts and easy methods of procedure are welcome. Drawing on squared paper will save much time and provide easily translatable proportional relationships of objects and their parts. Impressions and sugges-

*The Shop At Sly Corner.* The shop of the dealer in antiques can contain countless stage property items.

tions are usually more valuable than skilfully executed mechanical drawings. Saving time on detail work of this nature will usually release more man hours for reproduction, dressing and lighting of the setting.

In planning the setting the designer must go beyond the mere mechanics of selecting flats of the correct size, putting them together, painting them and setting them up on the stage. He must be able to project the spirit of the play. To do this he employs the elements of form, balance, proportion, symmetry, texture and color. Through his setting he attempts to capture, by means of these elements, the mood and atmosphere of the production. He tries to capture the underlying spirit, translate it into design elements and then spread this on the surface of his scenery.

In setting *You Can't Take It With You* the designer captures the wild weirdness of a family where each member lives in his own hilarious private world; the immaculate cleanliness of *Craig's Wife* is stiff, cold, formal and precise; characters of *The School for Scandal* live in a beautiful, brittle, artificial, fragile, pastel world; the violence and greediness of *Macbeth* is heavy, primitive and full of raw passion; anything may happen in *The Skin of Our Teeth* with its indefiniteness, yet the institution of marriage is shown as a great stabilizing force; *Peter Pan* has a light airiness that is exemplified by Peter in free flight. There is the usual clutter of a pawn shop about *Shop at Sly Corner,* yet there is also a feeling of foreboding and one feels that all is not as it should be. Maxwell Anderson inspires the designer of *Knickerbocker Holiday* with the jail cell where an inmate cut notches "bigger and bigger and bigger" until he cut one so big that he got out. *Star Wagon* dreams the actors back to a real past. *The Adding Machine* forcibly casts the hero into a horrible mechanical world.

Subtle or exaggerated elements of the design may suggest the spirit or general feeling. Gay, bright colors can be used to make one think of fun and laughter. Brilliant, rolling forms might suggest boisterous, noisy, farcical action. Deep values, heavy shadows and massive forms might suggest sombreness. Huge overpowering masses with upward lines might

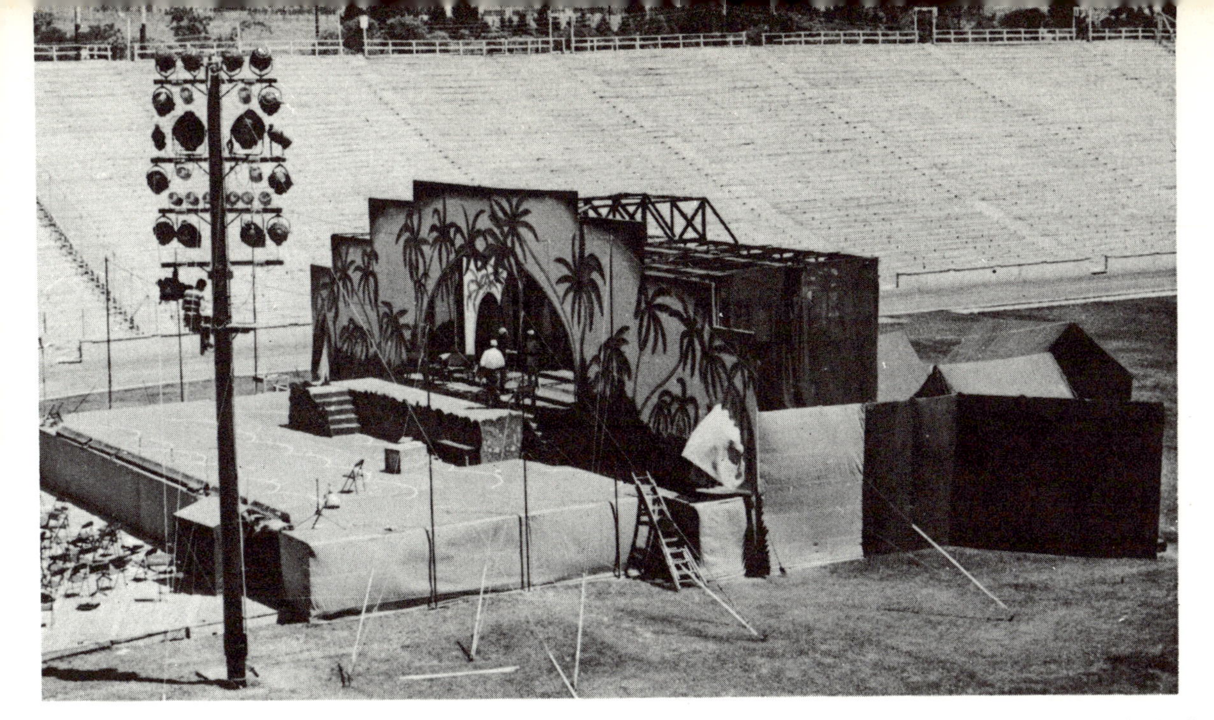

University of Utah portable stadium stage made of tubular scaffolding units and a plank deck. Basic setting for *South Pacific.* Sliding shutters used to make the many scene changes required. One of the spotlight poles is visible.

help to accent the basic idea of the smallness of man. Walls leaning inward, low ceilings and heavy colors pulling downward will help project the feeling of closing in, of claustrophobia. Green, dank colors, water stains and moss forms can suggest a stuffy, disease ridden atmosphere. Sharp, jagged lines and sharp contrasty colors can give a feeling of torture and constant injury. Unbalanced forms ready to collapse can help project the feeling of instability. The inevitable doom of depth, darkness and disaster can be projected through form and color. Warm colors and high light reflective surfaces combined with a feeling of openness can project warmth and sunshine. Hot, heavy colors will help develop the atmosphere of laziness and inertia. Cool colors, and stable forms can project a feeling of complacency and coolness. And sharpness of color, cleanness of line and simplicity of detail may suggest sophistication while careless use of the same elements suggests slovenliness.

The designer will need to learn the nationality, historical period and socio-economic level of the characters who inhabit the setting. A thorough investigation and correlation of these factors will enable him to check the desired architectural features. It may be that a slight oriental flavor overlaid on American architectural units is needed to suggest that the setting is the home of a Nesei, an American born Japanese; whereas the home of native Japanese will have the clean, simple, uncluttered lines of Nipponese architecture. One from another land might be expected to bring along some minor decorative details so that regardless of where he lives there will be a slight mixture of the two styles of design: the homeland and the land of residence.

The character of architectural features, including doorways, windows, moldings, furniture, room proportions, materials, decoration and furnishings, should be studied carefully and then the salient features chosen for the setting being designed. The degree of conformity to exact details will depend upon the individual play and the style of the scenery and production. A single beautiful and appropriately styled column may suffice to suggest that this scene is in Greece, Rome or Egypt. An archway may be designed to set the scene as China, Tudor England, Rome, Egypt, Peru or the land of Mohammed. Spaciousness and elaborateness of design will probably be used

Architectural units designed to project the mood and atmosphere of a time and place. *Romeo and Juliet.*

to suggest wealth and elevated position, while extreme simplification and space restriction will suggest modesty of means and position. The clutter and overabundance of color, furniture and knick knacks is typical of poor taste whereas the furbelows, tassels, heavy colors and gimcrack of Victorianism has a slightly different yet essentially the same flavor.

The designer will probably collect much material. In fact, he will probably scrounge around for old books, periodicals and pictures and will use them not only for the design of the moment, but will file them away for future reference. But for any one design he will gather much more material than is needed. He will sketch details and general designs from it, absorb the flavor and try to digest it rather completely. Then he will try to tell his story with a few essential lines and details, allowing for clutter only when and where it is actually needed. Robert Edmond Jones, the eminent American designer, had great facility for this. His settings were economical of detail but wealthy in projectivity. Any designer would do well to study his work carefully.

### Special Design Problems

There are theatrical productions that differ somewhat from the regular play performance. These should be discussed briefly. If the designer is working with Ballet, Opera, Arena Theatre, Outdoor Theatre or Touring Theatre he may encounter some special problems even though he is still working with basically the same materials, can design in the same styles and will have the same space and moving problems that he has with any other performance.

BALLET PRODUCTIONS—Ballet productions usually take place on one level. For traditional ballet, at least, the desire is for a large, unencumbered dancing space with several openings off right and left for entrances. Clear wing space for the dancers to begin and end their movement, beyond sight lines, is also desirable. The openings must be fairly wide to allow for the movement of the dancers and their costumes which are at times quite full. The choreographer will be the major consultant regarding the problems and requirements of a given program. He may be a traditionalist, in which case it will be necessary to refer to the original designs of the particular ballet. However, even if the original production is used as the basis for the design there is still a chance for considerable creativity.

In the main, because of its character and physical needs, ballet scenery will consist mostly of wings, backdrops and overhead borders. These needn't be conventional but may depart as far as possible from the traditional forms used for such stage elements. Basically the stage area should be clear and flat. It may be thought of as a lighted space occupied by the changing compositional forms of the moving dancers.

Opera Productions — The production of opera presents its problems too and they differ somewhat from ballet requirements. However, there are frequently ballet interludes so the ballet requirements must also be borne in mind during the course of the designing process. Opera can be exciting or static and stodgy. Usually there is a conflict between the stage director who is intent upon movement and stage pictures and the musical director who is conscious only of singing perfection, and complete attention of the singer upon his direction during the performance which often results in a static performance. In order to please all of the people involved it is necessary somehow to plan a flat space for dancing and still have levels for manipulation of choruses near the musical director in the orchestra pit. Ordinarily there are relatively large choruses that must be herded onto the stage quickly and ushered out just as fast. This, of course, means that there should be an adequate number and arrangement of fairly large openings. It will be necessary to know all of the factors about the individual opera before any definite plans can be made. Then design in terms of the number of people in an individual scene, the amount of space needed for dancing, entrances and exits needed to get crowds on and off stage, and level arrangements needed to place choruses in positions where they can see the musical director.

Arena Staging—The arena stage is usually viewed from all sides. It is therefore necessary to design so that sight lines are not impaired regardless of where the individual member of the audience is seated. Furniture, hand properties and low platform levels become extremely important factors in the design. Suspended scenery, low partial walls, scrim walls, diaphonous architectural units, forms made of wire, rope and string, lights and shadows, and painted floor cloths are all elements that the designer in the round may work with. If the room is so contrived it may be possible to shift the stage into different positions from play to play. It may be possible to run the stage along one end of the room, in a corner, diagonally from corner to corner, in two or more corners, at both ends, or almost any otherwhere. The variations are all but limitless in a flexible arena type theatre, but usually the audience is seated on more than one side so sight lines become a problem of more importance than in the regular proscenium theatre. Audience visibility, masking of off stage areas and doorways, baffles to prevent off stage light spill, careful arrangements of forms and lights to provide for fluid movement in all directions and careful execution of the scenic units so that they can pass close scrutiny of the audience that sits practically within the setting are all factors that must be considered very carefully.

Outdoor Productions—Outdoor productions also provide many headaches. Actually outdoor stages differ so greatly from one another that each has its own problems. The general problems of wind and rain, cold and heat, are more universal than any others and must be coped with in any outdoor theatre from the moment the project begins until it closes.

The wind is one unpredictable element that cannot be passed over lightly. A solid, permanent structure is the best solution to the problem. The next best solution is to have a solid framework to which movable flat scenery may be securely lashed. The third solution provides anchoring devices in the floor and floor that is securely anchored to the ground. Other convenient anchors are trees, permanent columns, pillars and pilons, stakes securely driven into the ground and heavily weighted wagons. Even slight gusts of wind can easily cause irreparable damage. A beautiful towering scenic structure can in a moment become a mess of tangled boards and shredded muslin. So when

Rectory scene for *Shadow and Substance.* Many items of stage dressing will aid in projecting the correct mood and atmosphere.

out of doors it is advisable to never leave scenery unanchored unless it is lying flat on the ground—and even then the wind can whip it around.

Rain causes less damage than wind, ordinarily, although it can cause glue to loosen, paint to run, and muslin to pull completely out of shape. Water proof glue, paint that is at least somewhat impervious to rain and heavier covering material than muslin may partially solve these problems. Rubber latex, casein, or glue base paint (sprayed with a solution of formaldehyde after it has dried) may solve the paint problem.

When the weather is calm and beautiful outdoor theatre seems ideal, but as soon as the weather takes a turn so do the technicians' ulcers.

TOURING THEATRE — Touring theatre can be fun for those dreaming nostalgically about it after it is over. But in the here and now there are packing, loading and unloading, repairing, setting up and striking problems that sometimes become almost insurmountable.

Scenery for touring should be built new if possible so that the flats will be relatively light in weight, sturdy and as impervious to rough treatment as possible. The size and proportions of the touring conveyance should be known so that scenery and props may be designed to pack and fit into the space neatly and easily. This will facilitate packing and unpacking. This will also aid in cutting down on the damage and repairs to the setting.

Scenery for the touring production should be designed so that it will brace and support itself as much as possible to hasten the setting up and striking process. The travelling company should attempt to be self sufficient scenery and property-wise so that there is adequate assurance that every item can be found when and where it is needed.

It is usually advisable to allow for expansion and contraction of the stage setting. Extra flats, a wide and narrow jog to fit in the same spot, places where portions of walls may be overlapped, are all devices that may prove handy when the set is being assembled on any one given stage.

If the problems of variable sized stages, loading and unloading, setting up and striking are considered carefully the touring problems may be cut to a minimum.

### Conclusion

In general it is advisable for the stage designer to plan his scenery in terms of the specific problem at hand. The designer is planning one special production of a play. When that play was first designed it was treated in the same way, as a special production. The first designer had problems just as every designer who followed him in planning the staging for that particular play.

Most designers in the nonprofessional theatre plan their settings for one theatre, or one set of theatres. They must of necessity carefully study the facilities available on their stage and make these facilities work for them. They must also know the other raw materials of their art; know the characteristics of lumber, paint, muslin; know how to manipulate flats, drops, dimmers; in short know how to stage a play. The material of this text may point in the right direction by describing a few of the basic materials that have been used in the past and that are being used at present. The theatre technician should always be willing not only to use these materials and ideas but also willing to try new and unusual materials and ideas. New products are constantly coming into the building trades, display, lighting, and television market that can be adapted to stage use. Materials need not be specifically manufactured for stage use before they can be taken into a theatre. In fact products that are manufactured for large commercial markets are usually mass produced and more economical to buy and use. The sealed beam PAR 38 and R 40 lamps are a good example of this economy. They were originally designed for window display use. Widely used they are produced in great quantities, hence are inexpensive. Many other products might be cited that have similar history. The designer should become aware of some of these products and try to use them, if they fit into his plan and design.

Ideally a designer should allow nothing to hamper his creative imagination. He should design freely in terms only of the total end result. Unfortunately however, there are other factors that enter in to hamper this freedom of design. The confines of the stage, the budget, the practicability of reproduction, the physical facilities of the stage, sight lines, and all of the other many and varied elements discussed previously must be carefully considered even before the creative work can begin. However, as stated from time to time, the conventions of the past and the shortcomings of the theatre plant should be employed rather than ignored and the designer's imagination should be tempered by thoughts of flats, paint and spotlights.

* * * *

The stage designer is not an easel artist; he isn't a carpenter or cabinetmaker; he isn't a draper or upholsterer; he isn't an interior decorator nor lighting artist; he isn't an electrician nor a magician; he isn't a special effects man, pawnbroker or thief. Indeed not!!! He is all of them rolled into one. He has the aesthetic feeling of an artist and the scrounging ability of a junkman.

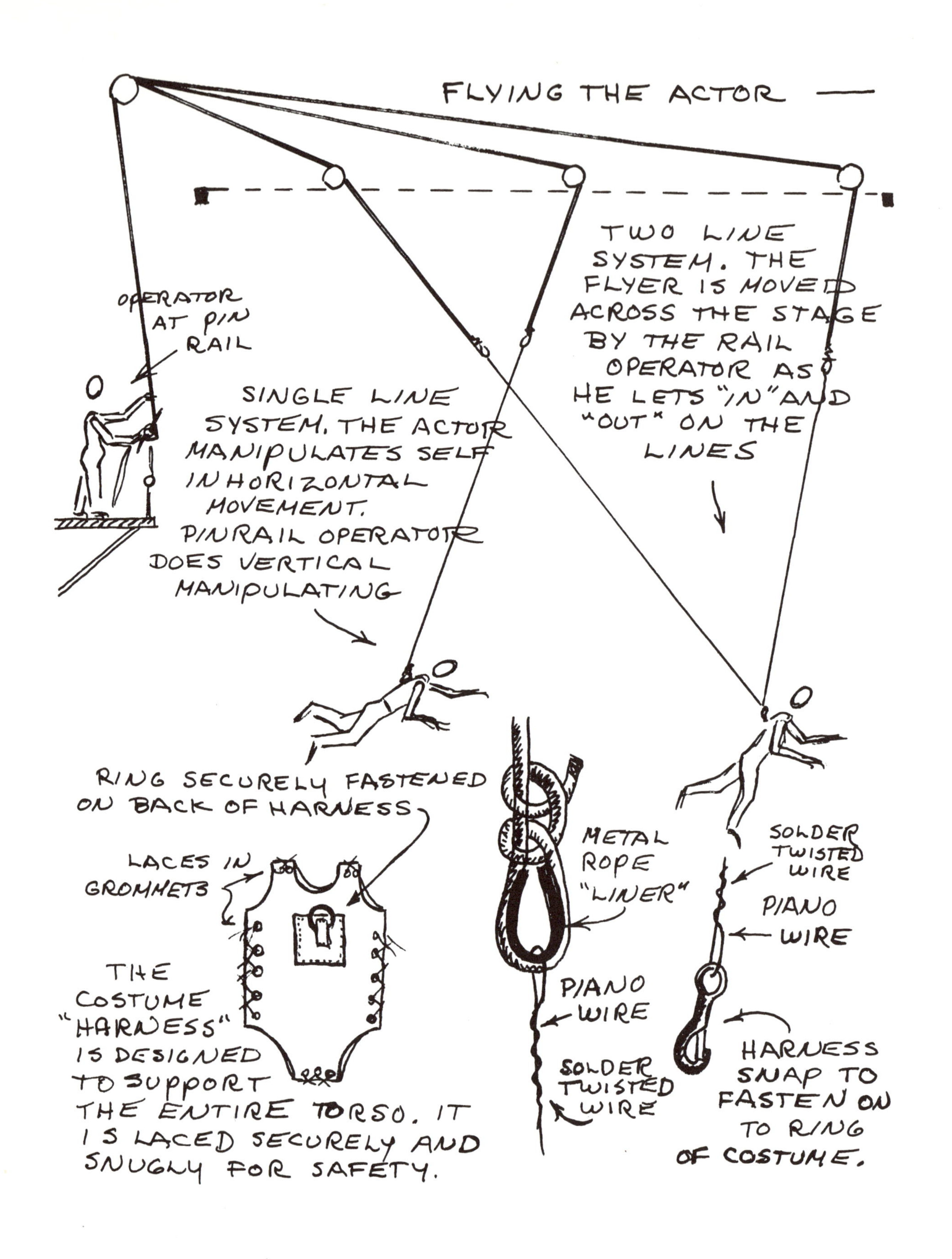

FLYING THE ACTOR —
TWO LINE SYSTEM. THE FLYER IS MOVED ACROSS THE STAGE BY THE RAIL OPERATOR AS HE LETS "IN" AND "OUT" ON THE LINES
OPERATOR AT PIN RAIL
SINGLE LINE SYSTEM. THE ACTOR MANIPULATES SELF IN HORIZONTAL MOVEMENT. PINRAIL OPERATOR DOES VERTICAL MANIPULATING
RING SECURELY FASTENED ON BACK OF HARNESS
LACES IN GROMMETS
THE COSTUME "HARNESS" IS DESIGNED TO SUPPORT THE ENTIRE TORSO. IT IS LACED SECURELY AND SNUGLY FOR SAFETY.
METAL ROPE "LINER"
PIANO WIRE
SOLDER TWISTED WIRE
SOLDER TWISTED WIRE
PIANO WIRE
HARNESS SNAP TO FASTEN ON TO RING OF COSTUME.

# *Appendices*

# SELECTED BIBLIOGRAPHY OF REFERENCE BOOKS.


## ARCHITECTURE

FLETCHER, BANISTER. *A History of Architecture.* New York: Charles Scribner and Sons. 14th Ed., 1948

HAMLIN, TALBOT. *Architecture Through the Ages.* G. P. Putnams Sons. New York, 1940

RICHARDSON, A. E. and CONFIATO, HECTOR O. *The Art of Architecture.* London: English University Press, 1952

SEWALL, JOHN IVES. *A History of Western Art.* New York: Henry Holt and Co., 1953

STATHAM, H. HEATHCOTE. *A History of Architecture.* London, New York: B. T. Batsford Ltd., 1950

STURGIS, RUSSELL. *A Dictionary of Architecture and Building.* New York: MacMillan Co., 1901 (3 Vol.)

THORNE, MRS. JAMES WARD. *Handbook to American Rooms in Miniature.* Chicago: Art Institute of Chicago, 1941

THORNE, MRS. JAMES WARD. *Handbook of European Rooms in Miniature.* Chicago: Art Institute of Chicago, 1943

VIOLLET, E. *Dictionnaire Raisonne de L'Architecture Francaise.* Paris: Libraires Imprimieries Reunies. (10 Vol.)

WASMUTH, GUNTHER, Ed. *Wasmuth Lexikon Der Baukunst.* Berlin: Verlag Ernst Wasmuth A. G., 1929 (5 Vol.)

## ARMS AND ARMOR

. . . *Handbook of the Severance Collection of Arms and Armor.* Cleveland, Cleveland Museum of Art, 1925

. . . *Catalog of Military Goods.* Bannerman's, 501 Broadway, New York

BOUTELL, CHARLES. *Arms and Armament in Antiquity and the Middle Ages.* London, Reeves and Turner, 1905

HAYWARD, J. F. *European Firearms.* New York: Philosophical Library, 1955

LELOIR, MAURICE. *Dictionnaire du Costume et de ses Accessoires des Armes et des Etoffes des Origines a nos jours.* Paris: Libraire Grund, 1951

WILKINSON, FREDERICK. *Arms and Armour.* London: Hamlyn, 1971

YAMAGAMI, HATIRO. *Japan's Ancient Armour.* Board of Tourist Industry, Japanese Government Railways. 1940

## ART COMPOSITION AND DESIGN

BISHOP, A. THORNTON. *Composition and Rendering.* New York: John Wiley and Sons Inc., 1933

CLARK, ARTHUR BRIDGMAN. *Perspective. A Textbook and Manual for Artists.* New York: Bridgman Publishers Inc., 1936

EMERSON, SYBIL. *Design, A Creative Approach.* New York: International Textbook Co., 1953

FELDSTED, C. J. *Design Fundamentals.* New York: Pitman Publishing Co., 1950

GUPTIL, ARTHUR L. *Color in Sketching and Rendering.* New York: Reinhold Publishing Corp., 1935

KAUFMANN, EDGAR. *What Is Modern Design.* New York: Museum of Modern Art. Simon and Schuster, 1950

POZZO, ANDREA. *Rules and Examples of Perspective* (Reissue from 1707 Ed.) New York: Blom, 1971

RASMUSSEN, HENRY. *Art Structure.* New York: McGraw Hill Co., 1950

RATHBONE, RICHARD ADAMS. *Introduction to the Fundamentals of Design.* New York: McGraw Hill Co., 1950

SCOTT, ROBERT. *Design Fundamentals.* New York: McGraw Hill Co., 1951

WIGGINS, RICHARD G. *Composing In Space.* Dubuque, Iowa. Wm. C. Brown Co., 1949

## COLOR

BIRREN, FABER. *Color, A Survey in Words and Pictures.* New Hyde Park, New York: University Books, Inc., 1963

BIRREN, FABER. *Color Form and Space.* New York: Reinhold Publishing Corp., 1961

BIRREN, FABER. *Creative Color.* New York: Reinhold Publishing Corp., 1961

BIRREN, FABER. *History of Color in Painting.* New York: Reinhold Publishing Corp., 1965

BOND, FRED. *Color, How to See and Use It.* San Francisco: Camera Craft Publishing Corp., 1954

BURRIS-MEYER, ELIZABETH. *Color and Design in the Decorative Arts.* New York: Prentice Hall Inc., 1935

BURRIS-MEYER, ELIZABETH. *Contemporary Color Guide.* New York: W. Helburn, 1947

BUSTANOBY, JACQUES HENRI. *Principles of Color and Color Mixing.* New York: McGraw Hill, 1947

GRAVES, MAITLAND E. *The Art of Color and Design*. New York: McGraw Hill Co., 1951
GRAVES, MAITLAND E. *Color Fundamentals*. New York: McGraw Hill Co., 1952
GUPTIL, ARTHUR L. *Color in Sketching and Rendering*. New York: Reinhold Publishing Corp., 1935
JACOBS, MICHEL. *The Art of Colour*. New York: Doubleday Doran and Co., 1931
JACOBSEN, EGBERT. *Basic Color*. Chicago: Paul Theobald, 1948
KING, JOHN L. *The Art of Using Color*. New York: The World Publishing Co., 1938
LUCKIESH, M. *The Language of Color*. New York: Dodd Mead, 1930
LUCKIESH, M. *Color and Colors*. New York: D. Van Nostrand Co., Inc., 1938
MAERZ, A. and REA, PAUL M. *Dictionary of Color*. New York: McGraw Hill Co., 1930
MUNSELL, A. H. *Munsell Book of Color*. Vol. 1 and Vol. 2. Baltimore: Munsell Color Co., Inc., 1929, 1942

## COSTUMING

BARTON, LUCY. *Historic Costumes for the Stage*. New York: Walter H. Baker Co., 1935
BOUCHER, FRANCOIS. *20,000 Years of Fashion*. New York: Harry N. Abrams, 1966
BROOKE, IRIS. *Footwear: A Short History of European and American Shoes*. New York: Theatre Arts Books, 1972
BROOKE, IRIS. *A History of English Costume*. New York: Theatre Arts Books, 1972
BROOKE, IRIS. *Mediaeval Theatre Costume*. New York: Theatre Arts Books, 1968
BROOKE, IRIS. *Western European Costume*. Vols. 1 and 2. New York: Theatre Arts Books, 1964, 1966
CHALMERS, HELENA. *Clothes On and Off the Stage*. New York: D. Appleton-Century Co., 1930
EVANS, MARY. *Costume Throughout the Ages*. Philadelphia: J. B. Lippincott Co., 1930
KOHLER, CARL and SICHART, EMMA VON. *A History of Costume*. Philadelphia: David McKay Co., 1928
KOMISARJEVSKY, THEODORE. *The Costume of the Theatre*. New York: Henry Holt and Co., 1932
LELOIR, MAURICE. *Dictionnaire Du Costume et de ses Accessoires des Armes et des Etoffes des Origines a nos jours*. Paris: Libraire Grund, 1951
MOTLEY, *Designing and Making Stage Costumes*. New York: Watson-Guptil, 1974
MOULTON, BERTHA. *Garment Cutting and Tailoring for Students*. New York: Theatre Arts Books, 1968
MOULTON, BERTHA. *Simplified Tailoring*. New York: Theatre Arts Books, 1969
PAYNE, BLANCHE. *History of Costumes from the Ancient Egyptians to the Twentieth Century*. New York: Harper, 1965
RACINET, ALBERT C. A. *Le Costume Historique*. Paris: Firmin-Didot, 1888
WALKUP, FAIRFAX PROUDFIT. *Dressing the Part*. New York: F. S. Crofts, 1939
WILCOX, R. TURNER. *The Mode In Costume*. New York: Charles Scribner, 1947
WILCOX, R. TURNER. *The Mode In Footwear*. New York: Charles Scribner, 1948
WILCOX, R. TURNER. *The Mode In Hats and Headdresses*. New York: Charles Scribner, 1946

## CRAFT PROCESSES

BOWMAN, NED. *A Handbook of Technical Practice for the Performing Arts*. Wilkinsburg, Penn: Scenographic Media, 1972
BRYSON, NICHOLAS. *Scenery for Theatre: Vol. 1. Vacuum Forming*. New York: Drama Book Specialists, 1972
BUTZCO, *Plastic Sheet Forming*. New York: Reinhold Publishing Co., 1958
CONWAY, JOHN O. *Plastics*. Homewood, Ill.: Goodheart–Wilcox Co., 1966
GRISWOLD, LESTER. *Handicraft, Simplified Procedure and Projects*. Prentice Hall Inc., 1951, 1952
PLYMOUTH CORDAGE CO. *How to Put Rope to Work*. Plymouth, Mass.: Plymouth Cordage Co., 1948
SUMNER PECKER. *Introduction to Plastic Stage Properties*. New Haven: Yale University, MFA Thesis
STEPHENSON, JESSIE BANE. *From Old Stencils to Silk Screening*. New York: Charles Scribner's Sons, 1953

## DESIGN MOTIVES

BOSSERT, HELMUTH. *An Encyclopedia of Color Decoration*. New York: E. Weyhe, 1928 (Color Plates)
ESTRIN, MICHAEL. *2,000 Designs, Forms, and Ornaments*. Philadelphia: Wm. Penn Publishing Corp., 1947
GLAZIER, RICHARD. *A Manual of Historic Ornament*. London, New York: Batsford Ltd., 6th Ed., 1948
HORNUNG, CLARENCE PEARSON. *Handbook of Designs and Devices*. New York: Dover Publications, 1946

JONES, OWEN. *The Grammar of Ornament*. London: Bernard Quaritch, 1910

SPELTZ, ALEXANDER. *Styles of Ornament*. New York: Grosset and Dunlap, 1936

## FURNITURE

ARONSON, JOSEPH. *The Book of Furniture and Decoration*. New York: Crown Publishers, 1941

ARONSON, JOSEPH. *The Encyclopedia of Furniture*. New York: Crown Publishers, 1938

BOSSERT, VON Prof. Dr. H. TH. *Geschichte Des Kunstgewerkes*. Vol. 6. Berlin: Verlag Ernst Wasmuth. Gm. B. H., 1935

BRUDER, KARL C. *Properties and Dressing the Stage*. New York: Richards Rosens Press, 1969

DERIEUX, MARY and STEVENSON, ISABELLE. *The Complete Book of Interior Decorating*. New York: Greystone Press, 1954

DYER, WALTER ALDEN. *Handbook of Furniture Styles*. New York: The Century Co., 1918

JACQUEMART, ALBERT. *A History of Furniture*. London: Chapman and Hall, 1878

KENTON, WARREN. *Stage Properties and How to Make Them*. London: Pitman, 1964

LEE, RUTH W. and BOLENDER, LOUISE T. *Fashions In Furnishings*. New York: McGraw Hill Co., 1948

MEILACH, DONA Z. *Creating With Plaster*. Chicago: Reilly and Lee, 1966

MILLER, EDGAR G., Jr. *The Standard Book of American Antique Furniture*. New York: Greystone Press, 1950

THORNE, Mrs. JAMES WARD. *Handbook to American Rooms in Miniature*. Chicago: Chicago Art Institute, 1941

THORNE, Mrs. JAMES WARD. *Handbook to European Rooms in Miniature*. Chicago: Chicago Art Institute, 1943

WHITON, SHERRILL. *Elements of Interior Design and Decoration*. Philadelphia: Lippincott, 1963

## SCENE DESIGN

BAY, HOWARD. *Stage Design Throughout the World*. New York: Drama Book specialists, 1974

BEAUMONT, CYRIL W. *Design for the Ballet*. London: The Studio Ltd., 1937

BURDICK, ELIZABETH B., HANSEN, PEGGY C., and ZENGER, BRENDA (Eds.). *Contemporary Stage Design U.S.A.* Middletown, Conn.: Wesleyan Univ. Press, 1974

BURIAN, JARKA. *The Scenography of Joseph Svoboda*. Middletown, Conn.: Wesleyan Univ. Press, 1971

CHENEY, SHELDON. *Stage Decoration*. New York: John Day Co., 1928

COREY, IRENE. *The Mask of Reality, An Approach to Design for the Theatre*. New Orleans: Anchorage Press, 1968

CRAIG, EDWARD GORDON. *Scene*. Oxford: Oxford University Press, 1923

FRIEDERICH, WILLARD J. and FRASER, JOHN H. *Scenery Design for the Amateur Stage*. New York: The Macmillan Co., 1950

FUERST, WALTER RENE and HUME, SAMUEL. *Twentieth Century Stage Decoration*. 2 vols. London: Aldred Knopf, Inc., 1928

GILLETTE, ARNOLD. *An Introduction to Scene Design*. New York, Harper, 1967

HAINAUX, RENE (Ed.). *Scene Design Throughout the World Since 1935*. New York: Theatre Arts Books, 1964

HAINAUX, RENE (Ed.). *Scene Design Throughout the World Since 1950*. New York: Theatre Arts Books, 1964

HAINAUX, RENE (Ed.). *Scene Design Throughout the World Since 1960*. New York: Theatre Arts Books, 1972

HELVENSTON, HAROLD. *Scenery*. California: Stanford University Press, 1931

JONES, ROBERT EDMOND. *Drawings for the Theatre*. New York: Theatre Arts, 1925

JONES, ROBERT EDMOND. *The Dramatic Imagination*. New York: Duell Sloane and Pearce, Inc., 1941

KOMISARJEVSKY, THEODORE and SIMONSON, LEE. *Settings and Costumes of the Modern Theatre*. New York: Studio Publications, Inc., 1933

LAVER, JAMES. *Drama, Its Costumes and Decor*. London: The Studio Publications, 1951

MacGOWAN, KENNETH and JONES, ROBERT E. *Continental Stagecraft*. New York: Harcourt Brace and Co., 1922

MIELZINER, JO. *Designing for the Theatre*. New York: Potter, 1970

MIELZINGER, JO. *The Shape of Our Theatre*. New York: Crown, 1970

MILLER, JAMES HULL. *Self Standing Scenery*. New Orleans, Anchorage Press, 1975

NICOLL, ALLARDYCE. *The Development of the Theatre*. New York: Harcourt Brace and Co., 1948

OENSLAGER, DONALD. *Scenery Then and Now*. New York: 1936

PARKER, W. OREN and SMITH, HARVEY K. *Scene Design and Stage Lighting*. New York: Holt Rinehart, Winston, 1968

PECTAL, LYNN. *Designing and Painting for the Theatre*. New York: Rinehart Winston, 1975

SHERINGHAM, GEORGE and LAVER, JAMES. *Design in the Theatre*. London: The Studio, Ltd., 1927
SIMONSON, LEE. *The Stage Is Set*. New York: Harcourt Brace and Co., 1932
SIMONSON, LEE. *Part of a Lifetime*. New York: Duell, Sloane and Pearce, 1943
SIMONSON, LEE. *The Art of Scenic Design*. New York: Harper and Bros., 1950
. . . *Theatre Arts Prints*. New York: Theatre Arts Magazine
ZINKEISEN, DORIS. *Designing for the Stage*. London: Studio Publications, 1938; rev. 1945

## STAGECRAFT

ASHWORTH, BRADFORD. *Notes on Scene Painting*. New Haven, Conn.: Whitlock's, 1952
BARBER, PHILIP. *The Scene Technician's Handbook*. New Haven: Whitlock's Book Store, 1928 (This is now a part of: GASSNER, JOHN. *Producing the Play*. New York: The Dryden Press, 1941)
BUERKI, FRED A. *Stagecraft for Non-Professionals*. Madison: University of Wisconsin Press, 1945
BURRIS-MEYER, HAROLD and COLE, EDWARD C. *Scenery for the Theatre*. Boston: Little Brown and Co., 1938
CORNBERG, SOL and GEBAUER, EMANUEL L. *A Stagecrew Handbook*. New York: Harper and Bros., 1941
GILLETTE, A. S. *Stage Scenery: Its Construction and Rigging*. New York: Harper, 1959
HAKE, HERBERT V. *Here's How*. Chicago: Row Peterson Co., 1942
KRANITCH, FRIEDRICH. *Buhnentechnik der Gegenwart*. Verlag von R. Oldenbourg. Munchen, 1929
NELMS, HENNING. *A Primer of Stagecraft*. New York: Dramatists Play Service, 1941
PHILIPPI, HERBERT. *Stagecraft and Scene Design*. Boston: Houghton Mifflin Co., 1953
SELDEN, SAMUEL and REZZUTO, TOM. *Essentials of Stage Scenery*. New York: Appleton Century Crofts, 1972
SELDEN, SAMUEL and SELLMAN, HUNTON D. *Stage Scenery and Lighting*. New York: F. S. Crofts and Co., 1930, rev. 1936
WADE, ROBERT J. *Staging TV Programs and Commercials*. New York: Hastings House, 1954
WADE, ROBERT J. *Designing for TV*. New York: Farr and Straus
WHITING, FRANK M. *An Introduction to the Theatre*. New York: Harper and Bros., 4th ed., 1978

## STAGE LIGHTING

BELLMAN, WILLARD F. *Lighting the Stage: Art and Practice*. Scranton, Penn.: Chandler Publishing Co., 1967
BENTHAM, FREDERICK. *Stage Lighting*. London: Sir Isaac Pitman and Sons Ltd., 1950
FUCHS, THEODORE. *Stage Lighting*. New York: Benjamin Blom, 1969
GILLETTE, MICHAEL. *Designing With Light*. Palo Alto, Cal.: Mayfield Publishing Co., 1978
McCANDLESS, STANLEY. *A Method of Lighting the Stage*. New York: Theatre Arts, 1958
NELMS, HENNING. *Lighting the Amateur Stage*. New York: Theatre Arts Inc., 1931
PARKER, W. OREN and SMITH, HARVEY K. *Scene Design and Stage Lighting*. New York: Holt Rinehart and Winston, 1974
RIDGE, C. HAROLD and ALDRED, F. S. *Stage Lighting*. London: Pitman and Co., 1935
ROSENTHAL, JEAN and WERTENBAKER, LAEL. *The Magic of Light*. Boston: Little Brown, 1972
RUBIN, JOEL E. and WATSON, LELAND H. *Theatrical Lighting Practice*. New York: Theatre Arts, 1955
SELDEN, S. and SELLMAN, H. *Stage Scenery and Lighting*. New York: F. S. Crofts and Co., 1941
SELLMAN, HUNTON D. *Essentials of Stage Lighting*. New York: Appleton Century Crofts, 1972
WILFRED, THOMAS. *Projected Scenery*. New York: Drama Book Specialists, 1965
WILLIAMS, ROLLO G. *The Technique of Stage Lighting*. New York: Pitman, 1958

*NOTE:* The list of suppliers of theatrical equipment, and the bibliography of books related to various technical theatre subject matter are not intended to be complete or exhaustive. For additional lists consult the directories and tabulated lists that are provided in other text books on the various subject matter. An extensive listing may also be found in *Simon's Directory of Theatrical Materials, Services and Information* available from Package Publicity Service, Inc., 1501 Broadway, New York, N.Y. 10036.

# THEATRICAL EQUIPMENT DEALERS

## COSTUMES

| | |
|---|---|
| Brooks–Van Horn | 117 West 17th St., New York, N.Y. 10011 |
| Colorado Costumes | 2100 Broadway, Denver, Colo. 89205 |
| Costume Armour, Inc. | 481 Canal Place, The Bronx, N.Y. 10451 |
| Eaves Costume Co., Inc. | 423 West 55th St., New York, N.Y. 10019 |
| Northwestern Costumes | N. Highway 100, Minneapolis, Minn. 55422 |
| Salt Lake Costume Co. | 1701 S. 11th East, Salt Lake City, Utah 84115 |
| Western Costume Co. | 5335 Melrose Ave., Hollywood, Cal. 90038 |
| Krause Costume Co. | 2445 Superior Ave., Cleveland, Ohio 44114 |

## FABRICS

| | |
|---|---|
| American Canvas Co. | 2232 Lawrence St., Denver, Colo. 80205 |
| Art Drapery Studios | 2766 North Lincoln Ave., Chicago, Ill. 60614 |
| Astrup Co | 39 Walker St., New York, N.Y. 10013 |
| Central Shippee, Inc. | Bloomingdale, N.J. 07403 |
| Dazians | 40 East 29th St., New York, N.Y. 10016 |
| Evans Supply Inc. | 509 W. 300 N., Salt Lake City, Utah 84116 |
| Frankel Assoc., Inc. | 1122 Broadway, New York, N.Y. 10010 |
| Home Silk Shop | 330 S. La Cienega Blvd., Los Angeles, Cal. 90048 |
| Hollywood Fancy Feathers | 512 S. Broadway, Los Angeles, Cal. 50013 |
| International Silks and Woolens | 8347 Beverly Blvd., Los Angeles, Cal. 90048 |
| Paramount Textile Mills | 34 Walker St., New York, N.Y. 10013 |
| Rose Brand Textile Fabrics | 517-527 West 35th St., New York, N.Y. 10001 |
| Theatre Production Service | 26 S. Highland, Ossining, N.Y.10562 |

## GENERAL THEATRE EQUIPMENT

| | |
|---|---|
| Alcone Co., Inc. (Paramount) | 575 8th Ave., New York, N.Y. 10018 |
| American Scenic Company | 11 Andrews St., Greenville, S.C. 29602 |
| Art Theatre Equipment Co. | 155 West 46th St., New York, N.Y. 10020 |
| Grand Stage Lighting Company | 630 W. Lake St., Chicago, Ill. 60606 |
| Oleson | 1535 Ivar Ave., Los Angeles, Cal. 90028 |
| Theatre Production Service | 26 S. Highland, Ossining, N.Y. 10562 |
| Theatrical Service and Supplies, Inc. | 205 C Brook Ave., Deer Park, N.Y. 11729 |

## LIGHTING, COLOR MEDIA

| | |
|---|---|
| Alcone Co., Inc. (Paramount) | 575 8th Ave., New York, N.Y. 10018 |
| Berkey Colortran | 1050 Chestnut St., Burbank, Cal. 91502 |
| Belden Communications & Co. (Lee) | 534 West 25th St., New York, N.Y. 10001 |
| Brigham Sheet Gelatine Co. | 5 Prospect Ave., Randolph, Ver. 05060 |
| Oleson Co. | 1535 Ivar, Hollywood, Cal. 90028 |
| Rosco Labs | 36 Bush Ave., Point Chester, N.Y. 10573 |
| | 1135 North Highland Ave., Los Angeles, Cal. 90038 |

## LIGHTING EQUIPMENT

| | |
|---|---|
| American Stage Lighting, Inc. | 1331 C North Ave., New Rochelle, N.Y. 10804 |
| Berkey Colortran | 1015 Chestnut St., Burbank, Cal. 91502 |
| Electro Controls | 2975 South 300 West, Salt Lake City, Utah 84115 |
| Electronics Diversified, Inc. | 1675 N.W., 216th St., Hillsboro, Ore. 97123 |
| Four Star Stage Lighting | 3935 M Mission, Los Angeles, Cal. 90031 |
| Hub Electric Co. | 940 Indiana Drive, Elmhurst, Ill. 60126 |
| Kliegl Brothers, Inc. | 32 48th Ave., Long Island City, N.Y. 11101 |
| Oleson Co. | 1535 Ivar Ave., Hollywood, Cal., 90028 |
| Mole-Richardson Co. | 937 N. Sycamore Ave., Hollywood, Cal. 90038 |
| Skirpan Lighting Control Corp. | 61st St. and 32nd Ave., Woodside, N.Y. 11377 |

Strong Electric Corp. . . . . . . . . . . . . . . . . . . . . . . . . 94 City Park Ave., Toledo, Ohio 43601
Superior Electric Co. . . . . . . . . . . . . . . . . . . . . . . . . . .383 Middle St., Bristol, Conn. 06010
Theatre Techniques Inc. . . . . . . . . . . . . . . . . . . 60 Connally Parkway, Hamden, Conn. 06514
Ward Leonard Electric Co. . . . . . . . . . . . . . . . . . . . .31 South, Mount Vernon, N.Y. 10550

## PAINT

American Scenic Colors . . . . . . . . . . . . . . . . . . . . . 11 Andrews St., Greenville, S.C. 29602
Aljo Mfg. Co. . . . . . . . . . . . . . . . . . . . . . . . . . . . . . .116 Prince, New York, N.Y. 10012
Bennet Paint and Glass Co. . . . . . . . . . . . . . 65 West 100 South, Salt Lake City, Utah 84101
M. Epstein's Sons . . . . . . . . . . . . . . . . . . . . . . . . . . 809 9th Ave., New York, N.Y. 10019
Gothic Color Co., Inc. . . . . . . . . . . . . . . . . . . . .727 Washington St., New York, N.Y. 10014
Playhouse Colors. . . . . . . . . . . . . . . . . . . . . . . . . . . 771 9th Ave., New York, N.Y. 10019
Rosco Labs. . . . . . . . . . . . . . . . . . . . . . . . . . . .36 Bush Ave., Point Chester, N.Y. 10573
(Also check out nationwide paint distributors such as: Devoe, Luminall, Iddings, Etc.)

## SOUND EFFECTS

Carroll Sound, Inc. . . . . . . . . . . . . . . . . . . . . . . . 351 West 41st St., New York, N.Y. 10036
Ross-Gaffney, Inc. . . . . . . . . . . . . . . . . . . . . . . . . .21 West 46th St., New York, N.Y. 10019
Theatresound, Inc.. . . . . . . . . . . . . . . . . . . . . . . . . . . .585 Gerard Ave., Bronx, N.Y. 10451
Thomas J. Valentino, Inc. . . . . . . . . . . . . . . . . . . .150 West 46th St., New York, N.Y. 10019

## STAGE HARDWARE

Channon Corp. . . . . . . . . . . . . . . . . . . . . . . . . . . . . 1345 W. Argyle, Chicago, Ill. 60640
J. R. Clancy Co. . . . . . . . . . . . . . . . . . . . . . . . .1010 West Belden Ave., Syracuse, N.Y. 13204
Evans Supply, Inc.. . . . . . . . . . . . . . . . . . . .509 West 300 North, Salt Lake City, Utah 84116
L & M Stagecraft, Inc. . . . . . . . . . . . . . . . . . . . . 2110 Superior Ave., Cleveland, Ohio 44114
Macton Corp. . . . . . . . . . . . . . . . . . . . . . . . . . . . .On-The-Airport, Danbury, Conn. 06810
Mutual Hdw. Corp. . . . . . . . . . . . . . . . . . . . . . .5-45 49th Ave., Long Island City, N.Y. 11001
Oleson. . . . . . . . . . . . . . . . . . . . . . . . . . . . . . . . . .1535 Ivar Ave., Los Angeles, Cal. 90028
Peter Albrecht Corp. . . . . . . . . . . . . . . . . . . . . . 325 E. Chicago St., Milwaukee, Wisc. 53202
Stagecraft Industries . . . . . . . . . . . . . . . . . . . . 1302 N.W. Kearney St., Portland, Ore. 97005
Theatrical Rigging Systems . . . . . . . . . . . . . 4090 West Broadway, Minneapolis, Minn. 55422

## PLASTIC MATERIALS

American Cyanimide ("Acrylite") . . . . . . . . . . . . . . . . . . . .Berden Ave., Wayne, N.J. 07470
American-Hoechst Corp. ("Pollyrubber"). . . . . . . . . . 202-206 North, Somerville, N.J. 08876
Atlantic Richfield, Plastics Div. ("Dylan", "Dylene"). . . . . . . . . . . . . . . . . . . . . . . . . .
3702 East Columbus Dr., Tampa, Fla. 33605
Borden Co., Adhesives (Elmer's Glue) . . . . . . . . . . . 180 E. Broad St., Columbus, Ohio 43215
Celanese Corp. ("Celanese", "Celanar"). . . 1211 Ave. of the Americas, New York, N.Y. 10036
Degussa Corp. ("Celastic") . . . . . . . . . . . . .Route 46 at Hollister Rd., Teterboro, N.J. 07608
Dow Chemical ("Dorvon", "Etha Foam", "Styron", "Styrofoam", "Silastic") . . . . . . . . . . . .
2020 Dow Center, Midland, Mich. 48640
E. I. DuPont de Nemours & Co., Inc. ("Dacron", "Mylar", "Lucite") . . . . . . . . . . . . . . . . .
1007 Market St., Wilmington, Del. 19898
Eastman Chemical Products, Inc. . . . . . . . . . . . . . . . .P.O. Box 431, Kingsport, Tenn. 37662
General Electric, Plastics Bus. Div. ("Textolite", "RTV 662") . . . . . . . . . . . . . . . . . . . . . .
One Plastics Ave., Pittsfield, Mass. 01201
General Tire and Rubber Co. ("Genthane", "Polyfoam") . One General St., Akron, Ohio 44329
Goodyear Tire and Rubber Co. ("Pliovic"). . . . . . . 1144 East Market St., Akron, Ohio 44316
Imex Polymers ("PVC") . . . . . . . . . . . . 129 John Vertente Blvd., New Bedford, Mass. 02740
Monsanto Plastics and Resins Co. ("Opalan", "Vyram", "Kydex", "Lustrex") . . . . . . . . . . . .
800 North Lindbergh Blvd., St. Louis, Mo. 63166
Owens Corning Fiberglass Corp. ("Fiberglass") . . . . . . .Fiberglass Tower, Toledo, Ohio 43659
Polymer Materials Inc. ("Polysar") . . . . . . . . . . . 100 Adams Blvd., Farmingdale, N.Y. 11735
Rohm-Haas Co. ("Plexiglass", "Kydex") . . Independence Mall West, Philadelphia, Penn. 19105
Stauffer Chemical Co. ("Everlon", "Restfoam") . . . . . . . . . . . . . . . . Westport, Conn. 06880
Union Carbide ("Bakelite"). . . . . . . . . . . . . . . . . . . .270 Park Ave., New York, N.Y. 10017

# MISCELLANEOUS INFORMATION

*Table of Equivalents.*

144 square inches equals 1 square foot
9 square feet equals 1 square yard
12 dozen, or 144 equals 1 gross
100 pounds equals one cwt. (hundred weight)
1 board foot (bd. ft.) is a piece of board 1′-0″ x 1′-0″ x 1″
2 cups equals 1 pint
2 pints equal 1 quart
4 quarts equal one gallon
16 ounces (avoirdupois) equals 1 pound
8 ounces equals 1 cup
1 tablespoonful equals ½ fluid ounce
1 teaspoonful equals 1/3 tablespoonful

* * *

*Stagecraft Material Measurements.*

Fabric or textiles are listed by the lineal or linear yard, which means a running yard of the specified width.
Pipe is measured by its inside diameter.
Aluminum conduit is measured by its outside diameter.
Lumber is measured by the board foot.
Plywood, Masonite, and other wall board materials are listed by the square foot.
Nails are listed by the penny which originally referred to price per hundred but which now refers to the length of the nail. The sizes of nails increase with the numbers.
Screws are listed by diameter, length, head shape, and finish.
Bolts are listed by diameter, type, length.
Casters are listed by diameter of the wheel.
Hammers are listed by their weight and the type of head and claws.
Saws are listed by the type, length and number of teeth per inch.

* * *

*Strength of Rope and Cable Material.* The safe working load is listed.

Piano Wire.
13 gauge ........ 250 lbs.
20 gauge ........ 500 lbs.
26 gauge ........ 900 lbs.
Manilla Rope.
¼″ ........ 130 lbs.
½″ ........ 400 lbs.
¾″ ........ 600 lbs.
1″ ........ 1450 lbs.
Braided cotton (sash cord), breaking strength.
¼″ ........ 450 lbs.
½″ ........ 1000 lbs.
¾″ ........ 1400 lbs.

*Thinners and Solvents.*

Water paint . . . use plain water or soap and water.
Oil paint . . . turpentine, kerosene, gasoline, brush cleaner.
Shellac . . . alcohol.
Lacquer . . . banana oil, lacquer thinner ,acetone.
Bronzing liquid . . . same as for oil paint.

* * *

Casting plaster and plaster of paris may be used for the same purposes. Casting plaster is usually obtainable in all lumber yards and is less expensive than plaster of paris.

* * *

Sal Ammoniac and Ammonium Chloride are the same.

* * *

*Blocks and Block and Tackle Systems.* Heavy objects may be lifted more easily if a block and tackle is employed. The mechanical advantage gained by using such a device is determined by the number and arrangement of the pulleys in relation to the load, the anchor and the operator. The friction of the rope moving on the wheel of the pulley and the friction of the pulley wheel on its axle are factors that cannot be ignored. Below is a list of the approximate advantages gained by using pulley or block systems.

1. One pulley. The rope is anchored above, passes up and over the pulley, (which is anchored above), and then back down to the operator. More energy is required to lift the object than when it is lifted without the pulley. Overcoming the friction adds to the strain of lifting. A 1000 pound weight would require a force of 1100 or 1200 pounds to lift it. The pull is downward, and there is no mechanical advantage.

2. One pulley. The rope is anchored above, passes down through the pulley and then back up again. The weight is fastened under the pulley and the operator pulls from above. A 1000 pound weight will require only half as much force as in the previous one pulley system. A 550 to 560 pound force will elevate the 1000 pound weight. The pull is upward and the M.A. is 2.

3. Two pulleys. The rope is anchored to the base of the top pulley, passes down around a lower pulley, and then up and over the top pulley and then down again. The pull is downward and the 100 pound weight will require a force of about 620 pounds to

lift it and overcome the friction of the pulleys. The pull is downward, and the M.A. is 2.

4. Two pulleys. The rope is anchored to the top of the lower block, passes up and over the top block, passes under the pulley of the lower block and then extends upward. 1000 pounds now seems to weigh about 420 pounds. The pull is upward and the M.A. is 3.

5. Double Block above and single block below. The rope is anchored to the top of the lower pulley, passes up around one of the upper pulleys, down around the lower pulley and back up around the other upper one, then back down. The apparent weight of the object is about 460 pounds. The pull is downward and the M.A. is 3.

6. Double block below and single block above. The apparent weight in this instance is about 350 pounds. The pull is upward and the M.A. is 4.

7. Two double blocks. The rope is anchored above, passed downward first and then in rotation through all of the pulleys. The apparent weight is about 385 pounds. The pull is downward and the M.A. is 4.

8. Two double blocks. The position of the lines is changed with the rope first anchored to the lower block. The apparent weight is only about 300 pounds. The pull is upward and the M.A. is 5.

It is to be noted that a loss of movement accompanies the gaining of the mechanical advantage. As the M.A. increases so also does the amount of rope that must be pulled so the movement on the object end decreases as the M.A. increases.

* * *

*The Steel Square or Framing Square.* The framing square is a valuable tool that has many uses. Ordinarily it is used for squaring the corners of flats or making marks across boards. It may be used for countless other computing tasks. Here are just a few of them.

1. Marking angles. Use the 12″ mark on one arm of the square and the number indicated on the other arm. Place the square with the 12″ mark at the junction of the lines of the angle and with the right angle of the square extending away from the point of convergence. The angle may be computed by drawing a line from the 12″ point toward the mark indicated on the other arm of the square.

12″ on one side and $1\frac{1}{16}$″ on the other for 5°
12″ on one side and $2\frac{1}{8}$″ on the other for 10°
12″ on one side and $3\frac{7}{32}$″ on the other for 15°
12″ on one side and $4\frac{3}{8}$″ on the other for 20°
12″ on one side and $5\frac{5}{8}$″ on the other for 25°
12″ on one side and $6\frac{15}{16}$″ on the other for 30°
12″ on one side and $8\frac{7}{16}$″ on the other for 35°
12″ on one side and $10\frac{3}{32}$ on the other for 40°
12″ on one side and 12″ on the other for 45°

2. In order to draw the mitre angles for the corners of various frames use the following measurements on the square:

12″ on one side and $20\frac{21}{32}$″ on the other for a triangle (equilateral)
12″ on one side and 12″ on the other for a square
12″ on one side and $8\frac{23}{32}$″ on the other for a pentagon
12″ on one side and $6\frac{15}{16}$″ on the other for a hexagon
12″ on one side and $5\frac{25}{32}$″ on the other for a heptagon
12″ on one side and $4\frac{31}{32}$″ on the other for an octagon
12″ on one side and $4\frac{3}{8}$″ on the other for a nonagon
12″ on one side and $3\frac{7}{8}$″ on the other for a decagon

3. In order to find the center of a circle with a framing square lay the square on the circle with the corner at the circumference. Mark where the outer edge of the blades cut the circle and draw a line connecting these points. This line is the diameter. Draw another line in a like manner and the point of intersection will be the center of the circle.

* * *

The mitre cut angle for corners of various figures may be determined by dividing the number of sides into 180. This holds only if the sides are equal in length.

A triangle has 3 sides so the boards for mitres are cut at 60°
A square has 4 sides so the boards for mitres are cut at 45°
A hexagon has 6 sides so the boards for mitres cut at 30°

* * *

The hypotenuse of a right triangle is equal to the square root of the sum of the square of the two sides.

# INDEX TO SUBJECT MATTER



ff: and following

**Professor Vern Adix**, B.A., University of Iowa, 1937, M.A., University of Minnesota, 1945, learned theatre fundamentals working with Sydney Spayde, E. C. Mabie, Hunton D. Sellman, Charles Elson, and Arnold Gillette. He began his professional career working under C. Lowell Lees, and Frank M. Whiting. Except for four summers early in his career, two serving as designer and technical director at the Cain Park Theatre in Cleveland, Ohio, and two as teacher and designer in the School of Fine Arts, Banff, Alberta, Canada, he has spent his entire career since 1943 with the Theatre Department, and various theatrical producing organizations of the University of Utah, in Salt Lake City, Utah. He has designed and/or staged in the neighborhood of 300 productions including plays, operas, musicals, and ballets; has adapted 15 plays for children's theatre; and has directed and/or supervised school time and summer time Theatre productions for, by, and with children. His general philosophy of staging and scenic design tends to be in favor of simplification: do not use a paragraph to describe a scenic locale if it can be done with a sentence, a phrase, or a word.